Measure, Integral, Probability & Processes

A CONCISE INTRODUCTION TO PROBABILITY AND RANDOM PROCESSES.

PROBAB(ILISTICAL)LY THE THEORETICAL MINIMUM

BY

RENÉ L. SCHILLING

TECHNISCHE UNIVERSITÄT DRESDEN

Mathematics Subject Classification 2020
Primary 60-01, 28-01.
Secondary 28A12; 28A25; 60A10; 60E10; 60Fxx; 60Gxx; 60J10; 60J65.

Author

Prof. Dr. René L. Schilling
Technische Universität Dresden
Institut für Mathematische Stochastik
01062 Dresden
Germany

rene.schilling@tu-dresden.de

ISBN-13: 979-8-5991-0488-9

Typeset by the author using LaTeX and the kpfonts package

Cover art: Dotstudio / Freepik (for the background)

Printed and bound: Amazon Fulfillment

Preface

The material is a honest rendering of my lecture notes of the courses »Measure and Integration« (3 contact hours §§1–19), »Introduction to probability« (4 contact hours, §§20–34, 38–40), »Further probability theory: Discrete random processes« (3–4 contact hours, §§35, 47–57) and »Probability with martingales« (3–4 contact hours, §§35–37, 41–46, 58–64) at TU Dresden. It is an introduction to measure and integration – suitable both for analysts and probabilists – and to probability theory & random processes up to the construction and first properties of Brownian motion.

The text is suitable for BSc students who have had a rigorous course in linear algebra and ϵ-δ-analysis. Some basic knowledge of functional analysis is helpful. The textbooks by Lang [16] (for linear algebra) and Rudin [27] (for analysis), [28, Chapters 4, 5] (for functional analysis) should be more than sufficient. For additional reference, I recommend Alt [2] (functional analysis) and the wonderful first chapter *Operator theory in finite-dimensional vector spaces* in Kato's book [14].

These notes contain the bare necessities. A more thorough treatment and plenty of exercises can be found in my books *Measures, Integrals and Martingales* [MIMS] and *Counterexamples in Measure and Integration* (with F. Kühn) [CEX] and my German-language textbooks *Maß und Integral* [MI], *Wahrscheinlichkeit* [WT] and *Martingale und Prozesse* [MaPs]. More on Brownian motion and stochastic (Itô) calculus is in *Brownian Motion. An introduction to stochastic processes* [BM] (with L. Partzsch). I tried to be as close as possible to the original lectures. Some essential material, which is usually set as (guided) exercise in the problem classes, is added as »starred items« like **Theorem***. Handouts are either set in small print (if they appear in the running text) or contained in chapter appendices.

The subtitle »the theoretical minimum« alludes to the physicist Lev Landau, who *developed a comprehensive exam called the "Theoretical Minimum" which students were expected to pass before admission to the school* `https://en.wikipedia.org/wiki/Kharkiv_Theoretical_Physics_School` (accessed 30/Oct/2020).

I would like to thank my friends, students and collaborators – David Berger,

Wojciech Cygan, Paolo Di Tella, Victoria Knopova, Franziska Kühn and Cailing Li – for their help when compiling these notes.

Dresden, summer of 2020
René L. Schilling

Contents

Abbreviations and symbols

Here I list the most common abbreviations and symbolic notation used throughout the text.

✎	this indicates that you should check it by yourself
⚠	this indicates a warning
💬	this indicates further information
a.a.	almost all
a.e.	almost every(where)
a.s.	almost surely
bdd	bounded
BL, cBL	(conditional) Beppo Levi theorem
BM	Brownian motion
b/o	because of
c.f.	characteristic function
cf.	confer, see
CLT	central limit theorem
d-convergence	convergence in distribution
DCT, cDCT	(conditional) domitated convergence theorem
e.g.	exempli gratia, for example
fdd	finite dimensional distributions
iid	independent identically distributed
LLN	law of large numbers
mble	measurable
MC	Markov chain
MCT	monotone class theorem *or* martingale convergence theorem
mg	martingale
MP	Markov property
$\mathbb{P}$-convergence	convergence in probability
PP, cPP	Poisson process, compound poisson process
rv, rvs	random variable, random variables

RW	random walk
SLLN	strong law of large numbers
SMP	strong Markov property
SRW	simple random walk
ui	uniformly integrable
WLLN	weak law of large numbers
wlog	without loss of generality
positive	always used in the sense »$\geqslant 0$«
negative	always used in the sense »$\leqslant 0$«
$\mathbb{N}$	natural numbers $1, 2, 3, \ldots$
$\mathbb{N}_0$	natural numbers $0, 1, 2, 3, \ldots$
$X \sim \mu$	the rv X is distributed like μ
$X \sim Y$	the rv X is distributed like the rv Y
$X \perp\!\!\!\perp Y$	X and Y are independent
$x \gg 1, \epsilon \ll 1$	x is sufficiently large, $\epsilon \in (0,1)$ is sufficiently small
$\mathcal{L}^p(\ldots)$, $L^p(\ldots)$	Lebesgue spaces of integrable functions $1 \leqslant p \leqslant \infty$
$\mathcal{L}^0(\mathscr{A})$	$\mathscr{A}$-measurable functions
$\mathcal{E}(\mathscr{A})$	$\mathscr{A}$-measurable simple (»step«) functions

A »+« as sub- or superscript, such as $\mathcal{E}^+$ or $\mathcal{L}^p_+$ means the positive ($\geqslant 0$) elements of $\mathcal{E}$ or $\mathcal{L}^p$ etc.

I

Measures

1 Introduction

The aim of measure theory is to assign all, or at least many, sets $A \subset \Omega$ of an abstract set Ω a **measure** $\mu(A) \in [0,\infty]$. The quantity $\mu(A)$ turns out to be a natural generalization of

- the length of an interval or a line $[a,b] \subset \mathbb{R}$;
- the area of a set $F \subset \mathbb{R}^2$;
- the volume of a body $V \subset \mathbb{R}^3$;
- the number of elements $\#A = |A|$ of a set $A \subset \Omega$.
- the increment $\mu([a,b)) := \phi(b) - \phi(a)$ of an increasing function $\phi(t)$ on a half-open interval $[a,b)$;
- the integral $\int_\Omega f(x)\,dx$ of a positive function $f(x) \geqslant 0$ over a set $\Omega \subset \mathbb{R}^d$. We interpret the integral as the »area« $\{(x,y) \mid x \in \Omega,\ 0 \leqslant y \leqslant f(x)\} \subset \mathbb{R}^{d+1}$, i.e. the geometric »volume« of the sub-graph.

The last example shows the intimate connection of measure and integral.

Measure theory was originally developed in connection with the theory of real functions; but soon it found applications in many other mathematical disciplines, e.g. functional analysis, probability theory and dynamical systems.

As we have seen, a **measure** is a function defined on a family of subsets of a base space Ω

$$\mu : \mathscr{F} \to [0,\infty], \quad \mathscr{F} \subset \mathscr{P}(\Omega) = \{A \mid A \subset \Omega\}.$$

⚠ $\mathscr{F} \subset \mathscr{P}(\Omega)$ means that $\mathscr{F}$ is a family of sets. Thus, the symbol »$\subset$« is consistently used at two levels, i.e. at the level of points $A \subset \Omega$ and the level of sets $\mathscr{F} \subset \mathscr{P}(\Omega)$.

💬 We will use curly letters $\mathscr{A}, \mathscr{B}, \mathscr{F}, \dots$ to indicate families of sets.

💬 We will use greek letters $\mu, \nu, \rho, \dots$ to denote measures.

Although we will later on consider very abstract spaces Ω, the basic principles of a measure can be explained using the measurement of area. Assume that $F \subset \mathbb{R}^2$ is some plane set and let μ be the usual notion of area. If F is simple, we can work out $\mu(F)$ directly, see Fig. 1.1. It is clear that **non-overlapping areas**

Fig. 1.1. Area of a plane rectancle $\mu(F) = g \cdot h$ and $\mu(F) = \mu(F_1) + \mu(F_2) + \mu(F_3)$.

add and, in the situation of Fig. 1.1, it is easy to check that

$$\mu(F_1) + \mu(F_2) + \mu(F_3) = g' \cdot h + g'' \cdot h'' + g'' \cdot h' = \cdots = g \cdot h = \mu(F).$$

Although it is not immediately obvious in the case of a rectangle, it is better to split up the area in triangles. Consider, for example an irregular quadrangle or an even more complicated geometric shape as in Fig. 1.2. This will allow us, if

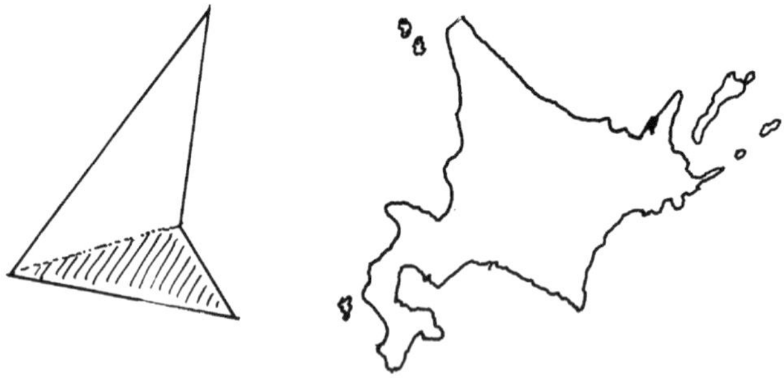

Fig. 1.2. A triangulation of a quadrangle and the island of Hokkaido, Japan.

we can exhaust F by triangles, to calculate very irregular shapes. It is important

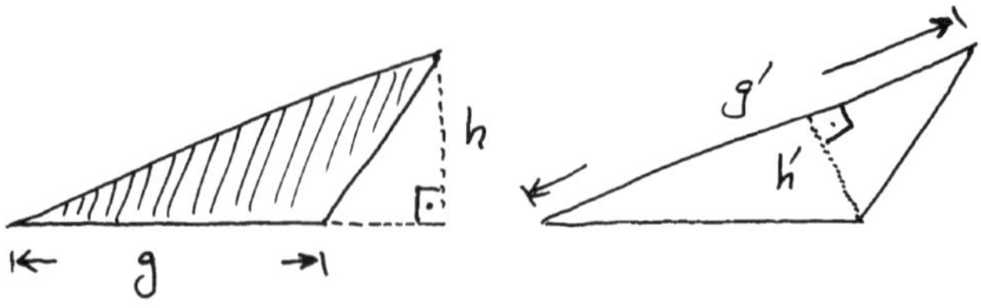

Fig. 1.3. The area of a triangle Δ.

that the function $\Delta \mapsto \mu(\Delta)$ is well-defined, i.e. it does not depend on the way how we determine it, e.g. (see Fig. 1.3)

$$\mu(\Delta) = \frac{1}{2} g \cdot h = \frac{1}{2} g' \cdot h'.$$

Let us collect the properties of μ which we have used so far.

a) μ maps subsets of Ω to positive real numbers $[0,\infty]$ and $\mu(\emptyset)=0$
b) μ is additive: $A\cap B=\emptyset \implies \mu(A\mathbin{\dot\cup} B)=\mu(A)+\mu(B)$
c) (only for volumes in $\mathbb{R}^d$): Congruent sets have the same measure. 💬 We will see that this property even characterizes d-dimensional volumes once we fix the volume of the unit cube $[0,1]^d$.

In general, the properties a) and b) are not enough to determine the area of complicated sets like »Hokkaido«. In that case we will have an infinite triangulation and the area will be an expression like

$$F=\dot{\bigcup_{n\in\mathbb{N}}}\Delta_n \quad\text{and}\quad \mu(F)=\sum_{n\in\mathbb{N}}\mu(\Delta_n); \tag{1.1}$$

(the symbol $\dot\bigcup$ indicates that the sets Δ_n are mutually disjont: $\Delta_n\cap\Delta_m=\emptyset$ if $n\neq m$). But there are problems

- ❓ Is there always a triangulation with **countably many** triangles?
- ❓ Does additivity carry over to countably many sets, i.e. is (1.1) valid?
- ❓ Do different triangulations give the same value $\mu(F)$?

In fact, the validity of (1.1) must be **assumed**. We will see, cf. Examples 1.1 and 1.2 below, that it is essentially a notion of continuity of the function $F\mapsto\mu(F)$.

d) μ is σ-additive: For disjoint $(A_n)_{n\in\mathbb{N}}$ we have $\mu\left(\dot\bigcup_n A_n\right)=\sum_n\mu(A_n)$

1.1 Example (area of a line). A line segment in $\mathbb{R}^2$ can be seen as a degenerate rectangle, i.e. a rectangle with zero width, say, $L=\{a\}\times[0,1]$. We have

$$\mu(L)=\mu\Bigg(\underbrace{\bigcap_{n=1}^{\infty}\left[a-\tfrac{1}{n},a+\tfrac{1}{n}\right]\times[0,1]}_{=L}\Bigg)=\lim_{n\to\infty}\mu\left(\left[a-\tfrac{1}{n},a+\tfrac{1}{n}\right]\times[0,1]\right)=\lim_{n\to\infty}\tfrac{2}{n}\times 1=0.$$

⚠ But it is not clear why we may pull out the »set-limit« $\bigcap_{n=1}^{\infty}$ of the measure function μ.

1.2 Example (area of a circle). Let $B_r(0)\subset\mathbb{R}^2$ be the open ball with centre 0 and radius $r>0$. By $U_n\subset B_r(0)\subset V_n$ we denote the open inscribed and circumscribed regular n-gons. We have

$$\bigcup_{n\in\mathbb{N}}U_n\subset B_r(0)\subset\overline{B_r(0)}\subset\bigcap_{n\in\mathbb{N}}\overline{V}_n$$

and each polygon can be split into n congruent triangles A_n and B_n, see Fig. 1.4 (right panel). Thus,

$$\mu(U_n)=n\mu(A_n)=n\tfrac{1}{2}\cdot 2r\sin\tfrac{\pi}{n}\cdot r\cos\tfrac{\pi}{n}=r^2\pi\frac{\sin(\pi/n)}{\pi/n}\cos\tfrac{\pi}{n},$$

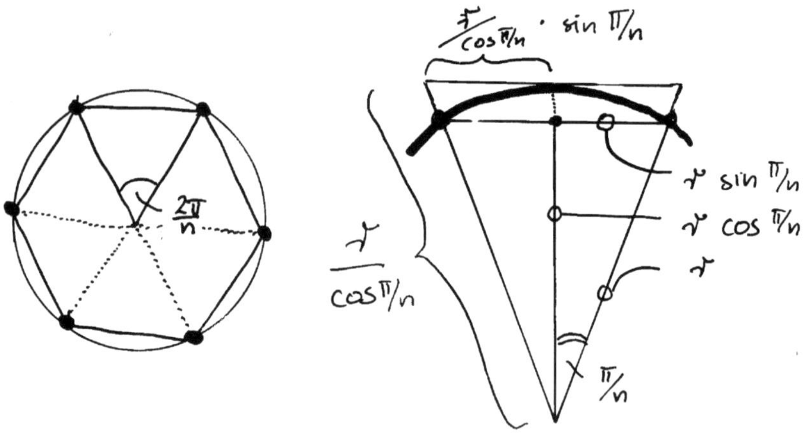

Fig. 1.4. The inscribed and circumscribed regular polygons U_n and V_n

and since $U_n \uparrow B_r(0)$, we conclude

$$\lim_{n\to\infty} \mu(U_n) \overset{\triangle}{=} \lim_{n\to\infty} r^2\pi \frac{\sin(\pi/n)}{\pi/n} \cos\tfrac{\pi}{n} = r^2\pi \cdot 1 \cdot 1.$$

Because of Example 1.1, the boundary of U_n has measure zero. Similarly,

$$\lim_{n\to\infty} \mu(\overline{V}_n) = r^2\pi$$

and we get $\mu(B_r(0)) = \mu(\overline{B_r(0)}) = r^2\pi$. In particular, the boundary $\mu(\partial B_r(0)) = 0$ has zero **area** (but, of course, not zero length).

We want to be able to determine the area of much more general objects such as fractals. It is interesting to note that the properties a), b) and d) give a rich enough theory to do this. For deep set-theoretic reasons, a measure with the properties a), b), c) and d) cannot be defined on **all** sets of $\mathbb{R}^d$ with $d \geqslant 3$, and we have to restrict ourselves, in general, to a proper subfamily $\mathscr{F} \subsetneqq \mathscr{P}(\Omega)$.

2 Sigma-algebras

Let Ω be an arbitrary (abstract) base set and $\mathscr{F} \subset \mathscr{P}(\Omega)$ be a family of sets. We have seen in Chapter 1 that a measure is a map $\mu : \mathscr{F} \to [0,\infty]$ satisfying the properties a), b) and d); this means, in particular, that the family $\mathscr{F}$ must be stable under certain set operations such as (countable) unions of its members.

2.1 Definition. A **σ-algebra** on a set $\Omega \neq \emptyset$ is a family $\mathscr{A} \subset \mathscr{P}(\Omega)$ satisfying

$$\Omega \in \mathscr{A}, \tag{Σ_1}$$

$$A \in \mathscr{A} \implies A^c := \Omega \setminus A \in \mathscr{A}, \tag{Σ_2}$$

$$(A_n)_{n\in\mathbb{N}} \subset \mathscr{A} \implies \bigcup_{n\in\mathbb{N}} A_n \in \mathscr{A}. \tag{Σ_3}$$

The elements $A \in \mathscr{A}$ are called **measurable sets.**

2.2 Properties (of a σ-algebra). Let $\mathscr{A}$ be a σ-algebra on Ω.

a) $\emptyset \in \mathscr{A}$, since $\emptyset = \Omega^c \in \mathscr{A}$ (use Σ_1, Σ_2).

b) $A, B \in \mathscr{A} \implies A \cup B \in \mathscr{A}$. If we take $A_1 := A$, $A_2 := B$ and $A_3 = A_4 = \ldots = \emptyset$, then

$$\overset{\Sigma_3}{\implies} A \cup B = A_1 \cup A_2 \cup A_3 \cup \ldots \in \mathscr{A}.$$

c) $(A_n)_{n\in\mathbb{N}} \subset \mathscr{A} \implies \bigcap_{n\in\mathbb{N}} A_n \in \mathscr{A}$. Indeed, we have

$$A_n \in \mathscr{A} \overset{\Sigma_2}{\implies} A_n^c \in \mathscr{A} \overset{\Sigma_3}{\implies} \bigcup_{n\in\mathbb{N}} A_n^c \in \mathscr{A}$$

$$\overset{\Sigma_2}{\implies} \bigcap_{n\in\mathbb{N}} A_n = \left(\bigcup_{n\in\mathbb{N}} A_n^c\right)^c \in \mathscr{A}.$$

d) $A, B \in \mathscr{A} \implies A \cap B \in \mathscr{A}$. To see this, argue as in b), where c) takes the role of (Σ_3).

e) $A, B \in \mathscr{A} \implies A \setminus B \in \mathscr{A}$. From d) and Σ_2 we infer

$$A \setminus B = A \cap (B^c) \in \mathscr{A}.$$

- Thus, a σ-algebra is stable under all set-operations – union, intersection, difference and complements – as long as we perform them at most countably often.
- If (Σ_3) is replaced by »$A \cup B \ \forall A, B \in \mathscr{A}$« we speak of a (**Boolean**) **algebra**. An algebra is stable under finite $\cap, \cup, \setminus$ and complements.
- A (**Boolean**) **ring** contains $\emptyset$ and it is stable under finitely many $\cap, \cup, \setminus$ of its members.

We will mostly deal with σ-algebras.

2.3 Example. Let $\Omega \neq \emptyset$ be an arbitrary base set and $A, B \subset \Omega$.

a) $\mathscr{P}(\Omega)$ is the largest (or finest) σ-algebra in Ω.
b) $\{\emptyset, \Omega\}$ is the smallest (or coarsest) σ-algebra in Ω.
c) $\{\emptyset, A, A^c, \Omega\}$ is a σ-algebra.
d) $\{\emptyset, B, \Omega\}$ is a σ-algebra if, and only if, $B = \emptyset$ or $B = \Omega$.
e) If Ω is not countable, then $\mathscr{A} := \left\{A \subset \Omega \mid \#A \leqslant \#\mathbb{N} \text{ or } \#A^c \leqslant \#\mathbb{N}\right\}$ is a σ-algebra. Indeed,

Σ_1: $\Omega^c = \emptyset$ is countable, so $\Omega \in \mathscr{A}$.

Σ_2: $A \in \mathscr{A}$ means that either A or A^c is countable. Since $A = (A^c)^c$, we see that $A^c \in \mathscr{A}$.

Σ_3: Let $(A_n)_{n\in\mathbb{N}} \subset \mathscr{A}$ and set $A = \bigcup_{n\in\mathbb{N}} A_n$. There are two possibilities:

- All A_n are countable, then A is the union of countably many countable sets, hence countable, and so $A \in \mathscr{A}$.
- There is some n_0 such that A_{n_0} is not countable, i.e. $A_{n_0}^c$ is countable. So,

$$A^c = \left(\bigcup_{n\in\mathbb{N}} A_n\right)^c = \bigcap_{n\in\mathbb{N}} A_n^c \subset A_{n_0}^c \implies A^c \text{ is countable}$$

which shows that $A \in \mathscr{A}$.

f) (**trace σ-algebra**) Let $E \subset \Omega$ be any set and $\mathscr{A}$ a σ-algebra in Ω. The family $\mathscr{A}|_E := \{E \cap A \mid A \in \mathscr{A}\}$ is a σ-algebra in E.
Note that (Σ_2) now reads $A' \in \mathscr{A}|_E \implies E \setminus A' \in \mathscr{A}|_E$.

g) (**pre-image of a σ-algebra**) Let $f : \Omega \to \Omega'$ be any map and $\mathscr{A}'$ a σ-algebra in Ω'. Since f^{-1} commutes with all (countable or uncountable) set-operations, $\mathscr{A} := \left\{f^{-1}(A') \mid A' \in \mathscr{A}'\right\}$ is a σ-algebra in Ω.

Notation. Let $\mathscr{A}_i$, $i \in I$, be arbitrarily many families of sets. We write

$$\bigcap_{i\in I} \mathscr{A}_i = \{A \mid \forall i \in I : A \in \mathscr{A}_i\} \quad \text{and} \quad \bigcup_{i\in I} \mathscr{A}_i = \{A \mid \exists i \in I : A \in \mathscr{A}_i\}$$

for those sets $A \subset \Omega$ which are contained in **all**, resp., **some** $\mathscr{A}_i$. This is consistent with the usual use of »∩« and »∪« for point sets – just note that sets now play the role of points.

2.4 Theorem. a) *The intersection $\mathscr{A} := \bigcap_{i\in I} \mathscr{A}_i$ of arbitrarily many σ-algebras in Ω is a σ-algebra.*

b) *For any family $\mathscr{G} \subset \mathscr{P}(\Omega)$ there is a smallest (also: minimal) σ-algebra $\mathscr{A}$ such that $\mathscr{G} \subset \mathscr{A}$. $\mathscr{A}$ is the **σ-algebra generated by** $\mathscr{G}$ and $\mathscr{G}$ is the **generator** of $\mathscr{A}$. One writes $\mathscr{A} = \sigma(\mathscr{G})$.*

Proof. a) Let us verify (Σ_1)–(Σ_3).

Σ_1: $\forall i \in I : \emptyset \in \mathscr{A}_i \implies \emptyset \in \bigcap_{i\in I} \mathscr{A}_i$.

Σ_2: $\forall i : A \in \mathscr{A}_i \implies \forall i : A^c \in \mathscr{A}_i \implies A^c \in \bigcap_{i\in I} \mathscr{A}_i$.

Σ_3: $\forall i : (A_k)_{k\in\mathbb{N}} \subset \mathscr{A}_i \implies \forall i : \bigcup_{k\in\mathbb{N}} A_k \in \mathscr{A}_i \implies \bigcup_{k\in\mathbb{N}} A_k \in \bigcap_{i\in I} \mathscr{A}_i$.

b) Part a) shows that the intersection

$$\mathscr{A} := \bigcap_{\substack{\mathscr{F} \text{ σ-algebra} \\ \mathscr{G} \subset \mathscr{F}}} \mathscr{F} \tag{2.1}$$

is a σ-algebra. Since $\mathscr{F} = \mathscr{P}(\Omega)$ appears in the intersection (2.1), $\mathscr{A}$ is well-defined; in particular it is a σ-algebra which contains $\mathscr{G}$.

Now let $\mathscr{A}'$ be any σ-algebra such that $\mathscr{G} \subset \mathscr{A}'$. Obviously, $\mathscr{F} = \mathscr{A}'$ appears in the intersection (2.1), and so $\mathscr{A}' \subset \mathscr{A}$. This shows that $\mathscr{A}$ is the smallest σ-algebra satisfying $\mathscr{G} \subset \mathscr{A}$. This means that the notation $\sigma(\mathscr{G})$ makes sense. □

2.5 Remark. a) If $\mathscr{A}$ is a σ-algebra, then $\sigma(\mathscr{A}) = \mathscr{A}$.
b) If $A \subset \Omega$, then $\sigma(\{A\}) = \{\emptyset, A, A^c, \Omega\}$.
c) If $\mathscr{G} \subset \mathscr{H}$, then $\sigma(\mathscr{G}) \subset \sigma(\mathscr{H})$. Indeed, $\mathscr{G} \subset \mathscr{H} \subset \sigma(\mathscr{H})$. Since $\sigma(\mathscr{H})$ is a σ-algebra, we see by the minimality of $\sigma(\mathscr{G})$ that $\sigma(\mathscr{G}) \subset \sigma(\mathscr{H})$.

In $\mathbb{R}^d$ (or in general topological spaces) the σ-algebra generated by the open sets plays a special role. Recall that

$$U \subset \mathbb{R}^d \text{ is open} \iff \forall x \in U\ \exists \epsilon > 0 : B_\epsilon(x) \subset U$$

($B_\epsilon(x)$ is the open ball with centre $x \in \mathbb{R}^d$ and radius $\epsilon > 0$). The family of open sets $\mathscr{O} = \mathscr{O}(\mathbb{R}^d)$ is the **topology**. It has the following properties

$$\emptyset, \Omega \in \mathscr{O}, \tag{$\mathscr{O}_1$}$$

$$U, V \in \mathscr{O} \implies U \cap V \in \mathscr{O}, \tag{$\mathscr{O}_2$}$$

$$U_i \in \mathscr{O},\ i \in I \text{ (arbitrary)} \implies \bigcup_{i \in I} U_i \in \mathscr{O}. \tag{$\mathscr{O}_3$}$$

In general, the properties ($\mathscr{O}_1$)–($\mathscr{O}_3$) are used to define a topology.

⚠ The intersection $\bigcap_{n \in \mathbb{N}} U_n$ of countably many open sets need not be an open set, consider e.g. $U_n = \left(-\frac{1}{n}, \frac{1}{n}\right) \subset \mathbb{R}$.

2.6 Definition. The σ-algebra generated by the open sets $\mathscr{O}$ in $\mathbb{R}^d$ is called the **Borel σ-algebra**. Notation: $\mathscr{B}(\mathbb{R}^d)$. The sets $B \in \mathscr{B}(\mathbb{R}^d)$ are called **Borel sets** or **Borel measurable sets.**

2.7 Theorem. *Let $\mathscr{O}$, $\mathscr{C}$, and $\mathscr{K}$ denote the open, closed and compact sets of $\mathbb{R}^d$. Then*

$$\mathscr{B}(\mathbb{R}^d) = \sigma(\mathscr{O}) = \sigma(\mathscr{C}) = \sigma(\mathscr{K}).$$

Proof. This is a simple exercise along the lines of the proof of the next theorem. □

The Borel sets are generated by several families of sets. For our investigations, the following generators are particularly important.

$$\begin{aligned}
\mathscr{I}^o_{[\mathrm{rat}]} &= \left\{(a_1, b_1) \times \cdots \times (a_d, b_d) \mid a_n, b_n \in \mathbb{R}\ [\in \mathbb{Q}]\right\} \\
&= \text{open »rectangles« [with rational vertices]} \\
\mathscr{I}_{[\mathrm{rat}]} &= \left\{[a_1, b_1) \times \cdots \times [a_d, b_d) \mid a_n, b_n \in \mathbb{R}\ [\in \mathbb{Q}]\right\} \\
&= \text{half-open »rectangles« [with rational vertices]}
\end{aligned}$$

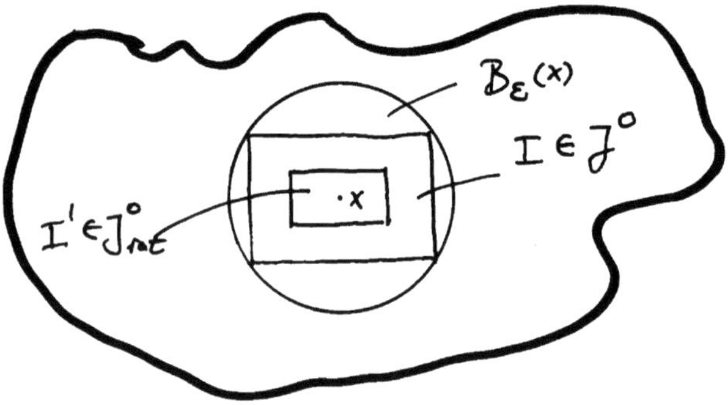

Fig. 2.1. U can be exhausted by rectangles with rational vertices.

⚠ $b < a \implies (a, b) = \emptyset$ and $A \times \cdots \times \emptyset \times \cdots \times Z = \emptyset$.

2.8 Theorem. *The Borel sets of $\mathbb{R}^d$ can be obtained as*

$$\sigma(\mathcal{O}) = \sigma(\mathcal{J}) = \sigma(\mathcal{J}^o) = \sigma(\mathcal{J}_{rat}) = \sigma(\mathcal{J}^o_{rat}).$$

Proof. 1° Since any open rectangle is an open set ✍, we have

$$\mathcal{J}^o_{\text{rat}} \subset \mathcal{J}^o \subset \mathcal{O} \implies \sigma(\mathcal{J}^o_{\text{rat}}) \subset \sigma(\mathcal{J}^o) \subset \sigma(\mathcal{O}).$$

2° Every open set $U \in \mathcal{O}$ can be written in the form

$$U = \bigcup_{I' \in \mathcal{J}^o_{\text{rat}}, I' \subset U} I'. \tag{2.2}$$

Indeed, the inclusion »⊃« in (2.2) is trivial. In order to see »⊂«, we note (cf. Fig. 2.1)

$$\forall x \in U \quad \exists \epsilon > 0 : B_\epsilon(x) \subset U.$$

Inside $B_\epsilon(x)$ there is some $I \in \mathcal{J}^o$, $x \in I$. Since $\mathbb{Q}^d$ is dense in $\mathbb{R}^d$, we can »squeeze« this rectangle and get some $I' \in \mathcal{J}^o_{\text{rat}}$ such that $x \in I' \subset I$.

Since an axis-parallel rectangle is uniquely determined through the vertices of its main diagonal, the intersecton in (2.2) is countable:

$$\#\mathcal{J}^o_{\text{rat}} = \#(\mathbb{Q}^d \times \mathbb{Q}^d) = \#\mathbb{N},$$

i.e. $U \in \sigma(\mathcal{J}^o_{\text{rat}}$. Thus, $\mathcal{O} \subset \sigma(\mathcal{J}^o_{\text{rat}}) \subset \sigma(\mathcal{J}^o) \subset \sigma(\mathcal{O})$; from this we conclude that $\sigma(\mathcal{O}) \subset \sigma(\mathcal{J}^o_{\text{rat}}) \subset \sigma(\mathcal{J}^o) \subset \sigma(\mathcal{O})$, i.e. all inclusions are actually equalities.

3° The following two identities

$$(a_1, b_1) \times \ldots \times (a_d, b_d) = \bigcup_{n \in \mathbb{N}} \left[a_1 + \tfrac{1}{n}, b_1\right) \times \ldots \times \left[a_d + \tfrac{1}{n}, b_d\right)$$

$$[\alpha_1, \beta_1) \times \ldots \times [\alpha_d, \beta_d) = \bigcap_{k \in \mathbb{N}} \left(\alpha_1 - \tfrac{1}{k}, \beta_1\right) \times \ldots \times \left(\alpha_d - \tfrac{1}{k}, \beta_d\right)$$

imply that

$$\mathcal{I}^o_{[\mathrm{rat}]} \subset \sigma(\mathcal{I}_{[\mathrm{rat}]}) \quad \text{and} \quad \mathcal{I}_{[\mathrm{rat}]} \subset \sigma(\mathcal{I}^o_{[\mathrm{rat}]})$$

and, therefore,

$$\sigma(\mathcal{I}^o) = \sigma(\mathcal{I}) \quad \text{and} \quad \sigma(\mathcal{I}^o_{\mathrm{rat}}) = \sigma(\mathcal{I}_{\mathrm{rat}}).$$

Since we already know from step 2° that $\sigma(\mathcal{I}^o) = \sigma(\mathcal{I}^o_{\mathrm{rat}})$, the proof is complete. □

2.9 Remark. a) An argument along the lines of Theorem 2.8 can be used to show that the Borel σ-algebra $\mathcal{B}(\mathbb{R})$ is generated by any of the following families ($D \subset \mathbb{R}$ is any dense subset):

$$\{(-\infty, a) \mid a \in D\}, \quad \{(-\infty, b] \mid b \in D\}, \quad \{(c, \infty) \mid c \in D\}, \quad \{[d, \infty) \mid d \in D\}.$$

b) In general, it is not possible to construct $\sigma(\mathcal{G})$ explicitly (adding, iteratively, complements, countable unions, countable intersections etc.) and any such argument requires transfinite induction or the axiom of choice. In Chapter 4 we introduce **Dynkin systems** which will allow us to reduce assertions on $\sigma(\mathcal{G})$ to properties of $\mathcal{G}$.

c) There are non-Borel measurable sets. Again, their construction needs the axiom of choice, see e.g. [MIMS, Appendix G].

Definition 2.6 still makes sense in a general topological space $(\Omega, \mathcal{O})$, and we call $\mathcal{B}(\Omega) = \sigma(\mathcal{O})$ Borel or **topological σ-algebra**.

Recall that the topology on a subset $A \subset \mathbb{R}^d$ are the **relatively open sets**, i.e. sets of the form $\mathcal{O}|_A = \{A \cap U \mid U \in \mathcal{O}\}$. The Borel sets of $(A, \mathcal{O}|_A)$ are defined as $\mathcal{B}(A) := \sigma(\mathcal{O}|_A)$. The following lemma shows that the Borel σ-algebra in A is the trace of the Borel σ-algebra (cf. Example 2.3.f)).

2.10 Lemma★. *Let* $A \subset \mathbb{R}^d$. *Then* $\mathcal{B}(A) = \mathcal{B}(\mathbb{R}^d)|_A$.

Proof. Clearly, $\mathcal{O}|_A \subset \mathcal{B}(\mathbb{R}^d)|_A$, hence $\mathcal{B}(A) = \sigma(\mathcal{O}|_A) \subset \mathcal{B}(\mathbb{R}^d)|_A$.

Conversely, define $\Sigma := \{B \subset \mathbb{R}^d \mid A \cap B \in \sigma(\mathcal{O}|_A)\}$. A simple calculation ✎ reveals that Σ is itself a σ-algebra.

Since $\mathcal{O} \subset \Sigma$, we get $\mathcal{B}(\mathbb{R}^d) = \sigma(\mathcal{O}) \subset \Sigma$. From the very definition of Σ, we conclude that $\mathcal{B}(\mathbb{R}^d)|_A \subset \sigma(\mathcal{O}|_A) = \mathcal{B}(A)$. □

3 Measures

Throughout this chapter, $\Omega \neq \emptyset$ is an arbitrary set. We are now going to introduce set functions along the lines explained in Chapter 1.

3.1 Definition. Let $\mathcal{F}$ be a family of sets in Ω such that $\emptyset \in \mathcal{F}$. A (positive)

measure relative to $\mathscr{F}$ is a map $\mu : \mathscr{F} \to [0,\infty]$ such that

$$\mu(\emptyset) = 0 \tag{M_1}$$

$$\left.\begin{array}{l}(A_n)_{n\in\mathbb{N}} \subset \mathscr{F} \text{ pairwise} \\ \text{disjoint, } \bigcup_{n\in\mathbb{N}} A_n \in \mathscr{F}\end{array}\right\} \Longrightarrow \mu\Big(\bigcup_{n\in\mathbb{N}}^{\boldsymbol{\cdot}} A_n\Big) = \sum_{n\in\mathbb{N}} \mu(A_n). \tag{M_2}$$

A **measure** is a measure relative to a σ-algebra.

▲ On a σ-algebra the condition $\bigcup_{n\in\mathbb{N}} A_n \in \mathscr{F}$ appearing in (M_2) is always satisfied.

💬 (M_2) is called **σ-additivity**.

💬 Sometimes, measures relative to a family are called »pre-measures«.

We will need the following useful notation:

$$A_n \uparrow A \iff A_1 \subset A_2 \subset A_3 \subset \ldots \quad \text{and} \quad \bigcup_{n\in\mathbb{N}} A_n = A,$$

$$B_n \downarrow B \iff B_1 \supset B_2 \supset B_3 \supset \ldots \quad \text{and} \quad \bigcap_{n\in\mathbb{N}} B_n = B.$$

3.2 Definition. Let $\mathscr{A}$ be a σ-algebra in Ω and μ a measure. The pair $(\Omega,\mathscr{A})$ is called a **measurable space** and the triplet $(\Omega,\mathscr{A},\mu)$ is called a **measure space**. A measure μ is said to be a

finite measure, if $\mu(\Omega) < \infty$,
σ-finite measure, if $\exists (A_n)_{n\in\mathbb{N}} \subset \mathscr{A},\ \mu(A_n) < \infty,\ A_n \uparrow \Omega$,
probability measure, if $\mu(\Omega) = 1$.

Accordingly, we speak of a **finite** and **σ-finite measure space** and a **probability space**.

The next theorem contains all elementary properties of measures.

3.3 Theorem. *Let* μ *be a measure on* $(\Omega,\mathscr{A})$ *and* $A, B, A_n, B_n \in \mathscr{A}$, $n \in \mathbb{N}$.

a) $A \cap B = \emptyset \Longrightarrow \mu(A \cup B) = \mu(A) + \mu(B)$ *(additive)*
b) $A \subset B \Longrightarrow \mu(A) \leqslant \mu(B)$ *(monotone)*
c) $A \subset B,\ \mu(A) < \infty \Longrightarrow \mu(B \setminus A) = \mu(B) - \mu(A)$
d) $\mu(A \cup B) + \mu(A \cap B) = \mu(A) + \mu(B)$ *(strongly additive)*
e) $\mu(A \cup B) \leqslant \mu(A) + \mu(B)$ *(subadditive)*
f) $A_n \uparrow A \Longrightarrow \mu(A) = \sup_{n\in\mathbb{N}} \mu(A_n) = \lim_{n\to\infty} \mu(A_n)$ *(continuous from below)*
g) $B_n \downarrow B,\ \mu(B_1) < \infty \Longrightarrow \mu(B) = \inf_{n\in\mathbb{N}} \mu(B_n) = \lim_{n\to\infty} \mu(B_n)$ *(continuous from above)*
h) $\mu\Big(\bigcup_{n\in\mathbb{N}} A_n\Big) \leqslant \sum_{n\in\mathbb{N}} \mu(A_n)$ *(σ-subadditive)*

Proof. a) We use the same trick as in the proof of the Property 2.2.b) of a σ-algebra:

$$\mu(A \mathbin{\dot\cup} B) = \mu(A \mathbin{\dot\cup} B \mathbin{\dot\cup} \emptyset \mathbin{\dot\cup} \emptyset \mathbin{\dot\cup} \dots) \overset{(M_2)}{=} \mu(A) + \mu(B) + \mu(\emptyset) + \dots \overset{(M_1)}{=} \mu(A) + \mu(B).$$

b) If $A \subset B$, then we have $B = A \mathbin{\dot\cup} (B \setminus A) = A \mathbin{\dot\cup} (B \setminus (A \cap B))$. By Part a)

$$\begin{aligned} \mu(B) &= \mu(A \mathbin{\dot\cup} (B \setminus A)) = \mu(A) + \mu(B \setminus A) \\ &\geqslant \mu(A). \end{aligned} \tag{3.1}$$

c) This follows, if we subtract $\mu(A) < \infty$ on both sides of (3.1).

d) Splitting $A \cup B$ as shown in Fig. 3.1, we see

$$\begin{aligned} &\mu(A \cup B) + \mu(A \cap B) \\ &= \mu(A \mathbin{\dot\cup} (B \setminus (A \cap B))) + \mu(A \cap B) \\ &\overset{a)}{=} \mu(A) + \underbrace{\mu(B \setminus (A \cap B)) + \mu(A \cap B)}_{=\mu(B) \text{ because of (3.1)}} \\ &= \mu(A) + \mu(B) \end{aligned}$$

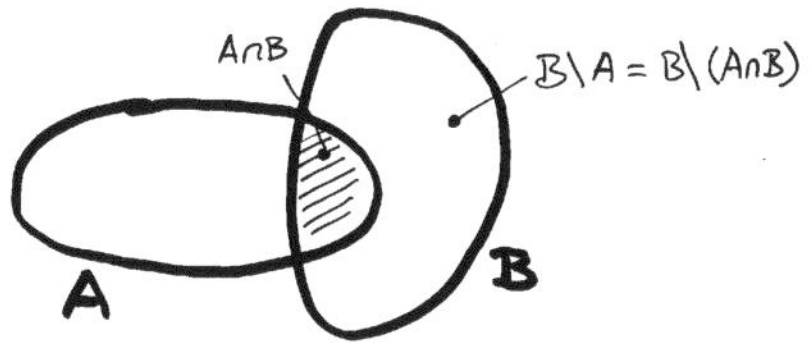

Fig. 3.1. Representing $A \cup B$ as disjoint union $A \cup (B \setminus (A \cap B))$

e) This follows from d), since $\mu(A \cap B) \geqslant 0$.

f) The sets $F_1 := A_1$, $F_2 := A_2 \setminus A_1, \dots, F_{n+1} := A_{n+1} \setminus A_n$ are pairwise disjoint and we have

$$A_n = \dot{\bigcup_{i=1}^{n}} F_i \xrightarrow{n\to\infty} A = \dot{\bigcup_{i=1}^{\infty}} F_i = \bigcup_{n=1}^{\infty} A_n,$$

see Fig. 3.2. Using σ-additivity (M_2) yields

$$\mu(A) = \mu\left(\dot{\bigcup_{i=1}^{\infty}} F_i\right) \overset{(M_2)}{=} \sum_{i=1}^{\infty} \mu(F_i) = \lim_{n\to\infty} \sum_{i=1}^{n} \mu(F_i) \overset{a)}{=} \lim_{n\to\infty} \mu(A_n).$$

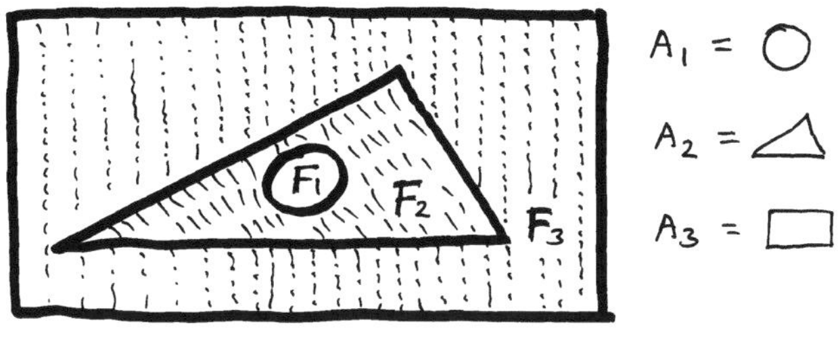

Fig. 3.2. Making a sequence of increasing sets A_n disjoint.

g) If $B_n \downarrow B$, then $B_1 \setminus B_n \uparrow B_1 \setminus B$. As $\mu(B_1) < \infty$, we can use f), c) and find

$$\mu(B_1) - \mu(B) \overset{c)}{=} \mu(B_1 \setminus B) = \lim_{n\to\infty} \mu(B_1 \setminus B_n) \overset{c)}{=} \lim_{n\to\infty} \big(\mu(B_1) - \mu(B_n)\big)$$
$$\overset{f)}{=} \mu(B_1) - \lim_{n\to\infty} \mu(B_n).$$

h) From f), g) we get

$$\mu\Big(\bigcup_{n=1}^{\infty} A_n\Big) \overset{f)}{=} \lim_{n\to\infty} \mu(A_1 \cup \ldots \cup A_n)$$
$$\overset{e)}{\leqslant} \lim_{n\to\infty} \big(\mu(A_1) + \ldots + \mu(A_n)\big) = \sum_{n=1}^{\infty} \mu(A_n).$$ □

3.4 Remark. Theorem 3.3 remains valid for measures relative to an **algebra** $\mathscr{F}$. Note that f), h) also needs that $\bigcup_{n\in\mathbb{N}} A_n \in \mathscr{F}$ and g) that $\bigcap_{n\in\mathbb{N}} B_n \in \mathscr{F}$.

At this stage we have only a few (and rather trivial) examples of measures. One reason for this is the problem that we have to assign a value $\mu(A)$ to all measurable sets $A \in \mathscr{A}$, while it is, in general, not possible to obtain all measurable sets $A \in \mathscr{A}$ constructively – at least not for »interesting« σ-algebras like the Borel sets. As a rule of thumb one can say that on a »big« σ-algebra there can only be a relatively »simple« measure.

3.5 Example. a) (**Dirac measure, δ-function, point mass**). Let $(\Omega, \mathscr{A})$ be any measurable space and $x \in \Omega$ some fixed point. The set function

$$\delta_x : \mathscr{A} \to \{0,1\}, \quad \delta_x(A) := \begin{cases} 0, & x \notin A \\ 1, & x \in A \end{cases}$$

is a measure. If $\Omega = \mathbb{Z}$, $\mathscr{A} = \mathscr{P}(\mathbb{Z})$ and $x = i$, then $\delta_i(\{k\}) = \delta_{ik}$ is the well-known **Kronecker's delta**

b) Let $\Omega = \mathbb{R}$ and $\mathscr{A}$ as in Example 2.3.e), i.e. $A \in \mathscr{A} \iff A$ or A^c is countable. The set function

$$\gamma(A) := \begin{cases} 0, & A \text{ countable} \\ 1, & A^c \text{ countable} \end{cases}$$

is a measure.

c) Let $(\Omega, \mathscr{A})$ be any measure space. The set function

$$|A| := \begin{cases} \#A, & \text{if A is a finite set} \\ +\infty, & \text{if A is an infinite set} \end{cases}$$

is a measure.

d) (**sum of measures**) Let $(a_n)_{n\in\mathbb{N}}$ be a sequence of positive numbers and $(\mu_n)_{n\in\mathbb{N}}$ a sequence of measures. Then $\mu := \sum_{n\in\mathbb{N}} a_n \mu_n$ is a measure.

e) (**discrete probability measure**) Let $\Omega = \{\omega_1, \omega_2, \dots\}$ be a countable set, let $\mathscr{A} = \mathscr{P}(\Omega)$ and $(p_n)_{n\in\mathbb{N}} \subset [0,1]$ be a sequence such that $\sum_{n=1}^{\infty} p_n = 1$. Then

$$\mathbb{P}(A) := \sum_{n,\,\omega_n\in A} p_n = \sum_{n\in\mathbb{N}} p_n \delta_{\omega_n}(A), \quad A \subset \Omega$$

is a probability measure. $(\Omega, \mathscr{P}(\Omega), \mathbb{P})$ is called a **discrete probability space**.

Observe that $\mathbb{P}(\{\omega_n\}) = p_n$; thus,

$$\mathbb{P}(A) = \mathbb{P}\Big(\bigcup_{\omega\in A}\{\omega\}\Big) = \sum_{\omega\in A} \mathbb{P}(\{\omega\}).$$

This shows that all discrete probability measures are of this form.

f) (**trivial measures**) Let $(\Omega, \mathscr{A})$ be any measurable space. The set functions

$$\mu(A) := \begin{cases} 0, & A = \emptyset, \\ \infty, & A \neq \emptyset, \end{cases} \quad \text{and} \quad \nu(A) = 0, \qquad A \in \mathscr{A},$$

are measures.

3.6 Definition. d-dimensional **Lebesgue measure** is a measure λ^d on the space $(\mathbb{R}^d, \mathscr{B}(\mathbb{R}^d))$ which assigns every rectangle $\times_{n=1}^{d} [a_n, b_n) \in \mathscr{I}$, $a_n \leqslant b_n$, the value

$$\lambda^d\Big(\mathop{\times}_{n=1}^{d} [a_n, b_n)\Big) = \prod_{n=1}^{d} (b_n - a_n).$$

There are several problems with this definition:

- λ^d is not a measure – it is not even defined on all of $\mathscr{B}(\mathbb{R}^d)$.
- can we extend λ^d to a measure?
  – The good news is that $\sigma(\mathscr{I}) = \mathscr{B}(\mathbb{R}^d)$.
  – The bad news is that we don't know if λ^d is a measure relative to $\mathscr{I}$
- if so, is the extension unique?

Anticipating the proof, let us give the answer right now.

3.7 Theorem. *Lebesgue measure λ^d exists as a measure on $(\mathbb{R}^d, \mathscr{B}(\mathbb{R}^d))$ and it is uniquely characterized by its restriction to $\mathscr{I}$. For $B \in \mathscr{B}(\mathbb{R}^d)$ it holds that*

a) *λ^d is invariant under translations*[(1)]*: $\lambda^d(x + B) = \lambda^d(B)$.*
b) *λ^d is invariant under rigid motions*[(2)] *R: $\lambda^d(R^{-1}(B)) = \lambda^d(B)$.*
c) *$\lambda^d(M^{-1}(B)) = |\det M|^{-1}\lambda^d(B)$ for all invertible matrices $M \in \mathbb{R}^{d\times d}$.*

(1) $x + B := \{x + y \mid y \in B\}$
(2) A **rigid motion** $R : \mathbb{R}^d \to \mathbb{R}^d$ is any combination of translations, rotations and reflections.

⚠ a)–c) only make sense, if

$$B \in \mathscr{B}(\mathbb{R}^d) \implies x + B,\ R^{-1}(B),\ M^{-1}(B) \in \mathscr{B}(\mathbb{R}^d).$$

We close this chapter by showing that the continuity properties f), g) from Theorem 3.3 are equivalent to σ-additivity.

3.8 Lemma. *Let $(\Omega, \mathscr{A})$ be a measurable space and $\mu : \mathscr{A} \to [0, \infty]$ be an additive set function such that $\mu(\emptyset) = 0$ and $\mu(\Omega) < \infty$; μ is a measure if, and only if, one of the following continuity properties holds:*

a) *μ is continuous from below (as in Theorem 3.3.f))*
b) *μ is continuous from above (as in Theorem 3.3.g))*
c) *μ is continous at $\emptyset$ (as in Theorem 3.3.g) for $B = \emptyset$)*

Proof. In the proof of Theorem 3.3 we have already seen that a)⇒b)⇒c). Let us assume c) and show that (M_2) holds. Let $(A_n)_{n\in\mathbb{N}} \subset \mathscr{A}$ be pairwise disjoint sets and set $A = \dot{\bigcup}_{n\in\mathbb{N}} A_n$.

Obviously, $B_n := A \setminus (A_1 \,\dot\cup \ldots \dot\cup\, A_n) \downarrow \emptyset$, and so

$$\begin{aligned}
\mu(A) &\overset{\text{additive}}{=} \mu(A \setminus (A_1 \,\dot\cup \ldots \dot\cup\, A_n)) + \mu(A_1 \,\dot\cup \ldots \dot\cup\, A_n) \\
&\overset{\text{additive}}{=} \mu(B_n) + \sum_{n=1}^{n} \mu(A_n) \\
&\xrightarrow[n\to\infty]{\text{c)}} \sum_{n=1}^{\infty} \mu(A_n),
\end{aligned}$$

thus we have (M_2). □

⚠ In the last step of the proof of Lemma 3.8 we use that $\mathscr{A}$ is stable under the formation of finite unions and differences.

💬 Lemma 3.8 holds for measures relative to an algebra, see Remark 3.4.

4 Uniqueness of measures

Before we turn to the problem of extending λ^d and more general measures on $(\Omega, \mathscr{A})$, let us see if it is enough to define measures on some generator $\mathscr{G}$ of the σ-algebra $\mathscr{A}$ – recall that we defined Lebesgue measure explicitly only for rectancles $\mathscr{I}$ but not for Borel sets. A direct approach using the properties of μ seems impossible since, in general, there is no constructive extension of $\mathscr{G}$ to $\sigma(\mathscr{G})$. One way out is the following auxiliary construction:

4.1 Definition. A family of sets $\mathscr{D} \subset \mathscr{P}(\Omega)$ is a **Dynkin system**, if

$$\Omega \in \mathscr{D}, \tag{D_1}$$

$$D \in \mathscr{D} \implies D^c = \Omega \setminus D \in \mathscr{D}, \tag{D_2}$$

$$(D_n)_{n\in\mathbb{N}} \subset \mathscr{D} \text{ pairwise disjoint} \implies \dot{\bigcup_{n\in\mathbb{N}}} D_n \in \mathscr{D}. \tag{D_3}$$

4.2 Remark. Since (D_3) is weaker than (Σ_3), every σ-algebra is a Dynkin system, but not vice versa. Exactly as in §2.2.a) and b) one can show that

$$\emptyset \in \mathscr{D} \quad \text{and} \quad A, B \in \mathscr{D},\ A \cap B = \emptyset \implies A \mathbin{\dot\cup} B \in \mathscr{D}.$$

4.3 Theorem. a) *For every family $\mathscr{G} \subset \mathscr{P}(\Omega)$ there is a smallest (or minimal) Dynkin system such that $\mathscr{G} \subset \mathscr{D}$. $\mathscr{D}$ is the Dynkin system* ***generated by*** *$\mathscr{G}$. Notation: $\mathscr{D} = \delta(\mathscr{G})$.*
b) $\mathscr{G} \subset \delta(\mathscr{G}) \subset \sigma(\mathscr{G})$.

Proof. The proof of Part a) is essentially the same as the proof of Theorem 2.4.
In order to show b), we note that the σ-algebra $\sigma(\mathscr{G})$ is a Dynkin system such that $\mathscr{G} \subset \sigma(\mathscr{G})$. Since $\delta(\mathscr{G})$ is minimal for this property, we see that $\delta(\mathscr{G}) \subset \sigma(\mathscr{G})$. □

The next lemma shows the relation between Dynkin systems and σ-algebras.

4.4 Lemma. *A Dynkin system $\mathscr{D}$ is a σ-algebra if, and only if, $\mathscr{D}$ is ∩-stable, i.e.* $D, E \in \mathscr{D} \implies D \cap E \in \mathscr{D}$.

Proof. »⇒« If $\mathscr{D}$ is a σ-algebra, then (Σ_1)–(Σ_3) imply (D_1)–(D_3), and ∩-stability follows from §2.2.c).

»⇐« We have to show that $(D_n)_{n\in\mathbb{N}} \subset \mathscr{D} \implies \bigcup_{n\in\mathbb{N}} D_n \in \mathscr{D}$, i.e. ($\Sigma_3$).

$$\text{Set} \qquad E_{n+1} := (D_{n+1} \setminus D_n) \setminus D_{n-1} \setminus \cdots \setminus D_1, \qquad (E_1 := D_1)$$
$$= \underbrace{D_{n+1} \cap D_n^c \cap D_{n-1}^c \cap \cdots \cap D_1^c}_{\in \mathscr{D} \text{ b/o } \cap\text{-stability and } (D_2)}$$

The sets E_n are, by definition, mutually disjoint. Therefore, (D_3) implies that $D = \bigcup_{n=1}^{\infty} D_n = \dot{\bigcup}_{n=1}^{\infty} E_n \in \mathscr{D}$. □

Generators are usually smaller and easier to handle than the systems generated by them. Therefore, the following theorem is of great importance.

4.5 Theorem. *If $\mathscr{G} \subset \mathscr{P}(\Omega)$ is ∩-stable, then $\delta(\mathscr{G}) = \sigma(\mathscr{G})$.*

Proof. We split the proof in several steps. Claims are highlighted.

1° Obviously, $\delta(\mathscr{G}) \subset \sigma(\mathscr{G})$.

2° If we can show that $\delta(\mathscr{G})$ is a σ-algebra, then we get that $\sigma(\mathscr{G}) \subset \delta(\mathscr{G})$, since $\sigma(\mathscr{G})$ is the smallest σ-algebra such that $\mathscr{G} \subset \sigma(\mathscr{G})$. Consequently, $\delta(\mathscr{G}) = \sigma(\mathscr{G})$.

3° If $\delta(\mathscr{G})$ is $\cap$-stable, then Lemma 4.4 shows that $\delta(\mathscr{G})$ is a σ-algebra and we are done.

4° Fix $D \in \delta(\mathscr{G})$. $\mathscr{D}_D := \{Q \subset \Omega \mid Q \cap D \in \delta(\mathscr{G})\}$ is a Dynkin system.

(D$_1$) is clear.

(D$_2$) For every $Q \in \mathscr{D}_D$ we have

$$Q^c \cap D = (Q^c \cup D^c) \cap D = (Q \cap D)^c \cap D = \Big(\underbrace{(Q \cap D)}_{\in \delta(\mathscr{G})} \mathbin{\dot\cup} \underbrace{D^c}_{\in \delta(\mathscr{G})}\Big)^c \in \delta(\mathscr{G})$$

which shows that $Q^c \in \mathscr{D}_D$.

(D$_3$) If $(Q_n)_{n\in\mathbb{N}} \subset \mathscr{D}_D$ are mutually disjoint sets, then $(Q_n \cap D)_{n\in\mathbb{N}} \subset \delta(\mathscr{G})$ (this follows from the very definition of $\mathscr{D}_D$) and the sets are still disjoint. Thus, by (D$_3$),

$$\delta(\mathscr{G}) \ni (Q_n \cap D) = \Big(\dot\bigcup_{n\in\mathbb{N}} Q_n\Big) \cap D \implies \dot\bigcup_{n\in\mathbb{N}} Q_n \in \mathscr{D}_D.$$

5° $\delta(\mathscr{G}) \subset \mathscr{D}_D$ for any $D \in \delta(\mathscr{G})$. Because of the definition of $\mathscr{D}_D$, and since D is arbitrary, we get from this the $\cap$-stabiliy of $\delta(\mathscr{G})$

$$\forall D \in \delta(\mathscr{G}) \quad \forall Q \in \delta(\mathscr{G}) \subset \mathscr{D}_D : D \cap Q \in \delta(\mathscr{G}).$$

Fix $D \in \delta(\mathscr{G})$ and let us show that $\delta(\mathscr{G}) \subset \mathscr{D}_D$. By construction, $\mathscr{G} \subset \delta(\mathscr{G})$. Since $\mathscr{G}$ is $\cap$-stable, we have

$$\begin{aligned}
& \mathscr{G} \subset \mathscr{D}_G && \forall G \in \mathscr{G} && \\
\implies\; & \delta(\mathscr{G}) \subset \mathscr{D}_G && \forall G \in \mathscr{G} && (\mathscr{D}_G \text{ is a Dynkin system}) \\
\implies\; & G \cap D \in \delta(\mathscr{G}) && \forall G \in \mathscr{G},\ \forall D \in \delta(\mathscr{G}) && (\text{definition of } \mathscr{D}_G) \\
\implies\; & G \in \mathscr{D}_D && \forall G \in \mathscr{G},\ \forall D \in \delta(\mathscr{G}) && \\
\implies\; & \mathscr{G} \subset \mathscr{D}_D && \forall D \in \delta(\mathscr{G}) && \\
\implies\; & \delta(\mathscr{G}) \subset \mathscr{D}_D && \forall D \in \delta(\mathscr{G}) && (\mathscr{D}_D \text{ is a Dynkin system}).
\end{aligned}$$

Looking once again at the definition of $\mathscr{D}_D$, the last condition, $\delta(\mathscr{G}) \subset \mathscr{D}_D$ for all $D \in \delta(\mathscr{G})$, means that $\delta(\mathscr{G})$ is $\cap$-stable. □

Theorem 4.5 is often used in the literature in the following guise.

4.6 Corollary★ (monotone class theorem – MCT). *Assume that $\mathscr{M}$ is a family of sets such that*

a) $\Omega \in \mathscr{M}$;
b) $A, B \in \mathscr{M},\ A \subset B \implies B \setminus A \in \mathscr{M}$;

c) $(A_n)_{n\in\mathbb{N}} \subset \mathscr{M}$, $A_n \uparrow A \implies A \in \mathscr{M}$.

If $\mathscr{G} \subset \mathscr{M}$ is $\cap$-stable, then $\sigma(\mathscr{G}) \subset \mathscr{M}$.

Proof. Taking $B = X$ in the »stability under proper differences«, we see that $\mathscr{M}$ enjoys (D_2). The identity $A \cup B = (A^c \setminus B)^c$ (note that $B \subset A^c$) together with the »stability under increasing unions«, shows that $\mathscr{M}$ enjoys (D_3), i.e. $\mathscr{M}$ is a Dynkin system. Thus, $\mathscr{M} \supset \delta(\mathscr{G})$ and $\delta(\mathscr{G}) = \sigma(\mathscr{G})$ since $\mathscr{G}$ is $\cap$-stable. □

There are other versions, e.g. if $\mathscr{G}$ is an algebra, $\mathscr{G} \subset \mathscr{M}$ and $\mathscr{M}$ is stable under increasing and decreasing limits of sets, then $\sigma(\mathscr{G}) \subset \mathscr{M}$, see e.g. [8, p. 18] and [MIMS, Problem 3.14].

Theorem 4.5 and the technique used in the proof are important in probability theory.

4.7 Theorem (uniqueness of measures). *Let μ and ν be measures on the measurable space $(\Omega, \mathscr{A})$ and assume that $\mathscr{A} = \sigma(\mathscr{G})$ where*

a) *$\mathscr{G}$ is $\cap$-stable,*
b) *$\mathscr{G}$ contains an exhausting sequence, i.e. $\exists\, (G_n)_{n\in\mathbb{N}} \subset \mathscr{G}$, $G_n \uparrow \Omega$, such that $\mu(G_n), \nu(G_n) < \infty$ for all $n \in \mathbb{N}$.*

If $\mu(G) = \nu(G)$ for all $G \in \mathscr{G}$, then $\mu(A) = \nu(A)$ for all $A \in \mathscr{A}$.(1)

4.8 Remark. If μ and ν are probability measures, we do not need the conditon 4.7.b). This follows from the observation that $\mu(\Omega) = \nu(\Omega) = 1$. Now use $\mathscr{G} \cup \{\Omega\}$ instead of $\mathscr{G}$ and pick $G_n = \Omega$.

Proof of Theorem 4.7. Claims are higlighted. Define

$$\mathscr{D}_n := \{A \in \mathscr{A} \mid \mu(G_n \cap A) = \nu(G_n \cap A)\}, \quad n \in \mathbb{N}.$$

1° $\mathscr{D}_n$ is a Dynkin system.

(D_1) is obvious.

(D_2) If $A \in \mathscr{D}_n$, then $\mu(G_n \cap A) = \nu(G_n \cap A) < \infty$, and

$$\begin{aligned}\mu(G_n \cap A^c) = \mu(G_n \setminus A) &= \mu(G_n) - \mu(G_n \cap A)\\ &= \nu(G_n) - \nu(G_n \cap A)\\ &= \nu(G_n \setminus A) = \nu(G_n \cap A^c)\end{aligned}$$

shows that $A^c \in \mathscr{D}_n$.

(1) This is often expressed as $\mu|_{\mathscr{G}} = \nu|_{\mathscr{G}} \implies \mu = \nu$.

(D₃) Let $(A_k)_{k\in\mathbb{N}} \subset \mathscr{D}_n$ be mutually disjoint. This property is inherited by the sequence $(A_k \cap G_n)_{k\in\mathbb{N}}$, and so

$$\mu\Big(\dot{\bigcup_{k\in\mathbb{N}}} A_k \cap G_n\Big) = \mu\Big(\dot{\bigcup_{k\in\mathbb{N}}} (A_k \cap G_n)\Big) \overset{(M_2)}{=} \sum_{k\in\mathbb{N}} \mu(A_k \cap G_n)$$
$$\overset{A_k\in\mathscr{D}_n}{=} \sum_{k\in\mathbb{N}} \nu(A_k \cap G_n) = \ldots = \nu\Big(\dot{\bigcup_{k\in\mathbb{N}}} A_k \cap G_n\Big)$$

which shows that $\dot{\bigcup}_{k\in\mathbb{N}} A_k \in \mathscr{D}_n$.

2° From $\mu|_{\mathscr{G}} = \nu|_{\mathscr{G}}$ we conclude that $\mathscr{G} \subset \mathscr{D}_n$. Since $\mathscr{G}$ is $\cap$-stable and $\mathscr{D}_n$ is a Dynkin system, we get $\sigma(\mathscr{G}) = \delta(\mathscr{G}) \subset \mathscr{D}_n$ for all $n \in \mathbb{N}$.

On the other hand, $\mathscr{A} = \sigma(\mathscr{G}) \subset \mathscr{D}_n \subset \mathscr{A}$, so $\mathscr{D}_n = \mathscr{A}$. In other words,

$$\mu(G_n \cap A) = \nu(G_n \cap A) \quad \forall n \in \mathbb{N},\ \forall A \in \mathscr{A}$$

We can now let $n \to \infty$ on both sides. Because of the continuity of measures, we find that $\mu(A) = \sup_{n\in\mathbb{N}} \mu(G_n \cap A) = \sup_{n\in\mathbb{N}} \nu(G_n \cap A) = \nu(A)$ for any $A \in \mathscr{A}$. □

Further applications of Dynkin systems

The next results exhibit a few very special properties of Lebesgue measure λ^d. We assume, tacitly, its existence. Recall that $x + B := \{x + y \mid y \in B\}$ denotes the set $B \in \mathscr{B}(\mathbb{R}^d)$ shifted by $x \in \mathbb{R}^d$.

4.9 Theorem. a) *λ^d is invariant w.r.t. translations, i.e. $\lambda^d(x + B) = \lambda^d(B)$ holds for all $x \in \mathbb{R}^d$ and $B \in \mathscr{B}(\mathbb{R}^d)$.*
b) *If μ is any translation invariant measure on $(\mathbb{R}^d, \mathscr{B}(\mathbb{R}^d))$ such that the unit cube has finite measure $\kappa = \mu([0,1)^d) < \infty$, then $\mu = \kappa \cdot \lambda^d$.*

Proof. We verify first that $x + B \in \mathscr{B}(\mathbb{R}^d)$ for all $B \in \mathscr{B}(\mathbb{R}^d)$ – otherwise a) would not make sense.

Fix $x \in \mathbb{R}^d$ and define $\mathscr{A}_x := \{B \in \mathscr{B}(\mathbb{R}^d) \mid x + B \in \mathscr{B}(\mathbb{R}^d)\}$. It is not hard to check ✎ that $\mathscr{A}_x$ is a σ-algebra. Since a shifted rectangle is again a rectangle,

$$I = \mathop{\times}_{n=1}^{d} [a_n, b_n) \implies x + I = \mathop{\times}_{n=1}^{d} [x_n + a_n, x_n + b_n),$$

we see that $\mathscr{I} \subset \mathscr{A}_x$. Therefore,

$$\mathscr{B}(\mathbb{R}^d) = \sigma(\mathscr{I}) \subset \sigma(\mathscr{A}_x) = \mathscr{A}_x \subset \mathscr{B}(\mathbb{R}^d)$$

which shows that $x + B$ is a Borel set.

Proof of a). Define $\nu(B) := \lambda^d(x + B)$ for $x \in \mathbb{R}^d$ as above and $B \in \mathscr{B}(\mathbb{R}^d)$. The set function ν is a measure on $\mathscr{B}(\mathbb{R}^d)$ ✎ whose value at $I = \times_{n=1}^{d} [a_n, b_n)$ is given by

$$\nu(I) = \lambda^d(x + I) = \prod_{n=1}^{d} (b_n + x_n - a_n - x_n) = \prod_{n=1}^{d} (b_n - a_n) = \lambda^d(I).$$

So, $\nu|_{\mathcal{J}} = \lambda^d|_{\mathcal{J}}$ and $\mathcal{J}$ is a $\cap$-stable generator of $\mathcal{B}(\mathbb{R}^d)$:

$$\bigtimes_{n=1}^{d}[a_n, b_n) \cap \bigtimes_{n=1}^{d}[\alpha_n, \beta_n) = \bigtimes_{n=1}^{d}\Big[\max\{a_n, \alpha_n\}, \min\{b_n, \beta_n\}\Big),$$

see Fig. 4.1. Since $[-k,k)^d \in \mathcal{J}$, $[-k,k)^d \uparrow \mathbb{R}^d$, $\lambda^d([-k,k)^d) = (2k)^d < \infty$ is an exhausting sequence, we can apply Thoerem 4.7 to get $\nu = \lambda^d$.

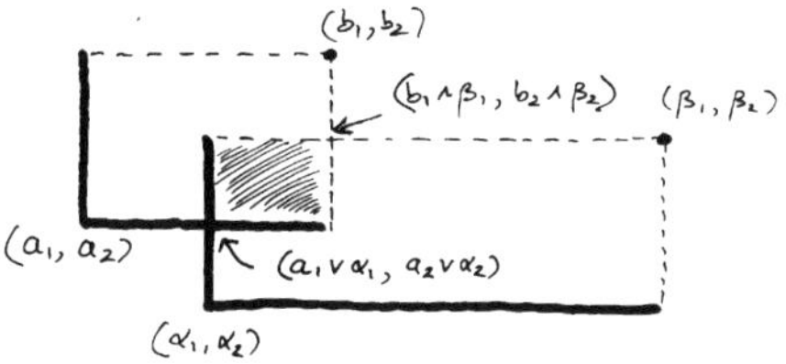

Fig. 4.1. The intersection of rectangles is again a rectangle.

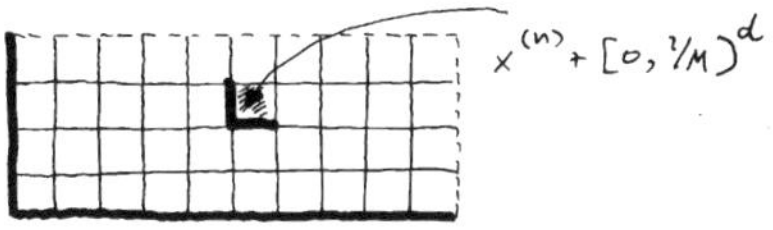

Fig. 4.2. Paving a rectangle with rational vertices with equal tiles.

Proof of b). For any $\mathrm{I} \in \mathcal{J}_{\mathrm{rat}}$ there are $\mathrm{M}, k(\mathrm{I}) \in \mathbb{N}$ and $x^{(n)} \in \mathbb{R}^d$ such that

$$\mathrm{I} = \biguplus_{n=1}^{k(\mathrm{I})} \Big(x^{(n)} + \big[0, \tfrac{1}{\mathrm{M}}\big)^d\Big),$$

see Fig. 4.2, i.e. we »pave« I with equal tiles $\big[0, \frac{1}{\mathrm{M}}\big)^d$ of side length $1/\mathrm{M}$ and lower left corner $x^{(n)}$. We may take M as the smallest common multiple of the denominators of the rational numbers a_n, b_n.

Since λ^d and μ are translation invariant, we have

$$\mu(\mathrm{I}) = k(\mathrm{I}) \cdot \mu\Big(\big[0, \tfrac{1}{\mathrm{M}}\big)^d\Big), \qquad \kappa = \mu\big([0,1)^d\big) = \mathrm{M}^d \mu\Big(\big[0, \tfrac{1}{\mathrm{M}}\big)^d\Big)$$

$$\lambda^d(\mathrm{I}) = k(\mathrm{I}) \cdot \lambda^d\Big(\big[0, \tfrac{1}{\mathrm{M}}\big)^d\Big), \qquad 1 = \lambda^d\big([0,1)^d\big) = \mathrm{M}^d \lambda^d\Big(\big[0, \tfrac{1}{\mathrm{M}}\big)^d\Big).$$

From the first and second line we see

$$\mu(\mathrm{I}) = \frac{k(\mathrm{I})}{\mathrm{M}^d} \cdot \kappa \quad \text{and} \quad \lambda^d(\mathrm{I}) = \frac{k(\mathrm{I})}{\mathrm{M}^d}$$

which finally gives $\mu(\mathrm{I}) = \kappa\lambda^d(\mathrm{I})$ for all $\mathrm{I} \in \mathcal{J}_{\mathrm{rat}}$, hence $\mu = \lambda^d$ on $\mathcal{B}(\mathbb{R}^d) = \sigma(\mathcal{J}_{\mathrm{rat}})$ because of the uniqueness theorem (Theorem 4.7). □

5 Extension and construction of measures

Let $\mathcal{G} \subset \mathcal{P}(\Omega)$ be some family of sets which is not a σ-algebra. We are interested how we can extend a set function μ defined on $\mathcal{G}$ to a measure on $\sigma(\mathcal{G})$. Obviously, the minimum requirement is that μ is a measure relative to $\mathcal{G}$. This is the typical situation for $\mu = \lambda^d$ which was explicitly defined on $\mathcal{G} = \mathcal{J}$ (Definition 3.6). The questions we have to address are

- how can we extend μ?
- is the extension unique?

Both questions are linked with properties of $\mathscr{G}$, size and structure. The following definition is inspired by the properties of rectangles.

5.1 Definition. A family of sets $\mathscr{S} \subset \mathscr{P}(\Omega)$ is a **semi-ring** in Ω, if

$$\emptyset \in \mathscr{S} \tag{S_1}$$

$$S, T \in \mathscr{S} \implies S \cap T \in \mathscr{S} \tag{S_2}$$

$$\forall S, T \in \mathscr{S} \quad \exists S_1, \ldots, S_m \in \mathscr{S}$$
$$m < \infty, \text{ disjoint, } S \setminus T = \dot{\bigcup_{n=1}^{m}} S_n \tag{S_3}$$

We can read off Fig. 4.1 and Fig. 5.1 that $\mathscr{I}$ is indeed a semi-ring

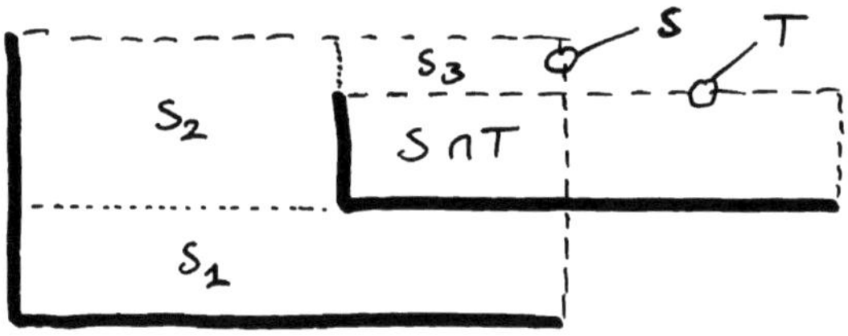

Fig. 5.1. The difference of two rectangles is the disjoint union of finitely many rectangles.

The **main theorem** of basic measure theory is Carathéodory's extension theorem.

5.2 Theorem (Carathéodory; extension of measures). *Let $\mathscr{S}$ be a semi-ring in Ω. Every measure μ relative to $\mathscr{S}$ has an extension to a measure on $\sigma(\mathscr{S})$.*

If $\mathscr{S}$ contains an exhausting sequence $(S_n)_{n \in \mathbb{N}}$, $S_n \uparrow \Omega$, $\mu(S_n) < \infty$, then the extension is unique.

5.3 Remark. If we want to construct a measure μ, Theorem 5.2 says that it is enough to assign values to μ on a semiring $\mathscr{S}$ in such a way that μ is σ-additive relative to $\mathscr{S}$. This shows the strength of Theorem 5.2: σ-finiteness on $\sigma(\mathscr{S})$ is automatically inherited by the extension.

The proof of Theorem 5.2 is quite technical and should be omitted at first reading. More important is to understand the idea behind the proof. Therefore, we split the proof into two parts, first explaining strategy and ingredients – this you should read right away – and then continuing with the gory details – keep this for a second reading.

Outline of the proof of Theorem 5.2. The **main problem** is the question how we can extend the set function μ. The **key concepts** are the notions of **outer measure** and **measurability** with respect to an outer measure.

I. We begin with an extension of μ **for every set** $A \subset \Omega$:

$$\mu^* : \mathscr{P}(\Omega) \to [0,\infty], \quad \mu^*(A) := \inf\left\{ \sum_{i=1}^{\infty} \mu(S_i) \mid (S_i)_{i\in\mathbb{N}} \in \mathcal{C}(A) \right\}. \tag{5.1}$$

By $\mathcal{C}(A)$ we denote all $\mathscr{S}$-covers of the set A, i.e.

$$\mathcal{C}(A) := \Big\{ (S_i)_{i\in\mathbb{N}} \subset \mathscr{S} \mid \underbrace{\textstyle\bigcup_{i\in\mathbb{N}} S_i}_{\substack{\text{not necessarily disjoint,}\\ \bigcup_i S_i \text{ need not be in } \mathscr{S}}} \supset A \Big\}$$

If there is no $\mathscr{S}$-cover, $\mathcal{C}(A) = \emptyset$, and we set $\inf \emptyset = \infty$.

II. μ^* is an **outer measure**, i.e. a set function $\mu^* : \mathscr{P}(\Omega) \to [0,\infty]$ such that

$$\mu^*(\emptyset) = 0 \tag{OM_1}$$

$$A \subset B \implies \mu^*(A) \leqslant \mu^*(B) \tag{OM_2}$$

$$\mu^*\Big(\bigcup_{i\in\mathbb{N}} A_i\Big) \leqslant \sum_{i\in\mathbb{N}} \mu^*(A_i) \tag{OM_3}$$

III. μ^* extends μ, i.e. $\mu^*|\mathscr{S} = \mu$

IV. μ^*-measurable sets

$$\mathscr{A}^* := \Big\{ A \subset \Omega \mid \forall Q \subset \Omega : \mu^*(Q) = \mu^*(Q \cap A) + \mu^*(Q \setminus A) \Big\}. \tag{5.2}$$

The family $\mathscr{A}^*$ is constructed in such a way that μ^* is additive on $\mathscr{A}^*$. With some extra work, this can be extended to σ-additivity. We have

- $\mathscr{A}^*$ is a σ-algebra
- $\mathscr{S} \subset \mathscr{A}^* \implies \sigma(\mathscr{S}) \subset \mathscr{A}^*$
- $\mu^*|_{\mathscr{A}^*}$ and $\mu^*|_{\sigma(\mathscr{S})}$ are measures which extend μ.

V. Uniqueness follows from Theorem 4.7. □

Proof of Theorem 5.2. Claims are highlighted. Let μ^* be the set function defined in (5.1).

1° μ^* is an **outer measure.**

(OM_1) Cover $\emptyset$ by $(\emptyset)_{i\in\mathbb{N}} \in \mathcal{C}(\emptyset)$.

(OM_2) If $B \supset A$, then every $\mathscr{S}$-covering of B is also a covering of A: $\mathcal{C}(B) \subset \mathcal{C}(A)$. Thus,

$$\mu^*(A) = \inf\left\{ \sum_{i\in\mathbb{N}} \mu(S_i) \mid (S_i)_{i\in\mathbb{N}} \in \mathcal{C}(A) \right\} \leqslant \inf\left\{ \sum_{k\in\mathbb{N}} \mu(T_k) \mid (T_k)_{k\in\mathbb{N}} \in \mathcal{C}(B) \right\} = \mu^*(B).$$

(OM_3) Without loss of generality we may assume that $\mu^*(A_i) < \infty$ for all $i \in \mathbb{N}$, so that $\mathcal{C}(A_i) \neq \emptyset$. Fix $\epsilon > 0$; because of the definition of the infimum, for every A_i there is a covering $(S_k^i)_{k\in\mathbb{N}} \in \mathcal{C}(A_i)$ such that

$$\sum_{k\in\mathbb{N}} \mu(S_k^i) \leqslant \mu^*(A_i) + \frac{\epsilon}{2^i}, \quad i \in \mathbb{N}. \tag{5.3}$$

The doubly-indexed sequence $(S_k^i)_{i,k\in\mathbb{N}}$ is an $\mathcal{S}$-covering of $A := \bigcup_{i\in\mathbb{N}} A_i$, and

$$\mu^*(A) \leqslant \sum\sum_{(i,k)\in\mathbb{N}\times\mathbb{N}} \mu(S_k^i) = \sum_{i\in\mathbb{N}}\sum_{k\in\mathbb{N}} \mu(S_k^i) \overset{(5.3)}{\leqslant} \sum_{i\in\mathbb{N}}\left(\mu^*(A_i) + \frac{\epsilon}{2^i}\right) = \sum_{i\in\mathbb{N}} \mu^*(A_i) + \epsilon.$$

Letting $\epsilon \to 0$ proves (OM_3).

2° Extension of μ onto $\mathcal{S}_\cup := \{S_1 \cup \ldots \cup S_M \mid M \in \mathbb{N},\ S_i \in \mathcal{S}\}$. Each additive set function on $\mathcal{S}_\cup$ which extends μ, $\bar\mu|\mathcal{S} = \mu$, **must** satisfy

$$\bar\mu(S_1 \cup \ldots \cup S_M) = \sum_{i=1}^{M} \mu(S_i). \tag{5.4}$$

If we can show that (5.4) does not depend on the representation of the set in $\mathcal{S}_\cup$, then we may use (5.4) as the – necessarily unique – extension of μ. Assume that

$$S_1 \cup \ldots \cup S_M = T_1 \cup \ldots \cup T_N, \quad M, N \in \mathbb{N},\ S_i, T_k \in \mathcal{S}.$$

Since

$$S_i = S_i \cap (T_1 \cup \ldots \cup T_N) = \dot{\bigcup_{k=1}^{N}} (S_i \cap T_k),$$

we can use the additivity of μ on $\mathcal{S}$ and the $\cap$-stability of $\mathcal{S}$ to see

$$\mu(S_i) = \sum_{k=1}^{N} \mu(S_i \cap T_k).$$

If we sum over $i = 1, 2, \ldots, M$, and interchange the order of the sums in i and k, we arrive at

$$\sum_{i=1}^{M} \mu(S_i) = \sum_{i=1}^{M}\sum_{k=1}^{N} \mu(S_i \cap T_k) = \sum_{k=1}^{N} \mu(T_k).$$

This shows that (5.4) does not depend on the particular representation. By definition, $\mathcal{S}_\cup$ is stable under the formation of finite unions of **disjoint** sets. If $S, T \in \mathcal{S}_\cup$ are arbitrary, then we have for suitable sets $S_i, T_k \in \mathcal{S}$

$$S \cap T = (S_1 \cup \ldots \cup S_M) \cap (T_1 \cup \ldots \cup T_N) = \dot{\bigcup_{i=1}^{M}} \dot{\bigcup_{k=1}^{N}} \underbrace{(S_i \cap T_k)}_{\in \mathcal{S}} \in \mathcal{S}_\cup.$$

Because of (S_3), $S_i \setminus T_k \in \mathcal{S} \subset \mathcal{S}_\cup$, and so

$$S \setminus T = (S_1 \cup \ldots \cup S_M) \setminus (T_1 \cup \ldots \cup T_N) = \dot{\bigcup_{i=1}^{M}} \bigcap_{k=1}^{N} \left(S_i \cap T_k^c\right) = \overbrace{\dot{\bigcup_{i=1}^{M}} \bigcap_{k=1}^{N} \underbrace{S_i \setminus T_k}_{\in \mathcal{S}_\cup}}^{\in \mathcal{S}_\cup} \in \mathcal{S}_\cup.$$

Here we use that $\mathcal{S}_\cup$ is stable under finite intersections and disjoint unions. Therefore,[1]

$$S \cup T = \left(S \setminus T\right) \cup \left(S \cap T\right) \cup \left(T \setminus S\right) \in \mathcal{S}_\cup,$$

and we can use (5.4) to extend μ onto finite unions of sets from $\mathcal{S}$.

[1] Incidentally, this shows that $\mathcal{S}_\cup$ is the Boolean ring generated by $\mathcal{S}$, i.e. the smallest ring containing $\mathcal{S}$.

3° $\bar\mu$ is a measure relative to $\mathscr{S}_\cup$. We only have to show σ-additivity relative to $\mathscr{S}_\cup$. Let $(T_k)_{k\in\mathbb{N}} \subset \mathscr{S}_\cup$ be a sequence of disjoint sets such that $T := \dot\bigcup_{k\in\mathbb{N}} T_k \in \mathscr{S}_\cup$.

By definition, there is a sequence of mutually disjoint sets $(S_i)_{i\in\mathbb{N}} \subset \mathscr{S}$ and integers $0 = n(0) < n(1) < n(2) < \dots$ such that for all $k \in \mathbb{N}$

$$T_k = S_{n(k-1)+1} \,\dot\cup \dots \dot\cup\, S_{n(k)} \quad \text{and} \quad T = U_1 \,\dot\cup \cdots \dot\cup\, U_N$$

where $U_\ell = \dot\bigcup_{i\in J(\ell)} S_i \in \mathscr{S}$ with mutually disjoint sets $J(1) \dot\cup J(2) \dot\cup \cdots \dot\cup J(N) = \mathbb{N}$. Thus, using the σ-additivity of μ relative to $\mathscr{S}$,

$$\bar\mu(T) \stackrel{\text{def}}{=} \sum_{\ell=1}^{N} \mu(U_\ell) \underset{\text{additive}}{\overset{\mu\text{ is }\sigma\text{-}}{=}} \sum_{\ell=1}^{N}\sum_{i\in J_\ell} \mu(S_i) = \sum_{k\in\mathbb{N}} \sum_{i=n(k-1)+1}^{n(k)} \mu(S_i) \stackrel{\text{def}}{=} \sum_{k\in\mathbb{N}} \bar\mu(T_k).$$

4° $\mu^*|_{\mathscr{S}} = \mu$. The measure $\bar\mu$ is σ-subadditive on $\mathscr{S}_\cup$. For any $S \in \mathscr{S}$ and any covering $(S_i)_{i\in\mathbb{N}} \in \mathcal{C}(S)$ we have

$$\mu(S) = \bar\mu(S) = \bar\mu\Big(\bigcup_{i\in\mathbb{N}} S_i \cap S\Big) \leqslant \sum_{i\in\mathbb{N}} \bar\mu(S_i \cap S) = \sum_{i\in\mathbb{N}} \mu(S_i \cap S) \leqslant \sum_{i\in\mathbb{N}} \mu(S_i).$$

Taking the infimum of all coverings $\mathcal{C}(S)$ yields $\mu(S) \leqslant \mu^*(S)$. The particular covering $(S,\emptyset,\emptyset,\dots) \in \mathcal{C}(S)$ shows $\mu^*(S) \leqslant \mu(S)$, and so $\mu(S) = \mu^*(S)$ for all $S \in \mathscr{S}$.

5° $\mathscr{A}^*$ is a σ-algebra and μ^* is a measure on $\mathscr{A}^*$.

(Σ_1) obviously, $\emptyset \in \mathscr{A}^*$.
(Σ_2) since the definition of $\mathscr{A}^*$ is symmetric in A and A^c, $A \in \mathscr{A}^* \iff A^c \in \mathscr{A}^*$.
(Σ_3) first we claim that $A, A' \in \mathscr{A}^* \implies A \cup A' \in \mathscr{A}^*$. For any $P \subset \Omega$ and $A, A' \in \mathscr{A}^*$,

$$\begin{aligned}
&\mu^*(P\cap(A\cup A')) + \mu^*(P\setminus(A\cup A')) \\
&\qquad = \mu^*(P\cap(A\cup[A'\setminus A])) + \mu^*(P\setminus(A\cup A')) \\
&\qquad \stackrel{(OM_3)}{\leqslant} \mu^*(P\cap A) + \mu^*(P\cap(A'\setminus A)) + \mu^*(P\setminus(A\cup A')) \\
&\qquad = \mu^*(P\cap A) + \mu^*((P\setminus A)\cap A') + \mu^*((P\setminus A)\setminus A') \\
&\qquad \stackrel{(5.2)}{=} \mu^*(P\cap A) + \mu^*(P\setminus A) \stackrel{(5.2)}{=} \mu^*(P).
\end{aligned} \tag{5.5}$$

In the last two steps we use the definition of $\mathscr{A}^*$ with $Q \rightsquigarrow P\setminus A$ and $A' \in \mathscr{A}$ and $Q \rightsquigarrow P$ and $A \in \mathscr{A}^*$. The converse inequality »⩾« follows from (OM_3). This shows that $A \cup A' \in \mathscr{A}^*$.

If $A \cap A' = \emptyset$, then we can use the equality (5.5) with $P := (A \,\dot\cup\, A') \cap Q$, $Q \subset \Omega$, to get

$$\mu^*(Q\cap(A\,\dot\cup\,A')) = \mu^*(Q\cap A) + \mu^*(Q\cap A') \quad \forall Q \subset \Omega.$$

Iterating this step yields, for mutually disjoint sets $A_1, A_2, \dots, A_M \in \mathscr{A}^*$,

$$\mu^*(Q\cap(A_1 \,\dot\cup \dots \dot\cup\, A_M)) = \sum_{i=1}^{M} \mu^*(Q\cap A_i) \quad \forall Q \subset \Omega. \tag{5.6}$$

$\mathscr{A}^*$ is a Dynkin system. We can now show (D_3) for $\mathscr{A}^*$. Let $A = \dot\bigcup_{i\in\mathbb{N}} A_i$ for some

mutually disjoint sets $(A_i)_{i\in\mathbb{N}} \subset \mathscr{A}^*$. Since $A_1 \cup \ldots \cup A_M \in \mathscr{A}^*$, we get with (OM_2) and (5.6)

$$\begin{aligned}\mu^*(Q) &= \mu^*(Q \cap (A_1 \cup \ldots \cup A_M)) + \mu^*(Q \setminus (A_1 \cup \ldots \cup A_M)) \\ &\geqslant \mu^*(Q \cap (A_1 \cup \ldots \cup A_M)) + \mu^*(Q \setminus A) \\ &= \sum_{i=1}^{M} \mu^*(Q \cap A_i) + \mu^*(Q \setminus A).\end{aligned}$$

The left-hand side does not depend on M. Thus, the limit $M \to \infty$ yields

$$\mu^*(Q) \geqslant \sum_{i=1}^{\infty} \mu^*(Q \cap A_i) + \mu^*(Q \setminus A) \overset{(OM_3)}{\geqslant} \mu^*(Q \cap A) + \mu^*(Q \setminus A). \tag{5.7}$$

The converse inequality $\mu^*(Q) \leqslant \mu^*(Q \cap A) + \mu^*(Q \setminus A)$ follows at once with the subadditivity of μ^*. This proves equality everywhere in (5.7), hence $A \in \mathscr{A}^*$.

μ^* **is a measure relative to** $\mathscr{A}^*$. Take $Q := A$ in (5.7) to get the σ-additivity of μ^* on $\mathscr{A}^*$.

$\mathscr{A}^*$ **is a σ-algebra.** We have already seen that $\mathscr{A}^*$ is a (∪-stable) Dynkin system. Since $A \cap B = (A^c \cup B^c)^c$, $\mathscr{A}^*$ is ∩-stable, hence a σ-algebra by Lemma 4.4. This finishes the proof of (Σ_3).

6° $\mathscr{S} \subset \mathscr{A}^*$. Let $S, T \in \mathscr{S}$. In view of (S_3), we have

$$T = (S \cap T) \mathbin{\dot\cup} (T \setminus S) = (S \cap T) \mathbin{\dot\cup} \dot{\bigcup_{i=1}^{M}} S_i, \quad S_i \in \mathscr{S},\ i = 1, 2, \ldots, M.$$

Now we use that μ^* is (σ-)subadditive, extends μ, and the additivity of μ, to get

$$\mu^*(S \cap T) + \mu^*(T \setminus S) \leqslant \mu(S \cap T) + \sum_{i=1}^{M} \mu(S_i) = \mu(T). \tag{5.8}$$

Let $B \subset \Omega$ and $(T_i)_{i\in\mathbb{N}} \in \mathcal{C}(B)$. Using $T = T_i$ in (5.8) and summing over $i \in \mathbb{N}$ gives

$$\sum_{i\in\mathbb{N}} \mu^*(T_i \setminus S) + \sum_{i\in\mathbb{N}} \mu^*(T_i \cap S) \leqslant \sum_{i\in\mathbb{N}} \mu(T_i).$$

Since $B \subset \bigcup_{i\in\mathbb{N}} T_i$, we see

$$\begin{aligned}\mu^*(B \setminus S) + \mu^*(B \cap S) &\overset{(OM_2)}{\leqslant} \mu^*\Big(\bigcup_{i\in\mathbb{N}} T_i \setminus S\Big) + \mu^*\Big(\bigcup_{i\in\mathbb{N}} T_i \cap S\Big) \\ &\overset{(OM_3)}{\leqslant} \sum_{i\in\mathbb{N}} \mu^*(T_i \setminus S) + \sum_{i\in\mathbb{N}} \mu^*(T_i \cap S) \\ &\leqslant \sum_{i\in\mathbb{N}} \mu(T_i).\end{aligned}$$

The infimum over all coverings from $\mathcal{C}(B)$ gives

$$\mu^*(B \setminus S) + \mu^*(B \cap S) \leqslant \mu^*(B) \quad \forall B \subset \Omega,\ S \in \mathscr{S}.$$

The reverse inequality »⩾« is always true because of the (σ-)subadditivity (OM_3) von μ^*. Thus, $S \in \mathscr{A}^*$ and $\mathscr{S} \subset \mathscr{A}^*$ follows.

7° μ^* **is a measure on $\sigma(\mathscr{S})$ which extends μ.** Step 6° shows that $\mathscr{S} \subset \mathscr{A}^*$ and since $\mathscr{A}^*$ is a σ-algebra (see 5°), we have $\sigma(\mathscr{S}) \subset \mathscr{A}^*$. Again by 5° , $\mu^*|_{\sigma(\mathscr{S})}$ is a measure which, see Step 4° , extends μ.

8° **$\mu^*|_{\sigma(\mathscr{S})}$ is unique.** This follows from Theorem 4.7. □

▲ In general, $\mathscr{A} = \sigma(\mathscr{S}) \subsetneqq \mathscr{A}^*$. The standard example is Lebesgue measure $\mu = \lambda^d$ where $\mathscr{S} = \mathscr{I}$ and $\sigma(\mathscr{I}) = \mathscr{A} = \mathscr{B}(\mathbb{R}^d)$; however, $\mathscr{A}^*$ are the **Lebesgue sets**, i.e.

$$\mathscr{B}^*(\mathbb{R}^d) = \sigma\left(\mathscr{B}(\mathbb{R}^d), \widetilde{\mathscr{N}}\right)$$

where $\widetilde{\mathscr{N}}$ is the family of all $\widetilde{N}$ such that $\widetilde{N} \subset N$ for some measurable null set $\lambda^d(N) = 0$, see Theorem 5.7 below.

▲ In general, $\mathscr{A}^* \subsetneqq \mathscr{P}(\Omega)$. The standard example is, again, Lebesgue measure. If $d \geqslant 3$, it is not possible to define a translation invariant measure μ on $\mathscr{P}(\Omega)$ such that $\mu|_{\mathscr{I}} = \lambda^d|_{\mathscr{I}}$, see the discussion in [CEX, §7.30] or [31]. The reason is the **Banach–Tarski paradox** which asserts that we can decompose $B_1(0) \subset \mathbb{R}^d$ ($d \geqslant 3$) into finitely many (non-measurable) disjoint pieces and re-assemble them to become $B_2(0)$.

💬 »Simple« measures like Dirac's δ_x can be defined for all sets $A \in \mathscr{P}(\mathbb{R}^d)$.

Construction of Lebesgue measure

Let us now continue our programme started in Theorem 3.7 and construct Lebesgue measure on $\mathbb{R}$.

5.4 Proposition. *λ^1 is a measure relative to $\mathscr{I} = \{[a,b) \subset \mathbb{R} \mid a \leqslant b\}$.*

Proof. We write $\lambda := \lambda^1$. If $[a,a'), [b',b) \in \mathscr{I}$ such that $a < b' \leqslant a' < b$, then

$$\lambda([a,a') \cup [b',b)) = \lambda[a,b) = b - a \leqslant (b - b') + (a' - a) = \lambda[b',b) + \lambda[a,a').$$

This shows that λ is subadditive on $\mathscr{I}$. Taking $a' = b'$ in this calculation gives »=« throughout, i.e. λ is additive.

Let $I_i = [a_i, b_i)$ be mutually disjoint such that $\dot{\bigcup}_{i\in\mathbb{N}} I_i = [a,b) \in \mathscr{I}$. Define $I_i^\epsilon := [a_i - 2^{-i}\epsilon, b_i)$ for some $0 < \epsilon < b - a$. The open intervals $\mathring{I}_i^\epsilon$ cover the compact set $[a, b-\epsilon]$; by compactness, there is a finite subcover

$$[a, b-\epsilon) \subset [a, b-\epsilon] \subset \bigcup_{i=1}^{N} \mathring{I}_i^\epsilon \subset \bigcup_{i=1}^{N} I_i^\epsilon$$

for a suitable $N \in \mathbb{N}$. From $\lambda(I_i) = \lambda(I_i^\epsilon) - \epsilon 2^{-i}$ we get

$$\lambda[a,b) - \sum_{i=1}^{N} \lambda(I_i) = \epsilon + \underbrace{\lambda[a, b-\epsilon) - \sum_{i=1}^{N} \lambda(I_i^\epsilon)}_{\leqslant 0 \text{ b/o subadditivity}} + \sum_{i=1}^{N} \frac{\epsilon}{2^i} \leqslant 2\epsilon$$

and conclude that

$$\lambda[a,b) \leqslant \sum_{i=1}^{\infty} \lambda(\mathrm{I}_i) + 2\epsilon.$$

On the other hand, we may enlarge $\mathrm{I}_i \subset \mathrm{I}_i' \in \mathscr{I}$ to exhaust $[a,b)$, i.e.

$$\forall \mathrm{N} \in \mathbb{N}: \quad \dot{\bigcup_{i=1}^{\mathrm{N}}} \mathrm{I}_i \subset [a,b) \quad \xRightarrow{\exists \mathrm{I}_1',\dots,\mathrm{I}_\mathrm{N}'} \quad \dot{\bigcup_{i=1}^{\mathrm{N}}} \mathrm{I}_i \subset \dot{\bigcup_{i=1}^{\mathrm{N}}} \mathrm{I}_i' = [a,b),$$

therefore,

$$\sum_{i=1}^{\mathrm{N}} \lambda(\mathrm{I}_i) \leqslant \sum_{i=1}^{\mathrm{N}} \lambda(\mathrm{I}_i') = \lambda\left(\dot{\bigcup_{i=1}^{\mathrm{N}}} \mathrm{I}_i'\right) \stackrel{\text{additive}}{=} \lambda[a,b).$$

Letting first $\mathrm{N} \to \infty$ and then $\epsilon \to 0$, we get $\sum_{i\in\mathbb{N}} \lambda(\mathrm{I}_i) = \lambda[a,b)$. Thus, λ is σ-additive relative to $\mathscr{I}$. □

The following corollary is now an immediate consequence of Theorem 5.2.

5.5 Corollary. *λ^1 is a measure on the Borel sets $\sigma(\mathscr{I}) = \mathscr{B}(\mathbb{R})$. It is the unique measure satisfying $\lambda[a,b) = b - a$.*

- The uniqueness in Corollary 5.5 (and Theorem 3.7) shows that the familiar geometric volume (length, area, volume) is unique.
- We will show the existence and uniqueness of d-dimensional Lebesgue measure later, using integration and product measures, cf. Corollary 15.8.

*Completion of measure spaces**

We will now discuss another extension of measures which will be important in probability theory.

5.6 Definition*. Let $(\Omega, \mathscr{A}, \mu)$ be a measure space.

a) $\mathscr{N}_\mu = \{\mathrm{N} \in \mathscr{A} \mid \mu(\mathrm{N}) = 0\}$ are the (measurable) **μ-null sets.**

b) $\widetilde{\mathscr{N}_\mu} = \left\{\tilde{\mathrm{N}} \subset \Omega \mid \exists \mathrm{N} \in \mathscr{N}_\mu, \tilde{\mathrm{N}} \subset \mathrm{N}\right\}$ are the (subsets of the measurable) **μ-null sets.**

c) A measure space is **complete** if $\widetilde{\mathscr{N}_\mu} \subset \mathscr{A}$.

⚠ In a complete measure space $\mathscr{N}_\mu = \widetilde{\mathscr{N}_\mu}$, i.e. all μ-null sets are measurable.

The following theorem shows that we may assume »without loss of generality« that a measure space is complete.

5.7 Theorem*. *Every measure space $(\Omega, \mathscr{A}, \mu)$ has a complete extension, i.e. there is a complete measure space $(\Omega, \widetilde{\mathscr{A}}, \widetilde{\mu})$ satisfying $\mathscr{A} \subset \widetilde{\mathscr{A}}$ and $\widetilde{\mu}|_{\mathscr{A}} = \mu$. The smallest complete extension is called the* **completion**, *given by*

$$\widetilde{\mathscr{A}} = \left\{\widetilde{A} \subset \Omega \mid \exists A, B \in \mathscr{A},\ A \subset \widetilde{A} \subset B,\ \mu(B \setminus A) = 0\right\} \tag{5.9}$$

$$\widetilde{\mu}(\widetilde{A}) := \frac{1}{2}(\mu(A) + \mu(B)), \quad \widetilde{A} \in \widetilde{\mathscr{A}}. \tag{5.10}$$

Moreover, $\widetilde{\mathscr{A}} = \sigma(\mathscr{A}, \widetilde{\mathscr{N}_\mu})$ and $\widetilde{\mathscr{N}_\mu} = \mathscr{N}_{\widetilde{\mu}}$.

Proof. It is enough to show that (5.9), (5.10) is the smallest complete extension. Throughout the proof, the sets $A, B, A_i, B_i \in \mathscr{A}$ correspond to $\widetilde{A}, \widetilde{A}_i \in \widetilde{\mathscr{A}}$ as indicated in (5.9).

1° Minimality of $\widetilde{\mathscr{A}}$. This follows once we know that $\widetilde{\mathscr{A}} = \sigma(\mathscr{A}, \widetilde{\mathscr{N}_\mu})$. Assume, for a moment, that $\widetilde{\mathscr{A}}$ defined by (5.9) is a σ-algebra. The definition of $\widetilde{\mathscr{A}}$ shows that $\mathscr{A} \subset \widetilde{\mathscr{A}}$ and $\widetilde{\mathscr{N}_\mu} \subset \widetilde{\mathscr{A}}$, and so we have the inclusion $\sigma(\mathscr{A}, \widetilde{\mathscr{N}_\mu}) \subset \widetilde{\mathscr{A}}$.

Conversely, if $\widetilde{A} \in \widetilde{\mathscr{A}}$, then $\widetilde{A} = A \cup (\widetilde{A} \setminus A)$ for some suitable set $A \in \mathscr{A}$. Since $\widetilde{A} \setminus A \subset B \setminus A \in \mathscr{N}_\mu$, we see that $\widetilde{A} \setminus A \in \widetilde{\mathscr{N}_\mu}$. Thus, $\widetilde{A} \in \sigma(\mathscr{A}, \widetilde{\mathscr{N}_\mu})$, and we get $\widetilde{\mathscr{A}} \subset \sigma(\mathscr{A}, \widetilde{\mathscr{N}_\mu})$.

2° $\widetilde{\mu}$ is well-defined, i.e. independent of the representation. Let $\widetilde{A} \in \widetilde{\mathscr{A}}$ such that $A_i \subset \widetilde{A} \subset B_i$ for $i = 1, 2$. Since $\mu(B_i) = \mu(A_i)$,

$$\mu(B_1 \cap B_2) \leqslant \mu(B_i) = \mu(A_i) \leqslant \mu(A_1 \cup A_2) \leqslant \mu(B_1 \cap B_2), \quad i = 1, 2,$$

and we see that $\mu(B_1) = \mu(B_2) = \mu(A_1) = \mu(A_2)$. Thus, $\widetilde{\mu}$ is well-defined. It is clear that $\widetilde{\mu}$ is a measure if we know that $\widetilde{\mathscr{A}}$ is a σ-algebra.

3° $\widetilde{\mathscr{A}}$ is a σ-algebra.

(Σ_1) $\emptyset \in \widetilde{\mathscr{A}}$ is obvious.

(Σ_2) If $\widetilde{A} \in \widetilde{\mathscr{A}}$, then $A \subset \widetilde{A} \subset B$ as in (5.9). However, $A^c \supset \widetilde{A}^c \supset B^c$ with $A^c, B^c \in \mathscr{A}$ and

$$B^c \setminus A^c = B^c \cap (A^c)^c = A \cap B^c = A \setminus B \in \mathscr{N}_\mu,$$

showing that $\widetilde{A}^c \in \widetilde{\mathscr{A}}$.

(Σ_3) Let $(\widetilde{A}_i)_{i \in \mathbb{N}} \subset \widetilde{\mathscr{A}}$ and $A_i \subset \widetilde{A} \subset B_i$ as in (5.9). Clearly,

$$A := \bigcup_{i \in \mathbb{N}} A_i \subset \bigcup_{i \in \mathbb{N}} \widetilde{A}_i \subset \bigcup_{i \in \mathbb{N}} B_i =: B,$$

and so, $A, B \in \mathscr{A}$, as well as

$$B \setminus A = \bigcup_{i \in \mathbb{N}} (B_i \setminus A) \subset \bigcup_{i \in \mathbb{N}} (B_i \setminus A_i) \in \mathscr{N}_\mu.$$

The latter follows from the σ-subadditivity of μ (Theorem 3.3.h)) as

$$\mu\left(\bigcup_{i\in\mathbb{N}} B_i \setminus A_i\right) \leqslant \sum_{i\in\mathbb{N}} \mu(B_i \setminus A_i) = 0.$$

This shows that $\bigcup_{i\in\mathbb{N}} \widetilde{A}_i \in \widetilde{\mathscr{A}}$. □

💬 The last lines in Step 3° of the above proof show that »countable unions of null sets are null sets«.

II
Integration

6 Measurable maps

Throughout this chapter we denote by $(\Omega, \mathscr{A})$ and $(\Omega', \mathscr{A}')$ measurable spaces and $X : \Omega \to \Omega'$ is some map between Ω and Ω'. We are interested in mappings which are compatible with the σ-algebras $\mathscr{A}$ and $\mathscr{A}'$ – just in the way continuous maps respect (and preserve) open sets: »pre-images of open sets are open«. We have already met this kind of problem at the beginning of the proof of Theorem 4.9 where we show

$$B \in \mathscr{B}(\mathbb{R}^d),\ x \in \mathbb{R}^d \overset{?!}{\Longrightarrow} x + B \in \mathscr{B}(\mathbb{R}^d).$$

6.1 Definition. A map $X : \Omega \to \Omega'$ is said to be **$\mathscr{A}/\mathscr{A}'$-measurable** (briefly: **measurable**), if

$$\forall A' \in \mathscr{A}' : X^{-1}(A') \in \mathscr{A}. \tag{6.1}$$

💬 Often (6.1) is written as $X^{-1}(\mathscr{A}') \subset \mathscr{A}$.
💬 We use $X : (\Omega, \mathscr{A}) \to (\Omega', \mathscr{A}')$ to indicate $\mathscr{A}/\mathscr{A}'$-measurability.

Based on Definition 6.1 we can re-write the above mentioned problem of Theorem 4.9 (and Theorem 3.7) in the following way: Using the shift operators

$$\begin{aligned} \tau_x : \mathbb{R}^d &\to \mathbb{R}^d & \qquad \tau_x^{-1} = \tau_{-x} : \mathbb{R}^d &\to \mathbb{R}^d \\ y &\mapsto y - x & y &\mapsto y + x \end{aligned}$$

we can rewrite $x + B$ as $x + B = \tau_{-x}(B) = \tau_x^{-1}(B)$, and so have for any Borel set $B \in \mathscr{B}(\mathbb{R}^d)$

$$\forall B \in \mathscr{B}(\mathbb{R}^d) : x + B \in \mathscr{B}(\mathbb{R}^d) \iff \tau_x \text{ is } \underbrace{\mathscr{B}(\mathbb{R}^d)/\mathscr{B}(\mathbb{R}^d)\text{-measurable}}_{\text{»Borel measurable«}}$$

In Theorem 4.9 we have used an »ad hoc« argument to show that τ_x is measurable, using the generator $\mathscr{I}$ of $\mathscr{B}(\mathbb{R}^d)$. This trick always works.

6.2 Lemma. *Assume that $\mathscr{A}' = \sigma(\mathscr{G}')$. A map $X : \Omega \to \Omega'$ is $\mathscr{A}/\mathscr{A}'$-measurable if, and only if, $X^{-1}(\mathscr{G}') \subset \mathscr{A}$, i.e.*

$$\forall G' \in \mathscr{G}' : X^{-1}(G') \in \mathscr{A}. \tag{6.2}$$

Proof. Since $\mathscr{G}' \subset \mathscr{A}'$, it is clear that (6.1)⇒(6.2).

Conversely, assume that (6.2) holds. Set $\Sigma' := \left\{A' \subset \Omega' \mid X^{-1}(A') \in \mathscr{A}\right\}$. It is not hard to see ✍ that Σ' is a σ-algebra. The condition (6.2) ensures that $\mathscr{G}' \subset \Sigma'$, and so

$$\sigma(\mathscr{G}') \subset \Sigma' \implies \mathscr{A}' \subset \Sigma' \implies (6.1). \qquad \square$$

The notion of **measurability** resembles the concept of (global) **continuity**. Recall that

$$\begin{aligned} & f : \mathbb{R}^d \to \mathbb{R}^n \text{ is continuous} \\ \iff & \forall x\, \forall \epsilon > 0 \quad \exists \delta = \delta_{\epsilon,x} > 0 \quad \forall |x-y| < \delta : \quad |f(x) - f(y)| < \epsilon; \\ \iff & \forall x\, \forall \epsilon > 0 \quad \exists \delta = \delta_{\epsilon,x} > 0 : \quad f(B_\delta(x)) \subset B_\epsilon(f(x)); \\ \iff & \forall U' \subset \mathbb{R}^n \text{ open} : \quad f^{-1}(U') \subset \mathbb{R}^d \text{ is open.} \end{aligned}$$

In general topological spaces $(\Omega, \mathscr{O})$, $(\Omega', \mathscr{O}')$ one uses the last equivalence as definition of (global) continuity.

$$f : \Omega \to \Omega' \text{ is continuous} \iff f^{-1}(\mathscr{O}') \subset \mathscr{O}. \tag{6.3}$$

6.3 Example. Every continuous map $X : \mathbb{R}^d \to \mathbb{R}^n$ is Borel measurable. This follows easily from the criterion from Lemma 6.2: $\mathscr{B}(\mathbb{R}^n) = \sigma(\mathscr{O}_{\mathbb{R}^n})$ where $\mathscr{O}_{\mathbb{R}^n}$ are the open sets of $\mathbb{R}^n$ and

$$\forall U' \in \mathscr{O}_{\mathbb{R}^n} : f^{-1}(U') \overset{\text{cont.}}{\in} \mathscr{O}_{\mathbb{R}^d} \subset \mathscr{B}(\mathbb{R}^d) \overset{\S 6.2}{\implies} f \text{ measurable.}$$

⚠ continuous ⇒ Borel masurable.

⚠ Borel measurable ⇏ continuous. Consider, e.g., $f : \mathbb{R} \to \mathbb{R}$, $f(x) := \mathbb{1}_{[-1,1]}(x)$. This function is clearly not (everywhere) continuous, but it is measurable:

$$f^{-1}(B) = \begin{cases} \emptyset, & 0, 1 \notin B; \\ [-1,1], & 1 \in B,\ 0 \notin B; \\ [-1,1]^c, & 1 \notin B,\ 0 \in B; \\ \mathbb{R} & 1 \in B,\ 0 \in B \end{cases} \implies f^{-1}(B) \in \mathscr{B}(\mathbb{R}).$$

6.4 Theorem. *Let $(\Omega_i, \mathscr{A}_i)$, $i = 1, 2, 3$, measurable spaces. If*

$$\begin{aligned} X : \Omega_1 \to \Omega_2 &\text{ is } \mathscr{A}_1/\mathscr{A}_2\text{-measurable} \\ Y : \Omega_2 \to \Omega_3 &\text{ is } \mathscr{A}_2/\mathscr{A}_3\text{-measurable} \\ \text{then } Y \circ X : \Omega_1 \to \Omega_3 &\text{ is } \mathscr{A}_1/\mathscr{A}_3\text{-measurable.} \end{aligned}$$

Proof. $\forall A_3 \in \mathscr{A}_3 : (Y \circ X)^{-1}(A_3) = X^{-1} \circ \underbrace{Y^{-1}(A_3)}_{\in \mathscr{A}_2} \in X^{-1}(\mathscr{A}_2) \subset \mathscr{A}_1.$ □

Sometimes we have a map $X : \Omega \to \Omega'$ and a σ-algebra $\mathscr{A}'$ in Ω' and we **ask** for some σ-algebra $\mathscr{A}$ in Ω, such that X becomes a measurable map. Here is a typical example from probability theory:

- $(\Omega', \mathscr{A}') = (\mathbb{R}^d, \mathscr{B}(\mathbb{R}^d))$ is the »real world« space where we can make observations. We want to embed these observations into a mathematical model Ω where we can define a probability measure $\mu = \mathbb{P}$.
- The map $X : \Omega \to \mathbb{R}^d$ linking the model with the real world is a **random variable**: $X^{-1}(B) = \{X \in B\} = \{\omega \in \Omega \mid X(\omega) \in B\}$ maps the observed values B to our model.
- In order to assign a probability to $X^{-1}(B)$, we need that $X^{-1}(B)$ is a measurable set in Ω, in other words: We need that $X : \Omega \to \mathbb{R}^d$ becomes measurable.

6.5 Lemma (and Definition). *Let $(X_i)_{i\in I}$ be arbitrarily many maps $X_i : \Omega \to \Omega_i$ which are defined on the measurable spaces $(\Omega_i, \mathscr{A}_i)$, $i \in I$. The σ-algebra*

$$\sigma(X_i \mid i \in I) := \sigma\Big(\bigcup_{i\in I} X_i^{-1}(\mathscr{A}_i)\Big) = \sigma\big(\{A \mid \exists i \in I,\ A \in X_i^{-1}(\mathscr{A}_i)\}\big)$$

is the smallest σ-algebra in Ω, such that all maps X_i become measurable. $\sigma(X_i : i \in I)$ is the σ-algebra ***generated by the maps*** $(X_i)_{i\in I}$.

Proof. By definition, the map $X_i : \Omega \to \Omega_i$ is $\mathscr{A}/\mathscr{A}_i$-measurable if, and only if, $X_i^{-1}(\mathscr{A}_i) \subset \mathscr{A}$. Thus

$$\text{all } X_i \text{ are measurable if, and only if, } \bigcup_{i\in I} X_i^{-1}(\mathscr{A}_i) \subset \mathscr{A}.$$

Since the σ-operation is minimal, $\mathscr{A} := \sigma\Big(\bigcup_{i\in I} X_i^{-1}(\mathscr{A}_i)\Big)$ is the smallest σ-algebra in Ω making all X_i measurable. □

⚠ Each $X_i^{-1}(\mathscr{A}_i)$ is a σ-algebra (Example 2.3.g)) but, in general, a union of (finitely or infinitely many) σ-algebras $\bigcup_{i\in I} X_i^{-1}(\mathscr{A}_i)$ is not a σ-algebra, see the discussion in [CEX, Example 4.7].

We will use measurable maps to transport »push forward« a measure ν on $(\Omega, \mathscr{A})$ to $(\Omega', \mathscr{A}')$.

6.6 Theorem (image measure). *Let $X : (\Omega, \mathscr{A}) \to (\Omega', \mathscr{A}')$ be a measurable map. For every measure ν on $(\Omega, \mathscr{A})$ the set function*

$$\nu'(A') := \nu(\underbrace{X^{-1}(A')}_{\in \mathscr{A}}), \quad \forall A' \in \mathscr{A}', \tag{6.4}$$

yields a measure on $(\Omega', \mathscr{A}')$. ν' is called the ***image measure*** *or* ***push forward*** *of ν under X, and it is frequently denoted by $X(\nu)$, $\nu \circ X^{-1}$ or $X_*\nu$.*

Proof. Since $X^{-1}(\mathscr{A}') \subset \mathscr{A}$ is a σ-algebra, the definition (6.4) makes sense.

(M₁) $A' = \emptyset \implies X^{-1}(\emptyset) = \emptyset \in \mathscr{A} \implies \nu'(\emptyset) = \nu(\emptyset) = 0.$

(M₂) $(A'_n)_{n\in\mathbb{N}} \subset \mathscr{A}'$ disjoint $\implies X^{-1}(A'_n) \in \mathscr{A}$ disjoint[(1)], and the σ-additivity of ν gives

$$\nu'\Big(\dot{\bigcup_{n\in\mathbb{N}}} A'_n\Big) = \nu\Big(X^{-1}\Big(\dot{\bigcup_{n\in\mathbb{N}}} A'_n\Big)\Big) = \nu\Big(\dot{\bigcup_{n\in\mathbb{N}}} X^{-1}(A'_n)\Big)$$
$$\overset{(M_2)}{=} \sum_{n\in\mathbb{N}} \nu(X^{-1}(A'_n)) = \sum_{n\in\mathbb{N}} \nu'(A'_n). \qquad \square$$

6.7 Example. a) $\lambda^d(x+B) = \lambda^d(\tau_x^{-1}(B)) = \tau_x(\lambda^d)(B)$ for all $B \in \mathscr{B}(\mathbb{R}^d)$.

b) Let $(\Omega, \mathscr{A}, \mathbb{P})$ be a probability space. A **random variable** is a measurable map

$$X : (\Omega, \mathscr{A}) \xrightarrow{\text{measurable}} (\mathbb{R}^d, \mathscr{B}(\mathbb{R}^d))$$

The image measure of $\mathbb{P}$ under X is a measure on $(\mathbb{R}^d, \mathscr{B}(\mathbb{R}^d))$

$$X(\mathbb{P})(B) = \mathbb{P} \circ X^{-1}(B) = \mathbb{P}(\underbrace{X \in B}_{\{X\in B\}=X^{-1}(B)}), \quad B \in \mathscr{B}(\mathbb{R}^d)$$

which is the (**probability**) **distribution** of X. Here is a concrete example: Casting two dice.

$$\Omega = \{(i,k) \mid 1 \leqslant i,k \leqslant 6\}, \quad \mathscr{A} = \mathscr{P}(\Omega), \quad \mathbb{P}(\{(i,k)\}) = \frac{1}{36}$$
$$X : \Omega \to \{2,3,\dots,12\}, \quad \xi((i,k)) := i+k.$$

The distribution (image measure) of the random variable X »sum of pips« is

n	2	3	4	5	6	7	8	9	10	11	12
$\mathbb{P}(X=n)$	$\frac{1}{36}$	$\frac{1}{18}$	$\frac{1}{12}$	$\frac{1}{9}$	$\frac{5}{36}$	$\frac{1}{6}$	$\frac{5}{36}$	$\frac{1}{9}$	$\frac{1}{12}$	$\frac{1}{18}$	$\frac{1}{36}$

Table 6.1. Probability distribution of the sum of pips

▲ $X^{-1}(B) \in \mathscr{P}(\Omega)$ is always true, that is, if $\mathscr{A} = \mathscr{P}(\Omega)$, then **every** mapping $X : (\Omega, \mathscr{P}(\Omega)) \to (\star, \star)$ is measurable. This is the reason why there are not measurability considerations in so-called discrete probability spaces, cf. also Example 3.5.e).

6.8 Theorem. *Let* $T \in O(\mathbb{R}^d)$ *be an orthogonal matrix, i.e.* $T \in \mathbb{R}^{d\times d}$, $T^\top T = \mathrm{id}_{\mathbb{R}^d}$. *Then* $T(\lambda^d) = \lambda^d$.

Proof. An orthogonal matrix $T \in O(\mathbb{R}^d)$ preserves distances, and so

$$\|Tx - Ty\| = \|x - y\| \quad \forall x, y \in \mathbb{R}^d.$$

(1) Indeed: $X^{-1}(C \cap D) = X^{-1}(C) \cap X^{-1}(D)$

This means that the map $x \mapsto \mathrm{T}x$ is (Lipschitz) continuous, hence measurable. Therefore, the image measure $\mu(\mathrm{B}) = \lambda^d(\mathrm{T}^{-1}(\mathrm{B}))$ is well-defined; since T^{-1} is linear, we have for any $\mathrm{B} \in \mathscr{B}(\mathbb{R}^d)$ and $x \in \mathbb{R}^d$, $y := \mathrm{T}^{-1}x$

$$\mu(x+\mathrm{B}) = \lambda^d(\mathrm{T}^{-1}(x+\mathrm{B})) = \lambda^d(y+\mathrm{T}^{-1}(\mathrm{B})) \overset{\S 3.7.a)}{=} \lambda^d(\mathrm{T}^{-1}(\mathrm{B})) = \mu(\mathrm{B}).$$

From Theorem 4.9 we know that in this case $\mu(\mathrm{B}) = \kappa\lambda^d(\mathrm{B})$ for all Borel sets $\mathrm{B} \in \mathscr{B}(\mathbb{R}^d)$. In order to identify the constant κ, we use $\mathrm{B} = \mathrm{B}_1(0)$ and we observe that $\mathrm{B}_1(0) = \{x \mid \|x\| < 1\} = \{x \mid \|\mathrm{T}x\| < 1\} = \mathrm{T}^{-1}(\mathrm{B}_1(0))$. Thus,

$$\lambda^d(\mathrm{B}_1(0)) = \lambda^d(\mathrm{T}^{-1}(\mathrm{B}_1(0))) \overset{\text{def}}{=} \mu(\mathrm{B}_1(0)) = \kappa\lambda^d(\mathrm{B}_1(0))$$

and we see that $\kappa = 1$. □

6.9 Theorem. *If* $\mathrm{M} \in \mathbb{R}^{d\times d}$ *is invertible, then*

$$\mathrm{M}(\lambda^d) = |\det \mathrm{M}^{-1}| \cdot \lambda^d = |\det \mathrm{M}|^{-1} \cdot \lambda^d. \tag{6.5}$$

Proof. In a finite-dimensional space, every linear map $x \mapsto \mathrm{M}x$ is continuous, hence Borel measurable (Example 6.3). As in the proof of Theorem 6.7 we find for $\mathrm{B} \in \mathscr{B}(\mathbb{R}^d)$

$$\mu(\mathrm{B}) := \lambda^d(\mathrm{M}^{-1}(\mathrm{B})) \implies \mu(x+\mathrm{B}) = \mu(\mathrm{B})$$

and we conclude from Theorem 3.7 that

$$\mu(\mathrm{B}) = \mu([0,1)^d) \cdot \lambda^d(\mathrm{B}) = \lambda^d(\mathrm{M}^{-1}[0,1)^d) \cdot \lambda^d(\mathrm{B}).$$

Since $\mathrm{M}^{-1}[0,1)^d$ is a parallelepiped spanned by $\mathrm{M}^{-1}e_k$, $k = 1,\dots,d$, (e_k is the kth unit vector of $\mathbb{R}^d$) its geometric volume, hence its Lebesgue measure (due to the uniqueness of Lebesgue measure), is $|\det \mathrm{M}|^{-1}$.(2) □

Since rigid motions consist of translations τ_x and rotations and reflections ($\mathrm{M} \in \mathbb{R}^{d\times d}$ with $\det \mathrm{M} = \pm 1$), we get

6.10 Corollary. λ^d *is invariant under rigid motions.*

Appendix: The volume of a parallelepiped

6.11 Theorem. $\lambda^d\left(\mathrm{A}[0,1)^d\right) = |\det \mathrm{A}|$ *for all invertible matrices* $\mathrm{A} \in \mathbb{R}^{d\times d}$.

For the proof of Theorem 6.11 we need two auxiliary results.

(2) This should be known from linear algebra. A proof is given in the appendix to this chapter.

6.12 Lemma. *If* $D = \operatorname{diag}[\lambda_1, \dots, \lambda_d]$, $\lambda_i > 0$, *is a* $d \times d$ *diagonal matrix, then* $\lambda^d(D(B)) = \det D \cdot \lambda^d(B)$ *for all* $B \in \mathscr{B}(\mathbb{R}^d)$.

Proof. Since both D and D^{-1} are continuous maps, D(B) is a Borel set if $B \in \mathscr{B}(\mathbb{R}^d)$. In view of the uniqueness theorem for measures (Theorem 4.7) it is enough to prove the lemma for half-open rectangles $\times_{i=1}^d [a_i, b_i)$, $a_i, b_i \in \mathbb{R}$. Obviously,

$$D\left(\mathop{\times}_{i=1}^{d} [a_i, b_i)\right) = \mathop{\times}_{i=1}^{d} [\lambda_i a_i, \lambda_i b_i),$$

and

$$\lambda^d\left(D\left(\mathop{\times}_{i=1}^{d} [a_i, b_i)\right)\right) = \prod_{i=1}^{d} (\lambda_i b_i - \lambda_i a_i) = \lambda_1 \cdots \lambda_d \prod_{i=1}^{d} (b_i - a_i)$$
$$= \det D \cdot \lambda^d\left(\mathop{\times}_{i=1}^{d} [a_i, b_i)\right). \qquad \square$$

6.13 Lemma. *Every invertible* $A \in \mathbb{R}^{d \times d}$ *can be written as* $A = SDT$, *where* $S, T \in O(\mathbb{R}^d)$ *are orthogonal matrices and* $D = \operatorname{diag}[\lambda_1, \dots, \lambda_d]$ *is a diagonal matrix with positive entries* $\lambda_i > 0$.

Proof. The matrix $A^\top A$ is symmetric, and so we can find some $U \in O(\mathbb{R}^d)$ such that

$$U^\top (A^\top A) U = \widetilde{D} = \operatorname{diag}[\mu_1, \dots, \mu_d].$$

Since for the ith unit vector e_i and the Euclidean norm $\|\cdot\|$

$$\mu_i = e_i^\top \widetilde{D} e_i = (e_i^\top U^\top A^\top)(AUe_i) = \|AUe_i\|^2 > 0,$$

we can define $D := \sqrt{\widetilde{D}} = \operatorname{diag}[\lambda_1, \dots, \lambda_d]$ where $\lambda_i := \sqrt{\mu_i}$. Thus,

$$D^{-1} U^\top A^\top A U D^{-1} = \mathrm{id}_d,$$

and this proves that $S := AUD^{-1} \in O(\mathbb{R}^d)$. Since $T := U^\top \in O(\mathbb{R}^d)$, we easily see that

$$SDT = (AUD^{-1}) D U^\top = A. \qquad \square$$

Proof of Theorem 6.11. Combining Lemma 6.12 and 6.13 we get for any invertible matrix $A \in \mathbb{R}^{d \times d}$

$$\lambda^d\left(A[0,1)^d\right) = \lambda^d\left(SDT[0,1)^d\right) = \lambda^d\left(DT[0,1)^d\right) = \det D \cdot \lambda^d\left(T[0,1)^d\right)$$
$$= \det D \cdot \lambda^d\left([0,1)^d\right).$$

Since $S, T \in O(\mathbb{R}^d)$, their determinants are either +1 or −1, and we conclude that $|\det A| = |\det(SDT)| = |\det S| \cdot |\det D| \cdot |\det T| = \det D$. $\square$

7 Borel functions

We continue our study of measurable maps. As before, $(\Omega, \mathscr{A})$ is an arbitrary measure space, but $(\Omega', \mathscr{A}')$ will now be a $(\mathbb{R}, \mathscr{B}(\mathbb{R}))$. We want to consider functions with values in $\overline{\mathbb{R}} = [-\infty, +\infty]$. Therefore, we extend the rules for addition and multiplication to include the »numbers« $\pm\infty$.

$$x + \infty = \infty + x = \infty, \qquad x - \infty = -\infty + x = -\infty,$$
$$\infty + \infty = \infty, \qquad (-\infty) + (-\infty) = -\infty,$$
$$(\pm x) \cdot \infty = \infty \cdot (\pm x) = \pm\infty \quad \forall x > 0,$$
$$(\pm x) \cdot (-\infty) = (-\infty) \cdot (\pm x) = \mp\infty \quad \forall x > 0,$$
$$\frac{1}{\pm\infty} = 0, \qquad 0 \cdot (\pm\infty) = (\pm\infty) \cdot 0 = 0. \text{ (1)}$$

⚠ The expressions »$\infty - \infty$« and »$\frac{\pm\infty}{\pm\infty}$« **are not defined.**

7.1 Definition. The Borel σ-algebra $\mathscr{B}(\overline{\mathbb{R}})$ comprises all sets $B^* = B \cup S$ where $B \in \mathscr{B}(\mathbb{R})$ and $S \in \big\{\emptyset, \{+\infty\}, \{-\infty\}, \{-\infty, +\infty\}\big\}$.

Check yourself ✍ that $\mathscr{B}(\overline{\mathbb{R}})$ is a σ-algebra.

7.2 Lemma. $\mathscr{B}(\overline{\mathbb{R}}) = \sigma([a, \infty], a \in \mathbb{Q})$. *We may replace* $[a, \infty]$ *by any of the following intervals:* $(a, \infty]$, $[-\infty, a)$ *or* $[-\infty, a]$.

Proof. Set $\Sigma := \sigma\big([a, \infty] \mid a \in \mathbb{Q}\big)$ and let $a, b \in \mathbb{Q}$. On the one hand,

$$[a, \infty] = [a, \infty) \cup \{\infty\} \in \mathscr{B}(\overline{\mathbb{R}}) \implies \Sigma \subset \mathscr{B}(\overline{\mathbb{R}});$$

on the other hand,

$$[a, b) = [a, \infty] \setminus [b, \infty] \in \Sigma \implies \mathscr{B}(\mathbb{R}) \subset \Sigma.$$

Since $\{+\infty\} = \bigcap_{n \in \mathbb{N}} [n, \infty] \in \Sigma$ and $\{-\infty\} = \bigcap_{n \in \mathbb{N}} [-n, \infty]^c \in \Sigma$, we conclude that the sets $B, B \cup \{\infty\}, B \cup \{-\infty\}, B \cup \{\pm\infty\} \in \Sigma$ for any $B \in \mathscr{B}(\mathbb{R})$. Therefore, $\mathscr{B}(\overline{\mathbb{R}}) \subset \Sigma$.

The other cases are similar. □

The following remark justifies our Definition 7.1.

(1) This is a special convention which is particular to measure and probability theory. You should interpret »0« as a true zero, while »±∞« comes from a limit, e.g. $\lim_n a_n = \pm\infty$. In this sense, $0 \cdot \pm\infty = 0 \cdot \lim_n a_n = \lim_n 0 \cdot a_n = 0$. The rationale of this convention will become clear when we discuss null sets in Chapter 10.

7.3 Remark★. $[-\infty,\infty]$ is the two-point compactification of $(-\infty,\infty)$, the open neighbourhoods of $\pm\infty$ are sets of the form $[-\infty,a)$ resp. $(b,\infty]$. Therefore, the open sets in $\overline{\mathbb{R}}=[-\infty,\infty]$ are given by

$$\mathscr{O}_{\overline{\mathbb{R}}}=\Big\{U,\ U\cup[-\infty,a),\ U\cup(b,\infty],\ U\cup[-\infty,a)\cup(b,\infty]\ \Big|\ U\in\mathscr{O}_{\mathbb{R}},\ a,b\in\mathbb{R}\Big\};$$

thus, $U\in\mathscr{O}_{\mathbb{R}}\iff U=V\cap\mathbb{R}$ for some $V\in\mathscr{O}_{\overline{\mathbb{R}}}$.

Lemma★. *$\mathscr{B}(\overline{\mathbb{R}})=\sigma(\mathscr{O}_{\overline{\mathbb{R}}})$ and $\mathscr{B}(\mathbb{R})=\mathscr{B}(\overline{\mathbb{R}})|_{\mathbb{R}}$ is the trace σ-algebra.*

Proof. Clearly, $\mathscr{O}_{\overline{\mathbb{R}}}\subset\mathscr{B}(\overline{\mathbb{R}})$, i.e. $\sigma(\mathscr{O}_{\overline{\mathbb{R}}})\subset\mathscr{B}(\overline{\mathbb{R}})$. Since the sets $(a,\infty]\in\mathscr{O}_{\overline{\mathbb{R}}}$ generate $\mathscr{B}(\overline{\mathbb{R}})$ (Lemma 7.2), we see that $\mathscr{B}(\overline{\mathbb{R}})=\sigma((a,\infty],a\in\mathbb{R})\subset\sigma(\mathscr{O}_{\overline{\mathbb{R}}})$.

The second assertion is proved like Lemma 2.10. □

7.4 Definition. A measurable map $u:(\Omega,\mathscr{A})\to(\mathbb{R},\mathscr{B}(\mathbb{R}))$, resp., $(\overline{\mathbb{R}},\mathscr{B}(\overline{\mathbb{R}}))$ is called an ($\overline{\mathbb{R}}$-valued) **Borel function.**

Since a real-valued function is also $\overline{\mathbb{R}}$-valued, and $\{x\mid u(x)=\pm\infty\}=\emptyset$ for $\mathbb{R}$-valued functions, we have

$$u \text{ Borel function} \iff u^{-1}(B)\in\mathscr{A}\quad\forall B\in\mathscr{B}(\overline{\mathbb{R}})$$
$$\overset{\S 6.2}{\iff} u^{-1}(G)\in\mathscr{A}\quad\forall G\in\mathscr{G}\ \text{ if }\ \sigma(\mathscr{G})=\mathscr{B}(\overline{\mathbb{R}}).$$

In particular, we may use for $\mathscr{G}$ intervals of the form

$$[a,\infty],\quad (b,\infty],\quad [-\infty,c),\quad [-\infty,d],\quad a,b,c,d\in\mathbb{R}\text{ or }\mathbb{Q}$$

cf. Remark 2.9 and Lemma 7.2. We will need the following **useful notation** in connection with functions $u,w:\Omega\to\overline{\mathbb{R}}$.

- 💬 $\{u\in B\}:=\{x\in\Omega\mid u(x)\in B\}=u^{-1}(B)$;
- 💬 $\{u\geqslant w\}:=\{x\in\Omega\mid u(x)\geqslant w(x)\}$ (similarly for $>,<,\leqslant,=,\neq$);
- 💬 in particular, $\{u\geqslant a\}=\{u\in[a,\infty]\}=u^{-1}[a,\infty]$ etc. (For $\mathbb{R}$-valued functions we may use $[a,\infty)$ etc.)

7.5 Lemma. *A function $u:(\Omega,\mathscr{A})\to(\mathbb{R},\mathscr{B}(\mathbb{R}))$, resp., $(\overline{\mathbb{R}},\mathscr{B}(\overline{\mathbb{R}}))$ is measurable if, and only if, for all $a\in\mathbb{R}$ or $a\in\mathbb{Q}$ one of the following conditions holds:*

a) $\{u\geqslant a\}\in\mathscr{A}$; c) $\{u\leqslant a\}\in\mathscr{A}$;

b) $\{u>a\}\in\mathscr{A}$; d) $\{u<a\}\in\mathscr{A}$.

7.6 Definition. $\mathcal{L}^0=\mathcal{L}^0(\mathscr{A})$, resp., $\mathcal{L}^0_{\overline{\mathbb{R}}}=\mathcal{L}^0_{\overline{\mathbb{R}}}(\mathscr{A})$ denote the families of measurable (Borel) functions

$$u : (\Omega, \mathscr{A}) \to (\mathbb{R}, \mathscr{B}(\mathbb{R})) \text{ resp. } (\overline{\mathbb{R}}, \mathscr{B}(\overline{\mathbb{R}})).$$

7.7 Example. a) $\mathbb{1}_A(x)$ is measurable if, and only if, $A \in \mathscr{A}$, i.e. if A is a measurable set.

Indeed, see also Fig. 7.1, $\{\mathbb{1}_A > \lambda\} = \begin{cases} \emptyset, & \lambda \geqslant 1, \\ A & 0 \leqslant \lambda < 1, \\ \Omega & \lambda < 0. \end{cases}$

b) Let $A_1, \dots, A_m \in \mathscr{A}$ be mutually disjoint and, $y_1, \dots, y_m \in \mathbb{R}$. The step function

$$g(x) := \sum_{i=1}^{m} y_i \mathbb{1}_{A_i} \quad \text{is measurable.}$$

Indeed, see Fig. 7.1, $\{g > \lambda\} = \dot{\bigcup}_{i, y_i > \lambda} A_i \in \mathscr{A}$.

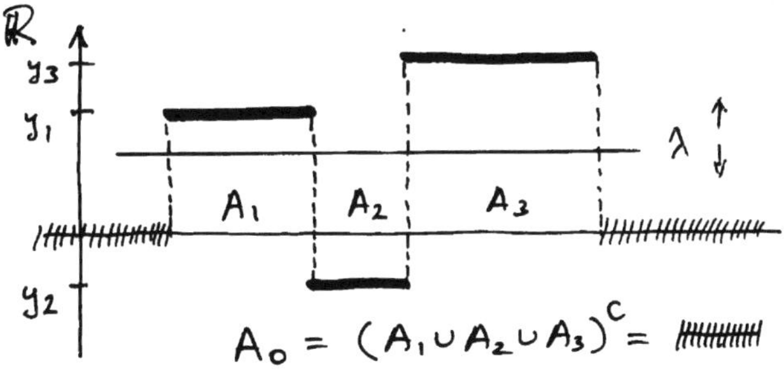

Fig. 7.1. Measurability of a simple function ($m = 3$); $\{u > \lambda\} = A_1 \,\dot{\cup}\, A_3$.

Set $A_0 := \Omega \setminus (A_1 \,\dot{\cup} \cdots \dot{\cup}\, A_m)$ and $y_0 := 0$. Clearly, $A_0 \in \mathscr{A}$ and

$$g(x) = \sum_{i=0}^{m} y_i \mathbb{1}_{A_i}(x) \quad \text{and} \quad \Omega = \dot{\bigcup_{i=0}^{m}} A_i.$$

7.8 Definition. A **simple function** on $(\Omega, \mathscr{A})$ is a step function of the form

$$g(x) = \sum_{i=0}^{m} y_0 \mathbb{1}_{A_i}(x), \quad m \in \mathbb{N},\ y_i \in \mathbb{R},\ A_i \in \mathscr{A} \text{ disjoint.} \tag{7.1}$$

If the sets $A_0, \dots, A_m$ are a tiling of Ω, $A_0 \,\dot{\cup} \cdots \dot{\cup}\, A_m = \Omega$, then (7.1) is called a **standard representation**.

The family of all simple functions is denoted by $\mathcal{E} = \mathcal{E}(\mathscr{A})$.

⚠ The representation (7.1) is not unique.

7.9 Example. a) Any $h \in \mathcal{L}^0(\mathscr{A})$ with finite range $h(\Omega) = \{y_1, \dots, y_m\}$ is a simple function. Indeed

$$h(x) = \sum_{z \in h(\Omega)} \mathbb{1}_{\{x \mid h(x)=z\}}(x)$$

is a standard representation. Note that the sets

$$A_z = \{x \mid h(x) = z\} = \{h = z\} = \{h \leqslant z\} \setminus \{h < z\} \in \mathscr{A}$$

are measurable and mutually disjoint.

Consquence: Every $h \in \mathcal{E}(\mathscr{A})$ has a standard representation.

b) $f, g \in \mathcal{E}(\mathscr{A}) \implies f \pm g,\ f \cdot g \in \mathcal{E}(\mathscr{A})$.

Using the standard representations $f = \sum_{i=0}^{m} y_i \mathbb{1}_{A_i}$ and $g = \sum_{k=0}^{n} z_k \mathbb{1}_{B_k}$ we get

$$f \pm g = \sum_{i=0}^{m} \sum_{k=0}^{n} (y_i \pm z_k) \mathbb{1}_{A_i \cap B_k},$$

$$f \cdot g = \sum_{i=0}^{m} \sum_{k=0}^{n} y_i z_k \mathbb{1}_{A_i \cap B_k}.$$

The family $(A_i \cap B_k)_{i,k}$ is the **common refinement** of the tilings $(A_i)_i$ and $(B_k)_k$. It is again a tiling, since

$$\dot{\bigcup_{i=0}^{m}} \dot{\bigcup_{k=0}^{n}} A_i \cap B_k = \dot{\bigcup_{i=0}^{m}} A_i \cap \Omega = \dot{\bigcup_{i=0}^{m}} A_i = \Omega.$$

The sets $A_i \cap B_k$ are mutually disjoint – many of them might be empty (never mind), the others inherit disjointness from $(A_i)_i$ or (B_k). Note that the functions f and g are constant on the sets $A_i \cap B_k$.

c) $f \in \mathcal{E}(\mathscr{A}) \implies f^+, f^-, |f| \in \mathcal{E}(\mathscr{A})$.

Example 7.9.c) uses the following definition.

7.10 Definition. The **positive part** u^+ and the **negative part** u^- of a function $u : \Omega \to \overline{\mathbb{R}}$ are defined as

$$u^+(x) := \max\{u(x), 0\} \qquad u^-(x) := -\min\{u(x), 0\}$$
$$= u(x) \vee 0, \qquad = -\big(u(x) \wedge 0\big).$$

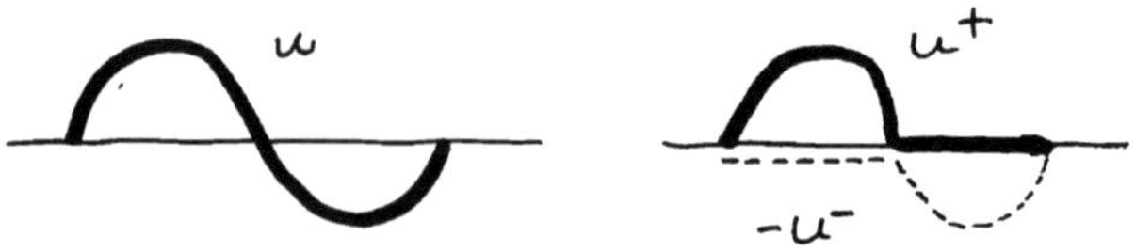

Fig. 7.2. The positive and negative part of a function u.

⚠ $u^+ \geqslant 0$, $u^- \geqslant 0$, i.e. also the negative part is a positive function.
💬 $u = u^+ - u^-$ and $|u| = u^+ + u^-$ and $u^+ \cdot u^- = 0$.

By definition, $\mathcal{E}(\mathscr{A}) \subset \mathcal{L}^0(\mathscr{A})$. We will now show that any $u \in \mathcal{L}^0(\mathscr{A})$ is the pointwise limit of simple functions.

7.11 Theorem (sombrero[2] lemma). *Every positive $\mathscr{A}/\mathscr{B}(\overline{\mathbb{R}})$-measurable function $u : \Omega \to [0,\infty]$ is the limit of an increasing sequence of simple functions $f_n \in \mathcal{E}(\mathscr{A})$, $f_n \geqslant 0$, i.e.*

$$u(x) = \sup_{n\in\mathbb{N}} f_n(x), \quad f_1 \leqslant f_2 \leqslant f_3 \leqslant \dots$$

Amendment 1: *The approximation is uniform in x, if u is bounded.*
Amendment 2: *Every [bounded] $\mathscr{A}/\mathscr{B}(\overline{\mathbb{R}})$-measurable $u : \Omega \to \overline{\mathbb{R}}$ is the [uniform] limit of simple functions $f_n \in \mathcal{E}(\mathscr{A})$ such that $|f_n| \leqslant |u|$.*

Proof. Define $f_n(x) := \sum_{k=0}^{n2^n} k2^{-n}\mathbb{1}_{A_k^n}(x)$ where

$$A_k^n = \begin{cases} \{k2^{-n} \leqslant u < (k+1)2^{-n}\}, & k = 0,1,2,\dots n2^n - 1, \\ \{u \geqslant n\}, & k = n2^n. \end{cases}$$

see Fig. 7.3. Clearly – do look at the figure – we have

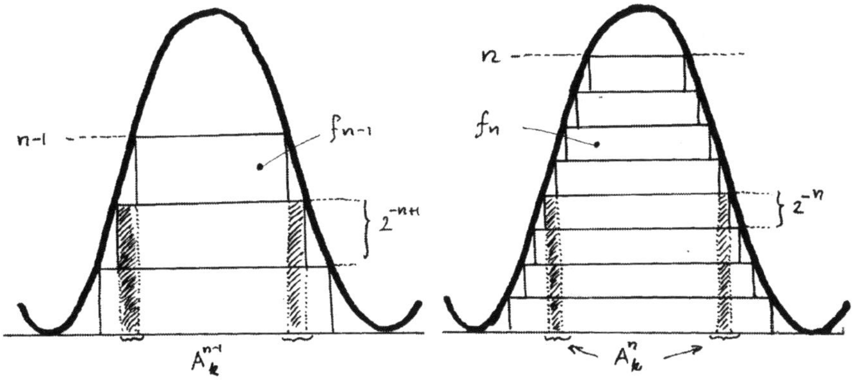

Fig. 7.3. The range of the measurable function u determines the tiling A_k^n of the domain. The picture shows two functions, f_{n-1} (left) and f_n (right).

i) $|f_n(x) - u(x)| \leqslant 2^{-n} \quad \forall x \in \{u < n\}$;
ii) $A_k^n = \{k2^{-n} \leqslant u\} \cap \{u < (k+1)2^{-n}\} \in \mathscr{A}$ and $\{u \geqslant n\} \in \mathscr{A} \implies f_n \in \mathcal{E}(\mathscr{A})$;
iii) $0 \leqslant f_n \leqslant f_{n+1} \leqslant u$ and $f_n \uparrow u$.

(2) A »sombrero« is a Mexican hat. If you look at Fig. 7.3 you'll understand the name of the game.

Proof of the first amendment. If $n > \sup_x u(x)$, then i) shows uniform convergence.

Proof of the second amendment. Note that

$$u \in \mathcal{L}^0(\mathscr{A}) \implies \{u^+ > \lambda\} = \begin{cases} \{u > \lambda\}, & \lambda \geqslant 0 \\ \Omega, & \lambda < 0 \end{cases} \in \mathscr{A} \implies u^+ \in \mathcal{L}^0(\mathscr{A}).$$

In the same way we see that $u^- \in \mathcal{L}^0(\mathscr{A})$. Using the first part of the proof, we can construct increasing sequences $f_n \uparrow u^+$ and $g_n \uparrow u^-$. Obviously,

$$f_n - g_n \to u \quad \text{and} \quad |f_n - g_n| \leqslant f_n + g_n \leqslant u^- + u^+ = |u|. \qquad \square$$

We will need the concepts of upper and lower limits which should be known from analysis. Recall that

$$\limsup_{n\to\infty} u_n(x) = \inf_{k\in\mathbb{N}} \sup_{n\geqslant k} u_n(x) \quad \text{and} \quad \liminf_{n\to\infty} u_n(x) = \sup_{k\in\mathbb{N}} \inf_{n\geqslant k} u_n(x);$$

The definition shows that $\limsup_{n\to\infty} u_n(x) = -\liminf_{n\to\infty}(-u_n(x))$. As the lower [upper] limit is the smallest [largest] subsequential limit, we arrive at the following important observation:

$$\underbrace{\exists \lim_{n\to\infty} u_n(x)}_{\in\mathbb{R} \text{ resp. } \overline{\mathbb{R}}} \iff \underbrace{\liminf_{n\to\infty} u_n(x) = \limsup_{n\to\infty} u_n(x)}_{\in\mathbb{R} \text{ resp. } \overline{\mathbb{R}}}.$$

7.12 Corollary. *If $(u_n)_{n\in\mathbb{N}} \subset \mathcal{L}^0_{\overline{\mathbb{R}}}(\mathscr{A})$ is a sequence of Borel functions, then*

$$\sup_{n\in\mathbb{N}} u_n, \quad \inf_{n\in\mathbb{N}} u_n, \quad \limsup_{n\to\infty} u_n, \quad \liminf_{n\to\infty} u_n \in \mathcal{L}^0_{\overline{\mathbb{R}}}(\mathscr{A}).$$

If the limit exists, then $\lim_{n\to\infty} u_n \in \mathcal{L}^0_{\overline{\mathbb{R}}}(\mathscr{A})$.

💬 $\lim_{n\to\infty} u_n$ is the **pointwise defined function** $x \mapsto \lim_{n\to\infty} u_n(x)$ (similarly $\inf_n$, $\sup_n$, etc.)

Proof. 1° $\sup_{n\in\mathbb{N}} u_n \in \mathcal{L}^0_{\overline{\mathbb{R}}}(\mathscr{A})$. We will show that

$$\{\sup_{n\in\mathbb{N}} u_n > \lambda\} = \bigcup_{n\in\mathbb{N}} \underbrace{\{u_n > \lambda\}}_{\in\mathscr{A}} \in \mathscr{A}.$$

We have

$$x \in \bigcup_{n\in\mathbb{N}} \{u_n > \lambda\} \implies \exists n_0 : \lambda < u_{n_0}(x) \leqslant \sup_{n\in\mathbb{N}} u_n(x)$$
$$\implies x \in \{\sup_{n\in\mathbb{N}} u_n(x) > \lambda\}.$$

Conversely, pick $x \in \{\sup_{n\in\mathbb{N}} u_n(x) > \lambda\}$ and assume that $u_n(x) \leqslant \lambda$ for all $n \in \mathbb{N}$. Then $\sup_{n\in\mathbb{N}} u_n(x) \leqslant \lambda$, which is impossible. This shows that we have equivalences in the above chain of implications.

2° $-u_n \in \mathcal{L}^0_{\overline{\mathbb{R}}}(\mathscr{A})$ This follows from $\{-u_n > \lambda\} = \{u_n < -\lambda\} \in \mathscr{A}$.

3° Step 1° and 2° imply that $\inf_{n\in\mathbb{N}} u_n = -\sup_{n\in\mathbb{N}}(-u_n) \in \mathcal{L}^0_{\overline{\mathbb{R}}}(\mathscr{A})$. Since lim inf and lim sup can be written in terms of inf and sup, we get that $\liminf_{n\to\infty} u_n$ and $\limsup_{n\to\infty} u_n$ are in $\mathcal{L}^0_{\overline{\mathbb{R}}}(\mathscr{A})$.

4° If $\lim_{n\to\infty} u_n$ exists, then $\lim_{n\to\infty} u_n = \liminf_{n\to\infty} u_n \in \mathcal{L}^0_{\overline{\mathbb{R}}}(\mathscr{A})$. □

7.13 Corollary. *If $u, v : \Omega \to \overline{\mathbb{R}}$ are measurable functions, then*

$$u \pm w, \quad u \cdot w, \quad u \vee w = \max\{u, w\}, \quad u \wedge w = \min\{u, w\}$$

are measurable – provided that these functions are everywhere defined.

⚠ Recall that we have the »$\infty - \infty$«, »$\frac{\infty}{\infty}$«-problem in $\overline{\mathbb{R}}$.

Proof. From Theorem 7.11 we know that there are simple functions such that $f_n \to u$ and $g_n \to w$. Note that

$$f_n \pm g_n, \quad f_n \cdot g_n, \quad f_n \vee g_n, \quad f_n \wedge g_n \in \mathcal{E}(\mathscr{A}).$$

These sequences converge to $u \pm w$, $u \cdot w$, $u \vee w$, and $u \wedge w$ (whenever these functions are defined), and so the claim follows from Corollary 7.12. □

7.14 Corollary. $u \in \mathcal{L}^0_{\overline{\mathbb{R}}}(\mathscr{A}) \iff u^+, u^- \in \mathcal{L}^0_{\overline{\mathbb{R}}}(\mathscr{A})$.

7.15 Corollary. *If $u, w \in \mathcal{L}^0_{\overline{\mathbb{R}}}(\mathscr{A})$, then*

$$\{u < w\}, \quad \{u \leqslant w\}, \quad \{u = w\}, \quad \{u \neq w\} \in \mathscr{A}.$$

Proof. This follows immediately from

$$\begin{aligned} \{u \leqslant w\} &= \Big(\{u \leqslant w\} \cap \{u = \infty\}\Big) \cup \Big(\{0 \leqslant w - u\} \cap \{u < \infty\}\Big) \\ &= \Big(\{w \geqslant \infty\} \cap \{u \geqslant \infty\}\Big) \cup \Big(\{0 \leqslant w - u\} \cap \{u < \infty\}\Big) \in \mathscr{A} \\ \{u = w\} &= \{u \leqslant w\} \cap \{u \geqslant w\}, \quad \{u \neq w\} = \{u = w\}^c, \quad \text{etc.} \end{aligned}$$ □

Let us close this chapter with two results which will be important in probability theory.

7.16 Lemma (factorization lemma). *Let $X : (\Omega, \mathscr{A}) \to (\Omega', \mathscr{A}')$ be a measurable map.*

$$\left.\begin{matrix} u : \Omega \to \overline{\mathbb{R}} \text{ is} \\ \sigma(X)/\mathscr{B}(\overline{\mathbb{R}})\text{-mble} \end{matrix}\right\} \iff \left\{\begin{matrix} \exists\, w : \Omega' \to \overline{\mathbb{R}}, \text{ which is} \\ \mathscr{A}'/\mathscr{B}(\overline{\mathbb{R}})\text{-mble and } u = w \circ X \end{matrix}\right.$$

Proof. »⇐« By the definition of $\sigma(X)$, the map X is $\sigma(X)$-measurable. Therefore, if w is $\mathscr{B}(\overline{\mathbb{R}})/\mathscr{A}'$-measurable, then

$$w \circ X : (\Omega, \sigma(X)) \xrightarrow[\text{mble}]{X} (\Omega', \mathscr{A}') \xrightarrow[\text{mble}]{w} (\overline{\mathbb{R}}, \mathscr{B}(\overline{\mathbb{R}})).$$

»⇒« Assume that u is $\sigma(X)$-measurable. We have to find a suitable w.

1° Assume that $u = \mathbb{1}_A$ is $\sigma(X)/\mathscr{B}(\overline{\mathbb{R}})$-measurable. In this case, we have

$$\begin{aligned} A \in \sigma(X) &\iff \exists A' \in \mathscr{A}' : A = X^{-1}(A') \\ &\implies u = \mathbb{1}_A = \mathbb{1}_{X^{-1}(A')} = \mathbb{1}_{A'} \circ X, \end{aligned}$$

and we can use $w = \mathbb{1}_{A'}$.

2° Assume that $u \in \mathcal{E}(\sigma(X))$. Applying the result of Step 1° to the terms of a standard representation of u shows that there is some $w \in \mathcal{E}(\mathscr{A})$ such that $u = w \circ X$.

3° Assume now that $u \in \mathcal{L}^0_{\overline{\mathbb{R}}}(\sigma(X))$. Using the sombrero lemma (Theorem 7.11, $\mathscr{A}$ is replaced by $\sigma(X)$), there is a sequence $f_n \in \mathscr{E}(\sigma(X))$ such that $f_n \to u$.
Step 2° shows that $f_n = w_n \circ X$ for some $w \in \mathcal{E}(\mathscr{A}')$.
Finally, the function $w := \liminf_{n\to\infty} w_n \in \mathcal{L}^0_{\overline{\mathbb{R}}}(\mathscr{A}')$ satisfies

$$w \circ X = (\liminf_{n\to\infty} w_n) \circ X = \lim_{n\to\infty} \underbrace{(w_n \circ X)}_{=f_n} = u. \qquad \square$$

▲ Note that $\lim_{n\to\infty} w_n$ may not exist, although $\lim_{n\to\infty}(w_n \circ X)$ exists.

Finally, here is a useful version of the monotone class theorem (MCT, Corollary 4.6) for functions.

7.17 Theorem* (MCT functional form). *Let $\mathscr{G} \subset \mathscr{P}(\Omega)$ be a $\cap$-stable family and $\mathcal{V}$ be a vector space of [bounded] functions $u : \Omega \to \mathbb{R}$ such that*

i) *$\mathbb{1} \in \mathcal{V}$ and $\mathbb{1}_G \in \mathcal{V}$ for all $G \in \mathscr{G}$;*
ii) *if $0 \leqslant u_1 \leqslant u_2 \leqslant \dots$, $u_n \in \mathcal{V}$, such that $u(x) := \sup_{n\in\mathbb{N}} u_n(x)$ is finite [bounded] for all $x \in \Omega$, then $u \in \mathcal{V}$.*

Under these assumptions $\mathcal{L}^0(\sigma(\mathscr{G})) \subset \mathcal{V}$ resp. $[\mathcal{L}^0_b(\sigma(\mathscr{G})) \subset \mathcal{V}]$.(3)

Proof. The family $\mathscr{D} := \{A \subset \Omega : \mathbb{1}_A \in \mathcal{V}\}$ is a Dynkin system. Indeed,

(D_1) We have $\Omega \in \mathscr{D}$ since $\mathbb{1} = \mathbb{1}_\Omega \in \mathcal{V}$.
(D_2) If $A \in \mathscr{D}$, then $\mathbb{1}_A \in \mathcal{V}$. Since $\mathcal{V}$ is a vector space containing $\mathbb{1}$, we get $\mathbb{1}_{A^c} = \mathbb{1} - \mathbb{1}_A \in \mathcal{V}$, and so $A^c \in \mathscr{D}$.
(D_3) Let $(A_n)_{n\in\mathbb{N}} \subset \mathscr{D}$ be a sequence of mutually disjoint sets. By assumption, $\mathbb{1}_{A_n} \in \mathcal{V}$ and $u_n := \mathbb{1}_{A_1} + \dots + \mathbb{1}_{A_n}$ is an increasing sequence of positive functions in $\mathcal{V}$. Because of ii), $\sup_{n\in\mathbb{N}} u_n = \mathbb{1}_{\bigcupdot_{n\in\mathbb{N}} A_n} \in \mathcal{V}$, i.e. $\bigcupdot_{n\in\mathbb{N}} A_n \in \mathscr{D}$.

(3) The subscript »b« in $\mathcal{L}^0_b$ denotes the bounded elements.

Since $\mathscr{G} \subset \mathscr{D}$ is $\cap$-stable, we know that $\sigma(\mathscr{G}) \subset \mathscr{D}$. This means, in particular, that all simple functions $\mathcal{E}(\sigma(\mathscr{G}))$ are contained in $\mathcal{V}$.

The sombrero lemma (Theorem 7.11) shows that every [bounded] positive, real-valued $u \in \mathcal{L}^{0,+}(\sigma(\mathscr{G}))$ can be approximated by an increasing sequence of $\sigma(\mathscr{G})$-measurable simple functions. Because of ii), we see that $\mathcal{L}^{0,+}_{[b]}(\sigma(\mathscr{G})) \subset \mathcal{V}$. Since $\mathcal{V}$ is a vector space, we conclude that every [bounded] measurable function $u = u^+ - u^- \in \mathcal{L}^0_{[b]}(\sigma(\mathscr{G}))$ is contained in $\mathcal{V}$. □

8 The integral of positive measurable functions

Throughout this chapter $(\Omega, \mathscr{A}, \mu)$ is a measure space and $\mathcal{E}(\mathscr{A})$, $\mathcal{L}^0(\mathscr{A})$ and $\mathcal{L}^0_{\overline{\mathbb{R}}}(\mathscr{A})$ denote the simple functions and the ($\overline{\mathbb{R}}$-valued) Borel functions. The superscript »+« as in $\mathcal{E}^+(\mathscr{A})$ or $\mathcal{L}^{0,+}(\mathscr{A})$ means that we consider only the positive members ($f \geqslant 0$) of the respective family.

The underlying **idea** of the integral is that it represents the »area below the graph« of a function. Consider, for example (see Fig. 8.1)

$$0 \leqslant f = \underbrace{\sum_{i=0}^{m} y_i \mathbb{1}_{A_i}}_{\text{standard representation}} \overset{\text{integral}}{\rightsquigarrow} I_\mu(f) = \sum_{i=0}^{m} y_i \mu(A_i) \qquad (\blacktriangle\ 0 \cdot \infty = 0)$$

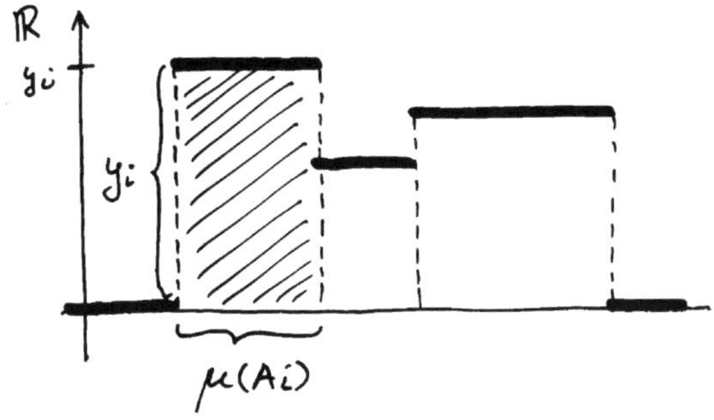

Fig. 8.1. Integral of a positive simple function

The **problem** is whether the sum appearing on the right-hand side is well-defined, i.e. independent of the chosen representation of f.

8.1 Lemma. *Assume that $f = \sum_{i=0}^m y_i \mathbb{1}_{A_i} = \sum_{k=0}^n z_k \mathbb{1}_{B_k}$ are two standard representations of the positive simple function $f \in \mathcal{E}^+(\mathscr{A})$. Then,*

$$\sum_{i=0}^{m} y_i \mu(A_i) = \sum_{k=0}^{n} z_k \mu(B_k).$$

Proof. Since $\Omega = A_0 \,\dot\cup \cdots \dot\cup\, A_m = B_0 \,\dot\cup \cdots \dot\cup\, B_n$, we have

$$A_i = \dot{\bigcup_{k=0}^{n}}(A_i \cap B_k) \quad \text{and} \quad B_k = \dot{\bigcup_{i=0}^{m}}(A_i \cap B_k).$$

Moreover, $y_i = z_k$ if $A_i \cap B_k \neq \emptyset$, and so

$$\begin{aligned}\sum_{i=0}^{m} y_i \mu(A_i) &= \sum_{i=0}^{m} y_i \sum_{k=0}^{n} \mu(A_i \cap B_k) \\ &= \sum_{i=0}^{m}\sum_{k=0}^{n} \underbrace{y_i \mu(A_i \cap B_k)}_{=z_k \mu(A_i \cap B_k)} \\ &= \sum_{k=0}^{n}\sum_{i=0}^{m} z_k \mu(A_i \cap B_k) \\ &= \sum_{k=0}^{n} z_k \mu(B_k).\end{aligned}$$ □

Since every positive simple function admits a standard representation, the following definition makes sense.

8.2 Definition. Let $f = \sum_{i=0}^{m} y_i \mathbb{1}_{A_i} \in \mathcal{E}^+(\mathscr{A})$ be given in standard representation. The (μ-)**integral** of f is defined as

$$I_\mu(f) := \sum_{i=0}^{m} y_i \mu(A_i) \in [0, \infty].$$

Let us briefly collect the main properties of the μ-integral.

8.3 Lemma. *Let $f, g \in \mathcal{E}^+(\mathscr{A})$ be positive simple functions.*

a) $I_\mu(\mathbb{1}_A) = \mu(A) \quad \forall A \in \mathscr{A}$

b) $I_\mu(\lambda f) = \lambda I_\mu(f) \quad \forall \lambda \geqslant 0$ *(positively homogeneous)*

c) $I_\mu(f + g) = I_\mu(f) + I_\mu(g)$ *(additive)*

d) $I_\mu(f) \leqslant I_\mu(g) \quad \forall f \leqslant g$ *(monotone)*

Proof. The properties a) and b) are obvious from the definition.

c) If $f = \sum_{i=0}^{m} y_i \mathbb{1}_{A_i}$ and $g = \sum_{k=0}^{n} z_k \mathbb{1}_{B_k}$ are standard representations, then

$$f + g = \sum_{i=0}^{m}\sum_{k=0}^{n} (y_i + z_k)\mathbb{1}_{A_i \cap B_k} \in \mathcal{E}^+(\mathscr{A})$$

is also a standard representation (Example 7.9.b), hence

$$\begin{aligned}
I_\mu(f+g) &= \sum_{i=0}^{m}\sum_{k=0}^{n}(y_i+z_k)\mu(A_i\cap B_k) \\
&= \sum_{i=0}^{m}\sum_{k=0}^{n} y_i\mu(A_i\cap B_k) + \sum_{i=0}^{m}\sum_{k=0}^{n} z_k\mu(A_i\cap B_k) \\
&= \sum_{i=0}^{m} y_i \sum_{k=0}^{n}\mu(A_i\cap B_k) + \sum_{k=0}^{n} z_k \sum_{i=0}^{m}\mu(A_i\cap B_k) \\
&= \sum_{i=0}^{m} y_i\mu(A_i) + \sum_{k=0}^{n} z_k\mu(B_k) = I_\mu(f)+I_\mu(g).
\end{aligned}$$

d) If $f \leqslant g$, then $g = f + (g-f)$ and $g-f$ is a positive simple function. Thus, by Part c),

$$I_\mu(g) = I_\mu(f) + \overbrace{I_\mu(g-f)}^{\geqslant 0} \geqslant I_\mu(f). \qquad \square$$

Using the sombrero lemma (Theorem 7.11) we can approximate any positive Borel function $u \in \mathcal{L}^{0,+}_{\overline{\mathbb{R}}}(\mathscr{A})$ with an increasing sequence of functions from $\mathcal{E}^+(\mathscr{A})$. This motivates our next definition.

8.4 Definition. The **(μ-)integral** of a positive measurable function (a Borel function) $u \in \mathcal{L}^{0,+}_{\overline{\mathbb{R}}}(\mathscr{A})$ is defined as

$$\int u \, d\mu := \sup\left\{I_\mu(g) \mid g \in \mathcal{E}(\mathscr{A}),\ 0 \leqslant g \leqslant u\right\} \in [0,\infty]. \tag{8.1}$$

⚠ The integral $\int u\,d\mu$ may be $+\infty$.

💬 Since the supremum ranges over **all** $\mathcal{E}^+ \ni f \leqslant u$, there is no well-definedness issue in (8.1).

💬 In order to emphasize the integration variable, we also write

$$\int u(x)\,\mu(dx) \quad \text{or} \quad \int u(x)\,d\mu(x).$$

8.5 Lemma. *The integral* $\int \cdots d\mu$ *extends* $I_\mu(\cdot)$, *i.e.*

$$\forall f \in \mathcal{E}^+(\mathscr{A}) : \int f\,d\mu = I_\mu(f).$$

Proof. If $f \in \mathcal{E}^+(\mathscr{A})$, then $g := f \leqslant f$ is an admissible function in the sup in (8.1), i.e.

$$I_\mu(f) \leqslant \sup\left\{I_\mu(g) \mid g \in \mathcal{E}(\mathscr{A}),\ 0 \leqslant g \leqslant f\right\} = \int f\,d\mu.$$

Conversely, if $\mathcal{E}^+(\mathscr{A}) \ni g \leqslant f$, then $\mathrm{I}_\mu(g) \leqslant \mathrm{I}_\mu(f)$, and we get

$$\int f \,\mathrm{d}\mu = \sup_{g \leqslant f,\, g \in \mathcal{E}^+} \mathrm{I}_\mu(g) \leqslant \mathrm{I}_\mu(f). \qquad \square$$

The nex result is the first in a series of so-called **convergence theorems** which deal with the problem to interchange $\int$ and $\lim_n$.

8.6 Theorem (Beppo Levi – »BL«). *Let $(u_n)_{n\in\mathbb{N}} \subset \mathcal{L}^{0,+}_{\overline{\mathbb{R}}}(\mathscr{A})$ be an increasing sequence of positive measurable functions: $0 \leqslant u_1 \leqslant u_2 \leqslant u_3 \leqslant \dots$. The limit $u := \sup_{n\in\mathbb{N}} u_n$ always exists in $[0,\infty]$, is measurable $u \in \mathcal{L}^{0,+}_{\overline{\mathbb{R}}}(\mathscr{A})$, and satisfies*

$$\int \sup_{n\in\mathbb{N}} u_n \,\mathrm{d}\mu = \sup_{n\in\mathbb{N}} \int u_n \,\mathrm{d}\mu. \tag{8.2}$$

💬 $u_n \leqslant u_{n+1}$ means $\forall x \in \Omega : u_n(x) \leqslant u_{n+1}(x)$.
💬 $u_n \uparrow u$ is short for $u_n \leqslant u_{n+1} \leqslant \dots$ and $\lim_{n\to\infty} u_n = u$.
💬 if $u_n \uparrow u$, then $\lim_{n\to\infty} u_n = \sup_{n\in\mathbb{N}} u_n$ is an increasing limit.

Proof. Since the sequence $(u_n(x))_{n\in\mathbb{N}}$ is increasing, the limit always exists in $[0,\infty]$. Corollary 7.12 shows that $u = \sup_{n\in\mathbb{N}} u_n$ is measurable. We will now proceed in several steps; claims are highlighted. Throughout this proof f, g denote positive simple functions.

1° $u, w \in \mathcal{L}^{0,+}_{\overline{\mathbb{R}}}(\mathscr{A}),\ u \leqslant w \implies \int u \,\mathrm{d}\mu \leqslant \int w \,\mathrm{d}\mu.$ Since $f \leqslant u$ entails $f \leqslant w$, the sets appearing under the sup in (8.1) increase with u, and so

$$\begin{aligned} \int u \,\mathrm{d}\mu &= \sup\left\{\mathrm{I}_\mu(f) \mid f \leqslant u,\ f \in \mathcal{E}^+\right\} \\ &\leqslant \sup\left\{\mathrm{I}_\mu(g) \mid g \leqslant w,\ g \in \mathcal{E}^+\right\} = \int w \,\mathrm{d}\mu. \end{aligned}$$

2° $\sup_{n\in\mathbb{N}} \int u_n \,\mathrm{d}\mu \leqslant \int \sup_{n\in\mathbb{N}} u_n \,\mathrm{d}\mu.$ Obviously, $u_k \leqslant \sup_{n\in\mathbb{N}} u_n$, and by Step 1°

$$\forall k : \int u_k \,\mathrm{d}\mu \leqslant \underbrace{\int \sup_{n\in\mathbb{N}} u_n \,\mathrm{d}\mu}_{\text{no dependence on } k} \implies \sup_{k\in\mathbb{N}} \int u_k \,\mathrm{d}\mu \leqslant \int \sup_{n\in\mathbb{N}} u_n \,\mathrm{d}\mu.$$

3° $f \leqslant u \implies \mathrm{I}_\mu(f) \leqslant \sup_{n\in\mathbb{N}} \int u_n \,\mathrm{d}\mu.$ Let $f \leqslant u$, and fix $\alpha \in (0,1)$. From the definition of $\sup_{n\in\mathbb{N}} u_n(x)$ we get

$$\forall x \quad \exists \mathrm{N}(x,\alpha) \in \mathbb{N} \quad \forall n \geqslant \mathrm{N}(x,\alpha) : \quad \alpha f(x) \leqslant u_n(x).$$

Since $f \leqslant u$ and $u_n \uparrow u$, we see that the sets $\mathrm{B}_n := \{x \mid \alpha f(x) \leqslant u_n(x)\}$ are measurable and $\mathrm{B}_n \uparrow \Omega$. By the definition of B_n, $\alpha f \leqslant u_n$ on B_n, and so

$$\alpha \mathbb{1}_{\mathrm{B}_n} \cdot f \overset{\text{def.}}{\leqslant} \mathbb{1}_{\mathrm{B}_n} \cdot u_n \overset{\mathbb{1}_{\mathrm{B}_n} \leqslant 1}{\leqslant} u_n.$$

If $f = \sum_{i=0}^{m} y_i \mathbb{1}_{A_i}$ is any standard representation, we get from $B_n \uparrow \Omega$

$$\alpha \underbrace{\sum_{i=0}^{m} y_i \overbrace{\mu(B_n \cap A_i)}^{\xrightarrow{B_n\uparrow\Omega} \mu(A_i)}}_{\xrightarrow[n\to\infty]{} \alpha\sum_{i=0}^{m} y_i\mu(A_i)} = I_\mu(\alpha \mathbb{1}_{B_n} \cdot f) \leqslant \int u_n \,d\mu \leqslant \underbrace{\sup_{i\in\mathbb{N}} \int u_i \,d\mu}_{\text{no dependence on } n}.$$

Thus, $\alpha I_\mu(f) \leqslant \sup_{i\in\mathbb{N}} \int u_i \,d\mu$, and the claim follows upon letting $\alpha \uparrow 1$.

4° From Step 3° we get for all simple functions satisfying $f \leqslant u$

$$I_\mu(f) \overset{\S 8.5}{=} \int f \,d\mu \leqslant \sup_{n\in\mathbb{N}} \int u_n \,d\mu$$

$$\implies \int u \,d\mu = \sup_{f\leqslant u} I_\mu(f) \leqslant \sup_{n\in\mathbb{N}} \int u_n \,d\mu$$

which is the reverse of the inequality from Step 2° . □

Often, the following consequence of the BL theorem for series is useful. In fact, it is even equivalent to BL ✎.

8.7 Corollary★. *If* $(w_i)_{i\in\mathbb{N}} \subset \mathcal{L}^{0,+}_{\overline{\mathbb{R}}}(\mathscr{A})$*, then* $\sum_{i=1}^{\infty} w_i \in \mathcal{L}^{0,+}_{\overline{\mathbb{R}}}(\mathscr{A})$ *and*

$$\sum_{i=1}^{\infty} \int w_i \,d\mu = \int \sum_{i=1}^{\infty} w_i \,d\mu \in [0,\infty].$$

Proof. Use $u_n := \sum_{i=1}^{n} w_i \uparrow \sum_{i=1}^{\infty} w_i =: u$ and Theorem 8.6. □

The following corollary is an important special case of Theorem 8.6.

8.8 Corollary. *Let* $u \in \mathcal{L}^{0,+}_{\overline{\mathbb{R}}}(\mathscr{A})$*. For any (!) sequence* $(f_n)_{n\in\mathbb{N}} \subset \mathcal{E}^+(\mathscr{A})$ *such that* $f_n \uparrow u$ *one has*

$$\int u \,d\mu = \lim_{n\to\infty} \int f_n \,d\mu. \tag{8.3}$$

💬 The equality (8.3) is often used to define the integral $\int u \,d\mu$. The problem is to show that $\int u \,d\mu$ is well-defined, i.e. independent of the approximating sequence. Our approach avoids this. Nevertheless, (8.3) is important as it shows that the sup in (8.1) is actually a(n increasing) limit, and $u \mapsto \int u \,d\mu$ is linear.

Combining Corollary 8.8 with the sombrero lemma (Theorem 7.11) and Lemma 8.3 yields

8.9 Corollary. *Let $u, w \in \mathcal{L}^{0,+}_{\overline{\mathbb{R}}}(\mathscr{A})$ be positive measurable functions.*

a) $\int \mathbb{1}_A \, d\mu = \mu(A) \quad \forall A \in \mathscr{A}$;

b) $\int \alpha u \, d\mu = \alpha \int u \, d\mu \quad \forall \alpha \geqslant 0$; *(positively homogeneous)*

c) $\int (u + w) \, d\mu = \int u \, d\mu + \int w \, d\mu$; *(additive)*

d) $\int u \, d\mu \leqslant \int w \, d\mu \quad \forall u \leqslant w$. *(monotone)*

8.10 Example. a) Let $(\Omega, \mathscr{A})$ be an arbitrary measurable space and $\mu = \delta_y$ for some fixed $y \in \Omega$. For every $u \in \mathcal{L}^{0,+}_{\overline{\mathbb{R}}}(\mathscr{A})$ we have

$$\int u \, d\delta_y = \int u(x) \, \delta_y(dx) = u(y). \tag{8.4}$$

Indeed, if $f = \sum_{i=0}^{m} \phi_i \mathbb{1}_{A_i}$ is a positive simple function in standard representation, then there is a unique $i_0 \in \{0, 1, \dots, m\}$ such that the point $y \in A_{i_0}$. Thus,

$$\int f \, d\delta_y = \sum_{i=0}^{m} \phi_i \delta_y(A_i) = \phi_{i_0} = f(y).$$

By the sombrero lemma, there is a sequence $\mathcal{E}^+ \ni f_n \uparrow u$, and we get

$$u(y) = \lim_{n\to\infty} f_n(y) = \lim_{n\to\infty} \int f_n \, d\delta_y \overset{\text{BL}}{=} \int u \, d\delta_y. \qquad \square$$

b) Let $(\Omega, \mathscr{A}) = (\mathbb{N}, \mathscr{P}(\mathbb{N}))$ and $\mu = \sum_{i=1}^{\infty} \alpha_i \delta_i$ mit $\alpha_i \geqslant 0$. From Example 3.5.d) we know that μ is a measure on $\mathbb{N}$. Since $\mathscr{A}$ is the power set, **every** function $u : (\mathbb{N}, \mathscr{P}(\mathbb{N})) \to (\mathbb{R}^+, \mathscr{B}(\mathbb{R}^+))$ is measurable; from the (tautological) identity

$$u(k) = \sum_{i\in\mathbb{N}} u(i) \mathbb{1}_{\{i\}}(k)$$

we conclude that

$$\begin{aligned} \int u \, d\mu &= \int \sum_{i\in\mathbb{N}} u(i) \mathbb{1}_{\{i\}} \, d\mu \\ &\overset{\text{BL}}{=} \sum_{i\in\mathbb{N}} u(i) \int \mathbb{1}_{\{i\}} \, d\mu \\ &= \sum_{i\in\mathbb{N}} u(i) \mu(\{i\}) \\ &= \sum_{i\in\mathbb{N}} \alpha_i u(i). \qquad \square \end{aligned}$$

c) If we choose $\alpha_i = 1$ and identify the function u from b) with the sequence $(u_i)_{i\in\mathbb{N}}$, $u_i := u(i)$, then we see that every series can be written as an integral w.r.t. counting measure:

$$\sum_{i=1}^{\infty} u_i = \int u \, d\nu \quad \text{and} \quad \nu(A) = \sum_{i=1}^{\infty} \delta_i(A) = \#A. \qquad \square$$

8.11 Theorem (Fatou's lemma). *Let $(u_i)_{i\in\mathbb{N}} \subset \mathcal{L}^{0,+}_{\overline{\mathbb{R}}}(\mathscr{A})$ be a sequence of positive measurable functions $u_i \geqslant 0$. Then $u := \liminf_{n\to\infty} u_n \in \mathcal{L}^{0,+}_{\overline{\mathbb{R}}}(\mathscr{A})$ and*

$$\int \liminf_{n\to\infty} u_n \, d\mu \leqslant \liminf_{n\to\infty} \int u_n \, d\mu. \tag{8.5}$$

Proof. Recall that $\liminf_{n\to\infty} u_n := \sup_{k\in\mathbb{N}}\left(\inf_{n\geqslant k} u_n\right) \in \overline{\mathbb{R}}$ is measurable (Corollary 7.12). Since $\inf_{n\geqslant k} u_n \uparrow u$ as $k \uparrow \infty$, we get

$$\int \liminf_{n\to\infty} u_n \, d\mu \overset{\text{BL}}{=} \sup_{k\in\mathbb{N}} \underbrace{\int \inf_{n\geqslant k} u_n \, d\mu}_{\leqslant \int u_\ell \, d\mu \ \forall \ell\geqslant k} \overset{\S}{\leqslant} \sup_{k\in\mathbb{N}} \left(\inf_{\ell\geqslant k} \int u_\ell \, d\mu \right)$$

$$= \liminf_{\ell\to\infty} \int u_\ell \, d\mu. \qquad \square$$

💬 A close inspection of the following proof reveals the true nature of the assumption $u_n \geqslant 0$ (see the step marked by »§«): It ensures that $\int u_n \, d\mu > -\infty$, i.e. 0 is actually an »integrable minorant« for all u_n.

9 The integral of measurable functions

Throughout this chapter, $(\Omega, \mathscr{A}, \mu)$ is any measure space. Our aim is to extend the integral from $\mathcal{L}^{0,+}_{\overline{\mathbb{R}}}(\mathscr{A})$ to measurable functions with arbitrary sign. Since we want that the integral is **linear**, we will use **linearity** to extend it. Assume that $u \in \mathcal{L}^{0}_{\overline{\mathbb{R}}}(\mathscr{A})$ can be written as $u = f - g$ for some positive $f, g \in \mathcal{L}^{0,+}_{\overline{\mathbb{R}}}(\mathscr{A})$. Since we always have $u = u^+ - u^-$, we see

$$u^+ + g = f + u^- \implies \int u^+ \, d\mu + \int g \, d\mu = \int f \, d\mu + \int u^- \, d\mu$$
$$\implies \int u^+ \, d\mu - \int u^- \, d\mu = \int f \, d\mu - \int g \, d\mu$$

provided that we can avoid »$\infty - \infty$« in the last step. In particular, the following definition does not depend on the representation of u.

9.1 Definition. A function $u : \Omega \to \overline{\mathbb{R}}$ is said to be (μ-)**integrable**, if

$$u \in \mathcal{L}^{0}_{\overline{\mathbb{R}}}(\mathscr{A}) \quad \text{and} \quad \int u^+ \, d\mu < \infty, \quad \int u^- \, d\mu < \infty.$$

The (μ-)**integral** of u is given by

$$\int u \, d\mu := \int u^+ \, d\mu - \int u^- \, d\mu \in \mathbb{R}. \tag{9.1}$$

The (μ-)**integrable functions** are denoted by $\mathcal{L}^1(\mu)$, resp., $\mathcal{L}^1_{\overline{\mathbb{R}}}(\mu)$.[(1)]

9.2 Remark. a) $\int u \, d\mu = \int u(x)\,\mu(dx) = \int u(x)\, d\mu(x)$.

b) If $\mu = \lambda^d$ is Lebesgue measure, u is said to be **Lebesgue integrable** and we often write $\lambda^d(dx) = dx$ and $\int u(x)\,dx$ etc.

c) The requirement $\int u^{\pm}\, d\mu < \infty$ excludes the case »$\infty - \infty$«. For positive $u \geqslant 0$ we still allow $\int u\, d\mu = \infty$, but $u \in \mathcal{L}^1_{\overline{\mathbb{R}}}(\mu)$ holds only if $\int u \, d\mu < \infty$.

💬 In order to avoid »$\infty - \infty$« in (9.1), it is enough to assume that **one** of the integrals $\int u^+\, d\mu$ or $\int u^-\, d\mu$ is finite, so $\int u\, d\mu \in [-\infty, \infty)$ or $\int u \, d\mu \in (-\infty, \infty]$. This is sometimes useful, and some books do this, but we won't.

9.3 Theorem (integrability criterion). *Let $u \in \mathcal{L}^0_{\overline{\mathbb{R}}}(\mathscr{A})$ be a measurable function. The following are equivalent:*

a) $u \in \mathcal{L}^1_{\overline{\mathbb{R}}}(\mu)$;

b) $u^+, u^- \in \mathcal{L}^1_{\overline{\mathbb{R}}}(\mu)$;

c) $|u| \in \mathcal{L}^1_{\overline{\mathbb{R}}}(\mu)$;

d) $\exists w \in \mathcal{L}^1_{\overline{\mathbb{R}}}(\mu) : |u| \leqslant w$.

Proof. a)⇔b) This is just the definition of the integral.

b)⇒c) $|u| = u^+ + u^-$, and so $\int |u|\, d\mu = \int u^+\, d\mu + \int u^-\, d\mu < \infty$.

c)⇒d) $w := |u|$.

d)⇒b) $u^{\pm} \leqslant |u| \leqslant w$, hence $\int u^{\pm}\, d\mu \leqslant \int w \, d\mu < \infty$. □

The integral is defined in such a way that all properties of the integral of positive measurable functions $\mathcal{L}^{0,+}_{\overline{\mathbb{R}}}(\mathscr{A})$ extend to $\mathcal{L}^1_{\overline{\mathbb{R}}}(\mu)$.

9.4 Theorem. *Let $u, w \in \mathcal{L}^1_{\overline{\mathbb{R}}}(\mu)$ and $\alpha \in \mathbb{R}$.*

a) $\alpha u \in \mathcal{L}^1_{\overline{\mathbb{R}}}(\mu)$, $\int \alpha u \, d\mu = \alpha \int u \, d\mu$ *(homogeneous)*

b) $\underbrace{u + w}_{\text{if defined}} \in \mathcal{L}^1_{\overline{\mathbb{R}}}(\mu)$, $\int (u+w)\, d\mu = \int u\, d\mu + \int w \, d\mu$ *(additive)*

c) $u \vee w,\ u \wedge w \in \mathcal{L}^1_{\overline{\mathbb{R}}}(\mu)$ *(lattice)*

d) $\int u \, d\mu \leqslant \int w\, d\mu \quad u \leqslant w$ *(monotone)*

e) $\left|\int u\, d\mu\right| \leqslant \int |u|\, d\mu$ *(triangle inequality)*

Proof. Since αu, $u + w$ (if everywhere defined, i.e. excluding »$\infty - \infty$«), $u \vee w$, $u \wedge w$ are all measurable (Corollary 7.13), we have to check only the finiteness of the integral of the absolute values:

a) $|\alpha u| = |\alpha| \cdot |u| \in \mathcal{L}^1_{\overline{\mathbb{R}}}(\mu)$ b/o Theorem 9.3, Lemma 8.9.b).

b) $|u + w| \leqslant |u| + |w| \in \mathcal{L}^1_{\overline{\mathbb{R}}}(\mu)$ b/o Theorem 9.3, Lemma 8.9.c).

c) $|u \vee w|$, $|u \wedge w| \leqslant |u| + |w| \in \mathcal{L}^1_{\overline{\mathbb{R}}}(\mu)$ b/o Theorem 9.3, Lemma 8.9.c).

(1) If we want to stress the σ-algebra or the underlying space we also write $\mathcal{L}^1(\mathscr{A})$ or $\mathcal{L}^1(\Omega)$.

d) If $u \leqslant w$, then $u^+ \leqslant w^+$ and $w^- \leqslant u^-$, and so

$$\int u\,d\mu = \int u^+\,d\mu - \int u^-\,d\mu \leqslant \int w^+\,d\mu - \int w^-\,d\mu = \int w\,d\mu.$$

e) Since $\pm u \leqslant |u|$, we get from d)

$$\pm \int u\,d\mu = \int (\pm u)\,d\mu \leqslant \int |u|\,d\mu \implies \left|\int u\,d\mu\right| \leqslant \int |u|\,d\mu. \qquad \square$$

9.5 Remark. If $\alpha u + \beta w$ exists in $\overline{\mathbb{R}}$, i.e. if »$\infty - \infty$« can be excluded, then 9.4.a) and 9.4.b) can be combined into

$$\int (\alpha u + \beta w)\,d\mu = \alpha \int u\,d\mu + \beta \int w\,d\mu. \quad \text{(linearity)}$$

If u, w are real-valued, this is always true. In particular, $\mathcal{L}^1(\mu) = \mathcal{L}^1_{\mathbb{R}}(\mu)$ is an $\mathbb{R}$-vector space with addition $(u+w)(x) = u(x) + w(x)$ and scalar multiplication $(\alpha u)(x) = \alpha u(x)$ $(\alpha \in \mathbb{R})$. The mapping $\mathcal{L}^1(\mu) \ni u \mapsto \int u\,d\mu$ is a positive linear form.

9.6 Example (cf. Example 3.5, Example 8.10).

a) Let $(\Omega, \mathscr{A})$ be an arbitrary measurable space and δ_y the Dirac measure for a fixed $y \in \Omega$. Then $u \in \mathcal{L}^1_{\overline{\mathbb{R}}}(\delta_y)$ if, and only if, $u \in \mathcal{L}^0_{\overline{\mathbb{R}}}(\mathscr{A})$ and $|u(y)| < \infty$; if so, the integral is given by $\int u(x)\,\delta_y(dx) = u(y)$.

b) Consider on $(\mathbb{N}, \mathscr{P}(\mathbb{N}))$ the measure $\mu = \sum_{i=1}^\infty \alpha_i \delta_i$, $\alpha_i \geqslant 0$. Since **all** functions $u : \mathbb{N} \to \mathbb{R}$ are measurable, we know from Example 8.10

$$\int |u|\,d\mu = \sum_{i=1}^\infty \alpha_i |u(i)|.$$

Therefore, $u \in \mathcal{L}^1(\mu) \iff \sum_{i=1}^\infty \alpha_i |u(i)| < \infty$.

If $\alpha_1 = \alpha_2 = \ldots = 1$, then $\mathcal{L}^1(\mu)$ is the space of all (**absolutely**) **summable sequences**:

$$\ell^1(\mathbb{N}) = \mathcal{L}^1\left(\sum_{i\in\mathbb{N}} \delta_i\right) = \left\{(u_i)_{i\in\mathbb{N}} \mid \sum_{i=1}^\infty |u_i| < \infty\right\}.$$

c) Let $(\Omega, \mathscr{A}, \mathbb{P})$ be a probability space and $X : \Omega \to \mathbb{R}$ a random variable (i.e. a measurable function). Probabilists call the integral (if it exists)

$$\mathbb{E}X := \int X\,d\mathbb{P}$$

the **expectation** or the **expected value** of X.

Important special case: If X is **bounded**, i.e. $\sup_{\omega\in\Omega} |X(\omega)| < \infty$, then

$$\int |X|\,d\mathbb{P} \leqslant \int \sup_{\omega\in\Omega} |X(\omega)|\,\mathbb{P}(d\omega) = \sup_{\omega\in\Omega} |X(\omega)| \underbrace{\int 1\,\mathbb{P}(d\omega)}_{=\mathbb{P}(\Omega)=1} < \infty.$$

This means that **bounded random variables** on a probability space $(\Omega, \mathscr{A}, \mathbb{P})$ are integrable, i.e. $\mathbb{E}X$ exists.

⚠ The converse is false: Consider, e.g. $\left(\mathbb{N}, \mathscr{P}(\mathbb{N}), \sum_{i=1}^{\infty} 2^{-i}\delta_i\right)$ and the random variable $X(i) = i$.

9.7 Definition. For $u \in \mathcal{L}^1_{\overline{\mathbb{R}}}(\mu)$ or $u \in \mathcal{L}^{0,+}_{\overline{\mathbb{R}}}(\mathscr{A})$ we set

$$\int_A u \, d\mu := \int \underbrace{u \mathbb{1}_A}_{\text{measurable b/o §7.13}} d\mu = \int u(x) \mathbb{1}_A(x) \mu(dx) \quad A \in \mathscr{A}.$$

💬 $\int_\Omega u \, d\mu = \int u \, d\mu$, since $\mathbb{1}_\Omega \equiv 1$.

9.8 Lemma (and Definition). *If $u \in \mathcal{L}^{0,+}_{\overline{\mathbb{R}}}(\mathscr{A})$, then*

$$\nu(A) := \int_A u \, d\mu := \int u \mathbb{1}_A \, d\mu, \quad A \in \mathscr{A}$$

is a measure on $(\Omega, \mathscr{A})$. ν is called a ***measure with density*** *u (w.r.t. μ). Notation: $\nu = u \cdot \mu$.*

Proof. Immediate from Corollary 8.7. □

9.9 Remark. If $\nu = u \cdot \mu$, the density u is often denoted by $u = \frac{d\nu}{d\mu}$. This notation is inspired by the fundamental theorem of integral calculus:

$$U(x) - U(a) = \int_a^x u(y) \, dy \implies \frac{dU}{dx} = u.$$

Recall that $\lambda(dx) = dx$ where $\lambda = \lambda^1$ Lebesgue measure on $\mathbb{R}$. Thus,

$$\frac{d(u\lambda)}{d\lambda} = \frac{dU}{dx} = u.$$

If $u \geqslant 0$, then we get a measure ν via $\nu[a,b) = U(b) - U(a) = \int_{[a,b)} u \, d\lambda$.

This story will be continued in Chapter 18.

10 Null sets

In this chapter, $(\Omega, \mathscr{A}, \mu)$ is an arbitrary measurable space. Recall the definition of (measurable) null sets $\mathscr{N}_\mu$ and (not necessarily measurable) subsets of null sets $\widetilde{\mathscr{N}_\mu}$,

$$\mathscr{N}_\mu = \{N \in \mathscr{A} \mid \mu(N) = 0\} \quad \text{and} \quad \widetilde{\mathscr{N}_\mu} = \left\{M \subset \Omega \mid \exists N \in \mathscr{N}_\mu,\ M \subset N\right\}.$$

10.1 Definition. A property $\Pi(x)$, $x \in \Omega$, is said to hold (μ-)**almost everywhere** (μ-)**a.e.** or **for** (μ-)**almost all** x ((μ)-**a.a.**), if

$$\left\{x \in \Omega \mid \Pi(x) \text{ fails}\right\} \subset N \text{ for some } N \in \mathcal{N}_\mu.$$

⚠ Note that $\left\{x \in \Omega \mid \Pi(x) \text{ fails}\right\} \in \widetilde{\mathcal{N}_\mu}$ need not be measurable.

💬 $u = w$ a.e. $\iff \{u \neq w\} \subset N \in \mathcal{N}_\mu$.

💬 If u, w are measurable, then $u = w$ a.e. $\iff \{u \neq w\} \in \mathcal{N}_\mu$.

⚠ If $u = w$ a.e. it may happen that u is measurable while w isn't; this means that $\{u \neq w\} \notin \mathscr{A}$ is possible.

10.2 Theorem. *Let* $u \in \mathcal{L}^0_{\overline{\mathbb{R}}}(\mathscr{A})$.

a) $\int |u| \, d\mu = 0 \iff |u| = 0$ *a.e.* $\iff \mu\{u \neq 0\} = 0$.

b) $N \in \mathcal{N}_\mu \implies \int_N u \, d\mu = 0$.

Proof. Let us first check b). Since $|u| \wedge k \uparrow |u|$ as $k \uparrow \infty$, we can use BL (Theorem 8.6) and find

$$\left|\int_N u \, d\mu\right| = \left|\int u \mathbb{1}_N \, d\mu\right| \leqslant \int |u| \mathbb{1}_N \, d\mu \overset{\text{BL}}{=} \sup_{k \in \mathbb{N}} \underbrace{\int (|u| \wedge k) \mathbb{1}_N \, d\mu}_{\leqslant \int k \mathbb{1}_N \, d\mu = k\mu(N) = 0} = 0.$$

a) From $|u| = 0$ a.e. $\iff u = 0$ a.e. $\iff \mu\{u \neq 0\} = 0$ we get the second equivalence. The direction »$\Leftarrow$« of the first equivalence holds, since

$$\int |u| \, d\mu = \int_{\{u \neq 0\}} |u| \, d\mu + \int_{\{u = 0\}} |u| \, d\mu = \underbrace{\int_{\{u \neq 0\}} |u| \, d\mu}_{=0 \text{ b/o b)}} + \underbrace{\int_{\{u = 0\}} 0 \, d\mu}_{=0} = 0.$$

For the converse direction »$\Rightarrow$« we use the following **Markov inequality**: For all measurable sets $A \in \mathscr{A}$ and all constants $c > 0$ we have

$$\begin{aligned}\mu\Big(\{|u| \geqslant c\} \cap A\Big) &= \int \mathbb{1}_{\{|u| \geqslant c\} \cap A} \, d\mu \\ &= \int_A \frac{c}{c} \mathbb{1}_{\{|u| \geqslant c\}} \, d\mu \\ &\leqslant \int_A \frac{|u|}{c} \mathbb{1}_{\{|u| \geqslant c\}} \, d\mu \leqslant \frac{1}{c} \int_A |u| \, d\mu.\end{aligned}$$

In particular, if we take $A = \Omega$, then

$$\mu\{|u| > 0\} = \mu\Big(\bigcup_{n \in \mathbb{N}} \{|u| \geqslant \tfrac{1}{n}\}\Big) \leqslant \sum_{n \in \mathbb{N}} \mu\{|u| \geqslant \tfrac{1}{n}\} \leqslant \sum_{n \in \mathbb{N}} n \underbrace{\int |u| \, d\mu}_{=0} = 0.$$ □

10.3 Corollary. *Let $u, w \in \mathcal{L}^0_{\overline{\mathbb{R}}}(\mathscr{A})$ and assume that $u = w$ a.e.*

a) *If $u, w \geqslant 0$, then $\int u \, d\mu = \int w \, d\mu \in [0, \infty]$.*

b) *If $u \in \mathcal{L}^1_{\overline{\mathbb{R}}}(\mu)$, then $w \in \mathcal{L}^1_{\overline{\mathbb{R}}}(\mu)$ and $\int u \, d\mu = \int w \, d\mu$.*

Proof. Since the functions u, w are measurable, we have $\{u \neq w\} \in \mathcal{N}_\mu$. Therefore,

a)
$$\begin{aligned} \int u \, d\mu &= \int_{\{u=w\}} u \, d\mu + \int_{\{u \neq w\}} u \, d\mu \\ &= \int_{\{u=w\}} w \, d\mu + 0 \\ &= \int_{\{u=w\}} w \, d\mu + \int_{\{u \neq w\}} w \, d\mu = \int w \, d\mu. \end{aligned}$$

b) From $u = w$ a.e. we see that $u^\pm = w^\pm$ a.e., and Part a), applied to $u^\pm, w^\pm$ proves the claim. □

10.4 Corollary. *Assume that $u \in \mathcal{L}^0_{\overline{\mathbb{R}}}(\mathscr{A})$ and $w \in \mathcal{L}^1_{\overline{\mathbb{R}}}(\mu)$. If $|u| \leqslant w$ a.e., then $u \in \mathcal{L}^1_{\overline{\mathbb{R}}}(\mu)$.*

Proof. From

$$\int |u| \, d\mu = \int_{|u| \leqslant w} |u| \, d\mu + \int_{|u| > w} |u| \, d\mu \leqslant \int_{|u| \leqslant w} w \, d\mu + 0 \leqslant \int w \, d\mu$$

we conclude that $u \in \mathcal{L}^1_{\overline{\mathbb{R}}}(\mu)$. □

10.5 Corollary (Markov inequality). *For every $u \in \mathcal{L}^0_{\overline{\mathbb{R}}}(\mathscr{A})$ we have*

$$\mu\big(\{|u| \geqslant c\} \cap A\big) \leqslant \frac{1}{c} \int_A |u| \, d\mu \quad \forall c > 0,\ A \in \mathscr{A}, \tag{10.1}$$

$$\mu\{|u| \geqslant c\} \leqslant \frac{1}{c} \int |u| \, d\mu \quad \forall c > 0. \tag{10.2}$$

Proof. See the proof of Theorem 10.2.a), direction »⇒«. □

10.6 Corollary. *Every $u \in \mathcal{L}^1_{\overline{\mathbb{R}}}(\mu)$ is a.e. $\mathbb{R}$-valued. In particular, for each $u \in \mathcal{L}^1_{\overline{\mathbb{R}}}(\mu)$ there exists some $\widetilde{u} \in \mathcal{L}^1_{\mathbb{R}}(\mu)$ such that $u = \widetilde{u}$ a.e.*

Proof. We have $N := \{|u| = \infty\} = \bigcap_{n \in \mathbb{N}} \{|u| \geqslant n\} \in \mathscr{A}$. Using the continuity of measures (from above) and the Markov inequality yields

$$\mu(N) = \lim_{n \to \infty} \mu(|u| \geqslant n) \overset{(10.2)}{\leqslant} \lim_{n \to \infty} \frac{1}{n} \underbrace{\int |u| \, d\mu}_{<\infty} = 0.$$

Obviously, $\widetilde{u} := u\mathbb{1}_{N^c} : \Omega \to \mathbb{R}$[1] satisfies $\{u \neq \widetilde{u}\} = N \in \mathcal{N}_\mu$. □

10.7 Remark★. a) An important **consequence** of Corollary 10.6 is that we can – at the cost of a null set – always replace $\mathcal{L}^1_{\overline{\mathbb{R}}}$ by $\mathcal{L}^1 = \mathcal{L}^1_{\mathbb{R}}$, and we can forget about the »$\infty - \infty$« problem. For example, if we want to add $u, w \in \mathcal{L}_{\overline{\mathbb{R}}}$, the »$\infty - \infty$« problem appears only inside the null set $\{|u| = \infty\} \cup \{|w| = \infty\}$. Thus, using the notation of Corollary 10.6, $\widetilde{u} + \widetilde{w}$ is a good replacement for the not everywhere existing $u + w$; the set where $u + w$ is not defined or where $u + w$ and $\widetilde{u} + \widetilde{w}$ differ is at most $\{|u| = \infty\} \cup \{|w| = \infty\}$.

b) Let us make the previous remark a bit more formal. We introduce the following equivalence relation ✍ in $\mathcal{L}^1_{\overline{\mathbb{R}}} := \mathcal{L}^1_{\overline{\mathbb{R}}}(\mu)$:

$$\forall u, w \in \mathcal{L}^1_{\overline{\mathbb{R}}} : \qquad u \sim w \overset{\text{def}}{\iff} u = w \text{ a.e.} \iff \mu\{u \neq w\} = 0.$$

The corresponding equivalence classes are denoted by

$$[u] := \left\{w \in \mathcal{L}^1_{\overline{\mathbb{R}}} \mid w \sim u\right\} \quad \text{and} \quad \mathrm{L}^1(\mu) := \left\{[u] \mid u \in \mathcal{L}^1_{\overline{\mathbb{R}}}\right\}.$$

Corollary 10.6 says that every equivalence class $[u]$ has a nice real-valued representative $\widetilde{u} \in \mathcal{L}^1_{\mathbb{R}}$, and so $[u] = [\widetilde{u}]$. This also explains why we do not need the notation $\mathrm{L}^1_{\overline{\mathbb{R}}}$. Since ✍

$$\forall u, w \in \mathcal{L}^1_{\overline{\mathbb{R}}}, \alpha \in \mathbb{R} : \quad [u] + [w] = [u + w] \quad \text{and} \quad [\alpha u] = \alpha[u],$$

we see that $\mathrm{L}^1(\mu)$ is a vector space. It is customary to identify $[u]$ with a convenient representative, e.g. $\widetilde{u}$, and to write $u \in \mathrm{L}^1$ and to speak of L^1-»functions«. This is not entirely correct, though customary and useful, but you must keep the following points in mind:

- ⚠ if $u, w \in \mathrm{L}^1$, then $u = w$, $u < w$, $u \leqslant w$ (etc.) always mean $u = w$ a.e., $u < w$ a.e., $u \leqslant w$ a.e., (etc.), i.e. **up to a null set.**
- ⚠ the elements $u \in \mathrm{L}^1$ may be defined only **up to a null set.**
- ⚠ if $(u_i)_{i\in\mathbb{N}} \subset \mathcal{L}^1_{\overline{\mathbb{R}}}$ is such that $\sum_{i\in\mathbb{N}} u_i \in \mathcal{L}^1_{\overline{\mathbb{R}}}$, then $\left[\sum_i u_i\right] = \sum_i [u_i]$ – we have to deal with a countable union of null-sets, one from each u_i, and this is still a null set ✍. But be **careful** not to »explode« the null sets by involving more than countably many u's or more than countably many operations between the u's.

[1] Please remember the convention »$0 \cdot \infty = 0$«, i.e. $\widetilde{u}(x) = 0$ for all $x \in N$.

III

Important theorems for integrals

11 Convergence theorems

Throughout this chapter we assume that $(\Omega, \mathscr{A}, \mu)$ is an arbitrary measure space. We are interested in the problem to interchange limits and integrals. Since the integral is itself a limit (Definition 8.4) this is a further variation of an old topic from analysis »under which conditions can we swap two limits?«. We have already seen two theorems in this direction: Beppo Levi (Theorem 8.8) and Fatou's Lemma (Theorem 8.11).

Simple convergence theorems are one of the advantages of Lebesgue's integral over Riemann's integral. The **standard example** of a function, which is Lebesgue but not Riemann integrable, is **Dirichlet's function**

$$u(x) = \mathbb{1}_{\mathbb{Q}\cap[0,1]}(x) = \begin{cases} 1, & x \in [0,1] \cap \mathbb{Q} \\ 0, & \text{otherwise.} \end{cases}$$

Since neither $\mathbb{Q}$ nor $[0,1] \setminus \mathbb{Q}$ contains any non-trivial interval, the largest step-function below u is $f \equiv 0$, and the smallest step-function above u is $f = \mathbb{1}_{[0,1]}$. Thus the Riemann–Darboux upper and lower integrals cannot coincide

$$\int^* u = 1 \neq 0 = \int_* u,$$

and u cannot be Riemann integrable.

The problem is, of course, that the »Riemannian« step-functions have steps which are based on intervals. For a Lebesgue approach we observe first that $\mathbb{Q} \cap [0,1] \in \mathscr{B}(\mathbb{R})$, i.e. $u = \mathbb{1}_{\mathbb{Q}\cap[0,1]}$ is measurable. Moreover,

$$\int u \, d\lambda = \int \sup_{n\in\mathbb{N}} \mathbb{1}_{\{q_1,\dots,q_n\}}(x)\,\lambda(dx) \overset{\text{BL}}{=} \sup_{n\in\mathbb{N}} \underbrace{\int \mathbb{1}_{\{q_1,\dots,q_n\}}(x)\,\lambda(dx)}_{=\lambda\{q_1,\dots,q_n\}=\sum_{i=1}^n \lambda\{q_i\}=0} = 0$$

shows that $u \in \mathcal{L}^1(\mathbb{R})$ and $\int u \, d\lambda = 0$.

We start with an extension of Beppo Levi's theorem to sequences of functions with arbitrary sign.

11.1 Theorem (monotone convergence – MCT). a) *If $(u_n)_{n\in\mathbb{N}} \subset \mathcal{L}^1(\mu)$ is an increasing sequence $u_1 \leqslant u_2 \leqslant \dots$ of integrable functions, and $u = \sup_{n\in\mathbb{N}} u_n$, then*

$$u \in \mathcal{L}^1_{\overline{\mathbb{R}}}(\mu) \iff \sup_{n\in\mathbb{N}} \int u_n \,d\mu < \infty.$$

In this case, one has

$$\int u \,d\mu = \int \sup_{n\in\mathbb{N}} u_n \,d\mu = \sup_{n\in\mathbb{N}} \int u_n \,d\mu \in \mathbb{R}. \qquad (\sup_n = \lim_n)$$

b) *If $(w_n)_{n\in\mathbb{N}} \subset \mathcal{L}^1(\mu)$ is a decreasing sequence $w_1 \geqslant w_2 \geqslant \dots$ of integrable functions, and $w = \inf_{n\in\mathbb{N}} w_n$, then*

$$w \in \mathcal{L}^1_{\overline{\mathbb{R}}}(\mu) \iff \inf_{n\in\mathbb{N}} \int w_n \,d\mu > -\infty.$$

In this case, one has

$$\int w \,d\mu = \int \inf_{n\in\mathbb{N}} w_n \,d\mu = \inf_{n\in\mathbb{N}} \int w_n \,d\mu \in \mathbb{R}. \qquad (\inf_n = \lim_n)$$

Proof. If we set $u_n = (-w_n)$, then b) is reduced to a). Let us show a). Clearly, u exists in $\overline{\mathbb{R}}$ for all x, it is measurable and satisfies

$$0 \leqslant u_n - u_1 \in \mathcal{L}^1(\mu), \quad u_n - u_1 \uparrow u - u_1.$$

Ths means that we can use Beppo Levi's theorem and get

$$0 \leqslant \sup_{n\in\mathbb{N}} \int (u_n - u_1) \,d\mu = \int (u - u_1) \,d\mu. \tag{11.1}$$

If $u_1, u \in \mathcal{L}^1_{\overline{\mathbb{R}}}(\mu)$, then (11.1) shows

$$\begin{aligned}\sup_{n\in\mathbb{N}} \int u_n \,d\mu - \int u_1 \,d\mu = \sup_{n\in\mathbb{N}} \int (u_n - u_1) \,d\mu &\overset{(11.1)}{=} \int (u - u_1) \,d\mu \\ &= \int u \,d\mu - \int u_1 \,d\mu,\end{aligned}$$

and we get $\sup_{n\in\mathbb{N}} \int u_n \,d\mu = \int u \,d\mu < \infty$. Conversely, if $\sup_{n\in\mathbb{N}} \int u_n \,d\mu < \infty$, then we can use (11.1) to conclude that $u - u_1 \in \mathcal{L}^1_{\overline{\mathbb{R}}}(\mu)$. Since we can represent $u = (u - u_1) + u_1$ as the sum of integrable functions, we have $u \in \mathcal{L}^1_{\overline{\mathbb{R}}}(\mu)$, and the equality $\sup_{n\in\mathbb{N}} \int u_n \,d\mu = \int u \,d\mu$ follows as before. □

Probably the most important and versatile convergence theorem is Lebesgue's dominated convergence theorem.

11.2 Theorem (Lebesgue – DCT). *Let $(u_n)_{n\in\mathbb{N}} \in \mathcal{L}^1(\mu)$ be a sequence of integrable functions satisfying*

i) $u_n(x) \xrightarrow[n\to\infty]{} u(x) \quad \forall x \in \Omega$;

ii) $|u_n(x)| \leqslant w(x)$, $w \in \mathcal{L}^1_{\overline{\mathbb{R}}}(\mu) \quad \forall n \in \mathbb{N}\ \forall x \in \Omega$.

Then, $u \in \mathcal{L}^1_{\overline{\mathbb{R}}}(\mu)$ and the following limits exist

a) $\lim_{n\to\infty} \int |u - u_n| \, d\mu = 0$;

b) $\lim_{n\to\infty} \int u_n \, d\mu = \int \lim_{n\to\infty} u_n \, d\mu = \int u \, d\mu$.

💬 The function w appearing in 11.2.ii) is called a (integrable) **majorant**.

Proof. 1° Since u is the limit of measurable functions, it is measurable, and the integrability follows with Lemma 9.3.d) and

$$|u| = \Big|\lim_{n\to\infty} u_n\Big| = \lim_{n\to\infty} |u_n| \leqslant w.$$

2° a)⇒b): This follows from

$$\left|\int u \, d\mu - \int u_n \, d\mu\right| = \left|\int (u - u_n) \, d\mu\right| \overset{9.4.e)}{\leqslant} \int |u - u_n| \, d\mu \xrightarrow[n\to\infty]{a)} 0.$$

3° We will now show a): Since $|u_n - u| \leqslant |u_n| + |u| \leqslant 2w$, we see that the function $2w - |u_n - u| \geqslant 0$, and Fatou's lemma yields

$$\begin{aligned}
\int 2w \, d\mu &= \int \Big(2w - \underbrace{\lim_{n\to\infty} |u - u_n|}_{=0}\Big) \, d\mu \\
&= \int \lim_{n\to\infty}(\inf)\Big(2w - |u - u_n|\Big) \, d\mu \\
&\leqslant \liminf_{n\to\infty} \int \Big(2w - |u - u_n|\Big) \, d\mu \\
&= \int 2w \, d\mu - \limsup_{n\to\infty} \int |u - u_n| \, d\mu.
\end{aligned}$$

This shows

$$\limsup_{n\to\infty} \int |u - u_n| \, d\mu = 0 \implies \lim_{n\to\infty} \int |u - u_n| \, d\mu = 0.^{(1)}$$

□

(1) If $a_n \geqslant 0$ is a sequence of positive numbers such that the upper limit $\limsup_n a_n = 0$, then we have $0 \leqslant \liminf_n a_n \leqslant \limsup_n a_n = 0$, and »=« holds throughout. This shows that $\liminf_n a_n = \limsup_n a_n = \lim_n a_n = 0$.

11.3 Remark*. We may replace in Theorem 11.2.i),ii) the assumption »$\forall x \in \Omega$« with »for a.a. $x \in \Omega$«. Note that in this case

$$N = \left\{x \mid \lim_{n\to\infty} u_n(x) \text{ does not exist}\right\} \cup \bigcup_{n\in\mathbb{N}} \left\{x \mid |u_n(x)| > w(x)\right\}$$

is a (measurable) μ-null set, and we can replace $u \rightsquigarrow u\mathbb{1}_{N^c}$, $u_n \rightsquigarrow u_n\mathbb{1}_{N^c}$. The thus modified functions satisfy 11.2.i),ii) **for all** x. We can now apply Theorem 11.2 since $\int u_n \mathbb{1}_{N^c}\,d\mu = \int u_n\,d\mu$ etc.

11.4 Remark. The existence of a majorant in 11.2.ii) is essential.

⚠ **Standard counterexample:** On $(\Omega, \mathscr{A}, \mu) = ([0,1], \mathscr{B}[0,1], dx)$ we use the sequence

$$u_n(x) := n\mathbb{1}_{[0,\frac{1}{n}]}(x) \xrightarrow[n\to\infty]{} \infty\mathbb{1}_{\{0\}}(x) \overset{\text{a.e.}}{=} 0,$$

but we have

$$\int u_n(x)\,dx = n\cdot\lambda\left[0,\tfrac{1}{n}\right] \overset{\forall n}{=} 1 \neq 0 = \int \infty\cdot\mathbb{1}_{\{0\}}(x)\,dx.$$

12 Parameter-dependent integrals

As before, $(\Omega, \mathscr{A}, \mu)$ is an arbitrary measure space. In this and in the following chapter we discuss two of the most important applications of the convergence theorems. We begin with the question under which conditions continuity and differentiability (of the integrand w.r.t. a free parameter) is preserved under integration. More precisely,

$$\begin{array}{ccc} \mu\text{-integrable} & & \\ \downarrow & & \\ u(t,x) & \xRightarrow{\;??\;} & U(t) = \int u(t,x)\,\mu(dx). \\ \uparrow & & \uparrow \\ \text{continuous / diff'ble} & & \text{continuous? / diff'ble?} \end{array}$$

Clearly, this question is a again a variation of the tune »interchange of limits«.

12.1 Theorem (continuity lemma). *Let $u : (a,b)\times\Omega \to \mathbb{R}$, $(a,b) \subset \mathbb{R}$ is an open interval, and assume that*

i) $x \mapsto u(t,x)$ *is in* $\mathcal{L}^1(\mu)$ — $\forall t \in (a,b)$ *fixed;*

ii) $t \mapsto u(t,x)$ *is continuous* — $\forall x \in \Omega$ *fixed;*

iii) $|u(t,x)| \leqslant w(x)$ — $\forall (t,x) \in (a,b)\times\Omega$ *and **one** $w \in \mathcal{L}^1(\mu)$.*

Then $t \mapsto U(t) := \int u(t,x)\,\mu(dx)$ is continuous at every $t \in (a,b)$.

Proof. The integral defining $V(t)$ exists because of i). Here is the **strategy** of the proof: Pick any $t \in (a,b)$. The function $U(\cdot)$ is continuous in t if, and only if for any sequence $(t_n)_{n\in\mathbb{N}} \subset (a,b)$ such that $t_n \to t$ we have $V(t_n) \to V(t)$.[1]

Let $(t_n)_{n\in\mathbb{N}} \subset (a,b)$ be a sequence such that $t_n \to t$ and define a sequence of functions by $u_n(x) := u(t_n, x)$. We have

- $u_n \in \mathcal{L}^1(\mu)$ because of i);
- $u_n(x) \to u(t,x)$ because of ii);
- $|u_n(x)| \leqslant w(x)$ because of iii).

These three conditions allow us to use the DCT (Theorem 11.2), and we see

$$U(t_n) = \int u(t_n,\cdot)\,d\mu = \int u_n\,d\mu \xrightarrow[n\to\infty]{\text{DCT}} \int u(t,\cdot)\,d\mu \stackrel{\text{def}}{=} U(t). \qquad \square$$

A similar startegy, with a slightly more elaborate proof, gives differentiability.

12.2 Theorem (differentiability lemma). *Let $u : (a,b) \times \Omega \to \mathbb{R}$ where $(a,b) \subset \mathbb{R}$ is an open interval, and assume that*

$$\begin{array}{lll} \text{i)} & x \mapsto u(t,x) \text{ is in } \mathcal{L}^1(\mu) & \forall t \in (a,b) \text{ fixed;} \\ \text{ii)} & t \mapsto u(t,x) \text{ is diff'ble} & \forall x \in \Omega \text{ fixed;} \\ \text{iii)} & |\partial_t u(t,x)| \leqslant w(x) & \forall (t,x) \in (a,b)\times\Omega \\ & & \text{and one } w \in \mathcal{L}^1(\mu). \end{array}$$

Then the function $t \mapsto U(t) := \int u(t,x)\,\mu(dx)$ is differentiable in $t \in (a,b)$ and

$$\frac{d}{dt}U(t) = \frac{d}{dt}\int u(t,x)\,\mu(dx) = \int \frac{\partial}{\partial t} u(t,x)\,\mu(dx). \tag{12.1}$$

Proof. We show differentiability of U and the formula (12.1) for every $t \in (a,b)$.[2]

Pick some sequence $(t_n)_{n\in\mathbb{N}} \subset (a,b)$ such that $t_n \to t$, $t_n \neq t$, and define

$$u_n(x) := \frac{u(t_n,x) - u(t,x)}{t_n - t} \xrightarrow[n\to\infty]{\text{b/o ii)}} \partial_t u(t,x).$$

In particular, $x \mapsto \partial_t u(t,x)$ is the limit of measurable functions, hence measurable; therefore, the r.h.s. of (12.1) is well definded. From the mean value theorem we get for some intermediate value $t = \theta$

$$|u_n(x)| = \left|\partial_t u(t,x)\Big|_{t=\theta}\right| \stackrel{\text{iii)}}{\leqslant} w(x) \qquad \forall n \in \mathbb{N}\ \forall x \in \Omega.$$

(1) Note that continuity is a **local** property, i.e. only a small neighbourhood of the particular t is relevant for our considerations.

(2) Like continuity, differentiability is also a **local** property.

This means that $u_n \in \mathcal{L}^1(\mu)$ is uniformly (in n) dominated by an integrable function, and we can apply the DCT (Theorem 11.2):

$$\begin{aligned}
U'(t) = \lim_{n\to\infty} \frac{U(t_n)-U(t)}{t_n - t} &= \lim_{n\to\infty} \int \frac{u(t_n,x)-u(t,x)}{t_n-t}\,\mu(dx) \\
&= \lim_{n\to\infty} \int u_n(x)\,\mu(dx) \\
&\overset{\text{DCT}}{=} \int \lim_{n\to\infty} u_n(x)\,\mu(dx) \\
&= \int \partial_t u(t,x)\,\mu(dx). \qquad \square
\end{aligned}$$

In the next example we need the following **fact** which will be proved in Chapter 13:

If $u:(a,b)\to\mathbb{R}$ is Riemann integrable, then u is Lebesgue integrable and the Riemann integral $\int_a^b u(x)\,dx$ coincides with the Lebesgue integral $\int_{(a,b)} u(x)\,\lambda(dx)$.

In particular, we may use all techniques known from Riemann integration in order to evaluate integrals.(3)

12.3 Example. Let $f_\alpha(x) := x^\alpha$, $x>0$ and $\alpha\in\mathbb{R}$.

a) $f_\alpha \in \mathcal{L}^1((0,1),dx) \iff \alpha > -1$;
b) $f_\alpha \in \mathcal{L}^1((1,\infty),dx) \iff \alpha < -1$.

Proof. a) Since $x \mapsto x^\alpha \mathbb{1}_{(0,1)}(x)$ is continuous, it is Borel measurable, i.e. the integral exists. Moreover, $x^\alpha \mathbb{1}_{[1/n,1)}(x) x^\alpha$ is Riemann integrable. Thus,

$$\begin{aligned}
\int_{(0,1)} x^\alpha\,\lambda(dx) &\overset{\text{BL}}{=} \lim_{n\to\infty} \int \mathbb{1}_{[1/n,1)}(x)x^\alpha\,\lambda(dx) = \lim_{n\to\infty}\int_{1/n}^1 x^\alpha\,dx \\
&= \lim_{n\to\infty} \frac{x^{\alpha+1}}{\alpha+1}\bigg|_{1/n}^1 = \frac{1}{\alpha+1} < \infty \iff \alpha > -1.
\end{aligned}$$

b) has a very similar proof ✍. $\square$

12.4 Example. $f(x) := x^\alpha e^{-\beta x}$, $x>0$, is Lebesgue integrable on $(0,\infty)$ if $\alpha > -1$ and $\beta > 0$.

Proof. 1° Since f is continuous on $(0,\infty)$, it is a Borel function.

2° The exponential series shows that for all $x>0$, all $\beta \geqslant 0$ and any $N \in \mathbb{N}_0$

$$\frac{(\beta x)^N}{N!} \leqslant \sum_{n=0}^\infty \frac{(\beta x)^n}{n!} = e^{\beta x} \implies e^{-\beta x} \leqslant \frac{N!}{\beta^N} x^{-N}.$$

(3) One is tempted to say that we »do the theory à la Lebesgue, but calculate à la Riemann.«

If we apply this estimate to the function f, we see

$$f(x) = x^{\alpha}e^{-\beta x} \leqslant \underbrace{x^{\alpha}\mathbb{1}_{(0,1)}(x)}_{\in\mathcal{L}^1 \text{ if } \alpha>-1} + \frac{N!}{\beta^N} \underbrace{x^{\alpha-N}\mathbb{1}_{[1,\infty)}(x)}_{\in\mathcal{L}^1 \text{ if } \alpha-N<-1}.$$

Since $N \in \mathbb{N}_0$ is arbitrary, we can always achieve $\alpha - N < -1$. □

12.5 Example★ (Gamma function). The **Gamma function** (Euler's integral of the first kind) is defined as

$$\Gamma(t) = \int_{(0,\infty)} x^{t-1}e^{-x}\,dx, \quad t > 0. \tag{12.2}$$

It has the following properties

a) Γ is continuous in $(0,\infty)$.
b) Γ is arbitarily often differentiable in $(0,\infty)$.
c) $t\Gamma(t) = \Gamma(t+1)$, in particular $\Gamma(n+1) = n!$.
d) $\Gamma\left(\frac{1}{2}\right) = \sqrt{\pi}$.
e) $\log\Gamma(t)$ is convex.

Proof. We will only show a), b) (for the first derivative), and d), the rest is set as an exercise ✍, cf. also [MIMS, Solution to Problem 12.27].

Example 12.4 shows that $\Gamma(t)$, $t > 0$, is well defined. Since **continuity** and **differentiability** are **local properties**, we only need to check a), b) for each t in a suitable neighbourhood $t \in (a,b) \subset [a,b] \subset (0,\infty)$; the values a,b may, and will, depend on t, only in this way we can guarantee the existence of an integrable majorant for the continuity and differentiability lemmas.

a) We want to apply Theorem 12.1. Fix $t \in (a,b)$ and observe that

$$u(t,x) := x^{t-1}e^{-x}, \quad u(t,\cdot) \in \mathcal{L}^1((0,\infty), dx) \quad \text{(Example 12.4)}.$$

Moreover, $t \mapsto u(t,x)$ is continuous. The majorant is given by:

$$\begin{aligned} x^{t-1}e^{-x} &\leqslant x^{t-1}\mathbb{1}_{(0,1)}(x) + N!x^{t-1-N}\mathbb{1}_{[1,\infty)}(x) \\ &\leqslant x^{a-1}\mathbb{1}_{(0,1)}(x) + N!\underbrace{x^{b-1-N}}_{\leqslant x^{-2} \text{ if } N=N(b)\gg 1}\mathbb{1}_{[1,\infty)}(x). \end{aligned}$$

The right-hand side is an integrable majorant; note that it does not depend on t, its dependence on a,b, however, is fine since we apply Theorem 12.1 only (a,b), i.e. **locally**.

This shows that Γ is continuous for every $t \in (a,b)$. Letting $a \downarrow 0$ and $b \uparrow \infty$, we conclude that Γ is continuous on $(0,\infty)$.

b) We want to apply Theorem 12.2. Fix $t \in (a,b)$ and observe that

- $u(t,\cdot)$ is integrable;
- $u(\cdot,x)$ is differentiable;
- $\partial_t u(t,x) = \partial_t\left(x^{t-1}e^{-x}\right) = x^{t-1}e^{-x}\log x$.

It remains to find an integrable majorant. For every $x \in (1,\infty)$ we have

$$\log x \leqslant x \implies |\partial_t u(t,x)| \leqslant x^t e^{-x} \leqslant x^b e^{-x} \overset{\text{Ex. 12.4}}{\in} \mathcal{L}^1([1,\infty), dx),$$

while for $x \in (0,1)$ we can use $|\log x| = \log \frac{1}{x}$, and so

$$|\partial_t u(t,x)| = e^{-x} x^{t-1} \log \tfrac{1}{x} \leqslant x^{a-1} \log \tfrac{1}{x} = x^{a-1-\epsilon} x^{\epsilon} \log \tfrac{1}{x}.$$

Since $a > 0$ there is some $\epsilon > 0$ such that $a - 1 - \epsilon > -1$. We also have

$$M := \sup_{x\in(0,1)} x^{\epsilon} \log \tfrac{1}{x} < \infty.^{(4)}$$

This means that we have for every $t \in (a,b)$

$$|\partial_t u(t,x)| \leqslant M x^{a-1-\epsilon} \mathbb{1}_{(0,1)}(x) + x^b e^{-x} \mathbb{1}_{[1,\infty)}(x) \in \mathcal{L}^1((0,\infty), dx).$$

We can now apply Theorem 12.2 and see that $\Gamma'(t)$ exists on (a,b) and satisfies

$$\Gamma'(t) = \int_0^{\infty} x^{t-1} e^{-x} \log x \, dx.$$

As in the previous part, we can now let $a \downarrow 0$ and $b \uparrow \infty$ to get the assertion on $(0,\infty)$.

c) Use integration by parts for the Riemann integral (in the same way as we use integration by substitution in Part d)).

d) Using Beppo Levi's theorem and the fact that $x^{-1/2} e^{-x}$ is Riemann integrable on every compact interval of the form $[1/n, n]$, we see[5]

$$\int_{(0,\infty)} x^{-\frac{1}{2}} e^{-x} \lambda(dx) \overset{\text{BL}}{=} \sup_{n\in\mathbb{N}} \int_{[1/n,n]} x^{-\frac{1}{2}} e^{-x} \lambda(dx)$$
$$\overset{\S}{=} \sup_{n\in\mathbb{N}} \int_{1/n}^{n} x^{-\frac{1}{2}} e^{-x} \, dx = 2 \sup_{n\in\mathbb{N}} \int_{1/n}^{n} e^{-t^2} \, dt$$

where we use the substitution rule (for Riemann integrals, $x = t^2$, $dx = 2t\,dt$) in the last step; $\lambda(dx)$, as opposed to dx, indicates a Lebesgue integral. A similar argument now shows

$$\Gamma\left(\tfrac{1}{2}\right) = 2 \int_{(0,\infty)} e^{-t^2} \lambda(dt) = \int_{\mathbb{R}} e^{-t^2} \lambda(dt).$$

We will see later, in Example 16.3, that this integral has the value $\sqrt{\pi}$.

e) Use Hölder's inequality (Thm. 14.3 below) to see $\Gamma(at + (1-a)t) \leqslant \Gamma^a(t)\Gamma^{1-a}(t)$, or a direct attack showing that

$$\tfrac{d^2}{dt^2} \log \Gamma(t) = \left(\Gamma'(t)\Gamma''(t) - [\Gamma'(t)]^2\right) / \Gamma^2(t) \leqslant 0.$$ ☐

(4) The function under the supremum is continuous and satisfies $\lim_{x\to 0} x^{\epsilon} \log \frac{1}{x} \overset{x=e^{-t}}{=} \lim_{t\to\infty} e^{-\epsilon t} t = 0$ for any $\epsilon > 0$.

(5) In the step marked by »§« in the following calculation we use that Riemann integrals, if they exist, coincide with Lebesgue integrals. This step is included to show all details of the argument, but we routinely skip this step in the sequel.

13 Riemann vs. Lebesgue

This is also an application of the convergence theorems from Chapter 11. In this chapter we use $\int_a^b \dots \mathrm{d}x$ for Riemann integrals, while $\int_{[a,b]} \dots \lambda(\mathrm{d}x)$ denote Lebesgue integrals. Later on, we will – for good reason – not insist on this distinction and even mix notation.

We begin with a reminder how the Riemann integral is defined. Let $[a,b] \subset \mathbb{R}$ be a compact interval and $\Pi = \{a = t_0 < t_1 < \dots < t_k = b\}$ a **partition** of $[a,b]$. Darboux' upper S^Π and lower S_Π sums of a function u are given by

$$S^\Pi[u] := \sum_{i=1}^k M_i(t_i - t_{i-1}) \quad \text{where} \quad M_i := \sup_{[t_{i-1},t_i]} u,$$

$$S_\Pi[u] := \sum_{i=1}^k m_i(t_i - t_{i-1}) \quad \text{where} \quad m_i := \inf_{[t_{i-1},t_i]} u.$$

13.1 Definition. A bounded function $u : [a,b] \to \mathbb{R}$ is **Riemann integrable**, if the lower and upper integrals coincide:

$$\int_* u = \sup_\Pi S_\Pi[u] = \inf_\Pi S^\Pi[u] =: \int^* u \in \mathbb{R}.$$

Their common value is the **Riemann integral** $\int_a^b u(x)\,\mathrm{d}x$ of u.

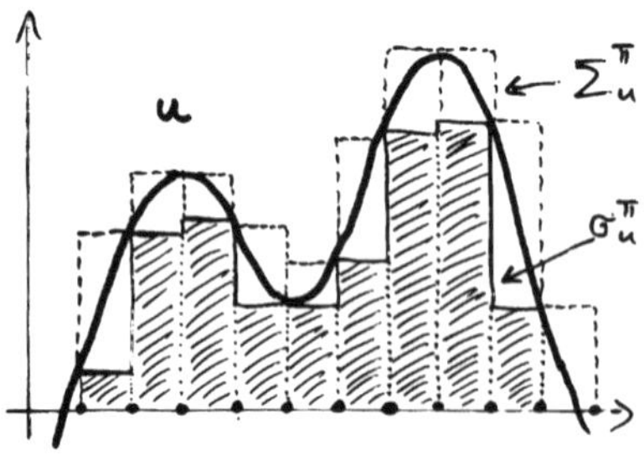

Fig. 13.1. Riemann: Fixed Π

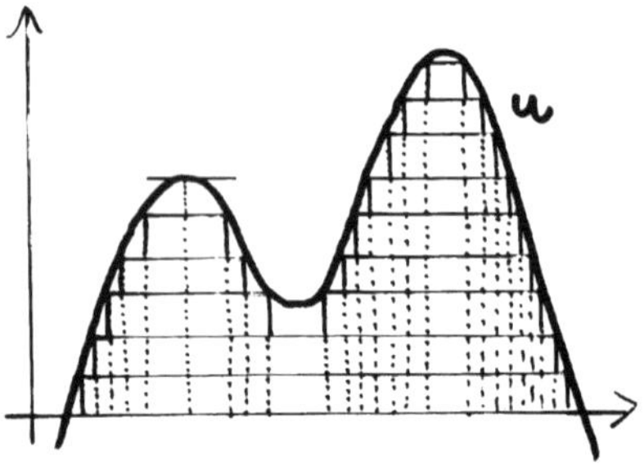

Fig. 13.2. Lebesgue: u determines Π

We can represent the lower (and upper) Darboux sums as Lebesgue integrals (see Fig. 13.1)

$$S_\Pi[u] = \sum_{i=1}^k m_i(t_i - t_{i-1}) = \sum_{i=1}^k m_i \int \mathbb{1}_{[t_{i-1},t_i)}\,\mathrm{d}\lambda = \int \underbrace{\sum_{i=1}^k m_i \mathbb{1}_{[t_{i-1},t_i)}}_{=:\, \sigma_u^\Pi \leqslant u,\ \sigma_u^\Pi \in \mathcal{E}} \mathrm{d}\lambda$$

and, in a similar fashion,

$$S^\Pi[u] = \int \Sigma_u^\Pi \,\mathrm{d}\lambda \quad \text{with} \quad u \leqslant \Sigma_u^\Pi = \sum_{i=1}^k M_i \mathbb{1}_{[t_{i-1},t_i)} \in \mathcal{E}.$$

For every refinement $\Pi' \supset \Pi$ we get

$$\sigma_u^{\Pi} \leqslant \sigma_u^{\Pi'} \leqslant \Sigma_u^{\Pi'} \leqslant \Sigma_u^{\Pi} \overset{\Pi\uparrow}{\Longrightarrow} \sigma_u^{\Pi}\uparrow, \quad \Sigma_u^{\Pi}\downarrow.$$

13.2 Theorem. *If $u : [a,b] \to \mathbb{R}$ is a Riemann integrable (hence, bounded) Borel function, then*

$$u \in \mathcal{L}^1(\lambda) \quad \textit{and} \quad \int_a^b u(x)\,dx = \int_{[a,b]} u(x)\,\lambda(dx).$$

Proof. If u is Riemann integrable, then there is a sequence of partitions Π_n of the interval $[a,b]$ such that

$$\lim_{n\to\infty} S_{\Pi_n}[u] = \int_* u = \int^* u = \lim_{n\to\infty} S^{\Pi_n}[u].$$

We may assume that the partitions are refining, $\Pi_n \subset \Pi_{n+1} \subset \cdots$, otherwise we could consider $\Pi_1 \cup \cdots \cup \Pi_n$ instead of Π_n. Since

$$\sigma_u := \sup_{n\in\mathbb{N}} \sigma_u^{\Pi_n} \leqslant u \leqslant \inf_{n\in\mathbb{N}} \Sigma_u^{\Pi_n} =: \Sigma_u,$$

we can use monotone convergence (Theorem 11.1) to see that

$$\begin{aligned}
\int_* u &\overset{\text{def}}{=} \lim_n S_{\Pi_n}[u] = \lim_n \int_{[a,b]} \sigma_u^{\Pi_n}\,d\lambda \overset{\text{MCT}}{=} \int_{[a,b]} \sigma_u\,d\lambda,\\
\int^* u &\overset{\text{def}}{=} \lim_n S^{\Pi_n}[u] = \lim_n \int_{[a,b]} \Sigma_u^{\Pi_n}\,d\lambda \overset{\text{MCT}}{=} \int_{[a,b]} \Sigma_u\,d\lambda.
\end{aligned} \tag{13.1}$$

By assumption, the upper and lower Riemann integrals coincide, and so

$$\int \underbrace{(\Sigma_u - \sigma_u)}_{\geqslant 0} d\lambda = 0 \Longrightarrow \Sigma_u \overset{\text{a.e.}}{=} \sigma_u \leqslant u \leqslant \Sigma_u.$$

Thus, $u = \Sigma_u$ Lebesgue a.e., and from (13.1) we get $\Sigma_u \in \mathcal{L}^1(\lambda)$. Since u is Borel measurable, we finally get $u \in \mathcal{L}^1(\lambda)$ (Corollary 10.3.b)). □

The proof of Theorem 13.2 shows a bit more: If u is Riemann integrable (but not necessarily Borel measurable), then there is some $\Sigma_u \in \mathcal{L}^1(\lambda)$ such that $u = \Sigma_u$ a.e., and $\int_a^b u(x)\,dx = \int \Sigma_u\,d\lambda$.

13.3 Theorem. *Let $u : [a,b] \to \mathbb{R}$ be a bounded function and denote by $D[u]$ the discontinuity points $\{x \in [a,b] \mid u \text{ not continuous in } x\}$. Then*

$$u \;\textit{Riemann integrable} \iff D[u] \textit{ is (subset of a) } \lambda\textit{-null set.}$$

💬 One can show, cf. Theorem 13.7, that the set of continuity points $C[u] = \{x \mid u \text{ continuous in } x\}$ of an arbitrary (!) function $u : [a,b] \to \mathbb{R}$ is a Borel set. Hence $D[u] = C[u]^c$ is a Borel set.

Proof. »⇒« Let u be Riemann integrable and let Π_n and σ_u, Σ_u as in the proof of Theorem 13.2. From the very definition of sup and inf we get

$$\forall \epsilon > 0 \quad \forall x \in [a,b] \quad \exists n_{\epsilon,x} \in \mathbb{N} \quad \exists t_{n0-1}, t_{n0} \in \Pi_{n_{\epsilon,x}} :$$
$$\text{a) } x \in [t_{n0-1}, t_{n0}]$$
$$\text{b) } \left|\sigma_u^{\Pi_n}(x) - \sigma_u(x)\right| + \left|\Sigma_u^{\Pi_n}(x) - \Sigma_u(x)\right| \leqslant \epsilon \quad \forall n \geqslant n_{\epsilon,x}.$$

Thus, for all $x \in [a,b] \setminus \bigcup_{n\in\mathbb{N}} \Pi_n$ and $y \in (t_{n0-1}, t_{n0})$, t_{n0-1}, t_{n0} as above,

$$|u(x) - u(y)| \leqslant M_{n_0} - m_{n_0} = \Sigma_u^{\Pi_{n_{\epsilon,x}}}(x) - \sigma_u^{\Pi_{n_{\epsilon,x}}}(x) \overset{\text{b)}}{\leqslant} \epsilon + |\Sigma_u(x) - \sigma_u(x)|.$$

From the proof of Theorem 13.2 we know that for a Riemann integrable u the set $\{\Sigma_u \neq \sigma_u\}$ is a null set. Thus,

$$D[u] = \bigcup_{n\in\mathbb{N}} \Pi_n \cup \{\Sigma_u \neq \sigma_u\} \in \mathscr{N}_\lambda$$

since it is the union of a countable set (which is a Lebesgue null set) and a Lebesgue null set.

»⇐« Conversely, assume that $D[u]$ is the subset of a measurable null set. For every $x \in [a,b] \setminus D[u] =: C[u]$ and every partition $\Pi \subset [a,b]$, there is an index $k = k(x,\Pi)$ and a partition interval such that $x \in [t_{k-1}, t_k]$. Therefore,

$$\Sigma_u(x) - \sigma_u(x) \leqslant M_k - m_k \xrightarrow[|\Pi|:=\max_i |t_i - t_{i-1}| \downarrow 0]{u \text{ continuous in } x} 0,$$

hence $\{\Sigma_u = \sigma_u\} \supset C[u]$ or $\{\Sigma_u \neq \sigma_u\} \subset D[u]$. Since $\{\Sigma_u \neq \sigma_u\}$ is measurable, we conclude that $\{\Sigma_u \neq \sigma_u\} \in \mathscr{N}_\lambda$. Therefore,

$$\int^* u \overset{\text{def}}{=} \int \Sigma_u \, d\lambda = \int \sigma_u \, d\lambda \overset{\text{def}}{=} \int_* u,$$

and we see that u is Riemann integrable. □

⚠ Theorem 13.2 and 13.3 do not hold for **improperly** Riemann integrable functions.

13.4 Example★. Consider the Borel function $u(x) = \sum_{n=1}^{\infty} (-1)^n \frac{1}{n} \mathbb{1}_{[n-1,n)}(x)$ on the half-line $([0,\infty), \mathscr{B}[0,\infty))$. On each interval $[0,N]$ the Riemann and Lebesgue integrals coincide, and the improper Riemann integral is

$$\lim_{N\to\infty} \int_{[0,N]} u \, d\lambda = \lim_{N\to\infty} \int_0^N u(x) \, dx = \sum_{n=1}^{\infty} \frac{(-1)^n}{n} = \log 2.$$

Since $\int_{[0,\infty)} |u| \, d\lambda = \sum_{n=1}^{\infty} \frac{1}{n} = \infty$ diverges, the Lebesgue integral $\int_{[0,\infty)} u \, d\lambda$ does not exist. A more interesting example of this kind is the sine integral, see Example 16.8.

Appendix: Measurability of continuity points

Let (Ω, d) be a metric space with metric $d(x,y)$, and denote by

$$B(r,x) := \{y \in \Omega \mid d(x,y) < r\}$$

the open ball (w.r.t. the metric d) with radius $r > 0$ and centre $x \in \Omega$. Recall that the Borel σ-algebra $\mathscr{B}(\Omega)$ is the σ-algebra generated by the open sets of the metric space (Ω, d). For any function $u : \Omega \to \mathbb{R}$ – we do not require f to be measurable – we define the **oscillation function**

$$w^u(x) = \inf_{r>0}\Big(\operatorname{diam} u(B(r,x))\Big) = \inf_{r>0}\Big(\sup_{y\in B(r,x)} u(y) - \inf_{y\in B(r,x)} u(y)\Big).$$

Note that $r \mapsto \operatorname{diam} u(B(r,x))$ decreases as r decreases, i.e. we can replace $\inf_{r>0}$ by $\inf_{0<r<\delta}$ for any fixed $\delta > 0$.

13.5 Lemma. *The function u is continuous at the point $x \in \Omega$ if, and only if, $w^u(x) = 0$.*

Proof. »⇒« If u is continuous at $x \in \Omega$, then

$$\forall \epsilon > 0 \quad \exists r_\epsilon > 0 \quad \forall r < r_\epsilon : \quad \Big(\sup_{y\in B(r,x)} u(y) - u(x)\Big) + \Big(u(x) - \inf_{y\in B(r,x)} u(y)\Big) < 2\epsilon.$$

Thus,

$$w^u(x) \leqslant \sup_{y\in B(r,x)} u(y) - \inf_{y\in B(r,x)} u(y) < 2\epsilon \xrightarrow{\epsilon\to 0} 0.$$

»⇐« We have for all $r > 0$ and $x, x' \in \Omega$ with $x' \in B(r,x)$ that

$$u(x) - u(x') \leqslant \sup_{y\in B(r,x)} u(y) - \inf_{y\in B(r,x)} u(y).$$

Changing the roles of x and x' gives

$$|u(x) - u(x')| \leqslant \sup_{y\in B(r,x)} u(y) - \inf_{y\in B(r,x)} u(y).$$

Assume now that $w^u(x) = 0$. Then we find for every $\epsilon > 0$ some r_ϵ such that for all $r < r_\epsilon$

$$|u(x) - u(x')| \leqslant \sup_{y\in B(r,x)} u(y) - \inf_{y\in B(r,x)} u(y) \leqslant \epsilon + w^u(x) = \epsilon \quad \forall x' \in B(r,x).$$

Letting $r \to 0$ (i.e. $x' \to x$), and then $\epsilon \to 0$, shows that u is continuous at x. □

13.6 Lemma. *The oscillation function w^u is upper semicontinuous, i.e. $\{w^u < \alpha\}$ is for all $\alpha > 0$ an open set.*

Proof. Let $x_0 \in \{w^u < \alpha\}$. There exists some $r = r(\alpha) > 0$ such that

$$\sup_{z\in B(r,x_0)} u(z) - \inf_{z\in B(r,x_0)} u(z) < \alpha.$$

Pick $y \in B(r/3, x_0)$, and observe that $B(r/3, y) \subset B(r, x_0)$. Thus,

$$\begin{aligned} w^u(y) &\leqslant \sup_{z\in B(r/3,y)} u(z) - \inf_{z\in B(r/3,y)} u(z) \\ &\leqslant \sup_{z\in B(r,x_0)} u(z) - \inf_{z\in B(r,x_0)} u(z) < \alpha. \end{aligned}$$

This shows that $y \in \{w^u < \alpha\}$, and so $B(r/3, x_0) \subset \{w^u < \alpha\}$. □

13.7 Theorem. *Let $u : \Omega \to \mathbb{R}$ be an arbitrary function on the metric space (Ω, d). The set of its continuity points $\mathrm{C}[u] := \{x \mid u \text{ continuous at } x\}$ is a Borel set.*

Proof. Lemma 13.5 shows that

$$\mathrm{C}[u] = \bigcap_{\delta>0} \left\{w^u < \delta\right\} = \bigcap_{n\in\mathbb{N}} \left\{w^u < \tfrac{1}{n}\right\},$$

and by Lemma 13.6 the sets $\left\{w^u < \frac{1}{n}\right\}$ are open, hence Borel measurable. Thus $\mathrm{C}[u]$ is the intersection of countably many Borel sets, hence Borel. □

14 The Lebesgue spaces $\mathcal{L}^p$ and L^p

Let $(\Omega, \mathscr{A}, \mu)$ be an arbitrary measure space. In analogy to the integrable functions

$$\mathcal{L}^1 = \mathcal{L}^1(\Omega, \mathscr{A}, \mu) \quad \text{and} \quad \mathrm{L}^1 = \mathrm{L}^1(\Omega, \mathscr{A}, \mu)$$

– depending on what we want to emphasize we also write $\mathcal{L}^1(\mu)$, $\mathcal{L}^1(\Omega)$, $\mathcal{L}^1(\mathscr{A})$ and $\mathrm{L}^1(\mu)$, $\mathrm{L}^1(\Omega)$, $\mathrm{L}^1(\mathscr{A})$ – we define the spaces of pth order integrable functions.

14.1 Definition. Let $p \in [1, \infty)$ and define for any measurable $u \in \mathcal{L}^0(\mathscr{A})$

$$\|u\|_p := \left(\int |u|^p \, d\mu\right)^{1/p} \qquad (1 \leqslant p < \infty)$$

$$\|u\|_\infty := \operatorname*{esssup}_{x\in\Omega} |u(x)| := \inf\left\{c > 0 \mid \mu\{|u| \geqslant c\} = 0\right\}. \qquad (p = \infty)$$

The space of **pth order integrable functions,** $1 \leqslant p \leqslant \infty$, is

$$\mathcal{L}^p(\mu) = \left\{u : \Omega \to \mathbb{R} \mid u \text{ measurable}, \|u\|_p < \infty\right\}.$$

The space of equivalence classes ($u \sim w \iff u = w$ a.e.) is denoted by $\mathrm{L}^p(\mu)$, see Remark 10.7.

The functional $u \mapsto \|u\|_p$, $p \in [1, \infty]$ is essentially a norm:

a) $\|u\|_p = 0 \iff u = 0$ a.e. (see Corollary 10.2.a) if $p < \infty$, for $p = \infty$ it is trivial). If we do not distinguish between a.e. coinciding functions (e.g. in L^p), we can even write $u = 0$.[(1)]
b) $\|\alpha u\|_p = |\alpha| \|u\|_p$, $\alpha \in \mathbb{R}$.

Only the triangle inequality for $\|\bullet\|_p$ is not obvious. For the proof we will need the following elementary inequality.

(1) Denote, for a moment, the elements of L^p, by $[u]$. Strictly speaking, the norm in L^p is $\|[u]\|_p := \inf_{w\in[u]} \|w\|_p$; since $\|w\|_p = \|v\|_p$ for all $v, w \in [u]$, we can identify $\|[u]\|_p$ with $\|u\|_p$.

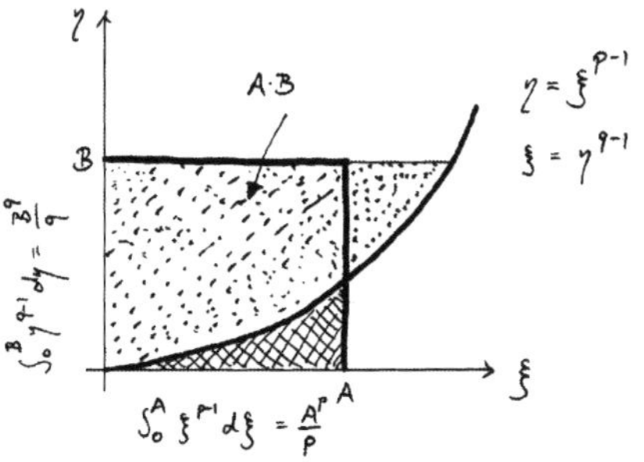

Fig. 14.1. Graphic proof of Young's inequality.

14.2 Lemma (Young's inequality). *Let $p,q \in (1,\infty)$ be conjugate indices, i.e. $p^{-1}+q^{-1}=1$ or $q=p/(p-1)$. Then*

$$AB \leqslant \frac{A^p}{p}+\frac{B^q}{q} \quad \forall A,B \geqslant 0. \tag{14.1}$$

Amendment: *Equality holds if, and only if, $B = A^{p-1}$.*

Proof without words. See Fig. 14.1. □

The next result is essential for the proof of the triangle inequality, but it is in itself interesting and one of the most important integral inequalities.

14.3 Theorem (Hölder's inequality). *Let $u \in \mathcal{L}^p(\mu)$ and $w \in \mathcal{L}^q(\mu)$ with conjugate indices $\frac{1}{p}+\frac{1}{q}=1$, $p,q \in [1,\infty]$.(2) Then $u \cdot w \in \mathcal{L}^1(\mu)$ and*

$$\left|\int uw\,d\mu\right| \leqslant \int |uw|\,d\mu \leqslant \|u\|_p\|w\|_q. \tag{14.2}$$

Proof. The first inequality in (14.2) is obvious. For the second inequality we assume first that $p,q \in (1,\infty)$. Let

$$A = \frac{|u(x)|}{\|u\|_p}, \qquad B = \frac{|w(x)|}{\|w\|_q}$$

and use (14.1) to obtain

$$\frac{|u(x)w(x)|}{\|u\|_p\|w\|_q} \leqslant \frac{|u(x)|^p}{p\|u\|_p^p}+\frac{|w(x)|^q}{q\|w\|_q^q}.$$

Integrating both sides of the inequality reveals

$$\int \frac{|u(x)w(x)|}{\|u\|_p\|w\|_q}\,\mu(dx) \leqslant \frac{\int |u(x)|^p\,\mu(dx)}{p\|u\|_p^p}+\frac{\int |w(x)|^q\,\mu(dx)}{q\|w\|_q^q} = \frac{1}{p}+\frac{1}{q}=1.$$

Now let $p=1$ and $q=\infty$. If $C > \|w\|_\infty$, then $C \geqslant |w|$ a.e., and so

$$\int |uw|\,d\mu \leqslant C\int |u|\,d\mu = C\|u\|_1.$$ □

14.4 Corollary (Cauchy–Schwarz inequality – CSI). *If $u,w \in \mathcal{L}^2(\mu)$, then the function $u \cdot w \in \mathcal{L}^1(\mu)$ and*

$$\int |uw|\,d\mu \leqslant \|u\|_2\|w\|_2. \tag{14.3}$$

(2) If $p=\infty$ we use $q=1$ and vice versa. This uses the convention that $\frac{1}{\infty}=0$.

Proof. Apply Theorem 14.3 for the conjugate indices $p = q = 2$. □

We can finally prove the triangle inequality for $u \mapsto \|u\|_p$.

14.5 Corollary (Minkowski's inequality). *If $u, w \in \mathcal{L}^p(\mu)$, $p \in [1, \infty]$, then the function $u + w \in \mathcal{L}^p(\mu)$ and*

$$\|u + w\|_p \leqslant \|u\|_p + \|w\|_p. \tag{14.4}$$

Proof. We leave the case $p = \infty$ to the reader ✍. Let $p \in [1, \infty)$. We show first p-integrability:

$$\begin{aligned} |u + w|^p &\leqslant (|u| + |w|)^p \leqslant (2\max\{|u|, |w|\})^p \\ &= 2^p \max\{|u|^p, |w|^p\} \\ &\leqslant 2^p(\underbrace{|u|^p}_{\in \mathcal{L}^1(\mu)} + \underbrace{|w|^p}_{\in \mathcal{L}^1(\mu)}) \in \mathcal{L}^1(\mu). \end{aligned}$$

From this we infer that $u + w \in \mathcal{L}^p(\mu)$. Moreover,

$$\begin{aligned} \int |u + w|^p \, d\mu &= \int |u + w| \cdot |u + w|^{p-1} \, d\mu \\ &\leqslant \int |u| \cdot |u + w|^{p-1} \, d\mu + \int |w| \cdot |u + w|^{p-1} \, d\mu \\ &\quad \text{(if } p = 1\text{, then we can stop here. } \square) \\ &\overset{\text{Hölder}}{\leqslant} \|u\|_p \cdot \| |u + w|^{p-1} \|_q + \|w\|_p \cdot \| |u + w|^{p-1} \|_q. \end{aligned}$$

Since $q = p/(p-1)$, we find

$$\| |u + w|^{p-1} \|_q = \left(\int |u + w|^{(p-1)q} \, d\mu \right)^{1/q} = \left(\int |u + w|^p \, d\mu \right)^{1-1/p}$$

and the claim follows upon dividing by $\| |u + w|^{p-1} \|_q$ (which is finite b/o the first part of our proof). □

14.6 Remark. a) We can consider $\mathcal{L}^p_{\overline{\mathbb{R}}}(\mu)$, but in view of Remark 10.7 and Corollary 10.6 we know that any $u \in \mathcal{L}^p_{\overline{\mathbb{R}}}(\mu)$ is μ-a.e. $\mathbb{R}$-valued.

b) $u \in \mathcal{L}^p(\mu) \iff u$ measurable and $|u|^p \in \mathcal{L}^1(\mu)$.

c) Corollary 14.5 shows, in particular, that $\mathcal{L}^p(\mu)$ is a vector space.

d) Since $\|u\|_p = 0 \implies u = 0$ a.e., $\mathcal{L}^p(\mu)$ is only a **pseudo-normed** linear space.

e) The space $L^p(\mu)$ is a normed linear space with norm $u \mapsto \|u\|_p$.

In the rest of this chapter, we study the normed spaces $L^p(\mu)$. All results also hold on $\mathcal{L}^p(\mu)$ if we add a few »μ-a.e.«'s at the obvious places.

14.7 Definition. Let $(u_n)_{n\in\mathbb{N}} \subset L^p(\mu)$, $u \in L^p(\mu)$ for some $p \in [1,\infty]$.

a) $u_n \xrightarrow[n\to\infty]{L^p} u$, $L^p\text{-}\lim_{n\to\infty} u_n = u \overset{\text{def}}{\iff} \lim_{n\to\infty} \|u_n - u\|_p = 0$.

b) $(u_n)_{n\in\mathbb{N}}$ is called an L^p **Cauchy sequence**, if

$$\forall \epsilon > 0 \quad \exists N = N_\epsilon \in \mathbb{N} \quad \forall n,k \geqslant N: \quad \|u_n - u_k\|_p \leqslant \epsilon.$$

14.8 Remark. Let $p \in [1,\infty]$ and $(u_n)_{n\in\mathbb{N}} \subset L^p(\mu)$ be a sequence in $L^p(\mu)$.

a) If $u_n \xrightarrow{L^p} u$, then $(u_n)_{n\in\mathbb{N}}$ is a Cauchy sequence. This follows from

$$\|u_n - u_k\|_p \overset{\text{Minkowski}}{\leqslant} \|u_n - u\|_p + \|u - u_k\|_p \xrightarrow[n\to\infty]{k\to\infty} 0.$$

b) ⚠ Pointwise convergence $u_n(x) \to u(x)$ for a.a. $x \in \Omega$, is not enough to guarantee $u_n \xrightarrow{L^p} u$. A **sufficient** condition is if **in addition** $|u_n| \leqslant w$ for all $n \in \mathbb{N}$ and some $w \in L^p(\mu)$ – cf. Theorem 11.2 or Theorem 14.12 below.

We are now going to show the converse of §14.8.a): $L^p(\mu)$, $p \in [1,\infty]$, is complete. For the proof we need the following auxiliary result.

14.9 Lemma. *Let $p \in [1,\infty]$ and $(u_n)_{n\in\mathbb{N}} \subset L^p(\mu)$, $u_n \geqslant 0$.*

$$\left\|\sum_{n=1}^\infty u_n\right\|_p \leqslant \sum_{n=1}^\infty \|u_n\|_p.$$

Proof. Applying Minkowski's inequality N times yields

$$\left\|\sum_{n=1}^N u_n\right\|_p \leqslant \sum_{n=1}^N \|u_n\|_p \leqslant \sum_{n=1}^\infty \|u_n\|_p.$$

Since $\sum_{n=1}^N u_n \uparrow \sum_{n=1}^\infty u_n$ as $N \uparrow \infty$, we can use Beppo Levi to get

$$\begin{aligned}\sup_{N\in\mathbb{N}} \left\|\sum_{n=1}^N u_n\right\|_p^p &= \sup_{N\in\mathbb{N}} \int \left(\sum_{n=1}^N u_n\right)^p d\mu = \int \left(\sup_{N\in\mathbb{N}} \sum_{n=1}^N u_n\right)^p d\mu \\ &= \int \left(\sum_{n=1}^\infty u_n\right)^p d\mu = \left\|\sum_{n=1}^\infty u_n\right\|_p^p.\end{aligned}$$

□

14.10 Theorem (Riesz–Fischer). *The spaces $L^p(\mu)$, $p \in [1,\infty]$, are **complete**, i.e. every Cauchy sequence $(u_n)_{n\in\mathbb{N}} \subset L^p(\mu)$ has a limit $u \in L^p(\mu)$.*

Proof. 1° The **main problem** is to identify the limit u. Since $(u_n)_{n\in\mathbb{N}}$ is a Cauchy sequence, there are integers

$$1 < n(1) < n(2) < \ldots < n(k) < \ldots \quad \text{such that} \quad \left\|u_n - u_{n(k)}\right\|_p < 2^{-k}$$

for all $n \geqslant n(k+1)$ and $k \in \mathbb{N}$. We can now express the limit u as a telescoping sum:

$$u_{n(k+1)} = \sum_{i=0}^{k} \left(u_{n(i+1)} - u_{n(i)}\right), \quad u_{n(0)} = 0.$$

If $u_n \xrightarrow{L^p} u$, then $u_{n(k+1)} \xrightarrow{L^p} u$, i.e. the series

$$u = \sum_{i=0}^{\infty} \left(u_{n(i+1)} - u_{n(i)}\right)$$

is a potential candidate for the limit.

2° u is well-defined. Using the estimate of Lemma 14.9 we get

$$\left\| \sum_{i=0}^{\infty} \left|u_{n(i+1)} - u_{n(i)}\right| \right\|_p \leqslant \sum_{i=0}^{\infty} \left\|u_{n(i+1)} - u_{n(i)}\right\|_p \leqslant \left\|u_{n(1)}\right\|_p + \sum_{i=1}^{\infty} 2^{-i}.$$

This shows that $\left(\sum_{i=0}^{\infty} \left|u_{n(i+1)} - u_{n(i)}\right|\right)^p < \infty$ a.e. Thus, $\sum_{i=0}^{\infty} \left(u_{n(i+1)} - u_{n(i)}\right)$ converges absolutely a.e., hence u is well-defined a.e. (we may set $u = 0$ if the series does not converge).

3° $u_{n(k)} \to u$ in L^p. Using again Lemma 14.9 we see that

$$\begin{aligned} \left\|u - u_{n(k)}\right\|_p &= \left\| \sum_{i=k}^{\infty} \left(u_{n(i+1)} - u_{n(i)}\right) \right\|_p \leqslant \left\| \sum_{i=k}^{\infty} \left|u_{n(i+1)} - u_{n(i)}\right| \right\|_p \\ &\leqslant \sum_{i=k}^{\infty} \left\|u_{n(i+1)} - u_{n(i)}\right\|_p \leqslant \sum_{i=k}^{\infty} 2^{-i} \xrightarrow[k\to\infty]{} 0. \end{aligned}$$

4° $u_n \to u$ in L^p. We have for all $n \geqslant n(k+1) > n(k)$

$$\begin{aligned} \left\|u - u_n\right\|_p &\leqslant \left\|u - u_{n(k)}\right\|_p + \left\|u_{n(k)} - u_n\right\|_p \\ &\leqslant \left\|u - u_{n(k)}\right\|_p + 2^{-k} \xrightarrow[k\to\infty]{} 0. \end{aligned}$$

□

The proof of Theorem 14.10 contains the following subsequence result.

14.11 Corollary. *If $(u_n)_{n\in\mathbb{N}} \subset L^p(\mu)$, $p \in [1,\infty]$, converges in L^p to u, then there is a subsequence $(u_{n(k)})_{k\in\mathbb{N}}$ such that $u_{n(k)}(x) \to u(x)$ for a.a. $x \in \Omega$.*

14.12 Theorem (L^p-DCT). *Let $(u_n)_{n\in\mathbb{N}} \subset L^p(\mu)$, $p \in [1,\infty)$ and assume that $|u_n| \leqslant w$ for all $n \in \mathbb{N}$ and some $w \in L^p(\mu)$. If $u_n(x) \to u(x)$ for a.a. $x \in \Omega$, then $u \in L^p(\mu)$ and*

a) $\|u - u_n\|_p \xrightarrow[n\to\infty]{} 0$; b) $\|u_n\|_p \xrightarrow[n\to\infty]{} \|u\|_p$.

Proof. ✎ Observe that for measurable functions $u \in \mathrm{L}^p \iff |u|^p \in \mathrm{L}^1$ and apply the »usual« dominated convergence theorem (Theorem 11.2) to the sequence $|u - u_n|^p \to 0$. □

⚠ Convergence in the L^p-norm **is strictly stronger than** convergence of the L^p-norms.

14.13 Theorem (F. Riesz). *Let $(u_n)_{n\in\mathbb{N}} \subset \mathrm{L}^p(\mu)$ for some $p \in [1,\infty)$. If the sequence $u_n(x) \to u(x)$ for a.a. $x \in \Omega$ and if $u \in \mathrm{L}^p(\mu)$, then*

$$\lim_{n\to\infty} \|u - u_n\|_p = 0 \iff \lim_{n\to\infty} \|u_n\|_p = \|u\|_p.$$

Proof. The direction »⇒« follows Minkowski's inequality

$$\|u_n\|_p = \|u_n - u + u\|_p \leqslant \|u_n - u\|_p + \|u\|_p,$$

and – changing the roles of u and u_n – the **lower triangle inequality**:

$$\big|\|u_n\|_p - \|u\|_p\big| \leqslant \|u_n - u\|_p. \tag{14.5}$$

For the converse direction »⇐« we observe that

$$|u_n - u|^p \leqslant 2^p(|u_n|^p + |u|^p).$$

Therefore, we can apply Fatou's Lemma (Theorem 8.11) to the sequence

$$2^p(|u_n|^p + |u|^p) - |u_n - u|^p \geqslant 0$$

and we see

$$\begin{aligned}
2^{p+1}\int |u|^p \,d\mu &= \int \liminf_{n\to\infty}\left\{2^p(|u_n|^p + |u|^p) - |u_n - u|^p\right\} d\mu \\
&\leqslant \liminf_{n\to\infty}\left(\int 2^p|u_n|^p \,d\mu + \int 2^p|u|^p \,d\mu - \int |u_n - u|^p \,d\mu\right) \\
&= \underbrace{2^p \int |u|^p \,d\mu + 2^p \int |u|^p \,d\mu}_{=2^{p+1}\int |u|^p \,d\mu} - \limsup_{n\to\infty} \int |u_n - u|^p \,d\mu.
\end{aligned}$$

Comparing both sides of the inequality shows that

$$\limsup_{n\to\infty} \int |u_n - u|^p \,d\mu = 0 \implies \lim_{n\to\infty} \int |u_n - u|^p \,d\mu = 0.^{(3)}$$ □

The following example continues the discussion of §§9.6, 8.10 where be started to study series as integrals.

(3) See the footnote on page 58.

14.14 Example. Let $\mu = \sum_{n=1}^{\infty} \delta_n$ be the counting measure on $(\mathbb{N}, \mathscr{P}(\mathbb{N}))$; as before we identify functions $u : \mathbb{N} \to \mathbb{R}$ with sequences $u = (u_n)_{n\in\mathbb{N}}$. Recall that **all** functions on $\mathbb{N}$ are measurable. As in Example 9.6 we see that

$$\|u\|_p = \sum_{n\in\mathbb{N}} |u_n|^p \quad (1 \leqslant p < \infty) \quad \text{and} \quad \|u\|_\infty = \sup_{n\in\mathbb{N}} |u_n|;$$

note that the only null set of the counting measure is the empty set. The spaces of p-**summable** resp. **bounded** sequences are

$$\ell^p(\mathbb{N}) = \mathrm{L}^p(\mu) = \left\{(u_n)_{n\in\mathbb{N}} \subset \mathbb{R} \mid \sum_{n=1}^{\infty} |u_n|^p < \infty\right\},$$

$$\ell^\infty(\mathbb{N}) = \mathrm{L}^\infty(\mu) = \left\{(u_n)_{n\in\mathbb{N}} \subset \mathbb{R} \mid \sup_{n\in\mathbb{N}} |u_n| < \infty\right\}.$$

The Minkowski and Hölder inequalities have the following form.

Hölder's inequality for series. Let $(a_n)_{n\in\mathbb{N}} \in \ell^p(\mathbb{N})$ and $(b_n)_{n\in\mathbb{N}} \in \ell^q(\mathbb{N})$.

$$\sum_{n=1}^{\infty} |a_n b_n| \leqslant \begin{cases} \left(\sum_{n=1}^{\infty} |a_n|^p\right)^{1/p} \left(\sum_{n=1}^{\infty} |b_n|^q\right)^{1/q}, & \dfrac{1}{p} + \dfrac{1}{q} = 1, \\ \sup_{n\in\mathbb{N}} |b_n| \sum_{n=1}^{\infty} |a_n|, & p = 1,\ q = \infty. \end{cases}$$

Minkowski's inequality for series. Let $(a_n)_{n\in\mathbb{N}}, (b_n)_{n\in\mathbb{N}} \in \ell^p(\mathbb{N})$.

$$\left(\sum_{n=1}^{\infty} |a_n + b_n|^p\right)^{1/p} \leqslant \left(\sum_{n=1}^{\infty} |a_n|^p\right)^{1/p} + \left(\sum_{n=1}^{\infty} |b_n|^p\right)^{1/p}, \qquad 1 \leqslant p < \infty,$$

$$\sup_{n\in\mathbb{N}} |a_n + b_n| \leqslant \sup_{n\in\mathbb{N}} |a_n| + \sup_{n\in\mathbb{N}} |b_n|, \qquad p = \infty.$$

We will now prove a useful convexity inequality. As preparation we need a result which is intuitive, but quite unpleasant to prove. Recall that a function $\mathrm{V} : (a,b) \to \mathbb{R}$, $-\infty \leqslant a < b \leqslant \infty$, is said to be **convex**, if

$$\forall x, y \in (a,b) \quad \forall t \in (0,1) : \quad \mathrm{V}(tx + (1-t)y) \leqslant t\mathrm{V}(x) + (1-t)\mathrm{V}(y)$$

see Fig. 14.2. A function $\Lambda : (a,b) \to \mathbb{R}$ is said to be **concave**, if $\mathrm{V} = -\Lambda$ is convex. The same definition works if we replace (a,b) by any other (closed, half-open) interval.

14.15 Lemma (see [MIMS, Lemma 13.12]). *Let $\mathrm{V} : (a,b) \to \mathbb{R}$ be a convex function and $-\infty \leqslant a < b \leqslant \infty$.*

a) $\mathrm{V}(x) = \sup\left\{\ell(x) \mid \ell \text{ affine-linear, } \ell(y) \leqslant \mathrm{V}(y)\ \forall y\right\}$ *(cf. Fig. 14.3).*
b) *V is continuous and admits one-sided derivatives at each $x \in (a,b)$.*

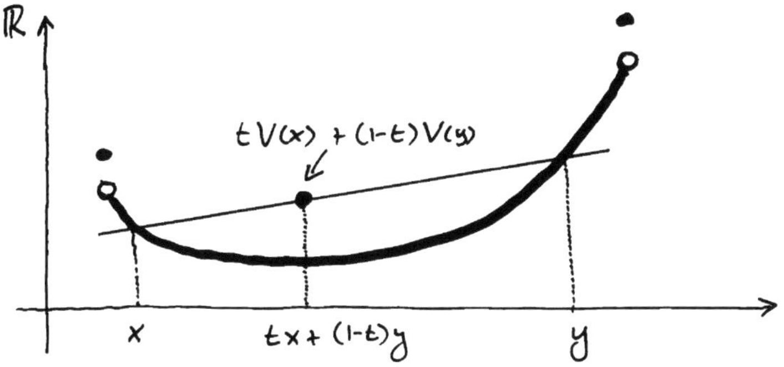

Fig. 14.2. A convex function is in the interior of its domain continuous. It looks like a »smile(y)«: Discontinuities can only appear at the boundary.

14.16 Theorem (Jensen's inequality). *Let* $V : [0,\infty) \to [0,\infty)$ *be a positive convex function and* μ *a probability measure on* $(\Omega, \mathscr{A})$. *Then*

$$V\left(\int u \, d\mu\right) \leqslant \int V(u) \, d\mu \quad \forall u \in \mathcal{L}^{0,+}(\mathscr{A}). \tag{14.6}$$

Proof. The function $V(u) = \mathbb{1}_{\{u=0\}}V(0) + \mathbb{1}_{\{u>0\}}V(u)$ is measurable, since $V|_{(0,\infty)}$ is continuous. Therefore, all integrals appearing in (14.6) exist in $[0,\infty]$. We extend V onto $[0,\infty]$ via $V(\infty) := \lim_{x\to\infty} V(x)$ – the limit exists b/o convexity. If $\int V(u)\,d\mu = \infty$, then there is nothing to be shown. Assume that $\int V(u)\,d\mu < \infty$.

If $\ell(x) = ax + b$ is an affine linear function, then we can use $\int b \, d\mu = b$ (since μ is a probability measure) and get

$$\ell\left(\int u \, d\mu\right) = a \int u \, d\mu + b = \int (au + b) \, d\mu = \int \ell(u) \, d\mu.$$

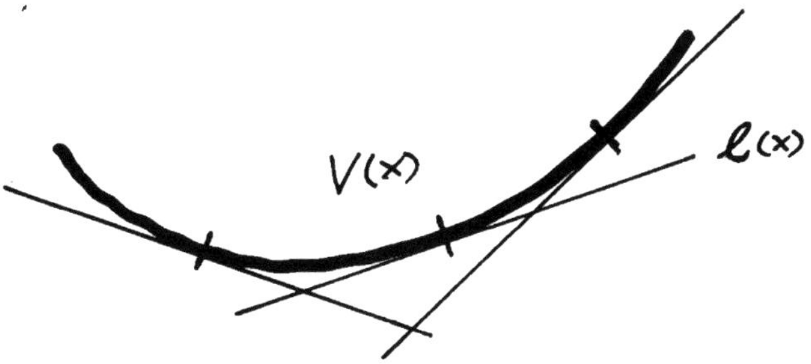

Fig. 14.3. A convex function is the (pointwise) upper envelope of all affine-linear function below its graph. At each point $(x, V(x))$, $x \in (a,b)$, there are not necessarily unique tangents from the left and right.

Therefore, by Lemma 14.15,

$$V\left(\int u\,d\mu\right) = \sup_{\substack{\ell \leqslant V,\\ \ell \text{ linear}}} \ell\left(\int u\,d\mu\right) = \sup_{\substack{\ell \leqslant V,\\ \ell \text{ linear}}} \int \underbrace{\ell(u)}_{\leqslant V(u)}\,d\mu \leqslant \int V(u)\,d\mu. \qquad \square$$

We want to emphasize a few important special cases of Jensen's inequality.

14.17 Example. Consider on $(\Omega, \mathscr{A})$ a probability measure μ and two further measures ν and ρ.

a) $\left|\int u\,d\mu\right|^p \leqslant \left(\int |u|\,d\mu\right)^p \leqslant \int |u|^p\,d\mu \qquad (p \geqslant 1)$

b) $\exp\left(\int u\,d\mu\right) \leqslant \int \exp(u)\,d\mu.$

c) If ν and ρ are not probability measures, you can use the following trick which still allows us to use Jensen's inequality:

If $\nu(\Omega) < \infty$, then $\mu := \nu/\nu(\Omega)$ is a probability measure.

If there is some $f \in L^1(\rho)$, $f \geqslant 0$, then $\mu := f\,d\rho \big/ \int f\,d\rho$ is a probability measure, and we have

$$\int u\,d\rho = \int \frac{u}{f} f\,d\rho = \int \frac{u}{f} \frac{f\,d\rho}{\int f\,d\rho} \cdot \int f\,d\rho.$$

d) The **concave Jensen's inequality**: If $u \in L^{1,+}(\mu)$ and $\Lambda : [0,\infty) \to [0,\infty)$ is concave (e.g. $\sqrt{x}$), then

$$\int \Lambda(u)\,d\mu \leqslant \Lambda\left(\int u\,d\mu\right).$$

IV

Products of measure spaces. Radon–Nikodým

15 Product measures

The idea behind product measures can be most easily explained by the method how we calculate the area of a rectangle:

$$\lambda^2\big([a_1,b_1)\times[a_2,b_2)\big) = (b_1-a_1)(b_2-a_2) = \lambda^1[a_1,b_1)\lambda^1[a_2,b_2),$$

We can interpret this equality in the following way

$$\begin{aligned}\lambda^2\big([a_1,b_1)\times[a_2,b_2)\big) &= \int_{[a_1,b_1)} \lambda^1[a_2,b_2)\,\lambda^1(dx)\\ &= \int\left(\int \mathbb{1}_{[a_2,b_2)}(y)\mathbb{1}_{[a_1,b_1)}(x)\,\lambda^1(dx)\right)\lambda^1(dy).\end{aligned}$$

This idea can also be used for more general sets $B \in \mathscr{B}(\mathbb{R}^2)$: We slice B vertically or horizontally, and add up (i.e. integrate) the resulting lower-dimensional measures. This is nothing but an infinitesimal version of

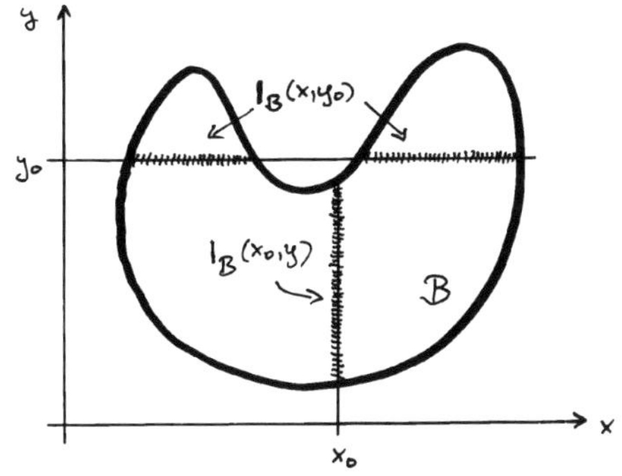

Fig. 15.1. Cavalieri's principle.

Cavalieri's principle

$$\begin{aligned}\lambda(B) &= \int_{\mathbb{R}^2} \mathbb{1}_B(x,y)\,\overbrace{\lambda^2(d(x,y))}^{=\lambda^2(dx\times dy)=\lambda^1(dx)\lambda^2(dy)}\\ &= \int_{\mathbb{R}}\int_{\mathbb{R}} \mathbb{1}_B(x,y)\,\lambda^1(dx)\,\lambda^1(dy)\\ &= \int_{\mathbb{R}}\left(\int_{\mathbb{R}} \mathbb{1}_B(x,y_0)\,\lambda^1(dx)\right)\lambda^1(dy_0).\end{aligned}$$

But there are **problems**: Our argument is heuristic and it is not clear at all that the »slices« of B are measurable. Moreover, we use implicitly that two-dimensional Lebesgue meassure λ^2 is uniquely determined by its one-dimensional marginals – in view of the uniqueness theorem for measures (Theorem 4.7) this amounts to showing that $\mathscr{B}(\mathbb{R}) \times \mathscr{B}(\mathbb{R})$ generates $\mathscr{B}(\mathbb{R} \times \mathbb{R})$.

15.1 Remark. Let us briefly review the rules for cartesian products of sets. Let $\mathscr{A}$, $\mathscr{B}$ be families of sets (not necessarily in the same basis set) and let $A, A', A_i \in \mathscr{A}$, $B, B', B_i \in \mathscr{B}$, $i \in I$ is an arbitrary index set. We have

$$\begin{aligned}
(\textstyle\bigcup_i A_i) \times B &= \textstyle\bigcup_i (A_i \times B)\\
(\textstyle\bigcap_i A_i) \times B &= \textstyle\bigcap_i (A_i \times B)\\
(A \times B) \cap (A' \times B') &= (A \cap A') \times (B \cap B')\\
A^c \times B &= \Omega \times B \setminus (A \times B)\\
A \times B \subset A' \times B' &\iff A \subset A' \text{ and } B \subset B'.
\end{aligned}$$

We use $\mathscr{A} \times \mathscr{B} = \{A \times B \mid A \in \mathscr{A},\ B \in \mathscr{B}\}$ to denote the (generalized) rectangles. ⚠ Note that $\mathscr{A} \times \mathscr{B}$ is, in general, not a σ-algebra.

15.2 Lemma. *Let $\mathscr{A}$ and $\mathscr{B}$ be semi-rings (see Def. 5.1) on the sets Ω and Θ, respectively. The family of rectangles $\mathscr{A} \times \mathscr{B}$ is again a semi-ring.*

Proof. Throughout we use $A, A' \in \mathscr{A}$ and $B, B' \in \mathscr{B}$.

(S_1) Clearly, $\emptyset = \emptyset \times \emptyset \in \mathscr{A} \times \mathscr{B}$.
(S_2) This follows from the identity

$$(A \times B) \cap (A' \times B') = (A \cap A') \times (B \cap B') \in \mathscr{A} \times \mathscr{B}. \tag{15.1}$$

(S_3) $$\begin{aligned}(A \times B)^c &= \big\{(x,y) \mid x \notin A,\ y \in B \text{ or } x \in A,\ y \notin B \text{ or } x \notin A,\ y \notin B\big\}\\ &= (A^c \times B) \uplus (A \times B^c) \uplus (A^c \times B^c)\end{aligned}$$

and with the help of (15.1) we see

$$\begin{aligned}(A \times B) \setminus (A' \times B') &= (A \times B) \cap (A' \times B')^c\\ &= \big[(A \setminus A') \times (B \cap B')\big] \uplus \big[(A \cap A') \times (B \setminus B')\big] \uplus \big[(A \setminus A') \times (B \setminus B')\big].\end{aligned}$$

By assumption, $A \setminus A'$ and $B \setminus B'$ can be written as finite unions of disjoint sets in $\mathscr{A}$ and $\mathscr{B}$, respectively. This proves the claim. □

15.3 Definition. Let $(\Omega, \mathscr{A})$ and $(\Theta, \mathscr{B})$ be measurable spaces. We call

$$\mathscr{A} \otimes \mathscr{B} := \sigma(\mathscr{A} \times \mathscr{B}) \qquad \textbf{product σ-algebra}$$

$$(\Omega, \mathscr{A}) \otimes (\Theta, \mathscr{B}) := (\Omega \times \Theta, \mathscr{A} \otimes \mathscr{B}) \qquad \textbf{product of measurable spaces}$$

We would like to reduce most assertions to the generator of $\mathscr{A} \otimes \mathscr{B}$. With a small caveat, the following lemma can be phrased in the following catchy way: »product of the generators = generator of the product«.

15.4 Lemma. *Let $\mathscr{A} = \sigma(\mathscr{F})$ and $\mathscr{B} = \sigma(\mathscr{G})$ be measurable spaces. If there exist exhausting sequences $(F_i)_{i\in\mathbb{N}} \subset \mathscr{F}$, $F_i \uparrow \Omega$ and $(G_k)_{k\in\mathbb{N}} \subset \mathscr{G}$, $G_k \uparrow \Theta$, then*

$$\sigma(\mathscr{F} \times \mathscr{G}) = \sigma(\mathscr{A} \times \mathscr{B}) \stackrel{def}{=} \mathscr{A} \otimes \mathscr{B}.$$

Proof. The inclusion »$\subset$« follows from $\mathscr{F} \times \mathscr{G} \subset \mathscr{A} \times \mathscr{B} \subset \mathscr{A} \otimes \mathscr{B}$ which implies that $\sigma(\mathscr{F} \times \mathscr{G}) \subset \mathscr{A} \otimes \mathscr{B}$. The proof of the converse inclusion »$\supset$« is split into several steps.

1° The family $\Sigma = \{A \in \mathscr{A} \mid A \times G \in \sigma(\mathscr{F} \times \mathscr{G})\ \forall G \in \mathscr{G}\}$ is a σ-algebra: Indeed, if $S, S_n \in \Sigma$ and $G \in \mathscr{G}$, then we have

(Σ_1) $\Omega \times G = \bigcup_{i\in\mathbb{N}} \underbrace{F_i \times G}_{\in\sigma(\mathscr{F}\times\mathscr{G})} \in \sigma(\mathscr{F} \times \mathscr{G})$ and so $\Omega \in \Sigma$.

(Σ_2) $S^c \times G = \underbrace{\Omega \times G}_{\in\sigma(\mathscr{F}\times\mathscr{G})} \setminus \underbrace{S \times G}_{\in\sigma(\mathscr{F}\times\mathscr{G})} \in \sigma(\mathscr{F} \times \mathscr{G})$ and so $S^c \in \Sigma$.

(Σ_3) $\left(\bigcup_{n\in\mathbb{N}} S_n\right) \times G = \bigcup_{n\in\mathbb{N}} \underbrace{(S_n \times G)}_{\in\sigma(\mathscr{F}\times\mathscr{G})} \in \sigma(\mathscr{F} \times \mathscr{G})$ and so $\bigcup_{n\in\mathbb{N}} S_n \in \Sigma$.

Obviously, $\mathscr{F} \subset \Sigma \subset \mathscr{A} = \sigma(\mathscr{F})$ and applying the σ-operation to this chain of inclusions, we get $\Sigma = \mathscr{A}$. From the definition of Σ we take it that $\mathscr{A} \times \mathscr{G} \subset \sigma(\mathscr{F} \times \mathscr{G})$.

2° As in Step 1° we see that $\mathscr{F} \times \mathscr{B} \subset \sigma(\mathscr{F} \times \mathscr{G})$.

3° Using the results of 1° and 2°, we see that $\mathscr{A} \times \mathscr{B} \subset \sigma(\mathscr{F} \times \mathscr{G})$:

$$A \times B = (A \times \Theta) \cap (\Omega \times B) = \bigcup_{i,k\in\mathbb{N}} \underbrace{(A \times G_k)}_{\in\mathscr{A}\times\mathscr{G}\subset\sigma(\mathscr{F}\times\mathscr{G})} \cap \overbrace{(F_i \times B)}^{\in\mathscr{F}\times\mathscr{B}\subset\sigma(\mathscr{F}\times\mathscr{G})} \in \sigma(\mathscr{F} \times \mathscr{G}).$$

4° Finally, $\mathscr{A} \otimes \mathscr{B} \stackrel{def}{=} \sigma(\mathscr{A} \times \mathscr{B}) \stackrel{3°}{\subset} \sigma(\mathscr{F} \times \mathscr{G})$. □

Let $(\Omega, \mathscr{A}, \mu)$ and $(\Theta, \mathscr{B}, \nu)$ be measure spaces. First we define a measure ρ relative to the rectangles $\mathscr{A} \times \mathscr{B}$ a measure ρ via

$$\rho(A \times B) := \mu(A)\nu(B), \qquad A \in \mathscr{A},\ B \in \mathscr{B}.$$

The main problem is to extend this measure onto $\sigma(\mathscr{A} \times \mathscr{B})$.

15.5 Theorem (uniqueness of product measure). *Let $(\Omega, \mathscr{A}, \mu)$ and $(\Theta, \mathscr{B}, \nu)$ be measure spaces such that $\mathscr{A} = \sigma(\mathscr{F})$, $\mathscr{B} = \sigma(\mathscr{G})$ and*

i) *$\mathscr{F}$, $\mathscr{G}$ are $\cap$-stable;*
ii) *$\exists F_i \in \mathscr{F}$, $G_k \in \mathscr{G}$, $F_i \uparrow \Omega$, $G_k \uparrow \Theta$ and $\mu(F_i) < \infty$, $\nu(G_k) < \infty$.*

Then there is at most one measure ρ on $(\Omega \times \Theta, \mathscr{A} \otimes \mathscr{B})$ such that

$$\rho(F \times G) = \mu(F)\nu(G) \quad \forall F \in \mathscr{F},\ G \in \mathscr{G}. \tag{15.2}$$

Proof. We reduce the assertion to the uniqueness theorem for measures (Theorem 4.7). To do so, note that

a) $\mathscr{F} \times \mathscr{G}$ is a generator of $\mathscr{A} \otimes \mathscr{B}$, cf. Lemma 15.4.

b) $\mathscr{F} \times \mathscr{G}$ is $\cap$-stable: $(F \times G) \cap (F' \times G') = (F \cap F') \times (G \cap G')$.

c) $F_i \times G_i \uparrow \Omega \times \Theta$ and $F_i \times G_i \in \mathscr{F} \times \mathscr{G}$ with $\rho(F_i \times G_i) = \mu(F_i)\nu(G_i) < \infty$.

Therefore, there is at most one measure satisfying (15.2). □

Recall that a measure space $(\Omega, \mathscr{A}, \mu)$ is σ-finite, if there is an exhausting sequence $(A_n)_{n\in\mathbb{N}} \subset \mathscr{A}$, $A_n \uparrow \Omega$ such that $\mu(A_n) < \infty$. Thus, Condition ii) of Theorem 15.5 implies that the measure spaces $(\Omega, \mathscr{A}, \mu)$ and $(\Theta, \mathscr{B}, \nu)$ are σ-finite. In most applications of Theorem 15.5, one has $\mathscr{F} = \mathscr{A}$ and $\mathscr{G} = \mathscr{B}$, i.e. ii) just means σ-finiteness.

We turn now to the existence problem.

15.6 Theorem (existence of prod. measure). *Let $(\Omega, \mathscr{A}, \mu)$ and $(\Theta, \mathscr{B}, \nu)$ be σ-finite measure spaces. The set function*

$$\rho : \mathscr{A} \times \mathscr{B} \to [0, \infty], \quad \rho(A \times B) := \mu(A)\nu(B)$$

has a unique extension – again denoted by ρ – which is a σ-finite measure on $\mathscr{A} \otimes \mathscr{B}$. For every $E \in \mathscr{A} \otimes \mathscr{B}$ one has

$$\rho(E) = \int_\Omega \left[\int_\Theta \mathbb{1}_E(x,y)\, \nu(dy) \right] \mu(dx) \tag{15.3}$$

$$= \int_\Theta \left[\int_\Omega \mathbb{1}_E(x,y)\, \mu(dx) \right] \nu(dy). \tag{15.4}$$

In particular, the functions

$$x \mapsto \mathbb{1}_E(x,y), \qquad y \mapsto \mathbb{1}_E(x,y),$$

$$y \mapsto \int_\Omega \mathbb{1}_E(x,y)\, \mu(dx), \qquad x \mapsto \int_\Theta \mathbb{1}_E(x,y)\, \nu(dy).$$

are $\mathscr{A}$-, resp., $\mathscr{B}$-measurable.

Proof. We split the proof in several parts. Claims are highlighted.

1° Uniqueness follows from Theorem 15.5 using $\mathscr{F} = \mathscr{A}$ and $\mathscr{G} = \mathscr{B}$.

2° Since μ and ν are σ-finite, there are sequences $(A_i)_{i\in\mathbb{N}} \subset \mathscr{A}$ and $(B_k)_{k\in\mathbb{N}} \subset \mathscr{B}$, such that $E_i := A_i \times B_i \uparrow \Omega \times \Theta$ and $\mu(A_i) < \infty$, $\nu(B_k) < \infty$.

For each $i \in \mathbb{N}$, we define a family $\mathscr{D}_i \subset \mathscr{A} \otimes \mathscr{B}$ as follows: $D \in \mathscr{D}_i$ if, and only if,

i) $x \mapsto \mathbb{1}_{D\cap E_i}(x,y)$ and $y \mapsto \mathbb{1}_{D\cap E_i}(x,y)$ are measurable;

ii) $y \mapsto \int \mathbb{1}_{D\cap E_i}(x,y)\, \mu(dx)$ and $x \mapsto \int \mathbb{1}_{D\cap E_i}(x,y)\, \nu(dy)$ are mble;

iii) $\iint \mathbb{1}_{D\cap E_i}(x,y)\, \mu(dx)\nu(dy) = \iint \mathbb{1}_{D\cap E_i}(x,y)\, \nu(dy)\mu(dx)$.

3° $\mathscr{A} \times \mathscr{B} \subset \mathscr{D}_i$. Take any set $A \times B \in \mathscr{A} \times \mathscr{B}$, and note that

$$\begin{aligned}\iint \mathbb{1}_{(A\times B)\cap E_i}\, d\mu\, d\nu &= \iint \mathbb{1}_{A\cap A_i} \mathbb{1}_{B\cap B_i}\, d\mu\, d\nu \\ &= \int \mu(A\cap A_i) \mathbb{1}_{B\cap B_i}\, d\nu \\ &= \mu(A\cap A_i)\nu(B\cap B_i) \\ &= \cdots = \iint \mathbb{1}_{(A\times B)\cap E_i}\, d\nu\, d\mu.\end{aligned}$$

This proves iii) for $D = A \times B$. The conditions i) and ii) are obvious because of the product structure:

$$\mathbb{1}_{A\times B}(x,y) = \mathbb{1}_A(x)\mathbb{1}_B(y), \qquad \int \mathbb{1}_{A\times B}(x,y)\,\mu(dx) = \mu(A)\mathbb{1}_B(y), \qquad \text{etc.}$$

4° $\mathscr{D}_i$ is a Dynkin system (see Def. 4.1):

(D_1) $\Omega \times \Theta \in \mathscr{D}_i$ is obvious.

(D_2) From 3° we know that $E_k = A_k \times B_k \in \mathscr{D}_i$ for any $k \in \mathbb{N}$. If $D \in \mathscr{D}_i$, then

$$\begin{aligned}\iint \mathbb{1}_{D^c\cap E_i}\, d\mu\, d\nu &= \int\left[\int\left(\mathbb{1}_{E_i} - \mathbb{1}_{D\cap E_i}\right) d\mu\right] d\nu \\ &= \iint \mathbb{1}_{E_i}\, d\mu\, d\nu - \iint \mathbb{1}_{D\cap E_i}\, d\mu\, d\nu \\ &\overset{\text{iii)}}{=} \iint \mathbb{1}_{E_i}\, d\nu\, d\mu - \iint \mathbb{1}_{D\cap E_i}\, d\nu\, d\mu \\ &= \iint \mathbb{1}_{D^c\cap E_i}\, d\nu\, d\mu.\end{aligned}$$

This shows that $D^c \in \mathscr{D}_i$, since the calculation implies – implicitly, as all (double) integrals in the course of the calculation are well-defined – that iii) along with i) and ii) hold for for D^c.

(D_3) Let $(D_k)_{k\in\mathbb{N}} \subset \mathscr{D}_i$ be a sequence of disjoint sets and set $D := \dot{\bigcup}_{k\in\mathbb{N}} D_k$. We have

$$\begin{aligned}\iint \mathbb{1}_{D\cap E_i}\, d\mu\, d\nu &= \iint \sum_{k=1}^{\infty} \mathbb{1}_{D_k\cap E_i}\, d\mu\, d\nu \\ &\overset{2\times\text{BL}}{=} \sum_{k=1}^{\infty} \iint \mathbb{1}_{D_k\cap E_i}\, d\mu\, d\nu \\ &\overset{D_k\in\mathscr{D}_i}{=} \sum_{k=1}^{\infty} \iint \mathbb{1}_{D_k\cap E_i}\, d\nu\, d\mu \\ &\overset{2\times\text{BL}}{=} \iint \mathbb{1}_{D\cap E_i}\, d\nu\, d\mu.\end{aligned}$$

This shows that $D \in \mathscr{D}_i$; notice that the conditions i), ii) for D are inherited from the D_k under the summation $\sum_{k=1}^{\infty}$.

5° Since $\mathscr{A} \times \mathscr{B}$ is $\cap$-stable and satisfies $\mathscr{A} \times \mathscr{B} \subset \mathscr{D}_i$, we see using Theorem 4.5 that $\sigma(\mathscr{A} \times \mathscr{B}) = \delta(\mathscr{A} \times \mathscr{B}) \subset \mathscr{D}_i$. Thus, all sets of the form $Q \cap E_i$ with $Q \in \mathscr{A} \otimes \mathscr{B}$ satisfy the conditions i)–iii). Finally, letting $E_i \uparrow \Omega \times \Theta$, yields i)–iii) for all $Q \in \mathscr{A} \otimes \mathscr{B}$.

6° The calculation for (D_3) in 5° with $E_k = \Omega \times \Theta$ shows that ρ is a measure. □

15.7 Definition. Let $(\Omega, \mathscr{A}, \mu)$, $(\Theta, \mathscr{B}, \nu)$ be σ-finite measure spaces. The measure $\rho := \mu \otimes \nu$ constructed in Theorem 15.6 is called the **product measure with marginals** μ, ν; the **product of measure spaces** is denoted by $(\Omega, \mathscr{A}, \mu) \otimes (\Theta, \mathscr{B}, \nu) := (\Omega \times \Theta, \mathscr{A} \otimes \mathscr{B}, \mu \otimes \nu)$.

In passing, we have (finally!) constructed d-dimensional Lebesgue measure λ^d. The programme started with Theorem 3.7 is thus complete.

15.8 Corollary. *d-dimensional Lebesgue measure exists and satisfies for $d = m+n$*

$$(\mathbb{R}^d, \mathscr{B}(\mathbb{R}^d), \lambda^d) = (\mathbb{R}^m, \mathscr{B}(\mathbb{R}^m), \lambda^m) \otimes (\mathbb{R}^n, \mathscr{B}(\mathbb{R}^n), \lambda^n)$$
$$= \bigotimes_{i=1}^{d} (\mathbb{R}, \mathscr{B}(\mathbb{R}), \lambda^1).$$

15.9 Example*. ⚠ The assumption of **σ-finiteness is extremely important** in Theorem 15.6. Consider on the measurable space $([0,1], \mathscr{B}[0,1])$ Lebesgue measure λ and the counting measure $\zeta := \#B$. Clearly, ζ is not σ-finite. If we try to define ρ with the Cavalieri formula (15.3) or (15.4) we get two different results if we take the diagonal $\Delta = \{(x,x) \mid x \in [0,1]\}$:

$$\int_0^1 \left[\int_0^1 \mathbb{1}_\Delta(x,y) \zeta(dy) \right] \lambda(dx) = \int_0^1 \underbrace{\zeta(\{x\})}_{=1} \lambda(dx) = 1,$$

whereas

$$\int_0^1 \left[\int_0^1 \mathbb{1}_\Delta(x,y) \lambda(dx) \right] \zeta(dy) = \int_0^1 \underbrace{\lambda(\{y\})}_{=0} \zeta(dy) = 0.$$

💬 Notice that in the above example the **diagonal is Borel measurable**, since $\mathscr{B}[0,1]$ is generated by $\mathscr{G} = \{[r,q) \mid 0 \leqslant r < q \leqslant 1,\ r,q \in \mathbb{Q}\}$ which is a countable family that separates points, i.e. for every $x, y \in [0,1]$ there are disjoint sets $G, G' \in \mathscr{G}$ with $x \in G$ and $y \in G'$. A moment's thought shows that

$$\Delta^c = ([0,1] \times [0,1]) \setminus \Delta = \bigcup_{G \in \mathscr{G}} (G \times G^c) \cup (G^c \times G).$$

16 The theorems of Fubini and Tonelli

As in the previous chapter, let $(\Omega, \mathscr{A}, \mu)$ and $(\Theta, \mathscr{B}, \nu)$ be two σ-finite measure spaces. The formulae (15.3), (15.4) defining the product measure can be seen as a change in the order of integration of a double integral of a one-step function, i.e.

$$\rho(E) = \int \mathbb{1}_E \, d\mu \otimes \nu = \iint \mathbb{1}_E \, d\mu \, d\nu = \iint \mathbb{1}_E \, d\nu \, d\mu.$$

With our standard techniques – linearity, sombrero Lemma and Beppo Levi – we can extend this formula to all positive measurable functions.

16.1 Theorem (Tonelli). *Let $(\Omega, \mathscr{A}, \mu)$ and $(\Theta, \mathscr{B}, \nu)$ be σ-finite measure spaces and $u : \Omega \times \Theta \to [0, \infty]$ an $\mathscr{A} \otimes \mathscr{B}$-measurable function.*

a) *$x \mapsto u(x,y)$ and $y \mapsto u(x,y)$ (y, resp., x is fixed) are $\mathscr{A}$- and $\mathscr{B}$-measurable, respectively.*
b) *$y \mapsto \int u(x,y)\,\mu(dx)$ and $x \mapsto \int u(x,y)\,\nu(dy)$ are $\mathscr{B}$- and $\mathscr{A}$-measurable, respectively.*
c) $$\int_{\Omega\times\Theta} u(x,y)\,\mu\otimes\nu(dx,dy) = \int_\Theta \left[\int_\Omega u(x,y)\,\mu(dx)\right]\nu(dy) = \int_\Omega \left[\int_\Theta u(x,y)\,\nu(dy)\right]\mu(dx) \in [0,\infty].$$

⚠ We can always interchange iterated integrals of positive measurable functions, if we allow the integrals to be $+\infty$. If one of the (iterated) integrals in §16.1.c) is finite, all are finite.

⚠ σ-finiteness is essential, see Example 15.9.

Proof. 1° If $u(x,y) = \mathbb{1}_E(x,y)$ for some $E \in \mathscr{A} \otimes \mathscr{B}$, then the claim follows from Theorem 15.6.

2° Let $u \in \mathcal{E}^+(\mathscr{A} \otimes \mathscr{B})$ be given in normal form, i.e.

$$u(x,y) = \sum_{k=0}^{n} \alpha_k \mathbb{1}_{E_k}(x,y), \quad \alpha_k \geqslant 0,\ E_k \in \mathscr{A} \otimes \mathscr{B}.$$

Since $\mathcal{L}^0(\mathscr{A} \otimes \mathscr{B})$ is a vector space, and since measurability is preserved under vector space operations, we obtain a), b) from the corresponding results for the one-step functions in 1° . As the (iterated) integral(s) are linear, this applies to c), too.

3° Let $u \in \mathcal{L}^{0,+}_{\overline{\mathbb{R}}}(\mathscr{A} \otimes \mathscr{B})$. Using the sombrero lemma (Theorem 7.11), there is a sequence of simple functions $(u_n)_{n\in\mathbb{N}} \subset \mathcal{E}^+(\mathscr{A} \otimes \mathscr{B})$ such that $u_n \uparrow u$. With Beppo Levi swe see that the claims a)–c) are preserved under increasing limits. ☐

16.2 Corollary (Fubini's theorem). *Let $(\Omega,\mathscr{A},\mu)$ and $(\Theta,\mathscr{B},\nu)$ be σ-finite measure spaces and $u:\Omega\times\Theta\to\overline{\mathbb{R}}$ a $\mathscr{A}\otimes\mathscr{B}$-measurable function. If one of the following three integral expressions is finite,*

$$\iint |u|\,d\mu\,d\nu,\quad \iint |u|\,d\nu\,d\mu,\quad \int |u|\,d(\mu\otimes\nu),$$

all integrals are finite, and we have

a) $u\in L^1(\mu\otimes\nu)$;
b) $x\mapsto u(x,y)\in L^1(\mu)$ *for ν-almost all y;*
c) $y\mapsto u(x,y)\in L^1(\nu)$ *for μ-almost all x;*
d) $y\mapsto \int_\Omega u(x,y)\,\mu(dx)\in L^1(\nu)$;
e) $x\mapsto \int_\Theta u(x,y)\,\nu(dy)\in L^1(\mu)$.
f) $$\int_{\Omega\times\Theta} u(x,y)\,\mu\otimes\nu(dx,dy)=\int_\Theta\left[\int_\Omega u(x,y)\,\mu(dx)\right]\nu(dy) = \int_\Omega\left[\int_\Theta u(x,y)\,\nu(dy)\right]\mu(dx)$$

»onion skin principle«: $\int\ldots d\nu$ acts like a pair of parentheses

💬 The integral on the l.h.S. of f) is well-defined because of a)!

Proof. Tonelli's theorem (Theorem 16.1) shows that

$$\int |u|\,d(\mu\otimes\nu)=\iint |u|\,d\mu\,d\nu=\iint |u|\,d\nu\,d\mu,$$

i.e. if one expression is finite, all are finite, and it follows that $u\in L^1(\mu\otimes\nu)$. Again by Tonelli's theorem we see that the functions

$$x\mapsto u^\pm(x,y)\quad\text{and}\quad y\mapsto\int u^\pm(x,y)\,\mu(dx)$$

are measurable. Since $u^\pm\leqslant|u|$, we see that

$$\int_\Omega u^\pm(x,y)\,\mu(dx)\leqslant\int_\Omega |u(x,y)|\,\mu(dx)<\infty\quad(\nu\text{-a.e. in } y);$$

here we use Corollary 10.6 (integrable functions are a.e. finite) and

$$\int_\Theta\int_\Omega u^\pm(x,y)\,\mu(dx)\,\nu(dy)\leqslant\int_\Theta\int_\Omega |u(x,y)|\,\mu(dx)\,\nu(dy)<\infty.$$

This proves b), d). The assertions c), e) are proved in the same way.
Finally, in order to see f), we use $u=u^+-u^-$, the linearity of the integral, and again Tonelli's theorem, applied to $u^\pm$.

16.3 Example* (normal distribution). Tonelli's theorem can be used to evaluate an integral which is of paramount importance in probability theory: It is the normalization for the normal distribution.

$$\int_{\mathbb{R}} e^{-x^2/2}\,dx = \sqrt{2\pi}. \tag{16.1}$$

Proof. We use a »variable doubling trick« to bring iterated integrals into play. Define $I := \int_{\mathbb{R}} e^{-x^2/2}\,dx$. We have

$$I^2 = \int_{\mathbb{R}} e^{-x^2/2}\,dx \int_{\mathbb{R}} e^{-y^2/2}\,dy = \int_{-\infty}^{\infty}\int_{-\infty}^{\infty} e^{-(x^2+y^2)/2}\,dy\,dx$$
$$= 4\int_0^{\infty}\int_0^{\infty} e^{-(x^2+y^2)/2}\,dy\,dx.$$

Since the integrands are continuous, we may treat these integrals as (improper) Riemann integrals. Changing variables in the inner integral according to $y = tx$, $dy = x\,dt$ yields

$$I^2 = 4\int_0^{\infty}\int_0^{\infty} xe^{-x^2(1+t^2)/2}\,dt\,dx = 4\int_0^{\infty}\int_0^{\infty} xe^{-x^2(1+t^2)/2}\,dx\,dt.$$

Since the inner function has a primitive, we see

$$I^2 = 4\int_0^{\infty}\left[-\frac{1}{1+t^2}e^{-x^2(1+t^2)/2}\right]_{x=0}^{\infty} dt = \int_0^{\infty}\frac{4\,dt}{1+t^2} = [4\arctan t]_0^{\infty} = 2\pi.$$

Distribution functions

16.4 Definition. Let $u \in \mathcal{L}^0(\mathscr{A})$. We call $t \mapsto \mu^u(t) := \mu\{u > t\}$, $t \in \mathbb{R}$, the **distribution function** of u with respect to μ.

- The function $t \mapsto \mu\{u > t\}$ is right-continuous and decreasing. If we replace »>« by »⩾«, we get a left-continuous function.
- ⚠ Probabilists call $\mu^u(t) := \mu\{u > t\}$ the **survival function** (of u, since u exceeds, i.e. »survives« t) or the **(right) tail**. The **distribution function** of u is usually $\mu_u(t) = \mu\{u \leqslant t\}$.

  If μ is a probability measure, then $\mu^u(t) + \mu_u(t) = \mu\{u > t\} + \mu\{u \leqslant t\} = 1$.

16.5 Theorem (layer-cake formula). *For any positive $u \in \mathcal{L}^{0,+}(\mathscr{A})$ one has*

$$\int u\,d\mu = \int_{(0,\infty)} \mu\{u > t\}\,\lambda(dt) \in [0,\infty].$$

Proof. The function $U(x,t) := (u(x),t)$ is $\mathscr{A}\otimes\mathscr{B}[0,\infty)$-measurable, since

$$U^{-1}(A\times I) = u^{-1}(A)\times I \in \mathscr{A}\otimes\mathscr{B}[0,\infty) \quad \forall A \in \mathscr{B}(\mathbb{R}),\ I = [a,b] \subset [0,\infty),$$

and so, the set $E = \{(x,t) \mid u(x) > t\} \in \mathscr{A} \otimes \mathscr{B}[0,\infty)$. Tonelli's theorem shows

$$\begin{aligned}\int u(x)\,\mu(dx) &= \iint_0^{u(x)} dt\,\mu(dx)\\ &= \iint \mathbb{1}_{(0,u(x))}(t)\,dt\,\mu(dx)\\ &= \iint_{\Omega\times(0,\infty)} \mathbb{1}_E(x,t)\,dt\,\mu(dx)\\ &\qquad \text{note: } \mathbb{1}_{(0,u(x))}(t) = \begin{cases} 0, & u(x) \leqslant t\\ 1, & u(x) > t\end{cases}\\ &= \int_{(0,\infty)} \underbrace{\left[\int_\Omega \mathbb{1}_E(x,t)\,\mu(dx)\right]}_{=\mu\{x \mid u(x)>t\}} dt.\end{aligned}$$ □

Below, $\int_a^b \ldots dx$ is a Riemann-, $\int_{(a,b]} \ldots \lambda(dx)$ a Lebesgue integral.

16.6 Corollary. *Let $\phi : [0,\infty) \to [0,\infty)$ be continuously differentiable, $\phi(0) = 0$ and $\phi' \geqslant 0$. For every measurable $u : \Omega \to [0,\infty)$ one has*

$$\int \phi \circ u \, d\mu = \int_0^\infty \phi'(s)\mu\{u > s\}\,ds \in [0,\infty].$$

In particular, $\phi \circ u \in L^1(\mu)$ if, and only if, $\phi'(s)\,\mu\{u > s\}$ is (improperly) Riemann-integrable.

💬 Most important special case: $\int_\Omega |u|^p \, d\mu = \int_0^\infty ps^{p-1}\,\mu\{|u| > s\}\,ds,\ p > 0.$

Proof. Since ϕ is continuous, $\phi \circ u$ is measurable. Therefore,

$$\begin{aligned}\int \phi \circ u \, d\mu \quad &\overset{\S 16.5}{\underset{\text{BL}}{=}} \quad \sup_{n\in\mathbb{N}} \int_{(n^{-1},n)} \mu\{\phi(u) > t\} \wedge n\,\lambda^1(dt)\\ &\overset{\S 13.3}{=} \quad \sup_{n\in\mathbb{N}} \int_{1/n}^n \mu\{\phi(u) > t\} \wedge n\,dt\\ &\overset{t=\phi(s)}{\underset{dt=\phi'(s)\,ds}{=}} \quad \sup_{n\in\mathbb{N}} \int_{1/n}^n \mu\{\phi(u) > \phi(s)\} \wedge n\,\phi'(s)\,ds\\ &= \quad \int_0^\infty \phi'(s)\mu\{u > s\}\,ds.\end{aligned}$$

⚠ In the above calculation, $\{u > s\} = \emptyset$ and $\{\phi(u) > t\} = \emptyset$ if $s \geqslant \sup u$ and $t \geqslant \phi(u)$, respectively.

In the step where we apply Theorem 13.3 we should make sure that the integrand $\mu\{\phi(u) > t\} \wedge n$ is Lebesgue-a.e. continuous. This follows easily from the fact that *every monotone function $m : \mathbb{R} \to \mathbb{R}$ has at most countably many jump discontinuities.*[1] □

(1) My favourite argument, the »bedside lamp lemma« (see Fig. 16.1): Since m is monotone, all

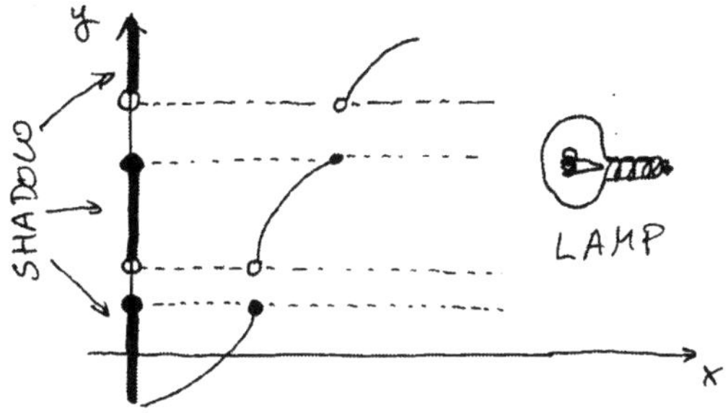

Fig. 16.1. The »bedside lamp lemma«: The jumps of a (wlog left-continuous) monotone function show up as disjoint »white« intervals on the wall (y-axis).

Integration by parts and the sine integral★

Fubini's theorem can be used to derive the classical integration by parts (IP) formula.

16.7 Example★ (parts – IP). Let $f, g \in \mathcal{L}^0(\mathbb{R})$ be Borel functions which are integrable over any compact set $K \subset \mathbb{R}$ (so-called **locally integrable** functions). Therefore, the primitives exist,

$$F(x) := \int_0^x f(t)\,dt := \begin{cases} \int f(t)\mathbb{1}_{[0,x]}(t)\,dt, & x \geqslant 0, \\ -\int f(t)\mathbb{1}_{[x,0]}(t)\,dt, & x < 0, \end{cases} \quad \text{and } G(x) := \int_0^x g(t)\,dt,$$

and we get for all $-\infty < a < b < \infty$

$$F(b)G(b) - F(a)G(a) = \int_a^b f(t)G(t)\,dt + \int_a^b F(t)g(t)\,dt. \tag{16.2}$$

Proof. Using Fubini's theorem (at the step marked by »§«) we get

$$\begin{aligned} \int_a^b f(t)(G(t) - G(a))\,dt &= \int_a^b f(t)\left\{\int_a^t g(s)\,ds\right\}dt \\ &= \int_a^b \int_a^b f(t)g(s)\mathbb{1}_{[a,t]}(s)\,ds\,dt \\ &\overset{\S}{=} \int_a^b \int_a^b f(t)g(s)\mathbb{1}_{[s,b]}(t)\,dt\,ds \\ &= \int_a^b \left(\int_s^b f(t)\,dt\right)g(s)\,ds \\ &= \int_a^b g(s)(F(b) - F(s))\,ds. \end{aligned}$$

discontinuities are jump discontinuities and we may assume that m is left-continuous and increasing. Draw the graph and imagine a light-source (a »bedside lamp«) to the right of the graph that projects the graph to the y-axis. Where there is a jump, we get a »white« interval of the form $(f(t), f(t+)] \neq \emptyset$ on the y-axis. Since these intervals are disjoint and contain at least one rational number, there are at most countably many of them.

A simple rearrangement yields (16.2). Note that Fubini may be applied since we do have

$$\iint_{[a,b]\times[a,b]} |f(t)g(s)|\,\mathrm{d}t\,\mathrm{d}s = \int_{[a,b]} |f(t)|\,\mathrm{d}t \cdot \int_{[a,b]} |g(s)|\,\mathrm{d}s < \infty. \qquad \square$$

The following example is an example of a function which is not Lebesgue integrable but has a convergent Riemann integral. It is important in connection with Fourier series and the inversion formula for the Fourier transform.

16.8 Example★ (sine integral). $\lim_{T\to\infty}\int_0^T \frac{\sin\xi}{\xi}\,\mathrm{d}\xi = \frac{1}{2}\pi$.(2)

Proof. Note that $\frac{1}{\xi} = \int_0^\infty e^{-t\xi}\,\mathrm{d}t$ and $\operatorname{Im} e^{i\xi} = \sin\xi$. Fubini's theorem shows

$$\begin{aligned}\int_0^T \frac{\sin\xi}{\xi}\,\mathrm{d}\xi &= \int_0^T\int_0^\infty e^{-t\xi}\sin\xi\,\mathrm{d}t\,\mathrm{d}\xi \\ &= \int_0^\infty\int_0^T e^{-t\xi}\operatorname{Im} e^{i\xi}\,\mathrm{d}\xi\,\mathrm{d}t = \int_0^\infty \operatorname{Im}\int_0^T e^{-(t-i)\xi}\,\mathrm{d}\xi\,\mathrm{d}t.\end{aligned}$$

The inner integral can be evaluated as follows

$$\begin{aligned}\operatorname{Im}\left[\int_0^T e^{-(t-i)\xi}\,\mathrm{d}\xi\right] &= \operatorname{Im}\left[\frac{e^{-(t-i)\xi}}{i-t}\right]_0^T \\ &= \operatorname{Im}\left(\frac{e^{(i-t)T}-1}{i-t}\right) = \operatorname{Im}\left(\frac{(e^{(i-t)T}-1)(-i-t)}{1+t^2}\right).\end{aligned}$$

Therefore, the dominated convergence theorem (Theorem 11.2) yields

$$\begin{aligned}\int_0^T \frac{\sin\xi}{\xi}\,\mathrm{d}\xi &= \int_0^\infty \frac{-te^{-tT}\sin T - e^{-tT}\cos T + 1}{1+t^2}\,\mathrm{d}t \\ &\overset{s=tT}{=} \int_0^\infty \frac{-se^{-s}\sin T}{T^2+s^2}\,\mathrm{d}s + \int_0^\infty \frac{-e^{-tT}\cos T + 1}{1+t^2}\,\mathrm{d}t \\ &\xrightarrow[T\to\infty]{\text{DCT}} \int_0^\infty \frac{1}{1+t^2}\,\mathrm{d}t = \arctan t\Big|_0^\infty = \frac{\pi}{2}.\end{aligned}$$

In order to see that the Lebesgue integral does not exist, we use the elementary estimate

$$\frac{|\sin x|}{x} \geqslant \frac{1}{(n+1)\pi}|\sin x|, \quad x \in [n\pi, (n+1)\pi)$$

to deduce for every $N \in \mathbb{N}$

$$\int_0^\infty \frac{|\sin x|}{x}\,\mathrm{d}x \geqslant \sum_{n=1}^N \frac{1}{(n+1)\pi}\int_{n\pi}^{(n+1)\pi} |\sin x|\,\mathrm{d}x = \frac{1}{\pi}\int_0^\pi \sin x\,\mathrm{d}x \cdot \sum_{n=1}^N \frac{1}{n+1}.$$

Since the harmonic series diverges, this shows that $\frac{1}{x}\sin x$ is not Lebesgue integrable. $\square$

(2) If you are not happy with the integral of $\mathbb{C}$-valued functions, please consult the appendix to this chapter.

More on the product of σ-algebras

Here we provide a further characterization of the product of two σ-algebras. Let $(\Omega_n, \mathscr{A}_n)$, $n = 1, 2, 3$, measurable spaces and denote by

$$\pi_n : \Omega_1 \times \Omega_2 \to \Omega_n, \quad (x_1, x_2) \mapsto x_n, \quad (n = 1, 2)$$

the canonical coordinate projections. According to Definition 6.5,

$$\sigma(\pi_1, \pi_2) = \sigma\Big(\pi_1^{-1}(\mathscr{A}_1),\ \pi_2^{-1}(\mathscr{A}_2)\Big)$$

is the smallest σ-algebra on $\Omega_1 \times \Omega_2$ such that π_1, π_2 are measurable.

16.9 Theorem. *Let $(\Omega_n, \mathscr{A}_n)$ and π_n, $n = 1, 2, 3$, be as above.*

a) $\mathscr{A}_1 \otimes \mathscr{A}_2 = \sigma(\pi_1, \pi_2)$;

b) $X : \Omega_3 \to \Omega_1 \times \Omega_2$ *is* $\mathscr{A}_3/\mathscr{A}_1 \otimes \mathscr{A}_2$*-measurable if, and only if* $\pi_n \circ X : \Omega_3 \to \Omega_n$ *is* $\mathscr{A}_3/\mathscr{A}_n$*-measurable for* $n = 1, 2$;

c) *If* $Y : \Omega_1 \times \Omega_2 \to \Omega_3$ *is* $\mathscr{A}_1 \otimes \mathscr{A}_2/\mathscr{A}_3$*-measurable, then* $Y(x_1, \bullet)$, $Y(\bullet, x_2)$ *are* $\mathscr{A}_2/\mathscr{A}_3$*-, resp.,* $\mathscr{A}_1/\mathscr{A}_3$*-measurable.*

Proof. a) Since $\pi_1^{-1}(\mathscr{A}_1) = \mathscr{A}_1 \times \Omega_2$, $\pi_2^{-1}(\mathscr{A}_2) = \Omega_1 \times \mathscr{A}_2$, we get

$$\sigma(\pi_1, \pi_2) = \sigma\Big(\{A_1 \times \Omega_2,\ \Omega_1 \times A_2 \mid A_1 \in \mathscr{A}_1,\ A_2 \in \mathscr{A}_2\}\Big)$$

and we conclude that $\mathscr{A}_1 \times \mathscr{A}_2 \subset \sigma(\pi_1, \pi_2) \subset \mathscr{A}_1 \otimes \mathscr{A}_2$, hence

$$\mathscr{A}_1 \otimes \mathscr{A}_2 \stackrel{\text{def}}{=} \sigma(\mathscr{A}_1 \times \mathscr{A}_2) \subset \sigma(\pi_1, \pi_2) \subset \mathscr{A}_1 \otimes \mathscr{A}_2.$$

b) Since $X : (\Omega_3, \mathscr{A}_3) \to (\Omega_1 \times \Omega_2, \mathscr{A}_1 \otimes \mathscr{A}_2)$ is measurable, so are the compositions $\pi_n \circ X$, $n = 1, 2$.

Conversely, assume that the maps $\pi_n \circ X : (\Omega_3, \mathscr{A}_3) \to (\Omega_n, \mathscr{A}_n)$ are measurable $n = 1, 2$. We have

$$\begin{aligned} X^{-1}(A_1 \times A_2) &= X^{-1}\Big(\pi_1^{-1}(A_1) \cap \pi_2^{-1}(A_2)\Big) \\ &= X^{-1}\Big(\pi_1^{-1}(A_1)\Big) \cap X^{-1}\Big(\pi_2^{-1}(A_2)\Big) \\ &= (\pi_1 \circ X)^{-1}(A_1) \cap (\pi_2 \circ X)^{-1}(A_2) \in \mathscr{A}_3. \end{aligned}$$

As $\mathscr{A}_1 \times \mathscr{A}_2$ generates the σ-algebra $\mathscr{A}_1 \otimes \mathscr{A}_2$, the claim follows.

c) Fix $x_1 \in \Omega_1$ and define the injection $\imath_{x_1} : \Omega_2 \to \Omega_1 \times \Omega_2$, $y \mapsto (x_1, y)$. The direct calculation

$$\imath_{x_1}^{-1}(A_1 \times A_2) = \begin{Bmatrix} \emptyset, & x_1 \notin A_1 \\ A_2, & x_1 \in A_1 \end{Bmatrix} \in \mathscr{A}_2$$

shows that $\imath_{x_1}$ is measurable, and so is $Y(x_1, \bullet) = Y \circ \imath_{x_1}$ as composition of measurable maps. The function $Y(\bullet, x_2)$ can be treated in a similar way. □

Appendix: Integrating complex-valued functions

Let $(\Omega, \mathscr{A}, \mu)$ be any measure space. The complex numbers $\mathbb{C}$ are a normed space, and so the notion of Borel sets $\mathscr{B}(\mathbb{C})$ makes sense. Let us show that $\mathscr{B}(\mathbb{C})$ can be identified with $\mathscr{B}(\mathbb{R}^2)$. Since the maps from $\mathbb{R}^2 \to \mathbb{C}$ and $\mathbb{C} \to \mathbb{R}^2$ defined by

$$(x,y) \mapsto z = x + \mathrm{i}y \quad \text{and} \quad z \mapsto (\operatorname{Re} z, \operatorname{Im} z) := \left(\tfrac{1}{2}(z+\bar{z}), \tfrac{1}{2\mathrm{i}}(z-\bar{z})\right)$$

are continuous and inverses of each other, we can identify $\mathscr{B}(\mathbb{C})$ and $\mathscr{B}(\mathbb{R}^2)$; in particular $f : \Omega \to \mathbb{C}$ is measurable if, and only if, $\operatorname{Re} f, \operatorname{Im} f : \Omega \to \mathbb{R}$ are measurable.

We want to extend the integral to complex-valued functions $f : \Omega \to \mathbb{C}$. In order to preserve the linearity of the integral, we extend it in a ($\mathbb{C}$-)linear way: Write $f = u + \mathrm{i}v$ where $u = \operatorname{Re} f$ and $v = \operatorname{Im} f$, and set

$$\int f \, d\mu := \int u \, d\mu + \mathrm{i} \int v \, d\mu = \int \operatorname{Re} f \, d\mu + \mathrm{i} \int \operatorname{Im} f \, d\mu.$$

We assume either that $u, v \geqslant 0$ or that $u^\pm, v^\pm \in L^1_{\mathbb{R}}(\mu)$. A simple calculation shows that $f \mapsto \int f \, d\mu$ is a $\mathbb{C}$-linear mapping ✎ which enjoys essentially the same properties as the integral for real-valued integrands. In addition, we have

$$\operatorname{Re} \int f \, d\mu = \int \operatorname{Re} f \, d\mu, \quad \operatorname{Im} \int f \, d\mu = \int \operatorname{Im} f \, d\mu,$$

$$\overline{\int f \, d\mu} = \int \bar{f} \, d\mu, \quad \left| \int f \, d\mu \right| \leqslant \int |f| \, d\mu.$$

Only the last inequality – the triangle inequality – is not immediately clear. To prove it, pick some $\theta \in [-\pi, \pi)$ such that the rotation of $\int f \, d\mu \in \mathbb{C}$ by the angle θ gives a positive number:

$$\left| \int f \, d\mu \right| = e^{\mathrm{i}\theta} \int f \, d\mu = \operatorname{Re}\left(e^{\mathrm{i}\theta} \int f \, d\mu \right) = \int \operatorname{Re}\left(e^{\mathrm{i}\theta} f \right) d\mu = \int \left| e^{\mathrm{i}\theta} f \right| d\mu = \int |f| \, d\mu.$$

The spaces $L^p_{\mathbb{C}}(\mu)$ are (equivalence classes w.r.t. μ-a.e. equality) of pth power integrable functions, i.e.

$$f \in L^p_{\mathbb{C}}(\mu) \iff f \in \mathcal{L}^0(\mathscr{B}(\mathbb{C})) \text{ and } |f| \in L^p_{\mathbb{R}}(\mu).$$

17 Integrals for image measures and convolutions

In this chapter, $(\Omega, \mathscr{A}, \mathbb{P})$ is a probability space, i.e. a measure space equipped with a probability measure $\mathbb{P}(\Omega) = 1$, and $(S, \mathscr{S})$ is a further measurable space. An (S-valued) **random variable (rv)** is a measurable map $X : (\Omega, \mathscr{A}) \to (S, \mathscr{S})$. The **distribution** or **(probability) law** of X is the image measure (Theorem 6.6) of $\mathbb{P}$ under X,

$$\mathbb{P}_X(B) := X(\mathbb{P})(B) = \mathbb{P}(X^{-1}(B)) = \mathbb{P}(X \in B), \quad B \in \mathscr{S}.$$

17.1 Theorem. *If $u : S \to \mathbb{R}$ is a measurable function and $X : \Omega \to S$ a random variable, then*

$$\int_\Omega u(X(\omega))\,\mathbb{P}(d\omega) = \int_S u(s)\,\mathbb{P}(X \in ds). \tag{17.1}$$

The formula (17.1) is understood in the following way: Either $u \geqslant 0$ and (17.1) holds in $[0,\infty]$, or $u \in L^1(\mathbb{P}_X)$ if, and only if, $u \circ X \in L^1(\mathbb{P})$.

Proof. 1° Assume that $u(x) = \mathbb{1}_B(s)$ for some $B \in \mathscr{S}$. We have

$$\int_S u\,d\mathbb{P}_X = \int_S \mathbb{1}_B\,d\mathbb{P}_X = \mathbb{P}_X(B) = \mathbb{P}(X \in B) = \int_\Omega \mathbb{1}_{\{X\in B\}}\,d\mathbb{P} = \int_\Omega \mathbb{1}_B \circ X\,d\mathbb{P}$$

where we use the identities $\mathbb{1}_{\{X\in B\}} = \mathbb{1}_{X^{-1}(B)} = \mathbb{1}_B \circ X$. This shows (17.1) and $u \circ X = \mathbb{1}_B \circ X \in L^1(\mathbb{P})$ if, and only if, $u \in L^1(\mathbb{P}_X)$.

2° Assume that $u(s) = \sum_{i=0}^n \alpha_i \mathbb{1}_{B_i}(s) \in \mathcal{E}^+(\mathscr{S})$. Combining 1° with the linearity of the integral proves the assertion for such functions.

3° Every $u \in \mathcal{L}^{0,+}(\mathscr{S})$ can be approximated with $u_n \in \mathcal{E}^+(\mathscr{S})$: $u_n \uparrow u$ and

$$\int u\,d\mathbb{P}_X \overset{\text{BL}}{=} \sup_{n\in\mathbb{N}} \int u_n\,d\mathbb{P}_X \overset{2^\circ}{=} \sup_{n\in\mathbb{N}} \int u_n \circ X\,d\mathbb{P} \overset{\text{BL}}{=} \int u \circ X\,d\mathbb{P}.$$

This shows that (17.1) holds in $[0,\infty]$.

4° If $u \in L^1(\mathbb{P}_X)$, then $u = u^+ - u^-$ and $u^\pm \in \mathcal{L}^{0,+}(\mathscr{S})$. We have

$$\begin{aligned}\int u\,d\mathbb{P}_X &= \int u^+\,d\mathbb{P}_X - \int u^-\,d\mathbb{P}_X \\ &\overset{3^\circ}{=} \underbrace{\int u^+ \circ X\,d\mathbb{P}}_{<\infty} - \underbrace{\int u^- \circ X\,d\mathbb{P}}_{<\infty} \\ &= \int u \circ X\,d\mathbb{P},\end{aligned}$$

i.e. we get (17.1) and $u \circ X \in L^1(\mathbb{P})$. Running the argument backwards, finishes the proof. □

Almost the same proof yields the following general result.

17.2 Corollary. *If $(\Omega, \mathscr{A}, \mu)$ is an arbitrary measure space, $(S, \mathscr{S})$ a measurable space and $X : \Omega \to S$ a measurable map, then*

$$\int_\Omega u(X(x))\,\mu(dx) = \int_S u(s)\,X(\mu)(ds). \tag{17.2}$$

The formula (17.2) is understood in the following way: Either $u \geqslant 0$ and (17.2) holds in $[0,\infty]$, or $u \in L^1(X(\mu))$ if, and only if, $u \circ X \in L^1(\mu)$.

17.3 Example. Let $(\Omega, \mathscr{A}, \mathbb{P})$ be a probability space and $X : \Omega \to \mathbb{R}$ a real-valued random variable.

a) $\mathbb{E}X := \int_\Omega X\,d\mathbb{P} = \int_{\mathbb{R}} x\,\mathbb{P}_X(dx) = \int_{\mathbb{R}} x\,\overbrace{\mathbb{P}(X \in dx)}^{\text{preferred notation in probability}}$;
b) $\mathbb{E}|X| = \int_{\mathbb{R}} |x|\,\mathbb{P}(X \in dx)$;
c) $\mathbb{V}X := \mathbb{E}\big[(X - \mathbb{E}X)^2\big] = \int_{\mathbb{R}} (x - \mathbb{E}X)^2\,\mathbb{P}(X \in dx)$;
d) Assume that (the distribution of) X has a density $f(x)$, i.e. $\mathbb{P}(X \in dx) = f(x)\,dx$. For any measurable $u : S \to \mathbb{R}$

$$\mathbb{E}(u \circ X) := \int u \circ X\,d\mathbb{P} = \int_{\mathbb{R}} u(y) f(y)\,dy;$$

e) Assume that the distribution of X is $\mathbb{P}_X = \sum_{i=1}^N p_i \delta_{x_i}$. For any measurable $u : S \to \mathbb{R}$

$$\mathbb{E}(u \circ X) := \int u \circ X\,d d\mathbb{P} = \int u(y)\,\mathbb{P}(X \in dy) = \sum_{i=1}^{N} p_i\,u(x_i).$$

17.4 Example (important). The functions

$$\begin{aligned} \sigma : \mathbb{R}^d &\to \mathbb{R}^d \\ y &\mapsto -y \end{aligned} \qquad\qquad \begin{aligned} \tau_z : \mathbb{R}^d &\to \mathbb{R}^d \\ y &\mapsto y - z \end{aligned}$$

are continuous, hence Borel measurable and they have Borel measurable inverse functions: $\sigma^{-1} = \sigma$ and $\tau_z^{-1} = \tau_{-z}$. Moreover,

$$\sigma(\lambda^d) = \lambda^d \quad\text{and}\quad \tau_z(\lambda^d) = \lambda^d.$$

At the level of integrals this leads to the following important formulae

$$\begin{aligned} \int u(-y)\,\lambda^d(dy) &= \int u(\sigma(y))\,\lambda^d(dy) \\ &= \int u(y)\,\sigma(\lambda^d)(dy) = \int u(y)\,\lambda^d(dy), \end{aligned}$$

and

$$\begin{aligned} \int u(y-z)\,\lambda^d(dy) &= \int u(\tau_z(y))\,\lambda^d(dy) \\ &= \int u(y)\,\tau_z(\lambda^d)(dy) = \int u(y)\,\lambda^d(dy). \end{aligned}$$

In particular, $u(-\bullet) \in L^1(\lambda^d) \iff u \in L^1(\lambda^d) \iff u(\bullet - z) \in L^1(\lambda^d)$.

17.5 Example★. Consider a linear map $\Phi(x) = Mx + b$ where $b \in \mathbb{R}^d$ and $M \in \mathbb{R}^{d\times d}$ is an invertible matrix. In Theorem 6.9 and 4.9.a) (also: Theorem 3.7.a),c)) we have seen that $M(\lambda^d) = \lambda^d \circ M^{-1} = |\det M^{-1}| \cdot \lambda^d$ and, since $\Phi(x) = \tau_{-b}(Mx)$, that $\Phi(\lambda^d) = \lambda^d \circ \Phi^{-1} = |\det M^{-1} \cdot \lambda^d|$. We can now apply Corollary 17.2 to, say, a measurable function $u \geqslant 0$

$$\begin{aligned}\int_{U} u(\Phi(x))|\det M| \cdot \lambda^{d}(dx) &\overset{\S}{=} \int_{\Phi(U)} u(y)\,\Phi(|\det M|\cdot\lambda^{d})(dy) \\ &= \int_{\Phi(U)} u(y)|\det M|\cdot\Phi(\lambda^{d})(dy) \\ &= \int_{\Phi(U)} u(y)|\det M|\cdot|\det M|^{-1}\lambda^{d}(dy) \\ &= \int_{\Phi(U)} u(y)\,\lambda^{d}(dy).\end{aligned}$$

In the step marked by »§«, we use first $\int_{U}\cdots = \int \mathbb{1}_{U}(x)\cdots = \int \mathbb{1}_{\Phi(U)}(\Phi(x))$, and then Corollary 17.2

Differentiating $\Phi(x)$, we get $D\Phi(x) = M$ and this explains the role of $|\det M| = |\det D\Phi(x)|$ on the left-hand side. We have thus shown

$$\int_{U} u(\Phi(x))|\det D\Phi(x)|\cdot\lambda^{d}(dx) = \int_{\Phi(U)} u(y)\,\lambda^{d}(dy). \tag{17.3}$$

Compare this to the well-known substitution rule for Riemann integrals:

$$\int_{a}^{b} u(\phi(x))\,\phi'(x)\,dx \xrightarrow[y=\phi(x)]{dy=\phi'(x)\,dx} \int_{\phi(a)}^{\phi(b)} u(y)\,dy$$

(there is no modulus around ϕ', since the Riemann integral knows the »direction«, i.e. we use $\int_{b}^{a} = -\int_{a}^{b}$, which is not the case for the Lebesgue integral as we integrate over $(a,b]$ and $\phi((a,b])$, respectively).

Formula (17.3) holds for all C^{1}-diffeomorphisms $\Phi : U \subset \mathbb{R}^{d} \to \Phi(U) \subset \mathbb{R}^{d}$. The proof is a bit involved (see e.g. [MIMS, Theorem 16.4]), but the essence of the proof is that we can approximate, by Taylor's formula, Φ locally by affine-linear functions of the form $\Phi(x) \approx \Phi(x_0) + D\Phi(x_0)x$, use (17.3) for small sets U and glue everything together in the end. Without proof we state

17.6 Theorem* (Jacobi's transformation theorem). *Let $U, V \subset \mathbb{R}^{d}$ be open sets and $\Phi : U \to V$ a C^{1}-diffeomorphism. For every positive measurable function u*

$$\int_{V} u(y)\,dy = \int_{U} u(\Phi(x))\,|\det D\Phi(x)|\,dx. \tag{17.4}$$

The formula also holds for integrable functions, and one has $u \in L^{1}(V,\lambda^{d})$ if, and only if, $|\det D\Phi|\, u\circ\Phi \in L^{1}(U,\lambda^{d})$.

Convolution

17.7 Definition. Let μ, ν be σ-finite measures on $(\mathbb{R}^{d}, \mathscr{B}(\mathbb{R}^{d}))$ and assume that $u, w : \mathbb{R}^{d} \to \mathbb{R}$ are measurable functions. We call

a) $u * w(x) = \int u(x-y)w(y)\,\lambda^d(dy), \quad x \in \mathbb{R}^d$;
b) $u * \mu(x) = \int u(x-y)\,\mu(dy), \quad x \in \mathbb{R}^d$;
c) $\nu * \mu(B) = \iint \mathbb{1}_B(x+y)\,\mu(dx)\nu(dy), \quad B \in \mathscr{B}(\mathbb{R}^d)$;

(provided that the integrals exist) the **convolution** or the convolution product of u, w or u, μ or ν, μ, respectively.

▲ The convolution of measures or the convolution of positive measurable functions always exist in $[0, \infty]$.

▲ The convolution of functions with arbitrary sign needs further conditions to ensure finiteness.

17.8 Properties. a) If the convolution exists, then $u * w = w * u$. Indeed:

$$\begin{aligned} u * w(x) &= \int u(x-y)w(y)\,\lambda^d(dy) \\ &= \int u(\sigma \circ \tau_x(y))w(\sigma \circ \tau_x \circ (\sigma \circ \tau_x)^{-1}(y))\,\lambda^d(dy) \\ &\overset{\text{Ex. 17.4}}{=} \int u(y)w(\underbrace{(\sigma \circ \tau_x)^{-1}(y)}_{=\tau_{-x}\circ\sigma(y)})\,\lambda^d(dy) \\ &= \int u(y)w(x-y)\,\lambda^d(dy) = w * u(x). \end{aligned}$$

b) $\mu * \nu = \nu * \mu$ because of Tonelli's theorem.

c) The convolution product is linear in each argument, for instance, if $a, b \in \mathbb{R}$,

$$(au + bv) * w = a(u * w) + b(v * w).$$

d) Let $\alpha : \mathbb{R}^d \times \mathbb{R}^d \to \mathbb{R}^d$, $(x, y) \mapsto \alpha(x, y) := x + y$. This map is continuous, hence Borel measurable and the calculation below shows that $\mu * \nu = \alpha(\mu \times \nu)$:

$$\begin{aligned} \mu * \nu(B) &= \int_{\mathbb{R}^d}\int_{\mathbb{R}^d} \mathbb{1}_B(x+y)\,\mu(dx)\nu(dy) \\ &= \int_{\mathbb{R}^d \times \mathbb{R}^d} \mathbb{1}_B(\alpha(x,y))\,(\mu \times \nu)(d(x,y)) \\ &= \int_{\mathbb{R}^d} \mathbb{1}_B(z)\,\alpha(\mu \times \nu)(dz). \end{aligned}$$

e) Note that $x + y \in B \iff x \in B - y$, and so $\mathbb{1}_B(x+y) = \mathbb{1}_{B-y}(x)$. Therefore, we have

$$\begin{aligned} \mu * \nu(B) &= \int_{\mathbb{R}^d}\int_{\mathbb{R}^d} \mathbb{1}_B(x+y)\,\mu(dx)\nu(dy) \\ &= \int_{\mathbb{R}^d}\left[\int_{\mathbb{R}^d} \mathbb{1}_{B-y}(x)\,\mu(dx)\right]\nu(dy) = \int_{\mathbb{R}^d} \mu(B-y)\,\nu(dy). \end{aligned}$$

In the same way we se, using Tonelli's theorem

$$\mu * \nu(B) = \int_{\mathbb{R}^d} \nu(B - x)\,\mu(dx).$$

f) If μ and ν are probability measures, then $\mu * \nu$ is again a probability measure:

$$\mu * \nu(\mathbb{R}^d) = \int_{\mathbb{R}^d} \nu(\mathbb{R}^d - x)\,\mu(dx) = \int_{\mathbb{R}^d} 1\,\mu(dx) = \mu(\mathbb{R}^d) = 1.$$

g) ✎ If $u, v \geqslant 0$, then the convolution of functions is compatible with the convolution of measures (with these functions as densities). To wit

$$\underbrace{(u\lambda^d) * (w\lambda^d)}_{\text{convolution of measures}} = \underbrace{(u * w)}_{\text{convolution of functions}} \lambda^d$$

It is not always possible to define $u * w$, since we need that $u(x - \bullet)w$ is either positive or integrable. This means that some of the calculation in §17.8 are **formal**. The next theorem provides a handy criterion for the existence of the convolution.

17.9 Theorem (Young's inequality). *If $u \in L^1(\lambda^d)$ and $w \in L^p(\lambda^d)$ for some $p \in [1, \infty]$, then $u * w$ exists as a function of $L^p(\lambda^d)$.*

Moreover, $u * w = w * u$ and

$$\|u * w\|_p \leqslant \|u\|_1 \cdot \|w\|_p. \tag{17.5}$$

Proof. If $u, w \geqslant 0$, then we can use §17.8.a) to see

$$\begin{aligned} u * w(x) &= \int u(x - y)w(y)\,\lambda^d(dy) \\ &= \int u(y)w(x - y)\,\lambda^d(dy) = w * u(x). \end{aligned}$$

In general, we have $u = u^+ - u^-$, $w = w^+ - w^-$ and we get by linearity that $u*w = w*u$ **provided that all expressions $u^\pm * w^\pm$ are λ^d-a.e. finite**. This follows, in particular, if the right-hand side of (17.5) is finite.

Case 1: $p = \infty$. We have $|w| \leqslant \|w\|_\infty$ a.e. Thus, we get for all x

$$\left|\int u(y)w(x - y)\,dy\right| \leqslant \int |u(y)||w(x - y)|\,dy \leqslant \int |u(y)|\|w\|_\infty\,dy = \|u\|_1\|w\|_\infty.$$

Case 2: $p \in [1,\infty)$. Here we have

$$\begin{aligned}
\|u * w\|_p^p &= \int \left| \int u(y) w(x-y)\, \lambda^d(dy) \right|^p \lambda^d(dx) \\
&= \int \left(\int |u(y)| \cdot |w(x-y)|\, \lambda^d(dy) \right)^p \lambda^d(dx) \\
&\leqslant \|u\|_1^p \int \left(\int |w(x-y)| \frac{|u(y)|}{\|u\|_1}\, \lambda^d(dy) \right)^p \lambda^d(dx) \\
&\overset{\text{Jensen}}{\leqslant} \|u\|_1^p \iint |w(x-y)|^p \frac{|u(y)|}{\|u\|_1}\, \lambda^d(dy)\, \lambda^d(dx) \\
&\overset{\text{Tonelli}}{=} \|u\|_1^p \int \underbrace{\left(\int |w(x-y)|^p\, \lambda^d(dx) \right)}_{=\|w\|_p^p \text{ cf. Example 17.4}} \frac{|u(y)|}{\|u\|_1}\, \lambda^d(dy) \\
&= \|u\|_1^p \|w\|_p^p.
\end{aligned}$$

□

The convolution of two functions is more regular (smoother) than its factors.

17.10 Lemma. *If $\phi \in C_c(\mathbb{R}^d)$ and $u \in L^1(\lambda^d)$, then $\phi * u$ is continuous.*

Proof. We have $|\phi| \leqslant M$ for the constant $M = \sup_{x\in\mathbb{R}^d} |\phi(x)|$. Therefore,

$$|u * \phi(x)| = \left| \int \phi(x-y) u(y)\, dy \right| \leqslant \int |\phi(x-y) u(y)|\, dy \leqslant M \int |u(y)|\, dy < \infty,$$

and it follows from the continuity lemma (Theorem 12.1, integrable majorant $Mu(y)$) that $x \mapsto u * \phi(x)$ is continuous. □

Lemma 17.10 has a far-reaching generalization. In order to prove it, we need the following result which will be proved in the appendix.

17.11 Theorem. *$C_c(\mathbb{R}^d)$ is is a dense subset of $L^p(\lambda^d)$, $p \in [1,\infty)$.*

That is, for every $u \in L^p(\lambda^d)$ there is a sequence $(u_n)_{n\in\mathbb{N}} \subset C_c(\mathbb{R}^d) \subset L^p(\mathbb{R}^d)$ such that $\lim_{n\to\infty} \|u - u_n\|_p = 0$.

17.12 Theorem. *Assume that $u \in L^p(\lambda^d)$ for some $p \in [1,\infty)$.*

a) $x \mapsto \int |u(x+y) - u(y)|^p\, \lambda^d(dy)$ *is a continuous function.*

b) *If $u \in L^1(\lambda^d)$ and $w \in L^\infty(\lambda^d)$, then $u * w$ is bounded and continuous.*

Proof. a) Assume first that $u = \phi \in C_c(\mathbb{R}^d)$ and $K = \operatorname{supp}\phi = \overline{\{\phi \neq 0\}}$. The continuity lemma (Theorem 12.1) implies that for all $x \in B_n(0)$ and some $n \in \mathbb{N}$

$$x \mapsto I(\phi; x) := \int \underbrace{|\phi(x+y) - \phi(y)|^p}_{\leqslant 2^p \|\phi\|_\infty \mathbb{1}_{K+B_n(0)}(y) \text{ – integrable majorant}} \lambda^d(dy) = \|\phi(x+\cdot) - \phi(\cdot)\|_p^p$$

is continuous. Since n is arbitrary, we get continuity on $\mathbb{R}^d$.

Using the density of $C_c(\mathbb{R}^d)$ in $\mathcal{L}^p(\lambda^d)$, we can find for every $u \in \mathcal{L}^p(\lambda^d)$ a sequence $(\phi_n)_{n\in\mathbb{N}} \subset C_c(\mathbb{R}^d)$ such that $\lim_{n\to\infty} \|u - \phi_n\|_p = 0$. Moreover, for every $x \in \mathbb{R}^d$

$$\begin{aligned} &\|(\phi_n(x+\cdot) - \phi_n(\cdot)) - (u(x+\cdot) - u(\cdot))\|_p \\ &\qquad \leqslant \|\phi_n(x+\cdot) - u(x+\cdot)\|_p + \|\phi_n - u\|_p \stackrel{\text{Ex. 17.4}}{=} 2\|\phi_n - u\|_p \end{aligned}$$

holds true and, therefore, $\lim_{n\to\infty} I(\phi_n;x) = I(u;x)$ uniformly for all x. Thus, $I(u;x)$ inherits the continuity of $I(\phi_n;x)$, which we have established in the first part of the proof.

b) For any $x, x' \in \mathbb{R}^d$ we have

$$\begin{aligned} \left|u * w(x) - u * w(x')\right| &\leqslant \int \left|w(y)u(x-y) - w(y)u(x'-y)\right| \lambda^d(dy) \\ &\leqslant \|w\|_\infty \|u(x-\cdot) - u(x'-\cdot)\|_1 \\ &\stackrel{\text{Ex. 17.4}}{=} \|w\|_\infty \|u(x-x'+\cdot) - u\|_1 \\ &\leqslant \|w\|_\infty (\|u(x-x'+\cdot)\|_1 + \|u\|_1) \\ &\leqslant 2\|w\|_\infty \|u\|_1. \end{aligned}$$

This proves boundedness, continuity follows from Part a). □

Appendix: Proof of Theorem 17.11★

In this appendix we want to show that the continuous functions with compact support, $C_c(\mathbb{R}^d)$, are a dense subset of $L^p(\lambda^d)$. Since any $\phi \in C_c^\infty$ is bounded by $\|\phi\|_\infty \mathbb{1}_{B_r(0)}$ where $r > 0$ is large enough to guarantee $\operatorname{supp}\phi \subset B_r(0)$, we have $\|\phi\|_p \leqslant \|\phi\|_\infty \lambda^d(B_r(0))^{1/p} < \infty$, i.e. $C_c(\mathbb{R}^d) \subset L^p$, i.e. **density** is the real issue.

We begin with an auxiliary result which is interesting in its own right. Recall that the **symmetric difference** of two sets $A, B \subset \Omega$ is $A \vartriangle B := (A \setminus B) \cup (B \setminus A)$.

17.13 Lemma★ (approximation by generators). *Let $(\Omega, \mathscr{A}, \mu)$ be a finite measure space such that $\mathscr{A} = \sigma(\mathscr{G})$ for some Boolean algebra $\mathscr{G}$ – i.e. $\mathscr{G}$ is stable under finitely many »∪«, »∩« and »\«, and $\Omega \in \mathscr{G}$. Then*

$$\forall A \in \mathscr{A} \quad \forall \epsilon > 0 \quad \exists G = G_{\epsilon,A} \in \mathscr{G} : \quad \mu(A \vartriangle G) < \epsilon. \tag{17.6}$$

Proof. If $A, B, C \subset \Omega$, then $A \vartriangle B = B \vartriangle A$ and $A \vartriangle (B \vartriangle C) = (A \vartriangle B) \vartriangle C$. The easiest way to show this is to use $\mathbb{1}_{A\vartriangle B} = \mathbb{1}_A + \mathbb{1}_B - 2\mathbb{1}_{A\cap B}$ and to show that $\mathbb{1}_{A\vartriangle(B\vartriangle C)}$ leads to an expression where A, B, C play symmetric roles ✍ (cf. [MIMS, Problem 2.6]). Moreover, note that

$$B \subset A \implies A \setminus B = A \vartriangle B \quad \text{and} \quad A \cap C = \emptyset \implies A \cup C = A \vartriangle C.$$

Define $\mathscr{D} := \{A \in \mathscr{A} : (17.6) \text{ holds for } A\}$. We claim that this family is a Dynkin system (Definition 4.1). Let $\epsilon > 0$ be fixed.

(D_1) Since $\Omega \in \mathscr{G}$, we also have $\Omega \in \mathscr{D}$.

(D_2) If $D \in \mathscr{D}$, then there is some $G \in \mathscr{G}$ such that $\mu(D \vartriangle G) < \epsilon$. Note that

$$D^c \vartriangle G^c = (D^c \setminus G^c) \cup (G^c \setminus D^c) = (D^c \cap G) \cup (G^c \cap D) \\ = (G \setminus D) \cup (D \setminus G) = D \vartriangle G,$$

so that $\mu(D^c \vartriangle G^c) = \mu(D \vartriangle G) < \epsilon$. Since $G^c \in \mathscr{G}$, we get $D^c \in \mathscr{D}$.

(D_3) Let $D_n \in \mathscr{D}$ be disjoint sets and set $D := \dot\bigcup_{n\in\mathbb{N}} D_n$ and $D^N = \dot\bigcup_{n\leqslant N} D_n$. Pick $N = N_\epsilon$ so large that $\mu\left(D \setminus D^N\right) < \epsilon$ – this is possible since $\mu(D) \leqslant \mu(\Omega) < \infty$. Since $D_n \in \mathscr{D}$, there is some $G_n \in \mathscr{G}$ such that $\mu(D_n \vartriangle G_n) < \epsilon$; define the set $G^N := G_1 \vartriangle G_2 \vartriangle \ldots \vartriangle G_N$. With the rules for symmetric differences, we get

$$D \vartriangle G^N = \left((D \setminus D^N) \,\dot\cup\, D^N\right) \vartriangle G^N = (D \setminus D^N) \vartriangle D^N \vartriangle G^N \\ = (D \setminus D^N) \vartriangle (D_1 \vartriangle G_1) \vartriangle \ldots \vartriangle (D_N \vartriangle G_N).$$

Since $\mu(A \vartriangle B) = \mu(A \setminus B) + \mu(B \setminus A) \leqslant \mu(A) + \mu(B)$, we see

$$\mu(D \vartriangle G^N) \leqslant \mu(D \setminus D^N) + \sum_{n=1}^{N} \mu(D_n \vartriangle G_n) = (N+1)\epsilon,$$

and since $G^N \in \mathscr{G}$ ($\mathscr{G}$ is an algebra) we get $D \in \mathscr{D}$.

By construction, $\mathscr{G} \subset \mathscr{D}$, and since $\mathscr{G}$ is $\cap$-stable, Theorem 4.5 proves

$$\mathscr{A} \overset{\text{def}}{=} \sigma(\mathscr{G}) \overset{\S 4.5}{=} \delta(\mathscr{G}) \subset \delta(\mathscr{D}) = \mathscr{D} \subset \mathscr{A}. \qquad \square$$

17.14 Lemma. *Let $(\Omega, \mathscr{A}, \mu)$ be a finite measure space such that $\mathscr{A} = \sigma(\mathscr{G})$ for some Boolean algebra $\mathscr{G}$. The functions of the form $\sum_{i=1}^N c_i \mathbb{1}_{G_i}$ ($N \in \mathbb{N}$, $c_i \in \mathbb{R}$, $G_i \in \mathscr{G}$) are dense in $L^p(\mathscr{A})$ for any $p \in [1,\infty)$.*

Proof. Fix $p \in [1,\infty)$, $\epsilon > 0$ and $u \in L^p(\mathscr{A})$. The second amendment of the sombrero lemma (Theorem 7.11) furnishes a sequence $\phi_n \in \mathcal{E}(\mathscr{A})$ such that

$$\phi_n(x) \xrightarrow[n\to\infty]{\forall x} u(x) \quad \text{and} \quad |\phi_n(x)| \leqslant |u(x)|.$$

Using $|u| \in L^p$ as integrable majorant, we can use dominated convergence (Thm. 14.12) to find some $\phi = \phi_\epsilon \in \mathcal{E}(\mathscr{A})$ such that

$$\|\phi - u\|_p < \epsilon.$$

By definition, $\phi = \sum_{i=1}^N c_i \mathbb{1}_{A_i}$; because of Lemma 17.13, we can approximate A_i with sets from the generator $G_i \in \mathscr{G}$ such that

$$\mu(A_i \vartriangle G_i) < \epsilon^p / (|c_1| + \cdots + |c_N|)^p.$$

Thus,

$$\left\|\sum_{i=1}^N c_i \mathbb{1}_{A_i} - \sum_{i=1}^N c_i \mathbb{1}_{G_i}\right\|_p \leqslant \sum_{i=1}^N |c_i| \cdot \left\|\mathbb{1}_{A_i} - \mathbb{1}_{G_i}\right\|_p = \sum_{i=1}^N |c_i| \cdot \mu(A_i \vartriangle G_i)^{1/p} < \epsilon,$$

and we get $\left\|u - \sum_{i=1}^N c_i \mathbb{1}_{G_i}\right\|_p < 2\epsilon$. $\square$

Proof of Theorem 17.11. Let $\mathscr{R}$ be the algebra generated by the rectangles $\times_{i=1}^d [a_i, b_i)$; we have

$$\mathscr{B}(\mathbb{R}^d) = \sigma(\mathscr{J}) \subset \sigma(\mathscr{R}) \subset \mathscr{B}(\mathbb{R}^d) \implies \sigma(\mathscr{R}) = \mathscr{B}(\mathbb{R}^d).$$

Throughout the proof, let $\epsilon > 0$ be fixed.

1° Let $J_m := [-m, m)^d \uparrow \mathbb{R}^d$. Because of dominated convergence,

$$u_m := u\mathbb{1}_{J_m} \to u \implies \forall m \geqslant m(\epsilon) : \|u - u_m\|_p < \epsilon.$$

2° We can interpret u_m as a function on the finite measure space $(J_m, \mathscr{B}(J_m), \lambda^d)$. By Lemma 17.14, there is some

$$\phi_m = \sum_{i=1}^{N} c_i \mathbb{1}_{R_i}, \quad N = N(m),\ c_i = c_i(m),\ R_i = R_i(m) \in \mathscr{R},$$

such that $\|u_m - \phi_m\|_p < \epsilon$.

3° We can represent the sets $R_i \subset J_m$ from Step 2° as union of finitely many rectangles. Thus,

$$\phi_m = \sum_{i=1}^{N'} c_i' \mathbb{1}_{Q_i}, \quad Q_i \in \mathscr{J}.$$

Consider in $\mathbb{R}$ and $\mathbb{R}^d$ the following (scaled) »tent« functions

$$h(t) = (1 - |t|)^+ \quad \text{and} \quad h_n(t) = nh(nt), \quad t \in \mathbb{R},$$

$$H_n(x_1, \dots, x_d) := \prod_{i=1}^{d} h_n(x_i), \quad x = (x_1, \dots, x_d) \in \mathbb{R}^d.$$

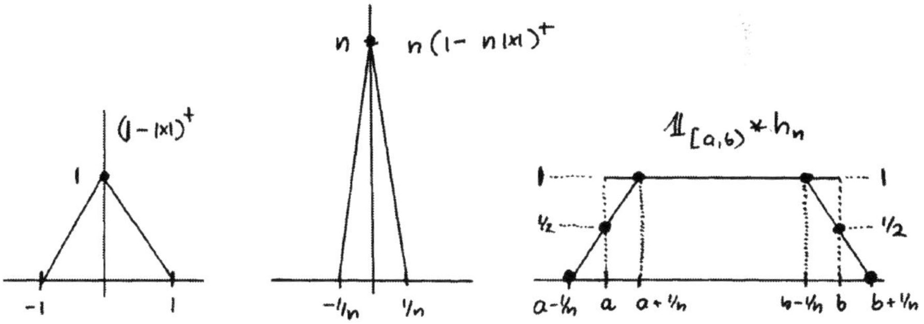

Fig. 17.1. The one-dimensional tent function $h(x) = (1 - |x|)^+$, the scaled tent function $h_n(x) = nh(nx) = n(1 - n|x|)^+$, and the convolution $\mathbb{1}_{[a,b)} * h_n(x)$

If $Q = \times_{i=1}^d [a_i, b_i)$, then

$$\mathbb{1}_Q * H_n(x) = \int \!\cdots\! \int \mathbb{1}_Q(y) H_n(x - y)\, dy = \prod_{i=1}^{d} \int_{a_i}^{b_i} h_n(x_i - y_i)\, dy_i,$$

and since everything factorizes, it is enough to consider the one-dimensional situation.

$$\int_a^b h_n(x - y)\, dy = \int_{a-x}^{b-x} nh(nz)\, dz \overset{t=nz}{\underset{dt=n\,dz}{=}} \int_{n(a-x)}^{n(b-x)} h(t)\, dt = \begin{cases} 0, & x \notin \left[a - \frac{1}{n}, b + \frac{1}{n}\right], \\ 1, & x \in \left(a + \frac{1}{n}, b - \frac{1}{n}\right), \\ \in [0,1], & \text{sonst.} \end{cases}$$

The boundary of the rectangle ∂Q is a Lebesgue null set, hence,

$$\mathbb{1}_{[a,b)} * h_n(t) \xrightarrow[n\to\infty]{\text{a.e.}} \mathbb{1}_{[a,b)}(t) \quad\text{and}\quad \mathbb{1}_Q * H_n(x) \xrightarrow[n\to\infty]{\text{a.e.}} \mathbb{1}_Q(x).$$

Observe that $|\mathbb{1}_Q * H_n(x)| \leqslant 1$ and $\|\mathbb{1}_Q * H\|_p \leqslant \|\mathbb{1}_Q\|_p\|H\|_1 = \|\mathbb{1}_Q\|_p$; see Theorem 17.9. Therefore, we can use dominated convergence (Theorem 14.12) to get

$$\forall n \gg 1 : \quad \|\mathbb{1}_Q - \mathbb{1}_Q * H_n\|_p < \epsilon^p/(|c_1| + \cdots + |c_N|)^p,$$

and we see, as in Lemma 17.14, that

$$\left\| \phi_m - \sum_{i=1}^{N} c_i \mathbb{1}_Q * H_n \right\|_p < \epsilon.$$

4° Since $\mathbb{1}_Q * H_n(x)$ is continuous and supported in $[-m-1/n, m+1/n]^d$, the claim follows from

$$\left\| u - \sum_{i=1}^{N} c_i \mathbb{1}_Q * H_n \right\|_p \leqslant \|u - u_m\|_p + \|u_m - \phi_m\|_p + \left\| \phi_m - \sum_{i=1}^{N} c_i \mathbb{1}_Q * H_n \right\|_p \leqslant 3\epsilon$$

provided that $m \gg 1$, $N = N(m) \gg 1$ and $n \gg 1$. □

18 The Radon-Nikodým theorem

Let $(\Omega, \mathscr{A}, \mu)$ be a measure space and $f : \Omega \to [0, \infty)$ a positive measurable function. We have seen in Lemma 9.8 that

$$\nu(A) := \int_A f \, d\mu, \quad A \in \mathscr{A},$$

is also a measure »with density f« on $(\Omega, \mathscr{A})$. Recall the notation $\nu = f \cdot \mu$.

From Theorem 10.2 we know that

$$\mu(N) = 0 \implies \nu(N) = \int_N f \, d\mu = 0$$

i.e. all (measurable) μ-null sets are ν-null sets: $\mathcal{N}_\mu \subset \mathcal{N}_\nu$.

18.1 Definition. Let μ and ν be measures on the same space $(\Omega, \mathscr{A})$. If

$$\forall N \in \mathscr{A} : \mu(N) = 0 \implies \nu(N) = 0,$$

then we call ν **absolutely continuous** with respect to μ. We write $\nu \ll \mu$.

Thus, if $\nu = f \cdot \mu$, then $f \cdot \mu \ll \mu$. The converse is also true.

18.2 Theorem (Radon-Nikodým). *Let μ, ν be finite measures on the same measurable space $(\Omega, \mathscr{A})$. The following assertions are equivalent:*

a) $\nu(A) = \int_A f(\omega) \, \mu(d\omega)$ *for some μ-a.e. unique* $f \in \mathcal{L}^{0,+}(\mathscr{A})$.

b) $\nu \ll \mu$.

The density $f := \dfrac{d\nu}{d\mu}$ *is called the* ***Radon-Nikodým derivative.***

Proof. We have already proved the direction a)⇒b). The proof of the converse b)⇒a) is split in several steps. Claims are highlighted.

1° It is enough to show that there is some $\mathscr{A}/\mathscr{B}(\mathbb{R})$-measurable function $f^* \geqslant 0$ such that

$$\int u \, d\nu = \int u f^* \, d\mu \quad \text{for all} \quad u \in \mathcal{L}^{0,+}(\mathscr{A}). \tag{18.1}$$

If we take $u = \mathbb{1}_A$, $A \in \mathscr{A}$, we get a).

2° Define a new measure $\rho := \mu + \nu$, a functional

$$\Phi(u) := \int u^2 \, d\mu + \int (1-u)^2 \, d\nu \quad \forall u \in L^2(\rho) = L^2(\mu) \cap L^2(\nu),$$

and set

$$d^2 := \inf_{u \in L^2(\rho)} \Phi(u).$$

Obviously, $d^2 \leqslant \Phi(0) = \int 1 \, d\nu = \nu(\Omega) < \infty$, as well as

$$\begin{aligned} &\Phi(u+w) + \Phi(u-w) \\ &= \int \left[(u+w)^2 + (u-w)^2\right] d\mu + \int \left[\big((1-u)-w\big)^2 + \big((1-u)+w\big)^2\right] d\nu \\ &= 2\int u^2 \, d\mu + 2\int w^2 \, d\mu + 2\int (1-u)^2 \, d\nu + 2\int w^2 \, d\nu \\ &= 2\Phi(u) + 2\|w\|^2_{L^2(\rho)}. \end{aligned} \tag{18.2}$$

3° There exists a unique minimizer $f \in L^2(\rho) : d^2 = \Phi(f)$. By the definition of the infimum there is a sequence $(f_i)_{i\in\mathbb{N}} \subset L^2(\rho)$ such that $d^2 = \lim_{i\to\infty} \Phi(f_i)$.

Now we use in (18.2) $u = \frac{1}{2}(f_i + f_k)$ and $w = \frac{1}{2}(f_i - f_k)$, and get

$$\begin{aligned} d^2 + \left\|\frac{f_i - f_k}{2}\right\|^2_{L^2(\rho)} &\overset{\inf}{\leqslant} \Phi\left(\frac{f_i + f_k}{2}\right) + \left\|\frac{f_i - f_k}{2}\right\|^2_{L^2(\rho)} \\ &\overset{(18.2)}{=} \frac{1}{2}\Phi(f_i) + \frac{1}{2}\Phi(f_k) \xrightarrow[i,k\to\infty]{} d^2. \end{aligned}$$

From this we conclude that $\lim_{i,k\to\infty} \|f_i - f_k\|_{L^2(\rho)} = 0$, i.e. $(f_i)_{i\in\mathbb{N}}$ is an $L^2(\rho)$ Cauchy sequence. Since $L^2(\rho)$ is complete (Theorem 14.10), the limit $f := L^2(\rho)\text{-}\lim_{i\to\infty} f_i$ exists.

Let us show that f is a minimizer: $d^2 = \Phi(f)$: Applying (18.2) with $u = f_i$ and $w = f - f_i$ yields

$$2d^2 \leqslant \Phi(f) + d^2 \leqslant \Phi(f) + \Phi(2f_i - f) \overset{(18.2)}{=} 2\Phi(f_i) + 2\|f - f_i\|^2_{L^2(\rho)} \xrightarrow[i\to\infty]{} 2d^2 + 0,$$

and we see that $\Phi(f) = d^2$.

f is unique: Assume that g is a further minimizer; using (18.2) for the functions $u = \frac{1}{2}(f+g)$ and $w = \frac{1}{2}(f-g)$ shows

$$d^2 + \left\|\frac{f-g}{2}\right\|^2_{L^2(\rho)} \leqslant \Phi\left(\frac{f+g}{2}\right) + \left\|\frac{f-g}{2}\right\|^2_{L^2(\rho)} \leqslant \frac{1}{2}\Phi(f) + \frac{1}{2}\Phi(g) = d^2.$$

Therefore, $\|f-g\|_{L^2(\rho)} = 0$ and we see that $f = g$ $(\mu+\nu)$-a.e.

4° The minimizer satisfies $0 \leqslant f \leqslant 1$. We note that

$$(0 \vee f \wedge 1)^2 = 0 \vee f^2 \wedge 1 \leqslant f^2$$
$$\text{and} \quad (0 \vee (1-f) \wedge 1)^2 = 0 \vee (1-f)^2 \wedge 1 \leqslant (1-f)^2,$$

and conclude $d^2 \leqslant \Phi(0 \vee f \wedge 1) \leqslant \Phi(f) = d^2$. Because of uniqueness of the minimizer, we get $f = 0 \vee f \wedge 1$, i.e. $0 \leqslant f \leqslant 1$.

5° For all $u \in \mathcal{L}^{0,+}(\mathscr{A})$, $u_n := u \wedge n$ and $t \in \mathbb{R}$ we have

$$\begin{aligned} 0 &\leqslant \Phi(f + tu_n) - \Phi(f) \\ &= \int \left[(f+tu_n)^2 - f^2\right] d\mu + \int \left[(1-f-tu_n)^2 - (1-f)^2\right] d\nu \\ &= t^2 \int u_n^2 \, d\mu + t^2 \int u_n^2 \, d\nu + 2t \int f u_n \, d\mu - 2t \int (1-f) u_n \, d\nu. \end{aligned}$$

If we divide both sides by t^2, we get

$$0 \leqslant \int u_n^2 \, d\mu + \int u_n^2 \, d\nu + \frac{2}{t}\left(\int f u_n \, d\mu - \int (1-f) u_n \, d\nu\right) \quad \forall t \in \mathbb{R},\ t \neq 0.$$

This shows – consider the limits $t \uparrow 0$ and $t \downarrow 0$ – that the expression in parentheses must be zero. With Beppo Levi, we find for all $u \in \mathcal{L}^{0,+}(\mathscr{A})$

$$\int f u \, d\mu = \sup_{n \in \mathbb{N}} \int f u_n \, d\mu = \sup_{n \in \mathbb{N}} \int (1-f) u_n \, d\nu = \int (1-f) u \, d\nu.$$

Taking $u = \mathbb{1}_N$ where $N = \{f = 1\}$, then $\mu(N) = \int_N 1 \, d\mu = \int_N 0 \, d\nu = 0$ and, using $\nu \ll \mu$, it follows that $\nu(N) = 0$.

Finally, if we take $f^* := f/(1-f)\mathbb{1}_{\{f \neq 1\}} \in \mathcal{L}^{0,+}(\mathscr{A})$, then we have

$$\begin{aligned} \int u f^* \, d\mu &= \int f \frac{u}{1-f} \mathbb{1}_{N^c} \, d\mu \\ &= \int (1-f) \frac{u}{1-f} \mathbb{1}_{N^c} \, d\nu \\ &= \int u \mathbb{1}_{N^c} \, d\nu \overset{\nu(N)=0}{=} \int u \, d\nu. \end{aligned}$$ □

18.3 Corollary. *Theorem 18.2 remains valid, if μ is a finite and ν an arbitrary measure.*

Proof. We only have to show the direction b)⇒a). Define the family of finite sets $\mathscr{F} := \{F \in \mathscr{A} \mid \nu(F) < \infty\}$ of the measure ν. Obviously, $\mathscr{F}$ is $\cup$-stable. Moreover, we can approximate

$$c := \sup_{F \in \mathscr{F}} \mu(F) \leqslant \mu(\Omega) < \infty$$

with an increasing sequence of sets $(F_n)_{n\in\mathbb{N}} \subset \mathscr{F}$, i.e.

$$F_\infty := \bigcup_{n\in\mathbb{N}} F_n \quad \text{and} \quad c = \mu(F_\infty) = \sup_{n\in\mathbb{N}} \mu(F_n).$$

By construction, the measure $\mathbb{1}_{F_\infty} \cdot \nu$ is σ-finite, and for every $A \subset F_\infty^c$, $A \in \mathscr{A}$, we have

$$\text{either} \quad \mu(A) = \nu(A) = 0 \qquad \text{or} \quad 0 < \mu(A) < \nu(A) = \infty. \tag{18.3}$$

Indeed: If $\nu(A) < \infty$, then $F_n \cup A \in \mathscr{F}$ for all $n \in \mathbb{N}$; thus,

$$c \geqslant \sup_{n\in\mathbb{N}} \mu(F_n \,\dot\cup\, A) = \sup_{n\in\mathbb{N}} (\mu(F_n) + \mu(A)) = \mu(F_\infty) + \mu(A) = c + \mu(A),$$

which shows that $\mu(A) = 0$; since $\nu \ll \mu$, we also get $\mu(A) = \nu(A) = 0$. If, on the other hand, $\nu(A) = \infty$, then we have necessarily (use again $\nu \ll \mu$) that $\mu(A) > 0$.

Now we define finite measures as follows:

$$\nu_n := \nu\big(\bullet \cap (F_n \setminus F_{n-1})\big), \quad \mu_n := \mu\big(\bullet \cap (F_n \setminus F_{n-1})\big), \qquad (F_0 := \emptyset).$$

Clearly, μ_n, ν_n are finite measures and $\nu_n \ll \mu_n$ for every $n \in \mathbb{N}$. Therefore, we can apply Theorem 18.2 and we see that $\nu_n = f_n \mu_n$ for a μ-a.e. unique density f_n. Set

$$f(x) := \begin{cases} f_n(x), & \text{if } x \in F_n \setminus F_{n-1}, \\ \infty, & \text{if } x \in F_\infty^c, \end{cases} \tag{18.4}$$

and observe that $\nu = f\mu$.

On the set F_∞, the density f inherits μ-a.e. uniqueness from the f_n. On F_∞^c we note that **every** density $\tilde{f}$ of ν w.r.t. μ satisfies the relation

$$\nu\big(\{\tilde{f} \leqslant n\} \cap F_\infty^c\big) = \int_{\{\tilde{f}\leqslant n\}\cap F_\infty^c} \tilde{f}\, d\mu \leqslant n\,\mu\big(\{\tilde{f} \leqslant n\} \cap F_\infty^c\big) < \infty.$$

Therefore, (18.3) shows that $\nu\big(\{\tilde{f} \leqslant n\} \cap F_\infty^c\big) = \mu\big(\{\tilde{f} \leqslant n\} \cap F_\infty^c\big) = 0$ for every $n \in \mathbb{N}$. From this we concldue that $\tilde{f}|_{F_\infty^c} = \infty$ which means that f, defined by (18.4), is also unique on F_∞^c. □

18.4 Corollary. *Theorem 18.2 remains valid, if μ is a σ-finite and ν an arbitrary measure.*

Proof. Since μ is σ-finite, there is an exhausting sequence $(A_n)_{n\in\mathbb{N}} \subset \mathscr{A}$ such that $A_n \uparrow \Omega$ and $\mu(A_n) < \infty$. Since for all $x \in \Omega$

$$h(x) := \sum_{n=1}^{\infty} \frac{2^{-n}}{1+\mu(A_n)} \mathbb{1}_{A_n}(x) > 0,$$

it is clear that the measures $h\cdot\mu$ and μ have the same (measurable) null sets. Therefore,

$$\nu \ll \mu \iff \nu \ll h\cdot\mu.$$

The Beppo Levi theorem (in the guise of Corollary 8.7) now shows that

$$\int_\Omega h \, d\mu = \sum_{n=1}^{\infty} \frac{2^{-n}}{1+\mu(A_n)} \int_\Omega \mathbb{1}_{A_n} \, d\mu = \sum_{n=1}^{\infty} \frac{\mu(A_n)}{1+\mu(A_n)} 2^{-n} \leqslant \sum_{n=1}^{\infty} 2^{-n} = 1,$$

Since $h\cdot\mu$ is a finite measure, Corollary 18.3 applies and yields

$$\nu = f\cdot(h\cdot\mu) \overset{\S}{=} (fh)\cdot\mu$$

for some density $f \in \mathcal{L}^{0,+}(\mathscr{A})$.

Let us quickly verify the equality marked by »§«: If $f = \sum_{n=0}^{M} y_n \mathbb{1}_{B_n} \geqslant 0$ is a positive simple function, then

$$\nu(A) = \int_A \sum_{n=0}^{M} y_n \mathbb{1}_{B_n} \, d(h\cdot\mu) = \sum_{n=0}^{M} y_n \int \mathbb{1}_{B_n \cap A} h \, d\mu = \int_A (fh) \, d\mu.$$

The general case $f \in \mathcal{L}^{0,+}(\mathscr{A})$ follows from the sombrero lemma and a Beppo Levi argument.

Uniqueness: We know that f is $(h\cdot\mu)$-a.e. unique, and so fh is μ-a.e. unique; here we use that $h > 0$ (everywhere) and that μ and $h\cdot\mu$ have the same null sets. □

We close with some kind of »opposite« of the absolute continuity.

18.5 Definition. Two measures μ, ν which are defined on the same measurable space $(\Omega, \mathscr{A})$ are called **(mutually) singular**, if

$$\exists N \in \mathscr{A} \; : \; \nu(N) = 0 = \mu(N^c).$$

In this case one writes $\mu \perp \nu$ or $\nu \perp \mu$.

18.6 Example. Consider Lebesgue measure λ^d on the space $(\mathbb{R}^d, \mathscr{B}(\mathbb{R}^d))$.

- $\delta_x \perp \lambda^n$ for any $x \in \mathbb{R}^n$.
- $f\cdot\lambda^d \perp g\cdot\lambda^d$ for all $f, g \in \mathcal{L}^{0,+}$ such that $\operatorname{supp} f \cap \operatorname{supp} g = \emptyset$.

If $\mu \perp \nu$, then the measures μ and ν charge disjoint sets of Ω.

18.7 Theorem (Lebesgue decomposition). *Let μ and ν be finite measures on $(\Omega, \mathscr{A})$. Up to null sets there is a unique decomposition*

$$\nu = \nu^\circ + \nu^\perp \quad \textit{such that} \quad \nu^\circ \ll \mu \quad \textit{and} \quad \nu^\perp \perp \mu.$$

Proof. **Existence:** $\rho := \mu + \nu$ is again a finite measure and Theorem 18.2 shows that there are densities

$$\mu \leqslant \rho \implies \mu \ll \rho \overset{18.2}{\implies} \mu = f_\mu \cdot \rho \quad \text{and} \quad \nu \leqslant \rho \implies \nu \ll \rho \overset{18.2}{\implies} \nu = f_\nu \cdot \rho.$$

From $\mu(\Omega), \nu(\Omega) < \infty$ we infer that $f_\mu, f_\nu \in L^1(\rho)$. Denote by

$$F := \{f_\mu \neq 0\} \quad \text{and} \quad f(x) := \frac{f_\nu(x)}{f_\mu(x)} \mathbb{1}_F(x)$$

and define for every $A \in \mathscr{A}$

$$\nu^\perp(A) := \nu(A \cap F^c) \quad \text{and} \quad \nu^\circ(A) := \int_A f \, d\mu = \nu(A \cap F).$$

The last equality follows from $f\,d\mu = \frac{f_\nu}{f_\mu} \mathbb{1}_F f_\mu \, d\rho = \mathbb{1}_F f_\nu \, d\rho = \mathbb{1}_F \, d\nu$.

Our construction yields $\nu^\circ + \nu^\perp = \nu$ with $\nu^\circ = f \cdot \mu \ll \mu$ and $\nu^\perp \perp \mu$; the latter follows from

$$\mu(F^c) = \int_{F^c} f_\mu \, d\rho \overset{\text{def. } F}{=} 0 \quad \text{and} \quad \nu^\perp(F) = \nu(F \cap F^c) = 0.$$

Uniqueness: Assume that $\nu = \nu_i^\circ + \nu_i^\perp$, $i = 1, 2$, are two decompositions such that $\nu_i^\circ \ll \mu$ and $\nu_i^\perp \perp \mu$. By assumption,

$$\exists N_i \in \mathscr{N}_\mu : \nu_i^\perp(A) = \nu_i^\perp(A \cap N_i). \tag{18.5}$$

Since $\nu_i^\circ \ll \mu$, we see that for the μ-null set $N := N_1 \cup N_2 \in \mathscr{N}_\mu$

$$\nu_i^\circ(A \cap N) \leqslant \nu_i^\circ(N) = 0.$$

This finally gives for every $A \in \mathscr{A}$

$$\nu(A \cap N) = \nu_i^\perp(A \cap N) \overset{(18.5)}{=} \nu_i^\perp(A \cap N \cap N_i) \overset{N_i \subset N}{=} \nu_i^\perp(A \cap N_i) \overset{(18.5)}{=} \nu_i^\perp(A),$$

and we conclude that $\nu_1^\perp = \nu_2^\perp$ as well as $\nu_1^\circ = \nu_2^\circ$. □

19 Products of infinitely many measure spaces

In this chapter we want to extend the product construction of Chapter 15 to infinitely many measures.

From the case of finite products we know that $\mu_1 \otimes \mu_2(\Omega_1 \times \Omega_2) = \mu_1(\Omega_1)\mu(\Omega_2)$, and this shows that it is reasonable to assume $\mu_i(\Omega_i) = 1$ to make sure that an infinite product $\prod_i \mu_i(\Omega_i)$ converges.

Throughout this chapter, I is an arbitrary (countable or uncountable) index set and $(\Omega_i, \mathscr{A}_i, \mathbb{P}_i)$, $i \in I$, are probability spaces. Let us introduce some essential notation.

19.1 Definition. Let $\emptyset \neq K \subset J \subset I$. The infinite product of the sets $(\Omega_i)_{i\in K}$ is defined as

$$\Omega_J := \bigtimes_{i\in J} \Omega_i := \left\{ f : J \to \bigcup_{i\in J} \Omega_i \mid f(i) \in \Omega_i \ \forall i \in J \right\};^{(1)}$$

$f(i)$ is called the i**th coordinate** of $f \in \Omega_J$. The **coordinate projection** from Ω_J onto Ω_K is denoted by

$$\pi^J_K : \Omega_J \to \Omega_K, \quad f \mapsto f|_K.$$

If $K = \{i\}$, we write $\pi^J_i(f) := \pi^J_{\{i\}}(f) = f(i)$.

For projections from Ω_I we use the shorthand $\pi_J := \pi^K_J$ and $\pi_i := \pi^I_i$. If we compose projections, we have

$$\pi^J_K = \pi^L_K \circ \pi^J_L \quad \forall K \subset L \subset J \subset I. \tag{19.1}$$

The **finite index sets** are denoted by $\mathcal{H} := \mathcal{H}(I) := \{K \subset I \mid \#K < \infty\}$.

19.2 Definition. The infinite product $\mathscr{A}_I := \bigotimes_{i\in I} \mathscr{A}_i$ of the σ-algebras $(\mathscr{A}_i)_{i\in I}$ is defined by

$$\mathscr{A}_I := \sigma(\pi_i, i \in I) = \sigma\big(\pi_i^{-1}(\mathscr{A}_i), i \in I\big). \tag{19.2}$$

19.3 Remark. a) Definition 19.2 is compatible with the earlier definition of the (finite) product of σ-algebras (Def. 15.3), see Theorem 16.9.

b) The projection $\pi_H : \Omega_I \to \Omega_H$, $H \in \mathcal{H}$, is $\mathscr{A}_I/\mathscr{A}_H$-measurable. In order to see this, we take a typical set from a generator of $\mathscr{A}_H$,

$$\bigtimes_{i\in H} A_i = \bigcap_{i\in H} (\pi^H_i)^{-1}(A_i), \quad A_i \in \mathscr{A}_i,$$

and we get for these sets

$$\pi_H^{-1}\left(\bigtimes_{i\in H} A_i\right) = \pi_H^{-1}\left(\bigcap_{i\in H} (\pi^H_i)^{-1}(A_i)\right) = \bigcap_{i\in H} \pi_H^{-1} \circ (\pi^H_i)^{-1}(A_i)$$
$$= \underbrace{\bigcap_{i\in H}}_{\text{finite}} \underbrace{\pi_i^{-1}(A_i)}_{\in \mathscr{A}_I} \in \mathscr{A}_I.$$

This proves measurability since it is enough to consider pre-images of generator sets.

(1) If $\Omega_i = \Omega$ for all $i \in I$, one also writes Ω^K instead of Ω_K. This is the usual notation for the family of all maps from $K \to \Omega$.

19.4 Theorem (Kolmogorov). *Let $(\Omega_i, \mathscr{A}_i, \mathbb{P}_i)$, $i \in I$, be arbitrarily many probability spaces. There is a unique probability measure $\mathbb{P} = \mathbb{P}_I$ on the product space $(\Omega_I, \mathscr{A}_I)$ such that*

$$\forall H \subset I,\ \#H < \infty: \quad \pi_H(\mathbb{P}) = \mathbb{P}_H = \bigotimes_{i \in H} \mathbb{P}_i. \tag{19.3}$$

The measure $\mathbb{P}$ is called the **(infinite) product** *of the measures $(\mathbb{P}_i)_{i \in I}$. Notation: $\mathbb{P} = \bigotimes_{i \in I} \mathbb{P}_i$.*

Proof. A set $C \in \mathscr{A}_I$ of the form

$$C = \pi_H^{-1}(A), \quad A \in \mathscr{A}_H = \bigotimes_{i \in H} \mathscr{A}_i, \quad H \in \mathcal{H},$$

is called a **cylinder set with basis** H (»H-cylinder«, for short).

The following picture – in clear abuse of notation – is helpful when thinking about cylinder sets: Think of it as a »tower« where the basis sets are »on top« of everything, i.e.

$$C = C_H \times \bigtimes_{i \notin H} \Omega_i, \quad \text{with } C_H \in \mathscr{A}_H.$$

Obviously, if $H \subset J \in \mathcal{H}$, then every H-cylinder is a J-cylinder. By

$$\mathscr{Z}_H = \pi_H^{-1}(\mathscr{A}_H) = \{\text{all H-cylinders}\} \quad \text{and} \quad \mathscr{Z} = \bigcup_{H \in \mathcal{H}} \mathscr{Z}_H$$

we denote the sets of all H-cylinders and all cylinder sets, respectively.

The proof is split into several steps. Claims are highlighted.

1° $\mathscr{A}_I = \sigma(\mathscr{Z})$. Since $\pi_H : \Omega_I \to \Omega_H$ is measurable, we see that $\mathscr{Z} \subset \mathscr{A}_I$, hence

$$\mathscr{A}_I \stackrel{\text{def}}{=} \sigma(\pi_i, i \in I) \subset \sigma(\pi_H, H \in \mathcal{H}) = \sigma(\mathscr{Z}) \subset \mathscr{A}_I.$$

2° $\mathscr{Z}$ is a Boolean algebra. Clearly, $\emptyset, \Omega_I \in \mathscr{Z}$ are cylinder sets. If $C_1, C_2 \in \mathscr{Z}$, then $C_i = \pi_{H_i}^{-1}(A_i)$ for suitable $H_i \in \mathcal{H}$ and $A_i \in \mathscr{A}_{H_i}$. Set $H := H_1 \cup H_2 \in \mathcal{H}$ and observe that $C_i \in \pi_H^{-1}(\mathscr{A}_H)$. Therefore,

$$C_1 \cap C_2,\ C_1 \setminus C_2,\ C_1 \cup C_2 \in \pi_H^{-1}(\mathscr{A}_H) \subset \mathscr{Z}.$$

⚠ $\mathscr{Z}$ is, in general, not a σ-algebra.

3° $\mathbb{P}_H := \bigtimes_{i \in H} \mathbb{P}_i$, $H \in \mathcal{H}$ is a projective family, i.e. it holds that

$$\forall H \subset J \in \mathcal{H}: \quad \mathbb{P}_H = \pi_H^J(\mathbb{P}_J).$$

Indeed, $\mathscr{A}_H$ is generated by rectangles $\bigtimes_{i \in H} A_i$, $A_i \in \mathscr{A}_i$. In view of the uniqueness of measures theorem (Theorem 4.7) it is enough to show this equality for rectangles. Since

$$(\pi_H^J)^{-1}\Big(\bigtimes_{i\in H} A_i\Big) = \bigtimes_{i\in J} A_i', \quad A_i' = \begin{cases} A_i, & i\in H,\\ \Omega_i, & i\in J\setminus H,\end{cases}$$

we see that

$$\begin{aligned}\mathbb{P}_H\Big(\bigtimes_{i\in H} A_i\Big) = \prod_{i\in H}\mathbb{P}_i(A_i) &= \prod_{i\in H}\mathbb{P}_i(A_i)\prod_{\ell\in J\setminus H}\mathbb{P}_\ell(\Omega_\ell)\\ &= \mathbb{P}_J\Big(\bigtimes_{i\in J} A_i'\Big)\\ &= \mathbb{P}_J\Big((\pi_H^J)^{-1}\Big(\bigtimes_{i\in H} A_i\Big)\Big).\end{aligned}$$

4° Construction of $\mathbb{P}|_{\mathscr{Z}}$. Since the measure $\mathbb{P}$ has to satisfy (19.3), we have necessarily that

$$\mathbb{P}(C) = \pi_H(\mathbb{P})(A) = \mathbb{P}_H(A) \quad \forall C = \pi_H^{-1}(A) \in \mathscr{Z}. \tag{19.4}$$

If we want to use (19.4) in order to define $\mathbb{P}|_{\mathscr{Z}}$, we must show that (19.4) only depends on C, but not in its representation as cylinder set.

Assume that $C = \pi_H^{-1}(A) = \pi_J^{-1}(B)$ for any two $H, J \in \mathcal{H}$ and $A \in \mathscr{A}_H$, $B \in \mathscr{A}_J$. We define $K := H \cup J$. Since C is also a K-cylinder, there is some $D \in \mathscr{A}_K$ such that $C = \pi_K^{-1}(D)$. Moreover,

$$(\pi_H^K)^{-1}(A) = \pi_K \circ \pi_H^{-1}(A) = \pi_K(C) = D$$

and – because of the projectivity property, cf. Step 3° – this implies that

$$\mathbb{P}_H(A) \overset{3^\circ}{=} \mathbb{P}_K\big((\pi_H^K)^{-1}(A)\big) = \mathbb{P}_K(D) = \overset{\text{similar}}{\cdots\cdots} = \mathbb{P}_J(B).$$

This means that we may define $\mathbb{P}$ for $C \in \mathscr{Z}$ by (19.4) using **any** cylinder representation.

5° $\mathbb{P}|_{\mathscr{Z}}$ is finitely additive and $\mathbb{P}(\emptyset) = 0$. This is obvious: All assertions can be reduced to the measure space $(\Omega_H, \mathscr{Z}_H, \mathbb{P}_H)$ for some sufficiently large $H \in \mathcal{H}$.

6° $\mathbb{P}|_{\mathscr{Z}}$ is $\emptyset$-continuous (part 1). In order to simplify notation, we will use the following notation: $\omega_H \in \Omega_H$, $H \subset I$, and $\omega = \omega_I \in \Omega_I$. If $K, L \subset I$ are disjoint index sets, we use

$$(\omega_K; \omega_L) \quad \text{as shorthand for} \quad \omega_{K\cup L},$$

and, in particular, $\omega = (\omega_K; \omega_{I\setminus K})$ for any $K \subset I$.

If $C \in \mathscr{Z}$ and $\bar\omega_H \in \Omega_H$ for some $H \in \mathcal{H}$, then

$$C(\bar\omega_H) = \big\{\omega = (\omega_H; \omega_{I\setminus H}) \in \Omega_I \mid (\bar\omega_H; \pi_{I\setminus H}(\omega)) = (\bar\omega_H; \omega_{I\setminus H}) \in C\big\}.$$

We have

$$\mathbb{P}(C) = \int_{\Omega_H} \mathbb{P}(C(\bar{\omega}_H))\,\mathbb{P}_H(d\bar{\omega}_H).$$

Indeed, if $C = (\pi_J)^{-1}(A)$, $H \subset J \in \mathcal{H}$, then

$$\begin{aligned} \mathbb{P}(C) &= \mathbb{P}_J(A) = \int \mathbb{1}_A(\omega_J)\,\mathbb{P}_J(d\omega_J) \\ &\overset{\text{Fubini}}{=} \iint \mathbb{1}_A(\bar{\omega}_H;\omega_{J\setminus H})\,\mathbb{P}_{J\setminus H}(d\omega_{J\setminus H})\,\mathbb{P}_H(d\bar{\omega}_H). \end{aligned} \tag{19.5}$$

Define, in analogy to $C(\bar{\omega}_H)$, the sets

$$A(\bar{\omega}_H) = \left\{\omega_{J\setminus H} \in \Omega_{J\setminus H} \mid (\bar{\omega}_H;\omega_{J\setminus H}) \in A\right\}.$$

Clearly, $\pi_{J\setminus H}^{-1}(A(\bar{\omega}_H)) = C(\bar{\omega}_H)$, and we see with (19.5)

$$\begin{aligned} \mathbb{P}(C) &= \int \mathbb{P}_{J\setminus H}(A(\bar{\omega}_H))\,\mathbb{P}_H(d\bar{\omega}_H) \\ &= \int \mathbb{P}(C(\bar{\omega}_H))\,\mathbb{P}_H(d\bar{\omega}_H). \end{aligned}$$

7° $\mathbb{P}|_{\mathcal{Z}}$ is $\emptyset$-continous (part 2). Assume that $(C_n)_{n\in\mathbb{N}} \subset \mathcal{Z}$ such that $C_n \downarrow$ and $C_n = \pi_{H_n}^{-1}(A_n)$ and $\mathbb{P}(C_n) \geqslant \delta > 0$ for some $\delta > 0$.

We have to show that $\bigcap_{n\in\mathbb{N}} C_n \neq \emptyset$.[2] Without loss of generality, we can assume that $H_1 \subset H_2 \subset \dots$. Define

$$Q_n := \left\{\bar{\omega}_{H_1} \in \Omega_{H_1} \mid \mathbb{P}(C_n(\bar{\omega}_{H_1})) \geqslant \tfrac{1}{2}\delta\right\}.$$

From Step 6° and Fubini's theorem we see that $Q_n \in \mathscr{A}_{H_1}$. Moreover, $C_n \downarrow$ implies that $Q_n \downarrow$ and

$$\begin{aligned} \delta \leqslant \mathbb{P}(C_n) &= \int_{Q_n^c} \mathbb{P}(C_n(\bar{\omega}_{H_1}))\,\mathbb{P}_{H_1}(d\bar{\omega}_{H_1}) + \int_{Q_n} \mathbb{P}(C_n(\bar{\omega}_{H_1}))\,\mathbb{P}_{H_1}(d\bar{\omega}_{H_1}) \\ &\leqslant \frac{\delta}{2}\mathbb{P}_{H_1}(Q_n^c) + 1\cdot\mathbb{P}_{H_1}(Q_n) \leqslant \frac{\delta}{2} + \mathbb{P}_{H_1}(Q_n). \end{aligned}$$

If we rearrange this inequality and use the continuity of the measure $\mathbb{P}_{H_1}$, we get

$$\forall n \in \mathbb{N}: \quad \mathbb{P}_{H_1}(Q_n) \geqslant \frac{\delta}{2} \implies \mathbb{P}_{H_1}\left(\bigcap_{n\in\mathbb{N}} Q_n\right) \geqslant \frac{\delta}{2}.$$

This means that $\bigcap_{n\in\mathbb{N}} Q_n \neq \emptyset$ and we can pick some $\omega_1 \in \bigcap_{n\in\mathbb{N}} Q_n$. We have thus shown

$$\exists\, \omega_1 \in \Omega_{H_1} \quad \forall n \in \mathbb{N}: \mathbb{P}(C_n(\omega_1)) \geqslant \frac{1}{2}\delta.$$

(2) Let $C_n \downarrow$ be decreasing. By contraposition, »$\bigcap_n C_n = \emptyset \implies \mu(C_n) \downarrow 0$« is the same as to say that »$\exists \delta > 0\ \forall n: \mu(C_n) > \delta \implies \bigcap_n C_n \neq \emptyset$.«

8° Apply Step 7° in the following setting:

$$C_n \rightsquigarrow C_n(\omega_1), \quad H_1 \rightsquigarrow H_2 \setminus H_1, \quad \delta \rightsquigarrow \frac{1}{2}\delta.$$

In this way, we see that

$$\exists \omega_2 \in \Omega_{H_2 \setminus H_1} \quad \forall n \in \mathbb{N} : \mathbb{P}(C_n(\omega_1)(\omega_2)) \geqslant \frac{1}{2} \cdot \frac{\delta}{2}.$$

Define $(\omega_1;\omega_2) \in \Omega_{H_2}$ and $C(\omega_1;\omega_2) := C(\omega_1)(\omega_2)$.

9° Iteratively, we see that

$$\exists (\omega_1;\ldots;\omega_k) \in \Omega_{H_k} \quad \forall n \in \mathbb{N} : \quad \mathbb{P}(C_n(\omega_1;\ldots;\omega_k)) \geqslant \frac{\delta}{2^k}.$$

Taking, in particular, $n = k$ shows that $C_k(\omega_1;\ldots;\omega_k) \neq \emptyset$, and so

$$\exists \omega \in \Omega_I \quad \forall k \in \mathbb{N} : \quad (\omega_1;\ldots;\omega_k, \pi_{I \setminus H_k}(\omega)) \in C_k.$$

Since there is some $\omega_0 \in \Omega_I$ such that $\pi_{H_k}(\omega_0) = (\omega_1;\ldots;\omega_k)$ for all $k \in \mathbb{N}$, we have found some element $\omega_0 \in \bigcap_{k \in \mathbb{N}} C_k$, i.e. the intersection is not empty. □

The following characterization of $\mathscr{A}_I$ in terms of countably-based cylinder sets is often useful.

19.5 Lemma*. $\mathscr{A}_I = \bigcup\{\mathscr{Z}_K \mid K \subset I \text{ countable}\}$.

Proof. Denote the union appearing on the right-hand side by Σ. Since $\mathscr{Z}_K \subset \mathscr{A}_I$, the inclusion »$\supset$« is obvious.

For the converse observe that $\mathscr{Z}_i = \pi_i^{-1}(\mathscr{A}_i) \subset \Sigma$ for all $i \in I$. If we can show that Σ is a σ-algebra, we get $\mathscr{A}_I \subset \Sigma$, since $\mathscr{A}_I = \sigma(\pi_i, i \in I)$.

Since each $\mathscr{Z}_K$ is a σ-algebra, the properties »$\emptyset \in \Sigma$« and »$S \in \Sigma \implies S^c \in \Sigma$« are obvious. Let $(S_n)_{n \in \mathbb{N}} \subset \Sigma$. For every $n \in \mathbb{N}$ there is some countable $K_n \subset I$ such that $S_n \in \mathscr{Z}_{K_n}$. Since $K = \bigcup_{n \in \mathbb{N}} K_n$ is still countable, $(S_n)_{n \in \mathbb{N}} \subset \mathscr{Z}_K$, hence $\bigcup_{n \in \mathbb{N}} S_n \in \mathscr{Z}_K \subset \Sigma$. □

V
Elementary probability

20 Hasard, chance and probability

The entry on »probability« in the Oxford Companion to Philosophy [13] begins with

> Although there is a well-established mathematical calculus of probability, the nature of its subject-matter is still in dispute. Someone who asserts that it will probably rain is not asserting outright that it will: the question is how such guarded assertions relate to the facts.

In other words, even if randomness will remain an opaque notion, we can treat it in a mathematical way. As it turns out, this is not much different from other mathematical disciplines, say geometry, where one uses terms like »point« and »line« etc. in an idealized, axiomatic way; but it seems that the intuitive understanding of these notions as real-world objects is less disputed than that of randomness.

In everyday language »hazard« – possibly from the Arab word *az-zahr* for gaming die – refers to randomness, often danger and in any case unforseen and inaccessible events. Frequently, »chance« is used to quantify the hazard. Although chance is interchangeable with probability, we will use probability as a strictly mathematical notion. In order to get some feeling for the going-ons let us try to find some hallmarks of random events. Contrary to deterministic phenomena, a random phenomenon

- may, but need not happen,
- need not produce the same outcome if it happens repeatedly,
- is not predictable.

There are many origins of randomness, for instance

- external forces beyond our control (e.g. queues, quantum effects in physics),
- physcial impossibilities (e.g. measurement errors, imprecision),
- high complexity of a (principally deterministic) system (e.g. weather forecasts),
- lack of information and data (e.g. in economics and politics);

some of them are due to us (e.g. our incompetence to deal with complex systems) others may be incumbent in the model (e.g. the role of randomness in quantum physics) and beyond our influence or understanding.

By **probability** we mean any reasonable quantification of randomness. We have to distinguish between two types of probabilities:

objective probability – this is independent of any person and inherent to the phenomenon: If we throw an (ideal) coin, then – for symmetry reasons – chances are 50% that »heads« comes up. Another example is the exact location of electrons in an atom.

The following example shows that even an **objective probability** is not absolute: Of 1000 children born in Germany, there are 513 boys and 487 girls.[(1)] This indicates an »objective« probability of having a girl of 48.7%. Now imagine that a pregnant woman has her first prenatal check-up by ultrasound. Before the check-up, the doctor has to assume that the likelihood for a baby girl is about 49%. During the check-up – only the doctor sees the result – the probability (for the mother) is still 49%, but the doctor knows the baby's sex, i.e. his probability is either 0% or 100% (modulo some error). Once he tells the mother, the objective *a priori* probability changes to 0% or 100% (modulo error).

subjective probability – this depends on the person and reflects some »degree of belief«.

Subjective probabilities – even of one and the same person – may be contradictory. If the probabilities are **coherent** (this is a well-defined notion to to de Finetti), one can show that subjective probabilities may be used to construct a theory of probability which does not contradict Kolmogorov's axioms (see below). The most important source is L.J. Savage's *The Foundations of Statistics* [29], a very good discussion can be found in Diaconis & Skyrms *Ten Great Ideas About Chance* [10].

Even if we cannot give a definition of randomness or chances, we can develop rules how to calculate with probabilities. Let us illustrate this with a simple example.

20.1 Example. Denote by $\omega_1,\dots,\omega_n$ **outcomes** (or **sample points**, **elementary events**) of an experiment (e.g. casting a die $\omega_i = i$, $i = 1,\dots,6$) and assume that, (e.g. for symmetry reasons) each outcome is equally likely.

If we norm a probability by 1 or 100%, we see that the probability to observe ω_i will be $\frac{1}{n}$. We write $p(\omega_i) = \frac{1}{n}$. If we are interested in »ω_i or ω_k«[(2)], then we will have the following probabilities:

$$p(\omega_i) + p(\omega_k) = \frac{2}{n} \quad (i \neq k) \quad \text{resp.} \quad p(\omega_i) = \frac{1}{n} \quad (i = k).$$

In our notation, we can get rid of the tiresome distinction between the two cases if we consider sets of outcomes $\{\omega_i\}$, $\{\omega_i, \omega_k\}$ etc. and a measure $\mathbb{P}$ defined

(1) Source: Statistisches Bundesamt, based on the years 2013–2016.

(2) Recall that »or« in mathematics is always inclusive, i.e. »a or b« = »a or b or both of them«

on those sets. We have

$$\mathbb{P}(\{\omega_i\}) = \frac{1}{n} \quad \text{and} \quad \mathbb{P}(\{\omega_i, \omega_k\}) = \sum_{\omega \in \{\omega_i, \omega_k\}} \mathbb{P}(\{\omega_i\}) = \begin{cases} \frac{2}{n}, & i \neq k, \\ \frac{1}{n}, & i = k. \end{cases} \tag{20.1}$$

Sets of outcomes are called **events**. Clearly, the family of events should be stable under the usual set operations such as $\cup, \cap, \setminus$ and the formation of complements. The family of all possible outcomes is $\Omega = \{\omega_1, \ldots, \omega_n\}$; Ω is called **sample space** or **certain event**.

We may extend this consideration and the following **rules** for a probability set function $\mathbb{P}$ emerge:

I. $\mathbb{P} : \mathscr{A} \to [0,1]$, $\mathscr{A}$ is the family of events;
II. $\mathscr{A}$ contains Ω and is $\cup, \cap, \setminus$-stable, i.e. $\mathscr{A}$ is an algebra;
III. $\mathbb{P}(\Omega) = 1$, i.e. $\mathbb{P}$ is normed;
IV. $\forall A, B \in \mathscr{A},\ A \cap B = \emptyset\ :\ \mathbb{P}(A \cup B) = \mathbb{P}(A) + \mathbb{P}(B)$, i.e. $\mathbb{P}$ is (finitely) additive

If Ω is finite and the probabilities $\mathbb{P}(\{\omega_i\})$ of all outcomes are equal, then (20.1) becomes

$$\mathbb{P}(A) = \sum_{\omega \in A} \mathbb{P}(\{\omega\}) = \frac{|A|}{|\Omega|} = \frac{\text{number of »favourable« outcomes}}{\text{number of all possible outcomes}}. \tag{20.2}$$

This is Laplace's definition of probability [18]. Nowadays it is called the **uniform distribution** or **equidistribution**.

⚠ The formula (20.2) requires that every outcome is equally likely.

20.2 Example (*problème des parties*(3)). *Two contestants* A *and* B *play the following game, betting equal stakes: A fair coin (Head & Tail) is tossed repeatedly.*

- A *wins, as soon as* »H« *appears for the 3rd time (not necessarily in a row);*
- B *wins, as soon as* »T« *appears for the 3rd time (not necessarily in a row);*

For some reason, the game has to be stopped after the first toss (which produces a H*), and cannot be resumed. How should one divide the stakes?*

First of all, it is clear that the game would stop after at most 5 tosses. We continue the game *hypothetically* and write down the results in the form of a tree diagram (Fig. 20.1, left side).

⚠ The »solution« to count the branches in the left diagram of Fig. 20.1 and to divide the stake accordingly as A : B = 6 : 4 is **false**, since the branches do

(3) This is the famous »partition problem«, which was presented in summer 1654 by the Chévalier de Méré to Blaise Pascal. The correspondence between Pascal and Pierre de Fermat is often seen as the beginning of modern probability theory. The present solution is due to Fermat. The *problème des parties* is much older, one can trace it back to a manuscript (national library of Florence) of around 1400 and the writings of Luca Pacioli (Venice 1494), Girolamo Cardano (Milan 1539) or Nicolò Tartagilia (Venice 1556), but their solutions are all incorrect; see [30].

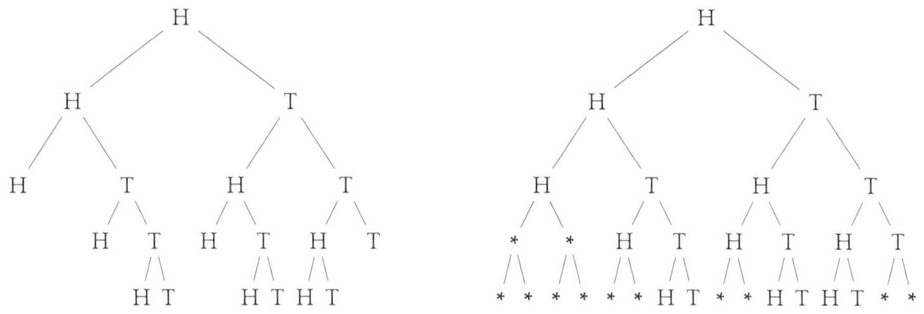

Fig. 20.1. **Left panel:** All possible outcomes of the game if we would continue after the first toss. Contestant A would win in 6 cases, B in 4. **Right panel:** The branches in the left diagram are not equally likely, as they stop after a different number of tosses. We have to extend all branches to 5 tosses to get equally likely outcomes: A wins in 11 cases, B in 5.

not have the same weight. We have to modify the diagram, as shown on the right panel, to get equally likely branches.

$$\text{A wins in 11 cases, B in 5 cases} \implies \text{fair distribution } 11:5$$

$$\text{probability that A wins (if we could finish the game): } \tfrac{11}{5+11} = \tfrac{11}{16},$$

$$\text{probability that B wins (if we could finish the game): } \tfrac{5}{5+11} = \tfrac{5}{16}.$$

It is striking that our probability consideration is based on events which we did not observe (but which one could have observed in principle).

The axioms I–IV are, in general, not sufficient to assign probabilities on an infinite sample space. Already the following situation is out of reach:

20.3 Example. *We draw, at random, a number $i \in \mathbb{N}$. What is the probability that i is an even number?* **Intuitively** we would expect $\frac{1}{2}$, since there are »as many« even numbers as odd numbers – use the bijections $2i \leftrightarrow i \leftrightarrow 2i-1$, $i \in \mathbb{N}$.

In order to justify this, we can argue as follows: $\Omega_n := \{1, 2, \dots, n\}$. We pick, at random, $i \in \Omega_n$. Clearly,

$$\mathbb{P}(\{i \leqslant n : \text{ is even}\}) = \frac{|\{2k : 2k \in \Omega_n\}|}{|\Omega_n|} = \begin{cases} \dfrac{n/2}{n} = \dfrac{1}{2}, & \text{if } n \text{ is even} \\[2ex] \dfrac{(n-1)/2}{n} = \dfrac{1}{2} - \dfrac{1}{2n}, & \text{if } n \text{ is odd} \end{cases}$$

and we see

$$\begin{aligned} \mathbb{P}(\{i \in \mathbb{N} : i \text{ is even}\}) &= \mathbb{P}\Big(\bigcup_{n \in \mathbb{N}} \{i \leqslant n : i \text{ is even}\}\Big) \\ &\overset{?!}{=} \lim_{n \to \infty} \mathbb{P}(\{i \leqslant n : i \text{ is even}\}) = \frac{1}{2}. \end{aligned}$$

The **problem** is whether we are allowed at the step marked by »?!« to pull out the limit $\mathbb{P}(\bigcup_1^\infty \dots) = \lim_n \mathbb{P}(\bigcup_1^n \dots)$; in fact, we have to **assume it** and to extend our axioms I–IV as follows:

II′. $\mathscr{A}$ is a σ-algebra
V. $\forall (A_n)_{n\in\mathbb{N}} \subset \mathscr{A},\ A_n \uparrow A = \bigcup_{n\in\mathbb{N}} A_n\ :\ \lim_{n\to\infty} \mathbb{P}(A_n) = \mathbb{P}(A)$, i.e. $\mathbb{P}$ is continuous from below.

We know from Theorem 3.3.f) and Lemma 3.7 that V is equivalent to

V′. $\forall (A_n)_{n\in\mathbb{N}} \subset \mathscr{A},\ A_n$ disjoint : $\mathbb{P}\left(\dot\bigcup_{n\in\mathbb{N}} A_n\right) = \sum_{n\in\mathbb{N}} \mathbb{P}(A_n) = \mathbb{P}(A)$, i.e. $\mathbb{P}$ is σ-additive.

The properties I, II′, III and V′ show that $\mathbb{P}$ is a **probability measure** in the sense of Definition 3.2. Although E. Borel (see e.g. [6]) already pioneered the use of measures within probability theory, it was A.N. Kolmogorov [15] in 1933 who formulated those axioms in the form we are still using today. This has finally established probability theory as an axiomatically founded mathematical discipline.

20.4 Definition (Kolmogorov 1933). A (mathematical) **probability** is a measure $\mathbb{P}$ on a measurable space $(\Omega, \mathscr{A})$ such that $\mathbb{P}(\Omega) = 1$. The following notation is commonly used:

$(\Omega, \mathscr{A}, \mathbb{P})$;	**probability space**
$\mathbb{P}$	**probability measure;**
$A \in \mathscr{A}$	**event;**
Ω	**certain event, sample space;**
$\emptyset$	**impossible event;**
$\omega \in \Omega$	**outcome, sample point, elementary event.**

There are, however, situations where the (naive) limit as in »?!« (from Example 20.3) does not exist.

20.5 Example (Benford's law). *We draw, at random, a number $i \in \mathbb{N}$. What is the probability that i starts with the digit »1«?* Let us argue as in Example 20.3:

$$\Omega_n := \{1, 2, \dots, n\} \quad \text{and} \quad A_n := \{i \in \Omega_n : i \text{ starts with the digit »1«}\}.$$

Clearly,

$$|A_{10}| = 2 \implies \frac{|A_{10}|}{|\Omega_{10}|} = \frac{2}{10}, \qquad |A_{200}| = 111 \implies \frac{|A_{200}|}{|\Omega_{200}|} = \frac{111}{200}$$

$$|A_{20}| = 11 \implies \frac{|A_{20}|}{|\Omega_{20}|} = \frac{11}{20}, \qquad |A_{1000}| = 112 \implies \frac{|A_{1000}|}{|\Omega_{1000}|} = \frac{112}{1000}$$

$$|A_{100}| = 12 \implies \frac{|A_{100}|}{|\Omega_{100}|} = \frac{12}{100}, \qquad |A_{2000}| = 1111 \implies \frac{|A_{2000}|}{|\Omega_{2000}|} = \frac{1111}{2000}$$

and this shows that the limit $\lim_{n\to\infty} |A_n|/|\Omega_n|$ **does not exist.**

In fact, we have chosen the wrong probability model since the integers with leading digit »1« are not uniformly distributed in $\mathbb{N}$! The correct answer is given by Benford's law, which tells us the probability distribution (»frequency«) of integers having leading digit $i \in \{1,\dots,9\}$:

leading digit	1	2	3	4	5	6	7	8	9
probability	0.301	0.176	0.125	0.097	0.079	0.067	0.058	0.051	0.046

Benford's law has interesting applications, e.g. in tax audits or election fraud, since one can detect in this way (naively) manipulated data. It is, however, important that the original data is sufficiently diverse and random.

In the next chapter we discuss a few combinatorial principles which will allow us – very much in the spirit of Laplace – to calculate certain probabilities.

21 (Very) basic combinatorics

In this chapter we discuss some important counting rules and combinatorial identities. As before, Ω is an arbitrary base set and $\#A = |A|$ denotes the cardinality of the set $A \subset \Omega$. The foundation of all combinatorial considerations are the following counting principles.

21.1 Counting Rules. *Let* $A, B \subset \Omega$ *be any subsets.*

a) *If there is a bijection* $A \leftrightarrow B$*, then* $|A| = |B|$.

b) *If* $|A|, |B| < \infty$ *are disjoint, i.e.* $A \cap B = \emptyset$*, then* $|A \cup B| = |A| + |B|$.

c) *If* $|A|, |B| < \infty$*, then* $|A \times B| = |A| \cdot |B|$.

Proof. The assertions a) and b) are obvious. In order to see Part c), we arrange the elements of the set $A \times B$ in a rectangle with $|A|$ rows and $|B|$ colums:

$$\begin{matrix} (a_1,b_1) & (a_1,b_2) & (a_1,b_3) & \dots & (a_1,b_{|B|}) \\ (a_2,b_1) & (a_2,b_2) & (a_2,b_3) & \dots & (a_2,b_{|B|}) \\ \vdots & \vdots & \vdots & & \vdots \\ (a_{|A|},b_1) & (a_{|A|},b_2) & (a_{|A|},b_3) & \dots & (a_{|A|},b_{|B|}) \end{matrix}$$

□

Arranging things – permutations

21.2 Lemma (permutations). *Let* $\mathcal{N}$ *be a set – i.e. all elements are distinguishable and appear only once – with cardinality* $|\mathcal{N}| = n$. *Then*

$$|\{(x_1,\dots,x_n) \mid x_i \in \mathcal{N},\ x_i \neq x_k\}| = n! = 1 \cdot 2 \cdot 3 \cdot \ldots \cdot n. \tag{21.1}$$

Proof. Define $\mathcal{N}_i := \mathcal{N} \setminus \{x_1, \dots, x_i\}$, $1 \leqslant i < n$. Plainly, we have for the selection of

$$\begin{aligned} x_1 &\in \mathcal{N} && \text{exactly } n \text{ choices,} \\ x_2 &\in \mathcal{N}_1 && \text{exactly } n-1 \text{ choices,} \\ x_3 &\in \mathcal{N}_2 && \text{exactly } n-2 \text{ choices,} \\ &\dots\dots && \\ x_n &\in \mathcal{N}_{n-1} && \text{exactly } 1 \text{ choice,} \end{aligned}$$

and the Counting Rule 21.1.c) proves the claim. □

21.3 Example (permutations of things which appear in several copies). Assume that we have m distinct things $x_1, \dots, x_m$, each of which has further indistinguishable copies, i.e.

$$\underbrace{\Big(\overbrace{(x_1, x_1, \dots, x_1)}^{k_1}, \overbrace{(x_2, x_2, \dots, x_2)}^{k_2}, \dots, \overbrace{(x_m, x_m, \dots, x_m)}^{k_m}\Big)}_{n = k_1 + \dots + k_m}$$

We can arrange these things in

$$\frac{n!}{k_1! \cdot k_2! \cdot \ldots \cdot k_m!} \quad (k_1 + \dots + k_m = n) \tag{21.2}$$

different ways.

Indeed: In a first step, we number the multiple copies so as to make them mutually distinct. From Lemma 21.2 we know that there are $n!$ different arrangements.

In each group $(x_{i,1}, \dots, x_{i,k_i})$ the numbering allows for $k_i!$ distinct arrangements. Thus, the Counting Rule 21.1.c) shows that we have to »divide out« these possibilities, since they are caused by the artificial numbering of the otherwise identical things of type x_i. This gives (21.2).

Letter jumble. How many different »words« (they may be linguistically senseless!) can we form from the letter ABRACADABRA? We have 11 letters which we make distinct by numbering,

$$\mathrm{A}_1, \mathrm{A}_2, \mathrm{A}_3, \mathrm{A}_4, \mathrm{A}_5, \quad \mathrm{B}_1, \mathrm{B}_2, \quad \mathrm{C}, \quad \mathrm{D}, \quad \mathrm{R}_1, \mathrm{R}_2.$$

This yields 11! different arrangements. Using the Counting Rules 21.1 we divide out multiplicities of the letters to get a total of

$$\frac{11!}{5!\,2!\,1!\,1!\,2!} = 83.160 \text{ different words.}$$

Selecting things – combinations, variations

If we want to select $k \in \mathbb{N}_0$ elements from a set $\mathcal{N} = \{1, 2, \dots, n\}$, then we have four possibilities depending on how we pick the elements (pick and put back:

»with replacement« – or pick and keep: »without replacement«) and whether we remember the order of our picks (»tuple«, »ordered sample«) or forget the

draw k numbers from $\mathcal{N} = \{1, 2, \ldots, n\}$	order important tuple $(\omega_1, \omega_2, \ldots, \omega_k)$ ordered sample	order not important set $\{\omega_i \mid i = 1, \ldots, k\}$ unordered sample	
with replacement repeated draws allowed	variations n^k $\Omega = \{(\omega_1, \ldots, \omega_k) \mid \omega_i \in \mathcal{N}\}$ 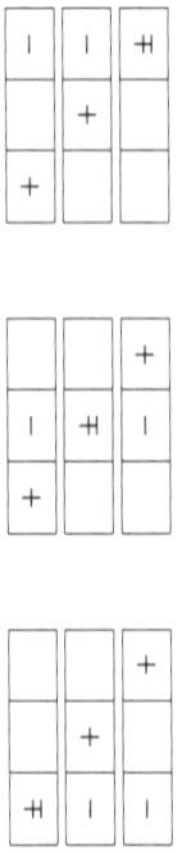	combinations $\binom{n+k-1}{k}$ $\Omega = \{(\omega_1, \ldots, \omega_k) \mid \omega_i \in \mathcal{N}, \omega_1 \le \omega_2 \le \cdots \le \omega_k\}$ 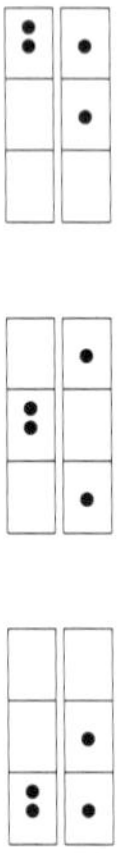	without exclusion no restrictions several things in a box
without replacement repeated draws not allowed	variations $\binom{n}{k} \cdot k!$ $\Omega = \{(\omega_1, \ldots, \omega_k) \mid \omega_i \in \mathcal{N}, \forall i \neq j : \omega_i \neq \omega_j\}$ 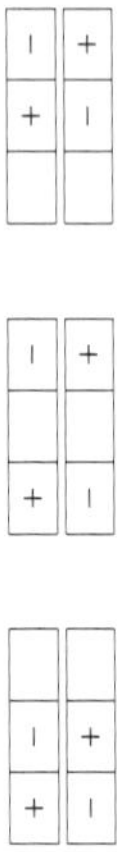	combinations $\binom{n}{k}$ $\Omega = \{(\omega_1, \ldots, \omega_k) \mid \omega_i \in \mathcal{N}, \omega_1 < \omega_2 < \cdots < \omega_k\}$ 	with exclusion with restrictions at most one thing in a box
	things are distinguishable (»people«)	things are indistinguishable (»peas«)	place k things in box no. $1, \ldots, n$

Fig. 21.1. Overview of the basic combinatorial formulae.

order (»set«, »unordered sample«) and arrange the things in, say, ascending order e.g. as in the presentation of the national lottery draws[(1)] – Fig. 21.1 gives an overview of these four possibilities.

We will now discuss these for cases for the urn model »select k things from n« and the box model »place k things in n boxes« – whichever is more intuitive.

21.4 Lemma (no replacement, no order). $\displaystyle\binom{n}{k} = \frac{n!}{k!(n-k)!}$

Proof. Define $\mathcal{N}_i = \mathcal{N} \setminus \{\omega_1,\dots,\omega_i\}$. Since we pick without replacement, the ith element is chosen from $\mathcal{N}_{i-1}$. Thus, the Counting Rule 21.1.c) shows that

$$|\mathcal{N}| \cdot |\mathcal{N}_1| \cdot \ldots \cdot |\mathcal{N}_{k-1}| = n(n-1) \cdot \ldots \cdot (n-k+1) = \frac{n!}{(n-k)!}.$$

The thus obtained tuples still keep the order of the actual selection. From 21.2 we know that there are $k!$ »clones« of these arrangements with the same elements (but in different places), i.e. we still have to divide by $k!$ to remove multiplicities. □

Interpretations. a) An urn contains n distinct (e.g. numbered) balls and we draw k balls which we put in a bag. This means that we do not replace the balls and we forget the order in which the numbers were drawn.

b) Distribute k peas (all green and undistinguishable) into n boxes and make sure that there is at most one pea per box.

c) $\Omega = \left\{(\omega_1,\omega_2,\dots,\omega_k) \in \mathcal{N}^k \mid \omega_1 < \omega_2 < \dots < \omega_k\right\}$, i.e. there are $\binom{n}{k}$ subsets of $\mathcal{N}$ containing k elements.

d) Number of strictly monotone functions $f : \{1,\dots,k\} \to \{1,\dots,n\}$. □

21.5 Lemma (no replacement, with order). $\displaystyle\binom{n}{k}k! = \frac{n!}{(n-k)!}.$

Proof. Similar to Lemma 21.4. □

Interpretations. a) An urn contains n distinct (e.g. numbered) balls and we draw k balls which we do not replace and arrange in a line according so as to remember the order of the draws.

b) Distribute k people (distinguishable, no Siamese twins allowed) onto n seats, and make sure that there is at most one person per seat.

c) $\Omega = \left\{(\omega_1,\omega_2,\dots,\omega_k) \in \mathcal{N}^k \mid \omega_i \neq \omega_j \ \forall i \neq j\right\}$, i.e. there are $\binom{n}{k}k!$ vectors of length k with mutually different entries from $\mathcal{N}$.

d) Number of injective functions $f : \{1,\dots,k\} \to \{1,\dots,n\}$. □

(1) It is a bit counterintuitive that »an unordered sample should be ordered«. Mathematically, this is just a good choice of a representative of the equivalence class of samples which differ by a permuation only.

21.6 Lemma (with replacement, with order). n^k.

Proof. Follows directly from the Counting Rule 21.1.c). □

Interpretations. a) An urn contains n distinguishable balls and we draw k balls. We note the number of the drawn ball and place it back into the urn. This means, we remember the order of the draws.

b) Distribute k people (distinguishable, no Siamese twins allowed) onto n sofas (which may seat several people).

c) $\Omega = \mathcal{N}^k = \left\{(\omega_1, \omega_2, \ldots, \omega_k) \in \mathcal{N}^k \mid \omega_i \in \mathcal{N}\right\}$, i.e. there are n^k vectors of length k with entries from $\mathcal{N}$.

d) Number of functions $f : \{1, \ldots, k\} \to \{1, \ldots, n\}$. □

21.7 Lemma (with replacement, no order). $\binom{n+k-1}{k}$.

Proof. Let us depict the situation (for $n = 6$ boxes and $k = 7$ peas »•«) as follows:

| ••• | | • | | •• | • |

We will only keep the $(n-1)$ separating walls between the boxes and the k peas

•••| |•| |••|•.

$k+(n-1)$ things

Now we can reformulate our enumeration problem as follows: Select $(n-1)$ separating walls from the $k + (n-1)$ things; in view of Lemma 21.4 we have

$$\binom{k+n-1}{n-1} = \binom{k+n-1}{k} \text{ different choices.}$$ □

Interpretations. a) An urn contains n distinguishable balls and we draw k balls. We note the number of the drawn ball and place it back into the urn. At the end we write the drawn balls in ascending order, i.e. we forget the actual order of the draws.

b) Distribute k peas (all look the same) into n boxes (which may contain more than one pea).

c) $\Omega = \left\{(\omega_1, \omega_2, \ldots, \omega_k) \in \mathcal{N}^k \mid \omega_1 \leqslant \omega_2 \leqslant \ldots \leqslant \omega_k\right\}$.

d) Number of monotone functions $f : \{1, \ldots, k\} \to \{1, \ldots, n\}$. □

Enumeration is best learned by solving as many problems as possible. Here are a few typical problems.

21.8 Example. a) **Lotto 6/49.**[2] How many different lotto draws are possible?

$$\text{pick 6 »good« numbers without replacement} \longrightarrow \binom{49}{6} = 13{,}983{,}816.$$

b) **Lotto 6/49.** How many possibilities are there for a »Match 3«? Using the counting rules we get

$$\underbrace{\overbrace{\binom{6}{3}}^{\text{pick »3 matching numbers« from 6 »good numbers«}} \cdot \binom{43}{3}}_{\text{pick 3 further numbers from 43 »bad numbers«}} = 20 \times 12{,}341 = 246{,}820.$$

c) **Lotto 6/49.** How many possibilities are there for a »Match 3 plus bonus number«? Using the counting rules we get

$$\overbrace{\binom{6}{3}}^{\text{pick »3 matching numbers« from 6 »good numbers«}} \cdot \underbrace{\binom{1}{1}}_{\text{1 number from 1 bonus number}} \cdot \underbrace{\binom{42}{2}}_{\text{pick 2 further numbers from the 42 »bad numbers«}} = 20 \times 1 \times 861 = 17{,}220.$$

d) An urn contains R red and W white balls. How many possibilities are there to select, without replacement, r red and w white balls?

$$\binom{R}{r} \cdot \binom{W}{w}$$

If we set $N = R + W$ and $n = r + w$, then the Counting Rule 21.1.b) shows that

$$\sum_{\substack{r+w=n\\ r\leqslant R, w\leqslant W}} \binom{R}{r}\binom{W}{w} = \binom{N}{n} \iff \sum_{r=0}^{n} \binom{R}{r}\binom{N-R}{n-r} = \binom{N}{n}.$$

💬 Please mind the **conventions:** $0! := 1$ and $\binom{n}{k} := 0$ if $k > n$.

e) **Multinomial coefficient.** We set

$$\binom{n}{k_1, k_2, \dots, k_m} := \frac{n!}{k_1! \cdot \ldots \cdot k_m!} \quad (m \leqslant n,\ k_1 + \dots + k_m = n).$$

The following generalization of the binomial formula holds:

$$(x_1 + \dots + x_m)^n = \sum_{\substack{k_1,\dots,k_m\\ k_1+\dots+k_m=n}} \binom{n}{k_1, k_2, \dots, k_m} x_1^{k_1} \cdot \ldots \cdot x_m^{k_m}.$$

Indeed, if we multiply out the product appearing on the left-hand side, we get a sum of expressions of the form

$$y_1 \cdot y_2 \cdot \ldots \cdot y_n, \quad y_i \in \{x_1, \dots, x_m\}.$$

[2] The UK national lottery's number pool is, since October 2015, 1,...,59. The German lottery still has 1,...,49. By 6/49 we mean that 6 numbers are drawn from 49 without replacement and your bet should match as many numbers as possible.

We arrange the sum according to powers and see that the product $x_1^{k_1} \cdot \ldots \cdot x_m^{k_m}$ appears in $\binom{n}{k_1, k_2, \ldots, k_m}$ copies.

f) **Counting functions.** Let $f : X \to Y$ be a function from the set $X = \{1, \ldots, k\}$ into the set $Y = \{1, \ldots, n\}$. If we want to enumerate functions with certain properties (see below), it is helpful to represent a function with the »box model«: The function f tells us how to place the elements $x \in X$ into the numbered boxes $y \in Y$, see Fig. 21.2. The boxes are arranged in increasing order of their number, the x are first thought of as indistinguishable »peas«.

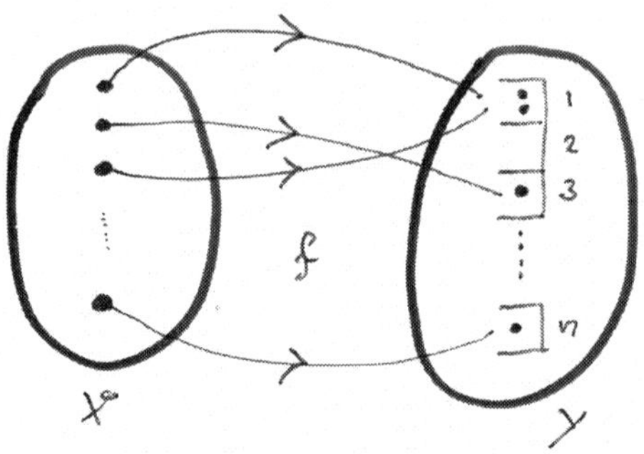

Fig. 21.2. Interpretation of a function $f : X \to Y$ in terms of the »box model« : X are »peas« or »people«, and Y repesent the »boxes«.

\# of all functions: n^k $(n, k$ are arbitrary). This is clear as every $x \in \{1, \ldots, k\}$ can be put in any of the n boxes, thus n^k.

Remark: Often the notation Y^X is used for the set $\{f \mid f : X \to Y\}$, i.e. we have the intuitively appealing formula $|Y^X| = |Y|^{|X|} = n^k$.

\# bijective functions: $n!$ $(n = k)$. We have to match each x with one y; this can be done in exactly one way. Then we number and permute the x and get $k! = n!$ functions.

\# injective functions: $\binom{n}{k} \cdot k!$ $(n \geqslant k)$. We select k boxes and place in each box a single x: $\binom{n}{k}$. All that remains is to number and permute the x, which gives an additional factor $k!$.

\# strictly monotone functions: $\binom{n}{k}$ $(n \geqslant k)$. Like »injections«, but we cannot permute the x, since this would destroy monotonicity.

\# monotone functions: $\binom{n+k-1}{k}$ $(n, k$ arbitrary). Since a value y in the range can be attained by several x – the graph of f can be »flat« –, we deal with the model »boxes with multiple occupation«: $\binom{n+k-1}{k}$. We think of x as »peas« (indistinguishable), place them into the ordered boxes such that several peas can be in one box. Then, we number the peas with $1,\dots,k$ in- or decreasingly to enforce monotonicity.

surjective functions: $T(k,n) := \sum_{\ell=0}^{n}(-1)^{\ell}\binom{n}{\ell}(n-\ell)^k \quad (n \leqslant k)$. We begin with enumerating all functions which **fail to be surjective**:

Write $A_y := \{f : X \to Y \setminus \{y\}\}$ for those functions which miss the value $y \in Y$. Clearly,

$$|A_y| = (n-1)^k \implies |A_{y(1)} \cap \dots \cap A_{y(\ell)}| = (n-\ell)^k, \quad (y(1) < \dots < y(\ell)).$$

Let us switch to indicator functions. Multiplying out we get

$$\begin{aligned}\prod_{\ell=1}^{n} \mathbb{1}_{A_\ell^c} &= \prod_{\ell=1}^{n}(1 - \mathbb{1}_{A_\ell}) \\ &= 1 + \sum_{\ell=1}^{n}(-1)^{\ell} \sum_{1\leqslant y(1)<\dots<y(\ell)\leqslant n} \mathbb{1}_{A_{y(1)}\cap\dots\cap A_{y(\ell)}}.\end{aligned} \tag{21.3}$$

If we integrate this identity w.r.t. counting measure, we see

$$\begin{aligned}\overbrace{|A_1^c \cap \dots \cap A_n^c|}^{\substack{\text{functions which do not leave out} \\ \text{any } y \in Y = \text{surjective functions}}} &= n^k + \sum_{\ell=1}^{n}(-1)^{\ell} \sum_{1\leqslant y(1)<\dots<y(\ell)\leqslant n} |A_{y(1)} \cap \dots \cap A_{y(\ell)}| \\ &= \sum_{\ell=0}^{n}(-1)^{\ell}\binom{n}{\ell}(n-\ell)^k.\end{aligned} \tag{21.4}$$

💬 $S(k,n) := k! \cdot T(k,n)$ is called the **Stirling number of the second kind.**

💬 The formulae (21.3) and (21.4) are called **inclusion–exclusion formulae.**

22 Discrete probability distributions

22.1 Definition. A probability space $(\Omega, \mathscr{A}, \mathbb{P})$ is **discrete**, if $\Omega = \{\omega_i : i \in I\}$, $I \subset \mathbb{N}$, is a finite or countably infinite set. In this case $\mathscr{A} = \mathscr{P}(\Omega)$.

22.2 Lemma. *If $(\Omega, \mathscr{A}, \mathbb{P})$ is a discrete probability space, then*

$$\forall A \subset \Omega : \mathbb{P}(A) = \sum_{\omega\in\Omega} \mathbb{P}(\{\omega\})\delta_\omega(A) \iff \mathbb{P}(\cdot) = \sum_{\omega\in\Omega} \mathbb{P}(\{\omega\})\delta_\omega(\cdot). \tag{22.1}$$

***Consequently**, any discrete probability measure is uniquely determined by the (finitely or countably many) values $\mathbb{P}(\{\omega\})$, $\omega \in \Omega$.*

Proof. Since $\mathscr{A} = \mathscr{P}(\Omega)$, the measure $\mathbb{P}$ is defined for all $A \subset \Omega$. Because of σ-additivity,

$$A = \dot{\bigcup_{\omega\in A}}\{\omega\} \implies \mathbb{P}(A) = \sum_{\omega\in A}\mathbb{P}(\{\omega\}) = \sum_{\omega\in\Omega}\mathbb{P}(\{\omega\})\delta_\omega(A). \qquad \square$$

22.3 Definition. Let $(\Omega,\mathscr{A},\mathbb{P})$ be a probability and $(E,\mathscr{E})$ a measurable space.

An E-**valued random variable** (rv, for short) is a measurable function $X : (\Omega,\mathscr{A}) \to (E,\mathscr{E})$. If $E = \mathbb{R}^d$ or $E = \mathbb{R}$, X is called a d-**dimensional** or **real random variable**, respectively.

If $f : E \to \mathbb{R}$ is a Borel function, then $f \circ X$ is measurable, i.e. a random variable. The **expectation** is the integral

$$\mathbb{E}f(X) := \int f(X)\,d\mathbb{P} = \int_\Omega f(X(\omega))\,\mathbb{P}(d\omega) \tag{22.2}$$

(if it exists). If $E = \mathbb{R}$, then

$$\mathbb{E}X := \int X\,d\mathbb{P} = \int_\Omega X(\omega)\,\mathbb{P}(d\omega). \tag{22.3}$$

22.4 Lemma. *On a discrete probability space* $(\Omega,\mathscr{A},\mathbb{P})$ ***all*** *mappings* $X : \Omega \to \mathbb{R}^d$ *are measurable, hence random variables. For* ***all*** $f : \mathbb{R}^d \to \mathbb{R}$ *one has*

$$\mathbb{E}f(X) = \sum_{\omega\in\Omega} f(X(\omega))\,\mathbb{P}(\{\omega\}) = \sum_{x\in X(\Omega)} f(x)\,\mathbb{P}(X = x), \tag{22.4}$$

if one of the series converges ***absolutely*** *or if all terms are positive; if* $d = 1$,

$$\mathbb{E}X = \sum_{\omega\in\Omega} X(\omega)\,\mathbb{P}(\{\omega\}) = \sum_{x\in X(\Omega)} x\,\mathbb{P}(X = x). \tag{22.5}$$

Proof. Since $\mathscr{A} = \mathscr{P}(\Omega)$ we see that $X^{-1}(B) = \{X \in B\} \in \mathscr{A}$ for all $B \in \mathscr{B}(\mathbb{R}^d)$, i.e. X is measurable, and so is $f(X) = f \circ X$. We will only show (22.4). We have

$$\begin{aligned}
\mathbb{E}f(X) &\overset{\text{def}}{=} \int_\Omega f(X(\theta))\,\mathbb{P}(d\theta) \\
&\overset{\text{def}}{=} \int_\Omega f(X(\theta)) \sum_{\omega\in\Omega}\mathbb{P}(\{\omega\})\delta_\omega(d\theta) \\
&\underset{\S 9.6.b)}{\overset{\S 8.10.b)}{=}} \sum_{\omega\in\Omega}\mathbb{P}(\{\omega\}) \int_\Omega f(X(\theta))\,\delta_\omega(d\theta) \\
&= \sum_{\omega\in\Omega}\mathbb{P}(\{\omega\}) f(X(\omega)),
\end{aligned}$$

and the integral exists, if $f \circ X \geqslant 0$ or if the series converges absolutely. The second equality in (22.4) follows from splitting $\Omega = \dot{\bigcup}_{x\in X(\Omega)}\{\omega : X(\omega) = x\}$ – i.e.

the random variables »re-arranges« the sample space – and

$$\sum_{\omega\in\Omega} \mathbb{P}(\{\omega\})f(X(\omega)) = \sum_{x\in X(\Omega)}\sum_{\omega\in\{X=x\}} \mathbb{P}(\{\omega\})f(X(\omega))$$
$$= \sum_{x\in X(\Omega)} \mathbb{P}(\{\omega \mid X(\omega)=x\})f(x)$$
$$= \sum_{x\in X(\Omega)} \mathbb{P}(X=x)f(x). \qquad \square$$

Recall the notation $\{X\in B\} = X^{-1}(B)$ and $\{X=x\} = X^{-1}(\{x\})$.

22.5 Definition. Let $X:\Omega\to\mathbb{R}^d$ be a random variable. The **(probability) distribution** or **(probability) law** of X is the image measure $\mathbb{P}_X(B) := \mathbb{P}\circ X^{-1}(B) = \mathbb{P}(X\in B)$, $B\in\mathscr{B}(\mathbb{R}^d)$.

22.6 Remark. a) $\mathbb{P}_X$ is an image measure in the sense of Theorem 6.6, see also Chapter 17. It is a probability measure on $(\mathbb{R}^d,\mathscr{B}(\mathbb{R}^d))$.

This makes $(\mathbb{R}^d,\mathscr{B}(\mathbb{R}^d),\mathbb{P}_X)$ a probability space.

b) If $\Omega = \{\omega_i : i\in I\}$ is a discrete sample space, then $X(\Omega)$ is again discrete: $X(\Omega) = \{x_i \mid x_i = X(\omega_i), i\in I\}$. In particular, $\mathbb{P}(X\in B)$ is determined by the values $\mathbb{P}(X=x_i), i\in I$:

$$B = \dot{\bigcup_{x\in B}}\{x\} \implies \mathbb{P}(X\in B) = \mathbb{P}\Big(\dot{\bigcup_{x\in B}}\{X=x\}\Big) = \sum_{x\in B}\mathbb{P}(X=x).$$

22.7 Example (coins and dice). A coin has two faces – obverse and reverse –, usually called

»**heads**« H = »success« = 1, with probability: p
»**tails**« T = »failure« = 0, with probability: $q = 1-p$

If $p = q = \frac{1}{2}$ we speak of a **ideal** or **fair** coin.

A standard die has sice faces, which are numbered by $1,2,\dots,6$ (the »pips«) such that two opposite faces add up to 7. A die is **ideal** or **fair**, if each side has probability $\frac{1}{6}$.

a) **simple coin toss**. The probability space is given by

$$\Omega = \{H,T\},\quad \mathscr{A} = \{\emptyset,\{H\},\{T\},\Omega\},\quad \mathbb{P}(\{H\}) = p,\quad \mathbb{P}(\{T\}) = q = 1-p.$$

An example of a random variable is $X(H)=1, X(T)=0$. We have

$$\mathbb{E}X = 1\cdot\mathbb{P}(X=1) + 0\cdot\mathbb{P}(X=0) = p.$$

The law $\mathbb{P}_X = p\delta_1 + (1-p)\delta_0$ is called the **Bernoulli distribution** and is denoted by $\mathsf{Bin}(p)$.

b) **casting a fair die**. The probability space is given by

$$\Omega = \{1,2,3,4,5,6\}, \quad \mathscr{A} = \mathscr{P}(\Omega), \quad \mathbb{P}(\{n\}) = \frac{1}{6}.$$

An example of a random variable is the number of pips $A(n) = n$, $n \in \Omega$. We have

$$\mathbb{E}A = \sum_{n=1}^{6} n\,\mathbb{P}(A = n) = \frac{1}{6}(1+2+\cdots+6) = \frac{7}{2}.$$

c) **casting two fair dice**. If we roll a die twice (or if we roll two distinguishable dice at the same time), then the probability space is

$$\Omega = \{1,2,3,4,5,6\}^2 = \{(n,m) \mid n,m = 1,\dots,6\},$$
$$\mathscr{A} = \mathscr{P}(\Omega), \quad \mathbb{P}(\{(n,m)\}) = \frac{1}{36}.$$

Consider the rv $S(n,m) = n + m$ which is the sum of pips. We could work out $\mathbb{E}S$ as in Part b), but here is an **alternative without further calculations:** Call $A_1(n,m) := n$ and $A_2(n,m) := m$ the pips of the first and second throw. Clearly, $S = A_1 + A_2$, and by the linearity of the expectation (the integral!)

$$\mathbb{E}S = \mathbb{E}(A_1 + A_2) = \mathbb{E}A_1 + \mathbb{E}A_2 \overset{A_1\sim A}{\underset{A_2\sim A}{=}} 2\mathbb{E}A \overset{b)}{=} 2 \cdot \frac{7}{2} = 7.$$

💬 We use the following **notation**

If X, Y are rv's: $X \sim Y \iff X, Y$ have the same law.

d) **Alternative point of view of c)**. The probability $\mathbb{P}$ in c) is the product of two throws:

$$\mathbb{P}(\{(m,n)\}) = \frac{1}{36} = \frac{1}{6} \cdot \frac{1}{6} = \mathbb{P}_1(\{m\}) \cdot \mathbb{P}_2(\{n\}), \quad m,n \in \{1,\dots,6\},$$

where $\mathbb{P}_i$ is the probability measure of the ith throw. Thus, $\mathbb{P} = \mathbb{P}_1 \otimes \mathbb{P}_2$ – this is the mathematical way to express that the two throws are **independent**.

e) ***n* successive coin tosses**. We toss the same coin (probability for »H« is p) n times. The probability space is

$$\Omega = \{H,T\}^n = \big\{(\omega_1,\dots,\omega_n) \mid \omega_i \in \{H,T\}\big\},$$
$$\mathscr{A} = \mathscr{P}(\Omega), \quad \mathbb{P}(\omega_1,\dots,\omega_n) = p^k q^{n-k},$$

where $k = |\{i : \omega_i = H\}|$ = number of H in the outcome $\omega = (\omega_1,\dots,\omega_n)$. As a function of ω, $k = k(\omega)$ is a random variable.

The following alternative modelling is a bit simpler. Identify $H \triangleq 1$ and $T \triangleq 0$. We have

$$\Omega = \{0,1\}^n = \big\{(\omega_1,\dots,\omega_n) \mid \omega_i \in \{0,1\}\big\}, \quad \mathscr{A} = \mathscr{P}(\Omega),$$
$$\mathbb{P}(\{(\omega_1,\dots,\omega_n)\}) = p^{\omega_1+\cdots+\omega_n} q^{n-\omega_1-\cdots-\omega_n},$$

where $\omega_1 + \cdots + \omega_n$ is the number of »1« or heads in the series of tosses.

Let us find $\mathbb{P}_S$ and $\mathbb{E}S$ for the rv $S(\omega_1,\dots,\omega_n) = \omega_1 + \cdots + \omega_n$.

We write $X_i(\omega_1,\dots,\omega_n) = \omega_i$ for the outcome of toss no. $i = 1,\dots,n$. As in c), we have $\mathbb{P}_{X_i} = \mathbb{P}_X$, where X is the rv appearing in a). By linearity,

$$\mathbb{E}S = \mathbb{E}X_1 + \dots + \mathbb{E}X_n \overset{X_i \sim X}{=} n\mathbb{E}X \overset{a)}{=} np.$$

In order to determine $\mathbb{P}_S$, we note that only the **number** but **not the position** of the »1« in the tuple $(\omega_1,\dots,\omega_n)$ determines the value of S. There are $\binom{n}{k}$ different positions for k ones in the tuple $(\omega_1,\dots,\omega_n)$, so

$$\mathbb{P}_S(\{k\}) = \mathbb{P}(S = k) = \sum_{\omega_1+\dots+\omega_n=k} p^k q^{n-k} = \binom{n}{k} p^k q^{n-k}, \quad k = 0,1,\dots,n;$$

this is the **binomial distribution** denoted by $\mathsf{Bin}(n,p)$.

f) **infinite coin toss**. Let us assume that $p = q = \frac{1}{2}$. The outcomes of this experiment are

$$\Omega = \{0,1\}^{\mathbb{N}} = \left\{\omega = (\omega_n)_{n\in\mathbb{N}} \mid \omega_n \in \{0,1\}\right\},$$

but this is no longer a countable set ✎. Naively, one would suspect that $\mathbb{P}(\{\omega\}) = \prod_{n=1}^\infty \frac{1}{2}$, but the infinite product diverges[(1)] as it has the value 0 ✎. In addition, it is not clear how we can define the probability of a set $A \subset \Omega$ as the formula $\mathbb{P}(A) = \sum_{\omega\in A} \mathbb{P}(\{\omega\})$ obviously fails.

⚠ At this point we need **infinite products** as in Chapter 19 and cylinder sets of the form

$$\exists m \in \mathbb{N} : Z = \{\omega_1\} \times \{\omega_2\} \times \dots \times \{\omega_m\} \times \{0,1\} \times \{0,1\} \times \dots.$$

Denote by $\mathbb{P}_n$ the probability measure on the nth copy of $\{0,1\}$ such that $\mathbb{P}_n(\{0\}) = \mathbb{P}_n(\{1\}) = \frac{1}{2}$, i.e. this measure models the nth coin toss. We have $\mathbb{P}(Z) = 2^{-m}$ with the probability measure $\mathbb{P} = \bigotimes_{n\in\mathbb{N}} \mathbb{P}_n$ on $\sigma(\mathscr{Z}) \subsetneq \mathscr{P}(\{0,1\}^{\mathbb{N}})$ (Theorem 19.4). **But this is not a discrete probability space!**

g) **coin toss up to the first success**. We toss a coin (success probability p) until »1« (heads = »success«) appears for the first time. The corresponding probability space is

$$\Omega = \{(1),(0,1),(0,0,1),\dots\}, \quad \mathscr{A} = \mathscr{P}(\Omega),$$
$$\mathbb{P}(\{\omega\}) = pq^n \quad \text{if} \quad \omega = (\underbrace{0,0,\dots,0}_{n \text{ times}},1).$$

A typical rv is the waiting time (i.e. the number of tosses needed) until the game stops: $T(\omega) = n+1$, where $n+1$ is the length of ω. We have

$$\mathbb{E}T = \sum_{n=0}^\infty (n+1)pq^n = p\sum_{n=0}^\infty (n+1)q^n = p\sum_{n=0}^\infty \frac{d}{dq} q^{n+1} = p\frac{d}{dq}\sum_{n=0}^\infty q^n$$

(1) An infinite product $\prod_1^\infty p_n = \lim_N \prod_1^N p_n$ is said to diverge, if the limit does not exist in $\mathbb{R}$ or if the limit is 0, see e.g. Whittaker & Watson [32, Chapter 2.7].

which gives $\mathbb{E}T = 1/p$ ✍. The law $\mathbb{P}_T(\{n\}) = \mathbb{P}(T = n) = pq^{n-1}$, $n = 1,2,\dots$ is called the **geometric distribution**.

22.8 Example (urn models). An **urn** is a container with coloured (but otherwise indistiguishable) balls. We can draw »with replacement« (i.e. we put the ball back into the urn) or »without replacement«. Thus,

»with replacement« means that the 2nd, 3rd etc. draw does not depend on the result of the 1st draw.

»without replacement« means that each draw depends on the result of the previous draws.

Variations of this scheme include coloured & numbered (hence, distinguishable) balls or adding/removing additional balls of the drawn colour.

a) **single draw**. We have an urn containing W white ($\triangleq 1$, »success«) and B black ($\triangleq 0$, »failure«) balls. The probability space is

$$\Omega = \{0,1\}, \quad \mathscr{A} = \{\emptyset, \{0\}, \{1\}, \Omega\}, \quad \mathbb{P}(\{0\}) = \frac{B}{B+W}, \quad \mathbb{P}(\{1\}) = \frac{W}{B+W}.$$

If $B = W$, this is the same as the toss of a fair coin.

b) *n* **draws with replacement**. We have an urn containing N_i balls of the colour $i = 0,1,2$ (e.g. black, white, red). Since we replace any drawn ball, the probability to draw a ball of colour i is $p_i = N_i/N$ where $N = N_0 + N_1 + N_2$. The probability space is

$$\Omega = \{0,1,2\}^n = \{\omega = (\omega_1,\dots,\omega_n) \mid \omega_i \in \{0,1,2\}\}, \quad \mathscr{A} = \mathscr{P}(\Omega),$$
$$\mathbb{P}(\{\omega\}) = p_0^{X_0(\omega)} p_1^{X_1(\omega)} p_2^{X_2(\omega)},$$

where $X_i(\omega)$ is the number of balls of colour i in the tuple $\omega = (\omega_1,\dots,\omega_n)$.
The distribution of the random variable $X = (X_0,X_1,X_2)$ is determined by

$$\mathbb{P}(X_0 = x_0, X_1 = x_1, X_2 = x_2) := \mathbb{P}(\{X_0 = x_0\} \cap \{X_1 = x_1\} \cap \{X_2 = x_2\});$$

as in Example 22.7.e), the position of the balls within the tuple ω does not matter. There are

$$\binom{n}{x_0, x_1, x_2} = \frac{n!}{x_0! \cdot x_1! \cdot x_2!}, \quad x_0, x_1, x_2 \in \mathbb{N}_0,\ x_0 + x_1 + x_2 = n,$$

possibilities to select positions, and so

$$\mathbb{P}(X_0 = x_0, X_1 = x_1, X_2 = x_2) = \binom{n}{x_0, x_1, x_2} p_0^{x_0} p_1^{x_1} p_2^{x_2}, \quad \begin{matrix} 0 \leqslant x_i \leqslant N_i, \\ x_0 + x_1 + x_2 = n \end{matrix};$$

this is the **multinomial distribution** for three classes (»colours«) of objects. If $N_2 = 0$ and $p = p_0$, $q = p_1$, we recover the binomial law $\mathsf{Bin}(n,p)$.

c) n **draws without replacement.** Consider an urn containing W white ($\triangleq 1$) and B black ($\triangleq 0$) balls, $N = W + B$. We perform n draws without replacing the drawn balls. The probability space is

$$\Omega = \{0,1\}^n = \{\omega = (\omega_1, \dots, \omega_n) \mid \omega_i \in \{0,1\}\}, \quad \mathscr{A} = \mathscr{P}(\Omega).$$

Since we do not return the drawn balls, we have

$$\begin{aligned}
\#\text{white balls before the } i\text{th draw} &\quad W(i) = W - \omega_1 - \dots - \omega_{i-1} \\
\#\text{all balls before the } i\text{th draw} &\quad N(i) = N - (i-1) \\
\#\text{black balls before the } i\text{th draw} &\quad B(i) = N(i) - W(i) \\
\text{probab. of 1 (white) in the } i\text{th draw} &\quad p_1(i) = \frac{W(i)}{N(i)} \\
\text{probab. of 0 (black) in the } i\text{th draw} &\quad p_0(i) = \frac{B(i)}{N(i)}
\end{aligned}$$

and so $\mathbb{P}(\{(\omega_1, \dots, \omega_n)\}) = p_{\omega_1}(1) \cdot \ldots \cdot p_{\omega_n}(n)$.

⚠ The probabilities of the ith draw depend on the results of **all** previous drawings: $\omega_1, \dots, \omega_{i-1}$.

Consider the special case $(\overbrace{1, \dots, 1}^{w}, \overbrace{0, \dots, 0}^{b})$ where we get first a run of w white balls and then a run of b black balls. The probability of this event is

$$\begin{aligned}
&\mathbb{P}(\{(1, \dots, 1, 0, \dots, 0)\}) \\
&= \frac{W \cdot (W-1) \cdot \ldots \cdot (W-w+1) \cdot B \cdot (B-1) \cdot \ldots \cdot (B-b+1)}{N \cdot (N-1) \cdot \ldots \cdot (N-n+1)} \\
&= \frac{\frac{W!}{(W-w)!} \frac{B!}{(B-b)!}}{\frac{N!}{(N-n)!}} = \frac{1}{\binom{n}{w}} \frac{\binom{W}{w}\binom{B}{b}}{\binom{N}{n}}.
\end{aligned}$$

This, however, is also the formula for the general case:

If we permute $\{(1, \dots, 1, 0, \dots, 0)\}$ to become $\{(\omega_1, \dots, \omega_n)\}$, the very same permutation leaves the product on the right-hand side of the above formula invariant.

Hypergeometric distribution $\mathsf{H}(N, W; n, w)$. Assume that an urn contains $N = W + B$ balls, B are black and W are white. We draw n balls without replacement. *What is the probability to obtain w white balls – and, a fortiori, $b = n - w$ black balls?* There are $\binom{n}{w}$ possibilities fot distribute w white balls within the n-tuple $(\omega_1, \dots, \omega_n)$. Therefore, the previous consideration shows

$$\mathsf{H}(N, W, n; w) = \frac{\binom{W}{w}\binom{B}{s}}{\binom{N}{n}} = \frac{\binom{W}{w}\binom{B}{s}}{\binom{B+W}{s+w}}, \quad \begin{matrix} 0 \leqslant s \leqslant B, \\ 0 \leqslant w \leqslant W. \end{matrix} \tag{22.6}$$

Remark★. It is quite difficult to calculate the expected value and the variance of the hypergeometric distributions. The following trick simplifies things. First we assign numbers to the white balls. The total number of white balls in the sample is $X = X_1 + \dots + X_W$ where $X_i = 1$ if, and only if, the white ball with number i is drawn;

Table 22.1. Important discrete probability laws

Nr.	name & symbol	parameter	support	»density«	$\mu = \mathbb{E}X$	$\sigma^2 = \mathbb{V}X$	$g_X(t) = \mathbb{E}t^X$	$\phi_X(\xi) = \mathbb{E}e^{i\xi X}$
0.	general discrete	$p_k \geqslant 0$, $\sum_k p_k = 1$	$\{x_k\}_k$	$\mathbb{P}(X = x_k) = p_k$	$\sum_k k p_k$	$\sum_k (k-\mu)^2 p_k$	$\sum_k t^{x_k} p_k$	$\sum_k e^{i\xi x_k} p_k$
1.	degenerate δ_c	$c \in \mathbb{R}$	$\{c\}$	$\mathbb{P}(X = c) = 1$, $\mathbb{P}(X \neq c) = 0$	c	0	t^c	$e^{i\xi c}$
2.	Bernoulli $\mathsf{Bin}(p)$	$p \in (0,1)$	$\{0,1\}$	$\mathbb{P}(X = 1) = p$, $\mathbb{P}(X = 0) = q$	p	pq	$q + tp$	$q + pe^{i\xi}$
3.	two-point	$p \in (0,1)$ $a, b \in \mathbb{R}$	$\{a,b\}$	$\mathbb{P}(X = a) = p$, $\mathbb{P}(X = b) = q$	$ap + bq$	$(a-b)^2 pq$	$pt^a + qt^b$	$pe^{ia\xi} + qe^{ib\xi}$
4.	symmetric two-point	$c > 0$	$\{\pm c\}$	$\mathbb{P}(X = \pm c) = \frac{1}{2}$,	0	c^2	$\dfrac{t^{2c}+1}{2t^c}$	$\cos(c\xi)$
5.	Binomial $\mathsf{Bin}(n,p)$	$p \in (0,1)$, $n \in \mathbb{N}$	$\{0,\dots,n\}$	$\binom{n}{k} p^k q^{n-k}$	np	npq	$(q+tp)^n$	$\left(q + pe^{i\xi}\right)^n$
6.	negative Binomial	$p \in (0,1)$, $n \in \mathbb{N}$	$\mathbb{N}_0$	$\binom{n+k-1}{n-1} p^n q^k$	$n\dfrac{q}{p}$	$n\dfrac{q}{p^2}$	$\dfrac{p^n}{(1-tq)^n}$	$\dfrac{p^n}{\left(1-qe^{i\xi}\right)^n}$
7.	geometric $\mathsf{g}(p)$	$p \in (0,1)$	$\mathbb{N}_0$	pq^k	$\dfrac{q}{p}$	$\dfrac{q}{p^2}$	$\dfrac{p}{1-tq}$	$\dfrac{p}{1-qe^{i\xi}}$
8.	hypergeometric $\mathsf{H}(N,M,n)$	$n, N, M \in \mathbb{N}_0$, $n \leqslant N, M \leqslant N$	$\{0,\dots,n\}$	$\dfrac{\binom{M}{k}\binom{N-M}{n-k}}{\binom{N}{n}}$	$\dfrac{nM}{N}$	$\dfrac{N-n}{N-1}\dfrac{nM}{N}\left(1-\dfrac{M}{N}\right)$	(1)	(2)
9.	Poisson $\mathsf{Poi}(\lambda)$	$\lambda > 0$	$\mathbb{N}_0$	$\dfrac{\lambda^k e^{-\lambda}}{k!}$	λ	λ	$\exp[\lambda(t-1)]$	$\exp\left[-\lambda\left(1-e^{i\xi}\right)\right]$

Remarks: $q := 1-p$. (**1**)$= {}_2F_1\left(-n,-M,-N,1-t\right)$. (2)$= {}_2F_1\left(-n,-M,-N,1-e^{i\xi}\right)$. ${}_2F_1$ is (Gauß') hypergeometric function [22, Chapter 15].

otherwise $X_i = 0$. Since every white ball has the same probability n/N to be among the drawn balls, we get

$$\mathbb{E}X_i = \mathbb{P}(X_i = 1) = \frac{n}{N} \implies \mathbb{E}X = W\mathbb{E}X_1 = W\frac{n}{N} = np.$$

Moreover,

$$X^2 = \sum_i X_i^2 + \sum_{i \neq k} X_i X_k = \sum_i X_i + \sum_{i \neq k} X_i X_k,$$

and we get

$$\mathbb{E}(X_i X_k) = \mathbb{P}(X_i = 1, X_k = 1) = \frac{n(n-1)}{N(N-1)}$$

and so

$$\mathbb{E}(X^2) = np + W(W-1)\frac{n(n-1)}{N(N-1)} = np + np(W-1)\frac{(n-1)}{(N-1)}.$$

With a few elementary calculations we arrive at the formula for the **variance**

$$\mathbb{V}X = \mathbb{E}(X^2) - (\mathbb{E}X)^2 = np + np(W-1)\frac{(n-1)}{(N-1)} - (np)^2 \cdots = npq\frac{N-n}{N-1}.$$

e) **Pólya's urn.** Assume that an urn contains N balls, of which W are white »1« and B are black »0«. We draw with replacement and add **in addition** $\delta \in \mathbb{N}$ balls of the drawn colour. Thus,

#white balls before the ith draw	$W(i) = W + (\omega_1 + \cdots + \omega_{i-1})\delta$
#all balls before the ith draw	$N(i) = N + (i-1)\delta$
#black balls before the ith draw	$B(i) = N(i) - W(i)$
probab. of 1 (white) in the ith draw	$p_1(i) = \frac{W(i)}{N(i)}$
probab. of 0 (black) in the ith draw	$p_0(i) = \frac{B(i)}{N(i)}$

As in c) we see that

$$\binom{w+b}{b} \prod_{i=0}^{w-1} \frac{W + i\delta}{N + i\delta} \prod_{k=0}^{b-1} \frac{B + k\delta}{N + w\delta + k\delta}, \quad b, w \in \mathbb{N}_0, \tag{22.7}$$

is the probability, to have w white and b black balls. This is **Pólya's distribution**.

23 Continuous probability distributions

Assume that $(\Omega, \mathscr{A}, \mathbb{P})$ is a general probability space and $(E, \mathscr{E})$ a measurable space. As in Definition 22.3, a measurable map $X : (\Omega, \mathscr{A}) \to (E, \mathscr{E})$ is said to be

Table 23.1. A list of important probability distributions with a density

Nr.	name & symbol	parameter	carrier I	density on I, else = 0	$\mu = \mathbb{E}X$	$\sigma^2 = \mathbb{V}X$	$\phi_X(\xi) = \mathbb{E}e^{i\xi X}$
0.	generic	—	$I \subset \mathbb{R}^d$	$p(x),\ \int_I p(x)\,dx = 1$	$\int_I x\,p(x)\,dx$	$\int_I (x-\mu)^2 p(x)\,dx$	$\int_I e^{i\xi x} p(x)\,dx$
1.	uniform, $\mathsf{U}[a,b]$	$a<b$, $a,b \in \mathbb{R}$	$[a,b]$	$\frac{1}{b-a}$	$\frac{a+b}{2}$	$\frac{(b-a)^2}{12}$	$\frac{e^{ib\xi}-e^{ia\xi}}{i\xi(b-a)}$
2.	uniform, $\mathsf{U}[-c,c]$	$c>0$	$[-c,c]$	$\frac{1}{2c}$	0	$\frac{c^2}{3}$	$\frac{\sin(c\xi)}{c\xi}$
3.	triangular	$a,b \in \mathbb{R}$	$[a,b]$	$\begin{cases} \frac{4(x-a)}{(b-a)^2}, & x \in \left(a, \frac{a+b}{2}\right] \\ \frac{4(b-x)}{(b-a)^2}, & x \in \left[\frac{a+b}{2}, b\right) \end{cases}$	$\frac{a+b}{2}$	$\frac{(b-a)^2}{24}$	$-\frac{4\left(e^{ia\xi/2}-e^{ib\xi/2}\right)^2}{(b-a)^2\xi^2}$
4.	symm. triangular	$a>0$	$[-a,a]$	$\frac{1}{a}\left(1-\frac{\lvert x\rvert}{a}\right)$	0	$\frac{a^2}{6}$	$\frac{4\sin^2(a\xi/2)}{a^2\xi^2}$
5.	inverse triangular	$a>0$	$\mathbb{R}$	$\frac{1-\cos ax}{\pi a x^2}$	$\nexists$	$\nexists$	$\left(1-\frac{\lvert\xi\rvert}{a}\right)\mathbb{1}_{[-a,a]}(\xi)$
6.	normal, Gaussian, $\mathsf{N}(\mu,\sigma^2)$	$\mu \in \mathbb{R}$, $\sigma>0$	$\mathbb{R}$	$\frac{1}{\sigma\sqrt{2\pi}} e^{-(x-\mu)^2/(2\sigma^2)}$	μ	σ^2	$e^{i\mu\xi-\sigma^2\xi^2/2}$
7.	normal, Gaussian, $\mathsf{N}(m,C)$	$m \in \mathbb{R}^d$, $C \in \mathbb{R}^{d\times d}$	$\mathbb{R}^d$	$\frac{e^{-\langle (x-m), C^{-1}(x-m)\rangle/2}}{\sqrt{(2\pi)^d \det C}}$	m	$\operatorname{Cov}(X_j,X_k) = c_{jk}$	$e^{i\langle m,\xi\rangle-\langle \xi, C\xi\rangle/2}$
8.	Cauchy, $\mathsf{C}(\lambda,\alpha)$	$\alpha \in \mathbb{R}$, $\lambda>0$	$\mathbb{R}$	$\frac{1}{\pi}\frac{\lambda}{\lambda^2+(x-\alpha)^2}$	$\nexists$	$\nexists$	$e^{i\alpha\xi-\lambda\lvert\xi\rvert}$

Table 23.1. continued

Nr.	name & symbol	parameter	carrier I	density on I, else = 0	$\mu = \mathbb{E}X$	$\sigma^2 = \mathbb{V}X$	$\phi_X(\xi) = \mathbb{E}e^{i\xi X}$
9.	Cauchy, $C(\lambda, a)$	$t > 0$, $a \in \mathbb{R}^d$	$\mathbb{R}^d$	$\dfrac{\Gamma\left(\frac{d+1}{2}\right)}{\pi^{\frac{d+1}{2}}} \dfrac{t}{\left(t^2 + \lvert x-a\rvert^2\right)^{\frac{d+1}{2}}}$	$\nexists$	$\nexists$	$e^{i\langle a,\xi\rangle - t\lvert\xi\rvert}$
10.	log-normal	$\mu \in \mathbb{R}$, $\sigma > 0$	$(0,\infty)$	$\dfrac{1}{x\sigma\sqrt{2\pi}} e^{-(\log x-\mu)^2/(2\sigma^2)}$	$e^{\mu+\sigma^2/2}$	$e^{2\mu+\sigma^2}(e^{\sigma^2}-1)$	unknown
11.	inverse normal	$\mu, \lambda > 0$	$(0,\infty)$	$\left(\dfrac{\lambda}{2\pi x^3}\right)^{\frac{1}{2}} e^{-\lambda(x-\mu)^2/(2\mu^2 x)}$	μ	$\dfrac{\mu^3}{\lambda}$	$\exp\left\{\dfrac{\lambda}{\mu}\left[1-\sqrt{1-\dfrac{2i\mu^2\xi}{\lambda}}\right]\right\}$
12.	Gamma (Erlang), $\Gamma_{\alpha,\beta}$	$\alpha, \beta > 0$	$(0,\infty)$	$\dfrac{1}{\Gamma(\alpha)\beta^\alpha} x^{\alpha-1} e^{-x/\beta}$	$\alpha\beta$	$\alpha\beta^2$	$(1-i\xi\beta)^{-\alpha}$
13.	exponential, $\mathsf{Exp}(\lambda)$	$\lambda > 0$	$(0,\infty)$	$\lambda e^{-\lambda x}$	$\dfrac{1}{\lambda}$	$\dfrac{1}{\lambda^2}$	$\lambda(\lambda - i\xi)^{-1}$
14.	two-sided exponential	$\mu \in \mathbb{R}$, $\sigma > 0$	$\mathbb{R}$	$\dfrac{1}{2\sigma} e^{-\lvert x-\mu\rvert/\sigma}$	μ	$2\sigma^2$	$\dfrac{e^{i\mu\xi}}{1+\sigma^2\xi^2}$
15.	double exponential	$q \in \mathbb{R}$, $\gamma > 0$	$\mathbb{R}$	$\dfrac{1}{\gamma} e^{-(x-q)/\gamma} \exp\left[-e^{-(x-q)/\gamma}\right]$	$\gamma C + q$ $C = 0.5772\ldots$	$\gamma^2, \dfrac{\pi^2}{6}$	$\Gamma(1-i\gamma\xi)e^{iq\xi}$
16.	χ^2 with r deg. of freedom, χ^2_r	$r > 0$	$(0,\infty)$	$\dfrac{1}{\Gamma(r/2)2^{r/2}} x^{r/2-1} e^{-x/2}$	r	$2r$	$(1-2i\xi)^{-r/2}$
17.	Beta, $B_{\alpha,\beta}$	$\alpha, \beta > 0$	$(0,1)$	$\dfrac{\Gamma(\alpha+\beta)}{\Gamma(\alpha)\Gamma(\beta)} x^{\alpha-1}(1-x)^{\beta-1}$	$\dfrac{\alpha}{\alpha+\beta}$	$\dfrac{\alpha\beta}{(\alpha+\beta+1)(\alpha+\beta)^2}$	${}_1F_1(\alpha,\beta,i\xi)$
18.	arc-sine	—	$(0,1)$	$\dfrac{1}{\pi\sqrt{x(1-x)}}$	$\dfrac{1}{2}$	$\dfrac{1}{8}$	$J_0\left(\frac{\xi}{2}\right)e^{i\xi/2}$

an E-valued (d-dimensional if $E = \mathbb{R}^d$ resp. real if $E = \mathbb{R}$) **random variable** (rv). The **distribution** or **law** of X is the image measure

$$\mathbb{P}_X(B) := \mathbb{P} \circ X^{-1}(B) = \mathbb{P}(X \in B), \quad B \in \mathscr{B}(\mathbb{R}^d).$$

In order to calculate the expectation of X, we use the integral w.r.t. the image measure (Theorem 17.1): For any measurable $f : E \to \mathbb{R}$ the composition $f \circ X$ is measurable (hence, a real random variable) and the **expectation** is given by

$$\mathbb{E}f(X) := \int f(X)\, d\mathbb{P} = \int_\Omega f(X(\omega))\, \mathbb{P}(d\omega) \overset{17.1}{=} \int_E f(x)\, \mathbb{P}(X \in dx), \tag{23.1}$$

provided that the integral exists. An important special case is $E = \mathbb{R}$ and $f = \mathrm{id}$

$$\mathbb{E}X = \int X\, d\mathbb{P} = \int_{\mathbb{R}} x\, \mathbb{P}(X \in dx).$$

23.1 Definition. The **distribution function** of a d-dimensional random variable $X = (X_1, \dots, X_d) : \Omega \to \mathbb{R}^d$ is

$$F_X(x_1, x_2, \dots, x_d) := \mathbb{P}(X_1 \leqslant x_1, X_2 \leqslant x_2, \dots, X_d \leqslant x_d), \; (x_1, x_2, \dots, x_d) \in \mathbb{R}^d.$$

The rv X (or the law $\mathbb{P}_X$) is said to be **continuous**, if $x \mapsto F_X(x)$ is a continuous function. If $\mathbb{P}_X$ has a density $p(x)$, i.e. $\mathbb{P}(X \in dx) = p(x)\,dx$, then we call X (or $\mathbb{P}_X$) **absolutely continuous.**

💬 Using dominated convergence one can show that absolutely continuous random variables are continuous.

23.2 Lemma. *The distribution function* $F(x) = \mathbb{P}(X \leqslant x)$ *of a real rv* X *has the following properties:*

i) $x \mapsto F(x)$ *is monotonically increasing;*
ii) $\lim_{x \to -\infty} F(x) = 0$ *and* $\lim_{x \to +\infty} F(x) = 1$;
iii) $x \mapsto F(x)$ *is right-continuous.*

Moreover, F *has at most countably many discontinuity points.*

If $F(x)$ *is (piecewise) continuously differentiable, then* X *is absolutely continuous with density* $p(x) = F'(x)$.

Proof. Clearly, $x \leqslant y$ implies $\{X \leqslant x\} \subset \{X \leqslant y\}$, and so

$$F(x) = \mathbb{P}(X \leqslant x) \leqslant \mathbb{P}(X \leqslant y) = F(y).$$

Right-continuity follows from the continuity of measures: If $x_n \downarrow x$, then

$$\{X \leqslant x_n\} \downarrow \{X \leqslant x\} \implies F(x_n) = \mathbb{P}(X \leqslant x_n) \downarrow \mathbb{P}(X \leqslant x) = F(x).$$

A similar argument with $x_n \uparrow \infty$ and $x_n \downarrow -\infty$ shows that we have $F(-\infty) = 0$ and $F(+\infty) = 1$ ✍.

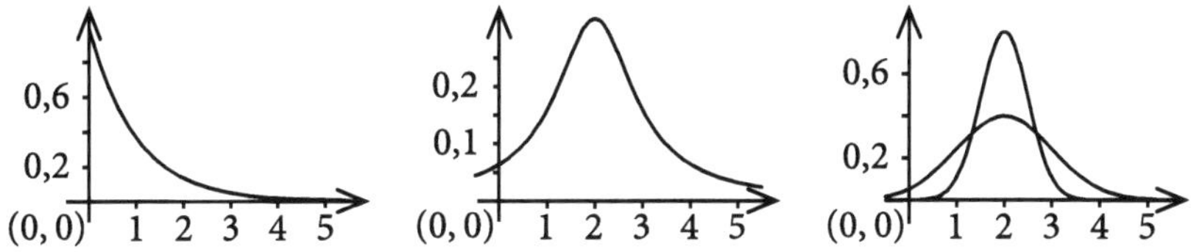

Fig. 23.1. **From left to right**: The densities of the distributions Exp(1), Cauchy(1, 2) and N(2, 1) (flat curve) and N(2, 1/4) (peaked curve).

Since F is monotone, it has at most countably many discontinuities, cf. the bedside lamp argument (Fig. 16.1) used in the proof of Corollary 16.6.

Finally, if $F'(x)$ is (piecewise) continuous, then $\mathbb{P}(X \leqslant x) = F(x) = \int_{-\infty}^{x} F'(t)\,dt$ exists as Riemann integral. From the uniqueness of measure theorem (Theorem 4.7) we know that the measure $B \mapsto \mathbb{P}(X \in B)$ is uniquely determined by the values on the generator $\{X \leqslant x\}, x \in \mathbb{R}$, and so $\mathbb{P}(X \in dx) = F'(x)\,dx$. □

We will see in Chapter 26 that the converse of Lemma 23.2 is also true: Every function $F : \mathbb{R} \to [0, 1]$ with the properties 23.2.i)–iii) is the distribution function of a real random variable. In particular, every positive function $p(x)$ on $\mathbb{R}$ such that $\int_{\mathbb{R}} p(x)\,dx = 1$ is the density of a(n absolutely continuous) random variable.

The following absolutely continuous probability distributions on $\mathbb{R}$ are particularly important.

23.3 Example. a) **Uniform distribution on an interval.** $X \sim U[a, b]$, i.e. X has the density $p(x) = (b-a)^{-1}\mathbb{1}_{[a,b]}(x)$.

Obviously, $\int \mathbb{1}_{[a,b]}(x)\,dx = b - a$, i.e. $p(x) \geqslant 0$ is indeed a probability density. Moreover,

$$\mathbb{E}X = \mu = \frac{1}{b-a}\int_a^b x\,dx = \frac{1}{2}(a+b),$$

$$\mathbb{V}X = \sigma^2 = \frac{1}{b-a}\int_a^b (x-\mu)^2\,dx = \frac{1}{12}(b-a)^2.$$

b) **Exponential law.** $X \sim \mathsf{Exp}(\lambda)$, $\lambda > 0$, has $p(x) = \lambda e^{-\lambda x}\mathbb{1}_{[0,\infty)}(x)$ as density. Note that $\int_0^\infty e^{-\lambda x}\,dx = 1/\lambda$, i.e. $p(x) \geqslant 0$ is a probability density such that

$$\mathbb{E}X = \mu = \int_0^\infty x\lambda e^{-\lambda x}\,dx = \frac{1}{\lambda},$$

$$\mathbb{V}X = \sigma^2 = \int_0^\infty (x-\mu)^2\lambda e^{-\lambda x}\,dx = \frac{1}{\lambda^2}.$$

The exponential distribution is often used to model waiting times, e.g. for radioactive decay, for waiting queues or survival times, and for growth processes. The exponential law is characterized by its lack-of-memory property, cf. Corollary 47.4 in Chapter 47.

c) **Cauchy distribution.** $X \sim C(\lambda, a)$ is a Cauchy rv, if $p(x) = \frac{1}{\pi}\frac{\lambda}{\lambda^2+(x-a)^2}$. From

$$\int_{-\infty}^{\infty} \frac{\lambda\,dx}{\lambda^2+(x-a)^2} \overset{\lambda y = x-a}{\underset{dy=dx/\lambda}{=}} \int_{-\infty}^{\infty} \frac{dy}{1+y^2} = \arctan y\Big|_{-\infty}^{\infty} = \pi$$

we see that $p(x)$ is a probability density. Since $|x|p(x) \approx c|x|^{-1}$ and $x^2 p(x) \approx c$ (as $|x| \to \infty$) neither $\mathbb{E}X$ nor $\mathbb{V}X$ exist.

d) **Normal or Gaussian distribution.** $X \sim N(\mu, \sigma^2)$, i.e. X has the density

$$g_{\mu,\sigma^2}(x) = \frac{1}{\sqrt{2\pi\sigma^2}} e^{-(x-\mu)^2/2\sigma^2}, \quad \mu \in \mathbb{R},\ \sigma^2 > 0,\ x \in \mathbb{R}.$$

The normalization $\int_{\mathbb{R}} g_{\mu,\sigma^2}(x)\,dx = 1$ is shown in Lemma 23.4, i.e. g_{μ,σ^2} is a probability density. The parameters μ and σ^2 are the expectation and variance of X. Changing variables according to $\sigma y = x - \mu$ and $dy = dx/\sigma$ shows

$$\mathbb{E}X = \frac{1}{\sqrt{2\pi\sigma^2}} \int_{-\infty}^{\infty} x e^{-(x-\mu)^2/2\sigma^2}\,dx = \frac{1}{\sqrt{2\pi}} \int_{-\infty}^{\infty} (\sigma y + \mu) e^{-y^2/2}\,dy = \mu,$$

$$\mathbb{V}X = \frac{1}{\sqrt{2\pi\sigma^2}} \int_{-\infty}^{\infty} (x-\mu)^2 e^{-(x-\mu)^2/2\sigma^2}\,dx = \frac{\sigma^2}{\sqrt{2\pi}} \int_{-\infty}^{\infty} y^2 e^{-y^2/2}\,dy = \sigma^2;$$

in the last equality we use integration by parts: $f = y$ and $g' = ye^{-y^2/2}$.

- 💬 $\sigma > 0$ is called the **standard deviation.**
- 💬 if $\mu = 0$, $\sigma = 1$, then $N(0,1)$ is the **standard** normal distribution.
- ⚠ if $\sigma = 0$, then $N(\mu, 0) = \delta_\mu$ is the **degenerate** normal distribution.
- ⚠ »normal« and »Gaussian« are used synonymously. If you meet French people, they'll insist on »Gauss–Laplace«.
- ⚠ $X \sim N(\mu, \sigma^2)$ is often called »normal« rv or »Gaussian« rv.

23.4 Lemma. $\displaystyle\int_{\mathbb{R}} e^{-(x-\mu)^2/2\sigma^2}\,dx = \sqrt{2\pi\sigma^2}.$

Proof. If $\mu = 0$ and $\sigma^2 = 1$, we have seen in Example 16.3 that the value of this integral is $\sqrt{2\pi}$. As we have noted there, we can calculate »in a Riemannian fashion« and change variables according to $y = (x-\mu)/\sigma$ and $dy = dx/\sigma$ to see that

$$\int_{\mathbb{R}} e^{-(x-\mu)^2/2\sigma^2}\,dx = \sigma \int_{\mathbb{R}} e^{-y^2/2}\,dy = \sqrt{2\pi\sigma^2}. \qquad \square$$

24 Conditional probability

Frequently, the probability of an event depends on some prior event(s), e.g. the second ball in the »6/49 lottery« depends on the outcome of the first draw. In this chapter we will investigate such »enchained probabilities«. As usual, $(\Omega, \mathscr{A}, \mathbb{P})$ is an arbitrary probability space.

24.1 Definition. Let $A, B \in \mathscr{A}$ be two events. The **conditional probability** of A given B is

$$\mathbb{P}(A \mid B) := \begin{cases} \dfrac{\mathbb{P}(A \cap B)}{\mathbb{P}(B)}, & \text{if } \mathbb{P}(B) > 0, \\ 0, & \text{if } \mathbb{P}(B) = 0. \end{cases} \tag{24.1}$$

$A \mapsto \mathbb{P}(A \mid B)$ is a probability measure on $(\Omega, \mathscr{A})$ or on $(B, \mathscr{A}|_B)$.

$B \mapsto \mathbb{P}(A \mid B)$ is not a measure.

24.2 Example. a) **Casting a die twice.** We take $\Omega = \{1, \dots, 6\}^2$ as sample space and the probability is $\mathbb{P}(\{(m,n)\}) = 1/36$. Consider the events

$$\begin{aligned} A &= \{6\} \times \{1, \dots, 6\} && \text{»first throw yields 6«} \\ B_\ell &:= \{(m,n) \mid m + n = \ell\} && \text{»sum of pips is } \ell = 2, \dots, 12\text{«} \\ B_{12} &= \{(6,6)\} \subset A \\ B_{11} &= \{(5,6), (6,5)\} \\ B_5 &= \{(1,4), (2,3), (3,2), (4,1)\}. \end{aligned}$$

Therefore, we get $\mathbb{P}(A) = \frac{1}{6}$, $\mathbb{P}(A \mid B_{12}) = 1$, $\mathbb{P}(A \mid B_{11}) = \frac{1}{2}$, $\mathbb{P}(A \mid B_5) = 0$.

Note that both expressions $\mathbb{P}(A \mid B)$ and $\mathbb{P}(B \mid A)$ make sense – and we cannot infer from it any temporal or causal dependence among A and B:

$$\begin{aligned} \mathbb{P}(B_{11} \mid A) &= \text{probability of } B_{11} \text{ if the first throw is a »6«.} \\ \mathbb{P}(A \mid B_{11}) &= \text{probability that the first throw is a »6«,} \\ & \text{ if we have observed } (5,6), (6,5). \end{aligned}$$

b) **2-stage experiment.** Assume that there are two urns u_1 and u_2, each of which contains b_i black and w_i white balls ($i = 1, 2$). First, we pick an urn – the probabilites are p and q – and then we draw from that urn a ball. The best way to visualize this is to draw a **tree diagram** (Fig. 24.1). An **edge** is the connection between two nodes, a **branch** is the succession of edges from the **root** (the first node) to the **leaves** (the top nodes without further edges). The edges are labeled with the conditional probabilities for the current stage of the experiment.

The probability space is given by $\Omega = \{u_1, u_2\} \times \{b, w\}$, and the probabilities of the outcomes

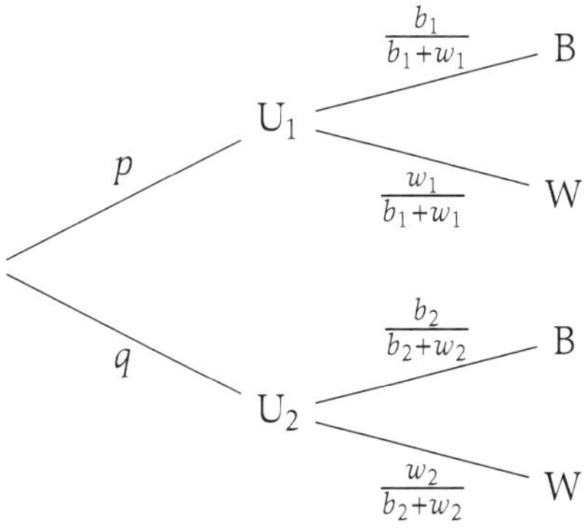

Fig. 24.1. A two-stage experiment.

$$\mathbb{P}(\{(u_1, b)\}) = p \cdot \frac{b_1}{b_1 + w_1} \quad \text{etc.}$$

are the probabilities of the branches where we multiply the probabilities of all edges of a branch. Consider the events

$$\begin{aligned} U_i &= \{\text{urn } u_i \text{ is chosen}\} = \{u_i\} \times \{r, s\} \\ B &= \{\text{a black ball is drawn}\} = \{u_1, u_2\} \times \{b\} \\ W &= \{\text{a white ball is drawn}\} = \{u_1, u_2\} \times \{w\}. \end{aligned}$$

With these sets, we can recover the conditional probabilities of the leaves

$$\mathbb{P}(W \mid U_i) = \frac{w_i}{w_i + b_i} \quad \text{and} \quad \mathbb{P}(B \mid U_i) = \frac{b_i}{w_i + b_i}.$$

c) Sometimes, it is useful to introduce random variables in order to describe the experiment b). Set

$$\begin{aligned} U : \Omega \to \{u_1, u_2\}, &\quad U(\omega) = \text{number of the urn}, \\ C : \Omega \to \{b, w\}, &\quad C(\omega) = \text{colour of the ball}. \end{aligned}$$

We have $U_i = \{U = u_i\}$ and $B = \{F = b\}$ or $W = \{F = w\}$, and we get

$$\mathbb{P}(W \mid U_i) = \mathbb{P}(F = w \mid U = u_i) \quad \text{etc.}$$

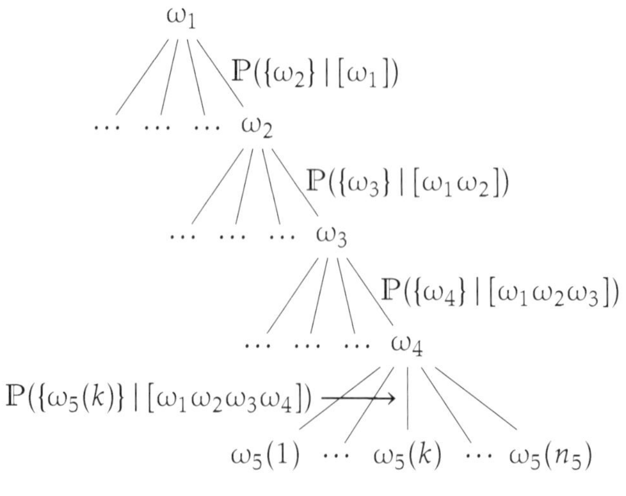

Fig. 24.2. The edges are labeled with $P(\{\omega_m(k)\} \mid [\omega_1\omega_2\cdots\omega_{m-1}])$, i.e. the conditional probabilities of the mth stage of the experiment. These are probability measures on Ω_m – which depend on the pre-history $\omega_1,\dots,\omega_{m-1}$ which represents the previous draws. Notice that, by definition, $\sum_{k=1}^{n_m} P(\{\omega_m(k)\} \mid [\omega_1\omega_2\cdots\omega_{m-1}]) = 1$.

d) Example b) nicely illustrates the principle how to construct a probability measure $\mathbb{P}$ from the conditional probabilities of each stage of the experiment. In reality, it is usually these probabilities which are known. We can model an n-stage experiment in the following way.

$$\Omega = \Omega_1 \times \cdots \times \Omega_n, \quad \omega = (\omega_1, \dots, \omega_n), \quad \omega_k \in \Omega_k = \{\omega_k(1), \dots, \omega_k(n_k)\}.$$

If $0 \leqslant m < n$, then we use the following notation to denote the »past draws«

$$\begin{aligned}[\omega_1\omega_2\cdots\omega_m] &:= \{\omega_1\}\times\cdots\times\{\omega_m\}\times\Omega_{m+1}\times\cdots\times\Omega_n\\ &:= \{(\omega_1,\dots,\omega_m)\}\times\Omega_{m+1}\times\cdots\times\Omega_n.\end{aligned}$$

We have

$$\mathbb{P}(\{\omega\}) = \mathrm{P}(\{\omega_1\})\prod_{m=2}^{n}\underbrace{\mathrm{P}(\{\omega_m\}\mid[\omega_1\omega_2\cdots\omega_{m-1}])}_{\substack{\text{probability, to get } \omega_m \text{ in stage } m, \text{ if we have}\\ \text{observed } \omega_1,\dots,\omega_{m-1} \text{ in the previous stages}}}.$$

It is fairly tedious (but straightforward ✍) to check that this is a discrete probability measure. As a hint, consider Fig. 24.2 and start from the leaves. Note that the sum of the edge probabilities emanating from any node adds up to 1.

Let us establish the main properties of conditional probabilities.

24.3 Theorem. *Assume that $A_m, B_n, A, B \in \mathscr{A}$ are events.*

a) $A \mapsto \mathbb{P}(A \mid B)$ *is a probability measure on* $(\Omega, \mathscr{A}, \mathbb{P})$ *if* $\mathbb{P}(B) > 0$.

b) ***Multiplication rule.***

$$\begin{aligned}\mathbb{P}\Big(\bigcap_{m=1}^{n} A_m\Big) &= \mathbb{P}(A_1)\prod_{m=2}^{n}\mathbb{P}(A_m \mid A_1\cap\cdots\cap A_{m-1}) \qquad (24.2)\\ &= \mathbb{P}(A_1)\mathbb{P}(A_2\mid A_1)\mathbb{P}(A_3\mid A_1\cap A_2)\cdots\mathbb{P}(A_n\mid A_1\cap\cdots\cap A_{n-1})\end{aligned}$$

c) ***Total probability.*** *If* $\bigcup_{n\in\mathbb{N}} B_n = \Omega$ *is a partitioning of* Ω, *then*

$$\mathbb{P}(A) = \sum_{n=1}^{\infty}\mathbb{P}(B_n)\mathbb{P}(A\mid B_n). \qquad (24.3)$$

d) ***Bayes's formula.*** *If* $\bigcup_{n\in\mathbb{N}} B_n = \Omega$ *is a partitioning of* Ω, *then*[1]

$$\mathbb{P}(B_m\mid A) = \frac{\mathbb{P}(B_m)\mathbb{P}(A\mid B_m)}{\sum_{n=1}^{\infty}\mathbb{P}(B_n)\mathbb{P}(A\mid B_n)}; \qquad (24.4)$$

in particular, if $n = 2$ *and* $\Omega = B \cup B^c$,

$$\mathbb{P}(B\mid A) = \frac{\mathbb{P}(B)\mathbb{P}(A\mid B)}{\mathbb{P}(B)\mathbb{P}(A\mid B) + \mathbb{P}(B^c)\mathbb{P}(A\mid B^c)}. \qquad (24.5)$$

⚠ If, in the above formulae, $\mathbb{P}(A) = 0$, then we agree that $\frac{0}{0} := 0$.

💬 Bayes's formula has a fascinating and controversial history, see [20].

Proof. a) follows straight from the definition.

b) Using the definition of the conditional probability recursively, we get

$$\begin{aligned}
&\mathbb{P}(A_1)\mathbb{P}(A_2 \mid A_1) &&= \mathbb{P}(A_1 \cap A_2)\\
\rightsquigarrow\ &\mathbb{P}(A_1)\mathbb{P}(A_2 \mid A_1)\cdot \mathbb{P}(A_3 \mid A_1 \cap A_2) &&= \mathbb{P}(A_1 \cap A_2)\cdot \mathbb{P}(A_3 \mid A_1 \cap A_2)\\
& &&= \mathbb{P}(A_1 \cap A_2 \cap A_3)
\end{aligned}$$

and so on.

c) Since $\Omega = \dot{\bigcup}_{n\in\mathbb{N}} B_n$, we get

$$\begin{aligned}
\mathbb{P}(A) = \mathbb{P}\Big(A \cap \dot{\bigcup}_{n\in\mathbb{N}} B_n\Big) &= \mathbb{P}\Big(\dot{\bigcup}_{n\in\mathbb{N}} A \cap B_n\Big)\\
&= \sum_{n\in\mathbb{N}} \mathbb{P}(A \cap B_n) = \sum_{n\in\mathbb{N}} \mathbb{P}(B_n)\mathbb{P}(A \mid B_n).
\end{aligned}$$

d) We have

$$\mathbb{P}(B_m \mid A) = \frac{\mathbb{P}(B_m \cap A)}{\mathbb{P}(A)} \overset{(24.3)}{=} \frac{\mathbb{P}(B_m)\mathbb{P}(A \mid B_m)}{\sum_{n=1}^{\infty} \mathbb{P}(B_n)\mathbb{P}(A \mid B_n)}. \qquad \square$$

24.4 Example. a) We draw from a deck of 52 cards (e.g. bridge, poker) 4 cards without replacement. What is the probability to get 4 aces?

Define $A_n = \{n\text{th draw yields an ace}\}$, and observe that

$$\begin{aligned}
\mathbb{P}(A_1) &= \frac{4}{52}, & \mathbb{P}(A_3 \mid A_1 \cap A_2) &= \frac{2}{50},\\
\mathbb{P}(A_2 \mid A_1) &= \frac{3}{51}, & \mathbb{P}(A_4 \mid A_1 \cap A_2 \cap A_3) &= \frac{1}{49}.
\end{aligned}$$

Therefore, $\mathbb{P}(A_1 \cap A_2 \cap A_3 \cap A_4) = \frac{4}{52}\cdot\frac{3}{51}\cdot\frac{2}{50}\cdot\frac{1}{49}$ because of the multiplication rule (24.2),

b) (Example 24.2.b) continued)

$$\mathbb{P}(B) = \mathbb{P}(U_1)\mathbb{P}(B \mid U_1) + \mathbb{P}(U_2)\mathbb{P}(B \mid U_2) = \frac{pb_1}{b_1 + w_1} + \frac{qb_2}{b_2 + w_2}.$$

In Fig. 24.1 this is the sum of the probabilities of the branches with black leaves.

c) (...continuation) Bayes's formula allows us to »reverse« the probabilities: What is the probability that we have selected u_1, if we have drawn a black ball?

$$\mathbb{P}(U_1 \mid B) = \frac{\mathbb{P}(U_1)\mathbb{P}(B \mid U_1)}{\mathbb{P}(U_1)\mathbb{P}(B \mid U_1) + \mathbb{P}(U_2)\mathbb{P}(B \mid U_2)} = \frac{\frac{pb_1}{b_1+w_1}}{\frac{pb_1}{b_1+w_1} + \frac{qb_2}{b_2+w_2}}.$$

d) (Kahnemann–Tversky paradox) In a town there are only green or blue taxi-cabs. A cab causes an accident and the driver escapes. The situation is observed by several witnesses. We set up a model by defining

$$B := \{\text{blue taxis}\}$$
$$G := \{\text{green taxis}\} = B^c.$$
$$\widehat{B} := \{\text{witness identifies the escaped taxi as blue}\}$$
$$\widehat{G} := \{\text{witness identifies the escaped taxi as green}\}.$$

The police runs experiments with witnesses under similar visibility conditions and finds that

$$\mathbb{P}(\widehat{B} \mid B) = 0.8 \quad \text{and} \quad \mathbb{P}(\widehat{G} \mid G) = 0.8.$$

These probabilities describe the reliability of the witnesses. How much can we trust the testimony in the accident case? That is, we ask for $\mathbb{P}(B \mid \widehat{B})$ or $\mathbb{P}(G \mid \widehat{G})$.

⚠ **Standard mistake:** $\mathbb{P}(B \mid \widehat{B}) = 0.8$.

💬 **Correct approach:** Use Bayes's formula

$$\mathbb{P}(B \mid \widehat{B}) = \frac{\mathbb{P}(\widehat{B} \mid B)\mathbb{P}(B)}{\mathbb{P}(\widehat{B} \mid B)\mathbb{P}(B) + \mathbb{P}(\widehat{B} \mid G)\mathbb{P}(G)}.$$

This means that we need the following **additional information on the taxi population in town**:

$$\mathbb{P}(B) = 0.15 \quad \text{and} \quad \mathbb{P}(G) = 0.85.$$

Thus, $\mathbb{P}(\widehat{B} \mid G) = 1 - \mathbb{P}(\widehat{G} \mid G) = 0.2 \quad \stackrel{\text{Bayes}}{\Longrightarrow} \quad \mathbb{P}(B \mid \widehat{B}) \approx 0.41.$

In comparison, $\mathbb{P}(G \mid \widehat{G}) \approx 0.96$.

💬 The source of the Khanemann–Tversky paradox is the rather unequal distribution in the taxi population which leads to grossly different credibilities for the testimonies.

e) (Medical screening) This is, essentially, a version of the Kahnemann–Tversky paradox. A company has developed a test to detect a certain illness. We write

$$P := \{\text{person is »positive« (= sick)}\}$$
$$N := \{\text{person is »negative« (= not sick)}\}$$
$$\widehat{P} := \{\text{test says person is »positive« (= sick)}\}$$
$$\widehat{N} := \{\text{test says person is »negative« (= not sick)}\}.$$

Every test has two kinds of problems (»type-I« and »type-II« errors):

$$\alpha = \mathbb{P}(\widehat{P} \mid N) = \text{false alarm} = \text{healthy person identified as sick}$$
$$\beta = \mathbb{P}(\widehat{N} \mid P) = \text{test failure} = \text{sick person identified as healthy.}$$

These probabilities are usually known. On the positive side, we call

$$1-\beta = \mathbb{P}(\widehat{\mathrm{P}} \mid \mathrm{P}) = \text{sensitivity} = \text{true positive rate}$$
$$1-\alpha = \mathbb{P}(\widehat{\mathrm{N}} \mid \mathrm{N}) = \text{specificity} = \text{true negative rate}.$$

We are interested in the probability that a person is ill if it is diagnosed as such:

$$\mathbb{P}\big(\mathrm{P} \mid \widehat{\mathrm{P}}\big) = \frac{\mathbb{P}\big(\widehat{\mathrm{P}} \mid \mathrm{P}\big)\mathbb{P}(\mathrm{P})}{\mathbb{P}\big(\widehat{\mathrm{P}} \mid \mathrm{P}\big)\mathbb{P}(\mathrm{P}) + \big(1-\mathbb{P}\big(\widehat{\mathrm{N}} \mid \mathrm{N}\big)\big)\big(1-\mathbb{P}(\mathrm{P})\big)}.$$

If an illness is relatively rare, i.e. $\mathbb{P}(\mathrm{P}) \ll 1$, a randomly selected person which is positively tested may have a very small likelihood to be ill! This is an disturbing effect in mass screenings, e.g. for breast cancer. It is known[2] that

$$\mathbb{P}(\widehat{\mathrm{P}} \mid \mathrm{P}) = 0.869, \quad \mathbb{P}(\widehat{\mathrm{N}} \mid \mathrm{N}) = 0.889, \quad \mathbb{P}(\mathrm{P}) = 0.040,$$

which shows that $\mathbb{P}(\mathrm{P} \mid \widehat{\mathrm{P}}) \approx 0.246$, i.e. only one of four randomly chosen (!) females with a positive screening is in fact ill. The story is different, if we sample from a population, where breast cancer is more prevalent,[3] i.e. with larger $\mathbb{P}(\mathrm{P})$; this follows from $\mathbb{P}\big(\mathrm{P} \mid \widehat{\mathrm{P}}\big) \uparrow 1$ as $\mathbb{P}(\mathrm{P}) \uparrow 1$.

24.5 Remark. Often it is better, to write (24.5) as **relative quantity**. In our Example 24.4.d) we would have

$$\frac{\mathbb{P}\big(\mathrm{B} \mid \widehat{\mathrm{B}}\big)}{\mathbb{P}\big(\mathrm{G} \mid \widehat{\mathrm{B}}\big)} = \frac{\mathbb{P}\big(\widehat{\mathrm{B}} \mid \mathrm{B}\big)}{\mathbb{P}\big(\widehat{\mathrm{B}} \mid \mathrm{G}\big)} \times \frac{\mathbb{P}\big(\mathrm{B}\big)}{\mathbb{P}\big(\mathrm{G}\big)}. \tag{24.5'}$$

If we translate this to a situation in a law-court, (24.5′) becomes

$$\frac{\mathbb{P}\big(\text{guilty} \mid \text{evidence}\big)}{\mathbb{P}\big(\text{not guilty} \mid \text{evidence}\big)} = \underbrace{\frac{\mathbb{P}\big(\text{evidence} \mid \text{guilty}\big)}{\mathbb{P}\big(\text{evidence} \mid \text{not guilty}\big)}}_{\text{e.g. expert assessment}} \times \overbrace{\frac{\mathbb{P}\big(\text{guilty}\big)}{\mathbb{P}\big(\text{not guilty}\big)}}^{\substack{\text{judge's opinion} \\ \textbf{prior}\text{ to evidence}}}.$$

24.6 Example (Monty Hall problem). A quiz show features three doors for you to choose. Behind one door, there is a car C; behind the others are goats G. The host knows what is behind the doors; he asks you to select one door. After you've told your selection, the host opens another door that shows a goat. You have now the possibility to change your selection. Should you?

Solution. We label the doors with $i = 1, 2, 3$, define events

$$\mathrm{C}_i = \{\text{»the car is behind door Nr. } i\text{«}\},$$
$$\mathrm{H}_i = \{\text{»the host opens door Nr. } i\text{«}\}.$$

and we get the tree diagram shown in Fig. 24.3.

(2) Data (2007–2013) of the US Breast Cancer Surveillance Consortium.
(3) e.g. if your doctor recommends that you should go to a non-routine screening

Suppose, you have chosen door Nr. 1. Then,

$$\mathbb{P}(H_1 \mid C_1) = 0, \qquad \mathbb{P}(H_2 \mid C_1) = \frac{1}{2}, \qquad \mathbb{P}(H_3 \mid C_1) = \frac{1}{2},$$

since the host chooses randomly door Nr. 2 or 3; he cannot choose door no. 1, since it's your door. For the other two branches we have

$$\begin{aligned} &\mathbb{P}(H_1 \mid C_2) = 0, \qquad \mathbb{P}(H_2 \mid C_2) = 0, \qquad \mathbb{P}(H_3 \mid C_2) = 1,\\ &\mathbb{P}(H_1 \mid C_3) = 0, \qquad \mathbb{P}(H_2 \mid C_3) = 1, \qquad \mathbb{P}(H_3 \mid C_3) = 0, \end{aligned}$$

Note that $\mathbb{P}(H_1 \mid C_i) = 0$ as door no. 1 is yours and $\mathbb{P}(H_i \mid C_i) = 0$, since the host cannot open the door with the car.

For symmetry reasons we may assume that the host opens door no. 3. With Bayes's formula we find

$$\mathbb{P}(C_1 \mid H_3) = \frac{\mathbb{P}(C_1)\mathbb{P}(H_3 \mid C_1)}{\sum_{i=1}^3 \mathbb{P}(C_i)\mathbb{P}(H_3 \mid C_i)} = \frac{\frac{1}{2}}{\frac{1}{2} + 1 + 0} = \frac{1}{3},$$

for the probability if we do not change our initial choice; if we change it, we get with a similar calculation

$$\mathbb{P}(C_2 \mid H_3) = \frac{\mathbb{P}(C_2)\mathbb{P}(H_3 \mid C_2)}{\sum_{i=1}^3 \mathbb{P}(C_i)\mathbb{P}(H_3 \mid C_i)} = \frac{1}{\frac{1}{2} + 1 + 0} = \frac{2}{3}.$$

Variants and a full discussion can be found in [WT, Chapter 4] and [26].

24.7 Example (Does the sun rise again tomorrow?). Let $S_n : \Omega \to \{0, 1\}$ be a $\mathsf{Bin}(n, p)$ distributed rv, but the parameter p is **not known**. We observe n successes in a row, i.e. $S_n = n$. *What is the probability that* $S_{n+1} = n + 1$*?*

This is a statistical problem: We know (or have at least a reasonable clue) the underlying probability mechanism »$\mathsf{Bin}(n, p)$« and we have collected data »n successes«, but we do not know the underlying »true« probability p. This is sometimes called an »inverse problem«, since we infer from observations to the probability. Thomas Bayes was the first mathematician who solved such a problem correctly [5] – as early as 1763!

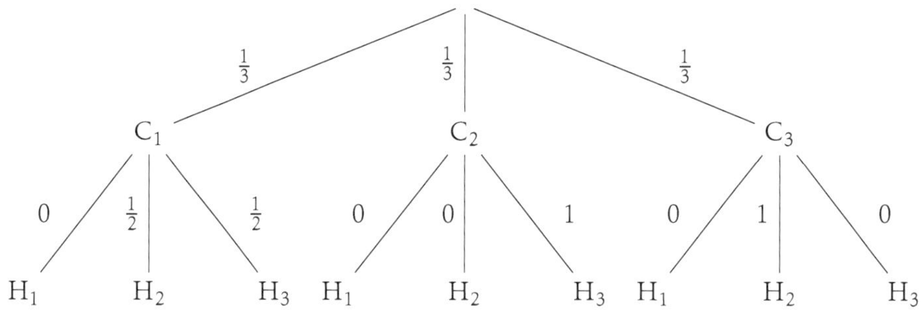

Fig. 24.3. You have chosen door no. 1; the host – he knows what is behind the doors – opens one door (randomly, if he has a choice) with a goat.

Laplace (1774) [17; 18] has generalized Bayes's solution and he's responsible for the »sunny header« of this example. He asked the (provocative) question: The sun rises every day since thousands of years, how likely is it that it will rise tomorrow, too?

If you are a die-hard frequentist, the answer is »for sure«, since your observations would tell you that $p = \frac{\#\{\text{days the sun rose}\}}{\#\{\text{all days}\}} = \frac{S_n}{n} = 1$.

But the former is pretty senseless, if n is small, say $n = 3$. You certainly would not assume that, after three rainy days, day number four is for sure rainy!

Bayesian solution (with uniform prior). We ask for

$$\begin{aligned}\mathbb{P}(S_{n+1} = n+1 \mid S_n = n) &= \frac{\mathbb{P}(\{S_{n+1} = n+1\} \cap \{S_n = n\})}{\mathbb{P}(\{S_n = n\})} \\ &= \frac{\mathbb{P}(\{S_{n+1} = n+1\})}{\mathbb{P}(\{S_n = n\})}.\end{aligned} \tag{24.6}$$

Since we do not know p, we assume that p is the outcome of a random variable $P \sim \mathsf{U}[0,1]$ – this is the so-called »**Bayesian prior**« – since every $p \in [0,1]$ should be equally likely. So,

$$\mathbb{P}(S_n = k) = \mathbb{E}\left[\binom{n}{k} P^k (1-P)^{n-k}\right] = \int_0^1 \binom{n}{k} p^k (1-p)^{n-k}\,\mathrm{d}p = \frac{1}{n+1}, \tag{24.7}$$

(the result is indeed independent of k), and we get

$$\mathbb{P}(S_{n+1} = n+1 \mid S_n = n) = \frac{\frac{1}{n+2}}{\frac{1}{n+1}} = \frac{n+1}{n+2}.$$

This is Laplace's famous »**rule of succession**«.

One can evaluate (24.7) by »brute force«, or with the following »**Bayes's billiard table argument**«. On a billiard table $[0,1] \times [0,\ell]$ you roll at random a pink ball and note its x-coordinate $P(\omega) = p \in [0,1]$. After that you roll n further white balls – the x-coordinates are denoted by $W_1(\omega), \dots, W_n(\omega)$ – and you ask for the probability that k of the balls are to the left of $P(\omega) = p$. This probability is given by a binomial formula

$$\mathbb{P}(\#\{i \mid W_i \leqslant P\} = k \mid P = p) = \binom{n}{k} p^k (1-p)^{n-k}.$$

Since we roll P »at random«, $P \sim \mathsf{U}[0,1]$, and we get

$$\mathbb{P}(\#\{i \mid W_i \leqslant P\} = k) = \mathbb{E}\left[\binom{n}{k} P^k (1-P)^{n-k}\right] = \int_0^1 \binom{n}{k} p^k (1-p)^{n-k}\,\mathrm{d}p.$$

We can, however, work out this probability straight away: Throw $n+1$ white billiard balls onto the table and then select one at random and colour it pink. Clearly, every configuration is equally likely, so

$$\mathbb{P}(\#\{i \mid W_i \leqslant P\} = k) = \frac{1}{n+1}.$$

In passing, we have established the following relation between Euler's Gamma- and Beta functions:

$$\mathrm{B}(x,y) = \int_0^1 p^{x-1}(1-p)^{y-1}\,\mathrm{d}p = \frac{(x-1)!(y-1)!}{(x+y-1)!} = \frac{\Gamma(x)\Gamma(y)}{\Gamma(x+y)}, \quad x,y \in \mathbb{N};$$

with some work, this can be extended to any $x, y > 1$.

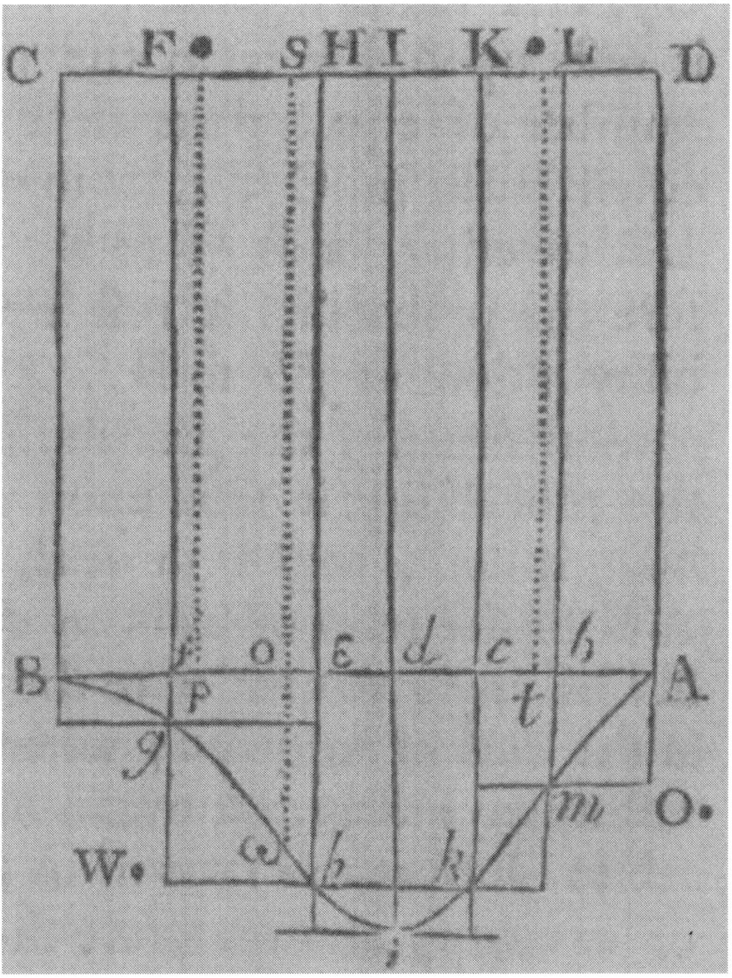

Fig. 24.4. Bayes's billiard table from the original paper [5, p. 385]. In modern coordinates, A = (0,0), B = (−1,0), and the base line $\overline{\mathrm{BA}}$ corresponds to the interval [0,1]; the pink ball is on the vertical line $\overline{os}$, $o = -p$. The graph below the line $\overline{\mathrm{BA}}$ shows the curve $\binom{n}{k}p^k(1-p)^{n-k}$.

VI
Independence

25 Independent events and random variables

Throughout this chapter, $(\Omega, \mathscr{A}, \mathbb{P})$ is an arbitrary probability space. Two events $A, B \in \mathscr{A}$ which do not influence each other (e.g. two consecutive rolls of the same die) are »independent«. On the other hand, two draws from a urn without replacement are »dependent«, since the second draw depends on the outcome of the first draw.

We can express this in a mathematically rigorous way in terms of conditional probabilities:

$$\begin{aligned} A, B \text{ independent} &\overset{\text{def}}{\iff} \mathbb{P}(A \mid B) = \mathbb{P}(A) \\ &\iff \mathbb{P}(A \cap B) = \mathbb{P}(A)\mathbb{P}(B). \end{aligned} \tag{25.1}$$

We use the latter equivalence for a more general definition.

25.1 Definition. Let I be any index set. The events $(A_i)_{i \in I} \subset \mathscr{A}$ are called **independent**, if

$$\mathbb{P}\Big(\bigcap_{i \in J} A_i\Big) = \prod_{i \in J} \mathbb{P}(A_i) \quad \forall J \subset I,\ |J| < \infty. \tag{25.2}$$

25.2 Remark. a) If $|I| = 2$, then (25.2) becomes (25.1).

b) We write $A \perp\!\!\!\perp B$ to indicate that A, B are independent.

c) Any sub-family $(A_i)_{i \in H} \subset (A_i)_{i \in I}$ of an independent family is independent.

d) Let $I = \{1, \dots, n\}$. The condition $\mathbb{P}\big(\bigcap_{i=1}^n A_i\big) = \prod_{i=1}^n \mathbb{P}(A_i)$ is **necessary** for the independence of $(A_i)_{i \in I}$.

⚠ If $n > 2$, it is **not sufficient** for independence. Here is the **standard counterexample:** $\Omega = \{1, 2, \dots, 6\}^2$, $\mathbb{P}(\{(i,k)\}) = 1/36$, $i, k \in \{1, \dots, 6\}$. Define the events

$$A = \{1, \dots, 6\} \times \{1, 2, 5\} \quad \text{and} \quad B = \{1, \dots, 6\} \times \{4, 5, 6\}$$

which have probabilities $\mathbb{P}(A) = \mathbb{P}(B) = \frac{1}{2}$, and the event

$$C = \{(i,k) \mid i+k=9\} = \{(3,6),(4,5),(5,4),(6,3)\}$$

which has probability $\mathbb{P}(C) = \frac{4}{36}$. Thus,

$$\begin{aligned}
\mathbb{P}(A \cap B) &= \frac{6}{36} \neq \mathbb{P}(A)\,\mathbb{P}(B),\\
\mathbb{P}(A \cap C) &= \frac{1}{36} \neq \mathbb{P}(A)\,\mathbb{P}(C),\\
\mathbb{P}(B \cap C) &= \frac{3}{36} \neq \mathbb{P}(B)\,\mathbb{P}(C),\\
\mathbb{P}(A \cap B \cap C) &= \frac{1}{36} = \mathbb{P}(A)\,\mathbb{P}(B)\,\mathbb{P}(C).
\end{aligned}$$

e) **Pairwise independence**, i.e.

$$\forall i,k \in I,\ i \neq k: \quad \mathbb{P}(A_i \cap A_k) = \mathbb{P}(A_i)\,\mathbb{P}(A_k)$$

is **necessary** for independence of the family $(A_i)_{i\in I}$.

⚠ If $n > 2$, pairwise independence is **not sufficient** for independence.
⚠ **Standard counterexample**: $\Omega = \{0,1\}^2$, $\mathbb{P}(\{(\omega_1,\omega_2)\}) = \frac{1}{4}$. Consider the events

$$A_1 = \{\omega_1 = 1\} = \left\{(\omega_1,\omega_2) \in \{0,1\}^2 \mid \omega_1 = 1\right\}$$
$$A_2 = \{\omega_2 = 1\} \quad \text{and} \quad A_3 = \{\omega_1 = \omega_2\}.$$

We have

$$\begin{aligned}
\mathbb{P}(A_i \cap A_k) &= \frac{1}{4} = \mathbb{P}(A_i)\,\mathbb{P}(A_k) \quad \forall i \neq k\\
\mathbb{P}(A_1 \cap A_2 \cap A_3) &= \frac{1}{4} \neq \frac{1}{8} = \mathbb{P}(A_1)\,\mathbb{P}(A_2)\,\mathbb{P}(A_3).
\end{aligned}$$

f) A further »real world« example showing that pairwise independence is different from independence is **Bernstein's die**: Paint three sides of a regular (hence, fair) tetrahedron in blue »B«, red »R« and white »W«, and mark the fourth side with all three colours. The result of a roll is the side where the tetrahedron comes to rest.

Since each colour appears on two sides, $\mathbb{P}(B) = \mathbb{P}(R) = \mathbb{P}(W) = \frac{1}{2}$. Since all colours appear together only on the fourth side, we have $\mathbb{P}(B \cap R) = \mathbb{P}(B \cap W) = \mathbb{P}(R \cap W) = \mathbb{P}(B \cap R \cap W) = \frac{1}{4}$. This shows that two colours are independent, but $\mathbb{P}(B \cap R \cap W) \neq \mathbb{P}(B)\mathbb{P}(R)\mathbb{P}(W)$.

The independence of both families of sets $\mathscr{F}_i \subset \mathscr{A}$ and of random variables $X_i : (\Omega, \mathscr{A}) \to (\mathbb{R}^{d_i}, \mathscr{B}(\mathbb{R}^{d_i}))$ can be reduced to Definition 25.1.

25.3 Definition. Let I be an arbitrary index set.

a) The families of sets $\mathscr{F}_i \subset \mathscr{A}$, $i \in I$, are called **independent**, if

$$\mathbb{P}\Big(\bigcap_{i\in J} A_i\Big) = \prod_{i\in J} \mathbb{P}(A_i) \quad \forall A_i \in \mathscr{F}_i,\ \forall J \subset I,\ |J| < \infty.$$

b) The random variables $X_i : (\Omega, \mathscr{A}) \to (\mathbb{R}^{d_i} \mathscr{B}(\mathbb{R}^{d_i}))$, $i \in I$, are independent, if the σ-algebas

$$\sigma(X_i) = X_i^{-1}(\mathscr{B}(\mathbb{R}^{d_i})) = \Big\{\{X_i \in F\} \mid F \in \mathscr{B}(\mathbb{R}^{d_i})\Big\}, \quad i \in I,$$

are independent.

💬 Definition 25.3.a) can be also stated as follows: $(\mathscr{F}_i)_{i\in I}$ are independent if all selections $(A_i)_{i\in I}$, $A_i \in \mathscr{F}_i$, are independent sets in the sense of Definition 25.1

If $A \in \mathscr{A}$, then $\mathbb{1}_A(\omega)$ is a random variable (and vice versa). From $\{\mathbb{1}_A = 1\} = A$ we see that $\sigma(\mathbb{1}_A) = \{\emptyset, A, A^c, \Omega\} = \sigma(A)$.

25.4 Lemma. *The following assertions are equivalent*

a) $A_1, \dots, A_n$ *are independent sets.*

b) $\sigma(A_1), \dots, \sigma(A_n)$ *are independent families of sets.*

c) $\mathbb{1}_{A_1}, \dots, \mathbb{1}_{A_n}$ *are independent random variables.*

Proof. The equivalence b)⇔c) is just the definition of the equivalence of random variables. The direction b)⇒a) is trivial. For the converse direction a)⇒b) we have to show that

$$A_1, A_2, \dots, A_n \text{ independent} \implies \begin{cases} B_1, B_2, \dots, B_n \text{ independent} \\ \text{for any } B_i \in \{\emptyset, A_i, A_i^c, \Omega\}. \end{cases}$$

Since we can repeat the following argument over and over again, it is enough to show that

$$A_1, A_2, \dots, A_n \text{ independent} \implies B_1, A_2, \dots, A_n \text{ indpendent}$$

for any $B_1 \in \{\emptyset, A_1, A_1^c, \Omega\}$. If $B_1 = \emptyset, A_1, \Omega$, there is nothing to show. Let us check (25.2) for $B_1 = A_1^c$.

Case 1: If $1 \notin J$, then (25.2) is trivial.

Case 2: If $1 \in J$, then we set $A := \bigcap_{i\in J, i\neq 1} A_i$.

$$\begin{aligned}\mathbb{P}(\underbrace{A_1^c \cap A}_{=A\setminus(A_1\cap A)}) &= \mathbb{P}(A) - \mathbb{P}(A_1 \cap A) = \prod_{1\neq i\in J} \mathbb{P}(A_i) - \mathbb{P}(A_1)\prod_{1\neq i\in J} \mathbb{P}(A_i) \\ &= (1 - \mathbb{P}(A_1))\prod_{1\neq i\in J} \mathbb{P}(A_i) = \mathbb{P}(A_1^c)\prod_{1\neq i\in J} \mathbb{P}(A_i).\end{aligned}$$

□

Let $\mathscr{F}_i \subset \mathscr{A}$, $i \in I$, be families of sets. Then ✍

$$\mathscr{F}_i,\ i \in I, \text{ independent} \iff \mathscr{F}_i \cup \{\Omega\},\ i \in I, \text{ independent}.$$

The direction »⇐« is obvious, »⇒« follows with a direct calculation.

25.5 Theorem. *Assume that $\mathscr{F}_i$, $i \in I$, are $\cap$-stable families (i.e. $F \cap G \in \mathscr{F}_i$ for all $F, G \in \mathscr{F}_i$). Then one has*

$$\mathscr{F}_i,\ i \in I \quad \textit{independent} \iff \sigma(\mathscr{F}_i),\ i \in I \quad \textit{independent}.$$

Proof. Without loss of generality, we may assume that $\Omega \in \mathscr{F}_i$ for all $i \in I$ (see the discussion before the theorem) and $I = \{i_1, \dots, i_n\}$ is any finite set (since independence is defined via finite families).

The direction »$\Leftarrow$« is clear, since we always can reduce the families without destroying independence. The converse is proved in several steps.

1° Fix arbitrary $F_i \in \mathscr{F}_i$, $i = i_2, i_3, \dots, i_n$, and define for any $F \in \sigma(\mathscr{F}_{i_1})$:

$$\mu(F) := \mathbb{P}\left(F \cap F_{i_2} \cap \dots \cap F_{i_n}\right),$$
$$\nu(F) := \mathbb{P}(F)\,\mathbb{P}\left(F_{i_2}\right) \cdots \mathbb{P}\left(F_{i_n}\right).$$

Obviously, μ and ν are finite measures on $\sigma(\mathscr{F}_{i_1})$.

2° As the families $\mathscr{F}_i$ are independent, we get $\mu|_{\mathscr{F}_{i_1}} = \nu|_{\mathscr{F}_{i_1}}$. The uniqueness theorem for measures (Theorem 4.7) shows that $\mu|_{\sigma(\mathscr{F}_{i_1})} = \nu|_{\sigma(\mathscr{F}_{i_1})}$, and thus

$$\mathbb{P}\left(F \cap F_{i_2} \cap \dots \cap F_{i_n}\right) = \mathbb{P}(F)\,\mathbb{P}\left(F_{i_2}\right) \cdots \mathbb{P}\left(F_{i_n}\right)$$
$$\forall F \in \sigma(\mathscr{F}_{i_1}),\ F_i \in \mathscr{F}_i,\ i = i_2, \dots, i_n.$$

Therefore, the families $\sigma(\mathscr{F}_{i_1}), \mathscr{F}_{i_2}, \dots, \mathscr{F}_{i_n}$ are independent: Here we use that $\Omega \in \mathscr{F}_i$ $\forall i$ entails that this product formula is also valid for any smaller index set $J \subset \{i_1, \dots, i_n\}$, i.e. we have verified Definition 25.3.a) with $I = \{i_1, \dots, i_n\}$.

3° Now we apply 1° and 2° to $\mathscr{F}_{i_2}, \mathscr{F}_{i_3}, \dots, \mathscr{F}_{i_n}, \sigma(\mathscr{F}_{i_1})$. This gives that

$$\sigma(\mathscr{F}_{i_2}), \mathscr{F}_{i_3}, \dots, \mathscr{F}_{i_n}, \sigma(\mathscr{F}_{i_1}) \text{ are independent,}$$

and we may iterate Step 3° to get, that all $\sigma(\mathscr{F}_i)$, $i \in I$, are independent.

4° Since the result of Step 3° holds for **any** finite index set I, and since independence of infinitely many families are defined via finite index sets, we have finally shown that $\sigma(\mathscr{F}_i)$, $i \in I$, are independent no matter what index set I we are considering. □

25.6 Corollary (first block lemma). *Let $\mathscr{F}_{ik} \subset \mathscr{A}$, $1 \leqslant i \leqslant m$, $1 \leqslant k \leqslant n(i)$ be independent $\cap$-stable families. The σ-algebras generated by the families in the ith row*

$$\mathscr{G}_i := \sigma(\mathscr{F}_{i1}, \dots, \mathscr{F}_{in(i)}), \quad 1 \leqslant i \leqslant m,$$

are again independent.

Proof. Without loss of generality we can assume that $\Omega \in \mathscr{F}_{ik}$ for all i, k. The following families of sets

$$\mathscr{F}_i^{\cap} := \left\{F_{i1} \cap \dots \cap F_{in(i)} \mid F_{ik} \in \mathscr{F}_{ik},\ 1 \leqslant k \leqslant n(i)\right\}, \quad 1 \leqslant i \leqslant m$$

are i) $\cap$-stable, ii) independent ✍ and iii) satisfy $\mathscr{F}_{ik} \subset \mathscr{F}_i^{\cap}$ for all k. Therefore, Theorem 25.5 shows that the families $\sigma\left(\mathscr{F}_i^{\cap}\right)$, $1 \leqslant i \leqslant m$, are independent. By construction,

$$\mathscr{F}_{i1},\dots,\mathscr{F}_{in(i)} \subset \mathscr{F}_i^{\cap} \subset \sigma(\mathscr{F}_{i1},\dots,\mathscr{F}_{in(i)}) \stackrel{\text{def}}{=} \mathscr{G}_i,$$

and if we apply the $\sigma(\dots)$-operation in this chain of inclusions, we get $\mathscr{G}_i \stackrel{\text{def}}{=} \sigma(\mathscr{F}_{i1},\dots,\mathscr{F}_{in(i)}) \subset \sigma(\mathscr{F}_i^{\cap}) \subset \mathscr{G}_i$; thus, the families $\mathscr{G}_i = \sigma(\mathscr{F}_i^{\cap})$ are independent. □

25.7 Corollary (second block lemma). *Let $X_{ik} : \Omega \to \mathbb{R}^d$, $1 \leqslant i \leqslant m$, $1 \leqslant k \leqslant n(i)$, be independent random variables and $f_i : \mathbb{R}^{d\times n(i)} \to \mathbb{R}$ Borel measurable functions. Then the random variables $f_i(X_{i1},\dots,X_{in(i)})$, $1 \leqslant i \leqslant m$, are independent.*

Proof. Set $\mathscr{F}_{ik} := \sigma(X_{ik})$ and $\mathscr{G}_i := \sigma(\mathscr{F}_{i1},\dots,\mathscr{F}_{in(i)})$. Each random variable

$$Z_i := f_i(X_{i1},\dots,X_{in(i)})$$

is $\mathscr{G}_i$-measurable, i.e. $\sigma(Z_i) \subset \mathscr{G}_i$. Since the $(\mathscr{G}_i)_{1\leqslant i\leqslant m}$ are independent (Corollary 25.6), so are the families $\sigma(Z_i)$ and the rv Z_i. □

25.8 Remark. Here are two typical applications of the Corollaries 25.6 and 25.7: Assume that $X_1,\dots,X_n$ are independent real random variables.

a) X_1 and $Y := X_2 \cdot X_3 \cdot \ldots \cdot X_n$ are independent
b) $S_n - S_m$ and $\mathbb{1}_{\{\max_{1\leqslant j\leqslant m} S_j > \alpha\}}$ are independent. ($S_k := X_1 + \dots + X_k$).

The following theorem contains a characterisation of independence based on the probability laws.

25.9 Theorem. *Let $X_1,\dots,X_n : \Omega \to \mathbb{R}^d$ be random variables. The following assertions are equivalent*

a) *$X_1,\dots,X_n$ are independent;*
b) $\mathbb{P}(X_1 \in A_1,\dots,X_n \in A_n) = \prod_{k=1}^n \mathbb{P}(X_k \in A_k) \quad \forall A_1,\dots,A_n \in \mathscr{B}(\mathbb{R}^d)$;
c) $\mathbb{P}(X_1 \in Q_1,\dots,X_n \in Q_n) = \prod_{k=1}^n \mathbb{P}(X_k \in Q_k) \quad \forall Q_1,\dots,Q_n \subset \mathbb{R}^d$ *half-open rectangles;*
d) $\mathbb{P}_{X_1,\dots,X_n} = \mathbb{P}_{X_1} \otimes \dots \otimes \mathbb{P}_{X_n}$;
e) $\mathbb{E}e^{i\sum_{k=1}^n \langle \xi_k, X_k\rangle} = \prod_{k=1}^n \mathbb{E}e^{i\langle \xi_k, X_k\rangle} \quad \forall \xi_1,\dots,\xi_n \in \mathbb{R}^d$.

💬 The law $\mathbb{P}_{X_1,\dots,X_n}$ of the vector $(X_1,\dots,X_n)$ is the **joint distribution**;
💬 $\mathbb{P}_{X_1} \otimes \dots \otimes \mathbb{P}_{X_n}$ is the product measure as in Chapter 15;
💬 The equivalence 25.9.a)⇔e) is often called **Kac's theorem**.

Proof of Theorem 25.9. a)⇒b) Let $A_1, \dots, A_n \in \mathscr{B}(\mathbb{R}^d)$ be arbitrary sets. We have

$$\begin{aligned}\mathbb{P}_{X_1,\dots,X_n}(A_1 \times \dots \times A_n) &\stackrel{\text{def}}{=} \mathbb{P}(X_1 \in A_1, \dots, X_n \in A_n) \stackrel{\text{def}}{=} \mathbb{P}\Big(\bigcap_{i=1}^n \{X_i \in A_i\}\Big) \\ &\stackrel{\perp\!\!\!\perp}{=} \prod_{i=1}^n \mathbb{P}(\{X_i \in A_i\}) = \prod_{i=1}^n \mathbb{P}_{X_i}(A_i) \\ &= \bigotimes_{i=1}^n \mathbb{P}_{X_i}(A_1 \times \dots \times A_n).\end{aligned}$$

b)⇒c) is obvious.

c)⇒d) Write $\mathscr{I}(\mathbb{R}^d)$ for the half-open rectangles. Clearly, $Q_1 \times \dots \times Q_n$ is again a half-open rectangle in $\mathscr{I}(\mathbb{R}^{nd})$, and the latter is a $\cap$-stable generator of the Borel sets $\mathscr{B}(\mathbb{R}^{nd})$. By assumption, the claimed identity holds on $\mathscr{I}(\mathbb{R}^{nd})$; with the help of the uniqueness theorem for measures (Theorem 4.7), we conclude that $\mathbb{P}_{X_1,\dots,X_n} = \bigotimes_{i=1}^n \mathbb{P}_{X_i}$ on $\mathscr{B}(\mathbb{R}^{nd}) = \mathscr{B}(\mathbb{R}^d)^{\otimes n}$.

d)⇒a) Let $J \subset \{1, \dots, n\}$ and define

$$A_i := \begin{cases} \text{any set from} \in \mathscr{B}(\mathbb{R}^d), & i \in J \\ \mathbb{R}^d, & i \notin J \end{cases}.$$

Then we have

$$\begin{array}{ccc} \mathbb{P}(X_i \in A_i, \ \forall i = 1, \dots, n) & = & \prod_{i=1}^n \mathbb{P}(X_i \in A_i) \\ \| & & \| \\ \mathbb{P}(X_i \in A_i, \ \forall i \in J) & & \prod_{i \in J} \mathbb{P}(X_i \in A_i). \end{array}$$

d)⇒e) We have for any $\xi_1, \dots, \xi_n \in \mathbb{R}^d$

$$\begin{aligned}\mathbb{E} e^{i \sum_{k=1}^n \langle \xi_k, X_k \rangle} &= \int_{\mathbb{R}^{nd}} e^{i\langle (\xi_1,\dots,\xi_n), (x_1,\dots,x_n)\rangle} \, \mathbb{P}_{X_1,\dots,X_n}(dx_1, \dots, dx_n) \\ &= \int_{\mathbb{R}^{nd}} e^{i\langle (\xi_1,\dots,\xi_n), (x_1,\dots,x_n)\rangle} \, \mathbb{P}_{X_1} \otimes \dots \otimes \mathbb{P}_{X_n}(dx_1, \dots, dx_n) \\ &= \int \!\cdots\! \int_{\mathbb{R}^d \times \dots \times \mathbb{R}^d} \prod_{k=1}^n e^{i\langle \xi_k, x_k \rangle} \, \mathbb{P}_{X_1}(dx_1) \dots \mathbb{P}_{X_n}(dx_n) \\ &\stackrel{\text{Fubini}}{=} \prod_{k=1}^n \mathbb{E} e^{i\langle \xi_k, X_k \rangle}.\end{aligned}$$

Finally, the direction e)⇒d) will be shown later in Corollary 27.8. □

The next result shows that we can treat expected values of independent random variables separately.

25.10 Corollary. *Let* $X, Y : \Omega \to \mathbb{R}^d$ *be independent random variables and let* $h : \mathbb{R}^{2d} \to \mathbb{R}$ *be a Borel function. If* $h \geqslant 0$ *or if* $h(X,Y) \in L^1(\mathbb{P})$*, then*

$$\begin{aligned}\mathbb{E}h(X,Y) &= \iint h(x,y)\,\mathbb{P}(X \in dx)\,\mathbb{P}(Y \in dy)\\ &= \mathbb{E}\int h(x,Y)\,\mathbb{P}(X \in dx)\\ &= \mathbb{E}\int h(X,y)\,\mathbb{P}(Y \in dy).\end{aligned}$$

Proof. Since $h(X,Y)$ is measurable, we see that

$$\begin{aligned}\mathbb{E}h(X,Y) &= \int_\Omega h(X,Y)\,d\mathbb{P}\\ &\overset{\S 17.1}{=} \int_{\mathbb{R}^{2d}} h(x,y)\,\mathbb{P}(X \in dx, Y \in dy)\\ &\overset{\S 17.1}{=} \int_{\mathbb{R}^{2d}} h(x,y)\,\mathbb{P}((X,Y) \in d(x,y))\\ &\overset{\S 25.9}{=} \int_{\mathbb{R}^{2d}} h(x,y)\,\mathbb{P}_X \otimes \mathbb{P}_Y(dx, dy)\\ &\underset{\text{Fubini}}{\overset{\text{Tonelli}}{=}} \int_{\mathbb{R}^d}\int_{\mathbb{R}^d} h(x,y)\,\mathbb{P}_X(dx)\,\mathbb{P}_Y(dy)\\ &\overset{\S 17.1}{=} \int_\Omega\int_{\mathbb{R}^d} h(x,Y)\,\mathbb{P}_X(dx)\,d\mathbb{P}\\ &= \mathbb{E}\int_{\mathbb{R}^d} h(x,Y)\,\mathbb{P}_X(dx).\end{aligned}$$

The remaining equality is shown in the same way. □

25.11 Corollary. *Let* $X, Y : \Omega \to \mathbb{R}^d$ *be independent random variables and let* $f, g : \mathbb{R}^d \to \mathbb{R}$ *be Borel functions. If* $f, g \geqslant 0$ *or if* $\mathbb{E}|f(X)|, \mathbb{E}|g(Y)| < \infty$*, then*

$$\mathbb{E}(f(X)g(Y)) = \mathbb{E}f(X)\,\mathbb{E}g(Y). \tag{25.3}$$

In particular,

$$X \perp\!\!\!\perp Y \;\&\; f(X) \in L^1(\mathbb{P}),\ g(Y) \in L^1(\mathbb{P}) \implies f(X)g(Y) \in L^1(\mathbb{P}).$$

Proof. Define $h(x,y) := f(x)g(y)$. If $f, g \geqslant 0$, then (25.3) follows directly from 25.10.

If $f, g : \mathbb{R} \to \mathbb{R}$ have arbitrary sign and $f(X), g(Y) \in L^1$, the first part of the proof shows

$$\mathbb{E}|f(X)g(Y)| = \mathbb{E}|f(X)|\,\mathbb{E}|g(Y)| < \infty$$

and so $f(X)g(Y) \in L^1$; (25.3) follows again from Corollary 25.10. □

Since we know from Remark 25.8 that

$$X_1, X_2, \dots, X_n \text{ independent} \implies X_1,\ Y := X_2 \cdot \ldots \cdot X_n \text{ independent},$$

the following result follows by iteration of Corollary 25.11.

25.12 Corollary. *Let $X_1, \dots, X_n : \Omega \to \mathbb{R}^d$ be independent random variables and $f_1, \dots, f_n : \mathbb{R}^d \to \mathbb{R}$ Borel functions. If all $f_i \geqslant 0$ or $f_i(X_i) \in L^1(\mathbb{P})$, then*

$$\mathbb{E}\left(\prod_{i=1}^n f_i(X_i)\right) = \prod_{i=1}^n \mathbb{E} f_i(X_i);$$

in particular, $f_i(X_i) \in L^1(\mathbb{P})\ \forall i \implies \prod_{i=1}^n f_i(X_i) \in L^1(\mathbb{P})$.

25.13 Example (independent ≠ uncorrelated). Two real random variables X, Y are called **uncorrelated**, if $\mathbb{E}(XY) = \mathbb{E}X\,\mathbb{E}Y$. This is necessary if $X \perp\!\!\!\perp Y$, but it is ⚠ **not sufficient**. Here is the **standard counterexample**: Consider rv X, Y with the following joint distribution

	$Y = 1$	$Y = 0$	$Y = -1$	$\mathbb{P}(X = \bullet)\ \Downarrow$
$X = 1$	0	a	0	$0 + a + 0$
$X = 0$	b	c	b	$b + c + a$
$X = -1$	0	a	0	$0 + a + 0$
$\mathbb{P}(Y = \bullet) \Rightarrow$	$0 + b + 0$	$a + c + a$	$0 + b + 0$	

Table 25.1. The table shows the joint distributions $\mathbb{P}(X = i, Y = k)$. On the right and bottom margins we can read off the distributions of X and Y, respectively.

We choose the parameters in such a way that

$$a, b > 0, \quad c \geqslant 0, \quad 2a + 2b + c = 1.$$

By construction, we have $X \cdot Y = 0$ and, by symmetry, $\mathbb{E}X = \mathbb{E}Y = 0$. Thus, $\mathbb{E}(XY) = 0 = \mathbb{E}X\,\mathbb{E}Y$. On the other hand,

$$\mathbb{P}(Y = 1, X = 1) = 0 \neq ab = \mathbb{P}(X = 1)\,\mathbb{P}(Y = 1),$$

i.e. X and Y are uncorrelated but not independent.

The law of a sum of independent random variables is given by the convolution (Definition 17.7).

25.14 Theorem. *Let $X, Y : \Omega \to \mathbb{R}^d$ be independent random variables.*

$$\mathbb{P}_{X+Y} = \mathbb{P}_X * \mathbb{P}_Y.$$

If $X \sim f(x)\,dx$ *and* $Y \sim g(y)\,dy$*, then* $X + Y \sim \int_{\mathbb{R}^d} f(z-y)g(y)\,dy\,dz$.

Proof. For any $B \in \mathscr{B}(\mathbb{R}^d)$ we have

$$\begin{aligned}\mathbb{P}(X+Y \in B) &= \int \underbrace{\mathbb{1}_B(X+Y)}_{=h(X,Y)}\,d\mathbb{P} \\ &\overset{25.10}{=} \iint \mathbb{1}_B(x+y)\,\mathbb{P}_X(dx)\mathbb{P}_Y(dy) \\ &\overset{17.7}{=} \int \mathbb{1}_B(z)\,\mathbb{P}_X * \mathbb{P}_Y(dz).\end{aligned}$$

The assertion on the densities follows from Remark 17.8. □

25.15 Example. We have the following identities, cf. Tables 22.1 and 23.1 for the notation.

- $\mathsf{N}(\mu,\sigma^2) * \mathsf{N}(m,s^2) = \mathsf{N}(\mu+m,\sigma^2+s^2)$ (normal law)
- $\gamma_{\alpha,\beta} * \gamma_{A,\beta} = \gamma_{\alpha+A,\beta}$ (Gamma law)
- $\mathsf{Bin}(m,p,\cdot) * \mathsf{Bin}(n,p,\cdot) = \mathsf{Bin}(m+n,p,\cdot)$ (Binomial law)
- $\mathsf{Poi}(\lambda) * \mathsf{Poi}(\mu) = \mathsf{Poi}(\lambda+\mu)$ (Poisson's law)

It is pretty tedious to check these identities by hand ✍. We will soon have characteristic functions at our disposal (Chapter 27 or Theorem 25.9.e)) which are a much better tool for this task.

25.16 Definition. Let $X, Y \in L^2(\mathbb{P})$ be real random variables. We call

$$\mathbb{V}X = \mathbb{E}\big((X - \mathbb{E}X)^2\big) \qquad \textbf{variance,}$$

$$\mathrm{Cov}(X,Y) = \mathbb{E}\big((X-\mathbb{E}X)(Y-\mathbb{E}Y)\big) \qquad \textbf{covariance.}$$

Let us collect some basic properties of the (co-)variance.

25.17 Lemma. *Let* $X, Y, Z \in L^2(\mathbb{P})$ *be real random variables. Then we have*

a) $\mathbb{V}X = \mathbb{E}(X^2) - (\mathbb{E}X)^2 \in [0,\infty)$;

b) $\mathbb{V}X \leqslant \mathbb{E}((X-a)^2)$, *i.e.* $a = \mathbb{E}X$ *is a minimiser of* $a \mapsto \mathbb{E}((X-a)^2)$.

c) $\mathbb{V}(aX+b) = a^2\mathbb{V}X$, $a, b \in \mathbb{R}$.

d) $\mathrm{Cov}(X,Y) = \mathbb{E}(XY) - \mathbb{E}X\mathbb{E}Y = \mathrm{Cov}(Y,X)$ *and* $\mathrm{Cov}(X,X) = \mathbb{V}X$.

e) $\mathrm{Cov}(aX+bY,Z) = a\,\mathrm{Cov}(X,Z) + b\,\mathrm{Cov}(Y,Z)$, $a, b \in \mathbb{R}$.

f) $\mathbb{V}(X+Y) - \mathbb{V}(X-Y) = 4\,\mathrm{Cov}(X,Y)$.

Proof. Since $\mathbb{E}|X| = \mathbb{E}(1\cdot|x|) \leqslant (\mathbb{E}(1)^2)^{1/2}(\mathbb{E}(X^2))^{1/2} < \infty$ etc., all expressions in the statement are finite. Therefore,

a), b): $(X-a)^2 = X^2 - 2aX + a^2$, and so

$$\mathbb{E}\big((X-a)^2\big) = \mathbb{E}\big(X^2\big) - 2a\mathbb{E}X + a^2.$$

Setting $a = \mathbb{E}X$, we get a), and b) follows if we minimize in a.

c): We have

$$aX + b - \mathbb{E}(aX + b) = aX - a\mathbb{E}X = a(X - \mathbb{E}X).$$

The claim follows if we take squares and then expectations.

d), e): Since $X, Y \in L^2(\mathbb{P}) \implies XY \in L^1(\mathbb{P})$, the covariance is well-defined. So,

$$(X - \mathbb{E}X)(Y - \mathbb{E}Y) = XY - X(\mathbb{E}Y) - Y(\mathbb{E}X) + (\mathbb{E}X)(\mathbb{E}Y),$$

and the claim follows if we take expectations on both sides.

The symmetry $\operatorname{Cov}(X, Y) = \operatorname{Cov}(Y, X)$ and linearity of $X \mapsto \operatorname{Cov}(X, Z)$ are clear.

f) follows if we check a) by brute force. A somewhat more intelligent approach is to use the bilinearity of the covariance and the familiar formulas for scalar products to get this polarization result. □

25.18 Theorem (Bienaymé). *Let $X_1, \dots, X_n : \Omega \to \mathbb{R}$ be pairwise independent, square integrable (i.e. $L^2(\mathbb{P})$) random variables. One has*

$$\mathbb{V}\left(\sum_{i=1}^n X_i\right) = \sum_{i=1}^n \mathbb{V}X_i.$$

Proof. Since $X \mapsto \operatorname{Cov}(X, Y)$ is linear and symmetric, the rules for scalar products also hold for the covariance. So,

$$\begin{aligned}\mathbb{V}(X_1 + \dots + X_n) &= \operatorname{Cov}(X_1 + \dots + X_n, X_1 + \dots + X_n)\\ &= \sum_{i=1}^n \sum_{k=1}^n \operatorname{Cov}(X_i, X_k) = \sum_{i=1}^n \operatorname{Cov}(X_i, X_i) = \sum_{i=1}^n \mathbb{V}X_i.\end{aligned}$$

In the penultimate equality we use pairwise independence which implies that $\operatorname{Cov}(X_i, X_k) = \mathbb{E}(X_i X_k) - \mathbb{E}(X_i)\mathbb{E}(X_k) = 0$ if $i \neq k$. □

💬 The proof of 25.18 remains valid, if we only assume that the X_i are **pairwise uncorrelated**, i.e. if $\operatorname{Cov}(X_i, X_k) = 0$ for $i \neq k$.

26 Construction of (independent) random variables

In this chapter we want to address the following three questions.

Problem 1: Is there for every probability measure μ on $(\mathbb{R}^d, \mathscr{B}(\mathbb{R}^d))$ a random variable such that $X \sim \mu$?

Problem 2: Let $X, Y : \Omega \to \mathbb{R}^d$ be two random variables. Can we »wlog« assume that $X \perp\!\!\!\perp Y$? **More precisely:** Assume that on $(\mathbb{R}^d, \mathscr{B}(\mathbb{R}^d))$ there are two probability measures μ_1, μ_2.

Is there **a single** probability space $(\Omega, \mathscr{A}, \mathbb{P})$ carrying two random variables X_1, X_2 such that $X_i \sim \mu_i$ and $X_1 \perp\!\!\!\perp X_2$?

Problem 3: Let $X_i : \Omega \to \mathbb{R}^d$, $i \in I$, be arbitrarily many random variables. Can we »wlog« assume that the rvs are independent (using, again, the **precise** wording as in Problem 2)?

Problem 1

The real problem is not so much the construction of the random variable, but the construction of a suitable probability space $(\Omega, \mathscr{A}, \mathbb{P})$.

Solution I: By assumption, μ is a probability measure on the measurable space $(\mathbb{R}^d, \mathscr{B}(\mathbb{R}^d))$. Therefore, $(\Omega, \mathscr{A}, \mathbb{P}) := (\mathbb{R}^d, \mathscr{B}(\mathbb{R}^d), \mu)$ is a probability space. For the measurable map $X : \underbrace{\Omega}_{=\mathbb{R}^d} \to \mathbb{R}^d$, $x \mapsto X(\omega) = \mathrm{id}(\omega) = \omega$, we have

$$\forall B \in \mathscr{B}(\mathbb{R}^d) : \quad \mathbb{P}(X \in B) = \mu(\mathrm{id}^{-1}(B)) = \mu(B) \implies X \sim \mu.$$

You may feel a bit uneasy about this solution which looks like a cheap trick, but it is a valid answer to our question and it nicely illustrates the fact that we have to construct both the probability space and the random variable. Nevertheless, we add a further solution.

Solution II: We restrict ourselves to $d = 1$, the case $d > 1$ is dealt with in the appendix. This solution can also be used for the numerical simulation of random variables.

Recall that $F(t) := \mu(-\infty, t]$ is the distribution function of the (probability) measure μ; we have seen in Lemma 23.2 that F is i) monotone increasing, ii) $F(-\infty) = 0$, $F(+\infty) = 1$ and iii) right-continuous. We will call every function enjoying the properties i)–iii) a distribution function. Our aim is to show that such an F is, in fact, the distribution function of a random variable.

We need the following auxiliary result which is interesting on its own.

26.1 Lemma. *Every distribution function* $F : \mathbb{R} \to [0,1]$ *has a unique increasing and right-continuous generalized inverse* $F^{-1} : [0,1] \to \mathbb{R}$

$$F^{-1}(s) = G(s) = \inf\{t \mid F(t) > s\}. \tag{26.1}$$

For all $s \in [0,1]$ *it holds that* $F(G(s)) \geqslant s$, *and if* $F(t)$ *is continuous at* $t = G(s)$, *then* $F(G(s)) = s$.

Proof. Intuitive argument: At the points where F is strictly increasing, there is nothing to show – see also Fig. 26.1. The remaining problem cases, where F has a jump or a flat stretch, are shown in the Figure:

Formal argument: The very definition of G shows that G is monotone increasing. Moreover,

$$\{t \mid F(t) > s\} = \bigcup_{\epsilon>0} \{t \mid F(t) > s + \epsilon\} \Rightarrow \inf\{t \mid F(t) > s\} = \inf_{\epsilon>0} \inf\{t \mid F(t) > s + \epsilon\}$$

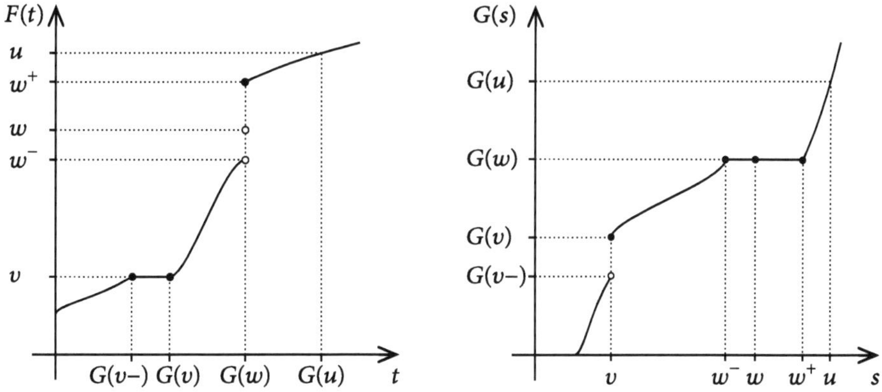

Fig. 26.1. If $F(t)$ has a jump discontinuity, then $t = G(w) = G(w^+) = G(w^-)$; if $F(t)$ flat, say on $[G(v-), G(v)]$, then we take the right end-point of the interval of constancy as value for G.

which shows that G is right-continuous. Again from the definition of G we get

$$\forall t > G(s): \quad F(t) \geqslant s \xRightarrow[\text{F right cts.}]{t \downarrow G(s)} F(G(s)) \geqslant s.$$

Finally, if $t = G(s)$ is a continuity point of F, then the definition of G as an infimum shows

$$\forall \epsilon > 0 \quad F(G(s) - \epsilon) \leqslant s \xRightarrow[\text{G(s) is cty. point of F}]{\epsilon \downarrow 0} F(G(s) \leqslant s,$$

and, together with the previous inequality, we see that $F(G(s)) = s$. □

26.2 Lemma. *For every distribution function* $F : \mathbb{R} \to [0,1]$ *there is a probability space* $(\Omega, \mathscr{A}, \mathbb{P})$ *and a random variable* $X : \Omega \to \mathbb{R}$ *such that* $\mathbb{P}(X \leqslant t) = F(t)$.

Proof. Consider the probability space $(\Omega, \mathscr{A}, \mathbb{P}) = ([0,1], \mathscr{B}[0,1], \lambda)$ and define $X = G$ where $G = F^{-1}$ is the generalized inverse from Lemma 26.1. The function G is monotone, hence measurable. Since F has at most countable many discontinuities, we see from Lemma 26.1 that the sets $\{s \mid G(s) \leqslant t\}$ and $\{s \mid s \leqslant F(t)\}$ differ by at most a countable set, and so

$$\begin{aligned}\mathbb{P}(\{\omega \in \Omega \mid X(\omega) \leqslant t\} &= \lambda(\{s \in [0,1] \mid G(s) \leqslant t\}) \\ &= \lambda(\{s \in [0,1] \mid s \leqslant F(t)\}) \\ &= \lambda([0, F(t)]) = F(t).\end{aligned}$$

□

Problem 2

Construct, as in Problem 1, probability spaces $(\Omega_i, \mathscr{A}_i, \mathbb{P}_i)$ and random variables $X_i \sim \mu_i$, $i = 1, 2$, and make a standard product construction.

$$(\Omega, \mathscr{A}, \mathbb{P}) := (\Omega_1 \times \Omega_2, \mathscr{A}_1 \otimes \mathscr{A}_2, \mathbb{P}_1 \otimes \mathbb{P}_2)$$

$$\begin{aligned} \widetilde{X}_1 : \Omega_1 \times \Omega_2 &\to \mathbb{R}^d & \qquad \widetilde{X}_2 : \Omega_1 \times \Omega_2 &\to \mathbb{R}^d \\ (\omega_1, \omega_2) &\mapsto X_1(\omega_1) & (\omega_1, \omega_2) &\mapsto X_2(\omega_2) \end{aligned}$$

We claim that

1° $\widetilde{X}_1 \sim X_1$, that is $\mathbb{P}_{\widetilde{X}_1} = (\mathbb{P}_1)_{X_1}$. Indeed, if $B \in \mathscr{B}(\mathbb{R}^d)$, then

$$\widetilde{X}_1^{-1}(B) = X_1^{-1}(B) \times \Omega_2 \in \mathscr{A}_1 \times \mathscr{A}_2 \subset \mathscr{A}_1 \otimes \mathscr{A}_2,$$

which means that $\widetilde{X}_1$ is $\mathscr{A}_1 \otimes \mathscr{A}_2/\mathscr{B}(\mathbb{R}^d)$-measurable, i.e. a random variable. Moreover,

$$\begin{aligned} \mathbb{P}(\widetilde{X}_1 \in B) = \mathbb{P}(\widetilde{X}_1^{-1}(B)) = \mathbb{P}(X_1^{-1}(B) \times \Omega_2) &= \mathbb{P}_1(X_1^{-1}(B)) \cdot \mathbb{P}_2(\Omega_2) \\ &= \mathbb{P}_1(X_1 \in B); \end{aligned}$$

a similar calculation works for $\widetilde{X}_2$.

2° $\widetilde{X}_1 \perp\!\!\!\perp \widetilde{X}_2$. In order to see this, pick any $A, B \in \mathscr{B}(\mathbb{R}^d)$. We have

$$\begin{aligned} \mathbb{P}(\widetilde{X}_1 \in A, \widetilde{X}_2 \in B) &= \mathbb{P}\left([X_1^{-1}(A) \times \Omega_2] \cap [\Omega_1 \times X_2^{-1}(B)]\right) \\ &= \mathbb{P}\left(X_1^{-1}(A) \times X_2^{-1}(B)\right) \\ &= \mathbb{P}_1\left(X_1^{-1}(A)\right) \cdot \mathbb{P}\left(X_2^{-1}(B)\right) \\ &= \mathbb{P}(X_1 \in A) \cdot \mathbb{P}(X_2 \in B). \end{aligned}$$

This standard construction applies, without problem, to **finitely many** random variables. In fact, the basic principle stays the same even in the case of arbitrarily many random variables, but this is exactly...

...Problem 3

We need to recall some notation from Chapter 19 on infinite products. Let I be an arbitrary index set and $(\Omega_i, \mathscr{A}_i, \mathbb{P}_i)$, $i \in I$, probability spaces. We denote by

$$(\Omega, \mathscr{A}, \mathbb{P}) = \left(\bigtimes_{i \in I} \Omega_i, \sigma(\pi_i, i \in I), \bigotimes_{i \in I} \mathbb{P}_i \right)$$

the (infinite) product space where $\pi_i : \Omega \to \mathbb{R}$, $\pi_i(\omega) := \omega(i)$, is the ith coordinate projection.

In order to solve **Problem 3**, we define – similar to the construction in Problem 2 – new random variables

$$\widetilde{X}_i : \Omega \to \mathbb{R}^d, \quad \widetilde{X}_i(\omega) := X_i \circ \pi_i(\omega) = X_i(\omega_i), \quad i \in I.$$

Exactly as in Problem 2 we see that $\widetilde{X}_i \sim X_i$. Moreover, for any finite index set $H \subset I$

$$\bigcap_{i\in H}\left\{\omega : \widetilde{X}_i(\omega)\in B_i\right\} = \mathop{\times}_{i\in H}\left\{\omega_i : X_i(\omega_i)\in B_i\right\}\times\mathop{\times}_{i\notin H}\Omega_i$$

(modulo some rearrangement, see our comments at the beginning of the proof of Theorem 19.4), and so

$$\begin{aligned}\mathbb{P}\Big(\bigcap_{i\in H}\left\{\omega : \widetilde{X}_i(\omega)\in B_i\right\}\Big) &= \mathbb{P}\Big(\mathop{\times}_{i\in H}\left\{\omega_i : X_i(\omega_i)\in B_i\right\}\times\mathop{\times}_{i\notin H}\Omega_i\Big)\\ &= \prod_{i\in H}\mathbb{P}_i(X_i\in B_i)\\ &\overset{X_i\sim\widetilde{X}_i}{=} \prod_{i\in H}\mathbb{P}(\widetilde{X}_i\in B_i).\end{aligned}$$

Since H is an arbitrary finite index set, we conclude that the family $(\widetilde{X}_i)_{i\in I}$ is independent. In fact, we get the following generalization of Theorem 25.9.d).

26.3 Corollary. *Let $X_i : \Omega \to \mathbb{R}^d$, $i \in I$, be random variables on the same space Ω.*

$$(X_i)_{i\in I} \textit{ independent} \iff \overbrace{\underbrace{\mathbb{P}_{(X_i)_{i\in I}}}_{\textit{joint distribution}} = \underbrace{\bigotimes_{i\in I}\mathbb{P}_{X_i}}_{\textit{product measure}}}^{\substack{\textit{probability measures on}\\ ((\mathbb{R}^d)_I,\,\mathscr{B}(\mathbb{R}^d)^{\otimes I})}}.$$

Proof. Let $H \subset I$ be a finite index set. We have

$$\underbrace{\forall H:\quad (X_i)_{i\in H}\text{ indep.}}\quad\overset{\text{Thm. 25.9}}{\iff}\quad\underbrace{\forall H:\quad \mathbb{P}_H := \mathbb{P}_{(X_i)_{i\in H}} = \bigotimes_{i\in H}\mathbb{P}_{X_i}}$$

$\Updownarrow$ def. $\qquad\qquad\qquad\qquad\qquad\qquad$ $\Updownarrow$

$(X_i)_{i\in I}$ independent $\qquad$ there is exactly one measure on $((\mathbb{R}^d)^I, \mathscr{B}(\mathbb{R}^d)^{\otimes I})$ such that the projections on the coordinates in H are of the form $\bigotimes_{i\in H}\mathbb{P}_{X_i}$, and this is the infinite product measure $\mathbb{P}_{(X_i)_{i\in I}} = \bigotimes_{i\in I}\mathbb{P}_{X_i}$.

□

We get finally the infinite-dimensional version of the Lemmas 25.6 and 25.7.

26.4 Corollary (third block lemma). *Let $X_i : (\Omega, \mathscr{A}) \to (\Omega_i, \mathscr{A}_i)$, $i \in I$, by arbitrarily many independent random variables, let $I = \biguplus_{k\in K} I_k$ be any partition of the* *index set I, and let*

$$f_k : \Big(\bigtimes_{i \in I_k} \Omega_i, \bigotimes_{i \in I_k} \mathscr{A}_i \Big) \to (\mathbb{R}^{d(k)}, \mathscr{B}(\mathbb{R}^{d(k)})), \quad k \in K,$$

be measurable functions. The random variables

$$f_k((X_i)_{i \in I_k}) : \Omega \to \mathbb{R}^{d(k)}, \quad k \in K, \quad \textit{are independent.}$$

Proof. Fix $k_1, \dots, k_N \in K$. We first claim that the families

$$\mathscr{G}_{k_n} = \bigcup_{i \in I_{k_n}} X_i^{-1}(\mathscr{A}_i) \quad \Big[\subset \mathscr{A} \Big], \quad n = 1, 2, \dots, N,$$

are independent. Indeed, for any choice $G_{k_n} \in \mathscr{G}_{k_n}$ – this includes $G_{k_n} = \Omega$ – and suitable Borel sets $A_{i_{k_n}}$ such that $G_{k_n} = X_{i_{k_n}}^{-1}(A_{i_{k_n}})$ we have

$$\mathbb{P}\Big(\bigcap_{n=1}^{N} G_{k_n} \Big) = \mathbb{P}\Big(\bigcap_{n=1}^{N} X_{i_{k_n}}^{-1}(A_{i_{k_n}}) \Big) = \prod_{n=1}^{N} \mathbb{P}\big(X_{i_{k_n}}^{-1}(A_{i_{k_n}}) \big) = \prod_{n=1}^{N} \mathbb{P}\big(G_{k_n} \big).$$

A similar, but notationally more involved, calculation shows that the $\cap$-stable families

$$\mathscr{G}_{k_n}^{\cap} := \Big\{ \text{finite intersection sof sets from } \mathscr{G}_{k_n} \Big\} \cup \{\Omega\}$$

are independent. According to Theorem 25.5, the σ-algebras generated by these families

$$\sigma\big(\mathscr{G}_{k_n} \big) = \sigma\big(\mathscr{G}_{k_n}^{\cap} \big) = \sigma(X_i \mid i \in I_{k_n}), \quad n = 1, 2, \dots, N,$$

inherit the independence of their generators. But then the families

$$\sigma\big(f_{k_n}((X_i)_{i \in I_{k_n}}) \big), \quad n = 1, 2, \dots N,$$

are independent, too, since we have the equality

$$\underbrace{\Big\{ f_{k_n}((X_i)_{i \in I_{k_n}}) \in \Gamma \Big\}}_{\substack{\text{if } \Gamma \text{ runs through all of} \\ \mathscr{B}(\mathbb{R}^{d(k_n)}), \text{ these sets} \\ \text{generate } \sigma(f_{k_n}(\cdots))}} = \underbrace{\Big\{ (X_i)_{i \in I_{k_n}} \in f_{k_n}^{-1}(\Gamma) \Big\}}_{\substack{\text{these sets are independent} \\ \text{for all } n = 1, \dots, N}}.$$

This shows that the blocks $f_k((X_i)_{i \in I_k})$, $k \in K$, are independent, since Definition 25.3 requires the independence of any finite sub-family. □

Appendix: An easy construction of countably many independent rvs★

Here is an alternative simple method how one can construct countably many independent random variables on the space $([0,1), \mathscr{B}[0,1), \lambda)$. Let $(\mu_i)_{i \in \mathbb{N}}$ be probability distributions on $(\mathbb{R}, \mathscr{B}(\mathbb{R}))$ and denote by $F_i(x) := \mu_i[0,x]$ their distribution functions. If we can construct iid uniformly distributed random variables $(U_i)_{i \in \mathbb{N}}$, $U_i \sim \mathsf{U}[0,1]$, on $([0,1), \mathscr{B}[0,1), \lambda)$, then Lemma 26.2 shows that the random variables $X_i := F_i^{-1}(U_i)$ are independent and $X_i \sim \mu_i$.

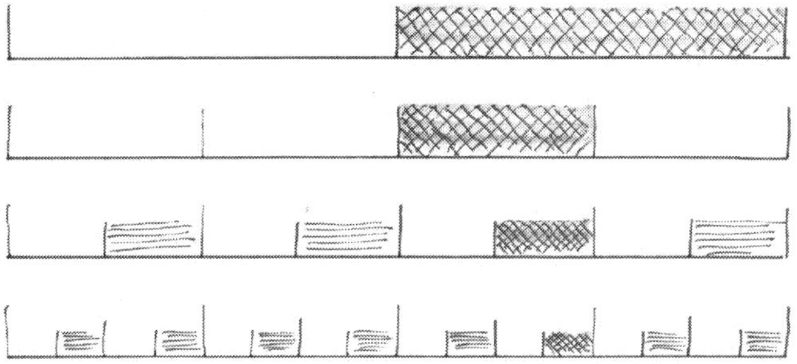

Fig. 26.2. In a **dyadic** ($g = 2$) expansion, the digit 0 corresponds to »take the left box« and 1 means »take the right box«. The figure shows the first few »decisions« of the number 0.1011... in criss-cross hatching. The horizontal hatch pattern in the last two lines shows the numbers $0.**1*...$ (the third digit is 1) and $0.***1*...$ (fourth digit is 1). Observe the periodicity.

Our starting point is the dyadic representation of a number $\omega \in [0,1)$:

$$\omega = 0.\epsilon_1(\omega)\epsilon_2(\omega)\epsilon_3(\omega)\cdots = \sum_{n=1}^{\infty} \frac{\epsilon_n(\omega)}{2^n}, \quad \epsilon_n(\omega) \in \{0,1\}.$$

In order to enforce uniqueness, we identify $0.\epsilon_1 \dots \epsilon_{n-1}0\overline{111}$ and $0.\epsilon_1 \dots \epsilon_{n-1}1\overline{000}$. Note that

$$\{\epsilon_n = \ell\} = \dot{\bigcup_{m=0}^{2^{n-1}-1}} \left[\frac{2m+\ell}{2^n}, \frac{2m+1+\ell}{2^n}\right)$$

which shows that the digits $\omega \mapsto \epsilon_n(\omega)$ are measurable, hence Bernoulli $\mathsf{Bin}(\frac{1}{2})$ random variables on $([0,1), \mathscr{B}[0,1), \lambda)$. Moreover, a moment's thought shows that prescribing a value $\epsilon_i(\omega) = 1$ or $\epsilon_i = 0$ in the dyadic representation of ω cuts the number of possible ω's by half, see also Fig. 26.2. Therefore,

$$\lambda\left(\left\{\omega \mid \epsilon_{n(i)} = \ell_i, i = 1,\dots,k\right\}\right) = \frac{1}{2^k} \prod_{i=1}^{k} \lambda\left(\left\{\omega \mid \epsilon_{n(i)} = \ell_i\right\}\right)$$

and this just means that the random variables $(\epsilon_n)_{n\in\mathbb{N}}$ are iid.

Now let $\phi : \mathbb{N} \times \mathbb{N} \to \mathbb{N}$ be a bijection and define new random variables $\phi_{i,k} := \epsilon_{\phi(i,k)}$. The random variables $(\phi_{i,k})_{i,k\in\mathbb{N}}$ are iid $\sim \mathsf{Bin}(\frac{1}{2})$ and, using the third block lemma (Corollary 26.4),

$$U_i := \sum_{k=1}^{\infty} \phi_{i,k} 2^{-k}$$

are independent. Since $(\phi_{i,k})_{k\in\mathbb{N}}$ runs through all 0-1 sequences, the U_i are iid $\sim \mathsf{U}[0,1)$.

Appendix: d-dimensional random variables via Borel isomorphisms★

In this appendix we show how one can generalize Solution II of Problem 1 to higher dimensions.

26.5 Definition. Let X and Y be topological spaces which are equipped with their Borel σ-algebras $\mathscr{B}(X)$, $\mathscr{B}(Y)$ which are generated by the open sets in X and Y, respectively. A bijection $f : X \to Y$ is said to be **bi-measurable [bi-continuous]** if both f and f^{-1} are Borel measurable [continuous]. We write $X \simeq Y$ and call X, Y **Borel isomorphic**, if there is a bi-measurable $f : X \to Y$.

We equip the space $\{0,1\}^{\mathbb{N}}$ with the following metric: $d(x,y) := \sum_{n\in\mathbb{N}} |x_n - y_n| 2^{-n}$.

26.6 Lemma. *The sets* $\{0,1\}^{\mathbb{N}}$ *and* $\left(\{0,1\}^{\mathbb{N}}\right)^d$ *are Borel isomorphic.*

Proof. Take any disjoint partition $\mathbb{N} = N_1 \cup \cdots \cup N_d$ such that all N_k are infinite. Write $x = (x_n)_{n\in\mathbb{N}}$ and $x|_{N_k} = (x_n)_{n\in N_k}$. The map

$$\{0,1\}^{\mathbb{N}} \ni x \mapsto (x|_{N_1}, \dots, x|_{N_d}) \in \left(\{0,1\}^{\mathbb{N}}\right)^d$$

is bijective and bi-continuous, hence bi-measurable. □

26.7 Lemma. *The sets* $\{0,1\}^{\mathbb{N}}$ *and* $[0,1]$ *are Borel isomorphic.*

Proof. Denote the dyadic numbers in $[0,1] \cap \mathbb{Q}$ by D – i.e. each $d \in D$ has either a **finite** dyadic representation or ends with the period $\overline{111}\dots$ – and write E for all sequences in $\{0,1\}^{\mathbb{N}}$ which are **finally constant** – i.e. sequences of the type $(***\overline{000}\dots)$ or $(***\overline{111}\dots)$. We will construct a bijection $f : [0,1] \to \{0,1\}^{\mathbb{N}}$.

1° On the set D we take any bijection $f|_D : D \to E$. This is possible since D and E have the same cardinality as $\mathbb{N}$.[1]

2° The elements of $[0,1] \setminus D$ can be written as a (truly) infinite series $x = \sum_{n\in\mathbb{N}} x_n 2^{-n}$, and we set

$$f|_{[0,1]\setminus D} : [0,1] \setminus D \to \{0,1\}^{\mathbb{N}} \setminus E, \quad x \mapsto (x_n)_{n\in\mathbb{N}}.$$

This is again a bijection. Since $d(x,y) \to 0 \iff \forall n \in \mathbb{N} : |x_n - y_n| \to 0$, the map $f|_{[0,1]\setminus D}$ is bi-continuous, hence bi-measurable.

3° Since the sets D and E are Borel sets of $[0,1]$ and $\{0,1\}^{\mathbb{N}}$, respectively, we can patch together f from $f|_D$ and $f|_{[0,1]\setminus D}$ and obtain a bi-measurable map $f : [0,1] \to \{0,1\}^{\mathbb{N}}$. □

26.8 Theorem. *The sets* $\overline{\mathbb{R}}$ *and* $\overline{\mathbb{R}}^d$ *are Borel isomorphic.*

Proof. We know from Lemma 26.7 that $\left(\{0,1\}^{\mathbb{N}}\right)^d \simeq [0,1]^d$. The Borel isomorphism $\overline{\mathbb{R}} \simeq [0,1]$ can be realized, for instance, with the map $x \mapsto \frac{1}{2} + \frac{1}{\pi}\arctan x$. Thus,

$$\overline{\mathbb{R}} \simeq [0,1] \overset{\S 26.7}{\simeq} \{0,1\}^{\mathbb{N}} \overset{\S 26.6}{\simeq} \left(\{0,1\}^{\mathbb{N}}\right)^d \overset{\S 26.7}{\simeq} [0,1]^d \simeq \overline{\mathbb{R}}^d.$$ □

(1) Note that we identify the dyadic numbers of the form $***\overline{111}$ and $***1\overline{000}$ whereas the corresponding sequences in $\{0,1\}^{\mathbb{N}}$ are different. This means that the bijection between E and D is really a (non-constructive) cardinality argument.

Let us return to the d-dimensional version of Solution II to Problem 1. Assume that μ is a probability measure on $(\mathbb{R}^d, \mathscr{B}(\mathbb{R}^d))$. We want to construct a probability space $(\Omega, \mathscr{A}, \mathbb{P})$ and a random variable $X_d : \Omega \to \mathbb{R}^d$ such that $X_d \sim \mu_d$. From Lemma 26.2 we know that this problem has in dimension $d = 1$ the following solution

$$(\Omega, \mathscr{A}, \mathbb{P}) = ([0,1], \mathscr{B}[0,1], \lambda) \quad \text{and} \quad X_1(t) = F_\mu(t) = \mu(-\infty, t]. \tag{26.2}$$

Consider now the Borel isomorphism $\phi : \overline{\mathbb{R}}^d \to \overline{\mathbb{R}}$ which exists due to Theorem 26.8. We may see μ_d as probability measure on $\overline{\mathbb{R}}^d$ – set $\mu_d(\overline{\mathbb{R}}^d \setminus \mathbb{R}^d) = 0$ – and define $\mu := \mu_d \circ \phi^{-1}$ and F_μ^{-1} as in (26.2). This shows that

$$\begin{array}{ccccc} [0,1] & \xrightarrow{\;F_\mu^{-1}\;} & \overline{\mathbb{R}} & \xrightarrow{\;\phi^{-1}\;} & \overline{\mathbb{R}}^d \\[1em] \lambda \overset{(26.2)}{=} \mu \circ F_\mu^{-1} & \longleftarrow & \mu := \mu_d \circ \phi^{-1} & \longleftarrow & \mu_d, \end{array}$$

and it is clear that $X_d := (F_\mu \circ \phi)^{-1}$ is a random variable on $([0,1], \mathscr{B}[0,1])$ taking values in $\overline{\mathbb{R}}^d$. Moreover, for all $B \in \mathscr{B}(\overline{\mathbb{R}}^d)$ we have

$$\mathbb{P}(X \in B) = \lambda(X^{-1}(B)) = \lambda(F_\mu \circ \phi(B)) = \lambda \circ F_\mu \circ \phi(B) = \mu \circ \phi(B) = \mu_d(B).$$

Since $\mu_d(\mathbb{R}^d) = 1$, it follows that $\mathbb{P}(X \in \mathbb{R}^d) = 1$, i.e. X is a.e. $\mathbb{R}^d$-valued; if necessary, we can set X to be 0 on that null set, i.e. X becomes an everywhere $\mathbb{R}^d$-valued random variable.

26.9 Remark. a) If we want to avoid working in $\overline{\mathbb{R}}$, we can argue in the following way. Clearly, $\mathbb{R} \simeq (0,1)$. Set $\bar{0} = (0,0,0\dots)$ and $\bar{1} = (1,1,1,\dots)$. If we remove from the set D in the proof of Lemma 26.7 the numbers 0 and 1, we get $\{0,1\}^{\mathbb{N}} \setminus \{\bar{0}, \bar{1}\} \simeq (0,1)$, and Lemma 26.6 can be adapted to show $\{0,1\}^{\mathbb{N}} \setminus \{\bar{0}, \bar{1}\} \simeq \left(\{0,1\}^{\mathbb{N}}\right)^d \setminus \{\bar{0}, \bar{1}\}$.

b) A simple modification of Lemma 26.6 shows that $\{0,1\}^{\mathbb{N}}$ and $\left(\{0,1\}^{\mathbb{N}}\right)^{\mathbb{N}}$ are Borel isomorphic. Therefore, we may apply the above technique to infinite-dimensional random variables taking values in $\mathbb{R}^{\mathbb{N}}$ or $[0,1]^{\mathbb{N}}$.

27 Characteristic functions

In this chapter we are going to see how we can characterize a random variable $X : \Omega \to \mathbb{R}^d$ or its probability law $X \sim \mu$ by a single non-random **function**.

27.1 Definition. Let $X : \Omega \to \mathbb{R}^d$ be a d-dimensional random variable. The function $\phi_X : \mathbb{R}^d \to \mathbb{C}$

$$\phi_X(\xi) := \mathbb{E} e^{i\langle \xi, X \rangle} = \int_{\mathbb{R}^d} e^{i\langle \xi, x \rangle}\, \mathbb{P}(X \in dx), \quad \xi \in \mathbb{R}^d, \tag{27.1}$$

is called the **characteristic function** (cf, for short) of X or μ.

In order to integrate complex-valued functions, we reduce things by linearity to the real and imaginary parts, i.e. if $f = u + iv$, then $\int f\, d\mu = \int u\, d\mu + i \int v\, d\mu$, and we have $f \in L^1_{\mathbb{C}}(\mu) \iff u, v \in L^1_{\mathbb{R}}(\mu)$. For more on this, see the appendix to Chapter 16 on page 90.

27.2 Example. Here are a few examples of characteristic functions of $\mathbb{R}$-valued random variables – see also the lists of probability distributions Table 22.1 and Table 23.1 on page 130 and 132*f.*

a) **Degenerate distribution.** $X \sim \delta_c$, i.e. $X \equiv c$ is a.s. constant.

$$\mathbb{E}e^{i\xi X} = \int_{\mathbb{R}} e^{i\xi x}\,\mathbb{P}(X \in dx) = \int_{\mathbb{R}} e^{i\xi x}\,\delta_c(dx) = e^{ic\xi}, \quad \xi \in \mathbb{R}.$$

b) **Bernoulli distribution.** $X \sim \mathsf{Bin}(p) = p\delta_1 + q\delta_0$. We have

$$\mathbb{E}e^{i\xi X} = pe^{i\xi\cdot 1} + qe^{i\xi\cdot 0} = q + pe^{i\xi}, \quad \xi \in \mathbb{R}.$$

c) **Two-point law.** $X \sim \frac{1}{2}\delta_a + \frac{1}{2}\delta_{-a}$, $a \in \mathbb{R}$. We have

$$\mathbb{E}e^{i\xi X} = \frac{1}{2}e^{ia\xi} + \frac{1}{2}e^{-ia\xi} = \cos(a\xi), \quad \xi \in \mathbb{R}.$$

d) **Poisson distribution.** $X \sim \mathsf{Poi}(\lambda)$, $\lambda > 0$, i.e. X takes values in the set $\mathbb{N}_0$ and $\mathbb{P}(X = n) = e^{-\lambda}\lambda^n/n!$. We have

$$\mathbb{E}e^{i\xi X} = \sum_{n=0}^{\infty} e^{i\xi n}\,\mathbb{P}(X = n) = \sum_{n=0}^{\infty} e^{i\xi n}\frac{\lambda^n}{n!}e^{-\lambda} = \sum_{n=0}^{\infty}\frac{[\lambda e^{i\xi}]^n}{n!}e^{-\lambda} = e^{\lambda e^{i\xi}}e^{-\lambda}.$$

e) **Uniform distribution on the interval** $[a,b]$. $X \sim \mathsf{U}[a,b]$, i.e. X has the density $(b-a)^{-1}\mathbb{1}_{[a,b]}(x)$. We have

$$\mathbb{E}e^{i\xi X} = \frac{1}{b-a}\int_a^b e^{i\xi x}\,dx = \frac{e^{ib\xi} - e^{ia\xi}}{i\xi(b-a)}, \quad \xi \in \mathbb{R}.$$

f) **Exponential distribution.** $X \sim \mathsf{Exp}(\lambda)$, $\lambda > 0$, i.e. the random variable X has the density $\lambda e^{-\lambda x}\mathbb{1}_{[0,\infty)}(x)$. We have

$$\mathbb{E}e^{i\xi X} = \lambda\int_0^{\infty} e^{i\xi x}e^{-\lambda x}\,dx = \frac{\lambda}{\lambda - i\xi}, \quad \xi \in \mathbb{R}.$$

One of the most important probability distributions is the normal distribution $\mathsf{N}(\mu,\sigma^2)$, cf. Example 23.3.d), whose density is

$$g_{\mu,\sigma^2}(x) = \frac{1}{\sqrt{2\pi\sigma^2}}e^{-(x-\mu)^2/2\sigma^2}, \quad x \in \mathbb{R}.$$

27.3 Theorem. *If* $G \sim \mathsf{N}(0,1)$ *is a standard normal random variable, then*

$$\mathbb{E}e^{i\xi G} = \frac{1}{\sqrt{2\pi}}\int_{\mathbb{R}} e^{i\xi x}e^{-x^2/2}\,dx = e^{-\xi^2/2}, \quad \xi \in \mathbb{R}. \tag{27.2}$$

Moreover, $\sigma G + \mu \sim \mathsf{N}(\mu,\sigma^2)$, *and*

$$\mathbb{E}e^{i\xi(\sigma G+\mu)} = \frac{1}{\sqrt{2\pi\sigma^2}}\int_{\mathbb{R}} e^{i\xi x}e^{-(x-\mu)^2/2\sigma^2}\,dx = e^{i\mu\xi - \sigma^2\xi^2/2}, \quad \xi \in \mathbb{R}. \tag{27.3}$$

Proof. Define $\phi(\xi) := \int_{\mathbb{R}} (2\pi)^{-1/2} e^{-x^2/2} e^{ix\xi} dx$. With the differentiability lemma (Theorem 12.1) and integration by parts we see that

$$\begin{aligned}\phi'(\xi) = \int_{\mathbb{R}} \frac{1}{\sqrt{2\pi}} e^{-x^2/2} \frac{d}{d\xi} e^{ix\xi} dx &= \frac{1}{\sqrt{2\pi}} \int_{\mathbb{R}} e^{-x^2/2} (ix) e^{ix\xi} dx \\ &= \frac{1}{\sqrt{2\pi}} \int_{\mathbb{R}} (-i) \left[\frac{d}{dx} e^{-x^2/2}\right] e^{ix\xi} dx \\ &= \frac{1}{\sqrt{2\pi}} \int_{\mathbb{R}} i e^{-x^2/2} \left[\frac{d}{dx} e^{ix\xi}\right] dx \\ &= \frac{1}{\sqrt{2\pi}} \int_{\mathbb{R}} i e^{-x^2/2} i\xi e^{ix\xi} dx \\ &= -\xi\phi(\xi).\end{aligned}$$

This leads to a differential equation which has a unique solution

$$\left.\begin{aligned}\phi'(\xi) &= -\xi\phi(\xi) \\ \phi(0) &= \mathbb{E}e^{i0\cdot X} = 1\end{aligned}\right\} \implies \phi(\xi) = \phi(0)e^{-\xi^2/2} = e^{-\xi^2/2}.$$

The random variable $\sigma G + \mu$ satisfies on the one hand

$$\mathbb{E}e^{i\xi(\sigma G+\mu)} = e^{i\xi\mu}\mathbb{E}e^{i(\xi\sigma)G} \stackrel{(27.2)}{=} e^{i\xi\mu}e^{-\sigma^2\xi^2/2},$$

while, on the other hand, we have for all $B \in \mathscr{B}(\mathbb{R})$

$$\begin{aligned}\mathbb{P}(\sigma G + \mu \in B) = \mathbb{P}(G \in (B-\mu)/\sigma) &= \frac{1}{\sqrt{2\pi}} \int_{(B-\mu)/\sigma} e^{-y^2/2} dy \\ &= \frac{1}{\sqrt{2\pi}\sigma} \int_B e^{-(x-\mu)^2/2\sigma^2} dx\end{aligned}$$

where we use the change of variables $y = (x-\mu)/\sigma$ and $dy = dx/\sigma$ in the last step. □

With the help of Fubini's theorem, we can get the multivariate version of Theorem 27.3 for Gaussian random variables with independent coordinates. General Gaussian random variables are discussed in Chapter 36. We use the following **standard notation**: $\langle x, \xi\rangle = \sum_{k=1}^d x_k\xi_k$ is the usual scalar product in $\mathbb{R}^d$, $|x|^2 = \sum_{k=1}^d x_k^2$ is the (square of the) Euclidean norm and $\mathrm{id} = \mathrm{id}_d \in \mathbb{R}^{d\times d}$ is the unit matrix.

27.4 Corollary. *Let $G_1, \dots, G_d \sim N(0,1)$ be independent standard normal random variables. The random vector $\sigma G + m$, $G = (G_1, \dots, G_d)^\top$, $\sigma > 0$, $m \in \mathbb{R}^d$, has the following probability density*

$$g_{m,\sigma^2 \mathrm{id}}(x) = \frac{1}{(2\pi\sigma^2)^{d/2}} e^{-|x-m|^2/2\sigma^2}, \quad x \in \mathbb{R}^d, \tag{27.4}$$

and the characteristic function

$$\mathbb{E}e^{i\langle\xi,\sigma G+m\rangle} = \frac{1}{(2\pi\sigma^2)^{d/2}}\int_{\mathbb{R}^d} e^{i\langle\xi,x\rangle}e^{-|x-m|^2/2\sigma^2}\,dx \\ = e^{i\langle m,\xi\rangle-\sigma^2|\xi|^2/2}, \quad \xi\in\mathbb{R}^d. \tag{27.5}$$

If we use $m = 0$ and $\sigma^2 = 1/t$ in (27.5) and change $\xi \rightsquigarrow -\xi$, then we get the following useful relation

$$\int_{\mathbb{R}^d} e^{-i\langle\xi,x\rangle}e^{-t|x|^2/2}\,dx = \left(\frac{2\pi}{t}\right)^{d/2} e^{-|\xi|^2/2t}, \quad \xi\in\mathbb{R}^d. \tag{27.6}$$

💬 The formulae (27.5) and (27.6) show, in particular, that there is a one-to-one correspondence

$$(2\pi\sigma^2)^{-d/2}e^{-|x-m|^2/2\sigma^2} \overset{\text{cf}}{\longleftrightarrow} e^{i\langle\xi,m\rangle-\sigma^2|\xi|^2/2} \quad\text{and}\quad (2\pi/t)^{-d/2}e^{-t|x|^2/2} \overset{\text{cf}}{\longleftrightarrow} e^{-|\xi|^2/2t},$$

i.e. we can recover the probability density from the characteristic function. This holds, in general, as we will see in a moment.

27.5 Theorem (properties of the cf). *Let $X:\Omega\to\mathbb{R}^d$ be a d-dimensional random variable. Its c.f. ϕ_X enjoys the following properties.*

a) $\xi\mapsto\phi_X(\xi)$ *is continuous.*

b) $|\phi_X(\xi)|\leqslant\phi_X(0)=1$.

c) $\phi_{-X}(\xi)=\phi_X(-\xi)=\overline{\phi_X(\xi)}$.

d) $\operatorname{Re}\phi_X(\xi)=\mathbb{E}[\cos\langle\xi,X\rangle]$ *is the c.f. of* $Y=\epsilon X$, *where* $\epsilon\sim\frac{1}{2}\delta_1+\frac{1}{2}\delta_{-1}$ *and* $X \perp\!\!\!\perp \epsilon$.

e) $|\phi_X|^2$ *is the c.f. of* $Y=X-\widetilde{X}$ *where* $X\perp\!\!\!\perp\widetilde{X}$ *and* $\widetilde{X}\sim X$.

f) *If* $T:\mathbb{R}^d\to\mathbb{R}^n$, $Tx:=\Sigma x+m$, *for some* $\Sigma\in\mathbb{R}^{n\times d}$ *and* $m\in\mathbb{R}^n$, *then*

$$\phi_{TX}(\xi)=e^{i\langle\xi,m\rangle}\phi_X(\Sigma^\top\xi), \quad \xi\in\mathbb{R}^n.$$

g) *If* $\mathbb{E}(|X|^n)<\infty$ *for some* $n\in\mathbb{N}$, *then* $\partial^\alpha\phi_X$ *exists for all multiindices* $\alpha\in\mathbb{N}_0^d$ *such that* $|\alpha|\leqslant n$. *In this case,*

$$\partial^\alpha\phi_X \quad \text{is continuous and} \quad \mathbb{E}(X^\alpha)=i^{-|\alpha|}\partial^\alpha\phi_X(0).$$

⚠ The converse direction in 27.5.g) is, in general, false. If $|\alpha|=2m$ is **even**, then one can show that the existence of $\partial^\alpha\phi_X(0)$ entails the existence of moments up to order $2m$, cf. [WT, Satz 7.6.f)].

💬 We use standard **multiindex notation**: If $\alpha=(\alpha_1,\dots,\alpha_d)\in\mathbb{N}_0^d$ and $x\in\mathbb{R}^d$, then

$$|\alpha|=\alpha_1+\dots+\alpha_d, \quad x^\alpha:=\prod_k x_k^{\alpha_k}, \quad\text{and}\quad \partial^\alpha=\partial_x^\alpha=\frac{\partial^{|\alpha|}}{\partial^{\alpha_1}x_1\dots\partial^{\alpha_d}x_d}.$$

These rules allow us to calculate with multiindices and vectors as if they were ordinary numbers.

Proof. a) Using the continuity lemma (Theorem 12.1) it is easy to see that $\xi \mapsto \phi_X(\xi) = \int_{\mathbb{R}^d} e^{i\langle \xi, x\rangle} \mathbb{P}(X \in dx)$ is continuous: $|e^{ix\xi}| = 1 \in L^1(\mathbb{P}_X)$ is the integrable majorant and the integrand $\xi \mapsto e^{i\langle \xi, x\rangle}$ is continuous.

b) &c) follow straight from the definition of the characteristic function.

d) Since $\operatorname{Re} e^{i\langle \xi, X\rangle} = \cos\langle \xi, X\rangle$, the first assertions follows from the linearity of the expectation. Moreover, let $\epsilon \perp\!\!\!\perp X$ as in the statement. Then

$$\mathbb{E}e^{i\langle \xi, \epsilon X\rangle} \underset{\S 25.10}{\overset{\epsilon \perp\!\!\!\perp X}{=}} \frac{1}{2}\mathbb{E}e^{i\langle \xi, X\rangle} + \frac{1}{2}\mathbb{E}e^{i\langle \xi, -X\rangle} \overset{c)}{=} \operatorname{Re}\mathbb{E}e^{i\langle \xi, X\rangle},$$

where we use $\operatorname{Re} z = \frac{1}{2}(z + \bar{z})$.

e) Let $\widetilde{X}$ be an independent and identically distributed (»iid«) copy of X. We have

$$\mathbb{E}e^{i\langle \xi, X-\widetilde{X}\rangle} \underset{\S 25.10}{\overset{X \perp\!\!\!\perp \widetilde{X}}{=}} \mathbb{E}e^{i\langle \xi, X\rangle} \cdot \mathbb{E}e^{-i\langle \xi, \widetilde{X}\rangle} \underset{c)}{\overset{\widetilde{X}\sim X}{=}} \phi_X(\xi)\cdot\overline{\phi_X(\xi)} = |\phi_X(\xi)|^2.$$

f) We have $\phi_{TX}(\xi) = \mathbb{E}e^{i\langle \xi, \Sigma X + m\rangle} = e^{i\langle \xi, m\rangle}\mathbb{E}e^{i\langle \xi, \Sigma X\rangle} = e^{i\langle \xi, m\rangle}\mathbb{E}e^{i\langle \Sigma^\top \xi, X\rangle}$
$$= e^{i\langle \xi, m\rangle}\phi_X(\Sigma^\top \xi).$$

g) Note that ✎ $|x^\alpha| \leqslant |x|^{|\alpha|}$ for all $x \in \mathbb{R}^d$ and $\alpha \in \mathbb{N}_0^d$. Therefore, it is enough to show that $\mathbb{E}|X^\alpha| < \infty$ guarantees the existence of $\partial^\alpha \phi_X$. This is done by repeated applications of the differentiability lemma (Theorem 12.2)

$$\begin{aligned}\partial^\alpha \phi_X(\xi) = \partial_\xi^\alpha \int e^{i\langle \xi, x\rangle}\mathbb{P}(X \in dx) &\overset{\S}{=} \int \partial_\xi^\alpha e^{i\langle \xi, x\rangle}\mathbb{P}(X \in dx)\\ &= \int (ix)^\alpha e^{i\langle \xi, x\rangle}\mathbb{P}(X \in dx)\\ &= i^{|\alpha|}\int x^\alpha e^{i\langle \xi, x\rangle}\mathbb{P}(X \in dx).\end{aligned}$$

In the step marked with § we use the integrable majroant

$$\left|\partial_\xi^\alpha e^{i\langle \xi, x\rangle}\right| = \left|(ix)^\alpha e^{i\langle \xi, x\rangle}\right| = \left|(ix)^\alpha\right| = |x^\alpha| \in L^1(\mathbb{P}_X).$$

Finally, the continuity lemma shows that $\partial^\alpha\phi_X(0) = i^{|\alpha|}\mathbb{E}(X^\alpha)$ for $\xi = 0$. □

Our next aim is to show that the c.f. ϕ_X characterizes the rv X.

27.6 Lemma. *Assume that $(\Omega, \mathscr{A}, \mathbb{P})$ and $(\Omega', \mathscr{A}', \mathbb{P}')$ are two probability spaces and $X : \Omega \to \mathbb{R}^d$ and $X' : \Omega' \to \mathbb{R}^d$ random variables.*

$$X \sim X' \iff \mathbb{E}u(X) = \mathbb{E}'u(X') \quad \forall u \in C_c(\mathbb{R}^d). \tag{27.7}$$

Proof. The direction »⇒« is clear. Conversely, »⇐« The distance of a point from a set $x \mapsto d(x, A) := \inf_{a\in A}|x - a|$, $A \subset \mathbb{R}^d$, is (Lipschitz) continuous,

$$\begin{aligned}|x - y| \geqslant |x - a| - |y - a| \geqslant \inf_{a\in A}|x - a| - |y - a| = d(x, A) - |y - a|\\ \implies |x - y| \geqslant d(x, A) - d(y, A).\end{aligned}$$

If we swap the roles of x and y, we see that $|d(x,\mathrm{A}) - d(y,\mathrm{A})| \leqslant |x-y|$.

Let $\mathrm{K} \subset \mathbb{R}^d$ be a compact set and $\mathrm{U}_n := \mathrm{K} + \mathrm{B}_{\frac{1}{n}}(0) = \left\{y \in \mathbb{R}^d \mid d(y,\mathrm{K}) < \frac{1}{n}\right\}$ its open $1/n$-enlargement. Clearly, ✍

$$u_n(x) := \frac{d(x,\mathrm{U}_n^c)}{d(x,\mathrm{U}_n^c) + d(x,\mathrm{K})} \in \mathrm{C}_c(\mathbb{R}^d) \quad \text{and} \quad u_n \downarrow \mathbb{1}_\mathrm{K}.$$

Using Beppo Levi, we see that

$$\mathbb{P}(\mathrm{X} \in \mathrm{K}) = \mathbb{E}\mathbb{1}_\mathrm{K}(\mathrm{X}) \stackrel{\mathrm{BL}}{=} \inf_{n\in\mathbb{N}} \mathbb{E}u_n(\mathrm{X}) = \inf_{n\in\mathbb{N}} \mathbb{E}' u_n(\mathrm{X}') = \cdots = \mathbb{P}'(\mathrm{X}' \in \mathrm{K}).$$

Since the family of all compact sets are $\cap$-stable generator of $\mathscr{B}(\mathbb{R}^d)$, and since $\mathrm{K}_n = \overline{\mathrm{B}_n(0)} \uparrow \mathbb{R}^d$, the uniqueness of measures theorem (Theorem 4.7) shows that $\mathbb{P}(\mathrm{X} \in \mathrm{B}) = \mathbb{P}'(\mathrm{X}' \in \mathrm{B})$ for all $\mathrm{B} \in \mathscr{B}(\mathbb{R}^d)$, hence $\mathrm{X} \sim \mathrm{X}'$. □

27.7 Theorem. *Assume that* $(\Omega, \mathscr{A}, \mathbb{P})$ *and* $(\Omega', \mathscr{A}', \mathbb{P}')$ *are probability spaces and* $\mathrm{X} : \Omega \to \mathbb{R}^d$, $\mathrm{X}' : \Omega' \to \mathbb{R}^d$ *random variables.*

$$\mathrm{X} \sim \mathrm{X}' \iff \phi_\mathrm{X} = \phi_{\mathrm{X}'}.$$

Proof. »$\Rightarrow$« : Since $x \mapsto e^{i\langle\xi, x\rangle}$ is a bounded continuous function, we see that $\mathrm{X} \sim \mathrm{X}'$ entails

$$\phi_\mathrm{X}(\xi) = \mathbb{E}e^{i\langle\xi,\mathrm{X}\rangle} = \mathbb{E}' e^{i\langle\xi,\mathrm{X}'\rangle} = \phi_{\mathrm{X}'}(\xi), \quad \xi \in \mathbb{R}^d.$$

Conversely, »$\Rightarrow$«, assume that $\phi_\mathrm{X} \equiv \phi_{\mathrm{X}'}$. We construct independent normal random variables $\mathrm{G}_k \sim \mathsf{N}(0,1)$, $k = 1,\dots,d$, which are also independent of the given random variable X.[1] Since $\mathrm{G} := (\mathrm{G}_1,\dots,\mathrm{G}_d)^\top \perp\!\!\!\perp \mathrm{X}$, we get for $u \in \mathrm{C}_c(\mathbb{R}^d)$

$$\begin{aligned}
\mathbb{E}u(\mathrm{X} + \sqrt{t}\mathrm{G}) &\stackrel{\perp\!\!\!\perp}{=} \int_{\mathbb{R}^d} \mathbb{E}u(\mathrm{X}+y) \underbrace{(2\pi t)^{-d/2} e^{-|y|^2/2t}\,dy}_{=\mathbb{P}(\sqrt{t}\mathrm{G} \in dy),\ (27.4)} \\
&\stackrel{z=\mathrm{X}+y}{=} \int_{\mathbb{R}^d} u(z)\,(2\pi t)^{-d/2}\,\mathbb{E}e^{-|\mathrm{X}-z|^2/2t}\,dz.
\end{aligned}$$

Now we use (27.6) with $x \rightsquigarrow \eta$ and $\xi \rightsquigarrow \mathrm{X} - z$ to infer

$$\begin{aligned}
\mathbb{E}u\left(\mathrm{X} + \sqrt{t}\mathrm{G}\right) &= \frac{1}{(2\pi)^d} \int_{\mathbb{R}^d} u(z)\,\mathbb{E}\int_{\mathbb{R}^d} e^{-i\langle\eta,\mathrm{X}-z\rangle} e^{-t|\eta|^2/2}\,d\eta\,dz \\
&= \frac{1}{(2\pi)^d} \int_{\mathbb{R}^d} u(z) \int_{\mathbb{R}^d} \mathbb{E}e^{-i\langle\eta,\mathrm{X}\rangle} e^{i\langle\eta,z\rangle} e^{-t|\eta|^2/2}\,d\eta\,dz \\
&= \frac{1}{(2\pi)^d} \int_{\mathbb{R}^d} u(z) \int_{\mathbb{R}^d} \phi_\mathrm{X}(-\eta) e^{i\langle\eta,z\rangle} e^{-t|\eta|^2/2}\,d\eta\,dz \\
&\stackrel{\phi_\mathrm{X}=\phi_{\mathrm{X}'}}{=} \frac{1}{(2\pi)^d} \int_{\mathbb{R}^d} u(z) \int_{\mathbb{R}^d} \phi_{\mathrm{X}'}(-\eta) e^{i\langle\eta,z\rangle} e^{-t|\eta|^2/2}\,d\eta\,dz \\
&= \dots = \mathbb{E}' u\left(\mathrm{X}' + \sqrt{t}\mathrm{G}'\right) \xrightarrow[t\to 0]{\mathrm{DCT}} \mathbb{E}' u(\mathrm{X}').
\end{aligned}$$

(1) This can be achieved by Problem 2 of Chapter 29.

(the random variable G' is similar to G, only that it lives in $(\Omega', \mathscr{A}', \mathbb{P}')$. Using Lemma 27.6, we conclude that $X \sim X'$. □

Theorem 27.7 would allow us to prove the still open implication e)⇒d) of Theorem 25.9. We will give an alternative proof.

27.8 Corollary (Kac). *Let* $X, Y : \Omega \to \mathbb{R}^d$ *be two random variables.*

$$X \perp\!\!\!\perp Y \iff \mathbb{E}e^{i\langle\xi, X\rangle + i\langle\eta, Y\rangle} = \mathbb{E}e^{i\langle\xi, X\rangle} \cdot \mathbb{E}e^{i\langle\eta, Y\rangle} \quad \forall \xi, \eta \in \mathbb{R}^d. \tag{27.8}$$

⚠ In order to ensure the independence of X and Y it is **not sufficient** to have (27.8) for all $\xi = \eta$. Consider, e.g. a bivariate rv (X, Y) with joint density $f_{X,Y}(x,y) = \frac{1}{4}(1 + xy(x^2 - y^2))\mathbb{1}^2_{[-1,1]}(x,y)$ ✍. X, Y do satisfy (27.8) for $\xi = \eta$, but they are not independent.

Proof. The direction »⇒« is Corollary 25.11. For the reverse implication »⇐« we note that

$$\mathbb{E}e^{i\langle\xi, X\rangle + i\langle\eta, Y\rangle} = \mathbb{E}e^{i\langle(\xi,\eta)^\top, (X,Y)^\top\rangle} = \phi_{(X,Y)^\top}(\xi, \eta).$$

Construct – e.g. using a product argument as in Chapter 26 – a new probability space $(\widetilde{\Omega}, \widetilde{\mathscr{A}}, \widetilde{\mathbb{P}})$ and random variables $\widetilde{X}$ and $\widetilde{Y}$ such that

$$\widetilde{X} \sim X, \quad \widetilde{Y} \sim Y \quad \text{and} \quad \widetilde{X} \perp\!\!\!\perp \widetilde{Y}.$$

We see for any $\xi, \eta \in \mathbb{R}^d$ that

$$\phi_{(X,Y)^\top}(\xi, \eta) = \phi_X(\xi)\phi_Y(\eta) \underset{\widetilde{Y}\sim Y}{\overset{\widetilde{X}\sim X}{=}} \phi_{\widetilde{X}}(\xi)\phi_{\widetilde{Y}}(\eta) \overset{\widetilde{X} \perp\!\!\!\perp \widetilde{Y}}{=} \phi_{(\widetilde{X},\widetilde{Y})^\top}(\xi, \eta).$$

From Theorem 27.7 we see that $(X, Y)^\top \sim (\widetilde{X}, \widetilde{Y})^\top$, and so

$$\mathbb{P}_{X,Y} = \widetilde{\mathbb{P}}_{\widetilde{X},\widetilde{Y}} \overset{\widetilde{X} \perp\!\!\!\perp \widetilde{Y}}{=} \widetilde{\mathbb{P}}_{\widetilde{X}} \otimes \widetilde{\mathbb{P}}_{\widetilde{Y}} \underset{\widetilde{Y}\sim Y}{\overset{\widetilde{X}\sim X}{=}} \mathbb{P}_X \otimes \mathbb{P}_Y.$$

In view of Theorem 25.9 this means that $X \perp\!\!\!\perp Y$. □

27.9 Example*. Let $X : \Omega \to \mathbb{R}^d$, $Y : \Omega \to \mathbb{R}^n$ be random variables and $\mathscr{F} \subset \mathscr{A}$.

a) $X \perp\!\!\!\perp \mathscr{F} \iff \forall \xi \in \mathbb{R}^d : \mathbb{E}\left(e^{i\langle\xi, X\rangle}\mathbb{1}_F\right) = \mathbb{E}e^{i\langle\xi, X\rangle}\mathbb{P}(F)$. We get »⇒« immediately from Corollary 25.11. For the converse »⇐« it is enough to show that $X \perp\!\!\!\perp Y$ for $Y := \mathbb{1}_F$ and every $F \in \mathscr{F}$. We want to apply Corollary 27.8. Fix $\xi \in \mathbb{R}^d$ and $\eta \in \mathbb{R}$. Observe that

$$e^{i\eta\mathbb{1}_F} = e^{i\eta}\mathbb{1}_F + \mathbb{1}_{F^c} = \left(e^{i\eta} - 1\right)\mathbb{1}_F + 1. \tag{27.9}$$

This shows that

$$\begin{aligned}\mathbb{E}\left(e^{i\langle\xi,X\rangle}e^{i\eta\mathbb{1}_F}\right) &\overset{(27.9)}{=} \left(e^{i\eta}-1\right)\mathbb{E}\left(e^{i\langle\xi,X\rangle}\mathbb{1}_F\right)+\mathbb{E}e^{i\langle\xi,X\rangle}\\ &\overset{\text{ass.}}{=} \left(e^{i\eta}-1\right)\mathbb{E}e^{i\langle\xi,X\rangle}\mathbb{P}(F)+\mathbb{E}e^{i\langle\xi,X\rangle}\\ &= \mathbb{E}e^{i\langle\xi,X\rangle}\left[\left(e^{i\eta}-1\right)\mathbb{P}(F)+1\right]\\ &\overset{(27.9)}{=} \mathbb{E}e^{i\langle\xi,X\rangle}\mathbb{E}e^{i\eta\mathbb{1}_F},\end{aligned}$$

and the assertion follows from Corollary 27.8.

b) $X \perp\!\!\!\perp Y, X \perp\!\!\!\perp Y, \mathscr{F}, Y \perp\!\!\!\perp \mathscr{F} \implies (X,Y) \perp\!\!\!\perp \mathscr{F}$. Pick $F \in \mathscr{F}$ and let $\xi \in \mathbb{R}^d$, $\eta \in \mathbb{R}^n$. We have

$$\begin{aligned}\mathbb{E}\left(e^{i(\langle\xi,X\rangle+\langle\eta,Y\rangle)}\mathbb{1}_F\right) &\overset{X\perp\!\!\!\perp Y,\mathscr{F}}{=} \mathbb{E}\left(e^{i\langle\xi,X\rangle}\right)\mathbb{E}\left(e^{i\langle\eta,Y\rangle}\mathbb{1}_F\right)\\ &\overset{Y\perp\!\!\!\perp\mathscr{F}}{=} \mathbb{E}\left(e^{i\langle\xi,X\rangle}\right)\mathbb{E}\left(e^{i\langle\eta,Y\rangle}\right)\mathbb{P}(F)\\ &\overset{X\perp\!\!\!\perp Y}{=} \mathbb{E}\left(e^{i(\langle\xi,X\rangle+\langle\eta,Y\rangle)}\right)\mathbb{P}(F).\end{aligned}$$

Since $\langle\xi,X\rangle+\langle\eta,Y\rangle = \langle(\xi,\eta),(X,Y)\rangle$, the claim follows from part a).

Without proof, we mention the following inversion formula for probability densities. A proof can be found in [WT, Satz 7.10].

27.10 Theorem★. *Let* $X:\Omega\to\mathbb{R}^d$ *be a random variable with c.f.* ϕ_X. *If* $\phi_X \in L^1(d\xi)$, *then* $X \sim p_X(x)\,dx$, *with*

$$p_X(x) = (2\pi)^{-d}\int_{\mathbb{R}^d}\phi_X(\xi)e^{-i\langle\xi,x\rangle}\,d\xi, \quad x\in\mathbb{R}^d. \tag{27.10}$$

Appendix: Inversion of the characteristic function★

If the c.f. ϕ_X is not integrable as in Theorem 27.10, we cannot expect to get a density and an inversion formula as in (27.10). On the other hand, $X \leftrightarrow \phi_X$ is a bijection, so some kind of inversion is to be expected. Indeed, we have the following result due to Paul Lévy. Here we discuss only the one-dimensional case, the multivariate case is essentially similar, see [MIMS, Corollary 19.7].

Assume that $X \sim \mu$ is a real random variable and write $\check{\mu}(\xi) = \mathbb{E}e^{i\xi X}$.

27.11 Theorem (Lévy). *Let* μ *be a probability measure on* $(\mathbb{R},\mathscr{B}(\mathbb{R}))$. *For all* $a<b$

$$\frac{1}{2}\mu\{a\}+\mu(a,b)+\frac{1}{2}\mu\{b\} = \lim_{T\to\infty}\frac{1}{2\pi}\int_{-T}^{T}\frac{e^{-i\xi a}-e^{-i\xi b}}{i\xi}\check{\mu}(\xi)\,d\xi. \tag{27.11}$$

Since a probability measure is uniquely determined by its values on, say, the open intervals (a,b), cf. Theorem 4.7, the formula (27.11) is indeed an inversion theorem for the characteristic function.

⚠ The limit $\lim_{T\to\infty}\int_{-T}^{T}\cdots$ appearing in (27.11) is a **Cauchy principal value integral** which is often denoted by $\mathrm{pv}\int_{-\infty}^{\infty}\ldots$ Because of the symmetric nature of the limit, there will be cancellations, and the limit cannot be expressed as a Lebesgue integral.

💬 Since $X \sim \mu$, we can re-write the lhS of (27.11) in the following way

$$\frac{1}{2}\mathbb{P}(X=a)+\mathbb{P}(a<X<b)+\frac{1}{2}\mathbb{P}(X=b)$$

or, using the distribution function $F(x)=F_X(x)=\mathbb{P}(X\leqslant x)$, as

$$\frac{1}{2}(F(b)+F(b-))-\frac{1}{2}(F(a)+F(a-)).$$

Proof of Theorem 27.11. 1° We have calculated in Example 16.8 the following sine integral

$$\lim_{T\to\infty}\int_0^T\frac{\sin\xi}{\xi}\,d\xi=\frac{\pi}{2}.$$

Using this formula we see that

$$\lim_{T\to\infty}\int_0^T\frac{\sin(\xi(x-a))}{\xi}\,d\xi=\lim_{T\to\infty}\int_0^{T(x-a)}\frac{\sin\eta}{\eta}\,d\eta=\begin{cases}\pi/2, & \text{if } x>a,\\ 0, & \text{if } x=a,\\ -\pi/2, & \text{if } x<a.\end{cases}$$

2° Now we use that $\operatorname{Im}e^{i\xi}=\sin\xi$ to get for $a<b$

$$\begin{aligned}\int_{-T}^{T}&\frac{e^{i\xi(x-a)}-e^{i\xi(x-b)}}{i\xi}\,d\xi\\ &=\int_{-T}^{0}\frac{e^{i\xi(x-a)}-e^{i\xi(x-b)}}{i\xi}\,d\xi+\int_0^T\frac{e^{i\xi(x-a)}-e^{i\xi(x-b)}}{i\xi}\,d\xi\\ &=-\int_0^T\frac{e^{-i\xi(x-a)}-e^{-i\xi(x-b)}}{i\xi}\,d\xi+\int_0^T\frac{e^{i\xi(x-a)}-e^{i\xi(x-b)}}{i\xi}\,d\xi\\ &=2\int_0^T\frac{\sin(\xi(x-a))}{\xi}\,d\xi-2\int_0^T\frac{\sin(\xi(x-b))}{\xi}\,d\xi.\end{aligned}$$

If we combine this with the result of Step 1° we get

$$\lim_{T\to\infty}\int_{-T}^{T}\frac{e^{i\xi(x-a)}-e^{i\xi(x-b)}}{i\xi}\,d\xi=\begin{cases}0, & \text{if } x<a \text{ or } x>b\\ \pi, & \text{if } x=a \text{ or } x=b\,.\\ 2\pi, & \text{if } a<x<b\end{cases}$$

3° Let us continue with a formal calculation. The steps marked with § and ‡ will be justified later on in 4° and 5° .

$$\begin{aligned}
\lim_{T\to\infty} \frac{1}{2\pi} \int_{-T}^{T} \frac{e^{-ia\xi} - e^{-ib\xi}}{i\xi} \check{\mu}(\xi)\, d\xi
&= \lim_{T\to\infty} \frac{1}{2\pi} \int_{-T}^{T} \frac{e^{-ia\xi} - e^{-ib\xi}}{i\xi} \int e^{ix\xi} \mu(dx)\, d\xi \\
&\overset{\S}{=} \lim_{T\to\infty} \frac{1}{2\pi} \int\int_{-T}^{T} \frac{e^{-ia\xi} - e^{-ib\xi}}{i\xi} e^{ix\xi}\, d\xi\, \mu(dx) \\
&= \lim_{T\to\infty} \frac{1}{2\pi} \int\int_{-T}^{T} \frac{e^{i(x-a)\xi} - e^{i(x-b)\xi}}{i\xi}\, d\xi\, \mu(dx) \\
&\overset{\ddagger}{=} \int \left[\frac{1}{2}\mathbb{1}_{\{a\}} + \mathbb{1}_{(a,b)} + \frac{1}{2}\mathbb{1}_{\{b\}}\right] d\mu \\
&= \tfrac{1}{2}\mu\{a\} + \mu(a,b) + \tfrac{1}{2}\mu\{b\}.
\end{aligned}$$

4° In § we use Fubini's theorem. This is admissible, since we conclude from the elementary inequality

$$|e^{it} - e^{is}| = \left|\int_s^t ie^{iu}\, du\right| \leqslant \int_s^t |ie^{iu}|\, du = |t - s|$$

that we have

$$\frac{1}{2\pi} \int_{-T}^{T} \left|\frac{e^{i\xi(x-a)} - e^{i\xi(x-b)}}{i\xi}\right| d\xi \leqslant \frac{1}{2\pi} \int_{-T}^{T} (b-a)\, d\xi = \frac{(b-a)T}{\pi}.$$

The rhS is μ-integrable, i.e. the integrability condition of Fubini's theorem is satisfied.

5° In ‡ we use the DCT. The integrand has the form $T \mapsto \int_0^T \frac{\sin u}{u}\, du$, and we have seen at the beginning of the proof that this is a bounded and continuous function. This allows us to use dominated convergence on the compact interval $[a, b]$. □

28 Three classic limit theorems

A classic and central topic of probability theory is the study of the limit behaviour of (sums of) indpendent random variables. We will use the following standard notation:

- $X_i : \Omega \to \mathbb{R}$, $i \in \mathbb{N}$, are iid (independent & identically distributed) rv.
- $S_n := X_1 + \cdots + X_n$ is the partial sum; we set $S_0 := 0$ »empty sum«.
- $X \sim Y$ and $X \sim \mu$ (μ is a probaility measure) means that $\mathbb{P}_X = \mathbb{P}_Y$ and $\mathbb{P}_X = \mu$, respectively.

Bortkiewicz' »law of small numbers« (1898)

Assume that the random variables $(X_i)_{i\in\mathbb{N}}$ are iid such that $X_i \sim \mathsf{Bin}(p)$, i.e. $\mathbb{P}(X_i = 1) = p$ and $\mathbb{P}(X_i = 0) = q = 1 - p$. We have

$$S_n = X_1 + \cdots + X_n \sim \mathsf{Bin}(n, p) \quad \text{i.e.} \quad \mathbb{P}(S_n = k) = \binom{n}{k} p^k q^{n-k}, \quad 0 \leqslant k \leqslant n.$$

Intuitive proof. $S_n = k$ means that we observe in n coin tosses exactly k heads and $n-k$ tails. This outcome has probability $p^k q^{n-k}$. Since we can place the heads in $\binom{n}{k}$ different positions, we get the total probability as asserted. □

Formal proof. We have $X_i \sim p\delta_1 + q\delta_0$. Using Theorem 25.14 and the fact that $\delta_a * \delta_b = \delta_{a+b}$, we get

$$\begin{aligned} X_1 + X_2 &\sim (p\delta_1 + q\delta_0) * (p\delta_1 + q\delta_0) \\ &= p^2 \underbrace{\delta_1 * \delta_1}_{=\delta_1^{*2}=\delta_2} + \underbrace{pq\delta_1 * \delta_0 + qp\delta_0 * \delta_1}_{=2pq\delta_1} + q^2 \underbrace{\delta_0 * \delta_0}_{=\delta_0}. \end{aligned}$$

This means that we can multiply out the convolution products like an ordinary binomial, and so

$$\begin{aligned} S_n = \sum_{i=1}^n X_i \sim (q\delta_0 + p\delta_1) * \cdots * (q\delta_0 + p\delta_1) &= \sum_{k=0}^n \binom{n}{k} p^k \delta_1^{*k} * q^{n-k} \delta_0^{*(n-k)} \\ &= \sum_{k=0}^n \binom{n}{k} p^k q^{n-k} \delta_k. \end{aligned}$$ □

Although $\mathsf{Bin}(n,p)$ is an elementary expression, the concrete evaluation is for large n complicated and some approximation is desirable. For large values of n and small p – i.e. for »rare events« – the **Poisson distribution** is a good approximation:

$$\mathsf{Bin}(n,p;k) = \binom{n}{k} p^k q^{n-k} \overset{\lambda=np}{\approx} e^{-\lambda}\frac{\lambda^k}{k!} = \mathsf{Poi}(\lambda;k), \quad k \in \mathbb{N}_0.$$

The following theorem contains the mathematically rigorous version of this fact.

28.1 Theorem (Poisson approximation). *Let $B_n \sim \mathsf{Bin}(n,p(n))$, $n \in \mathbb{N}$, be random variables on $(\Omega, \mathscr{A}, \mathbb{P})$. Then*

$$\lim_{n\to\infty} np(n) = \lambda \in [0,\infty) \implies \lim_{n\to\infty} \mathbb{P}(B_n = k) = e^{-\lambda}\frac{\lambda^k}{k!}. \tag{28.1}$$

Using Stirling's formula[(1)] one can check this limit relation directly, but we prefer the following more general argument since it gives a glimpse into the important coupling technique.

Proof. We show the following **auxiliary result:** *Let $X_i \sim \mathsf{Bin}(p_i)$, $1 \leqslant i \leqslant n$, be independent Bernoulli rv. Define $S_n := X_1 + \cdots + X_n$ and further independent rv $P_n \sim \mathsf{Poi}(\lambda_n)$ with $\lambda_n := p_1 + \cdots + p_n$. Then*

$$\sum_{k=0}^n \left|\mathbb{P}(S_n = k) - \mathbb{P}(P_n = k)\right| = \sum_{k=0}^n \left|\mathbb{P}(S_n = k) - e^{-\lambda_n}\frac{\lambda_n^k}{k!}\right| \leqslant 2\sum_{k=1}^n p_k^2. \tag{28.2}$$

(1) $n! = \sqrt{2\pi n}\, n^n e^{-n} e^{\theta(n)/12n}$ for all $n \in \mathbb{N}$ and some $0 < \theta(n) \leqslant 1$, cf. [3, Chapter 3] or [27, Chapter 8.22].

Let's take this claim for granted for now, and pick $p_1 = \cdots = p_n = p(n)$ (this means that we have $X_i = X_{i,n}$, $i = 1,\dots,n$) such that $\lambda_n = np(n) \to \lambda$. Since

$$2\sum_{k=1}^{n} p_k^2 = 2np(n)^2 = \underbrace{2np(n)}_{\to 2\lambda}\overbrace{p(n)}^{\to 0} \xrightarrow[n\to\infty]{} 0,$$

we see that Theorem 28.1 – and much more – follows from (28.2).

Proof of (28.2). Since (28.2) is an assertion on the distributions only, we can simplify the proof by a clever choice of the rv X_i and P_i.

1° Fix n and construct a probability space $(\Omega_n, \mathscr{A}_n, \mathbb{P}_n)$ and random variables $(X_1,\dots,X_n)$, $(Y_1,\dots,Y_n)$ with the following properties

- $X_i \sim \mathsf{Bin}(p_i)$, $i = 1,2,\dots,n$, are independent;
- $Y_i \sim \mathsf{Poi}(p_i)$, $i = 1,2,\dots,n$ are independent;
- the joint distribution of the bivariate rv (X_i, Y_i) is given in Table 28.1.

	$Y_i = 0$	$Y_i = y \in \mathbb{N}$	$\mathbb{P}_n(X_i = x) \Downarrow$
$X_i = 0$	$1 - p_i$	0	$1 - p_i$
$X_i = 1$	$e^{-p_i} - 1 + p_i$	$\frac{p_i^y}{y!} e^{-p_i}$	p_i
$\mathbb{P}_n(Y_i = y) \Rightarrow$	e^{-p_i}	$\frac{p_i^y}{y!} e^{-p_i}$	

Table 28.1. The table shows the joint probabilities $\mathbb{P}_n(X_i = x, Y_i = y)$. On the right and bottom margins we can read off the distributions of X_i, resp., Y_i. The joint distribution is the »coupling« of the marginals.

2° Now we define

$$S_n := X_1 + \cdots + X_n, \quad P_n := Y_1 + \cdots + Y_n, \quad \lambda_n := p_1 + \cdots + p_n,$$

and observe that $P_n \sim \mathsf{Poi}(\lambda_n)$, cf. Example 25.15. Therefore,

$$\begin{aligned}|\mathbb{P}_n(S_n = k) - \mathbb{P}_n(P_n = k)| &= |\mathbb{P}_n(S_n = k,\, P_n \neq S_n) + \mathbb{P}_n(S_n = k,\, P_n = S_n)\\ &\qquad - \mathbb{P}_n(P_n = k,\, S_n = P_n) - \mathbb{P}_n(P_n = k,\, S_n \neq P_n)|\\ &\leqslant \mathbb{P}_n(S_n = k,\, P_n \neq S_n) + \mathbb{P}_n(P_n = k,\, P_n \neq S_n).\end{aligned}$$

Summing this inequality over $k = 0,1,\dots,n$ yields

$$\sum_{k=0}^{n} \Big|\mathbb{P}_n(S_n = k) - \underbrace{\mathbb{P}_n(P_n = k)}_{= e^{-\lambda_n}\lambda_n^k/k!}\Big| \leqslant 2\mathbb{P}_n(S_n \neq P_n).$$

Let us estimate the right-hand side:

$$\mathbb{P}_n(S_n \neq P_n) \leqslant \mathbb{P}_n\left(\bigcup_{i=1}^n \{X_i \neq Y_i\}\right) \leqslant \sum_{i=1}^n \mathbb{P}_n(X_i \neq Y_i) = \sum_{i=1}^n (1 - \mathbb{P}_n(X_i = Y_i)),$$

and using the joint distribution from Table 28.1 gives

$$\mathbb{P}_n(S_n \neq P_n) \leqslant \sum_{i=1}^n \Big(1 - \overbrace{(1-p_i)}^{X_i=Y_i=0} - \overbrace{p_i e^{-p_i}}^{X_i=Y_i=1}\Big) = \sum_{i=1}^n p_i \underbrace{(1 - e^{-p_i})}_{\leqslant p_i} \leqslant \sum_{i=1}^n p_i^2. \qquad \square$$

Khintchine's weak law of large numbers (1929)

We will now show a first limit theorem for sums of independent (or uncorrelated) random variables.

28.2 Theorem (Khintchine; WLLN). *Assume that $X_i : \Omega \to \mathbb{R}$, $i \in \mathbb{N}$, are uncorrelated, square integrable ($X_i \in L^2(\mathbb{P})$) random variables. If*

$$\lim_{n\to\infty} \frac{1}{n^2} \sum_{i=1}^n \mathbb{V}X_i = 0,$$

then the $(X_i)_{i\in\mathbb{N}}$ satisfy the ***weak law of large numbers (WLLN)***

$$\forall \epsilon > 0 : \quad \lim_{n\to\infty} \mathbb{P}\left(\left|\frac{1}{n}\sum_{i=1}^n (X_i - \mathbb{E}X_i)\right| > \epsilon\right) = 0. \tag{28.3}$$

We will see in the next chapter that (28.3) means that $\frac{1}{n}\sum_{i=1}^n (X_i - \mathbb{E}X_i)$ converges in probability (or stochastically) to 0.

Proof. The Chebyshev-Markov inequality[(2)] shows that

$$\begin{aligned}\mathbb{P}\left(\left|\sum_{i=1}^n (X_i - \mathbb{E}X_i)\right| > n\epsilon\right) &\leqslant \frac{1}{(n\epsilon)^2}\mathbb{E}\left(\left|\sum_{i=1}^n (X_i - \mathbb{E}X_i)\right|^2\right) \\ &= \frac{1}{(n\epsilon)^2}\mathbb{V}\left(\sum_{i=1}^n X_i\right) \\ &\overset{\S 25.18}{=} \frac{1}{(n\epsilon)^2}\sum_{i=1}^n \mathbb{V}X_i \xrightarrow[n\to\infty]{} 0. \qquad \square\end{aligned}$$

(2) See Corollary 10.5. The Markov inequality is $\mathbb{P}(|X| > c) \leqslant c^{-1}\mathbb{E}|X|$. Taking $X \rightsquigarrow |X - \mathbb{E}X|^2$ gives the Chebyshev inequality

$$\mathbb{P}(|X - \mathbb{E}X| > c) = \mathbb{P}(|X - \mathbb{E}X|^2 > c^2) \leqslant c^{-2}\mathbb{E}[(X - \mathbb{E}X)^2] = c^{-2}\mathbb{V}X.$$

For iid Bernoulli random variables this theorem has been known much longer: Jacob Bernoulli knew this result as early as 1685/90 although it was published 8 years after his death in 1705 (cf. [7, Chapter 1]).

28.3 Corollary (Bernoulli 1713). *Let* $X_i : \Omega \to \mathbb{R}$, $i \in \mathbb{N}$, *be iid with* $X_1 \sim \mathsf{Bin}(p)$. *Then*

$$\forall \epsilon > 0 : \quad \lim_{n\to\infty} \mathbb{P}\left(\left|\frac{S_n}{n} - p\right| > \epsilon\right) = 0.$$

Proof. Since $|X_i| \leqslant 1$, we have $X_i \in L^2(\mathbb{P})$. The claim follows from Theorem 28.2 since

$$\frac{1}{n^2}\sum_{i=1}^n \mathbb{V}X_i = \frac{1}{n^2} n \mathbb{V}X_1 = \frac{1}{n}\mathbb{V}X_1 \xrightarrow[n\to\infty]{} 0. \qquad \square$$

28.4 Remark. The interpretation of Corollary 28.3 is very important: X_i is the result of the ith coin toss of a »p-q-coin«.

- $n^{-1}S_n$ is the **observed** relative frequency of »head«.
- p is the **theoretically** expected relative frequency of »head«.
- $\mathbb{P}\left(\left|n^{-1}S_n - p\right| > \epsilon\right)$ is the probability that theory and praxis differ by ϵ.

⚠ Corollary 28.3 **does not** mean that we have $n^{-1}S_n(\omega) \to p$ for some (or every) sequence of coin tosses ω.

⚠ Corollary 28.3 **does not** say that after observing a run of 1000 »tails«, a »head« is more likely to come next.

⚠ Corollary 28.3 **cannot be used** to argue that the relative frequency stabilizes and hence, the relative frequency can serve as **definition** of »probability«. This would be a circular conclusion: Corollary 28.3 requires, a priori, the probability (measure) $\mathbb{P}$.

A beautiful application of Bernoulli's WLLN is Bernstein's proof of the Weierstraß approximation theorem.

28.5 Theorem (Weierstraß; Bernstein). *Let* $u : [0,1] \to \mathbb{R}$ *be a continuous function. It holds that*

$$\lim_{n\to\infty} \sup_{0\leqslant x\leqslant 1} \Bigg| u(x) - \underbrace{\sum_{k=0}^n \binom{n}{k} u\left(\tfrac{k}{n}\right) x^k (1-x)^{n-k}}_{\text{so-called Bernstein polynomial}} \Bigg| = 0. \tag{28.4}$$

In particular, the polynomials on $[0,1]$ *are a dense subset of* $(C[0,1], \|\cdot\|_\infty)$.

Proof. Since $[0,1]$ is a compact set, u is uniformly continuous, i.e.

$$\forall \epsilon > 0 \quad \exists \delta = \delta(\epsilon) > 0 \quad \forall x, y \in [0,1],\ |x-y| < \delta : \quad |u(x) - u(y)| < \epsilon. \tag{28.5}$$

By definition, we have

$$\mathbb{E}u\left(\tfrac{S_n}{n}\right) = \sum_{k=0}^{n} u\left(\tfrac{k}{n}\right)\mathbb{P}(S_n = k) = \sum_{k=0}^{n}\binom{n}{k} u\left(\tfrac{k}{n}\right)p^k(1-p)^{n-k}.$$

This means that we have to estimate the following expression

$$\begin{aligned}
\left|\mathbb{E}u\left(\tfrac{S_n}{n}\right) - u(p)\right| &= \left|\mathbb{E}\left[u\left(\tfrac{S_n}{n}\right) - u(p)\right]\right| \\
&\leqslant \mathbb{E}\left|u\left(\tfrac{S_n}{n}\right) - u(p)\right| \\
&= \mathbb{E}\left[\left|u\left(\tfrac{S_n}{n}\right) - u(p)\right| \mathbb{1}_{\left\{\left|\frac{S_n}{n}-p\right|<\delta\right\}}\right] + \mathbb{E}\left[\left|u\left(\tfrac{S_n}{n}\right) - u(p)\right| \mathbb{1}_{\left\{\left|\frac{S_n}{n}-p\right|\geqslant\delta\right\}}\right] \\
&\overset{(28.5)}{\leqslant} \mathbb{E}\left[\epsilon \mathbb{1}_{\left\{\left|\frac{S_n}{n}-p\right|<\delta\right\}}\right] + \mathbb{E}\left[(\|u\|_\infty + \|u\|_\infty)\mathbb{1}_{\left\{\left|\frac{S_n}{n}-p\right|\geqslant\delta\right\}}\right] \\
&\leqslant \epsilon + 2\|u\|_\infty \mathbb{P}\left(\left|\tfrac{S_n}{n} - p\right| \geqslant \delta\right) \\
&\xrightarrow[n\to\infty]{\epsilon,\delta \text{ fixed}} \epsilon \xrightarrow[\epsilon\to 0]{} 0.
\end{aligned}$$

Since all limits are uniform in $p \in [0,1]$, the claim follows. □

The central limit theorem of DeMoivre–Laplace (1812)

The WLLN (Theorem 28.2) does not tell us anything about the limit of the distribution of S_n/n. If we assume that the rv X_i are iid, we can show the central limit theorem (CLT) which is of paramount importance in mathematical statistics.

28.6 Theorem (DeMoivre–Laplace; CLT). *Assume that $(X_i)_{i\in\mathbb{N}}$ are real-valued iid random variables such that $X_1 \in L^2(\mathbb{P})$, and let $S_n = X_1 + \cdots + X_n$, $\mu = \mathbb{E}X_1$ and $\sigma^2 = \mathbb{V}X_1 > 0$. Then one has for all $-\infty < a < b < \infty$*

$$\lim_{n\to\infty} \mathbb{P}\left(a < \frac{S_n - n\mu}{\sigma\sqrt{n}} \leqslant b\right) = \frac{1}{\sqrt{2\pi}}\int_a^b e^{-x^2/2}\,dx. \qquad (28.6)$$

For the proof we need an auxiliary result which we will prove in Chapter 30 (Theorem 30.3).

28.7 Theorem. *Let $S_n, S : \Omega \to \mathbb{R}$ be real random variables.*

$$\lim_{n\to\infty} \mathbb{E}e^{i\xi S_n} = \mathbb{E}e^{i\xi S} \implies \begin{cases} \lim_{n\to\infty} \mathbb{P}(S_n \leqslant x) = \mathbb{P}(S \leqslant x) \\ \forall \text{ continuity points } x \text{ of } x \mapsto \mathbb{P}(S \leqslant x) \end{cases} \qquad (28.7)$$

Proof of Theorem 28.6. Let us normalize X_i by setting $\widetilde{X}_i := (X_i - \mu)/\sigma$. Note that

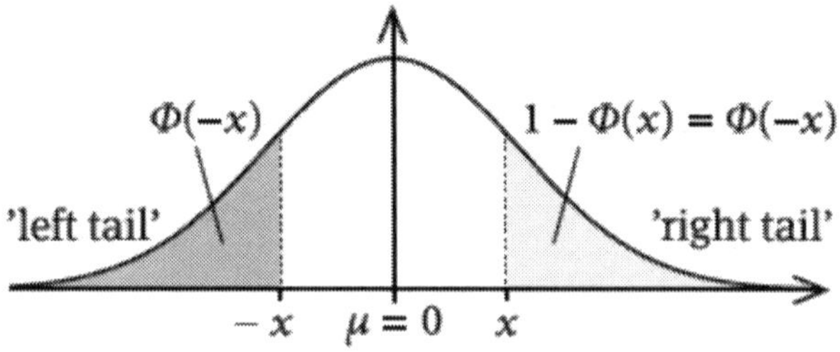

Fig. 28.1. The distribution function $\Phi(x)$ of the standard normal law $\mathsf{N}(0,1)$.

$\mathbb{E}\widetilde{X}_i = 0$ and $\mathbb{V}\widetilde{X}_i = 1$. The normalized rv $\bar{X}_i$ still are iid, and so

$$\mathbb{E}e^{i\xi\frac{\widetilde{S}_n}{\sqrt{n}}} = \mathbb{E}e^{i\frac{\xi}{\sqrt{n}}(\widetilde{X}_1+\cdots+\widetilde{X}_n)} \overset{\perp\!\!\!\perp}{=} \prod_{k=1}^{n} \mathbb{E}e^{i\frac{\xi}{\sqrt{n}}\widetilde{X}_k} \overset{\text{iid}}{=} \left(\mathbb{E}e^{i\frac{\xi}{\sqrt{n}}\widetilde{X}_1}\right)^n.$$

Denote by $\phi(\eta) := \mathbb{E}e^{i\eta\widetilde{X}_1}$ the characteristic function $\widetilde{X}_1$. Since $\widetilde{X}_1$ has two moments, we know from Theorem 27.5.g) that $\phi \in C^2$ and

$$\phi'(\eta) = i\mathbb{E}\left(\widetilde{X}_1 e^{i\eta\widetilde{X}_1}\right) \implies \phi'(0) = 0$$
$$\phi''(\eta) = -\mathbb{E}\left(\widetilde{X}_1^2 e^{i\eta\widetilde{X}_1}\right) \implies \phi'(0) = -1.$$

Using Taylor's theorem, we get

$$\phi(\eta) = 1 + 0 - \frac{1}{2}\eta^2 + \eta^2 R(\eta).$$

Since ϕ'' is continuous, the remainder terms satisfies $\lim_{\eta\to 0} R(\eta) = 0$. So,

$$\mathbb{E}e^{i\xi\frac{\widetilde{S}_n}{\sqrt{n}}} = \left\{\phi\left(\frac{\xi}{\sqrt{n}}\right)\right\}^n = \left\{1 - \frac{\xi^2}{2n} + \frac{\xi^2}{n}R\left(\frac{\xi}{\sqrt{n}}\right)\right\}^n \xrightarrow[n\to\infty]{} e^{-\frac{1}{2}\xi^2}$$

where we use the elementary relation

$$\lim_{n\to\infty}\left(1 - \frac{a_n}{n}\right)^n = e^{-\lim_{n\to\infty} a_n}.$$

This shows that for some normal rv $G \sim \mathsf{N}(0,1)$

$$\mathbb{E}e^{-i\xi\widetilde{S}_n/\sqrt{n}} \xrightarrow[n\to\infty]{} e^{-\xi^2/2} \overset{\S 27.3}{=} \mathbb{E}e^{i\xi G},$$

and Theorem 28.6 tells us that

$$\lim_{n\to\infty} \mathbb{P}\Bigg(a < \underbrace{\frac{S_n - n\mu}{\sigma\sqrt{n}}}_{=\widetilde{S}_n/\sqrt{n}} \leqslant b\Bigg) = \mathbb{P}(G \leqslant b) - \mathbb{P}(G \leqslant a) = \frac{1}{\sqrt{2\pi}}\int_a^b e^{-x^2/2}\,dx. \qquad \square$$

28.8 Example. A classic application of the CLT is, again, the approximation of $\mathsf{Bin}(n,p)$ – now, however, for **typical** and not for rare events. If $X_1 \sim \mathsf{Bin}(p)$, then

$$\mathbb{E}X_1 = 1\cdot\mathbb{P}(X=1) + 0\cdot\mathbb{P}(X=0) = p$$
$$\mathbb{V}X_1 = (1-p)^2\cdot\mathbb{P}(X=1) + (0-p)^2\cdot\mathbb{P}(X=0) = q^2p + p^2q = pq.$$

The sum $S_n = X_1 + \dots + X_n$ of iid $\mathsf{Bin}(p)$ random variables satisfies

$$\mathbb{E}S_n = \mathbb{E}(X_1+\dots+X_n) = \sum_{i=1}^n \mathbb{E}X_i \overset{\text{iid}}{=} np,$$

$$\mathbb{V}S_n = \mathbb{V}(X_1+\dots+X_n) \underset{§25.18}{\overset{\text{Bienaymé}}{=}} \sum_{i=1}^n \mathbb{V}X_i \overset{\text{iid}}{=} npq.$$

a) We toss a fair coin 100 times. What is the probability to obtain »at least 60 heads«? If X_i is the ith coin toss, then we are asking for

$$\mathbb{P}(S_{100} \geqslant 60) = \mathbb{P}\left(\frac{S_{100}-\mathbb{E}S_{100}}{\sqrt{\mathbb{V}S_{100}}} \geqslant \frac{60-50}{5} = 2\right)$$
$$\approx \frac{1}{\sqrt{2\pi}}\int_2^\infty e^{-x^2/2}\,dx = 1-\Phi(2) \approx 0,02.$$

b) Let $X_1, X_2, \dots$ be iid $\mathsf{Bin}(p)$-rv. The partial sum satisfies $S_n \sim \mathsf{Bin}(n,p)$ and we can normalize it

$$\frac{S_n - \mathbb{E}S_n}{\sqrt{\mathbb{V}S_n}} = \frac{S_n - np}{\sqrt{npq}} = \sqrt{\frac{n}{pq}}\left(\frac{S_n}{n} - p\right).$$

Therfore, we may interpret (28.6) as a relation between the relative frequency S_n/n and the theoretical probability p. We see

$$\mathbb{P}\left(a < \sqrt{\frac{n}{pq}}\left(\frac{S_n}{n}-p\right) < b\right) \approx \Phi(b) - \Phi(a). \tag{28.6'}$$

$$\mathbb{P}\left(a\sqrt{\frac{pq}{n}} < \frac{S_n}{n} - p < b\sqrt{\frac{pq}{n}}\right) \approx \Phi(b) - \Phi(a), \tag{28.6''}$$

and frequently one assumes that $b = -a > 0$, i.e.

$$\mathbb{P}\left(\left|\frac{S_n}{n} - p\right| < b\sqrt{\frac{pq}{n}}\right) \approx 2\Phi(b) - 1. \tag{28.6'''}$$

c) (election forecast 1) **After** the election we know that a political party P has got $p = 37,5\%$ of all votes. How many voters n **should we have polled before the election day**, to predict with 99.5% likelihood the true value p with an error of $\pm1\%$. If we have a random sample of voters, we can assume that $X_i \sim \mathsf{Bin}(p)$ and that $X_i = 1$ means »I vote for P«; thus, S_n/n is the fraction of voters who (claim that they would) vote for P. According to (28.6''') we get

$$\mathbb{P}\left(\left|\frac{S_n}{n} - 0.375\right| \leqslant \underbrace{\overbrace{0,01}^{\pm1\%}}_{=b\sqrt{\frac{pq}{n}}}\right) \geqslant \underbrace{0,995}_{=2\Phi(b)-1}.$$

Thus, we have to solve the following system of equations for n

$$b = \sqrt{\frac{n}{pq}} \times 0.01 = \frac{\sqrt{n} \times 0.01}{\sqrt{0.375 \times 0.625}} \quad \text{and} \quad 2\Phi(b) - 1 \geqslant 0.995.$$

Ths means that $\Phi(b) \geqslant \frac{1.995}{2}$, and using a table for the normal distribution gives $b \geqslant 2.81$. Inserting this and $p = 0.375$ into the first equation reveals that

$$\sqrt{n} \geqslant \frac{2.81}{0.01} \times \sqrt{pq} = \frac{2.81}{0.01} \times \sqrt{0.375 \times 0.625} \implies n \geqslant 18,507.$$

d) (election forecast 2) **Before** the election day we do not know p, i.e. the question asked in the previous example is a bit academic. Instead, we have to use the following »worst case« observation:

$$\max_{0 \leqslant p \leqslant 1} pq = \frac{1}{4} \implies \sqrt{pq} \leqslant \frac{1}{2}.$$

As before, we have

$$\sqrt{n} \geqslant \frac{2.81}{0.01} \times \sqrt{pq}$$

and we can guarantee this by the following lower estimate:

$$\sqrt{n} \geqslant \frac{2.81}{0.01} \times \frac{1}{2} \geqslant \frac{2.81}{0.01} \times \sqrt{pq} \quad \implies \quad n \geqslant \left(\frac{281}{2}\right)^2 \approx 19,741.$$

If we ask a fixed number of voters, this means that the forecast of a political party which is »large« $p \in (0.2, 0.8)$ is less precise than a forecast for a »small« $p \in (0, 0.1)$. See also the »package leaflet« of the *Deutschlandtrend* `www.tagesschau.de/inland/deutschlandtrend/` (accessed 26/01/2021).

29 Convergence of random variables

In Chapter 28 we have already seen limit theorems for sequences of random variables, e.g. the WLLN (weak law of large numbers) or the CLT (central limit theorem). We will add in the following chapters the SLLN (strong law of large numbers). The main difference between these results is the mode of convergence of the random variables. In this chapter we will systematically discuss the most common convergence modes for random variables.

29.1 Definition. Let the random variables $X, X_n : \Omega \to \mathbb{R}$, $n \in \mathbb{N}$, be defined on the same probability space $(\Omega, \mathscr{A}, \mathbb{P})$. We say that X_n **converges to** X **almost surely (a.s.)** ($\lim_{n\to\infty} X_n = X$ a.s., $X \xrightarrow{\text{a.s.}} X$), if

$$\mathbb{P}\Big(\Big\{\omega \in \Omega \mid \lim_{n\to\infty} X_n(\omega) = X(\omega)\Big\}\Big) = 1. \tag{29.1}$$

in L^p, $1 \leqslant p < \infty$ or in pth mean ($L^p\text{-}\lim_{n\to\infty} X_n = X$, $X_n \xrightarrow{L^p} X$), if

$$X_n, X \in L^p(\mathbb{P}) \quad \text{and} \quad \lim_{n\to\infty} \left[\mathbb{E}\left(|X_n - X|^p\right)\right]^{1/p} = 0. \tag{29.2}$$

in probability or **stochastically** ($\mathbb{P}\text{-}\lim_{n\to\infty} X_n = X$, $X_n \xrightarrow{\mathbb{P}} X$), if

$$\forall \epsilon > 0: \quad \lim_{n\to\infty} \mathbb{P}\left(|X_n - X| > \epsilon\right) = 0. \tag{29.3}$$

Frequently, a further type of convergence is important, but in this case the random variables may be defined on **different** probability spaces, as we are interested only in their probability distribution.

29.2 Definition (d-convergence). Let X, X_n be real random variables which are not necessarily defined on the same probability space. The sequence X_n converges to X **in distribution** or **weakly** ($X_n \xrightarrow{d} X$, $X_n \Rightarrow X$, $\mathbb{P}_{X_n} \xrightarrow{w} \mathbb{P}_X$), if

$$\forall f \in C_b(\mathbb{R}): \quad \lim_{n\to\infty} \mathbb{E} f(X_n) = \mathbb{E} f(X). \tag{29.4}$$

$$\left[\iff \forall f \in C_b(\mathbb{R}): \quad \lim_{n\to\infty} \int f(x)\,\mathbb{P}(X_n \in dx) = \int f(x)\,\mathbb{P}(X \in dx).\right]$$

29.3 Remark. a) The limits in Definition 29.1 are unique (hence, well-defined): $X_n \xrightarrow{\text{a.s.}/L^p/\mathbb{P}} X \;\;\&\;\; X_n \xrightarrow{\text{a.s.}/L^p/\mathbb{P}} Y \implies X \overset{\text{a.s.}}{=} Y$. Indeed,

- **a.s. convergence:** We have

$$\{X = Y\} = \left\{\lim_{n\to\infty} X_n = X\right\} \cap \left\{\lim_{n\to\infty} X_n = Y\right\},$$

and the intersection of sets of measure 1 is again a set of measure 1, so $\mathbb{P}(X = Y) = 1$.

- **L^p-convergence:** Using the Minkowski inequality yields

$$\begin{aligned}(\mathbb{E}|X - Y|^p)^{1/p} &= (\mathbb{E}|X - X_n + X_n - Y|^p)^{1/p} \\ &\leqslant (\mathbb{E}|X - X_n|^p)^{1/p} + (\mathbb{E}|X_n - Y|^p)^{1/p}.\end{aligned}$$

Each of the terms on the right-hand side converges to 0 as $n \to \infty$, hence $\mathbb{E}|X - Y|^p = 0$ and $X = Y$ a.s.

- **$\mathbb{P}$-convergence:** From the triangle inequality we find for every $\epsilon > 0$

$$\{|X - Y| > 2\epsilon\} \subset \{|X - X_n| > \epsilon\} \cup \{|X_n - Y| > \epsilon\}$$

Indeed, if we had both $|X - X_n| \leqslant \epsilon$ and $|X_n - Y| \leqslant \epsilon$, then we would get

$$|X - Y| \leqslant |X - X_n| + |X_n - Y| \leqslant 2\epsilon$$

which is a contradiction. Thus,

$$\mathbb{P}(|X - Y| > 2\epsilon) \leqslant \underbrace{\mathbb{P}(|X - X_n| > \epsilon)}_{\xrightarrow[n\to\infty]{} 0} + \underbrace{\mathbb{P}(|X_n - X| > \epsilon)}_{\xrightarrow[n\to\infty]{} 0}.$$

Now we take $\epsilon = 1/k$, $k \in \mathbb{N}$, and get

$$\mathbb{P}(X \neq Y) = \mathbb{P}(|X-Y|>0) = \mathbb{P}\Big(\bigcup_{k\in\mathbb{N}}\{|X-Y|>\tfrac{1}{k}\}\Big) \leqslant \sum_{k\in\mathbb{N}} \underbrace{\mathbb{P}\big(|X-Y|>\tfrac{1}{k}\big)}_{=0} = 0$$

which shows that $X = Y$ a.s.

b) Limits in distribution (Definition 29.2) are unique in the following sense: $X_n \xrightarrow{d} X$, $X_n \xrightarrow{d} Y \implies X \sim Y$. (Alternatively, we can express this for the probability laws in the following way: $\mathbb{P}_{X_n} \xrightarrow{w} \mathbb{P}_X$, $\mathbb{P}_{X_n} \xrightarrow{w} \mathbb{P}_Y \implies \mathbb{P}_X = \mathbb{P}_Y$ and this is the same as to say that $X \sim Y$.)

Indeed, as $e^{i\xi\bullet} \in C_b(\mathbb{R})$, we get

$$X_n \xrightarrow{d} X \implies \mathbb{E}e^{i\xi X_n} \to \mathbb{E}e^{i\xi X}$$

and the c.f. $\mathbb{E}e^{i\xi X}$ uniquely determines $\mathbb{P}_X$ and X (Theorem 27.7).

c) We have in Definition 29.1

$$X_n \xrightarrow{\text{a.s.}/L^p/\mathbb{P}} X \iff X_n - X \xrightarrow{\text{a.s.}/L^p/\mathbb{P}} 0.$$

⚠ this does not make sense for d-convergence: We cannot speak about differences $X_n - X_m$ if these rv are not defined on the same space!

d) Both L^p- and a.s. convergence is defined via complete norms, i.e. a sequence converges in L^p or a.s. if, and only if, the sequence is a Cauchy sequence; for a.s. convergence this should be known from anaylsis, for L^p convergence see Theorem 14.10. One can show that $\mathbb{P}$-convergence can be defined through a complete metric. The »easy« direction that a $\mathbb{P}$-convergent sequence $X_n \xrightarrow{\mathbb{P}} X$ is a $\mathbb{P}$-Cauchy sequence follows with the argument from a): For $n, k \in \mathbb{N}$ and $\epsilon > 0$ we have

$$\mathbb{P}(|X_n - X_k| > 2\epsilon) \leqslant \mathbb{P}(|X_n - X| > \epsilon) + \mathbb{P}(|X_k - X| > \epsilon) \xrightarrow{k,n\to\infty} 0.$$

For the converse, see the appendix to this chapter.

e) Definition 29.1 and 29.2 still hold for d-dimensional random variables ✍: If $X = (X^{(1)}, \dots, X^{(d)})$, $X_n = (X_n^{(1)}, \dots, X_n^{(d)}) \in \mathbb{R}^d$, then

$$X_n \xrightarrow{\text{a.s.}/L^p/\mathbb{P}} X \iff \begin{cases} X_n^{(k)} \xrightarrow{\text{a.s.}/L^p/\mathbb{P}} X^{(k)} \\ \forall k = 1, \dots, d. \end{cases}$$

This can be shown with a reasoning that resembles the one used in a) (if we think of $X \rightsquigarrow X^{(1)}$ and $Y \rightsquigarrow Y^{(2)}$ etc.).

29.4 Remark. The connection between the different modes of convergence is shown in Fig. 29.1.

a) $X_n \xrightarrow{L^p} X \implies X_n \xrightarrow{L^1} X$ for all $1 \leqslant p < \infty$. Let $\frac{1}{p} + \frac{1}{q} = 1$. An application of the Hölder inequality shows

$$\begin{aligned}\mathbb{E}|X_n - X| &= \int |X_n - X| \cdot 1 \, d\mathbb{P} \\ &\leqslant \left(\int |X_n - X|^p \, d\mathbb{P} \right)^{1/p} \left(\int 1^q \, d\mathbb{P} \right)^{1/q} \\ &= \Big(\mathbb{E}\big(|X_n - X|^p \big) \Big)^{1/p} \xrightarrow[n\to\infty]{} 0.\end{aligned}$$

This calculation shows, in particular, that $L^p(\mathbb{P}) \subset L^1(\mathbb{P})$.

b) $X_n \xrightarrow{L^1} X \implies X_n \xrightarrow{\mathbb{P}} X$. From the Chebyshev–Markov inequality we get

$$\forall \epsilon > 0 : \quad \mathbb{P}(|X_n - X| > \epsilon) \leqslant \frac{1}{\epsilon} \mathbb{E}|X_n - X| \xrightarrow[n\to\infty]{} 0.$$

c) $X_n \xrightarrow{\text{a.s.}} X \implies X_n \xrightarrow{\mathbb{P}} X$. We have for all $\epsilon > 0$

$$\mathbb{P}(|X_n - X| > \epsilon) = \mathbb{E} \mathbb{1}_{\{|X_n - X| > \epsilon\}} = \mathbb{E} \mathbb{1}_{[-\epsilon,\epsilon]^c}(X_n - X).$$

Since the random variables $Y_n := \mathbb{1}_{[-\epsilon,\epsilon]^c}(X_n - X) \leqslant 1 \in L^p(\mathbb{P})$ have an integ-

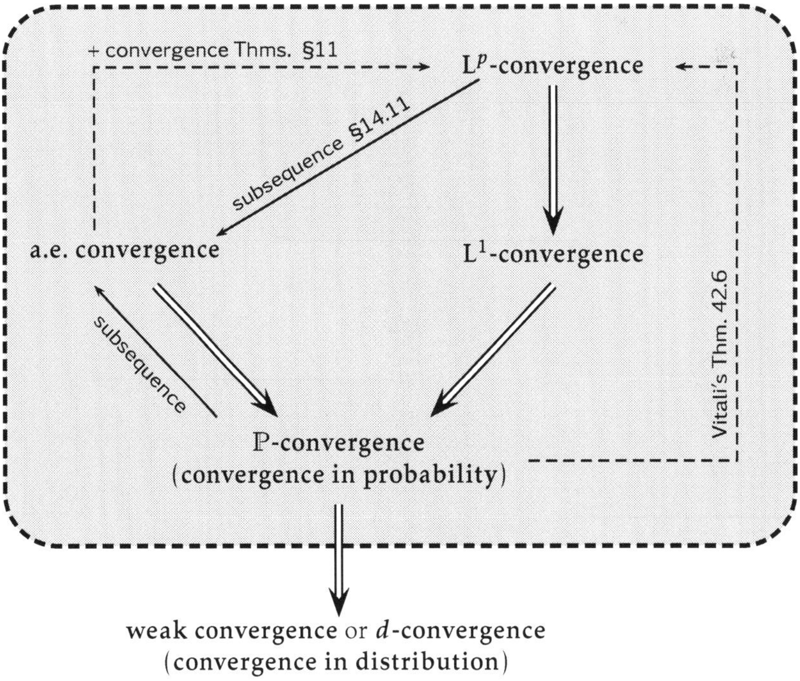

Fig. 29.1. Overview over the different modes of convergence.

rable majorant and since $Y_n \xrightarrow[n\to\infty]{} 0$ a.s., we can use the DCT to see

$$\forall \epsilon > 0 : \quad \mathbb{P}(|X_n - X| > \epsilon) \xrightarrow[n\to\infty]{} 0.$$

29.5 Theorem. *If* $X, X_n : \Omega \to \mathbb{R}$, $n \in \mathbb{N}$, *are random variables satisfying* $X_n \xrightarrow{\mathbb{P}} X$, *then* $X_n \xrightarrow{d} X$.

Amendment: *For all* $f \in C_b(\mathbb{R})$ *it holds that* $f(X_n) \xrightarrow{L^1} f(X)$.

Proof. It is enough to prove the amendment. Pick $f \in C_b(\mathbb{R})$ and $\epsilon > 0$.

1° $\exists N \quad \forall k \geqslant N : \mathbb{P}(|X| > k) < \epsilon$. Since $\{|X| > k\} \downarrow \emptyset$ if $k \uparrow \infty$, this assertion follows from the continuity of $\mathbb{P}$ at the empty set $\emptyset$.

2° Since $f \in C_b$, there is a constant M such that $|f| \leqslant M$ and $f|_{[-N-1,N+1]}$ is uniformly continuous. If we take N as in 1° , this means that

$$\exists \delta = \delta(\epsilon) > 0, \quad \forall |x|, |y| \leqslant N+1, \ |x-y| \leqslant \delta : \quad |f(x) - f(y)| \leqslant \epsilon.$$

Without loss of generality, we can assume that $\delta < 1$. Therefore,

$$\begin{aligned}
\left|\mathbb{E}f(X_n) - \mathbb{E}f(X)\right| &\leqslant \mathbb{E}|f(X_n) - f(X)| \\
&= \int\limits_{\substack{\{|X_n - X| \leqslant \delta\} \\ \cap \{|X| \leqslant N\}}} + \int\limits_{\substack{\{|X_n - X| \leqslant \delta\} \\ \cap \{|X| > N\}}} + \int\limits_{\{|X_n - X| > \delta\}} |f(X_n) - f(X)| \, d\mathbb{P} \\
&\leqslant \overbrace{\epsilon \mathbb{P}(|X_n - X| \leqslant \delta, |X| \leqslant N)}^{\text{uniform cont. } \& \ |X_n| \leqslant |X_n - X| + |X| \leqslant \delta + N \leqslant 1 + N} \\
&\qquad + \overbrace{2M\mathbb{P}(|X| > N) + 2M\mathbb{P}(|X - X_n| > \delta)}^{|f(x) - f(y)| \leqslant 2\|f\|_\infty \leqslant 2M} \\
&\leqslant \epsilon + 2M\epsilon + 2M\mathbb{P}(|X_n - X| > \delta) \\
&\xrightarrow[n\to\infty]{} (2M+1)\epsilon + 0 \xrightarrow[\epsilon\to 0]{} 0.
\end{aligned}$$

□

Let us now turn to the subsequential limits shown in Fig. 29.1. We need an auxiliary result which is due to Borel, it is the measure-theoretic »easy direction« of the Borel–Cantelli lemma.

29.6 Lemma (Borel–Cantelli; easy direction). *Let* $(A_n)_{n\in\mathbb{N}} \subset \mathscr{A}$.

$$\sum_{n=1}^{\infty} \mathbb{P}(A_n) < \infty \implies \mathbb{P}(\underbrace{A_n \text{ infinitely often}}_{= \limsup_{n\to\infty} A_n}) = 0.$$

Usually we write $\mathbb{P}(A_n$ **i.o.**$)$. What is **meant** by this assertion is that infinitely many of the events A_n occur – this is made precise in the proof below.

Proof. Recall the definition of the limsup of sets:

$$\omega \in \limsup_{n\to\infty} A_n = \bigcap_{k\in\mathbb{N}} \bigcup_{n\geqslant k} A_n \iff \forall k : \omega \in \bigcup_{n\geqslant k} A_n$$
$$\iff \omega \text{ is contained in } \infty\text{-many } A_n$$
$$\iff \sum_{n=1}^{\infty} \mathbb{1}_{A_n}(\omega) = \infty.$$

Together with our assumption, this shows that

$$\mathbb{E}\left(\sum_{n=1}^{\infty} \mathbb{1}_{A_n}\right) \overset{\text{Tonelli}}{=} \sum_{n=1}^{\infty} \mathbb{E}\mathbb{1}_{A_n} = \sum_{n=1}^{\infty} \mathbb{P}(A_n) < \infty$$

and we see that $\sum_{n=1}^{\infty} \mathbb{1}_{A_n} < \infty$ a.s., i.e. $\mathbb{P}(A_n \text{ i.o.}) = 0$. □

29.7 Lemma (»fast« $\mathbb{P}$-conv. $\Rightarrow$ a.s. conv.). *Assume that* $X_n \xrightarrow{\mathbb{P}} X$ *and that for some real sequence* $\epsilon_n \downarrow 0$ *it holds that* $\sum_{n=0}^{\infty} \mathbb{P}(|X_n - X| > \epsilon_n) < \infty$. *Then*

$$X_n \xrightarrow{a.s.} X.$$

💬 The convergence of the series with general term $\mathbb{P}(|X_n - X| > \epsilon_n)$ guarantees that $\mathbb{P}(|X_n - X| > \epsilon_n)$ converges to zero at least as fast as $\frac{1}{n}$ – in this sense we deal with »fast« $\mathbb{P}$-convergence.

Proof. Since the series converges, we know from the Borel–Cantelli Lemma 29.6 that $\mathbb{P}(|X_n - X| > \epsilon_n \text{ i.o.}) = 0$. We can re-write this as

$$\mathbb{P}\Big(|X_n - X| > \epsilon_n \text{ for finitely many } n\Big) = 1$$

and this is tantamount that there is some $\Omega' \subset \Omega$, $\mathbb{P}(\Omega') = 1$ with

$$\forall \omega \in \Omega' \quad \exists N(\omega) \quad \forall n \geqslant N(\omega) : \quad |X_n(\omega) - X(\omega)| \leqslant \epsilon_n.$$

Thus, $|X_n(\omega) - X(\omega)| \xrightarrow[n\to\infty]{} 0$ for all $\omega \in \Omega'$, hence a.s. □

29.8 Corollary. *If* $X_n \xrightarrow{\mathbb{P}} X$, *then there is a subsequence* $(X_{n(k)})_k \subset (X_n)_n$ *such that* $X_{n(k)} \xrightarrow[k\to\infty]{a.s.} X$.

Proof. By assumption, we have

$$\forall k \in \mathbb{N},\ \epsilon > 0 \quad \exists N(k,\epsilon) \quad \forall n \geqslant N(k,\epsilon) : \quad \mathbb{P}(|X_n - X| > \epsilon) \leqslant 2^{-k}.$$

If we pick $\epsilon = 2^{-k}$ and $n(k) := N(k, 2^{-k})$, then

$$\sum_{k=1}^{\infty} \mathbb{P}(|X_{n(k)} - X| > 2^{-k}) \leqslant \sum_{k=1}^{\infty} 2^{-k} < \infty.$$

The claim now follows from Lemma 29.7. □

We can characterize $\mathbb{P}$-convergence by subsequences.

29.9 Corollary* (subsequence principle). *Let $X_n, X : \Omega \to \mathbb{R}$. One has $X_n \xrightarrow{\mathbb{P}} X$ if, and only if, every subsequence $(X'_n)_{n\in\mathbb{N}} \subset (X_n)_{n\in\mathbb{N}}$ contains a further subsequence $(X''_n)_{n\in\mathbb{N}} \subset (X'_n)_{n\in\mathbb{N}}$ such that $X''_n \xrightarrow{a.s.} X$*

Proof. The direction »⇒« follows from Corollary 29.8.

Conversely, »⇐«, assume that every subsequence of $(X_n)_{n\in\mathbb{N}}$ has a further subsequence which converges a.s. to the same limit X. If X_n does not converge to X in probability, then there is some $\epsilon > 0$ and a subsequence $(X'_n)_{n\in\mathbb{N}} \subset (X_n)_{n\in\mathbb{N}}$ such that $\mathbb{P}(|X - X'_n| > \epsilon) \geqslant \epsilon$ for all $n \gg 1$ – but this subsequence cannot have a sub-subsequence which converges a.s. to X, and we have reached a contradiction. □

29.10 Lemma. *The convergence $X_n \xrightarrow{a.s.} X$ takes place if, and only if,*

$$\forall \epsilon > 0 : \quad \mathbb{P}(|X_n - X| > \epsilon \text{ i.o.}) = 0.$$

Proof. Fix $\epsilon > 0$ and define the set $N_\epsilon := \{|X_n - X| > \epsilon \text{ i.o.}\}$. We have

$$\mathbb{P}(N_\epsilon) = 0 \iff \mathbb{P}(N_\epsilon^c) = 1 \iff \mathbb{P}(|X_n - X| > \epsilon \text{ for finitely many } n) = 1$$

and this means that there is some $\Omega' \subset \Omega$, $\mathbb{P}(\Omega') = 1$ such that

$$\forall \omega \in \Omega' \quad \exists M(\omega) \in \mathbb{N} \quad \forall n \geqslant M(\omega) : \quad |X_n(\omega) - X(\omega)| \leqslant \epsilon$$

This shows that $\mathbb{P}(\limsup_{n\to\infty} |X_n - X| \leqslant \epsilon) = 1$. Now let $\epsilon = 1/k$, $k \in \mathbb{N}$, and note that

$$\mathbb{P}\Big(\bigcup_{k\in\mathbb{N}} N_{1/k}\Big) = 0 \iff \mathbb{P}\Big(\underbrace{\limsup_{n\to\infty} |X_n - X| \leqslant \tfrac{1}{k} \ \ \forall k}_{=\bigcap_{k\in\mathbb{N}}\{\limsup_n |X_n - X| \leqslant 1/k\} = \{\lim_n |X_n - X| = 0\}}\Big) = 1. \qquad \square$$

29.11 Lemma. *Assume that the random variables $X_n : \Omega \to \mathbb{R}$, $n \in \mathbb{N}$ are defined on the same probability space. If $X \equiv c$ a.s., then*

$$X_n \xrightarrow{\mathbb{P}} X \iff X_n \xrightarrow{d} X.$$

Proof. The direction »⇒« is always satisfied, even if X is not a.s. constant, see Theorem 29.5. For the converse, »⇐«, we pick a cut-out function $\chi_\epsilon \in C_b(\mathbb{R})$ such that $\chi_\epsilon(0) = 0$ and $\chi_\epsilon \geqslant \mathbb{1}_{[-\epsilon,\epsilon]^c}$, see Fig. 29.2. Since the shifted function

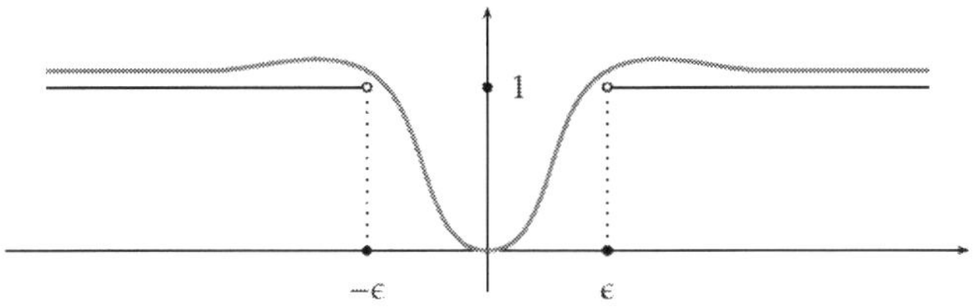

Fig. 29.2. The cut-out function $\chi_\epsilon \in C_b(\mathbb{R})$.

$\chi_\epsilon(\bullet - c) \in C_b(\mathbb{R})$, we get

$$\begin{aligned}\mathbb{P}(|X_n - X| > \epsilon) &\leqslant \int \chi_\epsilon(X_n - X)\,d\mathbb{P} \\ &\overset{X\equiv c \text{ a.s.}}{=} \int \chi_\epsilon(X_n - c)\,d\mathbb{P} \\ &\xrightarrow[\text{schw. Konv.}]{n\to\infty} \int \chi_\epsilon(\underbrace{X - c}_{=0 \text{ a.s.}})\,d\mathbb{P} = 0.\end{aligned}$$ □

The next few examples show a few standard counterexamples, indicating that the converse implications of the implications in Fig. 29.1 are, in general, false.

29.12 Example. We use the probability space $((0,1), \mathscr{B}(0,1), \lambda)$.

a) L^1-convergence $\not\Rightarrow$ L^p-convergence $(p > 1)$ »the chimney«:
$X_n(\omega) = \mathbb{1}_{[1/n,1)}(\omega)\,\omega^{-1/p}$ and $X(\omega) = \omega^{-1/p}$.

b) L^1-convergence $\not\Rightarrow$ a.s.-convergence »the moving hump«:
$X_{n,k}(\omega) = \mathbb{1}_{[k/n,(k+1)/n)}(\omega)$, $n \in \mathbb{N}$, $k = 0, 1, \dots, n-1$ (lexicographically ordered) and $X(\omega) \equiv 0$.

c) $\mathbb{P}$-convergence $\not\Rightarrow$ L^1-convergence »the rising tower«:
$X_n(\omega) = n\,\mathbb{1}_{(0,1/n]}(\omega)$ and $X(\omega) \equiv 0$.

d) $\mathbb{P}$-convergence $\not\Rightarrow$ a.s. convergence: see Example b).

e) d-convergence $\not\Rightarrow$ $\mathbb{P}$-convergence »the flip-flop sequence«: Consider the **Rademacher functions** $R_1, R_2, R_3, \dots$:

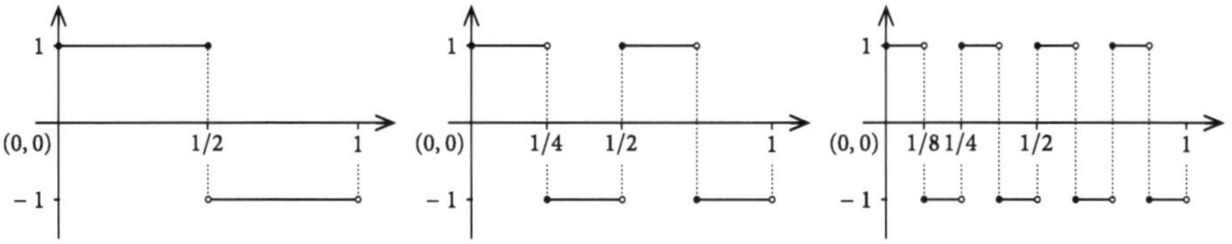

Fig. 29.3. The first three Rademacher functions R_1, R_2, R_3. Try to find a closed expression for R_n ✍.

We obtain R_{n+1} from R_n by halving the intervals of constancy of R_n and flipping the sign of the right half of $\{R_n = 1\}$ and the left half of $\{R_n = -1\}$.

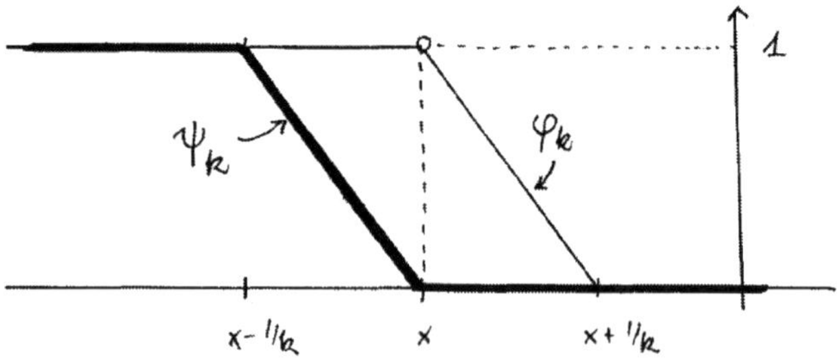

Fig. 29.4. The functions $\phi_k(t) = (1 - k(t-x)^+)^+$ and $\psi_k(t) = (1 - k(t - x + 1/k)^+)^+$ are uniformly continuous and they approximate $\mathbb{1}_{(-\infty,x]}(t)$ from below and above.

Clearly, $R_n \sim \frac{1}{2}\delta_1 + \frac{1}{2}\delta_{-1}$, since the intervals of constancy of R_n always have length $\frac{1}{2}$. Thus, $\mathbb{E}f(R_n) = \frac{1}{2}f(1) + \frac{1}{2}f(-1) \xrightarrow[n\to\infty]{} \mathbb{E}f(R_1)$, i.e. $R_n \xrightarrow{d} R_1$.

On the other hand, it is clear that the sequence $R_n(\omega)$ cannot converge for any fixed ω, since it alternates between ± 1:

$$\liminf_{n\to\infty} R_n(\omega) = -1 < +1 = \limsup_{n\to\infty} R_n(\omega) \quad \forall \omega \in (0,1).$$

In addition, we get for symmetry reasons that for $k \neq n$

$$R_n - R_k = \begin{cases} 0, & \{R_n = R_k\} \quad \to \frac{1}{2} \text{ of all cases} \\ +2, & \{R_n = 1, R_k = -1\} \quad \to \frac{1}{4} \text{ of all cases} \\ -2, & \{R_n = -1, R_k = 1\} \quad \to \frac{1}{4} \text{ of all cases} \end{cases}$$

and so we get $\mathbb{P}(|R_n - R_k| > \epsilon) = \frac{1}{2}$ for all $\epsilon < 2$. This means that $(R_n)_{n\in\mathbb{N}}$ cannot be a $\mathbb{P}$-Cauchy sequence, i.e. it cannot converge stochastically, see Remark 29.3.b).

Let us finally show an equivalent formulation of d-convergence which justifies the name **convergence in distribution**. Recall that we have introduced for real-valued rv the distribution function $F_Y(x) := \mathbb{P}(Y \leqslant x)$.

29.13 Theorem. *Let X_n, X be real random variables (not necessarily on the same probability space). One has*

$$X_n \xrightarrow{d} X \implies \forall x \in C^{F_X} : \quad F_{X_n}(x) \xrightarrow[n\to\infty]{} F_X(x)$$

where C^{F_X} are the continuity points of the limit distribution function F_X.

Proof. For every fixed $x \in \mathbb{R}$ we construct bounded, uniformly continuous functions $\psi_k, \phi_k \in C_b(\mathbb{R}) \cap C_u(\mathbb{R})$ such that (cf. Fig. 29.4)

$$\psi_k \leqslant \mathbb{1}_{(-\infty,x]} \leqslant \phi_k.$$

We have

$$F_{X_n}(x) = \mathbb{E}\left[\mathbb{1}_{(-\infty,x]}(X_n)\right] \leqslant \mathbb{E}\phi_k(X_n) \xrightarrow[n\to\infty]{d\text{-convergence}} \mathbb{E}\phi_k(X)$$

$$\implies \limsup_{n\to\infty} F_{X_n}(x) \leqslant \mathbb{E}\phi_k(X) \xrightarrow[k\to\infty]{\text{DCT}} \mathbb{E}\left[\mathbb{1}_{(-\infty,x]}(X)\right] = F_X(x).$$

Similarly, we get

$$F_{X_n}(x) = \mathbb{E}\left[\mathbb{1}_{(-\infty,x]}(X_n)\right] \geqslant \mathbb{E}\psi_k(X_n) \xrightarrow[n\to\infty]{d\text{-convergence}} \mathbb{E}\psi_k(X)$$

$$\implies \liminf_{n\to\infty} F_{X_n}(x) \geqslant \mathbb{E}\psi_k(X) \xrightarrow[k\to\infty]{\text{DCT}} \mathbb{E}\left[\mathbb{1}_{(-\infty,x)}(X)\right] = F_X(x-).$$

If x is a continuity point of F_X, then we see

$$\limsup_{n\to\infty} F_{X_n}(x) \leqslant F_X(x) = F_X(x-) \leqslant \liminf_{n\to\infty} F_{X_n}(x),$$

and the assertion follows. □

In order to prove the converse of Theorem 29.13, we need the following useful (and astonishing) theorem which connects d-convergence and a.s. convergence (on a new probability space).

29.14 Theorem. *Let X_n, X be real random variables such that $X_n \xrightarrow{d} X$. There exists a probability space and real rv Y_n, Y such that*

$$X_n \sim Y_n, \quad X \sim Y \quad \textit{and} \quad Y_n \xrightarrow{a.s.} Y.$$

Proof. As in Lemma 26.1 we use the generalized inverses of the distribution functions:

$$Y_n := F_{X_n}^{-1} \quad \text{and} \quad Y := F_X^{-1}$$

and interpret them as random variables on the space $([0,1], \mathscr{B}[0,1], dy)$.

It remains to show that $F_{X_n}^{-1}(y) \xrightarrow[n\to\infty]{} F_X^{-1}(y)$ for Lebesgue a.a. $y \in [0,1]$.

1° Identification of the exceptional set, see Figure 29.5. It is easy to see that $Y(y-) = \sup\{x \mid F_X(x) < y\}$ and $Y(y) = \inf\{x \mid F_X(x) > y\}$. We define the set Ω_0 as the set of all points x where F_X is strictly increasing – here F_X^{-1} is a proper inverse. The complement is of the form

$$\Omega_0^c = \{y \mid (Y(y-), Y(y)) \neq \emptyset\} = \text{»intervals of constancy« of } F_X.$$

Since the intervals $(Y(y-), Y(y))$ are disjoint and each contains a rational point, we see that

$$|\Omega_0^c| \leqslant |\mathbb{Q}| \implies \Omega_0^x \text{ is a Lebesgue null set.}$$

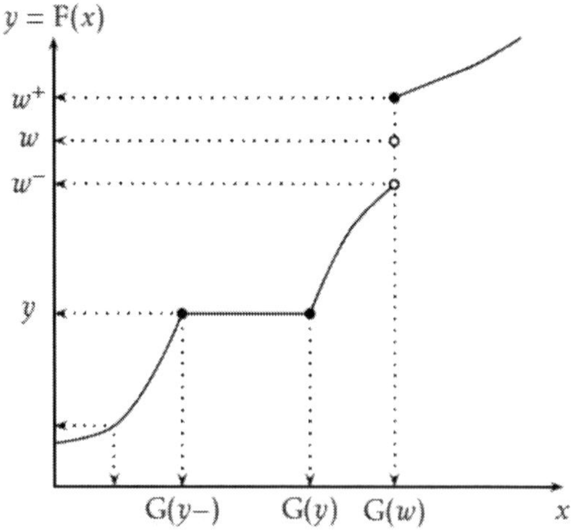

Fig. 29.5. Identification of the exceptional set.

2° $\forall y \in \Omega_0 : \liminf_{n\to\infty} F_{X_n}^{-1}(y) \geqslant F_X^{-1}(y)$. Indeed, if F_X is continuous at x, then

$$\begin{aligned} x < F_X^{-1}(y) &\implies F_X(x) < y \quad \text{as } y \in \Omega_0 \text{ is a point of strict growth} \\ &\overset{n\gg 1}{\implies} F_{X_n}(x) < y \quad \text{by Thm. 29.13: } F_{X_n} \to F_X \\ &\overset{n\gg 1}{\implies} x \leqslant F_{X_n}^{-1}(y) \quad \text{b/o defn. of } F_{X_n}^{-1} \\ &\implies x \leqslant \liminf_{n\to\infty} F_{X_n}^{-1}(y). \end{aligned}$$

Letting $x \uparrow F_X^{-1}(y)$ along a sequence of continuity points yields

$$\liminf_{n\to\infty} F_{X_n}^{-1}(y) \geqslant x \uparrow F_X^{-1}(y).$$

3° $\forall y \in \Omega_0 : \limsup_n F_{X_n}^{-1}(y) \leqslant F_X^{-1}(y)$. Use the argument of 2° .

4° Finally, we have

$$\liminf_{n\to\infty} F_{X_n}^{-1}(y) \leqslant \limsup_{n\to\infty} F_{X_n}^{-1}(y) \overset{3^\circ}{\leqslant} F_X^{-1}(y) \overset{2^\circ}{\leqslant} \liminf_{n\to\infty} F_{X_n}^{-1}(y). \qquad \square$$

The following corollary – which proves the converse of Theorem 29.13 – is a typical application of Theorem 29.14.

29.15 Theorem. *Let X_n, X be real random variables (not necessarily on the same* *probability space). One has*

$$X_n \xrightarrow{d} X \iff \forall x \in C^{F_X} : \quad F_{X_n}(x) \xrightarrow[n\to\infty]{} F_X(x)$$

where C^{F_X} are the continuity points of the limit distribution function F_X.

Proof. The direction »⇒« is Theorem 29.13. For the converse »⇐« let $f \in C_b(\mathbb{R})$. In view of Theorem 29.14, there is a probability space $(\Omega', \mathscr{A}', \mathbb{P}')$ and random variables $Y_n \sim X_n$, $Y \sim X$ such that $Y_n \xrightarrow{\mathbb{P}'\text{-a.s.}} Y$. Thus,

$$\mathbb{E} f(X_n) \overset{X_n \sim Y_n}{=} \mathbb{E}' f(Y_n) \xrightarrow[n\to\infty]{\text{DCT}} \mathbb{E}' f(Y) \overset{X\sim Y}{=} \mathbb{E} f(X). \qquad \square$$

💬 In many texts the equivalence in Theorem 29.15 is used as **definition** of convergence in distribution.

Appendix: Completeness of $\mathbb{P}$-convergence★

Let us show that convergence in probability is complete. We begin by showing that it is metrizable.

29.16 Definition (Ky Fan). Let $X, Y : \Omega \to \mathbb{R}^d$ be random variables. The **Ky Fan metric** is defined as

$$\rho(X,Y) = \inf\{\epsilon \geqslant 0 \mid \mathbb{P}(|X-Y| > \epsilon) \leqslant \epsilon\}. \tag{29.5}$$

29.17 Lemma. *$\rho(X,Y)$ is a metric on the space of d-dimensional random variables.*

Proof. Let $X, Y, Z : \Omega \to \mathbb{R}^d$. Clearly, $\rho(X,Y) \geqslant 0$, $\rho(X,Y) = \rho(Y,X)$ and

$$\rho(X,Y) = 0 \iff \forall k \in \mathbb{N} : \mathbb{P}(|X-Y| > 1/k) \leqslant 1/k \iff X = Y \text{ a.s.}$$

Only the triangle inequality is non-trivial. By the definition of $\rho(X,Y)$ we have that $|X(\omega) - Y(\omega)| \leqslant \rho(X,Y)$ for all ω outside a set of measure $\leqslant \rho(X,Y)$. Thus,

$$|X(\omega) - Z(\omega)| \leqslant |X(\omega) - Y(\omega)| + |Y(\omega) - Z(\omega)| \leqslant \rho(X,Y) + \rho(Y,Z)$$

outside an ω-set of probability of at most $\rho(X,Y) + \rho(Y,Z)$. Because of the definition of $\rho(X,Z)$, this shows $\rho(X,Z) \leqslant \rho(X,Y) + \rho(Y,Z)$. $\square$

29.18 Lemma. *Let X_n, X be d-dimensional random variables.*

$$X_n \xrightarrow[n\to\infty]{\mathbb{P}} X \iff \rho(X_n, X) \xrightarrow[n\to\infty]{} 0.$$

Proof. »⇒« If $X_n \xrightarrow{\mathbb{P}} X$, then $\mathbb{P}(|X_n - X| > 2^{-k}) \leqslant 2^{-k}$ for all $n \geqslant N(k)$. Thus, $\rho(X_n, X) \to 0$. »⇐« If $\rho(X_n, X) \to 0$, then $\rho(X_n, X) \leqslant \epsilon$ for all $n \geqslant N(\epsilon)$. Thus, $\mathbb{P}(|X_n - X| > \epsilon) \leqslant \epsilon$ for all $\epsilon > 0$, and so $X_n \xrightarrow{P} X$. $\square$

29.19 Theorem. *Let $(X_n)_{n\in\mathbb{N}}$ be a $\mathbb{P}$-Cauchy sequence of d-dimensional random variables, i.e.*

$$\forall\epsilon>0\quad \exists N(\epsilon)\quad \forall m,n\geqslant N(\epsilon):\quad \mathbb{P}(|X_n-X_m|>\epsilon)\leqslant\epsilon \tag{29.6}$$

$$\left[\iff \forall\epsilon>0:\quad \lim_{m,n\to\infty}\mathbb{P}(|X_n-X_m|>\epsilon)=0.\right]$$

Then there is a random variable X such that $X_n\xrightarrow{\mathbb{P}}X$.

Proof. From the condition (29.6) and the definition of the Ky-Fan metric ρ, we conclude that $\lim_{m,n\to\infty}\rho(X_n,X_m)=0$, i.e.

$$\forall k\in\mathbb{N}\quad \exists n(k)\quad \forall m\geqslant n(k):\quad \mathbb{P}\Big(|X_m-X_{n(k)}|>2^{-k}\Big)\leqslant 2^{-k}.$$

We may assume that $n(k)\uparrow\infty$ as $k\uparrow\infty$. Taking $m=n(\ell)$ for any $\ell>k$, we see that

$$\forall\ell>k\in\mathbb{N}:\quad \mathbb{P}\Big(|X_{n(\ell)}-X_{n(k)}|>2^{-k}\Big)\leqslant 2^{-k}.$$

Since $\sum_k 2^{-k}<\infty$, this is a »fast« $\mathbb{P}$-Cauchy sequence and, just as in Lemma 29.7, we see that $(X_{n(k)}(\omega))_{k\in\mathbb{N}}$ is an a.s. Cauchy sequence (outside an ω-set of $\mathbb{P}$-measure zero). Thus, $X_{n(k)}\to X$ exists a.s. – and we define $X:=0$ at all other points – qualifying X as a limit candidate for the $\mathbb{P}$-Cauchy sequence. Since a.s. convergence implies $\mathbb{P}$-convergence, we also have $X_{n(k)}\xrightarrow{\mathbb{P}}X$.

The rest is now a standard argument in metric spaces: We have Cauchy sequence and a convergent subsequence, so the full sequence has to converge (to the limit of the subsequence): We havde $\rho(X_n,X_m)\to 0$ and $\rho(X_{n(k)},X)\to 0$ as $m,n\to\infty$ and $k\to\infty$. For any $\epsilon>0$ there is some $N(\epsilon)$ such that

$$\forall n\geqslant n(k)\geqslant N(\epsilon):\quad \rho(X_n,X)\leqslant\rho(X_n,X_{n(k)})+\rho(X_{n(k)},X)\leqslant 2\epsilon.$$ □

30 Characteristic functions and convergence in distribution

We will state the following results for d-dimensional random variables, but we give only proofs for $d=1$. The extension to $d>1$ adds only notational complexity, complete proofs can be found in [WT, §§7.11, 9.17, 9.18].

30.1 Theorem (Lévy's truncation inequality). *Let $X=(X^{(1)},\dots,X^{(d)}):\Omega\to\mathbb{R}^d$ be a random variable with c.f. ϕ_X. Then*

$$\mathbb{P}\Big(\max_{1\leqslant k\leqslant d}|X^{(k)}|\geqslant r\Big)\leqslant 7r^d\int_0^{1/r}\dots\int_0^{1/r}(1-\operatorname{Re}\phi_X(\xi))\,d\xi. \tag{30.1}$$

Proof if $d=1$. We begin with the following elementary inequality, see Fig. 30.1.

$$\forall|t|\geqslant 1:\quad \frac{\sin t}{t}\leqslant\sin 1\overset{\text{symmetry}}{\iff}\forall t\in[1,\infty):\quad \sin t\leqslant t\cdot\sin 1.$$

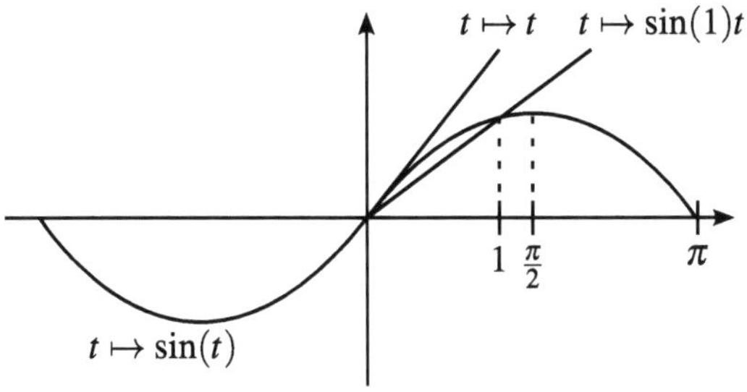

Fig. 30.1. $\sin t$ is a concave function on the interval $[0,\pi]$ and $\sin t \leqslant 1$. Therefore, we have $\sin t \leqslant t$ for all $t \geqslant 0$, as well as $\sin t \leqslant \sin(1)\cdot t$ if $t \geqslant 1$. Since $\sin t$ is an odd function, we get $|\sin t| \leqslant |t|$ and $|\sin t| \leqslant \sin(1)\cdot|t|$ for all $|t| \geqslant 1$.

Using this inequality in the calculation below yields

$$\begin{aligned}
r\int_0^{1/r}\left[1-\operatorname{Re}\mathbb{E}e^{iX\xi}\right]d\xi &= r\int_0^{1/r}\left[1-\mathbb{E}\cos(x\xi)\right]d\xi \\
&= r\int_0^{1/r}\int[1-\cos(x\xi)]\,\mathbb{P}(X\in dx)\,d\xi \\
&= r\int\int_0^{1/r}[1-\cos(x\xi)]\,d\xi\,\mathbb{P}(X\in dx) \\
&= \int\underbrace{\left[1-\frac{\sin(x/r)}{x/r}\right]}_{\geqslant 0}\mathbb{P}(X\in dx) \\
&\geqslant \int_{|x|>r}\underbrace{(1-\sin 1)}_{\geqslant 1/7}\,\mathbb{P}(X\in dx).
\end{aligned}$$ □

30.2 Corollary (tightness). *Let $X_n : \Omega \to \mathbb{R}^d$, $n \in \mathbb{N}$, be random variables. If the limit*

$$\mathbb{E}e^{i\langle\xi,X_n\rangle} = \phi_{X_n}(\xi) \xrightarrow[n\to\infty]{\forall\xi} \phi(\xi)$$

exists and if ϕ is continuous at $\xi = 0$, then

$$\forall\epsilon>0 \quad \exists r=r(\epsilon) \quad \forall R\geqslant r(\epsilon): \quad \sup_{n\in\mathbb{N}}\mathbb{P}(|X_n|>R)\leqslant\epsilon. \tag{30.2}$$

Proof if $d = 1$. Fix $\epsilon > 0$ and observe that $\phi(0) = \lim_{n\to\infty}\phi_{X_n}(0) = 1$. Since ϕ is continuous at 0,

$$\exists\delta=\delta(\epsilon) \quad \forall|\xi|\leqslant\delta: \quad |1-\operatorname{Re}\phi(\xi)|\leqslant\epsilon.$$

Therefore, we have

$$r\int_0^{1/r}\left(1-\operatorname{Re}\phi_{X_n}(\xi)\right)d\xi \xrightarrow[n\to\infty]{\text{DCT}} r\int_0^{1/r}(1-\operatorname{Re}\phi(\xi))\,d\xi.$$

If $n \geqslant N(\epsilon)$ is sufficiently large, and $r > 1/\delta = 1/\delta(\epsilon)$, then

$$\begin{aligned}\mathbb{P}(|X_n|>r) &\overset{(30.1)}{\leqslant} 7r\int_0^{1/r}\left(1-\operatorname{Re}\phi_{X_n}(\xi)\right)d\xi \\ &\overset{n\geqslant N(\epsilon)}{\leqslant} 7\epsilon + 7r\int_0^{1/r}\underbrace{(1-\operatorname{Re}\phi(\xi))}_{\leqslant\epsilon,\ \forall|\xi|\leqslant\delta}d\xi \\ &\overset{r>1/\delta}{\leqslant} 7(\epsilon+\epsilon) = 14\epsilon.\end{aligned}$$

Using the continuity of measures – if needed we can enlarge r – we see

$$\mathbb{P}(|X_n|>r)\leqslant 14\epsilon, \quad n=1,\dots,N(\epsilon)-1,$$

and the claim follows. □

We have already used the following result to prove the CLT §28.6.

30.3 Theorem (Lévy 1925; continuity theorem). *Let X, X_n be random variables with values in $\mathbb{R}^d$. Then*

$$X_n \xrightarrow[n\to\infty]{d} X \iff \forall\xi\in\mathbb{R}^d : \phi_{X_n}(\xi) \xrightarrow[n\to\infty]{} \phi_X(\xi).$$

Amendment: *The characteristic functions converge locally uniformly.*

💬 There is a stronger version of Theorem 30.3: If $\phi_{X_n} \to \phi$ where ϕ is continuous at 0, then ϕ is the c.f. of a random variable X – i.e. $\phi = \phi_X$ – and $X_n \xrightarrow{d} X$, see [WT, Satz 15.2].

Proof if $d = 1$. The direction »⇒« follows from Definition 29.2 and the fact that $e^{i\langle\xi,\bullet\rangle} \in C_b(\mathbb{R})$. For the converse »⇐« we pick

- $f \in C_b(\mathbb{R}^d)$, $\epsilon > 0$ and $R \geqslant r(\epsilon)$ as in (30.2);
- $G_t := \sqrt{t}G \sim N(0,t)$, $G_t \perp\!\!\!\perp (X_n)_{n\in\mathbb{N}}, X$.

1° Since f is uniformly continuous on compact sets, we have

$$\exists\delta=\delta(\epsilon)<1 : |f(X_n+G_t)-f(X_n)|\leqslant\epsilon \quad \text{on} \quad \{|X_n|\leqslant R\}\cap\{|G_t|\leqslant\delta\}.$$

For the function $f_R(x) := f(x)\mathbb{1}_{[0,R+1]}(|x|)$ and for all $t < \epsilon\delta^2$ it holds that

$$\begin{aligned}
|\mathbb{E} f_R(X_n + G_t) - \mathbb{E} f(X_n)| &\leqslant \mathbb{E}\left[|f(X_n + G_t) - f(X_n)|\mathbb{1}_{\{|X_n|\leqslant R\}\cap\{|G_t|\leqslant\delta\}}\right] \\
&\quad + \mathbb{E}\left[|f_R(X_n + G_t) - f(X_n)|\mathbb{1}_{\{|X_n|>R\}\cup\{|G_t|>\delta\}}\right] \\
&\leqslant \epsilon + 2\|f\|_{L^\infty}\left[\mathbb{P}(|X_n| > R) + \mathbb{P}(|G_t| > \delta)\right] \\
&\overset{(30.2)}{\leqslant} \epsilon + 2\|f\|_{L^\infty}\epsilon + 2\|f\|_{L^\infty}\frac{1}{\delta^2}\underbrace{\mathbb{E}\left[G_t^2\right]}_{=t} \leqslant c\epsilon.
\end{aligned}$$

2° As in the proof of Theorem 27.7 we find for f_R

$$\begin{aligned}
\mathbb{E} f_R(X_n + G_t) &= \frac{1}{(2\pi)^d}\int f_R(z)\int \phi_{X_n}(-\eta)e^{i\eta z}e^{-t|\eta|^2/2}\,d\eta\,dz \\
&\xrightarrow[n\to\infty]{\text{DCT}} \frac{1}{(2\pi)^d}\int f_R(z)\int \phi_X(-\eta)e^{i\eta z}e^{-t|\eta|^2/2}\,d\eta\,dz \\
&= \mathbb{E} f_R(X + G_t).
\end{aligned}$$

And, again using first d-covergence, and then twice the dominated convergence theorem,

$$\mathbb{E} f_R(X_n + G_t) \xrightarrow[n\to\infty]{} \mathbb{E} f_R(X + G_t) \xrightarrow[t\to0]{} \mathbb{E} f_R(X) \xrightarrow[R\to\infty]{} \mathbb{E} f(X).$$

3° In view of 1° we find for any $f \in C_b(\mathbb{R}^d)$, all $n \in \mathbb{N}$ and $t \leqslant \epsilon\delta^2$

$$\begin{aligned}
|\mathbb{E} f(X) - \mathbb{E} f(X_n)| &\leqslant |\mathbb{E} f(X) - \mathbb{E} f_R(X_n + G_t)| + |\mathbb{E} f_R(X_n + G_t) - \mathbb{E} f(X_n)| \\
&\leqslant |\mathbb{E} f(X) - \mathbb{E} f_R(X_n + G_t)| + c\epsilon \\
&\xrightarrow[n\to\infty,\ t\to0,\ R\to\infty]{\text{step } 2^\circ} 0 + c\epsilon \xrightarrow[\epsilon\to0]{} 0.
\end{aligned}$$

4° In order to see the **amendment**, we observe that the steps 1° –3° hold for $f(x) := e_\xi(x) := e^{i\xi x}$ uniformly for all $|\xi| \leqslant r$ and for any fixed $r > 0$. In particular, the limit $\phi_{X_n}(\xi) = \mathbb{E} e_\xi(X_n) \to \mathbb{E} e_\xi(X) = \phi_X(\xi)$ exists locally uniformly. □

We close with a technique which allows us to reduce assertions for $d > 1$ to the case where $d = 1$.

30.4 Corollary (Cramér–Wold trick). *If X, X_n are d-dimensional random variables, then*

$$X_n \xrightarrow[n\to\infty]{d} X \iff \forall \xi \in \mathbb{R}^d : \quad \langle \xi, X_n\rangle \xrightarrow[n\to\infty]{d} \langle \xi, X\rangle.$$

Proof. Note that $\mathbb{R}^d = \mathbb{R} \cdot \mathbb{R}^d = \{t\xi \mid t \in \mathbb{R},\ \xi \in \mathbb{R}^d\}$. So,

$$\begin{aligned}
X_n \xrightarrow{\ d\ } X &\overset{30.3}{\iff} \forall \xi \in \mathbb{R}^d \quad \forall t \in \mathbb{R}: \quad \lim_{n\to\infty} \phi_{X_n}(t\xi) = \phi_X(t\xi) \\
&\iff \forall \xi \in \mathbb{R}^d \quad \forall t \in \mathbb{R}: \quad \lim_{n\to\infty} \phi_{\langle \xi, X_n\rangle}(t) = \phi_{\langle \xi, X\rangle}(t) \\
&\overset{30.3}{\iff} \forall \xi \in \mathbb{R}^d: \quad \langle \xi, X_n\rangle \xrightarrow{\ d\ } \langle \xi, X\rangle.
\end{aligned}$$

□

31 Convergence of independent random variables

We will now discuss one of the central themes of probability theory: The convergence behaviour of (sums of) independent random variables. Recall from (the proof of) Lemma 29.6 that

i.o. = infinitely often = for infinitely many indices.

31.1 Theorem (Borel–Cantelli Lemma). *Let $(A_n)_{n\in\mathbb{N}} \subset \mathscr{A}$ be events.*

a) $\displaystyle\sum_{n=1}^{\infty} \mathbb{P}(A_n) < \infty \implies \mathbb{P}(\limsup_n A_n) = \mathbb{P}(A_n \text{ i.o.}) = 0.$

b) *If the events A_n are **pairwise independent**, then*
$$\sum_{n=1}^{\infty} \mathbb{P}(A_n) = \infty \implies \mathbb{P}(\limsup_n A_n) = \mathbb{P}(A_n \text{ i.o.}) = 1.$$

Proof. We have seen the »easy direction« a) already in Lemma 29.6. For the proof of b) we define

$$S_n := \sum_{i=1}^{n} \mathbb{1}_{A_i} \xrightarrow[n\to\infty]{} \sum_{i=1}^{\infty} \mathbb{1}_{A_i} =: S,$$

$$m_n := \mathbb{E}\sum_{i=1}^{n} \mathbb{1}_{A_i} = \sum_{i=1}^{n} \mathbb{P}(A_i) \xrightarrow[n\to\infty]{} \infty,$$

$$\mathbb{V}S_n \overset{\text{Bienaymé}}{=} \sum_{i=1}^{n} \mathbb{V}\mathbb{1}_{A_i} = \sum_{i=1}^{n} \left[\mathbb{E}(\mathbb{1}_{A_i}^2) - \left(\mathbb{E}\mathbb{1}_{A_i}\right)^2\right] \leqslant \sum_{i=1}^{n} \mathbb{E}\mathbb{1}_{A_i}^2 = m_n.$$

Obviously, we have $S_n \leqslant S$, and so $\left\{S \leqslant \frac{1}{2} m_n\right\} \subset \left\{S_n \leqslant \frac{1}{2} m_n\right\}$. This shows

$$\begin{aligned}
\mathbb{P}\left(S \leqslant \tfrac{1}{2} m_n\right) &\leqslant \mathbb{P}\left(S_n \leqslant \tfrac{1}{2} m_n\right) = \mathbb{P}\left(S_n - m_n \leqslant -\tfrac{1}{2} m_n\right) \\
&\overset{\S}{\leqslant} \mathbb{P}\left(|S_n - m_n| \geqslant \tfrac{1}{2} m_n\right) \overset{\ddagger}{\leqslant} \frac{4}{m_n^2} \mathbb{V}S_n \leqslant \frac{4}{m_n} \xrightarrow[n\to\infty]{} 0.
\end{aligned}$$

In the step marked by §, we use $S_n - m_n \leqslant -\frac{1}{2} m_n \implies |S_n - m_n| \geqslant \frac{1}{2} m_n$; in the step marked by ‡ we use the Chebyshev–Markov inequality. Thus,

$$\mathbb{P}(S < \infty) = \lim_{n\to\infty} \mathbb{P}\left(S \leqslant \tfrac{1}{2} m_n\right) = 0,$$

and the claim follows because of

$$\omega \in \limsup_{n\to\infty} A_n \iff \omega \text{ is contained in } \infty\text{-many } A_n \iff S(\omega) = \infty. \qquad \square$$

The following example and theorem contain typical applications of the Borel–Cantelli lemma.

31.2 Example. Let $(X_n)_{n\in\mathbb{N}}$ be iid Exp(1)-random variables, i.e.

$$\mathbb{P}(X_n \geqslant x) = \int_x^\infty e^{-t}\,dt = e^{-x}, \quad x \geqslant 0.$$

This shows that $\mathbb{P}(X_n > \alpha \log n) = n^{-\alpha}$ for any $\alpha > 0$, and we conclude from the Borel–Cantelli lemma

$$\mathbb{P}(X_n > \alpha \log n \text{ i.o.}) = \begin{cases} 0, & \alpha > 1, \\ 1, & \alpha \leqslant 1. \end{cases}$$

Let $L := \limsup_{n\to\infty} X_n/\log n$. Since $\alpha = 1$, we have

$$\mathbb{P}(L \geqslant 1) \geqslant \mathbb{P}(X_n > \log n \text{ i.o.}) = 1$$

and taking $\alpha = 1 + \frac{1}{k}$ we get

$$\mathbb{P}\left(L > 1 + \tfrac{1}{k}\right) \leqslant \mathbb{P}\left(X_n > \left(1 + \tfrac{1}{k}\right)\log n \text{ i.o.}\right) = 0.$$

Therefore, the set $\{L > 1\} = \bigcup_{k\in\mathbb{N}} \left\{L > 1 + \frac{1}{k}\right\}$ is a null set, and we get

$$\mathbb{P}(L = 1) = \mathbb{P}(\{L \geqslant 1\} \setminus \{L > 1\}) = 1 \implies L = \limsup_{n\to\infty} \frac{X_n}{\log n} = 1 \quad \text{a.s.}$$

Theorem 31.1 allows us to show a first version of the SLLN (strong law of large numbers).

31.3 Theorem (Cantelli; L^4-SLLN). *Let $(X_n)_{n\in\mathbb{N}} \subset L^4(\mathbb{P})$ be iid random variables and set $S_n := X_1 + \cdots + X_n$. Then*

$$\lim_{n\to\infty} \frac{S_n}{n} = \mathbb{E}X_1 = \mu \quad \textit{a.s.}$$

Proof. Since the random variables X_i are iid, we have

$$\mu = \mathbb{E}X_i \quad \text{and} \quad (X_i - \mu)_{i\in\mathbb{N}} \subset L^4, \text{ iid and } \mathbb{E}(X_i - \mu) = 0.$$

Without loss of generality we may, therefore, assume that $\mu = 0$. So,

$$\mathbb{E}\left[S_n^4\right] = \mathbb{E}\left[\left(\sum_{i=1}^n X_i\right)^4\right] = \mathbb{E}\left[\sum_{i,k,\ell,m} X_i X_k X_\ell X_m\right] = \sum_{i,k,\ell,m} \mathbb{E}\left[X_i X_k X_\ell X_m\right].$$

For the general term of the series, we have to distinguish between three cases: i)

$i=k=\ell=m$ (this happens in n cases), ii) $i=k$, $\ell=m$ but $k\neq\ell$ (this happens in $\binom{4}{2}n(n-1)$ cases) and iii) i is different from k,ℓ,m. Because of independence,

$$\mathbb{E}\left[X_iX_kX_\ell X_m\right]=\begin{cases}\mathbb{E}(X_i^4), & \text{in case i);}\\ \mathbb{E}(X_i^2)\mathbb{E}(X_\ell^2), & \text{in case ii);}\\ \underbrace{\mathbb{E}(X_i)}_{=0}\mathbb{E}(X_kX_\ell X_m)=0, & \text{in case iii).}\end{cases}$$

Therefore,

$$\mathbb{E}\left[S_n^4\right]\leqslant n\mathbb{E}\left[X_1^4\right]+6n(n-1)\left[\mathbb{E}(X_1^2)\right]^2\leqslant\kappa n^2$$

and, with the Markov inequality, we get for any $\epsilon>0$

$$\mathbb{P}(|S_n|>n\epsilon)\leqslant\frac{\mathbb{E}S_n^4}{n^4\epsilon^4}\leqslant\frac{\kappa}{n^2\epsilon^4}.$$

Therefore, the series $\sum_{n=1}^\infty\mathbb{P}(|S_n|>n\epsilon)$ converges and the Borel–Cantelli lemma shows that

$$\forall\epsilon>0:\mathbb{P}(|S_n|>n\epsilon\ \text{i.o.})=0\implies\forall\epsilon>0:\limsup_{n\to\infty}\frac{|S_n|}{n}\leqslant\epsilon\quad\text{a.s.}\tag{31.1}$$

This, however, means that $\limsup_n|S_n|/n=0$ a.s., and so $\lim_n|S_n|/n=0$ a.s. (recall the footnote on page 58).

⚠⚠ In the last conclusion, we should be very careful about the null sets. The exceptional set appearing in (31.1) can depend on ϵ: In fact, if

$$\Omega_\epsilon:=\left\{\limsup_{n\to\infty}\frac{|S_n|}{n}\leqslant\epsilon\right\},$$

then (31.1) tells us that $\mathbb{P}(\Omega_\epsilon)=1$. But we need only countably many ϵ's:

$$\left\{\limsup_{n\to\infty}\frac{|S_n|}{n}=0\right\}=\bigcap_{\epsilon>0}\Omega_\epsilon\overset{\text{key}}{=}\bigcap_{m\in\mathbb{N}}\Omega_{1/m}=:\Omega_0.$$

(this is the key observation). Therefore,

$$\mathbb{P}\left[\Omega_0^c\right]=\mathbb{P}\left[\bigcup_{m=1}^\infty\Omega_{1/m}^c\right]\leqslant\sum_{m=1}^\infty\mathbb{P}\left[\Omega_{1/m}^c\right]=0.$$ □

The integrability condition in the L^4-SLLN is quite strong (whereas the independence assumption is weak). The following theorem shows the how far we can go in reducing integrability.

31.4 Theorem. *Let $(X_n)_{n\in\mathbb{N}}$ be iid real random variables such that $\mathbb{E}|X_n|=\infty$. The partial sums $S_n=X_1+\cdots+X_n$ satisfy*

a) $\mathbb{P}\left(|X_n|\geqslant n\ \text{i.o.}\right)=1.$

b) $\mathbb{P}\left(\exists\lim_{n\to\infty}S_n/n\in\mathbb{R}\right)=0.$

Proof. a) Note that

$$\infty = \mathbb{E}|X_1| \overset{\S 16.5}{=} \int_0^\infty \mathbb{P}(|X_1| > t)\,dt = \sum_{n=0}^\infty \int_n^{n+1} \mathbb{P}(|X_1| > t)\,dt$$
$$\leqslant \sum_{n=0}^\infty \mathbb{P}(|X_1| \geqslant n) \overset{\text{iid}}{=} \sum_{n=0}^\infty \mathbb{P}(|X_n| \geqslant n).$$

In view of Theorem 31.1.b), we have

$$\mathbb{P}\Big(|X_n| \geqslant n \ \text{ i.o.}\Big) = 1.$$

b) Let $\Omega_0 := \{|X_n| \geqslant n \ \text{ i.o.}\}$ and $C := \{\lim_{n\to\infty} S_n/n \text{ exists in } \mathbb{R}\}$. For all $\omega \in C$ we have

$$\frac{X_{n+1}(\omega)}{n+1} = \frac{S_{n+1}(\omega)}{n+1} - \frac{S_n(\omega)}{n+1} = \underbrace{\frac{S_{n+1}(\omega)}{n+1}}_{\to a} - \underbrace{\frac{n}{n+1}}_{\to 1}\underbrace{\frac{S_n(\omega)}{n}}_{\to a}$$

This shows that $\lim_{n\to\infty} X_n(\omega)/n = 0$ for $\omega \in C$, and so $C \cap \Omega_0 = \emptyset$. Thus,

$$\mathbb{P}(C) = \mathbb{P}(C \cap \Omega_0) + \mathbb{P}(C \setminus \Omega_0) \leqslant \mathbb{P}(\Omega_0^c) = 0.$$ □

Let's return to the Borel–Cantelli lemma (Theorem 31.1). While the condition $\sum_{n=1}^\infty \mathbb{P}(A_n) = \infty$ or $< \infty$ is a dichotomy, this is not obvious for $\mathbb{P}(A) = 0$ or $= 1$ – that is $\mathbb{P}(A) \notin (0,1)$. Assertions of this kind are usually called **zero-one-laws**, and independence is often the reason for such phenomena.

31.5 Definition. Let $(\mathscr{A}_i)_{i\in\mathbb{N}}$ be a countably many families of events from $\mathscr{A}$ and define

$$\mathscr{T}_n := \sigma(\mathscr{A}_n, \mathscr{A}_{n+1}, \dots) = \sigma\Big(\bigcup_{i\geqslant n} \mathscr{A}_i\Big).$$

The family $\mathscr{T}_\infty := \bigcap_{n\in\mathbb{N}} \mathscr{T}_n$ is the **σ-algebra of terminal events** or **terminal σ-algebra**.

31.6 Example. Let $(A_i)_i \subset \mathscr{B}(\mathbb{R})$. Since the families $\mathscr{T}_n = \sigma(A_n, A_{n+1}, \dots)$ are decreasing, the event

$$\limsup_{i\to\infty} A_i = \bigcap_{n\geqslant 1}\underbrace{\bigcup_{i\geqslant n} A_i}_{\in \mathscr{T}_n} = \bigcap_{n\geqslant m}\underbrace{\bigcup_{i\geqslant n} A_i}_{\in \mathscr{T}_n \subset \mathscr{T}_m} \in \mathscr{T}_m$$

is for every $m \in \mathbb{N}$ in $\mathscr{T}_m$, hence in $\mathscr{T}_\infty$, hence terminal.

⚠ It is a common mistake to argue like »$\bigcap_{n=1}^\infty A_n \overset{?!}{\in} \bigcap_{n=1}^\infty \mathscr{T}_n = \mathscr{T}_\infty$ where we use at ?! the fact that $A_n \in \mathscr{T}_n$ for each n.« The **trouble** is that the intersection $\bigcap_{n=1}^\infty$ is (correctly!) used, but it has two different meanings, which are not compatible when looking at »$\in$«.

31.7 Example. Let $(X_i)_{i\in\mathbb{N}}$ be $\mathbb{R}$-valued random variables, $\mathcal{T}_n := \sigma(X_n, X_{n+1}, \dots)$ and $S_n := X_1 + \dots + X_n$.

a) We have for all $m \in \mathbb{N}$

$$\left\{\omega \mid \exists \lim_{n\to\infty} S_n(\omega)\right\} = \left\{\omega \mid \exists \lim_{n\to\infty} (X_m(\omega) + \dots + X_n(\omega))\right\} \in \mathcal{T}_m,$$

and so $\{\exists \lim_{n\to\infty} S_n\} \in \bigcap_{m=1}^{\infty} \mathcal{T}_m = \mathcal{T}_\infty$.

b) We have for all $m \in \mathbb{N}$

$$\left\{\limsup_{n\to\infty} \frac{S_n}{n} > x\right\} = \left\{\limsup_{n\to\infty} \frac{S_n - S_m}{n} + \underbrace{\lim_{n\to\infty} \frac{S_m}{n}}_{=0} > x\right\}$$

$$= \left\{\limsup_{n\to\infty} \frac{S_n - S_m}{n} > x\right\} \in \mathcal{T}_m,$$

and so $\{\limsup_{n\to\infty} S_n/n > x\} \in \bigcap_{m=1}^{\infty} \mathcal{T}_m = \mathcal{T}_\infty$.

c) The event $\{\limsup_{n\to\infty} S_n \geqslant 0\}$ is **not terminal**, since all X_i's may influence the value of the upper limit. For instance, if

$$X_1 = \mathbb{1}_A \quad \text{and} \quad X_i \equiv -2^{1-i} \quad (i \geqslant 2),$$

then we see

$$S_\infty = \lim_{n\to\infty} S_n = \mathbb{1}_A - \sum_{i=2}^{\infty} \frac{1}{2^{i-1}} = \mathbb{1}_A - 1 = -\mathbb{1}_{A^c},$$

which shows that $S_\infty \geqslant 0$ only depends on X_1. Thus, the event cannot be terminal.

31.8 Theorem (Kolmogorov's 0-1–law). *Let $(X_n)_{n\in\mathbb{N}}$ be independent random variables and let $A \in \mathcal{T}_\infty$ where*

$$\mathcal{T}_\infty = \bigcap_{n\in\mathbb{N}} \mathcal{T}_n \quad \textit{and} \quad \mathcal{T}_n = \sigma(X_n, X_{n+1}, \dots).$$

Then the event A is trivial, i.e. either $\mathbb{P}(A) = 0$ or $\mathbb{P}(A) = 1$.

Proof. We show that A is independent of itself. If so, then

$$\mathbb{P}(A) = \mathbb{P}(A \cap A) = \mathbb{P}(A)^2 \implies \mathbb{P}(A) = 0 \text{ or } \mathbb{P}(A) = 1.$$

Because of Lemma 26.4 we have

$$\sigma(X_1, \dots, X_k) \perp\!\!\!\perp \sigma(X_{k+1}, X_{k+2}, \dots) = \mathcal{T}_{k+1} \supset \mathcal{T}_\infty.$$

Thus, $\sigma(X_1, \dots, X_k) \perp\!\!\!\perp \mathcal{T}_\infty$ for all k, i.e. $\bigcup_{k=1}^{\infty} \sigma(X_1, \dots, X_k) \perp\!\!\!\perp \mathcal{T}_\infty$. Since the union extends over increasing families, it is $\cap$-stable, and so

$$\mathcal{T}_\infty \subset \sigma(X_1, X_2, \dots) \subset \sigma\left(\bigcup_k \sigma(X_1, \dots, X_k)\right) \perp\!\!\!\perp \mathcal{T}_\infty.$$

This proves that $\mathcal{T}_\infty \perp\!\!\!\perp \mathcal{T}_\infty$. □

Fluctuation of the simple random walk

A first application of the Kolmogorov zero-one law is the fluctuation behaviour of a symmetric **simple random walk** (SRW). Let $(X_i)_{i\in\mathbb{N}}$ be iid random variables such that $\mathbb{P}(X_i = \pm 1) = \frac{1}{2}$ and set $S_n = X_1 + \cdots + X_n$. The rv X_i are called **steps** of the simple random walk.

31.9 Theorem. *Let $(S_n)_{n\in\mathbb{N}}$ be a simple random walk with iid steps X_i such that $\mathbb{P}(X_i = \pm 1) = \frac{1}{2}$. Then*

$$\mathbb{P}\Big(\limsup_{n\to\infty} S_n = +\infty\Big) = \mathbb{P}\Big(\liminf_{n\to\infty} S_n = -\infty\Big) = 1.$$

Proof. Set $S = \limsup_{n\to\infty} S_n$ and observe that

$$S = \limsup_{n\to\infty} S_n = X_1 + \limsup_{n\to\infty}(X_2 + \cdots + X_n) =: X_1 + S'.$$

Since the steps $(X_i)_{i\in\mathbb{N}}$ are iid, we see that $S \sim S'$ – note that the sequences $(X_i)_{i\in\mathbb{N}}$ and $(X_{1+i})_{i\in\mathbb{N}}$ have the same distributions! – and $X_1 \perp\!\!\!\perp S'$. Moreover, from

$$\big\{S = \pm\infty\big\} = \big\{\limsup_{n\to\infty}(X_m + \cdots + X_n) = \pm\infty\big\} \in \mathscr{T}_m \quad \forall m \in \mathbb{N}$$

we conclude that $\{S = \pm\infty\}$ is terminal, i.e. $\mathbb{P}(S = \pm\infty)$ is 0 or 1 by Theorem 31.8.

Case 1: $\mathbb{P}(S = \pm\infty) = 1$. Since $X_i \sim -X_i$, we get

$$S = \limsup_{n\to\infty} S_n \sim \limsup_{n\to\infty}(-S_n) = -\liminf_{n\to\infty} S_n \tag{31.2}$$

and we conclude that $\mathbb{P}(S = +\infty) = 1$ and $\mathbb{P}(\liminf_n S_n = -\infty) = 1$.

Case 2: $\mathbb{P}(S = \pm\infty) = 0$. This means that S is a.s. finite. Therefore, we find for all $\xi \in \mathbb{R}$

$$\mathbb{E}e^{i\xi S} = \mathbb{E}e^{i\xi(X_1+S')} \overset{X_1 \perp\!\!\!\perp S'}{=} \mathbb{E}e^{i\xi S'}\mathbb{E}e^{i\xi X_1} \underset{\S 27.2.c)}{\overset{S\sim S'}{=}} \mathbb{E}e^{i\xi S}\cos\xi. \tag{31.3}$$

Since $\phi_S(\xi) = \mathbb{E}e^{i\xi S}$ is (uniformly) continuous in a neighbourhood of 0 and $\phi_S(0) = 1$, we see for sufficiently small $\epsilon > 0$ that $\phi_S|_{[-\epsilon,\epsilon]} \neq 0$, hence we have $\cos\xi = 1$ on $[-\epsilon,\epsilon]$ which is absurd. Therefore, $\mathbb{P}(S = \pm\infty) = 0$ cannot happen. □

A classic application: Borel's »normal numbers« (1909)

Consider the probability space $([0,1), \mathscr{B}[0,1), d\omega)$. We can write every $\omega \in [0,1)$ as a g-adic fraction with $g \in \mathbb{N}$, $g \geqslant 2$:

$$\omega = 0.t_1t_2t_2\cdots := \sum_{n=1}^{\infty} t_n g^{-n}$$

with »digits« $t_n \in \{0, 1, \ldots, g-1\}$.

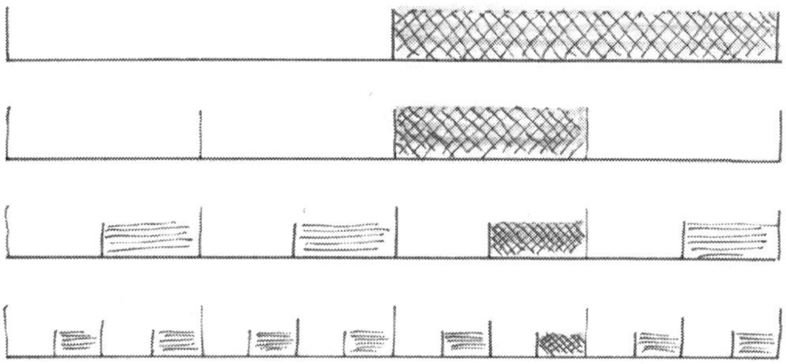

Fig. 31.1. In a **dyadic** ($g = 2$) expansion, the digit 0 corresponds to »take the left box« and 1 means »take the right box«. The figure shows the first few »decisions« of the number 0.1011... in criss-cross hatching. The horizontal hatch pattern in the last two lines shows the numbers $0.{*}{*}1{*}\ldots$ (the third digit is 1) and $0.{*}{*}{*}1{*}\ldots$ (fourth digit is 1). Observe the periodicity.

We identify $0.t_1\ldots t_n\overline{(g-1)}$ with $0.t_1\ldots t_{n-1}(t_n+1)\overline{000}$ to enforce the uniqueness of the representation. In general, we have (see Fig. 31.1)

$$\{t_n = \ell\} = \{\omega \mid \text{the } n\text{th digit} = \ell\} = \bigcup_{m=0}^{g^{n-1}-1} \left[\frac{mg+\ell}{g^n}, \frac{mg+\ell+1}{g^n}\right)$$

i.e. $t_n = t_n(\omega)$ is Borel measurable, hence a random variable. Since the numbers ω are uniformly distributed in $[0,1)$, we have

- $\mathbb{P}(t_n = \ell) = \frac{1}{g} \quad \forall \ell \in \{0,1,\ldots,g-1\}$;
- t_n, $n \in \mathbb{N}$, are independent ✍ random variables, see §29 Appendix.

We are interested in the relative frequency of the digits in a »typical« number $\omega \in [0,1)$. Define

$$X_n^\ell = \mathbb{1}_{\{t_n=\ell\}} \quad \text{and} \quad S_n^\ell = X_1^\ell + \cdots + X_n^\ell.$$

31.10 Definition. A number $\omega \in [0,1)$ is said to be a **normal number** (w.r.t. the basis g), if

$$\lim_{n\to\infty} \frac{S_n^\ell}{n} = \frac{1}{g} \quad \forall \ell = 0,1,2,\ldots,g-1, \quad \text{a.s.}$$

The number ω is **absolutely normal**, if ω is normal for all $g \in \mathbb{N}$, $g \geqslant 2$.

31.11 Theorem (Borel 1909). *Lebesgue-almost all $\omega \in [0,1)$ are absolutely normal.*

Proof. According to the L^4-SLLN, there is for every $g \in \mathbb{N}$, $g \geqslant 2$, a set $\Omega_g \subset [0,1)$

such that $\mathrm{Leb}(\Omega_g) = 1$ and

$$\lim_{n\to\infty} \frac{S_n^\ell(\omega)}{n} = \frac{1}{g} \qquad \forall\omega \in \Omega_g,\ \forall\ell = 0,\dots,g-1. \tag{31.4}$$

This proves that almost all $\omega \in [0,1)$ are normal. Since the countable intersection $\Omega_\infty := \bigcap_{g=2}^\infty \Omega_g$ still has Lebesgue measure 1 and since (31.4) holds on Ω_∞, all $\omega \in \Omega_\infty$ are absolutely normal. □

▲ Theorem 31.11 is a non-constructive existence result. For most concrete numbers e.g. $\frac{1}{2}\sqrt{2}$, $\frac{1}{4}\pi$, $\frac{1}{3}e$, $\ln 2$,... it is not known whether they are normal or not. Currently, only »trivial« (i.e. periodic) normal numbers are known.

32 The strong law of large numbers

Let us continue our investigation of the convergence of (sums of) independent random variables with the strong law of large numbers (SLLN). We have alerady seen a simple version, Cantelli's L^4-SLLN, in Theorem 31.3. On the other hand, integrability (L^1) is a necessary condition (Theorem 31.4.b)). Kolmogorov found in 1930/1933 the definitive L^1-version of the SLLN, using a truncation trick and the L^2-version of the SLLN (see below, Theorem 33.4 and 33.6); Etemadi [11] provided 50 years later, in 1981, a new elementary proof which reduces »independence« to »mutual independence«; both proofs can be found in [WT, Chapter 12]. Here we give another elementary proof which I learned from J.-F. Le Gall.

32.1 Theorem (Kolmogorov; L^1-SLLN). *Let $(X_i)_{i\in\mathbb{N}}$ be iid real random variables such that $X_i \in L^1$ and $S_n = X_1 + \cdots + X_n$.*

$$\lim_{n\to\infty} \frac{S_n}{n} = \mathbb{E}X_1 \quad a.s. \tag{32.1}$$

Proof. Pick $\lambda > \mathbb{E}X_1$ and define $M = \sup_{n\geqslant 0}(S_n - n\lambda) \in [0,\infty]$. Assume for a moment that $\mathbb{P}(M < \infty) = 1$. We have

$$S_n \leqslant n\lambda + M \text{ a.s.} \implies \limsup_{n\to\infty} \frac{S_n}{n} \leqslant \lambda \text{ a.s.}$$

Letting $\lambda \downarrow \mathbb{E}X_1$ along a countable sequence, we get $\limsup_{n\to\infty} \frac{S_n}{n} \leqslant \mathbb{E}X_1$.

Replacing in the above argument $X_i \rightsquigarrow -X_i$ gives $\liminf_{n\to\infty} \frac{S_n}{n} \geqslant \mathbb{E}X_1$, and (32.1) follows.

We still have to prove that $M < \infty$ a.s. The equality

$$\{M < \infty\} = \left\{\sup_{n\geqslant m}[(X_m + \cdots + X_n) - (n-m)\lambda] < \infty\right\} \in \mathscr{T}_m \quad \forall m \in \mathbb{N}$$

shows that $\{M < \infty\} \in \mathscr{T}_\infty$, hence $\mathbb{P}(M < \infty)$ is zero or one, cf. Theorem 31.8.

Define

$$\mathrm{M}_n := \sup_{0 \leqslant i \leqslant n} (\mathrm{S}_i - \lambda i) \quad \text{and} \quad \mathrm{M}'_n := \sup_{0 \leqslant i \leqslant n} (\mathrm{S}_{i+1} - \lambda i - \mathrm{X}_1).$$

Since the random variables X_i, $i \in \mathbb{N}$, are iid, we see that $\mathrm{M}_n \sim \mathrm{M}'_n$, thus

$$\mathrm{M} := \sup_{n \geqslant 0} \mathrm{M}_n \sim \sup_{n \geqslant 0} \mathrm{M}'_n = \mathrm{M}'.$$

Now we have

$$\begin{aligned}\mathrm{M}_{n+1} = 0 \vee \sup_{1 \leqslant i \leqslant n+1} (\mathrm{S}_i - \lambda i) &= 0 \vee (\mathrm{M}'_n + \mathrm{X}_1 - \lambda) \\ &= \mathrm{M}'_n - \mathrm{M}'_n \wedge (\lambda - \mathrm{X}_1).\end{aligned}$$

The last equality follows from the identity $0 \vee (a - b) = a - (a \wedge b)$ which can be easily checked. Using $\sim \mathrm{M}_n \leqslant \mathrm{M}_{n+1}$ and $\mathrm{M}'_n \sim \mathrm{M}_n$ shows

$$0 \leqslant \mathbb{E}\mathrm{M}_{n+1} - \mathbb{E}\mathrm{M}_n = \mathbb{E}\mathrm{M}_{n+1} - \mathbb{E}\mathrm{M}'_n = -\mathbb{E}\left[(\lambda - \mathrm{X}_1) \wedge \mathrm{M}'_n\right].$$

Since $|(\lambda - \mathrm{X}_1) \wedge \mathrm{M}'_n| \leqslant |\lambda - \mathrm{X}_1| \in \mathrm{L}^1(\mathbb{P})$, we can use the DCT to get

$$\mathbb{E}\left[(\lambda - \mathrm{X}_1) \wedge \mathrm{M}'\right] = \lim_{n \to \infty} \mathbb{E}\left[(\lambda - \mathrm{X}_1) \wedge \mathrm{M}'_n\right] \leqslant 0.$$

Assume that $\mathbb{P}(\mathrm{M}' = \infty) = 1$, then the last estimate yields

$$0 < \lambda - \mathbb{E}\mathrm{X}_1 = \mathbb{E}(\lambda - \mathrm{X}_1) \overset{\mathrm{M}'=\infty \text{ a.s.}}{=} \mathbb{E}\left[(\lambda - \mathrm{X}_1) \wedge \mathrm{M}'\right] \leqslant 0$$

which is impossible. Thus, $\mathbb{P}(\mathrm{M} = \infty) \overset{\mathrm{M} \sim \mathrm{M}'}{=} \mathbb{P}(\mathrm{M}' = \infty) = 0.$ □

The following converse of the SLLN shows again that the L^1-condition is best possible.

32.2 Theorem. *Let $(\mathrm{X}_i)_{i \in \mathbb{N}}$ be iid real random variables such that the limit*

$$\lim_{n \to \infty} \frac{\mathrm{X}_1(\omega) + \cdots + \mathrm{X}_n(\omega)}{n} = \mathrm{L}(\omega) \in \mathbb{R} \quad \textit{exists for a.a. } \omega.$$

In this case, $\mathrm{X}_1 \in \mathrm{L}^1(\mathbb{P})$ and $\mathrm{L} = \mathbb{E}\mathrm{X}_1$ a.s.

Proof. By definition, $\mathrm{L} = \mathrm{L}(\omega)$ is an a.s. finite random variable. Since $\mathrm{S}_n/n \to \mathrm{L}$, we see that

$$\frac{\mathrm{X}_n}{n} = \underbrace{\frac{\mathrm{S}_n}{n}}_{\to \mathrm{L}} - \underbrace{\frac{n-1}{n}}_{\to 1} \underbrace{\frac{\mathrm{S}_{n-1}}{n-1}}_{\to \mathrm{L}} \xrightarrow[n \to \infty]{} 0 \quad \text{a.s.}$$

Define $\mathrm{A}_n := \{|\mathrm{X}_n| > n\}$. For almost all ω, we know that $\omega \in \mathrm{A}_n$ happens for at most finitely many n. Since the sets A_n are independent, we know from the Borel–Cantelli lemma (Theorem 31.1) that

$$\sum_{n=1}^{\infty} \mathbb{P}(\mathrm{A}_n) < \infty.$$

Using the layer-cake formula (Theorem 16.5) we see that

$$\mathbb{E}|X_1| = \int_0^\infty \mathbb{P}(|X_1| > t)\,dt \leqslant \sum_{n=0}^\infty \int_n^{n+1} \mathbb{P}(|X_1| > t)\,dt$$
$$\leqslant 1 + \sum_{n=1}^\infty \mathbb{P}(|X_1| > n).$$

Now we use that $X_1 \sim X_n$ in each term under the sum, and so

$$\mathbb{E}|X_1| \leqslant 1 + \sum_{n=1}^\infty \mathbb{P}(|X_n| > n) = 1 + \sum_{n=1}^\infty \mathbb{P}(A_n) < \infty.$$

This shows that all assumptions of Theorem 32.1 hold, and we conclude that $L = \mathbb{E}X_1$ a.s. □

An application in numerical analysis: The Monte-Carlo method

We are interested in the numerical evaluation of integrals of the form $\int_a^b f(y)\,dy$.

32.3 Corollary. *Let $(X_n)_{n\in\mathbb{N}}$ be iid uniformly distributed rv, $X_i \sim \mathsf{U}[0,1]$, and $f : [0,1] \to \mathbb{R}$ a bounded measurable function. Then*

$$\lim_{n\to\infty} \frac{f(X_1) + \cdots + f(X_n)}{n} = \int_0^1 f(x)\,dx. \tag{32.2}$$

Proof. The rv ZV $f(X_n)$ are again iid and

$$\mathbb{E}f(X_1) = \int f(x)\mathbb{P}(X_1 \in dx) = \int f(x)\,\mathbb{1}_{[0,1]}(x)\,dx = \int_0^1 f(x)\,dx < \infty.$$

Therefore, the claim follows from the SLLN (Theorem 32.1). □

Extension 1: Different domains of integration.

Let $K \subset \mathbb{R}$ be any measurable set and $f : K \to \mathbb{R}$ be a bounded measurable function. We have

$$\int_K f(x)\,dx = \int_K \frac{f(x)}{p(x)}\,p(x)\,dx = \mathbb{E}\frac{f(Y_1)}{p(Y_1)} = \lim_{n\to\infty} \frac{\frac{f(Y_1)}{p(Y_1)} + \cdots + \frac{f(Y_n)}{p(Y_n)}}{n}$$

where the rv $Y_n \sim p(x)\,dx$ are iid and $p(x) > 0$ is any probability density on K, i.e. any measurable $p(x) > 0$ such that $\int_K p(x)\,dx = 1$. For instance, we can construct the rv Y_n as follows:

$$F(x) = \int_{(-\infty,x]} p(x)\,dx,\ \ X_n \sim \mathsf{U}[0,1] \text{ iid} \overset{\S 26.2}{\Longrightarrow} Y_n := F^{-1}(X_n) \sim p(x)\,dx \text{ iid}.$$

Extension 2: Higher dimensions $d > 1$.

Assume that $f : Q \to \mathbb{R}$ is a measurable and bounded function, defined on the cube $Q = [-a,a]\times\cdots\times[-a,a] \subset \mathbb{R}^d$. Again, we want to evaluate $\int_Q f(y)\,dy$.

We follow our previous strategy: Let Y_n be iid rv taking values in Q, e.g. $Y_n \sim (2a)^{-d}\mathbb{1}_Q(y)\,dy$. Because of the SLLN (Theorem 32.1) we have

$$\begin{aligned}\lim_{n\to\infty}\frac{f(Y_1)+\cdots+f(Y_n)}{n} = \mathbb{E}f(Y_1) &= \int f(y)\,\mathbb{P}(Y_1\in dy)\\ &= \int f(y)(2a)^{-d}\mathbb{1}_Q(y)\,dy\\ &= \frac{1}{(2a)^d}\int f(y)\mathbb{1}_Q(y)\,dy.\end{aligned}$$

The problem is how we can construct the rv Y_n.

As we need just **any** iid rv Y_n, we can use the following idea:

$$Y_n = (Y_{n,1},\dots,Y_{n,d}) \quad\text{and}\quad Y_{n,i} \sim \frac{1}{2a}\mathbb{1}_{[-a,a]}(y_i)\,dy_i,\quad i=1,\dots,d,\quad \text{iid}$$

and, if the coordinates are be independent, then we have (Theorem 25.9)

$$Y_n \sim \prod_{i=1}^{d}\frac{1}{2a}\mathbb{1}_{[-a,a]}(y_i)\,dy_i.$$

Thus, we only have to simulate nd iid, one-dimensional uniform random variables.

Extension 3: Error control. For this we can use the CLT. Recall that we have for iid $X_n \in L^2(\mathbb{P})$ with $\mathbb{E}X_1 = \mu$, $\mathbb{V}X_1 = \sigma^2$ (Theorem 28.6)

$$\mathbb{P}\left(a < \frac{S_n - n\mu}{\sigma\sqrt{n}} < b\right) \approx \frac{1}{\sqrt{2\pi}}\int_a^b e^{-t^2/2}\,dt = \Phi(b)-\Phi(a).$$

As in the second extension we use

$$X_i = f(Y_i),\quad \frac{S_n}{n} = \frac{f(Y_1)+\cdots+f(Y_n)}{n},\quad \mu = \mathbb{E}f(Y_1) = \frac{1}{|Q|}\int_Q f(y)\,dy$$

and set $a=-b$. A simple rearrangement gives

$$\mathbb{P}\left(\left|\frac{S_n}{n}-\mu\right| < \frac{b\sigma}{\sqrt{n}}\right) = \mathbb{P}\left(\frac{-b\sigma}{\sqrt{n}} < \frac{S_n}{n}-\mu < \frac{b\sigma}{\sqrt{n}}\right) \approx \Phi(b)-\Phi(-b).$$

Since Φ is symmetric (relative to $(0,\frac{1}{2})$) we get $\Phi(b)-\Phi(-b) = 2(1-\Phi(b))$. If we write $|Q|$ for the Volume (Lebesgue measure) of the set Q, we get

$$\mathbb{P}\left(\left|\frac{|Q|S_n}{n} - \int_Q f(x)\,dx\right| > \frac{b\sigma|Q|}{\sqrt{n}}\right) \approx 2(1-\Phi(b)).$$

Better error estimates can be obtained from the Berry–Esséen theorem, cf. [WT, p. 85] and Chung [8, Chapter 7.4].

⚠ The speed of approximation is $\sqrt{n}$ and it does **not depend on the dimension** d of the underlying Euclidean space. This means that Monte–Carlo methods are slow, but ideal for high dimensions.

An application in statistics: The Glivenko–Cantelli lemma★

Assume that $(X_i)_{i\in\mathbb{N}}$ are iid random variables. Their common distribution function will be denoted by

$$F(x) := \mathbb{P}(X_i \leqslant x).$$

The general problem in statistics is that we observe the outcomes $X_i(\omega) = x_i$ (x_i is a concrete value) and we can record their empirical frequencies. The true distribution function $F(x)$ is, however, unknown.

Typical problem: Find $F(x)$ from the »empirical values«, i.e. the »samples« $X_1(\omega) = x_1, X_2(\omega) = x_2, \dots$

Let n be fixed and let $(X_1(\omega),\dots,X_n(\omega)) = (x_1,\dots,x_n)$ be a sample which we order according to size:

$$X_{n:1}(\omega) \leqslant X_{n:2}(\omega) \leqslant \dots \leqslant X_{n:n}(\omega)$$

The **empirical distribution function** (from n observations) is defined as

$$F_n(x,\omega) := \begin{cases} 0, & \text{if } x < X_{n:1}(\omega), \\ \dfrac{k}{n}, & \text{if } X_{n:k}(\omega) \leqslant x < X_{n:k+1}(\omega),\ 1 \leqslant k \leqslant n-1, \\ 1 & \text{if } x \geqslant X_{n:n}(\omega). \end{cases}$$

In other words, $nF_n(x,\omega) = \#\{i \mid X_i(\omega) \leqslant x\}$ or $F_n(x,\omega)$ is the relative frequency of the observations $X_i(\omega) \leqslant x$. We have

$$F_n(x,\omega) = \frac{1}{n}\sum_{i=1}^{n} \mathbb{1}_{\{X_i\leqslant x\}}(\omega)$$

and this shows that $F_n(x,\cdot)$ is itself a random variable.

Since the X_i are independent, so are the functions $\xi_i(x,\cdot) := \mathbb{1}_{\{X_i\leqslant x\}}(\cdot)$; note that ξ_i are iid Bernoulli rv with $p = F(x)$ and $q = 1 - F(x)$. Thus, $\mathbb{E}\xi_i(x) = F(x)$, and the SLLN (Theorem 32.1) shows that

$$F_n(x,\omega) = \frac{1}{n}\sum_{i=1}^{n} \xi_i(x,\omega) \xrightarrow{\text{a.s.}} F(x). \tag{32.3}$$

⚠ The exceptional set appearing in (32.3) may depend on x, i.e. we may end up with more than countably many null sets!

Since we want to recover the whole function $F(x)$, we have to take care about the exceptional sets. In fact, we have the following fundamental theorem of mathematical statistics.

32.4 Theorem* (Glivenko, Cantelli). *The empirical distribution function satisfies*

$$\lim_{n\to\infty} \sup_{x\in\mathbb{R}} |F_n(x,\omega) - F(x)| = 0 \quad a.s.$$

Proof. Denote by $\mathbb{D}$ the discontinuity points of F. Since F is increasing, the beside-lamp lemma (Fig. 16.1 on page 87) shows that $\mathbb{D}$ is at most countable. Define

$$\eta_i(x,\omega) := \mathbb{1}_{\{X_i=x\}}(\omega) = \begin{cases} 1, & \text{if } X_i(\omega) = x \\ 0, & \text{if } X_i(\omega) \neq x \end{cases} \qquad \forall x \in \mathbb{D},$$

and note that $F_n(x+,\omega) - F_n(x-,\omega) = \frac{1}{n}\sum_{i=1}^{n} \eta_i(x,\omega)$. Using the SLLN we get, as above,

$$F_n(x+,\omega) - F_n(x-,\omega) \xrightarrow[n\to\infty]{} F(x+) - F(x-) = F(x) - F(x-)$$

for all $\omega \in \Omega_x$ from a set Ω_x of full probability $\mathbb{P}(\Omega_x) = 1$.

Define $\Omega' := \bigcap_{x\in\mathbb{Q}\cup\mathbb{D}} \Omega_x$. Since this is a countable intersection, we still have $\mathbb{P}(\Omega') = 1$, and for all $\omega \in \Omega'$ we find

$$\forall x \in \mathbb{D} : F_n(x+,\omega) - F_n(x-,\omega) \xrightarrow[n\to\infty]{} F(x) - F(x-)$$
$$\forall x \in \mathbb{Q}\cup\mathbb{D} : F_n(x,\omega) \xrightarrow[n\to\infty]{} F(x).$$

Since F is right-continuous, this gives uniform convergence for all $x \in \mathbb{R}$ – but this is a purely deterministic result which we defer to the following Lemma 32.5. □

The following lemma is a result from analysis which we only state in the form needed here. For a proof we refer to [WT, Lemma 12.11].

32.5 Lemma*. *Let* F_n, F *be right-continuous distribution functions and denote by* $\mathbb{D}$ *the discontinuity points of* F. *If*

$$\forall x \in \mathbb{D} : \quad F_n(x) - F_n(x-) \xrightarrow[n\to\infty]{} F(x) - F(x-),$$

and

$$\forall x \in \mathbb{Q}\cup\mathbb{D} : \quad F_n(x) \xrightarrow[n\to\infty]{} F(x),$$

then it holds that $\lim_{n\to\infty} \sup_{x\in\mathbb{R}} |F_n(x) - F(x)| = 0$.

33 Sums of independent random variables

We will conclude our investigations on the limits of sums of independent random variables. Throughout this chapter we assume that

- $X_1, X_2, \dots$ are real random variables on the same space $(\Omega, \mathscr{A}, \mathbb{P})$.
- $S_n = X_1 + \cdots + X_n$ is always the partial sum, and $S_0 = 0$.

We begin with a (surprisingly new) result in this very classical subject.

33.1 Theorem (maximal inequality; Etemadi 1985). *Let $X_1, \dots, X_n$ be independent real random variables. One has*

$$\mathbb{P}\Big(\max_{1\leqslant k\leqslant n} |S_k| \geqslant 3t\Big) \leqslant 3 \max_{1\leqslant k\leqslant n} \mathbb{P}\big(|S_k| \geqslant t\big) \quad \forall t \geqslant 0. \tag{33.1}$$

Proof. We define disjoint sets

$$B_k := \{\omega \mid |S_k(\omega)| \geqslant 3t \text{ and } |S_i(\omega)| < 3t \;\forall i < k\}.$$

which partition the set

$$\Big\{\max_{1\leqslant k\leqslant n} |S_k| \geqslant 3t\Big\} = \dot{\bigcup_{k=1}^{n}} B_k.$$

Therefore we find that

$$\begin{aligned}\mathbb{P}\Big(\max_{1\leqslant k\leqslant n} |S_k| \geqslant 3t\Big) &= \mathbb{P}\Big(\max_{1\leqslant k\leqslant n} |S_k| \geqslant 3t,\, |S_n| \geqslant t\Big) + \mathbb{P}\Big(\max_{1\leqslant k\leqslant n} |S_k| \geqslant 3t,\, |S_n| < t\Big) \\ &\leqslant \mathbb{P}(|S_n| \geqslant t) + \sum_{k=1}^{n} \mathbb{P}(B_k \cap \{|S_n| < t\}) \\ &\leqslant \mathbb{P}(|S_n| \geqslant t) + \sum_{k=1}^{n-1} \mathbb{P}(B_k \cap \{|S_n - S_k| > 2t\});\end{aligned}$$

here we use that $B_n \cap \{|S_n| < t\} = \emptyset$, reducing the upper bound of the summation $n \rightsquigarrow n-1$; moreover, on $B_k \cap \{|S_n| < t\}$ we have $|S_n - S_k| \geqslant |S_k| - |S_n| > 3t - t = 2t$ which brings the diffference of the partial sums into play. The latter allows us to use, in the next step, the independence $B_k \in \sigma(X_1, \dots, X_k) \perp\!\!\!\perp \sigma(X_{k+1}, \dots, X_n) \ni S_n - S_k$:

$$\begin{aligned}\mathbb{P}\Big(\max_{1\leqslant k\leqslant n} |S_k| \geqslant 3t\Big) &= \mathbb{P}(|S_n| \geqslant t) + \sum_{k=1}^{n-1} \mathbb{P}(B_k)\mathbb{P}(|S_n - S_k| > 2t) \\ &\leqslant \mathbb{P}(|S_n| \geqslant t) + \underbrace{\sum_{k=1}^{n-1} \mathbb{P}(B_k)}_{\leqslant 1} \max_{1\leqslant k\leqslant n} \mathbb{P}(|S_n - S_k| > 2t) \\ &\leqslant \mathbb{P}(|S_n| \geqslant t) + \max_{1\leqslant k\leqslant n} \mathbb{P}(|S_n - S_k| > 2t).\end{aligned}$$

Using the triangle inequality $|S_n - S_k| \leqslant |S_n| + |S_k|$, we see that on the set appearing on the right-hand side, it is not possible that both $|S_n| \leqslant t$ and $|S_k| \leqslant t$. Thus,

$$\mathbb{P}\left(\max_{1\leqslant k\leqslant n} |S_k| \geqslant 3t\right) \leqslant \mathbb{P}(|S_n| \geqslant t) + \max_{1\leqslant k\leqslant n} \mathbb{P}(|S_n| > t \text{ or } |S_k| > t)$$
$$\leqslant \mathbb{P}(|S_n| \geqslant t) + \mathbb{P}(|S_n| > t) + \max_{1\leqslant k\leqslant n} \mathbb{P}(|S_k| > t),$$

and the estimate follows. □

Etemadi's inequality has the following surprising consequence.

33.2 Theorem (Lévy). *Let X_n be independent real rv and $S_n = X_1 + \dots + X_n$.*

$$S_n \textit{ converges a.s.} \iff S_n \textit{ converges in probability.}$$

We can add another equivalence: $\iff$ S_n converges in distribution.

Here★ is the argument: Assume that $S_n \xrightarrow{d} S$. Since all partial sums live on the same probability space, the difference $S_n - S_m$ is well defined and we see from $S_m \perp\!\!\!\perp S_n - S_m$

$$\mathbb{E}e^{i\xi S_n} = \mathbb{E}e^{i\xi(S_n - S_m)} \cdot \mathbb{E}e^{i\xi S_m}.$$

Since $S_n \to S$ implies that $\mathbb{E}e^{i\xi S_n} \to \mathbb{E}e^{i\xi S}$, we conclude that

$$\lim_{m,n\to\infty} \mathbb{E}e^{i\xi(S_n - S_m)} = 1 \overset{\S 30.3}{\Longrightarrow} S_n - S_m \xrightarrow[m,n\to\infty]{} 0.$$

Since the limit is trivial, we see that $S_n - S_m \xrightarrow{\mathbb{P}} 0$, and by the completeness of convergence in probability (Theorem 29.19), $S_n \xrightarrow{\mathbb{P}} S$.
The converse that $\mathbb{P}$-convergence entails d-convergence is obvious, cf. Theorem 29.5.

Proof. The implicaton »$\Rightarrow$« holds without independence, cf. Remark 29.4.c).
For the converse »$\Leftarrow$« we do need independence. Set $Z := \mathbb{P}\text{-}\lim_{n\to\infty} S_n$. For any $\epsilon > 0$ and $m, n \in \mathbb{N}$, $m \leqslant n$ we have

$$\mathbb{P}\big(|S_n - Z| > 6\epsilon \text{ i.o.}\big) \leqslant \mathbb{P}\left(\sup_{n\geqslant m} |S_n - Z| > 6\epsilon\right)$$
$$\leqslant \mathbb{P}(|S_m - Z| > 3\epsilon) + \mathbb{P}\left(\sup_{n\geqslant m} |S_n - S_m| > 3\epsilon\right).$$

The last estimte uses the triangle inequality: $|S_n - Z| \leqslant |S_n - S_m| + |S_m - Z|$, and so it is impossible that both $|S_n - S_m| \leqslant 3\epsilon$ and $|S_m - Z| \leqslant 3\epsilon$. Thus,

$$\begin{aligned}
&\mathbb{P}\big(|S_n - Z| > 6\epsilon \text{ i.o.}\big)\\
&\overset{\text{Etemadi}}{\leqslant} \mathbb{P}(|S_m - Z| > 3\epsilon) + 3\sup_{n\geqslant m} \mathbb{P}(|S_n - S_m| > \epsilon)\\
&\leqslant \mathbb{P}(|S_m - Z| > 3\epsilon) + 3\left(\mathbb{P}(|Z - S_m| > \epsilon/2) + \sup_{n\geqslant m} \mathbb{P}(|S_n - Z| > \epsilon/2)\right)\\
&\leqslant 7\sup_{n\geqslant m} \mathbb{P}(|S_n - Z| > \epsilon/2).
\end{aligned}$$

The right-hand side converges to 0 as $m \to \infty$, since we have

$$\underbrace{\lim_{m\to\infty} \sup_{n\geqslant m}}_{=\limsup_n} \mathbb{P}(|S_n - Z| > \epsilon/2) = \lim_{m\to\infty} \mathbb{P}(|S_m - Z| > \epsilon/2) = 0.$$

Since $m \to \infty \implies n \to \infty$, we see that

$$\limsup_{n\to\infty} \mathbb{P}\Big(|S_n - Z| > 6\epsilon \ \text{i.o.}\Big) = 0,$$

and the claim follows from Lemma 29.10. □

33.3 Corollary (Kolmogorov). *Let* $(X_i)_{i\in\mathbb{N}} \subset L^2(\mathbb{P})$ *be independent real random variables.*

$$\sum_{i=1}^{\infty} \mathbb{V}X_i < \infty \implies \sum_{i=1}^{\infty} (X_i - \mathbb{E}X_i) \quad \textit{converges a.s.}$$

Proof. Since $\mu_i := \mathbb{E}X_i$ is a constant, it is clear that

$$\mathbb{V}X_i = \mathbb{V}(X_i - \mu_i) \quad \text{and} \quad (X_i - \mu_i)_{i\in\mathbb{N}} \text{ is independent, cf. §26.4.}$$

Therefore, we may assume that $\mathbb{E}X_i = 0$. For any $m < n$ we have

$$\mathbb{E}(S_n - S_m)^2 = \mathbb{V}(S_n - S_m) \overset{\text{Bienaymé}}{=} \sum_{m+1}^{n} \mathbb{V}(X_i) \xrightarrow{m,n\to\infty} 0,$$

and since $L^2(\mathbb{P})$ is complete, $S_n \xrightarrow{L^2} S$, hence in probability and, by Theorem 33.2, a.s. □

We can now show necessary and sufficient criteria for the convergence of a series of independent random variables. To do so, we need the following **truncation argument** which is due to Khintchine[(1)]. Let Z be a real random variable. We define

$$Z^K(\omega) := Z(\omega)\mathbb{1}_{\{|Z|\leqslant K\}}(\omega) = \begin{cases} Z(\omega) & \text{if } |Z(\omega)| \leqslant K; \\ 0 & \text{otherwise.} \end{cases}$$

33.4 Theorem (three series theorem; Kolmogorov). *Let* $(X_i)_{i\in\mathbb{N}}$ *be independent real random variables. The series* $\sum_{i=1}^{\infty} X_i$ *converges a.s. if, and only if, for some (or for all)* $K > 0$ *the following three series converge:*

$$\text{a)} \ \sum_{i=1}^{\infty} \mathbb{P}(|X_i| > K); \qquad \text{b)} \ \sum_{i=1}^{\infty} \mathbb{E}X_i^K; \qquad \text{c)} \ \sum_{i=1}^{\infty} \mathbb{V}X_i^K.$$

(1) Khintchine used this argument in order to reduce the L^2-assumption in Theorem 28.2 to L^1. His reasoning is similar to the one used in the following theorem.

Proof. We will only show that the conditions are sufficient – this is anyway more important for applications; necessity is proved in the appendix to this chapter.

Assume that the three series a)–c) converge for some fixed $K > 0$. So,

$$\sum_{i=1}^{\infty} \mathbb{P}\left(X_i \neq X_i^K\right) = \sum_{i=1}^{\infty} \mathbb{P}\left(|X_i| > K\right) < \infty.$$

The (easy direction of the) Borel–Cantelli lemma (Theorem 31.1.a)) tells us that $\mathbb{P}\left(X_i \neq X_i^K \text{ i.o.}\right) = 0$. This is the same as to say that there is an Ω_0 with $\mathbb{P}(\Omega_0) = 1$ such that

$$\forall \omega \in \Omega_0 \quad \exists N = N(\omega) \quad \forall i \geqslant N(\omega): \quad X_i(\omega) = X_i^K(\omega).$$

For fixed $\omega \in \Omega_0$ it is enough tho prove the convergence of the series

$$\sum_{i=1}^{\infty} X_i^K(\omega)$$

as only the tail $\sum_{N(\omega)}^{\infty} X_i(\omega) = \sum_{N(\omega)}^{\infty} X_i^K(\omega)$ decides on convergence or divergence. But now we have

$$\underbrace{\sum_{i=1}^{\infty} \mathbb{V}X_i^K < \infty}_{\text{from c)}} \overset{\S 33.3}{\Longrightarrow} \sum_{i=1}^{\infty} \left(X_i^K - \mathbb{E}X_i^K\right) \text{ converges}$$

$$\overset{\text{b)}}{\Longrightarrow} \sum_{i=1}^{\infty} X_i^K \text{ converges.} \qquad \square$$

Let us finally derive from Corollary 33.3 an L^2-SLLN which needs that the random variables are only independent, but not identically distributed. For this, we need a famous lemma from analysis.

33.5 Lemma*. *Let $(a_i)_{i\in\mathbb{N}_0} \subset (0,\infty)$ be an increasing sequence of positive numbers such that $a_i \uparrow \infty$.*

a) (**Cesàro**) *If the sequence $(v_i)_{i\in\mathbb{N}} \subset \mathbb{R}$ has a limit $v_\infty = \lim_{i\to\infty} v_i$, then*

$$\lim_{n\to\infty} \frac{1}{a_n} \sum_{i=1}^{n} (a_i - a_{i-1}) v_i = v_\infty.$$

b) (**Kronecker**) *If $(x_i)_{i\in\mathbb{N}} \subset \mathbb{R}$ is any sequence and $s_n := x_1 + x_2 + \cdots + x_n$, then*

$$\sum_{i=1}^{\infty} \frac{x_i}{a_i} \text{ converges} \quad \Longrightarrow \quad \lim_{n\to\infty} \frac{s_n}{a_n} = 0.$$

Proof. a) Since $v_\infty = \lim_{i\to\infty} v_i$, we have

$$\forall \epsilon > 0 \quad \exists N = N_\epsilon \in \mathbb{N} \quad \forall i \geqslant N: \quad v_\infty + \epsilon \geqslant v_i \geqslant v_\infty - \epsilon.$$

Therefore, we see for every natural number $n > N$ that

$$\frac{1}{a_n}\sum_{i=1}^{n}(a_i - a_{i-1})v_i \geqslant \frac{1}{a_n}\left(\sum_{i=1}^{N}(a_i - a_{i-1})v_i + \sum_{i=N+1}^{n}(a_i - a_{i-1})(v_\infty - \epsilon)\right)$$

$$= \underbrace{\frac{1}{a_n}\overbrace{\sum_{i=1}^{N}(a_i - a_{i-1})v_i}^{\text{does not depend on } n}}_{\xrightarrow[n\to\infty]{}0} + \underbrace{\frac{a_n - a_N}{a_n}}_{\xrightarrow[n\to\infty]{}1}(v_\infty - \epsilon).$$

This shows that $\liminf_{n\to\infty} a_n^{-1}\sum_{i=1}^{n}(a_i - a_{i-1})v_i \geqslant v_\infty - \epsilon$. A similar argument yields $\limsup_{n\to\infty} a_n^{-1}\sum_{i=1}^{n}(a_i - a_{i-1})v_i \leqslant v_\infty + \epsilon$. Since $\epsilon > 0$ is arbitrary, we conclude that the limit exists and has the value v_∞.

b) Define $u_n := \frac{x_1}{a_1} + \dots + \frac{x_n}{a_n}$ and $u_0 := 0$. Note that $u_\infty := \lim_{n\to\infty} u_n$ exists by our assumption. Since $u_i - u_{i-1} = x_i/a_i$, a direct calculation (expand the sums and re-group the terms – this is the so-called »Abel summation« or »summation by parts«) shows that

$$s_n = \sum_{i=1}^{n}(u_i - u_{i-1})a_i = a_n u_n - \sum_{i=1}^{n}(a_i - a_{i-1})u_{i-1}.$$

Now we can use Cesàro's lemma (Part a)) to conclude

$$\frac{s_n}{a_n} = u_n - \frac{1}{a_n}\sum_{i=1}^{n}(a_i - a_{i-1})u_{i-1} \xrightarrow[n\to\infty]{} u_\infty - u_\infty = 0. \qquad \square$$

Lemma 33.5.b) enables us to obtain from Corollary 33.3 the following L^2-SLLN.

33.6 Theorem* (L^2-SLLN; Kolmogorov). *Let $(\xi_i)_{i\in\mathbb{N}}$ be independent real random variables such that $\xi_i \in L^2(\mathbb{P})$. Then*

$$\sum_{i=1}^{\infty}\frac{\mathbb{V}\xi_i}{i^2} < \infty \implies \lim_{n\to\infty}\frac{1}{n}\sum_{i=1}^{n}(\xi_i - \mathbb{E}\xi_i) = 0 \quad \text{a.s.}$$

Proof. Since $\mathbb{V}\xi_i = \mathbb{V}(\xi_i - \mathbb{E}\xi_i)$, we can use Corollary 33.3 with $X_i = \xi_i/i$ to see that

$$\sum_{i=1}^{\infty}\frac{\xi_i - \mathbb{E}\xi_i}{i} \quad \text{converges a.s.}$$

Kronecker's Lemma 33.5.b) – take $a_i = i$ and $x_i = \xi_i(\omega) - \mathbb{E}\xi_i$ – shows that $\lim_{n\to\infty}\frac{1}{n}\sum_{i=1}^{n}(\xi_i - \mathbb{E}\xi_i) = 0$ a.s. $\square$

Appendix: Necessity in the three series theorem*

We will now complete the proof of Theorem 33.4, showing that the a.s. convergence of $\sum_{i=1}^{\infty} X_i$ entails the convergence of the three series a)–c). The key to the proof is the following (partial) converse to Theorem 33.3.

33.7 Theorem. *Let $(X_i)_{i\in\mathbb{N}}$ be independent rv such that $\mathbb{E}X_i = 0$ and $|X_i| \leqslant c < \infty$ for all $i \in \mathbb{N}$.*

$$\sum_{i=1}^{\infty} X_i \text{ converges a.s.} \Longrightarrow \sum_{i=1}^{\infty} \mathbb{V}X_i < \infty.$$

We will prove this theorem using L^2-martingales, see Theorem 41.5. The classical proof can be found in [WT, pp. 119, 120].

33.8 Corollary. *Let $(X_i)_{i\in\mathbb{N}}$ be independent random variables with $|X_i| \leqslant c < \infty$ for all $i \in \mathbb{N}$.*

$$\sum_{i=1}^{\infty} X_i \text{ converges a.s.} \Longrightarrow \sum_{i=1}^{\infty} \mathbb{E}X_i,\ \sum_{i=1}^{\infty} \mathbb{V}X_i \text{ converge.}$$

Proof. If we knew that $\mathbb{E}X_i = 0$, then we would be in the setting of Theorem 33.7. We will use a symmetrization argument to reduce things to this situation.

1° We construct two identical copies

$$\left(\Omega, \mathscr{A}, \mathbb{P}, (X_i)_{i\in\mathbb{N}}\right) \quad\text{and}\quad \left(\widetilde{\Omega}, \widetilde{\mathscr{A}}, \widetilde{\mathbb{P}}, (\widetilde{X}_i)_{i\in\mathbb{N}}\right)$$

and define on the product space$(\Omega \times \widetilde{\Omega}, \mathscr{A} \otimes \widetilde{\mathscr{A}}, \mathbb{P} \otimes \widetilde{\mathbb{P}})$ the random variables

$$Z_i(\omega, \widetilde{\omega}) := X_i(\omega) - \widetilde{X}_i(\widetilde{\omega}).$$

Because of the product construction, the random variables $X_1, \widetilde{X}_1, X_2, \widetilde{X}_2, \ldots$ are all independent, and this is inherited by the sequence $(Z_i)_{i\in\mathbb{N}}$.

2° Define $\sigma_i^2 := \mathbb{V}X_i = \mathbb{V}\widetilde{X}_i$. We have

$$\mathbb{E}Z_i = \mathbb{E}X_i - \mathbb{E}\widetilde{X}_i = 0,$$
$$\mathbb{V}Z_i = \mathbb{V}(X_i - \widetilde{X}_i) = \mathbb{V}X_i + \mathbb{V}\widetilde{X}_i = 2\sigma_i^2.$$

Moreover, we set

$$G := \left\{\omega \in \Omega \mid \sum_{i=1}^{\infty} X_i(\omega) \text{ converges}\right\} \quad\text{and}\quad \widetilde{G} := \left\{\widetilde{\omega} \in \widetilde{\Omega} \mid \sum_{i=1}^{\infty} \widetilde{X}_i(\widetilde{\omega}) \text{ converges}\right\}.$$

Clearly,

$$G^* := \left\{(\omega, \widetilde{\omega}) \in \Omega \times \widetilde{\Omega} \mid \sum_{i=1}^{\infty} Z_i(\omega, \widetilde{\omega}) \text{ converges}\right\} \supset G \times \widetilde{G},$$

and we conclude from this that $\mathbb{P} \otimes \widetilde{\mathbb{P}}(G^*) \geqslant \mathbb{P} \otimes \widetilde{\mathbb{P}}(G \times \widetilde{G}) = \mathbb{P}(G)\widetilde{\mathbb{P}}(\widetilde{G}) = 1$.

3° Now we can use Theorem 33.7 to conclude that $\sum_{i=1}^{\infty} 2\sigma_i^2 < \infty$.

Since the random variables $X_i - \mathbb{E}X_i$ are independent, have mean $t_i - t_{i-1}$, and variance $\mathbb{V}(X_i - \mathbb{E}X_i) = \mathbb{V}X_i = \sigma_i^2$, we see with the help of Corollary 33.3 that $\sum_{i=1}^{\infty}(X_i - \mathbb{E}X_i)$ converges a.s.

Finally, $\sum_{i=1}^{\infty} X_i$ a.s. converges by assumption, and this shows that $\sum_{i=1}^{\infty} \mathbb{E}X_i$ converges. □

Proof of »⇒« in Theorem 33.4. Fix any $K > 0$ and assume that $\sum_{i=1}^{\infty} X_i$ converges a.s. A necessary condition is that

$$\lim_{i\to\infty} X_i(\omega) = 0 \quad \text{a.s.}$$

If we had $\sum_{i=1}^{\infty} \mathbb{P}(|X_i| > K) = \infty$, then the (difficult direction of the) Borel–Cantelli Lemma 31.1 would entail that

$$\mathbb{P}(|X_i| > K \quad \text{i.o.}) = 1;$$

but this is impossible if $X_i(\omega) \to 0$ a.s. Thus, the series a) is convergent.

As in the first part of the proof of Theorem 33.4 we see now that there is a set $\Omega_0 \subset \Omega$ with $\mathbb{P}(\Omega_0) = 1$ such that

$$\forall \omega \in \Omega_0 \quad \exists N(\omega) \quad \forall i \geqslant N(\omega): \quad X_i(\omega) = X_i^K(\omega).$$

In particular, $\sum_{i=1}^{\infty} X_i^K$ converges a.s., and Corollary 33.8 shows that the series b), c) converge. □

VII
Conditioning

34 Conditional expectation

Let $(\Omega, \mathscr{A}, \mathbb{P})$ be a probability space. In Chapter 24 we have introduced the **conditional probability** given a set F as

$$\mathbb{P}(A \mid F) = \begin{cases} \dfrac{\mathbb{P}(A \cap F)}{\mathbb{P}(F)}, & \mathbb{P}(F) > 0, \\ 0, & \mathbb{P}(F) = 0, \end{cases} \qquad A, F \in \mathscr{A}. \tag{34.1}$$

We have seen that F contains some additional information which may influence the probability of the event A; if the knowledge of F does not influence the probability of A, i.e. if

$$\mathbb{P}(A \mid F) = \mathbb{P}(A) \iff \mathbb{P}(A \cap F) = \mathbb{P}(A)\mathbb{P}(F),$$

we call A and F independent.

Since $A \mapsto \mathbb{P}(A \mid F)$ is a probability measure, we can define the integral w.r.t. $\mathbb{P}(d\omega \mid F)$:

$$\mathbb{E}(X \mid F) = \int X(\omega)\mathbb{P}(d\omega \mid F) \overset{\S}{=} \begin{cases} \dfrac{\mathbb{E}(X\mathbb{1}_F)}{\mathbb{P}(F)}, & \mathbb{P}(F) > 0, \\ 0, & \mathbb{P}(F) = 0. \end{cases}$$

which we call the **conditional expectation** of $X \in L^1(\mathbb{P})$ given the event $F \in \mathscr{A}$. In order to see the equality marked by §, we argue in the usual way (wlog we may assume $\mathbb{P}(F) > 0$):

1° $X = \mathbb{1}_A \implies \mathbb{E}(\mathbb{1}_A \mid F) = \frac{\mathbb{P}(A \cap F)}{\mathbb{P}(F)} = \frac{\mathbb{E}(\mathbb{1}_A \mathbb{1}_F)}{\mathbb{P}(F)}$;

2° $X = \sum_i \alpha_i \mathbb{1}_{A_i} \implies \mathbb{E}(X \mid F) = \frac{1}{\mathbb{P}(F)}\mathbb{E}(X\mathbb{1}_F)$;

3° $X \in \mathcal{L}^{0,+}(\mathscr{A}) \implies \mathbb{E}(X \mid F) = \frac{1}{\mathbb{P}(F)}\mathbb{E}(X\mathbb{1}_F)$; (sombrero lemma, BL)

4° $X = X^+ - X^- \in L^1(\mathscr{A}) \implies \mathbb{E}(X \mid F) = \frac{1}{\mathbb{P}(F)}\mathbb{E}(X\mathbb{1}_F)$. (linearity)

We want to extend the notion of conditional probability and conditional expectation and define

$$\mathbb{P}(A \mid \mathscr{F}) \quad \text{and} \quad \mathbb{E}(X \mid \mathscr{F})$$

for a σ-algebra $\mathscr{F}$ rather than for a set F.

34.1 Definition. Let $\mathscr{F} \subset \mathscr{A}$ be a σ-algebra and X a real random variable such that $X \in \mathcal{L}^{0,+}(\mathscr{A})$ or $X \in L^1(\mathscr{A})$. The **conditional expectation of X w.r.t. $\mathscr{F}$** is any $\mathscr{F}$-measurable random variable $X^{\mathscr{F}}$ such that

$$\int_F X\,d\mathbb{P} = \int_F X^{\mathscr{F}}\,d\mathbb{P} \quad \forall F \in \mathscr{F}. \tag{34.2}$$

We denote $X^{\mathscr{F}}$ also as $\mathbb{E}(X \mid \mathscr{F})$.

The **conditional probability** of $A \in \mathscr{A}$ is $\mathbb{P}(A \mid \mathscr{F}) := \mathbb{E}(\mathbb{1}_A \mid \mathscr{F})$.

An alternative way to express (34.2) is

$$\mathbb{E}(X\mathbb{1}_F) = \mathbb{E}(X^{\mathscr{F}}\mathbb{1}_F) \quad \forall F \in \mathscr{F}. \tag{34.3}$$

Before we discuss the **existence** and **uniqueness** of conditional expectations, we want to explain the connection with the classical conditional probability from (34.1).

34.2 Example. Let $A, F \in \mathscr{A}$ and define $\mathscr{F} = \sigma(\{F\}) = \{\emptyset, F, F^c, \Omega\} \subset \mathscr{A}$ and

$$Z(\omega) := \mathbb{P}(A \mid F)\mathbb{1}_F(\omega) + \mathbb{P}(A \mid F^c)\mathbb{1}_{F^c}(\omega).$$

Obviously, Z is an $\mathscr{F}$-measurable random variable, and we have

$$\begin{aligned}\int_F Z\,d\mathbb{P} &= \mathbb{P}(A \mid F)\int_F \mathbb{1}_F(\omega)\mathbb{P}(d\omega) + \mathbb{P}(A \mid F^c)\int_F \mathbb{1}_{F^c}(\omega)\mathbb{P}(d\omega)\\ &= \mathbb{P}(A \mid F)\int_F \mathbb{1}_F(\omega)\mathbb{P}(d\omega) = \mathbb{P}(A \mid F)\mathbb{P}(F) = \mathbb{P}(A \cap F) = \int_F \mathbb{1}_A\,d\mathbb{P}.\end{aligned}$$

The same calculation with $G = F^c$, Ω or ∅ yields $\int_G Z\,d\mathbb{P} = \int_G \mathbb{1}_A\,d\mathbb{P}$. Therefore,

$$Z = \mathbb{E}(\mathbb{1}_A \mid \mathscr{F}) \overset{\text{def}}{=} \mathbb{P}(A \mid \mathscr{F}) \tag{34.4}$$

$$\text{and so,} \quad \mathbb{P}(A \mid \mathscr{F}) = \mathbb{P}(A \mid F)\mathbb{1}_F + \mathbb{P}(A \mid F^c)\mathbb{1}_{F^c} \tag{34.5}$$

34.3 Example. Example 34.2 can be extended to finitely many disjoint F's. Let $F_1 \cup \dots \cup F_n = \Omega$ be a partiton and define $\mathscr{F} := \sigma(F_1, \dots, F_n)$. Since we can obtain every $F \in \mathscr{F}$ as a union of finitely many of the F_i, the reasoning used in Example 34.2 yields

$$\mathbb{E}(\mathbb{1}_A \mid \mathscr{F}) = \sum_{i=1}^n \mathbb{P}(A \mid F_i)\mathbb{1}_{F_i} = \sum_{i=1}^n \frac{\mathbb{P}(A \cap F_i)}{\mathbb{P}(F_i)}\mathbb{1}_{F_i}. \tag{34.6}$$

If Ω is a square equipped with (normalized) Lebesgue measure, $A \subset \Omega$ is a black spot, and the F_i are rectangular cells (see Fig. 34.1), then we may see $\mathbb{E}(\mathbb{1}_A \mid \mathscr{F})$ as a greyscale image of the black spot where the shade of grey depends on the proportion of A in a cell F_i.

In the same way we may think of a random variable as a colour image (the

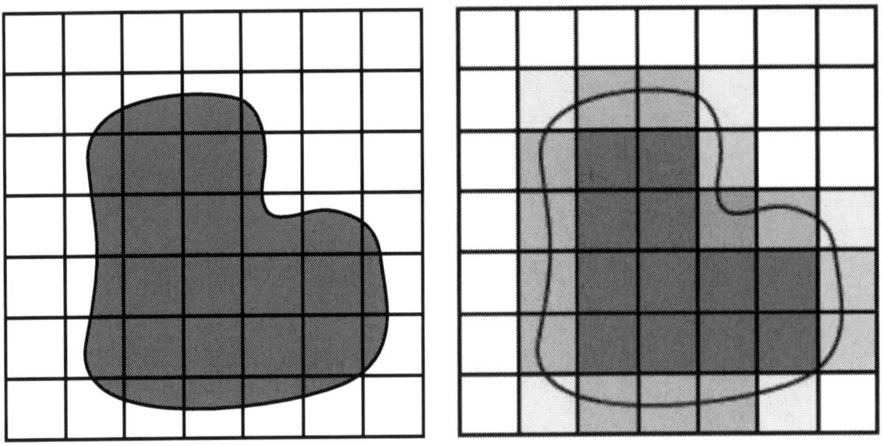

Fig. 34.1. The conditional expectation of a one-step function $\mathbb{1}_A$ (»b/w image«, black = 1, white = 0) relative to the pixels $F_1,\dots,F_n$ is an $\mathscr{F}$-step »greyscale« function.

value $X(\omega)$ is the shade of colour at the point ω) and $\mathbb{E}(X \mid \mathscr{F})$ is a picture with the »resolution« given by the »grid« of the F_i's.

$$\mathbb{E}(X \mid \mathscr{F}) = \sum_{i=1}^{n} \mathbb{E}(X \mid F_i)\mathbb{1}_{F_i} = \sum_{i=1}^{n} \frac{\mathbb{E}(X\mathbb{1}_{F_i})}{\mathbb{P}(F_i)}\mathbb{1}_{F_i}. \tag{34.7}$$

In order to prove the existence of conditional expectation we use the Radon–Nikodým theorem (Theorem 18.2): *If* $\mathbb{P}$ *and* $\mathbb{Q}$ *are probability measures on* $(\Omega, \mathscr{A})$ *such that* $\mathbb{Q} \ll \mathbb{P}$ *(i.e.* $\mathbb{P}(A) = 0 \implies \mathbb{Q}(A) = 0$*), then there is some positive density* $\xi \in L^1(\mathscr{A}, \mathbb{P})$, $\xi \geqslant 0$ *such that* $\mathbb{Q}(A) = \int_A \xi \, d\mathbb{P}$ *for all* $A \in \mathscr{A}$.

34.4 Theorem. *Let* $\mathscr{F} \subset \mathscr{A}$ *be a σ-algebra and* X *be a positive or integrable rv.*

a) *If* $X^{\mathscr{F}}, Y^{\mathscr{F}}$ *are* $\mathscr{F}$*-measurable and satisfy* (34.2), *then* $X^{\mathscr{F}} = Y^{\mathscr{F}}$ *a.s.*
b) *The conditional expectation* $\mathbb{E}(X \mid \mathscr{F})$ *exists.*

⚠ Note that the uniqueness in §34.4.a) justifies the notation $\mathbb{E}(X \mid \mathscr{F})$.
⚠ Conditional expectations are only determined up to sets of measure zero, so we call a concrete rv $X^{\mathscr{F}}$ a **version** of $\mathbb{E}(X \mid \mathscr{F})$.

Proof. a) Uniqueness: Let $X^{\mathscr{F}}, Y^{\mathscr{F}}$ be as in the statement of the theorem and set $F := F_k := \{k \geqslant X^{\mathscr{F}} > Y^{\mathscr{F}} > -k\} \in \mathscr{F}$.[1] Because of (34.2)

$$\int_F X^{\mathscr{F}} \, d\mathbb{P} = \int_F X \, d\mathbb{P} = \int_F Y^{\mathscr{F}} \, d\mathbb{P} \implies \int_{\{k \geqslant X^{\mathscr{F}} > Y^{\mathscr{F}} > -k\}} \left(X^{\mathscr{F}} - Y^{\mathscr{F}}\right) d\mathbb{P} = 0$$

(1) We use the bound k to ensure that the integrals below are finite. This is only needed if $X \geqslant 0$ is not integrable.

and this shows that $\mathbb{P}\left(k \geqslant X^{\mathscr{F}} > Y^{\mathscr{F}} \geqslant -k\right) = 0$. Using the continuity of measures, we can let $k \to \infty$ and get $\mathbb{P}(X^{\mathscr{F}} > Y^{\mathscr{F}}) = 0$. Now we change the roles of $X^{\mathscr{F}}$ and $Y^{\mathscr{F}}$ to get $\mathbb{P}(X^{\mathscr{F}} < Y^{\mathscr{F}}) = 0$, and so $\mathbb{P}(X^{\mathscr{F}} \neq Y^{\mathscr{F}}) = 0$.

b) Existence if $X \geqslant 0$: Define on $(\Omega, \mathscr{F})$ the measure

$$F \mapsto \mathbb{Q}(F) := \int_F X \, d\mathbb{P}, \quad F \in \mathscr{F}.$$

Obviously, $\mathbb{P}(F) = 0 \implies \mathbb{Q}(F) = 0$, i.e. $\mathbb{Q}|_{\mathscr{F}} \ll \mathbb{P}|_{\mathscr{F}}$ and the Radon–Nikodým theorem – used on the measurable space $(\Omega, \mathscr{F})$ – shows that there is an $\mathscr{F}$-measurable density $\xi \geqslant 0$ such that $\mathbb{Q}|_{\mathscr{F}} = \xi \cdot \mathbb{P}|_{\mathscr{F}}$. Thus,

$$\forall F \in \mathscr{F}: \quad \int_F X \, d\mathbb{P} \overset{\text{def}}{=} \mathbb{Q}(F) = \int_F d\mathbb{Q} \overset{\mathbb{Q} \ll \mathbb{P}}{=} \int_F \xi \, d\mathbb{P},$$

and so $X^{\mathscr{F}} := \xi$ is a version of $\mathbb{E}(X \mid \mathscr{F})$.

b) Existence if $X \in L^1(\mathbb{P})$: Write $X = X^+ - X^-$ and observe that $\int_F X^\pm \, d\mathbb{P} < \infty$. Therefore, we can use the previous step to infer

$$\int_F X \, d\mathbb{P} = \int_F X^+ \, d\mathbb{P} - \int_F X^- \, d\mathbb{P} = \int_F \xi_1 \, d\mathbb{P} - \int_F \xi_2 \, d\mathbb{P} = \int_F (\xi_1 - \xi_2) \, d\mathbb{P}$$

for all $F \in \mathscr{F}$. Thus, $\xi := \xi_1 - \xi_2$ is the a.s. unique choice for $\mathbb{E}(X \mid \mathscr{F})$. ☐

In many respects, the conditional expectation $X \mapsto \mathbb{E}(X \mid \mathscr{F})$ behaves like an ordinary expectation.

34.5 Proposition (basic properties). *Let* X, Y *be positive or integrable random variables,* $a, b, c \in \mathbb{R}$*, and* $\mathscr{F} \subset \mathscr{A}$ *a* σ*-algebra.*

a) $X \geqslant 0 \implies \mathbb{E}(X \mid \mathscr{F}) \geqslant 0$; *(positive)*
$X \equiv c \implies \mathbb{E}(X \mid \mathscr{F}) \equiv c$; *(conservative)*

b) $\mathbb{E}(X \mid \{\emptyset, \Omega\}) = \mathbb{E}X$;

c) $\mathbb{E}(\mathbb{E}(X \mid \mathscr{F})) = \mathbb{E}X$;

d) $\mathbb{E}(aX + bY \mid \mathscr{F}) = a\mathbb{E}(X \mid \mathscr{F}) + b\mathbb{E}(Y \mid \mathscr{F})$; *(linear)*

e) $X \geqslant Y \implies \mathbb{E}(X \mid \mathscr{F}) \geqslant \mathbb{E}(Y \mid \mathscr{F})$; *(monotone)*

f) $|\mathbb{E}(X \mid \mathscr{F})| \leqslant \mathbb{E}(|X| \mid \mathscr{F})$. *(triangle ineq.)*

Proof. a) Positivity follows directly from the proof of Theorem 34.4.b). Since the constant rv $\omega \mapsto c$ is $\mathscr{F}$-measurable and $\int_F X \, d\mathbb{P} = \int_F c \, d\mathbb{P}$ for all $F \in \mathscr{F}$, we see that $\mathbb{E}(X \mid \mathscr{F}) = c$.

b) Note that $\omega \mapsto \mathbb{E}X$ is $\mathscr{F}$-measurable and

$$\int_F X \, d\mathbb{P} = \begin{Bmatrix} 0, & F = \emptyset \\ \mathbb{E}X, & F = \Omega \end{Bmatrix} = \int_F \mathbb{E}X \, d\mathbb{P}.$$

c) We have $\mathbb{E}(\mathbb{E}(X \mid \mathscr{F})) = \int_\Omega \mathbb{E}(X \mid \mathscr{F}) \, d\mathbb{P} \overset{\Omega \in \mathscr{F}}{=} \int_\Omega X \, d\mathbb{P} = \mathbb{E}X$.

d) We have for every $F \in \mathscr{F}$

$$\begin{aligned}\int_F (aX + bY)\,d\mathbb{P} &= a\int_F X\,d\mathbb{P} + b\int_F Y\,d\mathbb{P}\\ &= a\int_F \mathbb{E}(X \mid \mathscr{F})\,d\mathbb{P} + b\int_F \mathbb{E}(Y \mid \mathscr{F})\,d\mathbb{P}\\ &= \int_F \Big(a\mathbb{E}(X \mid \mathscr{F}) + b\mathbb{E}(Y \mid \mathscr{F})\Big)\,d\mathbb{P}.\end{aligned}$$

Since $a\mathbb{E}(X \mid \mathscr{F}) + b\mathbb{E}(Y \mid \mathscr{F})$ is $\mathscr{F}$-measurable, it is a candidate for the conditional expectation of $aX + bY$, so

$$\mathbb{E}(aX + bY \mid \mathscr{F}) = a\mathbb{E}(X \mid \mathscr{F}) + b\mathbb{E}(Y \mid \mathscr{F}).$$

e) $X \geqslant Y \implies X - Y \geqslant 0$, and because of d) and a)

$$\mathbb{E}(X \mid \mathscr{F}) - \mathbb{E}(Y \mid \mathscr{F}) = \mathbb{E}(X - Y \mid \mathscr{F}) \geqslant 0.$$

This proves $\mathbb{E}(X \mid \mathscr{F}) \geqslant \mathbb{E}(Y \mid \mathscr{F})$.

f) follows from e) if we take $Y \rightsquigarrow \pm X$ and $X \rightsquigarrow |X|$. ☐

As for the ordinary expectation, we have the following convergence theorems for conditional expectations.

34.6 Proposition (convergence theorems). *Let $X, X_n : \Omega \to \mathbb{R}$ be random variables and $\mathscr{F} \subset \mathscr{A}$ a σ-algebra.*

a) (cBL – *conditional Beppo Levi) If $X_n \geqslant 0$ and $X_n \uparrow X$, then*

$$\mathbb{E}(X_n \mid \mathscr{F}) \uparrow \mathbb{E}(X \mid \mathscr{F}).$$

b) (cFATOU – *conditonal Fatou) If $X_n \geqslant 0$, then*

$$\mathbb{E}\Big(\liminf_n X_n \mid \mathscr{F}\Big) \leqslant \liminf_n \mathbb{E}\Big(X_n \mid \mathscr{F}\Big).$$

c) (cDCT – *conditional dominated convergence) If $X_n \to X$ a.s. such that $|X_n| \leqslant Y$ for some integrable majorant $Y \in L^1(\mathscr{A})$, then*

$$\mathbb{E}(X_n \mid \mathscr{F}) \to \mathbb{E}(X \mid \mathscr{F}) \text{ a.s. and in } L^1.$$

d) (cJENSEN – *conditional Jensen inequality) If $V : \mathbb{R} \to \mathbb{R}^+$ is a convex function and $X \in L^1(\mathscr{A})$ or $X \geqslant 0$, then*

$$V\Big(\mathbb{E}(X \mid \mathscr{F})\Big) \leqslant \mathbb{E}\Big(V(X) \mid \mathscr{F}\Big).$$

In particular, $V(x) = |x|^p, 1 \leqslant p < \infty$: $\|\mathbb{E}(X \mid \mathscr{F})\|_{L^p} \leqslant \|X\|_{L^p}$.

Proof. a) Because of §34.5.e) we have $\mathbb{E}(X_n \mid \mathscr{F}) \uparrow$, and so we get for all $F \in \mathscr{F}$

$$\begin{aligned}\int_F \sup_{n\in\mathbb{N}} \mathbb{E}(X_n \mid \mathscr{F})\,d\mathbb{P} &\overset{\text{BL}}{=} \sup_{n\in\mathbb{N}} \int_F \mathbb{E}(X_n \mid \mathscr{F})\,d\mathbb{P}\\ &= \sup_{n\in\mathbb{N}} \int_F X_n\,d\mathbb{P} \overset{\text{BL}}{=} \int_F \sup_{n\in\mathbb{N}} X_n\,d\mathbb{P}.\end{aligned}$$

Thus, $\mathbb{E}(X \mid \mathscr{F}) = \sup_{n\in\mathbb{N}} \mathbb{E}(X_n \mid \mathscr{F})$.

b), c): Mimic the proof of the usual Fatou and DCT theorems, but replace BL by cBL. In order to see L^1-convergence in c), observe that

$$\mathbb{E}\left|\mathbb{E}(X \mid \mathscr{F}) - \mathbb{E}(X_n \mid \mathscr{F})\right| \leqslant \mathbb{E}\left[\mathbb{E}(|X - X_n| \mid \mathscr{F})\right] = \mathbb{E}|X - X_n| \xrightarrow[n\to\infty]{\text{DCT}} 0.$$

d) can be shown as in the ordinary »un-conditional« setting. If we extend V onto $[-\infty, \infty]$ by $V(\pm\infty) := \infty$, we do not loose convexity. Using Lemma 14.15 we get

$$\ell(\mathbb{E}(X \mid \mathscr{F})) \overset{\text{linear}}{=} \mathbb{E}(\ell(X) \mid \mathscr{F}) \overset{\text{monotone}}{\leqslant} \mathbb{E}(V(X) \mid \mathscr{F}).$$

On the lhS we can take the sup over all affine-linear functions $\ell \leqslant V$ and obtain $V(\mathbb{E}(X \mid \mathscr{F}))$. □

The next few properties are particular to conditional expectations.

34.7 Proposition.

a) *(pull out) If $X \in L^1(\mathscr{A})$, Z is $\mathscr{F}$-measurable, and $XZ \in L^1(\mathscr{A})$, then*

$$\mathbb{E}(ZX \mid \mathscr{F}) = Z\mathbb{E}(X \mid \mathscr{F}).$$

b) *(tower property) If $\mathscr{G} \subset \mathscr{F} \subset \mathscr{A}$ are σ-algebras and $X \in L^1(\mathscr{A})$ or $X \geqslant 0$, then*

$$\mathbb{E}\Big[\mathbb{E}(X \mid \mathscr{F}) \,\Big|\, \mathscr{G}\Big] = \mathbb{E}(X \mid \mathscr{G}).$$

c) *(projection) If $X \in L^2(\mathbb{P})$ and $Y \in L^2(\mathscr{F})$, then*

$$\mathbb{E}\left[(X - \mathbb{E}(X \mid \mathscr{F}))^2\right] \leqslant \mathbb{E}\left[(X - Y)^2\right].$$

In particular: $\mathbb{E}[(X - \mathbb{E}(X \mid \mathscr{F}))(Y - \mathbb{E}(X \mid \mathscr{F}))] = 0$, i.e. $X - \mathbb{E}(X \mid \mathscr{F}) \perp Y$ for all $Y \in L^2(\mathscr{F})$.

- §34.7.c) means that $\Pi : L^2(\mathscr{A}) \to L^2(\mathscr{F})$, $\Pi(X) := \mathbb{E}(X \mid \mathscr{F})$, is the **orthogonal projection**, that is $\Pi \circ \Pi = \Pi$ & $\text{Image}(\Pi) = \text{Ker}(\Pi)^\perp$ (which is equivalent to $\Pi \circ \Pi = \Pi$ & Π self-adjoint).
- Frequently $\mathbb{E}(\bullet \mid \mathscr{F})$ is defined on $L^2(\mathscr{A})$ as the orthogonal projection onto $L^2(\mathscr{F})$, and then extended to $L^1(\mathscr{A})$, see [MaPs, Kapitel 2] or [MIMS, Ch. 27].

Proof. a) Assume, for now, that $Z = \mathbb{1}_G$ for some $G \in \mathscr{F}$. For all $F \in \mathscr{F}$ we find

$$\int_F Z\mathbb{E}(X \mid \mathscr{F})\,d\mathbb{P} = \int_{F\cap G} \mathbb{E}(X \mid \mathscr{F})\,d\mathbb{P} = \int_{F\cap G} X\,d\mathbb{P} = \int_F \mathbb{1}_G X\,d\mathbb{P} = \int_F ZX\,d\mathbb{P}.$$

By linearity, this shows the claim for all (positive) $\mathscr{F}$-measurable simple functions. Using (c)BL and the (second amendment of the) sombrero lemma, we get the assertion for all $Z \in L^\infty(\mathscr{F})$.

Now let Z be $\mathscr{F}$-measurable such that $\mathbb{E}|XZ| < \infty$. Clearly, the truncated random variables $Z_n := (-n) \vee Z \wedge n \in L^\infty(\mathscr{F})$, and so

$$\mathbb{E}(ZX \mid \mathscr{F}) \overset{\text{cDCT}}{=} \lim_{n\to\infty} \mathbb{E}(Z_n X \mid \mathscr{F}) = \lim_{n\to\infty} Z_n \mathbb{E}(X \mid \mathscr{F}) = Z\mathbb{E}(X \mid \mathscr{F}).$$

b) We find for all $G \in \mathscr{G}$

$$\int_G \underbrace{\mathbb{E}(X \mid \mathscr{F})}_{\text{original rv}} d\mathbb{P} = \int \mathbb{1}_G \mathbb{E}(X \mid \mathscr{F})\, d\mathbb{P} = \int \mathbb{E}(\mathbb{1}_G X \mid \mathscr{F})\, d\mathbb{P}$$
$$= \int \mathbb{1}_G X\, d\mathbb{P} = \int_G X\, d\mathbb{P} = \int_G \mathbb{E}(X \mid \mathscr{G})\, d\mathbb{P}.$$

And so, $\mathbb{E}(X \mid \mathscr{G})$ is a version of the $\mathscr{G}$-conditional expectation of the random variable $\mathbb{E}(X \mid \mathscr{F})$.

c) Set $X' = \mathbb{E}(X \mid \mathscr{F})$. From Proposition 34.6.d) we get that $X' \in L^2(\mathscr{F})$. Further,

$$\mathbb{E}\left[(X - X')(Y - X')\right] \overset{\text{tower}}{=} \mathbb{E}\Big[\mathbb{E}\{(X - X')(Y - X') \mid \mathscr{F}\}\Big]$$
$$\overset{\text{pull}}{=} \mathbb{E}\Big[(Y - X')\underbrace{\mathbb{E}\{(X - X') \mid \mathscr{F}\}}_{=\mathbb{E}(X|\mathscr{F}) - \mathbb{E}(X'|\mathscr{F}) = X' - X' = 0}\Big] = 0.$$

In particular, $X - X' \perp Y - X'$. If we replace $Y \rightsquigarrow Y + X'$, then we see $X - X' \perp Y$. Moreover,

$$\mathbb{E}[(X - Y)^2] = \mathbb{E}[(\{X - X'\} + \{X' - Y\})^2]$$
$$= \mathbb{E}[\{X - X'\}^2] + \underbrace{\mathbb{E}[\{X' - Y\}^2]}_{\geqslant 0} + \underbrace{\mathbb{E}[\{X - X'\}\{X' - Y\}]}_{=0}$$
$$\geqslant \mathbb{E}[\{X - X'\}^2].$$ □

34.8 Example (Example 34.2 reloaded). Let us use the rules for conditional expectations to derive, once again, the formula (34.5). Let

$$A, F \in \mathscr{A}, \quad \mathscr{F} = \{\emptyset, F, F^c, \Omega\}, \quad \mathbb{P}(A \mid F) \text{ classical conditional probab.}$$

We have

$$L^1(\mathscr{F}) = \{a\mathbb{1}_F + b\mathbb{1}_{F^c} \mid a, b \in \mathbb{R}\}$$

and for all $X \in L^1(\mathscr{A})$ we get

$$\mathbb{E}(X \mid \mathscr{F}) = \alpha\mathbb{1}_F + \beta\mathbb{1}_{F^c} \quad (\alpha, \beta \text{ suitable}).$$

Let us determine the constants α, β. First

$$\mathbb{E}(\mathbb{1}_F X \mid \mathscr{F}) \overset{\text{pull}}{=} \mathbb{1}_F \mathbb{E}(X \mid \mathscr{F}) = \mathbb{1}_F \cdot (\alpha\mathbb{1}_F + \beta\mathbb{1}_{F^c}) = \alpha\mathbb{1}_F$$

and taking on both sides the expectation yields

$$\mathbb{E}(\mathbb{1}_F X) = \mathbb{E}[\mathbb{E}(\mathbb{1}_F X \mid \mathscr{F})] = \mathbb{E}(\alpha\mathbb{1}_F) = \alpha\mathbb{P}(F)$$

This shows

$$\alpha = \frac{1}{\mathbb{P}(F)}\mathbb{E}(\mathbb{1}_F X) = \frac{1}{\mathbb{P}(F)}\int_F X\, d\mathbb{P} = \int X\, d\mathbb{P}(\bullet \mid F),$$

and with a similar calculation we get

$$\beta = \int X \, d\mathbb{P}(\bullet \mid F^c).$$

Together we have

$$\mathbb{E}(X \mid \mathscr{F}) = \alpha \mathbb{1}_F + \beta \mathbb{1}_{F^c} = \mathbb{1}_F \int X \, d\mathbb{P}(\bullet \mid F) + \mathbb{1}_{F^c} \int X \, d\mathbb{P}(\bullet \mid F^c).$$

Taking, in partiuclar, $X = \mathbb{1}_A$ and $\mathscr{F} = \{\emptyset, F, F^c, \Omega\}$ shows

$$\mathbb{P}(A \mid \mathscr{F}) = \mathbb{P}(A \mid F)\mathbb{1}_F + \mathbb{P}(A \mid F^c)\mathbb{1}_{F^c}.$$

The same argument for countably many, mutually disjoint sets $(F_i)_{i\in\mathbb{N}} \subset \mathscr{A}$ such that $\dot\bigcup_i F_i = \Omega$ yields for the σ-algebra $\mathscr{H} = \sigma(F_i,\ i \in \mathbb{N})$

$$\mathbb{P}(A \mid \mathscr{H}) = \sum_{i=1}^{\infty} \mathbb{P}(A \mid F_i)\mathbb{1}_{F_i}$$

$$\iff \forall i: \quad \mathbb{P}(A \mid \mathscr{H}) = \mathbb{P}(A \mid F_i) \quad \mathbb{P}\text{-a.s. on } F_i$$

For the next example, the following characterization of $\mathbb{E}(X \mid \mathscr{F})$ with a generator of $\mathscr{F}$ is useful.

34.9 Lemma. *Let $X, Y \in L^1(\mathscr{A})$ and $\mathscr{F} = \sigma(\mathscr{G}) \subset \mathscr{A}$. If $\mathscr{G}$ is $\cap$-stable and contains a sequence $(G_i)_{i\in\mathbb{N}} \subset \mathscr{G}$, $G_i \uparrow \Omega$, then the following assertions are equivalent:*

$$\int_F X \, d\mathbb{P} = \int_F Y \, d\mathbb{P} \quad \forall F \in \mathscr{F}, \tag{34.8}$$

$$\int_G X \, d\mathbb{P} = \int_G Y \, d\mathbb{P} \quad \forall G \in \mathscr{G}. \tag{34.9}$$

In particular, we only have to verify (34.2) on a »well-behaved« generator.

Proof. As $\mathscr{G} \subset \mathscr{F}$ it is enough to check that (34.9)$\Rightarrow$(34.8). Re-arrange (34.8) as follows:

$$\mu(F) := \int_F (X^+ + Y^-) \, d\mathbb{P} = \int_F (Y^+ + X^-) \, d\mathbb{P} = \nu(F).$$

Obviously, μ, ν are finite measures on $\mathscr{F}$; (34.9) shows that $\mu|_{\mathscr{G}} = \nu|_{\mathscr{G}}$. Since $\mathscr{G}$ satisfies the conditions of the uniqueness theorem for measures (Theorem 4.7), we get (34.8). □

34.10 Example. Let $\mathscr{H} = \sigma(F_1, F_2, \dots)$ where $\Omega = \dot\bigcup_{i\in\mathbb{N}} F_i$ and $\mathbb{P}(F_i) > 0$. We have for every rv $X \in L^1(\mathscr{A})$

$$\mathbb{E}(X \mid \mathscr{H}) = \sum_{i=1}^{\infty} \frac{\mathbb{E}(X\mathbb{1}_{F_i})}{\mathbb{P}(F_i)} \mathbb{1}_{F_i} \tag{34.10}$$

$$\iff \forall i: \quad \mathbb{E}(X \mid \mathscr{H})(\omega) = \frac{\mathbb{E}(X\mathbb{1}_{F_i})}{\mathbb{P}(F_i)} \quad \text{for } \mathbb{P} \text{ a.a. } \omega \in F_i. \tag{34.11}$$

Since we have for each k

$$\int_{F_k} \sum_{i=1}^{\infty} \frac{\mathbb{E}(X\mathbb{1}_{F_i})}{\mathbb{P}(F_i)} \mathbb{1}_{F_i}\,d\mathbb{P} = \sum_{i=1}^{\infty} \frac{\mathbb{E}(X\mathbb{1}_{F_i})}{\mathbb{P}(F_i)} \int_{F_k} \mathbb{1}_{F_i}\,d\mathbb{P} = \sum_{i=1}^{\infty} \frac{\mathbb{E}(X\mathbb{1}_{F_i})}{\mathbb{P}(F_i)} \mathbb{P}(F_k \cap F_i)$$
$$= \frac{\mathbb{E}(X\mathbb{1}_{F_k})}{\mathbb{P}(F_k)} \mathbb{P}(F_k) = \mathbb{E}(X\mathbb{1}_{F_k}) = \int_{F_k} X\,d\mathbb{P},$$

the claim follows from Lemma 34.9. □

34.11 Example. An electronic device has life-time $\zeta : \Omega \to [0, \infty)$. We assume that $\zeta \sim f_\zeta(z)\,dz$. We are interested in the **mean survival time** if the device has already been in use for a years, i.e.

$$\mathbb{E}(\zeta - a \mid \zeta \geqslant a).$$

Fix a and assume that $\mathbb{P}(\zeta \geqslant a) > 0$. If $\mathscr{H} = \sigma(\{\zeta \geqslant a\}, \{\zeta < a\})$, then

$$\mathbb{E}(\zeta - a \mid \zeta \geqslant a) = \frac{\mathbb{E}\big((\zeta - a)\mathbb{1}_{\{\zeta \geqslant a\}}\big)}{\mathbb{P}(\zeta \geqslant a)} = \frac{\int (\zeta - a)\mathbb{1}_{\{\zeta \geqslant a\}}\,d\mathbb{P}}{\mathbb{P}(\zeta \geqslant a)} = \frac{\int_a^\infty (z-a) f_\zeta(z)\,dz}{\int_a^\infty f_\zeta(z)\,dz}.$$

In many concrete situations, $f_\zeta(z) = \lambda \exp(-\lambda z)$, $z \geqslant 0$, i.e. ζ is an exponential random variable with mean $1/\lambda$. In this case we have

$$\mathbb{E}(\zeta \mid \zeta \geqslant 0) = \frac{1}{\lambda} \quad \text{and} \quad \forall a > 0 : \quad \mathbb{E}(\zeta - a \mid \zeta \geqslant a) = \frac{1}{\lambda}.$$

This means that the exponential law »has no memory«. One can show

$$\mathbb{P}(\zeta - a \leqslant x \mid \zeta \geqslant a) = \frac{\mathbb{P}(a \leqslant \zeta \leqslant a + x)}{\mathbb{P}(\zeta \geqslant a)} = \cdots = 1 - e^{-\lambda x} = \mathbb{P}(\zeta \leqslant x).$$

This, in fact, characterizes $\mathsf{Exp}(\lambda)$. For more on this topic, see Chapter 47, in particular Corollary 47.4.

Conditional expectations and independence

Recall that in the classical case $A \perp\!\!\!\perp F \iff \mathbb{P}(A \mid F) = \mathbb{P}(A)$. Similar formulae hold for abstract conditional expectations.

34.12 Proposition. *Let* $X, Y \in L^1(\mathscr{A})$ *and* $\mathscr{F}, \mathscr{G} \subset \mathscr{A}$ *σ-algebras.*

a) *If* $X \perp\!\!\!\perp \mathscr{F}$, *then* $\mathbb{E}(X \mid \mathscr{F}) = \mathbb{E}X$.
b) *If* $X, \mathscr{G} \perp\!\!\!\perp \mathscr{F}$, *then* $\mathbb{E}(X \mid \mathscr{G}, \mathscr{F}) = \mathbb{E}(X \mid \mathscr{G})$.
c) *If* $X \perp\!\!\!\perp \mathscr{F}$, *if* Y *is* $\mathscr{F}$*-measurable, and if* $g : \mathbb{R} \times \mathbb{R} \to \mathbb{R}$ *bounded and measurable, then*

$$\mathbb{E}(g(X, Y) \mid \mathscr{F})(\omega) = \mathbb{E}g(X, t)\big|_{t = Y(\omega)}.$$

A word on notation: $X, \mathscr{G} \perp\!\!\!\perp \mathscr{F}$ is short for »the rv X and the family $\mathscr{G}$ are independent of $\mathscr{F}$«; i.e. $\sigma(\sigma(X), \mathscr{G}) \perp\!\!\!\perp \mathscr{F}$. Similarly, $\mathbb{E}(\cdots \mid \mathscr{F}, \mathscr{G})$ is short for $\mathbb{E}(\cdots \mid \sigma(\mathscr{F}, \mathscr{G}))$.

Proof. a) We have

$$\int_F X\,d\mathbb{P} = \int \mathbb{1}_F X\,d\mathbb{P} \overset{\perp\!\!\!\perp}{=} \int \mathbb{1}_F\,d\mathbb{P} \int X\,d\mathbb{P} = \mathbb{P}(F)\mathbb{E}X = \int_F \mathbb{E}X\,d\mathbb{P},$$

i.e. $\mathbb{E}X$ is a version of the $\mathscr{F}$-conditional expectation of X.

b) Sets of the form $F \cap G$ where $F \in \mathscr{F}$ and $G \in \mathscr{G}$ generate $\sigma(\mathscr{F}, \mathscr{G})$. Therefore,

$$\begin{aligned}\int \mathbb{1}_F \mathbb{1}_G X\,d\mathbb{P} &\overset{\perp\!\!\!\perp}{=} \mathbb{P}(F) \int \mathbb{1}_G X\,d\mathbb{P}\\ &\overset{\text{tower}}{=} \mathbb{P}(F) \int \mathbb{E}(\mathbb{1}_G X \mid \mathscr{G})\,d\mathbb{P}\\ &\overset{\perp\!\!\!\perp}{=} \int \mathbb{1}_F \mathbb{E}(\mathbb{1}_G X \mid \mathscr{G})\,d\mathbb{P}\\ &\overset{\text{pull}}{=} \int \mathbb{1}_F \mathbb{1}_G \mathbb{E}(X \mid \mathscr{G})\,d\mathbb{P}\\ &= \int_{F \cap G} \mathbb{E}(X \mid \mathscr{G})\,d\mathbb{P},\end{aligned}$$

which shows that $\mathbb{E}(X \mid \mathscr{G})$ is a version of the $\sigma(\mathscr{G}, \mathscr{F})$-conditional expectation of X.

c) Step 1° : Assume that $g(x,y) = \mathbb{1}_A(x)\mathbb{1}_B(y)$ for some $A, B \in \mathscr{B}(\mathbb{R})$. Since Y, hence $\mathbb{1}_B(Y)$, is $\mathscr{F}$-measurable and $X \perp\!\!\!\perp \mathscr{F}$, we get

$$\begin{aligned}\mathbb{E}\big(g(X,Y) \mid \mathscr{F}\big) &= \mathbb{E}\big(\mathbb{1}_A(X)\mathbb{1}_B(Y) \mid \mathscr{F}\big) \overset{\text{pull}}{=} \mathbb{1}_B(Y)\mathbb{E}\big(\mathbb{1}_A(X) \mid \mathscr{F}\big)\\ &\overset{\perp\!\!\!\perp}{=} \mathbb{1}_B(Y)\mathbb{E}(\mathbb{1}_A(X)) = \mathbb{E}(\mathbb{1}_A(X)\mathbb{1}_B(t))\big|_{t=Y} = \mathbb{E}g(X,t)\big|_{t=Y}.\end{aligned}$$

By linearity, the claim follows for all step functions whose steps are rectangles of the type $A \times B$.

Step 2° : We can now use the monotone class theorem (Theorem 7.17) for the vector space $\mathcal{V}$ of all g satisfying the claim. Step 1° and cBL show that the conditions i) and ii) of Theorem 7.17 hold. □

*Conditional expectation and measure-changes**

Sometimes we need to replace the probability measure $\mathbb{P}$ on $(\Omega, \mathscr{A})$ by another probability measure $\mathbb{Q}$ which is of the form

$$\mathbb{Q}(A) = \int_A \beta(\omega)\,\mathbb{P}(d\omega), \quad A \in \mathscr{A},$$

for some integrable density $\beta \geqslant 0$ such that $\mathbb{E}\beta = 1$. We denote by

$$\mathbb{E}_{\mathbb{Q}} X := \int X\,d\mathbb{Q} \overset{\text{def}}{=} \int X\beta\,d\mathbb{P} = \mathbb{E}(X\beta)$$

the expected value with respect to the measure $\mathbb{Q}$. We want to relate the conditional expectations $\mathbb{E}_{\mathbb{Q}}(\cdots \mid \mathscr{F})$ and $\mathbb{E}(\cdots \mid \mathscr{F})$ for some sub-σ-algebra $\mathscr{F} \subset \mathscr{A}$.

34.13 Lemma. *Let $\mathbb{Q} = \beta\mathbb{P}$ be a further probability measure on the measure space $(\Omega, \mathscr{A}, \mathbb{P})$ and $\mathscr{F} \subset \mathscr{A}$ a σ-algebra. Then*

$$\mathbb{E}_{\mathbb{Q}}(X \mid \mathscr{F}) = \frac{\mathbb{E}(X\beta \mid \mathscr{F})}{\mathbb{E}(\beta \mid \mathscr{F})} \tag{34.12}$$

for all positive or $\mathbb{Q}$-integrable random variables X.

Proof. Fix $F \in \mathscr{F}$. Then we have

$$\begin{aligned}
\mathbb{E}_{\mathbb{Q}}(X\mathbb{1}_F) = \mathbb{E}(X\beta\mathbb{1}_F) &\overset{\text{tower}}{\underset{\text{pull}}{=}} \mathbb{E}\left[\mathbb{E}(X\beta \mid \mathscr{F})\mathbb{1}_F\right] \\
&= \mathbb{E}\Bigg[\underbrace{\frac{\mathbb{E}(X\beta \mid \mathscr{F})}{\mathbb{E}(\beta \mid \mathscr{F})}}_{\mathscr{F}\text{-mble}} \mathbb{E}\{\beta \mid \mathscr{F}\}\mathbb{1}_F\Bigg] \\
&\overset{\text{pull}}{=} \mathbb{E}\left[\mathbb{E}\left\{\frac{\mathbb{E}(X\beta \mid \mathscr{F})}{\mathbb{E}(\beta \mid \mathscr{F})}\beta\mathbb{1}_F \,\middle|\, \mathscr{F}\right\}\right] \\
&\overset{\text{tower}}{=} \mathbb{E}\left[\frac{\mathbb{E}(X\beta \mid \mathscr{F})}{\mathbb{E}(\beta \mid \mathscr{F})}\beta\mathbb{1}_F\right] \\
&= \mathbb{E}_{\mathbb{Q}}\left[\frac{\mathbb{E}(X\beta \mid \mathscr{F})}{\mathbb{E}(\beta \mid \mathscr{F})}\mathbb{1}_F\right]
\end{aligned}$$

and this proves that $\mathbb{E}_{\mathbb{Q}}(X \mid \mathscr{F}) = \mathbb{E}(X\beta \mid \mathscr{F})/\mathbb{E}(X \mid \mathscr{F})$. □

If the density β happens to be $\mathscr{F}$-measurable, then we see from (34.12) and a further pull-out argument that $\mathbb{E}_{\mathbb{Q}}(\cdots \mid \mathscr{F}) = \mathbb{E}(\cdots \mid \mathscr{F})$. One of the most interesting cases happens if $\beta = \mathbb{1}_G/\mathbb{P}(G)$ for some $G \in \mathscr{F}$ with $\mathbb{P}(G) > 0$, i.e. if we consider a classical conditional probability

$$\mathbb{Q}(A) = \mathbb{P}(A \mid G) = \frac{\mathbb{P}(A \cap G)}{\mathbb{P}(G)} = \int_A \frac{\mathbb{1}_G}{\mathbb{P}(G)}\, d\mathbb{P}.$$

34.14 Corollary. *Let $(\Omega, \mathscr{A}, \mathbb{P})$ be a probability space, $\mathscr{F} \subset \mathscr{A}$ a σ-algebra and $G \in \mathscr{F}$ such that $\mathbb{P}(G) > 0$. Define $Q(A) := \mathbb{P}(A \mid G)$. The conditional expectations $\mathbb{E}_{\mathbb{Q}}(\cdots \mid \mathscr{F}) = \mathbb{E}(\cdots \mid \mathscr{F})$ coincide.*

35 Conditioning on $\mathscr{F} = \sigma(Y)$

Let $(\Omega, \mathscr{A}, \mathbb{P})$ be an arbitrary probability space and $Y : \Omega \to E$ be a random variable taking values in a measurable space $(E, \mathscr{E})$. We are interested in the σ-algebra $\mathscr{F} = \sigma(Y) = Y^{-1}(\mathscr{E})$ induced by Y. We will follow the usual custom and use

$\mathbb{E}(X \mid Y)$, $\mathbb{E}(X \mid Y_1, Y_2, \dots)$ etc. as shorthand for
$\mathbb{E}(X \mid \sigma(Y))$, $\mathbb{E}(X \mid \sigma(Y_i,\, i \in \mathbb{N}))$ etc.

In the same spirit we use $\mathbb{P}(A \mid Y)$ and $\mathbb{P}(A \mid Y_1, Y_2, \dots)$ etc.

Recall the factorization theorem (Lemma 7.16) from measure & integration which says that a $\sigma(Y)$-measurable random variable $Z : \Omega \to \mathbb{R}$ can be expressed as $Z = g(Y)$ with a measurable $g : (E, \mathscr{E}) \to (\mathbb{R}, \mathscr{B}(\mathbb{R}))$. If $Z \geqslant 0$ or $\mathbb{E}|Z| < \infty$, then g is $\mathbb{P}_Y$-a.s. unique:

Indeed, if $g(Y) = f(Y) = Z$, then we find for $B_n := \{-n \leqslant g < f \leqslant n\}$, $n \in \mathbb{N}$, that

$$0 \leqslant \int_{B_n} (f(y) - g(y))\, \mathbb{P}(Y \in dy) = \int_{\{Y \in B_n\}} (f(Y) - g(Y))\, d\mathbb{P} = 0.$$

This shows that each B_n, hence $\bigcup_n B_n = \{g < f\}$, is a $\mathbb{P}_Y$ null set. If we change the roles of f and g, we get $\mathbb{P}_Y(f \neq g) = 0$.

These observations are the basis for the next definition.

35.1 Definition. Let $Y : \Omega \to E$, $X : \Omega \to \mathbb{R}$ random variables and $X \in L^1(\mathscr{A})$ or $X \geqslant 0$. We denote by $\mathbb{E}(X \mid Y = y)$ the ($\mathbb{P}_Y$-a.e. unique) function $g : E \to \mathbb{R}$ which satisfies $\mathbb{E}(X \mid Y) = g(Y)$ a.s.

35.2 Example. If $Y : \Omega \to \{y_1, y_2, \dots\}$ is a discrete random variable, then we can use Example 34.10, (34.11), and $\mathscr{H} = \sigma\big(\{Y = y_i\},\ i \in \mathbb{N}\big)$ to see

$$\mathbb{E}(X \mid Y = y) = \underbrace{\mathbb{E}(X \mid \{Y = y\})}_{\text{classical}} = \begin{cases} \dfrac{\mathbb{E}(X \mathbb{1}_{\{y\}}(Y))}{\mathbb{P}(Y = y)}, & \mathbb{P}(Y = y) > 0; \\ 0, & \mathbb{P}(Y = y) = 0. \end{cases}$$

This means, in particular, that the classical use and our use of conditional probabilities coincide: $\mathbb{P}(A \mid \{Y = y\}) = \mathbb{P}(A \mid Y = y)$.

⚠ Note that things are different if Y is not discrete; there might be »too many« y's such that $\mathbb{P}(Y = y) = 0$.

35.3 Theorem. *Let $Y : \Omega \to E$, $X : \Omega \to \mathbb{R}$ be random variables and $X \in L^1(\mathscr{A})$ or $X \geqslant 0$. Every measurable function $g : E \to \mathbb{R}$ such that $g(Y) = \mathbb{E}(X \mid Y)$ satisfies*

$$\int_B g(y)\, \mathbb{P}(Y \in dy) = \int_{\{Y \in B\}} X \, d\mathbb{P} \quad \forall B \in \mathscr{E}. \tag{35.1}$$

Conversely, every g which satisfies (35.1) also satisfies $g(Y) = \mathbb{E}(X \mid Y)$.

The function g is $\mathbb{P}_Y$-a.s. determined through the relation (35.1).

Proof. We begin with a general formula. For any $B \in \mathscr{E}$

$$\int_B g(y)\, \mathbb{P}(Y \in dy) = \int \mathbb{1}_B(Y) \cdot g(Y)\, d\mathbb{P} = \int_{\{Y \in B\}} g(Y)\, d\mathbb{P}. \tag{35.2}$$

1° Since $\mathbb{E}(X \mid Y)$ is $\sigma(Y)$-measurable, there is some function $g : E \to \mathbb{R}$ with $\mathbb{E}(X \mid Y) = g(Y)$. Thus,

$$\int_B g(y)\, \mathbb{P}(Y \in dy) \overset{(35.2)}{=} \int_{\{Y \in B\}} g(Y)\, d\mathbb{P} = \int_{\{Y \in B\}} \mathbb{E}(X \mid Y)\, d\mathbb{P} = \int_{\{Y \in B\}} X \, d\mathbb{P},$$

and we get (35.1).

2° Conversely, if (35.1) holds, then (35.2) shows for all $B \in \mathscr{E}$

$$\int_{\{Y\in B\}} g(Y)\,d\mathbb{P} = \int_{\{Y\in B\}} X\,d\mathbb{P},$$

and since $\{Y \in B\}$ is a generic element of $\sigma(Y)$, we get $g(Y) = \mathbb{E}(X \mid Y)$.

3° Uniqueness is clear from the considerations before Def. 35.1. □

Conditional densities

Assume that $X, Y : \Omega \to \mathbb{R}$ are random variables with a joint density $f_{X,Y}(x,y)$, i.e.

$$\mathbb{P}(X \leqslant a, Y \leqslant b) = \int_{(-\infty,a]}\int_{(-\infty,b]} f_{X,Y}(x,y)\,dy\,dx.$$

The marginal distributions are given by

$$\mathbb{P}(X \leqslant a) = \mathbb{P}(X \leqslant a, Y < \infty) = \int_{(-\infty,a]} \underbrace{\int_{\mathbb{R}} f_{X,Y}(x,y)\,dy}_{=:f_X(x)}\,dx.$$

i.e. $X \sim f_X(x)\,dx$ and $Y \sim f_Y(y)\,dy$ with

$$f_X(x) = \int_{\mathbb{R}} f_{X,Y}(x,y)\,dy, \quad f_Y(y) = \int_{\mathbb{R}} f_{X,Y}(x,y)\,dx.$$

35.4 Definition. Let $(X,Y) \sim f_{X,Y}(x,y)\,dx\,dy$. The function

$$f_{X|Y}(x \mid y) := \begin{cases} \dfrac{f_{X,Y}(x,y)}{f_Y(y)}, & f_Y(y) \neq 0, \\ 0, & \text{else}, \end{cases} \tag{35.3}$$

is the **conditional density of X given Y.**

We can use conditional densities in order to evaluate conditional expectations.

35.5 Theorem. *Let* $(X,Y) \sim f_{X,Y}(x,y)\,dx\,dy$ *and* $h : \mathbb{R}^2 \to \mathbb{R}$ *be a Borel function such that* $\mathbb{E}|h(X,Y)| < \infty$.

$$\mathbb{E}(h(X,Y) \mid Y = y) = \int_{\mathbb{R}} h(x,y) f_{X|Y}(x \mid y)\,dx.$$

Proof. We have to show that

$$\begin{aligned} &\mathbb{E}(h(X,Y) \mid Y) = g(Y) \quad \text{for the function} \\ &g(y) := \int_{\mathbb{R}} h(x,y) f_{X|Y}(x \mid y)\,dx. \end{aligned} \tag{35.4}$$

Since $A \in \sigma(Y) \iff A = \{Y \in B\}$ for some suitable Borel set $B \in \mathscr{B}(\mathbb{R})$, we have $\mathbb{1}_A = \mathbb{1}_B \circ Y$ and

$$\begin{aligned}\int_A h(X,Y)\,d\mathbb{P} &= \int \mathbb{1}_B(Y)h(X,Y)\,d\mathbb{P}\\ &= \int \mathbb{1}_B(y)h(x,y)\,\mathbb{P}(X \in dx, Y \in dy)\\ &= \iint \mathbb{1}_B(y)h(x,y)\,f_{X,Y}(x,y)\,dx\,dy.\end{aligned}$$

In the other hand, using the definition of g, we find

$$\begin{aligned}\int_A g(Y)\,d\mathbb{P} &= \int \mathbb{1}_B(Y)g(Y)\,d\mathbb{P}\\ &= \int \mathbb{1}_B(Y)\int h(x,Y)\,f_{X|Y}(x\mid Y)\,dx\,d\mathbb{P}\\ &= \int \mathbb{1}_B(y)\int h(x,y)\,f_{X|Y}(x\mid y)\,dx\,\mathbb{P}(Y \in dy)\\ &= \int \mathbb{1}_B(y)\int h(x,y)\,\underbrace{f_{X|Y}(x\mid y)\,f_Y(y)}_{=f_{X,Y}(x,y)}\,dx\,dy\\ &= \iint \mathbb{1}_B(y)h(x,y)\,f_{X,Y}(x,y)\,dx\,dy.\end{aligned}$$

□

35.6 Corollary. *Let* $(X,Y) \sim f_{X,Y}(x,y)\,dx\,dy$ *and* $B \in \mathscr{B}(\mathbb{R})$. *Then one has*

$$\mathbb{P}(X \in B \mid Y) = \int_B f_{X|Y}(x\mid Y)\,dx.$$

Proof. Since the set $\{X \in B\}$ is a generic element of $\sigma(X)$, the claim follows with Theorem 35.5 upon taking $h(X,Y) = \mathbb{1}_B(X) = \mathbb{1}_{\{X\in B\}}$. □

35.7 Remark. Observe the analogy of (35.3) with the classic formula for $\mathbb{P}(A \mid B)$

classical	$\mathbb{P}(B)\cdot\mathbb{P}(A\mid B) = \mathbb{P}(A\cap B) = \mathbb{P}(A)\cdot\mathbb{P}(B\mid A)$;
cond. density	$f_Y(y)\cdot f_{X|Y}(x\mid y) = f_{X,Y}(x,y) = f_X(x)\cdot f_{Y|X}(y\mid x)$.

Conditional distributions

We want to extend the assertion of Theorem 35.5. For this we need a further concept which will again become important in connection with Markov chains and stochastic processes.

35.8 Definition. Let $(E,\mathscr{E})$, $(F,\mathscr{F})$ be measurable spaces. A measure kernel $N : E \times \mathscr{F} \to [0,1]$ is said to be a **Markov kernel**, if

i) $\forall B \in \mathscr{F}: \quad y \mapsto N(y,B)$ is $\mathscr{E}$-measurable;
ii) $\forall y \in E: \quad B \mapsto N(y,B)$ is a probability measure on F.

35.9 Example. Let $E = \mathbb{R}^d$ and $F = \mathbb{R}^n$ with their Borel σ-algebras. If

- $f : \mathbb{R}^d \times \mathbb{R}^n \to [0,\infty)$ is measurable,
- $\mu(dx)$ is a probability measure on $\mathbb{R}^n$,
- $\int_{\mathbb{R}^d} f(y,x)\,\mu(dx) = 1$,

then $N(y,B) := \int_B f(y,x)\,\mu(dx)$ is a Markov kernel.

You should imagine Markov kernels as families of probability measures which depend measurably on a parameter.

The following theorem on regular conditional distributions is a quite deep result. Our simple-looking proof heavily depends – at least in the multivariate case – on abstract measurability considerations from the appendix to Ch. 26.

35.10 Theorem (regular conditional distribution). *Assume that* $X : \Omega \to \mathbb{R}^n$ *and* $Y : \Omega \to \mathbb{R}^d$ *are random variables. There exists a* $\mathbb{P}_Y$*-a.s. unique Markov kernel* N *on* $\mathbb{R}^d \times \mathscr{B}(\mathbb{R}^n)$ *such that*

$$\mathbb{E}(u(X) \mid Y) = \int_{\mathbb{R}^n} u(x)\,N(Y,dx) \quad \forall u \in \mathcal{L}_b^0(\mathscr{B}(\mathbb{R}^n)). \tag{35.5}$$

The kernel $N(y,dx)$ *is a (version of the)* ***conditional distribution*** *of* X *given* $Y = y$.

Proof. 1° Uniqueness of the kernel. If $M(x.dy)$ is a further Markov kernel satisfying (35.5), we can take $u(x) = \mathbb{1}_Q(x)$ where $Q \in \mathscr{J}^n_{\text{rat}}$ is any half-open rectangle with rational vertices. This shows $M(Y,Q) = N(Y,Q)$ $\mathbb{P}$ a.s., or $M(y,Q) = N(y,Q)$ $\mathbb{P}_Y$ a.s., with an exceptional set depending on Q. Since there are only countably many such rectangles, we can take the union of all exceptional sets, and this is still a null set. In view of the uniqueness theorem of measures (Theorem 4.7), the family $\mathscr{J}^n_{\text{rat}}$ determines $N(y,Q)$ $\mathbb{P}_Y$ a.s., hence $M(y,B) = N(y,B)$ $\mathbb{P}_Y$ a.s. for all $B \in \mathscr{B}(\mathbb{R}^n)$.

Assume first that $n = 1$, i.e. X is a real random variable

2° Existence of the measure $N(y,dx)$: Let $r \in \mathbb{Q}$. We pick an $\mathbb{P}_Y$-a.s. unique version of the distribution function

$$F_y(-\infty, r] := \mathbb{E}(\mathbb{1}_{(-\infty,r]} \mid Y = y).$$

Clearly, we have for all $r \leqslant r'$, $r, r' \in \mathbb{Q}$,

a) $F_y(-\infty,r] = \mathbb{E}(\mathbb{1}_{(-\infty,r]} \mid Y = y) \overset{\mathbb{P}_Y \text{ a.s.}}{\leqslant} \mathbb{E}(\mathbb{1}_{(-\infty,r']} \mid Y = y) = F_y(-\infty,r']$;

b) $F_y\big(-\infty, r+n^{-1}\big] = \mathbb{E}\big(\mathbb{1}_{(-\infty,r+n^{-1}]} \mid Y = y\big) \xrightarrow[n\to\infty]{\mathbb{P}_Y \text{ a.s.}} \mathbb{E}\big(\mathbb{1}_{(-\infty,r]} \mid Y = y\big)$;

c) $\lim_{n\to-\infty} F_y(-\infty,n] = 0$ and $\lim_{n\to\infty} F_y(-\infty,n] = 1$.

Since $r, r' \in \mathbb{Q}$, there are only countably many null sets appearing in a)–c); denoting their union by M, we still have $\mathbb{P}_Y(M) = 0$. Thus,

$$\forall y \in N^c : \quad \mathbb{Q} \ni r \mapsto F_y(-\infty, r] \text{ is a (right cts.) distribution function;}$$

because of monotonicity, it suffices to consider in b) for each $r \in \mathbb{Q}$ just **a single** sequence $r_n \uparrow r$ – otherwise we might have a problem with exceptional sets as there are more than countably many rational sequences with $r_n \uparrow r$. Therefore,

$$\forall y \in M^c \quad \forall x \in \mathbb{R} : \quad F_y(-\infty, x] := \inf_{r \geqslant x} F_y(-\infty, r]$$

extends F_y uniquely to become a right-continuous distribution function which induces a probability measure $N(y, dx)$ on $M^c \times \mathscr{B}(\mathbb{R})$.

3° Measurability of $y \mapsto N(y, B)$: Denote by $\mathscr{J} := \{(-\infty, r) \mid r \in \mathbb{Q}\}$. We define

$$\mathscr{D} := \left\{ B \in \mathscr{B}(\mathbb{R}) \mid y \mapsto N(y, B) \text{ mble } \& \ N(y, B) = \mathbb{E}(\mathbb{1}_B \mid Y = y) \ \mathbb{P}_Y \text{ a.s.} \right\}.$$

Using the properties of conditional expactation (linearity, cBL) is is easy to see ✍ that $\mathscr{D}$ is a Dynkin system. Thus,

$$\mathscr{D} \overset{\text{def}}{\subset} \mathscr{B}(\mathbb{R}) \overset{\S 2.9}{=} \sigma(\mathscr{J}) \underset{\mathscr{J} \cap\text{-stable}}{\overset{\S 4.5}{=}} \delta(\mathscr{J}) \subset \mathscr{D},$$

and we see that $\mathscr{D} = \mathscr{B}(\mathbb{R})$.

4° Proof of (35.5): The first two steps show that $N(Y(\omega), dx)$ is unique and defines a measure on $(\mathbb{R}, \mathscr{B}(\mathbb{R}))$. Therefore, (35.5) follows with the usual reasoning: First for one-step simple functions, then for positive simple functions (by linearity), then for positive measurable functions (by the sombrero lemma and (c)BL) and finally for bounded measurable functions (by linearity).

Now we consider $n > 1$

5° From Theorem 26.8 and Remark 26.9 we know that $\mathbb{R}$ and $\mathbb{R}^n$ are Borel isomorphic, i.e. there exists a bijective and bi-measurable mapping $\phi : \mathbb{R} \to \mathbb{R}^n$. Let $X : \Omega \to \mathbb{R}^n$ be a random variable, then $X' := \phi^{-1}(X)$ is a one-dimensional random variable. Since ϕ is bi-measurable, we can write every $u \in \mathcal{L}_b^0(\mathbb{R}^n)$ in the form $u = (u \circ \phi) \circ \phi^{-1}$ where $u \circ \phi \in \mathcal{L}_b^0(\mathbb{R})$. Using the results from the previous steps, there is a $\mathbb{P}_Y$-a.s. unique kernel $N'(y, dx')$ on $\mathbb{R}^d \times \mathscr{B}(\mathbb{R})$ such that

$$\begin{aligned} \mathbb{E}(u(X) \mid Y) = \mathbb{E}(u \circ \phi(X') \mid Y) &= \int_{\mathbb{R}} u \circ \phi(x') N'(y, dx') \\ &= \int_{\mathbb{R}^n} u(x) N'(y, \phi \in dx) \end{aligned}$$

where $N'(y, \phi \in dx)$ denotes the image measure of $N'(y, dx')$ under the map ϕ. This shows that $N(y, dx) := N'(y, \phi \in dx)$ is a Markov kernel on $\mathbb{R}^d \times \mathscr{B}(\mathbb{R}^n)$ which is the conditional distribution of X given Y. □

35.11 Remark. The proof of Theorem 35.10 can be extended to spaces $(E,\mathscr{E})$ and $(F,\mathscr{F})$ such that the σ-algebra $\mathscr{E}$ is countably generated and $(F,\mathscr{F})$ is Borel-isomorphic to $(\mathbb{R},\mathscr{B}(\mathbb{R}))$. Typical examples are Polish spaces $(E,\mathscr{E})$, $(F,\mathscr{F})$, i.e. separable, complete metric spaces, see also Parthasarathy [23, Chapter V.8] or, for a completely different proof, Bauer [4, Chapter 44].

35.12 Remark. Since $\sigma(Y) = Y^{-1}(\mathscr{B}(\mathbb{R}^d))$, the equality (35.5) is equivalent to

$$\int_{\{Y\in B\}} u(X)\,d\mathbb{P} = \iint_{\{Y\in B\}\times\mathbb{R}^n} u(x)\,N(Y,dx)\,d\mathbb{P} \quad \forall B\in\mathscr{B}(\mathbb{R}^d). \tag{35.6}$$

35.13 Theorem. *Let $X:\Omega\to\mathbb{R}^n$ and $Y:\Omega\to\mathbb{R}^d$ be random variables. The kernel $N(y,dx)$ is the conditional distribution of X given Y if, and only if. for all $h\in\mathcal{L}^0_b(\mathscr{B}(\mathbb{R}^n\times\mathbb{R}^d))$*

$$\mathbb{E}h(X,Y) = \iint_{\mathbb{R}^n\times\mathbb{R}^d} h(x,y)\,N(y,dx)\,\mathbb{P}(Y\in dy) \tag{35.7}$$

$$= \mathbb{E}\int_{\mathbb{R}^n} h(x,Y)\,N(Y,dx) \tag{35.8}$$

and, therefore,

$$\mathbb{E}(h(X,Y)\mid Y=y) = \int_{\mathbb{R}^n} h(x,y)\,N(y,dx) \quad \mathbb{P}(Y\in dy)\text{-a.s.} \tag{35.9}$$

Proof. Note that (35.7)⇒(35.8)⇒(35.9) follows directly from the definitions. We show that $N(x,dy)$ is a conditional distribution if, and only if, (35.7) holds.

»⇐« Assume that (35.7) holds. Take $h(x,y) = u(x)\mathbb{1}_B(y)$ where $B\in\mathscr{B}(\mathbb{R}^d)$ and $u\in\mathcal{L}^0_b(\mathscr{B}(\mathbb{R}^n))$. This gives (35.6) and (35.5), i.e. $N(x,dy)$ is a conditional distribution.

»⇒« Assume that $N(x,dy)$ is a conditional distribution. In order to show (35.7), we assume first that $h(x,y) = \mathbb{1}_A(x)\mathbb{1}_B(y)$ for some $A\in\mathscr{B}(\mathbb{R}^n)$ and $B\in\mathscr{B}(\mathbb{R}^d)$. Note that

$$\begin{aligned}
\mathbb{E}h(X,Y) &= \mathbb{E}\big(\mathbb{1}_A(X)\mathbb{1}_B(Y)\big) = \int_{\{Y\in B\}} \mathbb{1}_A(X)\,d\mathbb{P} \\
&\underset{u=\mathbb{1}_A}{\overset{(35.6)}{=}} \int_{\{Y\in B\}}\int \mathbb{1}_A(x)\,N(Y,dx)\,d\mathbb{P} \\
&= \iint \mathbb{1}_B(y)\mathbb{1}_A(x)\,N(y,dx)\,\mathbb{P}(Y\in dy) \\
&= \iint h(x,y)\,N(y,dx)\,\mathbb{P}(Y\in dy).
\end{aligned}$$

We want to use the monotone class theorem (Theorem 7.17). Clearly, the family $\mathcal{V}$ of measurable functions h satisfying (35.7) is a vector space. Our calculation shows that $\mathbb{1}_{A\times B}\in\mathcal{V}$, and a standard Beppo-Levi argument shows that $\mathcal{V}$ is stable under increasing limits. Finally, the rectangles $\mathscr{B}(\mathbb{R}^n) \times \mathscr{B}(\mathbb{R}^d)$ are a $\cap$-stable generator of $\mathscr{B}(\mathbb{R}^n \times \mathbb{R}^d)$. Thus the MCT shows that $\mathcal{L}_b^0(\mathscr{B}(\mathbb{R}^n \times \mathbb{R}^d)) \subset \mathcal{V}$. $\square$

The following result is a direct consequence of Theorem 35.5.

35.14 Corollary. *Let* $X : \Omega \to \mathbb{R}^n$ *and* $Y : \Omega \to \mathbb{R}^d$ *be random variables with a joint density* $(X, Y) \sim f_{X,Y}(x, y)\,dx\,dy$.

$$N(y, A) = \frac{\int_A f_{X,Y}(x,y)\,dx}{\int_{\mathbb{R}^n} f_{X,Y}(x,y)\,dx} = \int_A f_{X|Y}(x \mid y)\,dx.$$

VIII

Gaussian distributions and the Lindeberg-Lévy CLT

36 The multivariate normal law

In this chapter we study a multivariate version of the normal distribution

$$\nu_{\mu,\sigma^2}(\mathrm{d}x) = g_{\mu,\sigma^2}(x)\,\mathrm{d}x = \frac{1}{\sqrt{2\pi\sigma^2}} e^{(x-\mu)^2/2\sigma^2}\,\mathrm{d}x. \tag{36.1}$$

An important tool is the characteristic function (cf) of a rv X with values in $\mathbb{R}^d$

$$\phi_X(\xi) := \mathbb{E}e^{i\langle\xi, X\rangle}, \quad \langle\xi, X\rangle = \sum_{i=1}^{d} \xi_i X_i \tag{36.2}$$

For example, if $G \sim \nu_{\mu,\sigma^2}$, we have (Theorem 27.3)

$$\phi_G(\xi) = \frac{1}{\sqrt{2\pi\sigma^2}} \int_{\mathbb{R}} e^{ix\xi} e^{-(x-\mu)^2/2\sigma^2}\,\mathrm{d}x = e^{i\xi\mu} e^{-\frac{1}{2}\sigma^2\xi^2} \tag{36.3}$$

36.1 Definition. A rv $X : \Omega \to \mathbb{R}^d$ is [non-degnerate] **Gaussian** or **normal**, if $\langle \ell, X\rangle$ is for all $\ell \in \mathbb{R}^d$ a one-dimensional [non-degenerate] normal random variable.

36.2 Theorem. *The following assertions are equivalent.*

a) $X : \Omega \to \mathbb{R}^d$ *is a non-degenerate Gaussian random variable.*

b) *There exists some* $m \in \mathbb{R}^d$ *and a strictly positive definite* $C \in \mathbb{R}^{d\times d}$ *such that*

$$\mathbb{E}e^{i\langle\xi, X\rangle} = \exp\left(i\langle\xi, m\rangle - \tfrac{1}{2}\langle\xi, C\xi\rangle\right), \quad \xi \in \mathbb{R}^d.$$

c) $X \sim g(x)\,\mathrm{d}x$ *where* $m \in \mathbb{R}^d$, $C \in \mathbb{R}^{d\times d}$ *is strictly positive definite and*

$$g(x) = \left(\frac{1}{2\pi}\right)^{d/2} \frac{1}{\sqrt{\det C}} \exp\left(-\tfrac{1}{2}\langle x-m, C^{-1}(x-m)\rangle\right).$$

We denote d-dimensional Gaussian random variables by $X \sim \mathsf{N}(m, C)$.

- $m = \mathbb{E}X$, $m = (m_1, \dots, m_d)^\top$ is the mean of X. Note that the mean of a vector is defined as the vector of the means of its coordinates: $\mathbb{E}X := (\mathbb{E}X_1, \dots, \mathbb{E}X_d)^\top$
- $C = (c_{ij})_{ij}$ with $c_{ij} = \mathbb{E}(X_i - m_i)(X_j - m_j) = \operatorname{Cov}(X_i, X_j)$ is the covariance matrix. Note that the strict positive definiteness of C gives $\det C > 0$.

Proof. a)⇒b): By definition, $\langle \xi, X\rangle$ is Gaussian for every $\xi \in \mathbb{R}^d$. That is,

$$\langle \xi, X\rangle \sim \nu_{\mu(\xi),\sigma^2(\xi)} \implies \mathbb{E}e^{it\langle \xi, X\rangle} = e^{it\mu(\xi) - \frac{1}{2}t^2\sigma^2(\xi)}.$$

This allows us to determine $\mu(\xi)$ and $\sigma(\xi)$:

$$\begin{aligned}
\mu(\xi) &= \mathbb{E}\langle \xi, X\rangle = \langle \xi, \mathbb{E}X\rangle \\
\sigma^2(\xi) = \mathbb{V}\big(\langle \xi, X\rangle\big) &= \mathbb{V}\Big(\sum_{i=1}^d \xi_i X_i\Big) = \sum_{i,k=1}^d \operatorname{Cov}(\xi_i X_i, \xi_k X_k) \\
&= \sum_{i,k=1}^d \xi_i \xi_k \operatorname{Cov}(X_i, X_k) = \langle \xi, C\xi\rangle
\end{aligned}$$

with the covariance matrix $C = (\operatorname{Cov}(X_i, X_k))_{i,k}$. The calculation shows that C is strictly positive definite, as $\sigma^2(\xi) > 0$ for all $\xi \in \mathbb{R}^d$.

b)⇒c): Since the c.f. characterizes X uniquely, it is enough to show that the density g from c) satisfies $\int g(x) e^{i\langle \xi, x\rangle}\,dx = e^{i\langle \xi, m\rangle - \frac{1}{2}\langle \xi, C\xi\rangle}$. We have

$$\begin{aligned}
\int e^{i\langle \xi, x\rangle} g(x)\,dx &= \int e^{i\langle \xi, x\rangle} \frac{1}{\sqrt{\det C}} \left(\frac{1}{2\pi}\right)^{d/2} e^{-\frac{1}{2}\langle (x-m), C^{-1}(x-m)\rangle}\,dx \\
&= e^{i\langle m, \xi\rangle} \int e^{i\sqrt{C}\langle z, \xi\rangle} \left(\frac{1}{2\pi}\right)^{d/2} e^{-\frac{1}{2}\langle \sqrt{C}z, \sqrt{C}^{-1} z\rangle}\,dz,
\end{aligned}$$

and changing variables according to $\sqrt{C}z = x - m$, $dz = (\det\sqrt{C})^{-1}\,dx$ shows

$$\int e^{i\langle \xi, x\rangle} g(x)\,dx = e^{i\langle m, \xi\rangle} \int e^{i\langle \sqrt{C}\xi, z\rangle} \left(\frac{1}{2\pi}\right)^{d/2} e^{-\frac{1}{2}|z|^2}\,dz.$$

Since $\sqrt{C}$ is symmetric, $|\sqrt{C}\xi|^2 = \xi \cdot C\xi$, and so

$$\begin{aligned}
\int e^{i\langle \xi, x\rangle} g(x)\,dx &= e^{i\langle m, \xi\rangle} \prod_{k=1}^d \int_{\mathbb{R}} e^{iz_k(\sqrt{C}\xi)_k} e^{-\frac{1}{2}z_k^2} \frac{dz_k}{2\pi} \\
&= e^{i\langle m, \xi\rangle} \prod_{k=1}^d e^{-\frac{1}{2}(\sqrt{C}\xi)_k^2} \\
&\overset{(36.3)}{=} e^{i\langle m, \xi\rangle} e^{-\frac{1}{2}|\sqrt{C}\xi|^2} \\
&= e^{i\langle m, \xi\rangle} e^{-\frac{1}{2}\langle \xi, C\xi\rangle}.
\end{aligned}$$

c)⇒a): Fix $\ell \in \mathbb{R}^d$ and assume that $X \sim g(x)\,dx$. We have

$$\begin{aligned}\mathbb{P}(\langle \ell, X\rangle \leqslant t) &= \int_{\langle \ell, x\rangle \leqslant t} g(x)\,dx \\ &= \int_{\langle \ell, x\rangle \leqslant t} \frac{1}{(2\pi)^{d/2}\sqrt{\det C}}\, e^{-\frac{1}{2}\langle (x-m), C^{-1}(x-m)\rangle}\,dx \\ &= \int_{\langle \ell, \sqrt{C}y\rangle \leqslant t - \langle \ell, m\rangle} \frac{1}{(2\pi)^{d/2}}\, e^{-\frac{1}{2}|y|^2}\,dy;\end{aligned}$$

in the last line we use the symmetry of $\sqrt{C}$ and change variables according to $\sqrt{C}y = x - m$, $dy = dx/\det\sqrt{C} = dx/\sqrt{\det C}$. Again by symmetry, $\langle \ell, \sqrt{C}y\rangle = \langle \sqrt{C}\ell, y\rangle$. Pick a rotation $U \in \mathbb{R}^{d\times d}$ such that

$$U\sqrt{C}\ell = |\sqrt{C}\ell|\,\vec{e_1}, \quad \vec{e_1} = (1,0,\dots,0)$$

and set $y = U^\top z$, $dy = dz$, $|y| = |z|$. Set $\mu = \langle \ell, m\rangle$ and $\sigma = |\sqrt{C}\ell|$. Then

$$\begin{aligned}\mathbb{P}(\langle \ell, X\rangle \leqslant t) &= \int_{\langle \sqrt{C}\ell, U^\top z\rangle \leqslant t - \langle \ell, m\rangle} \frac{1}{(2\pi)^{d/2}}\, e^{-\frac{1}{2}|z|^2}\,dz \\ &= \int_{\mathbb{R}^{d-1}} \int_{-\infty}^{\frac{t-\mu}{\sigma}} e^{-\frac{1}{2}z_1^2}\frac{dz_1}{\sqrt{2\pi}}\, e^{-\frac{1}{2}(z_2^2+\dots+z_d^2)}\frac{d(z_2,\dots,z_d)}{\sqrt{2\pi}^{d-1}} \\ &= \int_{\mathbb{R}^{d-1}} \int_{-\infty}^{t} \exp\left[-\frac{1}{2}\left(\frac{s-\mu}{\sigma}\right)^2\right] ds\, e^{-\frac{1}{2}(z_2^2+\dots+z_d^2)}\frac{d(z_2,\dots,z_d)}{\sqrt{2\pi}^{d-1}} \\ &= \int_{-\infty}^{t} \exp\left[-\frac{1}{2}\left(\frac{s-\mu}{\sigma}\right)^2\right] ds\, \underbrace{\int_{\mathbb{R}^{d-1}} e^{-\frac{1}{2}(z_2^2+\dots+z_d^2)}\frac{d(z_2,\dots,z_d)}{\sqrt{2\pi}^{d-1}}}_{=1}.\end{aligned}$$

□

36.3 Corollary. *Let $X = (X_1,\dots,X_d) : \Omega \to \mathbb{R}^d$ be a Gaussian random variable.*

$$\left.\begin{array}{r} X_1,\dots,X_d \\ \textit{independent}\end{array}\right\} \iff \left(\operatorname{Cov}(X_i,X_k)\right)_{i,k=1}^d = \begin{pmatrix} c_1 & 0 & 0 & \dots & 0 \\ 0 & c_2 & 0 & \dots & 0 \\ 0 & 0 & c_3 & \dots & 0 \\ \vdots & & & \ddots & \vdots \\ 0 & 0 & 0 & \dots & c_d \end{pmatrix}$$

▲ Attention: Corollary 36.3 only holds if we assume that the vector X is Gaussian. Without this assumption, the assertion is false.

Proof of Corollary 36.3. From Kac's theorem (Corollary 27.8) we know that the independence of the rv's $X_1,\dots,X_d$ is equivalent to

$$\mathbb{E}e^{i\langle \xi, X\rangle} = \prod_{k=1}^d \mathbb{E}e^{i\xi_k X_k} \quad \forall \xi = (\xi_1,\dots,\xi_d) \in \mathbb{R}^d,$$

which is the same as to say that the covariance matrix is diagonal. □

Conditional expectations and Gaussian rv's

Let $X : \Omega \to \mathbb{R}$ and $Y = (Y_1, \dots, Y_d) : \Omega \to \mathbb{R}^d$ be random variables. We know from Proposition 34.7.c) that $\mathbb{E}(X \mid Y) = \mathbb{E}(X \mid \sigma(Y))$ is the best mean square approximation of X by an $\sigma(Y)$-measurable random variable. For Gaussian random variables the best approximation is, in fact, an affine-linear function

$$E(X \mid Y) = a + \sum_{k=1}^{d} b_k Y_k$$

with suitable coefficients $a, b_1, \dots, b_d \in \mathbb{R}$. This is an important tool in statistics (linear models).

36.4 Theorem. *Let $X : \Omega \to \mathbb{R}$ and $Y : \Omega \to \mathbb{R}^d$ be random variables such that the vector (X, Y) is a $(d+1)$-dimensional Gaussian random variable. There exist $a, b_1, \dots, b_d \in \mathbb{R}$ such that*

a) $\mathbb{E}(X \mid Y) = a + \sum_{k=1}^{d} b_k Y_k$;
b) $X = a + \sum_{k=1}^{d} b_k Y_k + Z$ *and* $Z \sim \mathsf{N}(0, \sigma^2)$, $Z \perp\!\!\!\perp Y$;
c) $\mathbb{P}(X \in dx \mid Y = y) = \mathsf{N}\Big(a + \sum_{k=1}^{d} b_k y_k, \sigma^2\Big)$.

Proof. a), b): The space

$$\mathcal{H} := \left\{ \alpha + \sum_{k=1}^{d} \beta_k Y_k \mid \alpha, \beta_1, \dots, \beta_d \in \mathbb{R} \right\} \subset L^2(\mathscr{A})$$

is a closed subspace of $L^2(\mathscr{A})$ – just note that it is finite-dimensional, i.e. $\simeq \mathbb{R}^{d+1}$. Denote by

$$P_{\mathcal{H}} : L^2(\mathscr{A}) \to \mathcal{H}$$

the orthogonal projection onto $\mathcal{H}$. Since $P_{\mathcal{H}} X \in \mathcal{H}$, there are coefficients $a, b_k \in \mathbb{R}$ such that $P_{\mathcal{H}} X = a + \sum_{k=1}^{d} b_k Y_k$. Moreover,

$$Z := X - P_{\mathcal{H}} X \perp \mathcal{H} \implies \begin{cases} Z \perp 1 & \implies \mathbb{E}Z = \mathbb{E}(Z1) = 0, \\ Z \perp Y_k \ \forall k & \implies \mathbb{E}(ZY_k) = 0 \ \forall k. \end{cases}$$

Since the vector (Z, Y) is a $(d+1)$-dimensional Gaussian rv ✎ (hint: we have $\zeta Z + \langle \eta, Y \rangle = \zeta a + \langle (\zeta, \zeta b + \eta), (X, Y) \rangle$) satisfying $\mathrm{Cov}(Z, Y_k) = 0$, we see that $Z \perp\!\!\!\perp Y$ ✎. Thus,

$$\mathbb{E}(Z \mid Y) = \mathbb{E}Z = 0$$

and

$$\mathbb{E}(X \mid Y) = \mathbb{E}\left(a + \sum_{k=1}^{d} b_k Y_k + Z \,\middle|\, Y \right) = a + \sum_{k=1}^{d} b_k Y_k + \mathbb{E}(Z \mid Y) = a + \sum_{k=1}^{d} b_k Y_k.$$

c) *First proof*: Setting $Y = (Y_1, \dots, Y_d) = (y_1, \dots, y_d)$ in b), we can read off the result with the help of Part a). This requires that $\mathbb{E}(h(U,W) \mid W = w) = h(U,w)$ if $U \perp\!\!\!\perp W$, see Theorem 34.12.c).

Direct proof: Let $g : \mathbb{R}^d \to \mathbb{R}$ be bounded and measurable. For all $\xi \in \mathbb{R}$

$$\begin{aligned}
\mathbb{E}\left(e^{i\xi X} g(Y)\right) &\overset{\text{tower}}{=} \mathbb{E}\left(\mathbb{E}\left[e^{i\xi X} g(Y) \mid Y\right]\right) \\
&\overset{\text{pull}}{=} \mathbb{E}\left(g(Y)\mathbb{E}\left[e^{i\xi X} \mid Y\right]\right) \\
&= \mathbb{E}\left(g(Y) \int e^{i\xi x}\, \mathbb{P}(X \in dx \mid Y)\right).
\end{aligned}$$

On the other hand,

$$\begin{aligned}
\mathbb{E}\left(e^{i\xi X} g(Y)\right) &= \mathbb{E}\left(g(Y) e^{i\xi\left(a + \sum_{k=1}^d b_k Y_k + Z\right)}\right) \\
&\overset{Z \perp\!\!\!\perp Y}{=} \int g(y) \mathbb{E}\left(e^{i\xi\left(a + \sum_{k=1}^d b_k y_k + Z\right)}\right) \mathbb{P}(Y \in dy).
\end{aligned}$$

Since g is arbitrary, we get

$$\begin{aligned}
\int e^{i\xi x}\, \mathbb{P}(X \in dx \mid Y = y) &= \mathbb{E}\left(e^{i\xi\left(a + \sum_{k=1}^d b_k y_k + Z\right)}\right) \\
&= e^{i\xi\left(a + \sum_{k=1}^d b_k y_k\right)} e^{-\frac{1}{2}\sigma^2\xi^2}.
\end{aligned}$$

□

37 The central limit theorem (CLT)

In Chapter 28 we have proved the following classical CLT of **de Moivre-Laplace**.

37.1 Theorem (DeMoivre–Laplace; CLT). *Let $(X_k)_{k\in\mathbb{N}}$ be real-valued iid rv such that $X_1 \in L^2(\mathbb{P})$; define $S_n = X_1 + \cdots + X_n$, $\mu = \mathbb{E}X_1$ and $\sigma^2 = \mathbb{V}X_1$. Then one has for all $-\infty < a < b < \infty$*

$$\lim_{n\to\infty} \mathbb{P}\left(a < \frac{S_n - n\mu}{\sigma\sqrt{n}} \leqslant b\right) = \frac{1}{\sqrt{2\pi}} \int_a^b e^{-x^2/2}\, dx. \tag{37.1}$$

We can express (37.1) in an equivalent way, namely

$$\frac{S_n - \mathbb{E}S_n}{\sqrt{\mathbb{V}S_n}} = \frac{X_1 + \cdots + X_n - n\mu}{\sigma\sqrt{n}} = \frac{X_1^* + \cdots + X_n^*}{\sqrt{n}} \xrightarrow[n\to\infty]{d} G \sim \mathsf{N}(0,1) \tag{37.2}$$

(an »asterisk« indicates normalization $X^* = (X-\mu)/\sigma$) and this formulation gives us a clue how we might prove this result: Recall that d-convergence is equivalent to pointwise convergence of the characteristic functions (Theorem 30.3), so it is enough to show that

$$\mathbb{E}\exp\left[i\frac{\xi}{\sqrt{n}} \sum_{k=1}^n X_k^*\right] \overset{\text{iid}}{=} \left(\mathbb{E}e^{i\frac{\xi}{\sqrt{n}}X_1^*}\right)^n \xrightarrow[n\to\infty]{\forall\xi\in\mathbb{R}} e^{-\frac{1}{2}\xi^2}. \tag{37.3}$$

This can be done by **Taylor's theorem**, see the proof of Theorem 28.6. Our aim is to show a more general version of the CLT. Let us, first of all, discuss the limitations of the CLT.

37.2 Definition. A sequence of real rv $(X_k)_{k\in\mathbb{N}}$ **satisfies the CLT**, if there exist sequences $(a_n)_{n\in\mathbb{N}}\subset\mathbb{R}$ and $(s_n)_n\subset(0,\infty)$ such that

$$\frac{X_1+\cdots+X_n-a_n}{s_n}\xrightarrow[n\to\infty]{d} G\sim N(0,1).$$

37.3 Example. Let $(C_k)_{k\in\mathbb{N}}$ be a sequence of iid Cauchy random variables, i.e. $C_k\sim\gamma_1(dx)=\pi^{-1}(1+x^2)^{-1}\,dx$. The c.f. is given by

$$\mathbb{E}e^{i\xi C_k}=e^{-|\xi|}.$$

Proof. We argue »backwards«, using the inversion formula for the characteristic function, Theorem 27.10.(1) We have

$$\begin{aligned}\frac{1}{2\pi}\int_{-\infty}^{\infty}e^{-|\xi|}e^{-ix\xi}\,d\xi&=\frac{1}{2\pi}\left(\int_0^\infty e^{-\xi}e^{-ix\xi}\,d\xi+\int_0^\infty e^{-\xi}e^{ix\xi}\,d\xi\right)\\&=\frac{1}{2\pi}\left(\frac{1}{1+ix}+\frac{1}{1-ix}\right)\\&=\frac{1}{\pi}\frac{1}{1+x^2}.\end{aligned}$$

In view of the inversion formula of the c.f. we see that C_1 has indeed the characteristic function $e^{-|\xi|}$. □

Thus,

$$\mathbb{E}\exp\left(i\xi\frac{1}{n}(C_1+\cdots+C_n)\right)\overset{\text{iid}}{=}\prod_{k=1}^n e^{-|\xi|/n}=e^{-|\xi|}=\widehat{\gamma}_1(\xi),$$

and we conclude that $\frac{1}{n}(C_1+\cdots+C_n)\xrightarrow{d}C\sim\gamma_1$, violating the CLT. In fact $(C_k)_{k\in\mathbb{N}}$ violates the CLT **for every choice of** $(a_n)_{n\in\mathbb{N}},(s_n)_{n\in\mathbb{N}}\subset\mathbb{R}$: Assume that $[(C_1+\cdots+C_n)-a_n]/s_n\xrightarrow[n\to\infty]{d}G\sim N(0,1)$. Then

$$\begin{aligned}e^{-|\xi|^2/2}&\overset{\text{CLT}}{=}\lim_{n\to\infty}\left|\mathbb{E}\exp\left(i\xi\frac{1}{s_n}(C_1+\cdots+C_n-a_n)\right)\right|\\&\underset{\text{iid}}{\overset{|e^{i\xi a_n/s_n}|=1}{=}}\lim_{n\to\infty}\left|\mathbb{E}\exp\left(i\xi\frac{1}{s_n}(C_1+\cdots+C_n)\right)\right|\\&=\lim_{n\to\infty}\left|\mathbb{E}\exp\left(i\left(\frac{n}{s_n}\xi\right)\frac{1}{n}(C_1+\cdots+C_n)\right)\right|\\&\overset{\text{iid}}{=}\lim_{n\to\infty}e^{-(n/s_n)|\xi|}.\end{aligned}$$

(1) The usual approach is via the residue theorem. A direct calculation, avoiding complex function theory, can be found in [WT, Example 9.5].

This means that $s = \lim_{n\to\infty} n/s_n$ exists and $e^{-|\xi|^2/2} = e^{-s|\xi|}$ for all $\xi \in \mathbb{R}$. This is, however, impossible.

⚠ Note that we have $\mathbb{E}|C_1| = \infty$, i.e. $C_1 \notin L^2(\mathbb{P})$.

37.4 Theorem (Lindeberg 1922; Lévy 1925; Feller 1935). *Let $(X_k)_{k\in\mathbb{N}} \subset L^2(\mathbb{P})$ be independent real-valued random variables such that*

$$\mathbb{E}X_k = 0, \quad \mathbb{V}X_k = \sigma_k^2 > 0 \quad X_k \sim \mu_k.$$

Define $s_n^2 := \sigma_1^2 + \cdots + \sigma_n^2$. If the **Lindeberg condition**

$$\forall \epsilon > 0: \quad \lim_{n\to\infty} \frac{1}{s_n^2} \sum_{k=1}^n \int_{|x|>\epsilon s_k} x^2 \,\mu_k(dx) = 0 \tag{L}$$

holds, then $s_n^{-1}(X_1 + \cdots + X_n) \xrightarrow{d} G \sim N(0,1)$ as $n \to \infty$.

Let us prepare the proof of Theorem 37.4 with a few auxiliary results.

37.5 Remark. a) By independence we have $s_n^2 = \mathbb{V}(X_1 + \cdots + X_n) = \mathbb{V}S_n$ (Bienaymé's theorem).

b) The **classical Lindeberg condition** is

$$\forall \epsilon > 0: \quad \lim_{n\to\infty} \frac{1}{s_n^2} \sum_{k=1}^n \int_{|x|>\epsilon\, s_n} x^2 \,\mu_k(dx) = 0. \tag{L'}$$

c) (L)⇒(L'). This follows immediately from

$$\forall k \leqslant n: \quad s_k^2 \leqslant s_n^2 \implies \{|x| > \epsilon s_n\} \subset \{|x| > \epsilon s_k\}.$$

d) (L')⇒(L). We have for all $\delta > 0$

$$\begin{aligned}
&\frac{1}{s_n^2} \sum_{k=1}^n \int_{|x|>\epsilon s_k} x^2 \,\mu_k(dx) \\
&= \frac{1}{s_n^2} \left\{ \sum_{k:\, s_k \leqslant \delta s_n} + \sum_{k:\, s_k > \delta s_n} \right\} \int_{|x|>\epsilon s_k} x^2 \,\mu_k(dx) \\
&\leqslant \underbrace{\frac{1}{s_n^2} \underbrace{\sum_{k:\, s_k \leqslant \delta s_n} \underbrace{\int x^2 \,\mu_k(dx)}_{=\sigma_k^2}}_{\leqslant \delta^2 s_n^2}}_{\leqslant \delta^2} + \underbrace{\frac{1}{s_n^2} \sum_{1 \leqslant k \leqslant n} \int_{|x|>\epsilon\delta s_n} x^2 \,\mu_k(dx)}_{\xrightarrow[n\to\infty]{\text{(L') for } \epsilon\delta} 0} \\
&\xrightarrow[n\to\infty]{\text{(L')}} \delta^2 \xrightarrow[\delta\to 0]{} 0.
\end{aligned}$$

37.6 Lemma (Feller). (L) *implies* ***Feller's condition***

$$\lim_{n\to\infty} \max_{1\leqslant k\leqslant n} \frac{\sigma_k}{s_n} = 0. \tag{F}$$

Proof. For $1 \leqslant k \leqslant n$ and $\epsilon > 0$ we have

$$\begin{aligned}\sigma_k^2 = \int x^2\,\mu_k(dx) &= \left\{\int_{|x|\leqslant\epsilon s_n} + \int_{|x|>\epsilon s_n}\right\} x^2\,\mu_k(dx)\\ &\leqslant \epsilon^2 s_n^2 + \int_{|x|>\epsilon s_n} x^2\,\mu_k(dx)\end{aligned}$$

and this shows that

$$\max_{1\leqslant k\leqslant n} \frac{\sigma_k^2}{s_n^2} \leqslant \epsilon^2 + \frac{1}{s_n^2}\sum_{k=1}^n \int_{|x|>\epsilon s_n} x^2\,\mu_k(dx) \xrightarrow[n\to\infty]{(L')} \epsilon^2 \xrightarrow{\epsilon\to 0} 0. \qquad \square$$

37.7 Lemma. *For all $a_1,\dots a_n, b_1,\dots,b_n \in \mathbb{C}$ such that $|a_k|, |b_k| \leqslant 1$ one has*

$$\left|\prod_{k=1}^n a_k - \prod_{k=1}^n b_k\right| \leqslant \sum_{k=1}^n |a_k - b_k|.$$

Proof. We use induction on n. The induction step $n \rightsquigarrow n+1$ is as follows: Set $A := \prod_{k=1}^n a_k$ and $B := \prod_{k=1}^n b_k$. Then

$$\begin{aligned}|Aa_{n+1} - Bb_{n+1}| &\leqslant |Aa_{n+1} - Ba_{n+1}| + |Ba_{n+1} - Bb_{n+1}|\\ &\leqslant |A-B|\cdot\underbrace{|a_{n+1}|}_{\leqslant 1} + \underbrace{|B|}_{\leqslant 1}\cdot|a_{n+1}-b_{n+1}|\\ &\leqslant \sum_{k=1}^n |a_k - b_k| + |a_{n+1} - b_{n+1}|.\end{aligned} \qquad \square$$

37.8 Lemma. *Let $X \in L^2(\mathbb{P})$ be a real rv with $\mathbb{E}X = 0$ and $\mathbb{V}X = \sigma^2 > 0$. Then*

$$\left|\mathbb{E}e^{i\xi X} - 1 - i\cdot 0\cdot\xi + \tfrac{1}{2}\sigma^2\xi^2\right| \leqslant \xi^2\,\mathbb{E}\left(|X|^2 \wedge \frac{|\xi|\cdot|X|^3}{6}\right).$$

Proof. This is a remainder estimate for the Taylor series. We have (use $\int_0^a = -\int_a^0$)

$$\left|e^{ix} - 1 - ix - \tfrac{1}{2}(ix)^2\right| = \left|\int_0^x \int_0^t (1 - e^{is})\,ds\,dt\right| \leqslant \int_0^{|x|}\int_0^{|t|} |1 - e^{is}|\,ds\,dt.$$

We can estimate the integrand of the last integral in two ways:

$$|1 - e^{is}| \to \begin{cases} \leqslant 2, \\ = \left|\int_0^s e^{iu}\,du\right| \leqslant |s|, \end{cases}$$

and this yields

$$\left|e^{ix} - 1 - ix - \tfrac{1}{2}(ix)^2\right| \leqslant x^2 \wedge \frac{|x|^3}{6}.$$

Now we insert $x = \xi X$, take expectations, and use $\mathbb{E}X^2 = \mathbb{V}X = \sigma^2$ to see

$$\begin{aligned}\left|\mathbb{E}e^{i\xi X} - 1 - i\xi\mathbb{E}X + \tfrac{1}{2}\xi^2\mathbb{E}X^2\right| &\leqslant \mathbb{E}\left|e^{i\xi X} - 1 - i\xi X - \tfrac{1}{2}(i\xi X)^2\right| \\ &\leqslant \mathbb{E}\left((\xi^2 X^2) \wedge \frac{|\xi X|^3}{6}\right).\end{aligned}$$

□

We are now ready for the proof of the CLT.

Proof of Theorem 37.4. Construct independent $G_k \sim \mathsf{N}(0, \sigma_k^2)$, $1 \leqslant k \leqslant n$. Note that the G_k are chosen in such a way that X_k and G_k have the same mean and variance: $\mathbb{E}G_k = \mathbb{E}X_k$ and $\mathbb{V}G_k = \mathbb{V}X_k$.

1° Using a Taylor expansion we get

$$\begin{aligned}\mathbb{E}e^{i\xi X_k/s_n} &= \mathbb{E}e^{iX_k\xi/s_n} = 1 - \frac{1}{2}\xi^2\frac{\sigma_k^2}{s_n^2} + R_1(\xi/s_n), \\ \mathbb{E}e^{i\xi G_k/s_n} &= \mathbb{E}e^{iG_k\xi/s_n} = 1 - \frac{1}{2}\xi^2\frac{\sigma_k^2}{s_n^2} + R_2(\xi/s_n),\end{aligned}$$

with remainder terms R_1 and R_2. Lemma 37.8 shows

$$\begin{aligned}\left|\mathbb{E}e^{i\xi X_k/s_n} - \mathbb{E}e^{i\xi G_k/s_n}\right| &\leqslant |R_1(\xi/s_n)| + |R_2(\xi/s_n)| \\ &\leqslant \frac{\xi^2}{s_n^2}\left[\mathbb{E}\left(|X_k|^2 \wedge \frac{1}{6}\frac{|\xi|}{s_n}|X_k|^3\right) + \mathbb{E}\left(|G_k|^2 \wedge \frac{1}{6}\frac{|\xi|}{s_n}|G_k|^3\right)\right].\end{aligned}$$

Let Z denote either X_k or G_k. We have

$$\begin{aligned}\mathbb{E}\left(|Z|^2 \wedge \frac{1}{6}\frac{|\xi|}{s_n}|Z|^3\right) &= \left\{\int_{|Z|>\epsilon s_n} + \int_{|Z|\leqslant\epsilon s_n}\right\}\left(|Z|^2 \wedge \frac{1}{6}\frac{|\xi|}{s_n}|Z|^3\right)d\mathbb{P} \\ &\leqslant \int_{|Z|>\epsilon s_n}|Z|^2\,d\mathbb{P} + \int_{|Z|\leqslant\epsilon s_n}\frac{1}{6}\frac{|\xi|}{s_n}|Z|^3\,d\mathbb{P} \\ &\leqslant \int_{|Z|>\epsilon s_n}|Z|^2\,d\mathbb{P} + \frac{1}{6}|\xi|\frac{\epsilon s_n}{s_n}\int_{|Z|\leqslant\epsilon s_n}|Z|^2\,d\mathbb{P} \\ &\leqslant \int_{|Z|>\epsilon s_n}|Z|^2\,d\mathbb{P} + \frac{\epsilon}{6}|\xi|\,\mathbb{V}Z.\end{aligned}$$

This implies

$$\begin{aligned}&\left|\mathbb{E}e^{i\xi X_k/s_n} - \mathbb{E}e^{i\xi G_k/s_n}\right| \\ &\quad\leqslant \frac{\xi^2}{s_n^2}\int_{|X_k|>\epsilon s_n}|X_k|^2\,d\mathbb{P} + \frac{\xi^2}{s_n^2}\int_{|G_k|>\epsilon s_n}|G_k|^2\,d\mathbb{P} + \frac{\epsilon}{3}|\xi|^3\frac{\sigma_k^2}{s_n^2}.\end{aligned} \tag{37.4}$$

2° Since the rv $G_1, \dots, G_n$ are independent, we see that

$$G := G_1 + \dots + G_n \sim \mathsf{N}(0, \sigma_1^2) * \dots * \mathsf{N}(0, \sigma_n^2) = \mathsf{N}(0, s_n^2),$$

and so $G/s_n \sim \mathsf{N}(0,1)$. The independence of $X_1,\dots,X_n$ and Lemma 37.7 show

$$\begin{aligned}\left|\mathbb{E}\exp\left[i\xi\frac{X_1+\dots+X_n}{s_n}\right]-\mathbb{E}e^{i\xi G/s_n}\right| &= \left|\prod_{k=1}^n \mathbb{E}e^{i\xi X_k/s_n}-\prod_{k=1}^n \mathbb{E}e^{i\xi G_k/s_n}\right| \\ &\leqslant \sum_{k=1}^n \left|\mathbb{E}e^{i\xi X_k/s_n}-\mathbb{E}e^{i\xi G_k/s_n}\right|.\end{aligned}$$

Now use the estimate (37.4) from 1° and get

$$\begin{aligned}&\left|\mathbb{E}\exp\left[i\xi\frac{X_1+\dots+X_n}{s_n}\right]-\mathbb{E}e^{i\xi G/s_n}\right| \\ &\leqslant \xi^2\Bigg(\underbrace{\frac{1}{s_n^2}\sum_{k=1}^n\int_{|X_k|>\epsilon s_n}|X_k|^2\,d\mathbb{P}}_{\to 0 \text{ b/o (L')}}+\underbrace{\frac{1}{s_n^2}\sum_{k=1}^n\int_{|G_k|>\epsilon s_n}|G_k|^2\,d\mathbb{P}}_{\to 0 \text{ see below } 3^\circ}+\underbrace{\sum_{k=1}^n\frac{\epsilon}{3}|\xi|\frac{\sigma_k^2}{s_n^2}}_{=\epsilon|\xi|/3}\Bigg) \\ &\xrightarrow[n\to\infty]{} \frac{1}{3}\epsilon|\xi| \xrightarrow[\epsilon\to 0]{} 0.\end{aligned}$$

3° We still have to show the convergence of the G_k-term. Since the G_k are normal random variables,

$$\begin{aligned}\frac{1}{s_n^2}\int_{|G_k|>\epsilon s_n}|G_k|^2\,d\mathbb{P} &= \int_{|x|>\epsilon s_n}\left|\frac{x}{s_n}\right|^2 e^{-x^2/2\sigma_k^2}\frac{dx}{\sigma_k\sqrt{2\pi}} \\ &= \frac{1}{\sqrt{2\pi}}\int_{|y|>\epsilon s_n/\sigma_k}\frac{\sigma_k^2}{s_n^2}y^2 e^{-y^2/2}\,dy \\ &\leqslant \frac{\sigma_k^2}{s_n^2}\frac{1}{\sqrt{2\pi}}\int_{|y|>\epsilon \min_{k\leqslant n} s_n/\sigma_k} y^2 e^{-y^2/2}\,dy.\end{aligned}$$

Summing over $k=1,2,\dots,n$ and using $\sum_{k=1}^n \sigma_k^2 = s_n^2$ gives

$$\frac{1}{s_n^2}\sum_{k=1}^n\int_{|G_k|>\epsilon s_n}|G_k|^2\,d\mathbb{P} \leqslant \frac{1}{\sqrt{2\pi}}\int_{|y|>\epsilon\min_{k\leqslant n}s_n/\sigma_k} y^2 e^{-y^2/2}\,dy \xrightarrow[n\to\infty]{\text{DCT}} 0.$$

In order to apply the DCT, we use (F):

$$\min_{1\leqslant k\leqslant n}\frac{s_n}{\sigma_k}\to\infty \implies \left\{|y|>\epsilon\min_{k\leqslant n}\frac{s_n}{\sigma_k}\right\}\xrightarrow[n\to\infty]{}\emptyset. \qquad \square$$

37.9 Corollary (Lyapounov 1901). *Let $(X_k)_{k\in\mathbb{N}} \subset L^{2+\delta}(\mathbb{P})$, $\delta>0$, be independent rv such that $\mathbb{E}X_k = 0$, $s_n^2 = \mathbb{V}(X_1+\dots+X_n)$, and satisfying the* ***Lyapounov condition***

$$\lim_{n\to\infty}\frac{1}{s_n^{2+\delta}}\sum_{k=1}^n \mathbb{E}\left(|X_k|^{2+\delta}\right) = 0.$$

Then the CLT holds:

$$\frac{X_1 + \cdots + X_n}{s_n} \xrightarrow{d} G \sim \mathsf{N}(0,1).$$

Proof. We check (L'). For every $\epsilon > 0$

$$\begin{aligned}\frac{1}{s_n^2}\sum_{k=1}^n \int_{|X_k|>\epsilon s_n} |X_k|^2 \, d\mathbb{P} &\leqslant \frac{1}{s_n^2}\sum_{k=1}^n \int_{|X_k|>\epsilon s_n} |X_k|^2 \left|\frac{X_k}{\epsilon s_n}\right|^\delta d\mathbb{P} \\ &\leqslant \frac{1}{\epsilon^\delta}\frac{1}{s_n^{2+\delta}}\sum_{k=1}^n \mathbb{E}\left(|X_k|^{2+\delta}\right) \xrightarrow[n\to\infty]{} 0.\end{aligned}$$

□

37.10 Theorem (Feller 1935). *Let* $(X_k)_k \subset L^2(\mathbb{P})$ *be independent rv such that* $\mathbb{E}X_k = 0$ *and* $\mathbb{V}X_k = \sigma_k^2 > 0$.

$$\text{(F) \& CLT} \iff \text{(L)}.$$

Proof. »⇐«: This is Theorem 37.4 and Lemma 37.6.

»⇒«: Assume that (F) and CLT hold:

$$\lim_{n\to\infty}\max_{1\leqslant k\leqslant n}\frac{\sigma_k}{s_n} = 0 \tag{F}$$

$$\frac{X_1 + \cdots + X_n}{s_n} \xrightarrow{d} G \sim \mathsf{N}(0,1) \tag{CLT}$$

1° We show that (F) implies the so-called **asymptotic negligibility**

$$\lim_{n\to\infty}\max_{1\leqslant k\leqslant n}\mathbb{P}(|X_k| > \epsilon s_n) = 0. \tag{A}$$

Indeed, by the Chebyshev–Markov inequality

$$\mathbb{P}(|X_k| > \epsilon s_n) \leqslant \frac{1}{\epsilon^2 s_n^2}\mathbb{E}\left(X_k^2\right) = \frac{1}{\epsilon^2}\frac{\sigma_k^2}{s_n^2} \leqslant \frac{1}{\epsilon^2}\max_{1\leqslant k\leqslant n}\frac{\sigma_k^2}{s_n^2} \xrightarrow[\text{(F)}]{n\to\infty} 0$$

uniformly for all $k = 1, 2, \dots n$.

2° Using the continuity of $z \mapsto |e^z - 1|$, $|e^0 - 1| = 0$ and (A) gives

$$\begin{aligned}\left|\mathbb{E}\left(e^{i\xi X_k/s_n} - 1\right)\right| &\leqslant \int_{|X_k|\leqslant\epsilon s_n} \underbrace{\left|e^{i\xi X_k/s_n} - 1\right|}_{\leqslant \eta\ \forall\epsilon\ll 1} d\mathbb{P} + \int_{|X_k|>\epsilon s_n} \underbrace{\left|e^{i\xi X_k/s_n} - 1\right|}_{\leqslant 2} d\mathbb{P} \\ &\leqslant \eta + 2\max_{1\leqslant k\leqslant n}\mathbb{P}(|X_k| > \epsilon s_n) \qquad (37.5) \\ &\xrightarrow[n\to\infty]{} \eta \xrightarrow[\epsilon\to 0]{\eta\to 0} 0\end{aligned}$$

uniformly for all $k = 1, 2, \dots, n$. Moreover, because of Lemma 37.8,

$$\left|\mathbb{E}\left(e^{i\xi X_k/s_n} - 1\right)\right| \leqslant \left|\mathbb{E}R_1(\xi \cdot X_k/s_n)\right| \leqslant \frac{3}{2}\xi^2\frac{1}{s_n^2}\mathbb{E}X_k^2 = \frac{3}{2}\xi^2\frac{\sigma_k^2}{s_n^2}. \tag{37.6}$$

3° We need an estimate from complex analysis: If $z \in \mathbb{C}$ with $|z| < \frac{1}{2}$, then

$$\begin{aligned}|\log(1+z) - z| &= \left|\int_{\overrightarrow{0z}} \left(\frac{1}{\zeta+1} - 1\right) d\zeta\right| \\ &\leqslant |z| \max_{\zeta \in \overrightarrow{0z}} \left|\frac{\zeta}{\zeta+1}\right| \\ &\leqslant |z| \max_{|\zeta| \leqslant 1/2} \frac{|\zeta|}{1-|\zeta|} \\ &\leqslant |z| \frac{|z|}{1-\frac{1}{2}} = 2|z|^2.\end{aligned}$$

4° (37.5) shows that $z = \mathbb{E}e^{i\xi X_k/s_n} - 1$ satisfies $|z| < \frac{1}{2}$ if $n \gg 1$. Thus

$$\begin{aligned}&\sum_{k=1}^{n} \left|\log \mathbb{E}\left(e^{i\xi X_k/s_n} - 1 + 1\right) - \mathbb{E}\left(e^{i\xi X_k/s_n}\right) + 1\right| \\ &\quad \overset{3^\circ}{\leqslant} 2 \sum_{k=1}^{n} \left|\mathbb{E}e^{i\xi X_k/s_n} - 1\right|^2 \\ &\quad \leqslant 2 \max_{1\leqslant k\leqslant n} \mathbb{E}\left|e^{i\xi X_k/s_n} - 1\right| \sum_{k=1}^{n} \mathbb{E}\left|e^{i\xi X_k/s_n} - 1\right| \qquad (37.7) \\ &\quad \overset{(37.5),(37.6)}{\leqslant} \epsilon_n \sum_{k=1}^{n} \xi^2 \frac{\sigma_k^2}{s_n^2} \\ &\quad = \xi^2 \epsilon_n \xrightarrow[n\to\infty]{} 0.\end{aligned}$$

5° Now we use that the CLT holds:

$$\prod_{k=1}^{n} \mathbb{E}e^{i\xi X_k/s_n} = \mathbb{E}\left(\exp\left[i\xi \frac{1}{s_n}(X_1 + \cdots + X_n)\right]\right) \xrightarrow[n\to\infty]{} e^{-\xi^2/2}.$$

If we take logarithms, we see that

$$-\frac{1}{2}\xi^2 = \sum_{k=1}^{n} \log \mathbb{E}e^{i\xi X_k/s_n} + \epsilon_n' \overset{(37.7)}{=} \sum_{k=1}^{n} \left(\mathbb{E}e^{i\xi X_k/s_n} - 1\right) + \epsilon_n''$$

for suitable error terms $\epsilon_n', \epsilon_n''$ which tend to 0 as $n \to \infty$.

This equality can be written as

$$\frac{\xi^2}{2} - \sum_{k=1}^{n} \int_{|X_k| \leqslant \epsilon s_n} \left(1 - e^{i\xi X_k/s_n}\right) d\mathbb{P} = \sum_{k=1}^{n} \int_{|X_k| > \epsilon s_n} \left(1 - e^{i\xi X_k/s_n}\right) d\mathbb{P} + \epsilon_n''.$$

Now we consider the real parts only

$$\begin{aligned}
\frac{\xi^2}{2} - \sum_{k=1}^n \int_{|X_k|\leqslant \epsilon s_n} \left(1-\cos\frac{\xi X_k}{s_n}\right) d\mathbb{P}
&= \sum_{k=1}^n \int_{|X_k|> \epsilon s_n} \left(1-\cos\frac{\xi X_k}{s_n}\right) d\mathbb{P} + \tilde{\epsilon}_n \\
&\leqslant \sum_{k=1}^n \int_{|X_k|> \epsilon s_n} 2\, d\mathbb{P} + \tilde{\epsilon}_n
\end{aligned}$$

and the Chebyshev inequality shows

$$\begin{aligned}
&\leqslant \sum_{k=1}^n 2\,\mathbb{E}\frac{X_k^2}{\epsilon^2 s_n^2} + \tilde{\epsilon}_n \\
&= \frac{2}{\epsilon^2} \underbrace{\frac{\sum_{k=1}^n \mathbb{E}X_k^2}{s_n^2}}_{=1,\ s_n^2=\sum_{k=1}^n \sigma_k^2} + \tilde{\epsilon}_n \\
&= \frac{2}{\epsilon^2} + \tilde{\epsilon}_n.
\end{aligned}$$

On the other hand, $1-\cos t \leqslant \frac{1}{2}t^2$; applying this with $t = \xi X_k/s_n$ on the left side of the above estimate, yields

$$\frac{1}{2}\xi^2 - \sum_{k=1}^n \int_{|X_k|\leqslant \epsilon s_n} \frac{\xi^2 X_k^2}{2 s_n^2}\, d\mathbb{P} \leqslant \frac{2}{\epsilon^2} + \tilde{\epsilon}_n$$

and this can be rearranged to become

$$1 - \frac{1}{s_n^2}\sum_{k=1}^n \int_{|X_k|\leqslant \epsilon s_n} X_k^2\, d\mathbb{P} \leqslant \frac{4}{\xi^2\epsilon^2} + \frac{2}{\xi^2}\tilde{\epsilon}_n \xrightarrow[n\to\infty]{} \frac{4}{\xi^2\epsilon^2} \xrightarrow[|\xi|\to\infty]{} 0.$$

Finally, using that $1 = s_n^2/s_n^2 = \sum_{k=1}^n \mathbb{E}X_k^2/s_n^2$ gives

$$1 - \frac{1}{s_n^2}\sum_{k=1}^n \int_{|X_k|\leqslant \epsilon s_n} X_k^2\, d\mathbb{P} = \frac{1}{s_n^2}\sum_{k=1}^n \int_{|X_k|> \epsilon s_n} X_k^2\, d\mathbb{P}.$$

This proves (L') which is, by §37.5 equivalent to (L). □

37.11 Remark. Combining all results from above, we have shown

$$(F) + \text{CLT} \overset{\S 37.10}{\Longrightarrow} (A) + \text{CLT} \overset{\S 37.10}{\Longrightarrow} (L) \begin{cases} \overset{\S 37.4}{\Longrightarrow} \text{CLT} \\ \overset{37.5}{\Longleftrightarrow} (L') \overset{37.6}{\Longrightarrow} (F) \end{cases}$$

37.12 Remark. Often the CLT is usef for so-called **triangular arrays**, i.e. families of real rvs $X_{n,k}$ of the following form

$$\begin{array}{l} X_{1,1}, X_{1,2}, X_{1,3}, \dots, X_{1,k(1)} \\ X_{2,1}, X_{2,2}, X_{2,3}, \dots,\dots X_{2,k(2)} \\ \vdots \\ X_{n,1}, X_{n,2}, X_{n,3}, \dots,\dots,\dots, X_{n,k(n)} \end{array}$$

where $k(n) \in \mathbb{N}$. It is assumed that the rv **in each row** are independent, but the rows may be dependent.

We have considered so far the following array:

$$\begin{array}{l} X_1 \\ X_1, X_2 \\ X_1, X_2, X_3 \\ \dots \end{array}$$

Our proof shows that need only **row-wise independence** (and not overall independence). Therefore, we can use the following »translation« table to extend our results to general **triangular arrays**:

37.13 Remark*. Let us sketch a simple version of the CLT for random vectors. Let $Y_n = (Y_n^{(1)}, \dots Y_n^{(d)})$ be a sequence of random vectors. The key observation is that

$$\begin{aligned} Y_n \xrightarrow[n\to\infty]{d} Y \iff \forall \xi \in \mathbb{R}^d : \quad & \mathbb{E}e^{i\langle \xi, Y_n\rangle} \xrightarrow[n\to\infty]{} \mathbb{E}e^{i\langle \xi, Y\rangle} \\ \iff \forall \xi \in \mathbb{R}^d : \quad & \langle \xi, Y_n\rangle \xrightarrow[n\to\infty]{d} \langle \xi, Y\rangle, \end{aligned}$$

see Corollary 30.4, i.e. we can reduce everything to a one-dimensional setting.

Let $(X_n)_{n\in\mathbb{N}}$ be a sequence of iid random vectors $X_n = (X_n^{(1)}, \dots X_n^{(d)})$ such that $X_1^{(i)} \in L^2(\mathbb{P})$, $\mathbb{E}X_1^{(i)} = 0$, $c_{ij} = \operatorname{Cov}(X_1^{(i)}, X_1^{(j)})$ and $C = (c_{ij})_{ij}$. If the covariance matrix C is invertible, then

$$\frac{X_1 + \dots + X_n}{\sqrt{n}} \xrightarrow[n\to\infty]{d} G \sim N(0, C).$$

In order to see this we show that for each $\xi \in \mathbb{R}^d$ the one-dimensional variables $\langle \xi, C^{-1/2}(X_1 + \dots + X_n)\rangle / \sqrt{n}$ converge to $\langle \xi, \Gamma\rangle$ where $\Gamma \sim N(0, \mathrm{id})$ is a d-dimensional Gaussian random vector. This proves the claim if we set $G = C^{1/2}\Gamma \sim N(0, C)$.

Observe that the random variables $\langle \xi, C^{-1/2}X_n\rangle \in L^2(\mathbb{P})$ are iid with mean zero

Here	General array
$X_k \sim \mu_k$	$X_{n,k} \sim \mu_{n,k}, \quad 1 \leqslant k \leqslant k(n)$
X_k independent $\forall k$	$X_{n,k}$ independent $\forall k = 1, 2, \dots, k(n)$
$\sigma_k^2 = \mathbb{V}X_k$	$\sigma_{n,k}^2 = \mathbb{V}X_{n,k}$
$s_n^2 = \sum_{k=1}^{n} \sigma_k^2$	$s_n^2 = \sum_{k=1}^{k(n)} \sigma_{n,k}$
$S_n = \sum_{k=1}^{n} X_n$	$S_n = \sum_{k=1}^{k(n)} X_{n,k}$
(L)	$\frac{1}{s_n^2} \sum_{k=1}^{k(n)} \int_{\lvert x\rvert > \epsilon s_k} x^2 \mu_{n,k}(dx) \xrightarrow[n\to\infty]{\forall \epsilon>0} 0$
(L')	$\frac{1}{s_n^2} \sum_{k=1}^{k(n)} \int_{\lvert x\rvert > \epsilon s_n} x^2 \mu_{n,k}(dx) \xrightarrow[n\to\infty]{\forall \epsilon>0} 0$
(F)	$\max_{1\leqslant k\leqslant k(n)} \frac{\sigma_{n,k}}{s_n} \xrightarrow[n\to\infty]{} 0$
(A)	$\max_{1\leqslant k\leqslant k(n)} \mathbb{P}\left(\lvert X_{n,k}\rvert > \epsilon s_n\right) \xrightarrow[n\to\infty]{\forall \epsilon>0} 0$
CLT	$\frac{S_n - \mathbb{E}S_n}{s_n} \xrightarrow{d} G \sim \mathsf{N}(0,1)$

Table 37.1. Passing from an iid sequence to a general triangular array with row-wise independent random variables.

and variance

$$\begin{aligned}\mathbb{V}\langle \xi, C^{-1/2}X_n\rangle = \mathbb{V}\langle C^{-1/2}\xi, X_n\rangle &= \sum_{i,j=1}^{d} (C^{-1/2}\xi)_i (C^{-1/2}\xi)_j c_{ij} \\ &= \langle C^{-1/2}\xi, CC^{-1/2}\xi\rangle = \lvert\xi\rvert^2.\end{aligned}$$

The CLT (in the de-Moivre–Laplace form §37.1) shows that

$$\frac{\langle \xi, C^{-1/2}(X_1 + \cdots + X_n)\rangle}{\lvert\xi\rvert\sqrt{n}} \xrightarrow[n\to\infty]{d} \Gamma' \sim \mathsf{N}(0,1).$$

Since $\langle \xi, \Gamma\rangle / \lvert\xi\rvert \sim \mathsf{N}(0,1)$, and since only the probability distribution is important, the proof is finished.

IX

Martingales

38 Discrete martingales

Let $(\Omega, \mathscr{A}, \mathbb{P})$ be a probability space. We begin with a classical example from gambling.

38.1 Example. Consider a wager where two people bet unit stakes and the winner takes it all. Using random variables, we can describe this wager as follows:

$X_i = \pm 1$ iid	gain / loss in the ith round;
$S_n = X_1 + \cdots + X_n$	net gain after n rounds, $S_0 := 0$;
$\mathscr{F}_n = \sigma(X_1, \ldots, X_n)$	information on win/loose in the rounds $1, 2, \ldots, n$;
$= \sigma(S_1, \ldots, S_n)$	information on net gain in the rounds $1, 2, \ldots, n$;
$\mathbb{E}(S_{n+k} \mid \mathscr{F}_n)$	prediction on future net gain based on the results of all previous rounds $1, 2, \ldots, n$.

We can work out the expected net gain explicitly

$$\begin{aligned}
\mathbb{E}(S_{n+k} \mid \mathscr{F}_n) &= \mathbb{E}(\underbrace{S_n}_{\mathscr{F}_n\text{-mble}} + \underbrace{X_{n+1} + \cdots + X_{n+k}}_{\perp\!\!\!\perp \mathscr{F}_n} \mid \mathscr{F}_n) \\
&= \mathbb{E}(\underbrace{S_n}_{\mathscr{F}_n\text{-mble}} \mid \mathscr{F}_n) + \sum_{i=1}^{k} \mathbb{E}(\underbrace{X_{n+i}}_{\perp\!\!\!\perp \mathscr{F}_n} \mid \mathscr{F}_n) \\
&= S_n + \mathbb{E}X_{n+1} + \cdots + \mathbb{E}X_{n+k}
\end{aligned}$$

Depending on $\mathbb{E}X_1 = 0$, $\mathbb{E}X_1 \geqslant 0$ or $\mathbb{E}X_1 \leqslant 0$, the game is fair, or you have an edge, or your opponent has an edge, resulting in $\mathbb{E}(S_{n+k} \mid \mathscr{F}_n) = S_n$, $\geqslant S_n$ or $\leqslant S_n$, respectively.

We can make the game more interesting by admitting variable stakes (hence, strategies):

$$e_{n+1} = e_{n+1}(X_1, \ldots, X_n) \geqslant 0 \quad \text{(your stake in round } n+1\text{)}.$$

Note that e_{n+1} depends only on $X_1, \ldots, X_n$, i.e. it is $\mathscr{F}_n$-measurable. This makes sense since your stake should depend on your wealth which depends on the previous rounds, and not many banks will lend money on the basis of the unknown outcome of the $(n+1)$st round. Among the many possible choices are $e_{n+1} = 2e_n$ if you have lost in the nth round[(1)] or $e_{n+1} = 0$ if $S_n > c$ (this would be a stopping rule: you stop playing once you've reached a certain income).

We call such random variables **predictable** or **previsible** (since the information from $\mathscr{F}_n$ is good enough to know what happens at the future time $n+1$). We have to modify our formula for the net gain

$$S_{n+1} = S_n + e_{n+1}X_{n+1}$$

and using the rules for conditional expectations we get

$$\mathbb{E}(S_{n+1} \mid \mathscr{F}_n) = S_n + \mathbb{E}(e_{n+1}X_{n+1} \mid \mathscr{F}_n) \overset{\text{pull}}{=} S_n + \underbrace{e_{n+1}}_{\geqslant 0}\underbrace{\mathbb{E}(X_{n+1} \mid \mathscr{F}_n)}_{=\mathbb{E}X_{n+1}}$$

which shows that we still have $\mathbb{E}(S_{n+1} \mid \mathscr{F}_n) = S_n, \geqslant S_n$ or $\leqslant S_n$, depending on the sign of $\mathbb{E}X_1$.

Variable stakes, hence any predictable strategy, does not change the basic character of the game! So don't play the martingale.[(2)]

Before we introduce some general notation, let us settle one technical detail.

38.2 Lemma. *If $X_1,\dots,X_n$ are random variables and $S_k = X_1 + \dots + X_k$, then $\sigma(X_1,\dots,X_n) = \sigma(S_1,\dots,S_n)$.*

Proof. Define $S_0 = 0$. For each $k = 1,\dots,n$, the partial sum

$$S_k = X_1 + \dots + X_k$$

is $\sigma(X_1,\dots,X_k)$-, hence $\sigma(X_1,\dots,X_n)$-measurable. Thus,

$$\sigma(S_1,\dots,S_n) \subset \sigma(X_1,\dots,X_n).$$

On the other hand, for each $k = 1, 2,\dots,n$

$$X_k = S_k - S_{k-1}$$

is $\sigma(S_1,\dots,S_k)$-, hence $\sigma(S_1,\dots,S_n)$-measurable. Thus,

$$\sigma(S_1,\dots,S_n) \supset \sigma(X_1,\dots,X_n). \qquad \square$$

💬 The key to the proof of Lemma 38.2 is the fact that we have a bi-measurable function $X_1,\dots,X_n \leftrightarrow S_1,\dots,S_n$ – try to find an explicit representation as an invertible matrix ✍.

We will now extend the notion of a fair game.

(1) This is the classical martingale strategy

(2) A hommage to William Makepeace Thackeray: *You have not played as yet? Do not do so; above all avoid a martingale, if you do.* (The Newcomes, Chapter XXVIII)

38.3 Definition. Let $I \subset \mathbb{Z}$ be an index set and $\mathscr{F}_n \subset \mathscr{A}$, $n \in I$, σ-algebras.

a) If $\mathscr{F}_m \subset \mathscr{F}_n$ for all $m \leqslant n$, then $(\mathscr{F}_n)_{n\in I}$ is called a **filtration**

b) $\mathscr{F}_\infty := \sigma(\bigcup_{n\in I} \mathscr{F}_n) = \sigma(\mathscr{F}_n : n \in I)$.

c) A **(stochastic) process** is a family of random variables $X = (X_n)_{n\in I}$.

d) A process $X = (X_n)_{n\in I}$ is **($\mathscr{F}_n$-)adapted**, if each random variable X_n is $\mathscr{F}_n$-measurable: $X_n^{-1}(\mathscr{B}(\mathbb{R})) \subset \mathscr{F}_n$ for all $n \in I$. Notation: $(X_n, \mathscr{F}_n)_{n\in I}$.

e) A process $(e_n)_{n\in\mathbb{N}}$ is **predictable** or **previsible** w.r.t. the filtration $(\mathscr{F}_n)_{n\in\mathbb{N}_0}$, if e_n is $\mathscr{F}_{n-1}$-measurable.

⚠ Often the following convention is used: $e_0 := 0$ and $\mathscr{F}_{-1} := \{\emptyset, \Omega\}$.

38.4 Definition. Let $X = (X_n)_{n\in I}$ be a real-valued process and $(\mathscr{F}_n)_{n\in I}$ a filtration. If

a) $(X_n)_{n\in I}$ is adapted,

b) $X_n \in L^1(\mathbb{P})$, $n \in I$,

c) $\mathbb{E}(X_n \mid \mathscr{F}_m) = X_m$ for all $m \leqslant n$, $m, n \in I$,

then X is called a **martingale** (mg). If we only know

c') $\mathbb{E}(X_n \mid \mathscr{F}_m) \geqslant X_m$, $\forall m \leqslant n$ – then we have a **sub-martingale.**

c") $\mathbb{E}(X_n \mid \mathscr{F}_m) \leqslant X_m$, $\forall m \leqslant n$ – then we have a **super-martingale.**

💬 My **favourite way to remember** this is the slightly counterintuitive: »sUB goes UP« or »a SUB-martingale goes UP, a SUPER-martingale goes DOWN«. For instance, a sub-mg satisfies

$$X_m \leqslant \mathbb{E}(X_n \mid \mathscr{F}_m),\ \ m \leqslant n \implies \mathbb{E}X_m \leqslant \mathbb{E}\Big[\mathbb{E}(X_n \mid \mathscr{F}_m)\Big] \overset{\text{tower}}{=} \mathbb{E}X_n.$$

38.5 Remark. It is enough to require §38.4.c), c') or c") for $m = n-1$. One direction is clear, the other follows from the tower property: If $m < n$, then

$$\begin{aligned}\mathbb{E}(X_n \mid \mathscr{F}_m) &\overset{\text{tower}}{=} \mathbb{E}(\mathbb{E}(X_n \mid \mathscr{F}_{n-1}) \mid \mathscr{F}_m) \overset{\text{ass'n}}{=} \mathbb{E}(X_{n-1} \mid \mathscr{F}_m)\\ &\overset{\text{tower}}{=} \mathbb{E}(\mathbb{E}(X_{n-1} \mid \mathscr{F}_{n-2}) \mid \mathscr{F}_m) \overset{\text{ass'n}}{=} \mathbb{E}(X_{n-2} \mid \mathscr{F}_m) = \cdots = X_m.\end{aligned}$$

38.6 Lemma. *$(X_n, \mathscr{F}_n)_{n\in I}$ is a sub-martingale if, and only if,*

$$\forall F \in \mathscr{F}_m,\ m \leqslant n :\quad \int_F X_n \, d\mathbb{P} \geqslant \int_F X_m \, d\mathbb{P}. \tag{38.1}$$

Proof. »⟹« Assume that $\mathbb{E}(X_n \mid \mathscr{F}_m) \geqslant X_m$. Then we have for all $F \in \mathscr{F}_m$

$$X_m \leqslant \mathbb{E}(X_n \mid \mathscr{F}_m) \implies \int_F X_m \, d\mathbb{P} \leqslant \int_F \mathbb{E}(X_n \mid \mathscr{F}_m) \, d\mathbb{P} \underset{\mathbb{E}(\cdots \mid \mathscr{F}_m)}{\overset{\text{def. of}}{=}} \int_F X_n \, d\mathbb{P}.$$

»⇐« Assume that (38.1) holds. In this case we have

$$\forall F \in \mathscr{F}_m: \quad \int_F (\mathbb{E}(X_n \mid \mathscr{F}_m) - X_m)\, d\mathbb{P} \overset{\text{def. of}}{\underset{\mathbb{E}(\cdots\mid\mathscr{F}_m)}{=}} \int_F (X_n - X_m)\, d\mathbb{P} \overset{(38.1)}{\geqslant} 0.$$

If $F := \{\mathbb{E}(X_n \mid \mathscr{F}_m) - X_m < 0\} \in \mathscr{F}_m$, then $\mathbb{P}(\mathbb{E}(X_n \mid \mathscr{F}_m) - X_m < 0) = 0$, and so $\mathbb{E}(X_n \mid \mathscr{F}_m) \geqslant X_m$ a.s. □

38.7 Example. a) Let $S_n = \sum_{i=1}^n X_i$, $X_i \in L^1(\mathbb{P})$ iid, $\mathscr{F}_n = \sigma(X_1,\dots,X_n)$. From Example 38.1 we know that

$$S_n \overset{\S 38.1}{=} \begin{cases} \text{martingale} & \iff \mathbb{E}X_1 = 0, \\ \text{sub-mg} & \iff \mathbb{E}X_1 \geqslant 0, \\ \text{super-mg} & \iff \mathbb{E}X_1 \leqslant 0. \end{cases}$$

b) $X_i \geqslant 0$ iid, $\mathbb{E}X_1 = 1$, $\mathscr{F}_n = \sigma(X_1,\dots,X_n)$. Set $M_n = \prod_{i=1}^n X_i$, $M_0 = 1$, $\mathscr{F}_0 = \{\emptyset, \Omega\}$. We have

$$\begin{aligned} \mathbb{E}(M_n \mid \mathscr{F}_{n-1}) &= \mathbb{E}(M_{n-1} \cdot X_n \mid \mathscr{F}_{n-1}) \\ &\overset{\text{pull}}{=} M_{n-1}\mathbb{E}(X_n \mid \mathscr{F}_{n-1}) \overset{X_n \perp\!\!\!\perp \mathscr{F}_{n-1}}{=} M_{n-1}\mathbb{E}X_n = M_{n-1}, \end{aligned}$$

hence $(M_n, \mathscr{F}_n)_{n\in\mathbb{N}_0}$ is a martingale; the fact $M_n \in L^1(\mathbb{P})$ is an exercise ✍.

c) (Lévy's martingale) Let $(\mathscr{F}_n)_{n\in\mathbb{N}_0}$ be any given filtration and $X \in L^1(\mathscr{A})$. Then $M_n := \mathbb{E}(X \mid \mathscr{F}_n)$ is a martingale. Indeed, »adapted« is clear. Integrability follows from

$$\mathbb{E}|M_n| = \mathbb{E}|\mathbb{E}(X \mid \mathscr{F}_n)| \leqslant \mathbb{E}(\mathbb{E}(|X| \mid \mathscr{F}_n)) = \mathbb{E}|X| < \infty.$$

Using the tower property,

$$\mathbb{E}(M_{n+k} \mid \mathscr{F}_n) = \underbrace{\mathbb{E}(X \mid \mathscr{F}_{n+k} \mid \mathscr{F}_n)}_{\text{short for: } \mathbb{E}[\mathbb{E}(X\mid\mathscr{F}_{n+k})\mid\mathscr{F}_n]} \overset{\text{tower}}{=} \mathbb{E}(X \mid \mathscr{F}_n) = M_n.$$

We will see in §43 that many martingales are actually of this type.

d) Further examples of martingales are Pólya's urn (Example 22.8.e)), branching processes (from biology) and the likelihood ratio (from statistics), see [MaPs, Beispiel 3.4].

38.8 Theorem. *All (sub-, super-) mg are relative to the filtration* $(\mathscr{F}_n)_{n\in I}$.

a) $(X_n)_n, (Y_n)_n$ *mg* $\implies (aX_n + bY_n)_n$ *mg* $(a, b \in \mathbb{R})$.

b) $(X_n)_n, (Y_n)_n$ *sub-mg* $\implies (aX_n + bY_n)_n$ *sub-mg* $(a, b \geqslant 0)$.

c) $(X_n)_n, (Y_n)_n$ *super-mg* $\implies (X_n \wedge Y_n)_n$ *super-mg.*

d) $(X_n)_n$ *sub-mg* $\implies (X_n^+)_n$ *sub-mg.*

e) $(X_n)_n$ *mg,* ϕ *convex,* $\phi(X_n) \in L^1(\mathbb{P}) \implies (\phi(X_n))_n$ *sub-mg.*

f) $(X_n)_n$ *sub-mg,* ϕ *increasing & convex,* $\phi(X_n) \in L^1(\mathbb{P}) \implies (\phi(X_n))_n$ *sub-mg.*

g) $(X_n)_n$ *mg* $\iff$ $(X_n)_n$ *sub- and super-mg.*

h) $(X_n)_n$ *sub-mg* $\iff$ $(-X_n)_n$ *super-mg.*

Proof. Since all proofs are quite similar, we will check c)–f) only.

c) Clearly, $X_n \wedge Y_n = \frac{1}{2}(X_n + Y_n - |X_n - Y_n|)$ is adapted and integrable. If $m \leqslant n$, then

$$\mathbb{E}(X_n \wedge Y_n \mid \mathscr{F}_m) \overset{\text{monotone}}{\leqslant} \begin{cases} \mathbb{E}(X_n \mid \mathscr{F}_m) \overset{\text{super-mg}}{\leqslant} X_m \\ \mathbb{E}(Y_n \mid \mathscr{F}_m) \overset{\text{super-mg}}{\leqslant} Y_m \end{cases}$$

and so $\mathbb{E}(X_n \wedge Y_n \mid \mathscr{F}_m) \leqslant X_m \wedge Y_m$.

f) By the cJensen inequality §34.6.d) we see that for a sub-mg

$$\mathbb{E}(\phi(X_n) \mid \mathscr{F}_m) \geqslant \phi\Big(\underbrace{\mathbb{E}(X_n \mid \mathscr{F}_m)}_{\geqslant X_m}\Big) \overset{\phi \text{ increasing}}{\geqslant} \phi(X_m);$$

the case of a martingale – e) – is even easier and does not need that ϕ is increasing

d) follows from f) with $\phi(x) = x^+ = \max\{x, 0\}$. □

The Doob decomposition

We want to establish a connection between martingales and submartingales.

38.9 Theorem (Doob decomposition). *If* $X = (X_n, \mathscr{F}_n)_{n\in\mathbb{N}_0} \subset L^1$ *is any adapted process, then*

$$X_n = X_0 + M_n + A_n \quad \forall n \in \mathbb{N}_0, \tag{38.2}$$

where $M = (M_n)_{n\in\mathbb{N}_0} \subset L^1$ *is a martingale,* $A = (A_n)_{n\in\mathbb{N}_0} \subset L^1$ *is a predicable process and* $M_0 = A_0 = 0$.

The representation (38.2) is (up to ***indistinguishability****) unique, i.e. if there is any other such decompositon* $X_n = X_0 + M_n' + A_n'$*, then*

$$\mathbb{P}(M_n = M_n',\ A_n = A_n' \ \forall n \in \mathbb{N}) = 1.$$

Amendment: $(X_n)_{n\in\mathbb{N}_0}$ *is a sub-mg if, and only if,* $A_n \leqslant A_{n+1} \leqslant \ldots$ *a.s.*

Proof. 1° We begin by deriving some **necessary properties** of the representation (38.2). Assume that X satisfies (38.2). Since A is predictable, i.e. A_i is $\mathscr{F}_{i-1}$-measurable, we have

$$\begin{aligned}\mathbb{E}(X_i - X_{i-1} \mid \mathscr{F}_{i-1}) &= \mathbb{E}(M_i - M_{i-1} \mid \mathscr{F}_{i-1}) + \mathbb{E}(A_i - A_{i-1} \mid \mathscr{F}_{i-1}) \\ &= \underbrace{0}_{\text{b/o mg}} + \underbrace{A_i - A_{i-1}}_{\text{b/o predictable}} .\end{aligned} \tag{38.3}$$

Now we sum over $i = 1, \dots, n$ and find

$$A_n = \sum_{i=1}^{n} \mathbb{E}(X_i - X_{i-1} \mid \mathscr{F}_{i-1}). \tag{38.4}$$

2° We will now use the result of 1° to construct the decomposition. Assume that X is any adapted, integrable process. We will use the formula (38.4) to define the process A and we set

$$M_n := X_n - X_0 - A_n. \tag{38.5}$$

By definition, A_n is $\mathscr{F}_{n-1}$-measurable, i.e. A is predictable; moreover,

$$\mathbb{E}(M_n - M_{n-1} \mid \mathscr{F}_{n-1}) \overset{(38.5)}{=} \mathbb{E}(X_n - X_{n-1} \mid \mathscr{F}_{n-1}) - \underbrace{\mathbb{E}(A_n - A_{n-1} \mid \mathscr{F}_{n-1})}_{=\mathbb{E}(X_n - X_{n-1} | \mathscr{F}_{n-1}) \text{ by } (38.4)} = 0,$$

and we see that $(M_n, \mathscr{F}_n)_{n\in\mathbb{N}_0}$ is a martingale.

3° Since (38.4) is a necessary property, the prescription (38.5) defines $(M_n)_{n\in\mathbb{N}_0}$ uniquely, i.e. for any other decomposition $X_n = X_0 + A'_n + M'_n$ we have

$$\mathbb{P}(M_n = M'_n) = \mathbb{P}(A_n = A'_n) = 1 \quad \forall n \in \mathbb{N}.$$

Since the intersection of countably many sets with full measure is again of full measure, we get

$$\mathbb{P}(M_n = M'_n,\ A_n = A'_n \quad \forall n \in \mathbb{N}) = \mathbb{P}\Bigg(\bigcap_{n\in\mathbb{N}} \{M_n = M'_n\} \cap \{A_n = A'_n\}\Bigg) = 1.$$

4° Finally, we can read off the »amendment« directly from (38.3):

$$(X_n)_{n\in\mathbb{N}_0} \text{ sub-mg} \iff A_n - A_{n-1} \geqslant 0 \text{ a.s.} \qquad \square$$

The compensator

We will now study a particularly important example of the Doob decomposition.

38.10 Definition. Let $X = (X_n, \mathscr{F}_n)_{n\in\mathbb{N}_0}$ be a martingale.

a) If $X_n \in L^2(\mathbb{P})$ for all $n \in \mathbb{N}_0$, we call X a **square integrable** or L^2-martingale.
b) $\mathcal{M}^2 = \mathcal{M}^2((\mathscr{F}_n)_n)$ denote all L^2-martingales with the same filtration.
c) The **compensator** of an L^2-martingale X is the unique, increasing process $\langle X\rangle := (\langle X\rangle_n)_{n\in\mathbb{N}_0}$ of the Doob decomposition of the sub-mg $(X_n^2)_{n\in\mathbb{N}_0}$ (cf. §38.8.e)).

⚠ The compensator of X »compensates« what lacks X^2 to become a mg: $X^2 - \langle X\rangle$ is a mg. $\langle X\rangle$ is often called the **angle bracket**.

38.11 Lemma. *If $(X_n, \mathscr{F}_n)_{n\in\mathbb{N}_0}$ is an L^2-mg with compensator $\langle X\rangle$, then for all $m \leqslant n$*

$$\mathbb{E}\left[(X_n - X_m)^2 \mid \mathscr{F}_m\right] = \mathbb{E}\left[(X_n^2 - X_m^2) \mid \mathscr{F}_m\right] = \mathbb{E}\left[\langle X\rangle_n - \langle X\rangle_m \mid \mathscr{F}_m\right] \tag{38.6}$$

and

$$\langle X\rangle_n = \sum_{i=1}^{n} \mathbb{E}\left[(X_i - X_{i-1})^2 \mid \mathscr{F}_{i-1}\right] = \sum_{i=1}^{n} \mathbb{E}\left[(X_i^2 - X_{i-1}^2) \mid \mathscr{F}_{i-1}\right]. \tag{38.7}$$

Proof. It is enough to show (38.6), since we obtain (38.7) from (38.6) if we set $n \rightsquigarrow i$, $m \rightsquigarrow i-1$ and sum over $i = 1, \dots, n$. Let $m \leqslant n$. We have

$$\begin{aligned}
\mathbb{E}\left[(X_n - X_m)^2 \mid \mathscr{F}_m\right] &= \mathbb{E}\left[X_n^2 - 2X_n X_m + X_m^2 \mid \mathscr{F}_m\right] \\
&= \mathbb{E}\left[X_n^2 \mid \mathscr{F}_m\right] - 2X_m \underbrace{\mathbb{E}\left[X_n \mid \mathscr{F}_m\right]}_{=X_m \text{ b/o mg}} + X_m^2 \\
&= \mathbb{E}\left[X_n^2 \mid \mathscr{F}_m\right] - X_m^2 = \mathbb{E}\left[X_n^2 - X_m^2 \mid \mathscr{F}_m\right].
\end{aligned}$$

Since $(X_n^2 - \langle X\rangle_n)_{n\in\mathbb{N}_0}$ is a martingale, we see that

$$\mathbb{E}\left[X_n^2 - \langle X\rangle_n \mid \mathscr{F}_m\right] = X_m^2 - \langle X\rangle_m$$

which we can rearrange to become

$$\mathbb{E}\left[X_n^2 - X_m^2 \mid \mathscr{F}_m\right] = \mathbb{E}\left[\langle X\rangle_n \mid \mathscr{F}_m\right] - \langle X\rangle_m = \mathbb{E}\left[\langle X\rangle_n - \langle X\rangle_m \mid \mathscr{F}_m\right]. \qquad \square$$

The martingale transformation

38.12 Definition. Let $(M_n, \mathscr{F}_n)_{n\in\mathbb{N}_0}$ be a martingale and $(C_n)_{n\in\mathbb{N}}$ a predictable process. The process $C \bullet M$,

$$C \bullet M_n := \sum_{i=1}^{n} C_i (M_i - M_{i-1}), \quad C \bullet M_0 := 0, \quad n \in \mathbb{N}, \tag{38.8}$$

is called a **martingale transform.**

Please **observe** the analogy of (38.8) with Riemann(–Stieltjes) sums and integrals: $C \simeq$ integrand, $M \simeq$ integrator, $dM_i = M_i - M_{i-1}$. »Predictability« means that on each interval $(i-1, i]$ the measurability of the integrand C_i is determined by the left end-point $i-1$ of the interval: C_i is $\mathscr{F}_{i-1}$-measurable.

38.13 Theorem. *Let $M = (M_n, \mathscr{F}_n)_{n\in\mathbb{N}_0}$ be adapted, $C = (C_n)_{n\in\mathbb{N}}$ predictable, and $C_n(M_n - M_{n-1}) \in L^1$.*

a) *If M is a mg, then $C \bullet M = (C \bullet M_n)_{n\in\mathbb{N}_0}$ is a mg.*

b) *If M is a sub-mg, and $C_n \geqslant 0$, then $C \bullet M$ is a sub-mg.*

💬 The condition $C_n(M_n - M_{n-1}) \in L^1$ typically holds, if $C_n \in L^p$ and $M_n \in L^q$ for $1 \leqslant p, q \leqslant \infty$ such that $1/p + 1/q = 1$.

Proof. We concentrate on b), since the proof of a) is similar.

$$
\begin{aligned}
\mathbb{E}\big(C \bullet M_n \,\big|\, \mathscr{F}_{n-1}\big) &= \mathbb{E}\Big(\sum_{i=1}^{n} C_i(M_i - M_{i-1}) \,\Big|\, \mathscr{F}_{n-1}\Big) \\
&= \underbrace{\sum_{i=1}^{n-1} C_i(M_i - M_{i-1})}_{\mathscr{F}_{n-1}\text{-mble}} + \mathbb{E}\big(\underbrace{C_n}_{\mathscr{F}_{n-1}\text{-mble, b/o predictable}}(M_n - M_{n-1}) \,\big|\, \mathscr{F}_{n-1}\big) \\
&\overset{\text{pull}}{=} \sum_{i=1}^{n-1} C_i(M_i - M_{i-1}) + C_n \underbrace{\mathbb{E}\big((M_n - M_{n-1}) \,\big|\, \mathscr{F}_{n-1}\big)}_{=\mathbb{E}(M_n|\mathscr{F}_{n-1})-M_{n-1}\geqslant 0 \text{ b/o sub-mg}} \\
&\geqslant \sum_{i=1}^{n-1} C_i(M_i - M_{i-1}) = C \bullet M_{n-1},
\end{aligned}
$$

(if $n = 1$, then we use the »empty sum convention« $\sum_1^0 = 0$), so we have a submartingale. □

You may want to go back to the beginning of this chapter to see that our »variable stakes wager« is actually a martingale transform $e \bullet X$. This example gives a nice interpretation of predictability.

38.14 Theorem. *Let $C = (C_n)_{n\in\mathbb{N}_0}$ be a bounded predictable process, i.e. $|C_n| \leqslant K$ for some fixed K, and $M \in \mathcal{M}^2$ an L^2-mg with compensator $\langle M\rangle$.*

a) $C\bullet : \mathcal{M}^2 \to \mathcal{M}^2$;

b) $M \mapsto C \bullet M$ *and* $C \mapsto C \bullet M$ *are linear;*

c) $\langle C \bullet M\rangle_n = C^2 \bullet \langle M\rangle_n := \sum_{i=1}^n C_{i-1}^2(\langle M\rangle_i - \langle M\rangle_{i-1})$;

d) $\mathbb{E}\big[(C \bullet M_n - C \bullet M_m)^2 \mid \mathscr{F}_m\big] = \mathbb{E}(\langle C \bullet M\rangle_n - \langle C \bullet M\rangle_m \mid \mathscr{F}_m)$, *$m \leqslant n$;*

e) $\mathbb{E}\big[(C \bullet M_n)^2\big] = \mathbb{E}\big[C^2 \bullet \langle M\rangle_n\big]$.

Proof. a) follows from 38.13.b): Since the random variables C_i are bounded, we get for all $0 \leqslant i, k \leqslant n$:

$$\mathbb{E}(|C_i M_k|^2) \leqslant K^2 \mathbb{E}(M_k^2) < \infty \implies C \bullet M_n \in L^2.$$

b) Trivial.

c) First, we apply Lemma 38.11 with the L^2-martingale $X = C \bullet M$, and then with the L^2-martingale $X = M$:

$$
\begin{aligned}
\langle C \bullet M\rangle_n &\overset{(38.7)}{=} \sum_{i=1}^{n} \mathbb{E}\big[(C \bullet M_i - C \bullet M_{i-1})^2 \,\big|\, \mathscr{F}_{i-1}\big] \\
&= \sum_{i=1}^{n} \mathbb{E}\big[C_i^2 (M_i - M_{i-1})^2 \,\big|\, \mathscr{F}_{i-1}\big] \qquad (38.9)\\
&= \sum_{i=1}^{n} C_i^2 \mathbb{E}\big[(M_i - M_{i-1})^2 \,\big|\, \mathscr{F}_{i-1}\big] \\
&\overset{(38.6)}{=} \sum_{i=1}^{n} C_i^2 \mathbb{E}\big[\underbrace{\langle M\rangle_i - \langle M\rangle_{i-1}}_{\text{predictable}} \,\big|\, \mathscr{F}_{i-1}\big] \\
&= \sum_{i=1}^{n} C_i^2 \big(\langle M\rangle_i - \langle M\rangle_{i-1}\big) \\
&= C^2 \bullet \langle M\rangle_n.
\end{aligned}
$$

d) This is (38.6) if we take $X = C \bullet M$.

e) Follows from Part d) if we use $m = 0$, Part c) and (38.9). □

39 Stopping

Throughout this chapter $(\Omega, \mathscr{A}, \mathbb{P})$ is a probability space and $(\mathscr{F}_n)_{n\in\mathbb{N}_0}$ a filtration.

We want to evaluate a process $(X_n)_{n\in\mathbb{N}_0}$ at a random time, e.g. we might want to stop a wager once we have reached a certain threshold of our losses (or gains):

$$T(\omega) = \inf\{n \geqslant 0 \mid X_n(\omega) \leqslant -100\}.$$

⚠ As usual $\inf\emptyset = \infty$, and $T(\omega) = \infty$ would mean that the event does not happen (in finite time).

39.1 Definition. Let $(\mathscr{F}_n)_{n\in\mathbb{N}_0}$ be a filtration. A **stopping time** is a random variable $T : \Omega \to \mathbb{N}_0 \cup \{\infty\}$ such that

$$\forall n \in \mathbb{N}_0 : \quad \{T \leqslant n\} \in \mathscr{F}_n.$$

39.2 Remark (properties of stopping times). a) T is a stopping time if, and only if, $\{T = n\} \in \mathscr{F}_n$ for all $n \geqslant 0$.

»⇒« $\{T = n\} = \underbrace{\{T \leqslant n\}}_{\in\mathscr{F}_n} \setminus \underbrace{\{T \leqslant n-1\}}_{\in\mathscr{F}_{n-1}\subset\mathscr{F}_n} \in \mathscr{F}_n.$

»⇐«: $\{T \leqslant n\} = \bigcup_{i=0}^{n} \underbrace{\{T = i\}}_{\in\mathscr{F}_i\subset\mathscr{F}_n} \in \mathscr{F}_n.$

b) T is a stopping time if, and only if, $\{T < n\} \in \mathscr{F}_{n-1}$ for all $n \geqslant 1$.
This follows immediately from $\{T < n\} = \{T \leqslant n-1\}$.

c) $\{T = \infty\} = \{T < \infty\}^c = \Omega \setminus \bigcup_{n\in\mathbb{N}}\{T = n\} \in \mathscr{F}_\infty$.

d) $T \equiv m$ is a stopping time.
Indeed: $\{T = n\} = \left\{ \begin{array}{ll} \emptyset, & m \neq n \\ \Omega, & m = n \end{array} \right\} \in \mathscr{F}_n$.

e) If S, T are stopping times, then $S \wedge T$, $S \vee T$ are stopping times.
Indeed: $\{S \wedge T \leqslant n\} = \{S \leqslant n\} \cup \{T \leqslant n\} \in \mathscr{F}_n$. The proof of $S \vee T$ is similar.

f) If $(T_i)_{i\in\mathbb{N}}$ are stopping times, then $\inf_i T_i$, $\sup_i T_i$ are stopping times.
Hint: We have $\{\sup_i T_i \leqslant n\} = \bigcap_i\{T_i \leqslant n\}$, the rest is ✍.

39.3 Definition. Let $(X_n, \mathscr{F}_n)_{n\in\mathbb{N}_0}$ be an adapted process and T a stopping time. We define

$$X_T(\omega) := X_{T(\omega)}(\omega) \quad \forall \omega \in \{T < \infty\}; \tag{39.1}$$

$$X_n^T(\omega) := X_{n\wedge T(\omega)}(\omega) \tag{39.2}$$

$$= X_T(\omega)\, \underbrace{\mathbb{1}_{[0,n]}(T(\omega))}_{=\mathbb{1}_{\{T\leqslant n\}}(\omega)} + X_n(\omega)\, \underbrace{\mathbb{1}_{(n,\infty)}(T(\omega))}_{=\mathbb{1}_{\{T>n\}}(\omega)} \tag{39.3}$$

⚠ Since rvs are only defined up to $\mathbb{P}$-null sets, X_T is again a random variable if $\mathbb{P}(T < \infty) = 1$. You may define X_T to be 0 on $\{T = \infty\}$. Measurability of X_T is clear since it is the composition of two measurable functions:

$$\{X_T \in B\} = \bigcup_{n\in\mathbb{N}_0} \{T = n\} \cap \{X_n \in B\} \cup \{T = \infty\} \cap \{0 \in B\} \in \mathscr{A}.$$

39.4 Theorem. *If $(X_n, \mathscr{F}_n)_{n\in\mathbb{N}_0}$ is a sub-martingale and T a stopping time, then $(X_n^T, \mathscr{F}_n)_{n\in\mathbb{N}_0}$ is again a sub-martingale; in particular, $\mathbb{E}X_{T\wedge n} \geqslant \mathbb{E}X_0$.*

Proof. We define $C_n := \mathbb{1}_{[0,T]}(n)$ and observe that

$$\{C_n = 1\} = \{T \geqslant n\} = \{T < n\}^c = \{T \leqslant n-1\}^c \in \mathscr{F}_{n-1};$$

(it is essential that T is integer-valued!). Therefore, C is predictable.

It is easy to see that $X_n^T - X_0 = \mathbb{1}_{[0,T]} \bullet X$, and so the claim follows from Theorem 38.13.b).

The addendum follows from $\mathbb{E}(X_{T\wedge n} \mid \mathscr{F}_0) \geqslant X_{T\wedge 0} = X_0$, if we take expectations on both sides. □

39.5 Example. a) A direct consequence of §39.4 is the observation

$$(M_n)_n \text{ mg, T stopping time.} \implies (M_{T\wedge n})_n \text{ mg} \implies \mathbb{E}M_{T\wedge n} = \mathbb{E}M_0.$$

b) **In general**, however, $\mathbb{E}M_T \neq \mathbb{E}M_0$.
⚠ standard counterexample: $M_n = X_1 + \cdots + X_n$, $X_i \sim \frac{1}{2}(\delta_{-1} + \delta_1)$ iid, and $M_0 = 0$; M is a symmetric simple random walk. We use the stopping time ✍ $T = \inf\{n \in \mathbb{N}_0 \mid M_n = 1\}$, i.e. the **first passage time** of 1.
From Theorem 31.9 – an alternative proof is given in Part c) – we know that $\mathbb{P}(T < \infty) = 1$, and so $M_T = 1$. Thus, $\mathbb{E}M_{T\wedge n} = \mathbb{E}M_0 = 0$, but $\mathbb{E}M_T = \mathbb{E}1 = 1$.

c) We continue with b). We want to show that $\mathbb{P}(T < \infty) = 1$.
It is easy to see that

$$N_0 := 1 \quad \text{and} \quad N_n := \frac{\exp[\theta M_n]}{(\mathbb{E}\exp[\theta X_1])^n}, \qquad n \in \mathbb{N},\ \theta > 0$$

is a martingale ✍, and from Part a) the stopped martingale

$$N_n^T = \frac{\exp[\theta M_{n\wedge T}]}{(\mathbb{E}\exp[\theta X_1])^{n\wedge T}}, \qquad n \in \mathbb{N}_0, \quad \text{is again a mg.}$$

Obviously, $\mathbb{E}\exp[\theta X_1] = \frac{1}{2}(e^{\theta} + e^{-\theta}) = \cosh\theta \geqslant 1$. Because of the definition of the stopping time T we see that

$$0 \leqslant \frac{\exp[\theta M_{n\wedge T}]}{\cosh^{n\wedge T}\theta} \leqslant \exp[\theta M_{n\wedge T}] \leqslant e^{\theta}, \qquad \theta > 0,$$

and on the set $\{T < \infty\}$ we have $M_T = 1$. Therefore,

$$\lim_{n\to\infty} \frac{\exp[\theta M_{n\wedge T}]}{\cosh^{n\wedge T}\theta} = \begin{cases} \dfrac{\exp[\theta M_T]}{\cosh^T\theta} = \dfrac{e^{\theta}}{\cosh^T\theta}, & \text{if } T < \infty, \\ 0, & \text{if } T = \infty. \end{cases}$$

Since a martingale has constant expectation, we get

$$1 = \lim_{n\to\infty} \mathbb{E}\frac{\exp[\theta M_{n\wedge T}]}{\cosh^{n\wedge T}\theta} \overset{\text{DCT}}{=} e^{\theta}\mathbb{E}\left[\frac{\mathbb{1}_{\{T<\infty\}}}{\cosh^T\theta}\right] \leqslant e^{\theta}\mathbb{E}\mathbb{1}_{\{T<\infty\}},$$

and $\theta \to 0$ shows that $\mathbb{P}(T < \infty) = 1$.

The next theorem contains conditions ensuring that $\mathbb{E}M_T = \mathbb{E}M_0$.

39.6 Theorem (Doob; optional sampling). *Let $(X_n, \mathscr{F}_n)_{n\in\mathbb{N}_0}$ be a [sub-] martingale and T a stopping time such that $T < \infty$ a.s. Then $X_T \in L^1$ and $\mathbb{E}X_T = \mathbb{E}X_0$ [resp. $\mathbb{E}X_T \geqslant \mathbb{E}X_0$] if one of the following conditions hold:*

a) *T is a.s. bounded, i.e.* $\exists N : \mathbb{P}(T \leqslant N) = 1$;
b) *X is a.s. bounded, i.e.* $\exists K : \mathbb{P}\left(\sup_{n\in\mathbb{N}} |X_n| \leqslant K\right) = 1$;
c) $\mathbb{E}T < \infty$ *and* $\exists K : \mathbb{P}(\sup_{n\in\mathbb{N}} |X_n - X_{n-1}| \leqslant K) = 1$.

Proof. It is enough to consider the sub-martingale case. From Theorem 39.4 we know that $X_{T\wedge n} \in L^1$ and $\mathbb{E}X_{T\wedge n} \geqslant \mathbb{E}X_0$.

a) Pick $n \geqslant N$.

b) Since $|X_{n\wedge T}| \leqslant K$ a.s. and $X_{n\wedge T} \xrightarrow[n\to\infty]{\text{a.s.}} X_T$, we can use dominated convergence to obtain

$$X_{n\wedge T} \xrightarrow[n\to\infty]{L^1} X_T \implies X_T \in L^1$$

and

$$\mathbb{E}X_T = \lim_{n\to\infty} \underbrace{\mathbb{E}X_{T\wedge n}}_{\geqslant \mathbb{E}X_0} \geqslant \mathbb{E}X_0.$$

c) We have

$$|X_{T\wedge n} - X_0| = \left|\sum_{i=1}^{T\wedge n}(X_i - X_{i-1})\right| \leqslant \sum_{i=1}^{T} |X_i - X_{i-1}| \leqslant T\cdot K \in L^1.$$

Therefore, we may use the DCT and conclude that $|X_{T\wedge n} - X_0| \xrightarrow{L^1} |X_T - X_0|$. In particular, $X_T \in L^1$ and

$$\mathbb{E}X_T = \lim_{n\to\infty} \mathbb{E}X_{T\wedge n} \geqslant \mathbb{E}X_0.$$ □

Stopped processes are important objects and we want to study them in-depth. In order to do so, we need the notion of a σ-algebra associated with a stopping time T.

39.7 Definition. Let T be a stopping time w.r.t. the filtration $(\mathscr{F}_n)_{n\in\mathbb{N}_0}$. The σ-algebra associated with T is

$$\mathscr{F}_T := \left\{A \in \mathscr{F}_\infty \mid A \cap \{T \leqslant i\} \in \mathscr{F}_i \quad \forall i \in \mathbb{N}_0\right\}. \tag{39.4}$$

39.8 Remark. Let S, T be two stopping times.

a) $\mathscr{F}_T$ is a σ-algebra ✎.

⚠ Although our notation may give a different impression, $\mathscr{F}_T$ is a **non-random** family of sets .

b) If $S \leqslant T$, then $\mathscr{F}_S \subset \mathscr{F}_T$.
Indeed: Let $A \in \mathscr{F}_S$. For every $i \in \mathbb{N}_0$ we have

$$A \cap \{T \leqslant i\} = A \cap \overbrace{\{S \leqslant T\}}^{=\Omega} \cap \{T \leqslant i\} = \underbrace{A \cap \{S \leqslant i\}}_{\in \mathscr{F}_i} \cap \underbrace{\{T \leqslant i\}}_{\in \mathscr{F}_i} \in \mathscr{F}_i,$$

and this means that $A \in \mathscr{F}_T$.

c) $\{S < T\} \in \mathscr{F}_S \cap \mathscr{F}_T$. ✎

d) $\mathscr{F}_S \cap \mathscr{F}_T = \mathscr{F}_{S\wedge T}$.
Indeed: As $S \wedge T \leqslant S, T$ we conclude from b) that $\mathscr{F}_{S\wedge T} \subset \mathscr{F}_S \cap \mathscr{F}_T$.
Conversely, if $F \in \mathscr{F}_S \cap \mathscr{F}_T$, then we get for all $i \in \mathbb{N}_0$

$$F \cap \{S \wedge T \leqslant i\} = F \cap \Big(\{S \leqslant i\} \cup \{T \leqslant i\}\Big) = \underbrace{\Big(F \cap \{S \leqslant i\}\Big)}_{\in \mathscr{F}_i} \cup \underbrace{\Big(F \cap \{T \leqslant i\}\Big)}_{\in \mathscr{F}_i} \in \mathscr{F}_i,$$

and this means that $F \in \mathscr{F}_{S\wedge T}$.

39.9 Theorem. *Let $(X_n, \mathscr{F}_n)_{n\in\mathbb{N}_0} \subset L^1$ be adapted and integrable. The following are equivalent:*

a) $(X_n)_{n\in\mathbb{N}_0}$ *is a sub-martingale;*
b) $\mathbb{E}X_S \leqslant \mathbb{E}X_T$ *for all a.s. bounded stopping times* $S \leqslant T$;
c) $\int_F X_S \, d\mathbb{P} \leqslant \int_F X_T \, d\mathbb{P} \quad \forall F \in \mathscr{F}_S$ *and all a.s. bdd stopping times* $S \leqslant T$;
d) $X_S \leqslant \mathbb{E}(X_T \mid \mathscr{F}_S)$ *for all a.s. bounded stopping times* $S \leqslant T$.

Proof. Let $S \leqslant T$ be a.s. bounded stopping times, i.e. $S \leqslant T \leqslant N \in \mathbb{N}$ a.s. Let us first check that $X_S, X_T \in L^1$:

$$X_T = \sum_{i\in\mathbb{N}_0\cup\{\infty\}} X_i \mathbb{1}_{\{T=i\}} \overset{\text{a.s.}}{=} \sum_{i=0}^{N} X_i \mathbb{1}_{\{T=i\}}$$

and so, X_T is $\mathscr{A}$-measurable, and

$$\mathbb{E}|X_T| = \sum_{i=1}^{N} \mathbb{E}(|X_i| \mathbb{1}_{\{T=i\}}) \leqslant \sum_{i=1}^{N} \mathbb{E}|X_i| < \infty.$$

A similar calculation works for X_S.

a)⇒b) Because of the Doob decomposition (Theorem 38.9) we know that

$$X_T - X_S = M_T + A_T - M_S - A_S \geqslant M_T - M_S$$

since $A_T \geqslant A_S$, and we get

$$\mathbb{E}(X_T - X_S) \geqslant \mathbb{E}(M_T - M_S) = \mathbb{E}M_T - \mathbb{E}M_S \overset{\S 39.6}{=} \mathbb{E}M_0 - \mathbb{E}M_0 = 0.$$

b)⇒c) Take any $F \in \mathscr{F}_S \subset \mathscr{F}_T$ and define $\rho := S\mathbb{1}_F + T\mathbb{1}_{F^c}$. For every $i \in \mathbb{N}_0$ it holds that

$$\{\rho \leqslant i\} = \{\rho \leqslant i\} \cap (F \cup F^c) = \underbrace{(F \cap \{S \leqslant i\})}_{\in \mathscr{F}_i} \cup \underbrace{(F^c \cap \{T \leqslant i\})}_{\in \mathscr{F}_i} \in \mathscr{F}_i,$$

and we conclude that ρ is a stopping time. Moreover, $\rho \leqslant T\mathbb{1}_F + T\mathbb{1}_{F^c} = T$, and so

$$\mathbb{E}X_\rho = \mathbb{E}(X_S \mathbb{1}_F + X_T \mathbb{1}_{F^c}) \overset{b)}{\leqslant} \mathbb{E}X_T.$$

Using $\mathbb{1}_F = 1 - \mathbb{1}_{F^c}$, we can rearrange this inequality to get

$$\mathbb{E}(X_S \mathbb{1}_F) \leqslant \mathbb{E}(X_T \mathbb{1}_F)$$

and c) follows.

c)⇔d) This can be shown with the argument used for Lemma 38.6.

c)⇒a) Use the (non-random) stopping times $S \equiv n-1$ and $T \equiv n$. □

40 The martingale convergence theorem

Our starting point is a purely non-random consideration. Let $(a_n)_n \subset \mathbb{R}$ be a sequence of real numbers. We have

$$
\begin{aligned}
&\lim_{n\to\infty} a_n \text{ does } \textbf{not} \text{ exist in } [-\infty,\infty] \\
&\iff -\infty \leqslant \liminf_{n\to\infty} a_n < \limsup_{n\to\infty} a_n \leqslant \infty \\
&\iff \exists\, a,b \in \mathbb{Q} : \liminf_{n\to\infty} a_n < a < b < \limsup_{n\to\infty} a_n \\
&\iff \exists\, a,b \in \mathbb{Q} : \text{infinitely many } a_n \text{ are »above«} \\
&\qquad\qquad\qquad \text{and »below« the strip } [a,b]\times\mathbb{N} \\
&\iff \exists\, a,b \in \mathbb{Q} : \text{the sequence } (a_n)_n \text{ »crosses«} \\
&\qquad\qquad\qquad [a,b]\times\mathbb{N} \text{ infinitely often}
\end{aligned}
$$

If we apply the above reasoning to a sequence of rv $X_n : \Omega \to \mathbb{R}$, we see

$$
\begin{aligned}
&\left\{\omega \mid \lim_{n\to\infty} X_n(\omega) \text{ does not exist in } [-\infty,\infty]\right\} \\
&= \bigcup_{a<b,a,b\in\mathbb{Q}} \left\{\omega \mid \liminf_{n\to\infty} X_n(\omega) < a < b < \limsup_{n\to\infty} X_n(\omega)\right\} \\
&= \bigcup_{a<b,a,b\in\mathbb{Q}} \left\{\omega \mid D([a,b],\omega) = \infty\right\}
\end{aligned}
$$

where

$$
D([a,b],\omega) = \#\left\{\text{downcrossings over } [a,b]\times\mathbb{N}\right\};
$$

a *downcrossing* happens if we start from above the strip $[a,b]\times\mathbb{N}_0$ and cross it in a downwards movement, cf. Fig. 40.1.

40.1 Definition. Let $(X_n, \mathscr{F}_n)_{n\in\mathbb{N}_0}$ be an adapted process, $a < b$ and $N \in \mathbb{N}$. The number of downcrossings across $[a,b]\times\{0,1,\dots N\}$ is

$$
D_N([a,b],\omega) := \max\left\{\begin{matrix} k \geqslant 0 \mid \exists\, 0 \leqslant \sigma_1 < \tau_1 < \dots < \sigma_k < \tau_k \leqslant N \\ \text{with } X_{\sigma_i}(\omega) > a,\ X_{\tau_i}(\omega) < b,\ 0 \leqslant i \leqslant k \end{matrix}\right\}
$$

($\max\emptyset = 0$). We define $D[a,b] := \sup_{N\in\mathbb{N}_0} D_N[a,b]$.

40.2 Lemma (Doob; downcrossing estimate). *Let $(X_n, \mathscr{F}_n)_{n\in\mathbb{N}_0}$ be a sub-martingale. For all $a < b$ and $N \in \mathbb{N}$ one has*

$$
(b-a)\mathbb{E}D_N[a,b] \leqslant \mathbb{E}(X_N - b)^+. \tag{40.1}
$$

⚠ Often **upcrossings** are considered. However, since there must be a downcrossing between two upcrossings and vice versa, the number of up- and downcrossings differs by at most 1. In other words, we get a similar estimate for upcrossings (in our proof below, just change C by $1 - C$ and see what happens).

Proof. In order to describe the downcrossings (marked in Fig. 40.1 with solid circles) we think of X as a game and we introduce a betting strategy C as follows:

1° Wait until X is above b (including this step);
2° Bet unit stakes until X is below a (including this step);
3° go to 1° ;

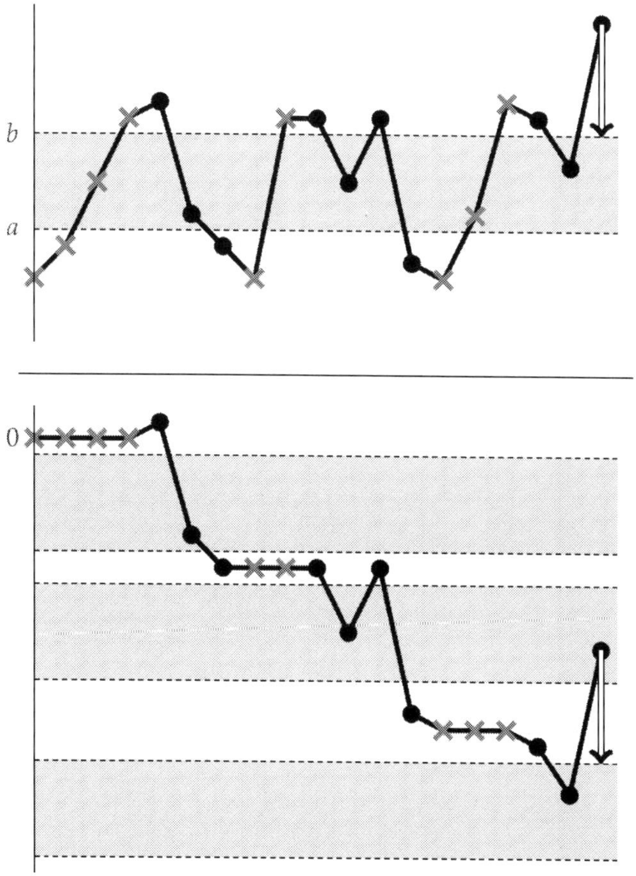

Fig. 40.1. Two full downcrossings of X over the strip $[a,b]\times\{0,1,\dots,N\}$ (top panel) and the mg transform $C\bullet X$ (bottom panel). Our picture shows the worst case of an unsuccessful last crossing attempt which terminates at $X_N > b$ resulting in a overshoot $(X_N - b)^+$.

»wait« and »bet« are shown in Fig. 40.1 as crosses and circles, respectively. A downcrossing corresponds to every »wait-bet-...-bet-wait« sequence. The fact that we start (stop) betting in the step after we went above b (below a) makes our strategy predictable. A bit more formally, we can write

$$C_1 := \begin{cases} 1, & \text{if } X_0 > b; \\ 0, & \text{otherwise;} \end{cases} \qquad C_n := \begin{cases} 1, & \text{if } C_{n-1} = 1 \ \& \ X_{n-1} > a; \\ 1, & \text{if } C_{n-1} = 0 \ \& \ X_{n-1} \leqslant b; \\ 0, & \text{otherwise.} \end{cases}$$

This shows that C_n depends only on C_{n-1} and X_{n-1}, hence (by recursion) on $X_0, \dots, X_{n-1}$. Thus it is predictable.

The martingale transform $C\bullet X_N$ does not change during an upcrossing, but in each step of a downcrossing it adds $(X_n - X_{n-1})$, and in each full downcrossing it decreases by at least $(b-a)$. If, in the end, $X_N \geqslant a$, there was a last, unsuccessful, downcrossing attempt. This might result in an increase of $C\bullet X$ by – in the worst case – $(X_N - b)^+$ (look at the picture, the worst case happens if $X_N > b$). Thus,

$$C \bullet X_N \leqslant -(b-a) D_N[a,b] + (X_N - b)^+.$$

Since $C \bullet X$ is a sub-martingale (Theorem 38.13), we have

$$0 = \mathbb{E} C \bullet X_0 \overset{\text{sub-mg}}{\leqslant} \mathbb{E} C \bullet X_N \leqslant -(b-a)\mathbb{E} D_N[a,b] + \mathbb{E}(X_N - b)^+,$$

and the claim follows by re-arranging this inequality. □

Doob's inequality 40.2 is the basis of all convergence results for martingales – MCT = martingale convergence theorem.

40.3 Theorem ((sub-)mg convergence theorem). *Let $(X_n, \mathscr{F}_n)_{n\in\mathbb{N}_0}$ be a sub-martingale. If $\sup_{n\in\mathbb{N}_0} \mathbb{E} X_n^+ < \infty$, then the a.s. limit $X_\infty := \lim_{n\to\infty} X_n$ exists, is finite and defines an $\mathscr{F}_\infty$-measurable random variable.*

💬 We may define X_∞ for all ω by $X_\infty := \limsup_{n\to\infty} X_n \in \overline{\mathbb{R}}$ or $X_\infty = \lim_{n\to\infty} X_n$ (if the limit exists) and $= 0$ (otherwise).

Proof. In view of our reasoning from the beginning of this chapter, it is enough to show that

$$\forall a, b \in \mathbb{Q},\ a < b : \quad \mathbb{P}(D[a,b] = \infty) = 0.$$

This entails immediately that

$$\mathbb{P}\Bigg(\bigcup_{a,b\in\mathbb{Q}, a<b} \{D[a,b] = \infty\} \Bigg) = 0.$$

Doob's estimate 40.2 shows that

$$\begin{aligned} \mathbb{E} D[a,b] \overset{\text{BL}}{=} \sup_{N\in\mathbb{N}} \mathbb{E} D_N[a,b] &\overset{\S 40.2}{\leqslant} \frac{1}{b-a} \sup_{N\in\mathbb{N}} \mathbb{E}(X_N - b)^+ \\ &\leqslant \frac{1}{b-a} \Big(\sup_{N\in\mathbb{N}} \mathbb{E} X_N^+ + |b| \Big) < \infty; \end{aligned}$$

(we use that $(a-b)^+ \leqslant a^+ + |b|$). Consequently, $\mathbb{P}(D[a,b]=\infty)=0$, i.e.

$$\lim_{n\to\infty} X_n(\omega) \quad \text{exists a.s. in } \overline{\mathbb{R}}.$$

Using Fatou's lemma and the elementary identity $\mathbb{E}|Y| = 2\mathbb{E}Y^+ - \mathbb{E}Y$ gives

$$\begin{aligned} \mathbb{E}\Big(\liminf_{n\to\infty}|X_n|\Big) &\overset{\text{Fatou}}{\leqslant} \liminf_{n\to\infty}\mathbb{E}|X_n| \\ &= \liminf_{n\to\infty}(2\mathbb{E}X_n^+ - \mathbb{E}X_n) \\ &\leqslant \underbrace{2\liminf_{n\to\infty}\mathbb{E}X_n^+}_{\leqslant 2\sup_{n\in\mathbb{N}}\mathbb{E}X_n^+} \underbrace{-\,\mathbb{E}X_1}_{\text{sub-mg}} < \infty. \end{aligned} \tag{40.2}$$

Thus, $|X_\infty| \overset{\text{a.s.}}{=} |\lim_{n\to\infty} X_n| = \lim_{n\to\infty}|X_n| \overset{\text{a.s.}}{=} \liminf_{n\to\infty}|X_n| < \infty$ a.s. □

40.4 Corollary. *Each of the following conditions ensures that* $\lim_{n\to\infty} X_n$ *exists a.s. in* $\mathbb{R}$*:*

a) $(X_n)_{n\in\mathbb{N}_0}$ *is a super-martingale and* $\sup_{n\in\mathbb{N}}\mathbb{E}X_n^- < \infty$*;*
b) $(X_n)_{n\in\mathbb{N}_0}$ *is a positive super-martingale:* $X_n \geqslant 0$*,* $n\in\mathbb{N}$*;*
c) $(X_n)_{n\in\mathbb{N}_0}$ *is a martingale and* $\sup_{n\in\mathbb{N}}\mathbb{E}|X_n| < \infty$*.*

Proof. For a), b) we use Theorem 40.3 for $-X_n$; c) follows immediately from §40.3. □

40.5 Example. ▲ In general, Theorem 40.3 and Corollary 40.4 do **not entail** L^1-convergence, i.e. we do **not have** $X_n \xrightarrow[n\to\infty]{L^1} X_\infty$. Here is the **standard counter-example**:

$$X_n = \xi_1 + \cdots + \xi_n, \quad X_0 = 0, \quad \xi_i \text{ iid} \sim \frac{1}{2}(\delta_1 + \delta_{-1}),$$

is a martingale. Consider the stopping time ✎ $S := \min\{n \mid X_n = -1\}$; the stopped process is $(X_{S\wedge n})_{n\in\mathbb{N}_0}$ is also a martingale and $Y_n = X_{S\wedge n} + 1$ is a positive (super-)martingale. From Corollary 40.4.b) we know that

$$Y_\infty = \lim_{n\to\infty} Y_n = X_S + 1 = 0 \text{ a.s.}$$

On the other hand, $\mathbb{E}Y_n = \mathbb{E}(X_{S\wedge n}+1) = \mathbb{E}(X_0+1) = 1$. If Y_n converges in L^1, there must be an a.s. convergent subsequence, and so $Y_\infty = 0$ is the only possible L^1-limit – but this is ruled out by the fact that $\mathbb{E}Y_n = 1$ for all n.

41 L^2-martingales

We have seen in Chapter 40 that L^1-**boundedness** $\sup_{n\in\mathbb{N}_0}\mathbb{E}|X_n| < \infty$ of a (sub-, super-)martingale gives a.s. convergence $X_\infty = \lim_{n\to\infty} X_n \in \mathbb{R}$, but it does not entail L^1-convergence, see Ex. 40.5. This changes if we have L^2-boundedness.

41.1 Definition. A sub/super-martingale $X = (X_n, \mathscr{F}_n)_{n\in\mathbb{N}_0}$ is **square-integrable** or an **L^2-sub/super-martingale** if

$$\forall n \in \mathbb{N}_0 : \quad \mathbb{E}(|X_n|^2) < \infty.$$

If $\sup_{n\in\mathbb{N}_0} \mathbb{E}(|X_n|^2) < \infty$, X is said to be **$L^2$-bounded.**

Recall that L^2 is a Hilbert space with scalar product $\langle X, Y\rangle_{L^2} := \mathbb{E}(XY)$; this means that we have a geometry in a Hilbert space: Two elements $X, Y \in L^2(\mathscr{A})$ are **orthogonal** $X \perp Y \iff \langle X, Y\rangle_{L^2} = 0$.

41.2 Lemma. *Let $(X_n, \mathscr{F}_n)_{n\in\mathbb{N}_0}$ be a square-integrable martingale. The* **increments** *satisfy*

$$X_n - X_m \perp L^2(\mathscr{F}_m) \quad \textit{for all } m \leqslant n.$$

In particular, $\langle X_k - X_i, X_n - X_m\rangle_{L^2} = 0$ for all $i \leqslant k \leqslant m \leqslant n$, i.e. non-overlapping increments are orthogonal.

Proof. For all $Z \in L^2(\mathscr{F}_m)$

$$\langle X_n - X_m, Z\rangle_{L^2} = \mathbb{E}((X_n - X_m)Z) \overset{\text{tower}}{=} \mathbb{E}\Big(\mathbb{E}[(X_n - X_m)Z \mid \mathscr{F}_m]\Big)$$

$$\overset{\text{pull}}{=} \mathbb{E}\Big(Z\underbrace{\mathbb{E}[X_n - X_m \mid \mathscr{F}_m]}_{=0 \text{ b/o mg}}\Big) = 0.$$

For the second part, note that $Z = X_k - X_i \in L^2(\mathscr{F}_k) \subset L^2(\mathscr{F}_m)$. □

Using a telescoping trick $X_n - X_0 = \sum_{i=1}^n (X_i - X_{i-1})$, Lemma 41.2 yields

41.3 Corollary (Pythagoras' theorem). *Let $(X_n, \mathscr{F}_n)_{n\in\mathbb{N}_0}$ be a square-integrable martingale.*

$$\mathbb{E}X_n^2 = \mathbb{E}X_0^2 + \sum_{i=1}^{n} \mathbb{E}\big((X_i - X_{i-1})^2\big). \tag{41.1}$$

41.4 Theorem (convergence of L^2-mg). *Let $(X_n, \mathscr{F}_n)_{n\in\mathbb{N}_0}$ be a square-integrable martingale.*

$$(X_n)_{n\in\mathbb{N}_0} \textit{ is } L^2\textit{-bounded} \iff \sum_{i=1}^{\infty} \mathbb{E}\big((X_i - X_{i-1})^2\big) < \infty. \tag{41.2}$$

In this case, there is some $X_\infty \in L^2(\mathscr{F}_\infty)$, such that

$$X_n \xrightarrow[n\to\infty]{} X_\infty \quad \textit{a.s. and in } L^2.$$

Amendment: *$(X_n, \mathscr{F}_n)_{n\in\mathbb{N}_0\cup\{\infty\}}$ is a martingale.*

We can use the random variable X_∞ to recover X_m using the filtration: We have $\mathbb{E}(X_\infty \mid \mathscr{F}_m) = X_m$. This means that the martingale $(X_n, \mathscr{F}_n)_{n\in\mathbb{N}_0}$ is a »Lévy martingale«, cf. Example 38.7.c). We say that the rv. X_∞ **closes the martingale to the right.**

Proof. The equivalence (41.2) is exactly Corollary 41.3. An L^2-bounded martingale is also L^1-bounded,

$$\mathbb{E}|X_n| \leqslant \sqrt{\mathbb{E}(|X_n|^2)};$$

therefore the MCT (Corollary 40.4) shows that $X_\infty = \lim_{n\to\infty} X_n$ exists a.s. for some $X_\infty \in L^1(\mathscr{F}_\infty)$.

For $m < n$ we see that

$$\begin{aligned}\mathbb{E}\left[(X_n - X_m)^2\right] &= \mathbb{E}\left[\left(\sum_{i=m+1}^{n}(X_i - X_{i-1})\right)^2\right] \\ &\overset{\S 41.2}{\underset{\S 41.3}{=}} \mathbb{E}\left[\sum_{i=m+1}^{n}(X_i - X_{i-1})^2\right] \xrightarrow[(41.2)]{m,n\to\infty} 0.\end{aligned}$$

So, $(X_n)_{n\in\mathbb{N}_0}$ is an L^2-Cauchy sequence, and we get

$$Z = L^2\text{-}\lim_{n\to\infty} X_n \quad \text{and, for a subsequence,} \quad Z = \lim_{i\to\infty} X_{n(i)} \text{ a.s.}$$

Since a.s. limits are unique, $X_\infty = Z$ a.s., hence $X_n \to X_\infty$ a.s. and in L^2.

In order to see the amendment, note that $X_n \xrightarrow{L^2} X_\infty$ implies $X_n \mathbb{1}_F \xrightarrow{L^1} X_\infty \mathbb{1}_F$ for all $F \in \mathscr{A}$. Fix $m \in \mathbb{N}$ and $F \in \mathscr{F}_m$. We get

$$\int_F X_\infty \, d\mathbb{P} \overset{L^1\text{-conv.}}{=} \lim_{n\to\infty} \int_F X_{n+m} \, d\mathbb{P} \overset{\text{mg}}{=} \lim_{n\to\infty} \int_F X_m \, d\mathbb{P} = \int_F X_m \, d\mathbb{P}$$

So, $\mathbb{E}(X_\infty \mid \mathscr{F}_m) = X_m$, and we see that $(X_n)_{n\in\mathbb{N}_0\cup\{\infty\}}$ is a martingale. □

The following result was used to prove Kolmogorov's three series theorem (Theorem 33.7). We have seen an alternative proof of the first (sufficiency) part of this result in Corollary 33.3, while the necessity part was postponed to this point. The following proof uses only martingales, and does not rely on results from §33, i.e. our argument is not circular.

41.5 Theorem. *Let $(X_i)_{i\in\mathbb{N}} \subset L^2(\mathbb{P})$ be independent random variables such that $\mathbb{E}X_i = 0$ and $\mathbb{V}X_i = \sigma_i^2 > 0$.*

$$\sum_{i=1}^{\infty} \mathbb{V}X_i < \infty \implies \sum_{i=1}^{\infty} X_i \text{ converges a.s.} \tag{41.3}$$

If $\sup_{i\in\mathbb{N}}|X_i| \leqslant \kappa < \infty$ *a.s., the converse assertion holds*

$$\sum_{i=1}^{\infty} \mathbb{V}X_i < \infty \Longleftarrow \sum_{i=1}^{\infty} X_i \text{ converges a.s.} \tag{41.4}$$

Proof. 1° The partial sums $M_n = X_1 + \cdots + X_n$, $M_0 = 0$ are a martingale for the natural filtration $\mathscr{F}_n = \sigma(X_1,\ldots,X_n)$. Because of $\mathbb{E}X_i = 0$, we have $\mathbb{V}X_i = \mathbb{E}X_i^2$, hence

$$\mathbb{E}(M_k - M_{k-1})^2 = \mathbb{E}X_k^2 = \sigma_k^2 \tag{41.5}$$

$$\mathbb{E}M_n^2 \overset{(41.1)}{=} \sum_{i=1}^{n} \sigma_i^2 =: A_n. \tag{41.6}$$

2° Proof of (41.3). Using the martingale convergence theorem (Theorem 41.4) we see

$$\sum_{i=1}^{\infty} \sigma_i^2 < \infty \overset{(41.6)}{\Longrightarrow} \sup_{n\in\mathbb{N}} \mathbb{E}M_n^2 < \infty \overset{\text{MCT}}{\Longrightarrow} (M_n)_{n\in\mathbb{N}} \text{ converges a.s.}$$

3° Proof of (41.4). We have

$$\begin{aligned}\mathbb{E}(M_n^2 \mid \mathscr{F}_{n-1}) &= \mathbb{E}((M_{n-1} + X_n)^2 \mid \mathscr{F}_{n-1})\\ &= \mathbb{E}(M_{n-1}^2 + 2M_{n-1}X_n + X_n^2 \mid \mathscr{F}_{n-1})\\ &= M_{n-1}^2 + 2M_{n-1}\mathbb{E}X_n + \mathbb{E}X_n^2 \;=\; M_{n-1}^2 + \mathbb{E}X_n^2.\end{aligned}$$

Thus, $Y_n := M_n^2 - A_n$, $A_n := \sum_{i=1}^n \mathbb{E}X_i^2$ is a martingale.(1)

Fix $c \in (0,\infty)$ and define stopping times ✎

$$\tau := \tau_c := \inf\{k \mid |M_k| > c\}.$$

By Theorem 39.9.d) – use $S = \tau \wedge m$, $T = \tau \wedge n$ – we see that the stopped process $Y^\tau := (Y_{\tau\wedge n}, \mathscr{F}_{\tau\wedge n})_{n\in\mathbb{N}_0}$ is a mg. Thus,

$$0 = \mathbb{E}Y_0^\tau = \mathbb{E}Y_n^\tau = \mathbb{E}M_{n\wedge\tau}^2 - \mathbb{E}A_{n\wedge\tau}. \tag{41.7}$$

But we also have

$$|M_{n\wedge\tau}| \leqslant \begin{Bmatrix} c, & \text{if } \tau > n\\ |M_\tau|, & \text{if } \tau \leqslant n\end{Bmatrix} = c\mathbb{1}_{\{\tau>n\}} + |M_\tau|\mathbb{1}_{\{n\geqslant\tau\}},$$

and

$$|M_{n\wedge\tau}| \leqslant c\mathbb{1}_{\{\tau>n\}} + \underbrace{|M_\tau - M_{\tau-1}|}_{=|X_\tau|\leqslant\kappa}\underbrace{\mathbb{1}_{\{\tau\leqslant n\}}}_{\leqslant 1} + \underbrace{|M_{\tau-1}|}_{\leqslant c}\mathbb{1}_{\{\tau\leqslant n\}} \leqslant c + \kappa.$$

If we use this estimate in (41.7) and rearrange the resulting inequality, we get $\mathbb{E}A_{n\wedge\tau} \leqslant (c+\kappa)^2$ for all $n \in \mathbb{N}$.

(1) In other words $A = \langle X\rangle$ is the compensator, cf. page 254.

By assumption $\sum_{i=1}^\infty X_i$ converges a.s. Thus, $\mathbb{P}(\tau := \tau_{c_0} = \infty) > 0$ for some constant c_0. So

$$\mathbb{P}(\tau = \infty) \cdot A_n = \mathbb{E}(\mathbb{1}_{\{\tau=\infty\}} A_n) = \mathbb{E}(\mathbb{1}_{\{\tau=\infty\}} A_{n\wedge\tau}) \leqslant \mathbb{E} A_{n\wedge\tau} \leqslant (c_0+\kappa)^2.$$

Divide by $\mathbb{P}(\tau = \infty) > 0$ to get

$$\sum_{i=1}^\infty \sigma_i^2 = \sup_{n\in\mathbb{N}} A_n \leqslant \frac{(c_0+\kappa)^2}{\mathbb{P}(\tau=\infty)} < \infty. \qquad \square$$

42 Uniform integrability

Let us collect the martingale convergence results which we have obtained so far. If a martingale $X = (X_n, \mathscr{F}_n)_{n\in\mathbb{N}_0}$ is...

- L^1-bounded $\implies X_n \xrightarrow{\text{a.s.}} X_\infty$ — MCT Theorem 40.3;
- L^1-bounded $\not\Rightarrow X_n \xrightarrow{L^1} X_\infty$ — Example 40.5;
- L^2-bounded $\implies X_n \xrightarrow{\text{a.s.}, L^2, L^1} X_\infty$ — L^2-MCT Theorem 41.4.

The reason as to why L^2-bounded martingales behave better than L^1-bounded martingales is **compactness**. The key notion in this connection is **uniform integrability**.

42.1 Definition. Let I be an arbitrary index set. The family of random variables $X_\lambda : \Omega \to \mathbb{R}$, $\lambda \in I$, is **uniformly integrable (ui)**, if

$$\lim_{R\to\infty} \sup_{\lambda\in I} \int_{\{|X_\lambda|>R\}} |X_\lambda| \, d\mathbb{P} = 0. \tag{42.1}$$

⚠ Uniform integrability is a condition that prevents that the X_λ shift too much mass to infinity. A typical »bad guy« is $I = \mathbb{N}$, $X_n \sim \delta_n$ i.e. $\mathbb{P}(X_n = n) = 1$. The »limit« X_∞ has only mass at infinity.

42.2 Lemma. *Let $X \in L^1(\mathscr{A})$ be an integrable random variable.*

a) *$\{X\}$ is uniformly integrable*

b) *$\{\mathbb{E}(X \mid \mathscr{F}) \mid \mathscr{F} \subset \mathscr{A}$ is a σ-algebra$\}$ is uniformly integrable*

Proof. a) Since X is integrable, $\mathbb{P}(|X| = \infty) = 0$, and so $\lim_{R\to\infty} \mathbb{1}_{\{|X|>R\}} = 0$ a.s. Using $|X|$ as integrable majorant, we can use dominated convergence to get

$$\int_{\{|X|>R\}} |X| \, d\mathbb{P} = \int \underbrace{\mathbb{1}_{\{|X|>R\}} \cdot |X|}_{\leqslant |X| \in L^1} \, d\mathbb{P} \xrightarrow[R\to\infty]{\text{DCT}} 0.$$

b) Let $\mathscr{F} \subset \mathscr{A}$ be a σ-algebra and set $Y := \mathbb{E}(X \mid \mathscr{F})$. From the triangle inequality we get

$$|Y| = |\mathbb{E}(X \mid \mathscr{F})| \leqslant \mathbb{E}(|X| \mid \mathscr{F}).$$

Since $\{|Y| > R\} \in \mathscr{F}$ we have for every $R > 0$

$$\begin{aligned}\int_{\{|Y|>R\}} |Y|\,d\mathbb{P} &\leqslant \int_{\{|Y|>R\}} \mathbb{E}(|X| \mid \mathscr{F})\,d\mathbb{P} \overset{\text{def}}{=} \int_{\{|Y|>R\}} |X|\,d\mathbb{P}\\ &= \int_{\{|Y|>R\}\cap\{|X|>R/2\}} |X|\,d\mathbb{P} + \int_{\{|Y|>R\}\cap\{|X|\leqslant R/2\}} |X|\,d\mathbb{P}\end{aligned}$$

and if we use that $|X| \leqslant \frac{1}{2}R < \frac{1}{2}|Y|$ on the set $\{|Y| > R\} \cap \{|X| \leqslant R/2\}$,

$$\begin{aligned}&\leqslant \int_{\{|Y|>R\}\cap\{|X|>R/2\}} |X|\,d\mathbb{P} + \int_{\{|Y|>R\}\cap\{|X|\leqslant R/2\}} \frac{1}{2}|Y|\,d\mathbb{P}\\ &\leqslant \int_{\{|X|>R/2\}} |X|\,d\mathbb{P} + \frac{1}{2}\int_{\{|Y|>R\}} |Y|\,d\mathbb{P}.\end{aligned}$$

Rearranging this inequality we get

$$\int_{\{|Y|>R\}} |Y|\,d\mathbb{P} \leqslant 2\int_{\{|X|>R/2\}} |X|\,d\mathbb{P} \xrightarrow[R\to\infty]{\text{DCT}} 0$$

uniformly for all $Y = \mathbb{E}(X \mid \mathscr{F})$ where $\mathscr{F} \subset \mathscr{A}$ is a σ-algebra. □

Usually, it is difficult to check uniform integrability directly from the definition. Here are some useful criteria.

42.3 Theorem. $(X_\lambda)_{\lambda\in I}$ *be a family of random variables. Eeach of the following conditions ensures that* $(X_\lambda)_{\lambda\in I}$ *is uniformly integrable:*

a) **L^p-boundedness** ($p > 1$): $\sup_{\lambda\in I} \mathbb{E}(|X_\lambda|^p) < \infty$.
b) **integrable majorant:** $\exists Y \in L^1 \quad \forall \lambda \in I: \quad |X_\lambda| \leqslant Y$.

Proof. a) Fix $R > 0$.

$$\begin{aligned}\int_{\{|X_\lambda|>R\}} |X_\lambda|\,d\mathbb{P} \leqslant \int_{\{|X_\lambda|>R\}} |X_\lambda| \frac{|X_\lambda|^{p-1}}{R^{p-1}}\,d\mathbb{P} &= R^{1-p}\int_{\{|X_\lambda|>R\}} |X_\lambda|^p\,d\mathbb{P}\\ &\leqslant \underbrace{R^{1-p}}_{\to 0}\underbrace{\sup_{\lambda\in I}\mathbb{E}(|X_\lambda|^p)}_{<\infty}\end{aligned}$$

uniformly for all $\lambda \in I$.

b) Since $|X_\lambda| \leqslant Y$ we get that $\{|X_\lambda| > R\} \subset \{Y > R\}$, and so

$$\int_{\{|X_\lambda|>R\}} |X_\lambda|\,d\mathbb{P} \leqslant \int_{\{|X_\lambda|>R\}} Y\,d\mathbb{P} \leqslant \int_{\{Y>R\}} Y\,d\mathbb{P} \xrightarrow[R\to\infty]{\text{DCT}} 0$$

uniformly for all $\lambda \in I$. □

42.4 Example. ⚠⚠Theorem 42.3.a) does not hold for $p = 1$. L^1-boundedness $\sup_{\lambda\in I} \mathbb{E}|X_\lambda| < \infty$ does not imply unform integrability.

Here is a **counterexample**: Consider on $((0,1], \mathscr{B}(0,1], dt)$ the random variables $X_n(t) := n\mathbb{1}_{(0,1/n]}(t)$, $n \in \mathbb{N}$. Then

- $(X_n)_{n\in\mathbb{N}}$ is not ui;
- $\sup_{n\in\mathbb{N}} \mathbb{E}|X_n| = 1 < \infty$;
- $\lim_{n\to\infty} X_n = 0$ a.s., but $\mathbb{E}X_n \not\to 0$.

The condition a) in Theorem 42.3 is a convexity condition and one has the following theorem. The proof of sufficiency is similar to §42.3.a), a proof of the converse can be found in [MIMS].

42.5 Theorem* (de la Vallée–Poussin). *A family $(X_\lambda)_{\lambda\in I}$ of random variables is uniformly integrable if, and only if, for some positive, increasing convex function $\Phi : [0,\infty) \to [0,\infty)$ such that $\lim_{R\to\infty} \Phi(R)/R$ the following condition holds:*

$$\sup_{\lambda\in I} \mathbb{E}\Phi(|X_\lambda|) < \infty.$$

We will now use uniform integrability to give an »optimal« version of the dominated convergence theorem (cf. Theorem 11.2 and 14.12). Recall the definition of **convergence in probability** or **$\mathbb{P}$-convergence**:

$$X_n \xrightarrow[n\to\infty]{\mathbb{P}} X \iff \forall \epsilon > 0 : \quad \lim_{n\to\infty} \mathbb{P}(|X_n - X| > \epsilon) = 0.$$

Moreover, see Theorem 29.5, $X_n \xrightarrow{\mathbb{P}} X$ also implies that $f(X_n) \xrightarrow{L^1} f(X)$ for all $f \in C_b(\mathbb{R})$.

The following theorem is an »optimal« version of the DCT.

42.6 Theorem (Vitali). *Let $(X_i)_{i\in\mathbb{N}} \subset L^p$, $p \geqslant 1$, be random variables such that $X_i \xrightarrow{\mathbb{P}} X$ for some random variable X. The following are equivalent*

a) *$(|X_i|^p)_{i\in\mathbb{N}}$ is uniformly integrable*

b) *$X_i \xrightarrow{L^p} X$; in particular: $X \in L^p$.*

c) *$\mathbb{E}|X_i|^p \xrightarrow{i\to\infty} \mathbb{E}|X|^p < \infty$.*

💬 Theorem 42.6 contains the dominated convergence theorem ✍.

Proof. a)⇒b): The function $\chi_n(x) := (-n) \vee (x \wedge n)$ is continuous and bounded (cf. Fig. 42.1) and satisfies

$$|x - \chi_n(x)| \leqslant |x|\mathbb{1}_{\{|x|\geqslant n\}}.$$

Use $(a+b+c)^p \leqslant 3^{p-1}(a^p+b^p+c^p)$ (✍– Hölder) to get for all $i,k \in \mathbb{N}$

$$
\begin{aligned}
|X_i - X_k|^p &\leqslant 3^{p-1}\Big(|X_i - \chi_n(X_i)|^p + |\chi_n(X_i) - \chi_n(X_k)|^p + |\chi_n(X_k) - X_k|^p\Big) \\
&\leqslant 3^{p-1}\Big(|X_i|^p \cdot \mathbb{1}_{\{|X_i|\geqslant n\}} + |\chi_n(X_i) - \chi_n(X_k)|^p + |X_k|^p \cdot \mathbb{1}_{\{|X_k|\geqslant n\}}\Big).
\end{aligned}
$$

If we take expectations on both sides, we see

$$
\begin{aligned}
\mathbb{E}|X_i - X_k|^p &\leqslant 2\cdot 3^{p-1} \sup_{l\in\mathbb{N}} \mathbb{E}\Big(|X_l|^p \cdot \mathbb{1}_{\{|X_l|\geqslant n\}}\Big) + 3^{p-1}\mathbb{E}|\chi_n(X_i) - \chi_n(X_k)|^p \\
&\xrightarrow[\S 29.5]{i,k\to\infty} \underbrace{2\cdot 3^{p-1} \sup_{l\in\mathbb{N}} \mathbb{E}\Big(|X_l|^p \cdot \mathbb{1}_{\{|X_l|\geqslant n\}}\Big)}_{\to 0 \text{ as } n\to\infty \text{ b/o ui}} + \underbrace{\mathbb{E}|\chi_n(X) - \chi_n(X)|^p}_{=0} \\
&\xrightarrow[n\to\infty]{} 0.
\end{aligned}
$$

This shows that $(X_i)_{i\in\mathbb{N}}$ is an L^p Cauchy sequence. Since L^p is complete (Theorem 14.10), there is a rv Y such that $X_i \to Y$ in L^p, hence in $\mathbb{P}$ (Remark 29.4). Since $\mathbb{P}$-limits are unique (Remark 29.3), we see that $X = Y$ a.s.

b)⇒c): By the (lower) triangle inequality for $\|X\|_{L^p} = (\mathbb{E}|X|^p)^{1/p}$ we get

$$
\big|\|X_i\|_{L^p} - \|X\|_{L^p}\big| \leqslant \|X_i - X\|_{L^p} \xrightarrow[i\to\infty]{\text{b)}} 0.
$$

c)⇒a): Fix $S > 0$, $\epsilon > 0$ and construct a continuous function ψ_S such that

$$
\mathbb{1}_{[-S+1,S-1]} \leqslant \psi_S \leqslant \mathbb{1}_{[-S,S]} \quad \text{and set} \quad f_S(x) := |x|\psi_S(x),
$$

as it is shown in Fig. 42.2.

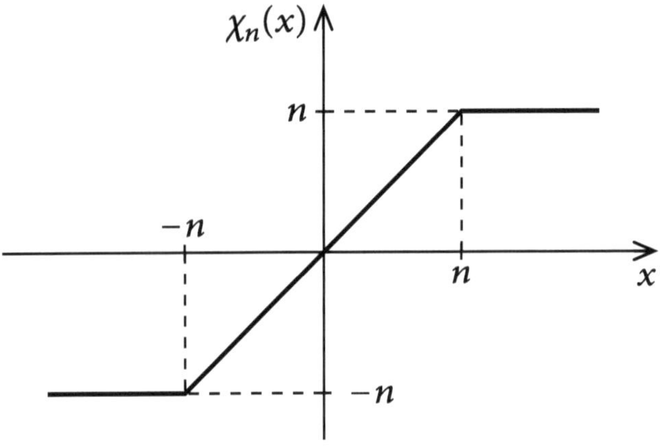

Fig. 42.1. The function $\chi_n(x) := (-n) \vee (x \wedge n)$.

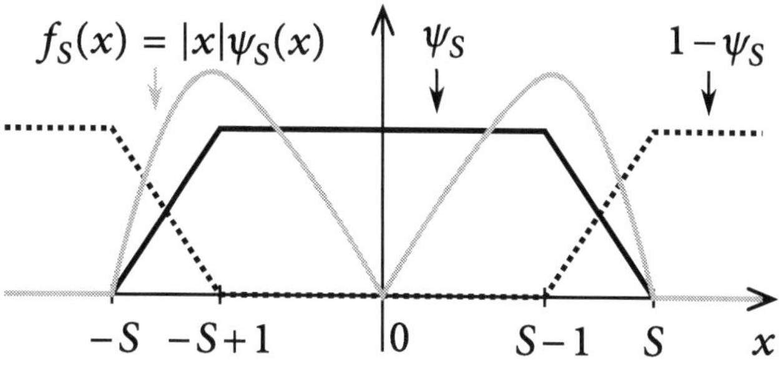

Fig. 42.2. The functions $\psi_S(x)$ and $f_S(x) = |x|\psi_S(x)$.

1° Since $f_S(x) \leqslant |x|$, we can use dominated convergence and get

$$\lim_{S\to\infty} \int f_S^p(X)\,d\mathbb{P} = \mathbb{E}|X|^p \implies 0 \leqslant \mathbb{E}|X|^p - \int f_S^p(X)\,d\mathbb{P} \leqslant \epsilon$$

for all $S \geqslant S_\epsilon$ and some $S_\epsilon > 0$.

2° Set $S = S_\epsilon$. f_S^p is bounded and continuous, and Theorem 29.5 shows

$$\lim_{i\to\infty} \int f_S^p(X_i)\,d\mathbb{P} = \int f_S^p(X)\,d\mathbb{P}.$$

Since, by assumption, $\lim_{i\to\infty} \mathbb{E}|X_i|^p = \mathbb{E}|X|^p$, we also have

$$\mathbb{E}|X_i|^p - \int f_S^p(X_i)\,d\mathbb{P} \leqslant \epsilon + \mathbb{E}|X|^p - \int f_S^p(X)\,d\mathbb{P} \leqslant 2\epsilon$$

for all $i \geqslant N_\epsilon$ and some $N_\epsilon = N(\epsilon, S_\epsilon) \in \mathbb{N}$.

3° The previous steps show that for all $R > S = S_\epsilon$ and all $i \geqslant N_\epsilon$ we have

$$\int_{\{|X_i|>R\}} |X_i|^p\,d\mathbb{P} \leqslant \int (1 - \psi_S^p(X_i))|X_i|^p\,d\mathbb{P} = \mathbb{E}|X_i|^p - \int f_S^p(X_i)\,d\mathbb{P} \leqslant 2\epsilon,$$

hence

$$\limsup_{R\to\infty} \sup_{i\geqslant N_\epsilon} \int_{\{|X_i|>R\}} |X_i|^p\,d\mathbb{P} \leqslant 2\epsilon.$$

4° Since $X_1, X_2, \dots, X_{N_\epsilon - 1} \in L^1(\mathbb{P})$, we get with dominated convergence

$$\lim_{R\to\infty} \max_{i<N_\epsilon} \int_{\{|X_i|>R\}} |X_i|^p\,d\mathbb{P} = 0 \overset{3^\circ}{\implies} \limsup_{R\to\infty} \sup_{i\in\mathbb{N}} \int_{\{|X_i|>R\}} |X_i|^p\,d\mathbb{P} \leqslant 2\epsilon.$$

5° If we combine the steps 3° and 4° , we get

$$\sup_{i\in\mathbb{N}} \int_{\{|X_i|>R\}} |X_i|^p\,d\mathbb{P} \leqslant 2\epsilon \implies \limsup_{R\to\infty} \sup_{i\in\mathbb{N}} \int_{\{|X_i|>R\}} |X_i|^p\,d\mathbb{P} = 0$$

since $\epsilon > 0$ is arbitrary. Thus, we have established a). □

43 Uniformly integrable martingales

We will now combine Vitali's convergence theorem 42.6 and the MCT.

43.1 Theorem (MCT for ui mg). *Let* $(X_n, \mathscr{F}_n)_{n\in\mathbb{N}_0} \subset L^p(\mathbb{P})$, $p \geqslant 1$, *be a [sub-]martingale. The following are equivalent:*

a) $(|X_n|^p)_{n\in\mathbb{N}_0}$ *is uniformly integrable;*

b) $X_n \xrightarrow[n\to\infty]{a.e., L^p} X_\infty$ *for some* $X_\infty \in L^p(\mathscr{F}_\infty)$.

In this case, we have

c) $(X_n, \mathscr{F}_n)_{n\in\mathbb{N}_0\cup\{\infty\}}$ *is a [sub-]martingale and* $\lim_{n\to\infty} \mathbb{E}X_n = \mathbb{E}X_\infty$.
In particular: $X_n \leqslant \mathbb{E}(X_\infty \mid \mathscr{F}_n)$, $n \in \mathbb{N}_0$ *(for a mg one has »=«).*

Proof. a)⇔b): Note that

$$(|X_n|^p)_n \text{ ui} \implies \sup_{n\in\mathbb{N}} \mathbb{E}|X_n|^p < \infty \implies \sup_{n\in\mathbb{N}} \mathbb{E}|X_n| < \infty,$$

$$X_n \xrightarrow[n\to\infty]{L^p} X \implies \sup_{n\in\mathbb{N}} \mathbb{E}|X_n|^p < \infty \implies \sup_{n\in\mathbb{N}} \mathbb{E}|X_n| < \infty,$$

and the MCT (Theorem 40.3) shows that there is some $\mathscr{F}_\infty$-measurable random variable X_∞ such that $X_\infty = \lim_{n\to\infty} X_n$ a.s.

Since a.s. convergence gives $\mathbb{P}$-convergence, the equivalence a)⇔b) follows from Vitali's Theorem 42.6.

b)⇒c): Fix $n \in \mathbb{N}$. For all $F \in \mathscr{F}_n$ and $N > n$

$$\int_F X_n \, d\mathbb{P} \overset{\text{sub-mg}}{\leqslant} \int_F X_N \, d\mathbb{P} \xrightarrow[N\to\infty]{} \int_F X_\infty \, d\mathbb{P},$$

with equality in the mg case. The convergence follows from $X_N \xrightarrow{L^1} X_\infty$ and

$$|\mathbb{E}[X_N \mathbb{1}_F] - \mathbb{E}[X_\infty \mathbb{1}_F]| \leqslant \mathbb{E}|X_N \mathbb{1}_F - X_\infty \mathbb{1}_F| \leqslant \mathbb{E}|X_N - X_\infty| \xrightarrow[N\to\infty]{} 0.$$

This gives the first part of c); pick $F = \Omega$ to get $\lim_{n\to\infty} \mathbb{E}X_n = \mathbb{E}X_\infty$. □

43.2 Corollary. *If* $(X_n, \mathscr{F}_n)_{n\in\mathbb{N}_0\cup\{\infty\}}$ *is a martingale, then it is ui.*

Proof. Use $X_n = \mathbb{E}(X_\infty \mid \mathscr{F}_n)$ and Lemma 42.2.b). □

43.3 Corollary. *If* $(X_n, \mathscr{F}_n)_{n\in\mathbb{N}_0\cup\{\infty\}}$ *is a sub-mg with* $\mathbb{E}X_n \to \mathbb{E}X_\infty$, *then it is ui*

Proof. From $\lim_{n\to\infty} \mathbb{E}X_n = \mathbb{E}X_\infty$ we get

$$\forall \epsilon > 0 \quad \exists m = m(\epsilon) \in \mathbb{N} \quad \forall n \geqslant m: \quad \mathbb{E}X_n \geqslant \mathbb{E}X_\infty - \epsilon. \tag{43.1}$$

Fix $R > 0$. For all $n \geqslant m = m(\epsilon)$

$$\begin{aligned}
\int_{\{|X_n|>R\}} |X_n|\,d\mathbb{P} &= \int_{\{X_n<-R\}} (-X_n)\,d\mathbb{P} + \int_{\{X_n>R\}} X_n\,d\mathbb{P} \\
&= \int_{\{X_n\geqslant -R\}} X_n\,d\mathbb{P} - \mathbb{E}X_n + \int_{\{X_n>R\}} X_n\,d\mathbb{P} \\
&\overset{\text{sub-mg}}{\underset{(43.1)}{\leqslant}} \int_{\{X_n\geqslant -R\}} X_\infty\,d\mathbb{P} - \mathbb{E}X_\infty + \epsilon + \int_{\{X_n>R\}} X_\infty\,d\mathbb{P} \\
&\leqslant \int_{\{|X_n|>R\}} |X_\infty|\,d\mathbb{P} + \epsilon \\
&= \int_{\{|X_n|>R\}\cap\{|X_\infty|>R/2\}} |X_\infty|\,d\mathbb{P} + \int_{\{|X_n|>R\}\cap\{|X_\infty|\leqslant R/2\}} |X_\infty|\,d\mathbb{P} + \epsilon \\
&\leqslant \int_{\{|X_\infty|>R/2\}} |X_\infty|\,d\mathbb{P} + \frac{1}{2}\int_{\{|X_n|>R\}} |X_n|\,d\mathbb{P} + \epsilon.
\end{aligned}$$

In the last estimate we use that the inequality $|X_\infty| \leqslant R/2 < |X_n|/2$ holds on the set $\{|X_n| > R\} \cap \{|X_\infty| \leqslant R/2\}$. If we rearrange this, we see

$$\sup_{n\geqslant m(\epsilon)} \int_{\{|X_n|>R\}} |X_n|\,d\mathbb{P} \leqslant 2\int_{\{|X_\infty|>R/2\}} |X_\infty|\,d\mathbb{P} + 2\epsilon.$$

By dominated convergence, we find some $R = R(\epsilon)$ such that

$$\int_{\{|X_\infty|>R/2\}} |X_\infty|\,d\mathbb{P} \leqslant \epsilon \quad \text{and} \quad \forall 0 \leqslant i < m(\epsilon): \quad \int_{\{|X_i|>R\}} |X_i|\,d\mathbb{P} \leqslant \epsilon.$$

Thus,

$$\sup_{n\in\mathbb{N}} \int_{\{|X_n|>R\}} |X_n|\,d\mathbb{P} \leqslant 4\epsilon \text{ for all } R \geqslant R_\epsilon.$$ □

Ui and optional stopping

Uniform integrability allows us to avoid the boundedness of the stopping times in the optional stopping theorem (Theorem 39.9).

43.4 Theorem (optional stopping). *Let $(X_n, \mathscr{F}_n)_{n\in\mathbb{N}_0}$ be a ui sub-mg and S, T stopping times with $\mathbb{P}(S \leqslant T \leqslant \infty) = 1$.*[(1)] *Then $X_S, X_T \in L^1$ and*

$$\mathbb{E}(X_T \mid \mathscr{F}_S) \geqslant X_S. \tag{43.2}$$

If $S = 0$, we have $\mathbb{E}X_T \geqslant \mathbb{E}X_0$. If $(X_n)_{n\in\mathbb{N}_0}$ is a ui mg, then we have everywhere »=« .

Proof. Note that $X_\infty = \lim_{n\to\infty} X_n$ exists, and that $(X_n)_{n\in\mathbb{N}_0\cup\{\infty\}}$ (Theorem 43.1) and $(X_n^+)_{n\in\mathbb{N}_0}$ are sub-mg.

From Theorem 39.9, we know that $(X_{n\wedge T}^+, \mathscr{F}_n)_{n\in\mathbb{N}_0}$ is a sub-mg. Moreover,

$$\mathbb{E}|X_{n\wedge T}| = 2\mathbb{E}X_{n\wedge T}^+ - \mathbb{E}X_{n\wedge T} \underset{\S 39.9}{\overset{\text{sub-mg}}{\leqslant}} 2\mathbb{E}X_n^+ - \mathbb{E}X_0 \leqslant 3\sup_{n\in\mathbb{N}_0} \mathbb{E}|X_n| \overset{\text{ui}}{<} \infty.$$

Using Fatou's lemma

$$\mathbb{E}|X_T| = \mathbb{E}\left(\liminf_{n\to\infty} |X_{n\wedge T}|\right) \leqslant \liminf_{n\to\infty} \mathbb{E}\left(|X_{n\wedge T}|\right) \leqslant 3\sup_{n\in\mathbb{N}_0} \mathbb{E}|X_n| < \infty,$$

and $X_T \in L^1(\mathbb{P})$ follows; similarly, $X_S \in L^1(\mathbb{P})$. The estimates

$$|X_{n\wedge T}| = |X_n \mathbb{1}_{\{T\geqslant n\}} + X_T \mathbb{1}_{\{T<n\}}| \leqslant |X_n| + |X_T| \quad \text{and} \quad |X_{n\wedge S}| \leqslant |X_n| + |X_S|$$

together with the uniform integrability of $(X_n)_{n\in\mathbb{N}_0}$ and $X_S, X_T \in L^1$ show that $(X_{n\wedge T})_{n\in\mathbb{N}_0}$ and $(X_{n\wedge S})_{n\in\mathbb{N}_0}$ are ui ✎.[2] Thus, by the ui-MCT, Theorem 43.1,

$$X_{n\wedge T} \xrightarrow{L^1} X_T \quad \text{and} \quad X_{n\wedge S} \xrightarrow{L^1} X_S. \tag{43.3}$$

Finally, for all $F \in \mathscr{F}_S$ and $k \leqslant n \in \mathbb{N}_0$ we have

$$F \cap \{S \leqslant k\} \in \mathscr{F}_k \subset \mathscr{F}_n \quad \text{and} \quad F \cap \{S \leqslant k\} \in \mathscr{F}_S.$$

Moreover, $F \cap \{S \leqslant k\} \in \mathscr{F}_S \cap \mathscr{F}_n \overset{\S 39.8.d)}{=} \mathscr{F}_{n\wedge S}$, and we can use Thm. 39.9.d):

$$\begin{aligned}\int_{F\cap\{S\leqslant k\}} X_T \, d\mathbb{P} &\overset{(43.3)}{=} \lim_{n\to\infty} \int_{F\cap\{S\leqslant k\}} X_{n\wedge T} \, d\mathbb{P} \overset{\S 39.9.d)}{\geqslant} \lim_{n\to\infty} \int_{F\cap\{S\leqslant k\}} X_{n\wedge S} \, d\mathbb{P} \\ &\overset{(43.3)}{=} \int_{F\cap\{S\leqslant k\}} X_S \, d\mathbb{P}.\end{aligned}$$

Letting $k \to \infty$ yields $\int_F X_T \, d\mathbb{P} \geqslant \int_F X_S \, d\mathbb{P}$ for all $F \in \mathscr{F}_S$. □

Ui and backwards martingales

A **backwards (sub-)martingale** is is a (sub-)mg whose index set »goes to the left«:

$$(\mathscr{F}_\nu)_{\nu\in-\mathbb{N}_0} \text{ filtration, i.e. } \mathscr{F}_{-\infty} := \bigcap_{\mu\in-\mathbb{N}_0} \mathscr{F}_\mu \subset \mathscr{F}_{\nu-1} \subset \mathscr{F}_\nu \subset \mathscr{F}_0, \quad \nu \in -\mathbb{N},$$

an adapted process $X_\nu \in L^1(\mathscr{F}_\nu)$ is a sub-mg, if

$$\forall \nu \in -\mathbb{N} : \mathbb{E}(X_\nu \mid \mathscr{F}_{\nu-1}) \geqslant X_{\nu-1} \iff \forall \nu, \mu \in -\mathbb{N}, \mu \leqslant \nu : \mathbb{E}(X_\nu \mid \mathscr{F}_\mu) \geqslant X_\mu.$$

(1) We will see that the ui assumption guarantees the existence of X_∞, i.e. X_S, X_T are everywhere defined.

(2) Hint: If $|X_n| \leqslant Y_n + Z$, then we have $\int_{|X_n|>R} |X_n| \, d\mathbb{P} \leqslant \int_{Y_n+Z>R} (Y_n + Z) \, d\mathbb{P} = \int_{Y_n+Z>R,\, Y_n\leqslant Z} (Y_n + Z) \, d\mathbb{P} + \int_{Y_n+Z>R,\, Y_n>Z} (Y_n + Z) \, d\mathbb{P} \leqslant \int_{2Z>R} (2Z) \, d\mathbb{P} + \int_{2Y_n>R} (2Y_n) \, d\mathbb{P}$.

⚠ This is exactly our Definition 38.4, so the new notion »backwards« is actually not needed – but it is often used in the literature

⚠ Often, backwards sub-mg $(X_\nu, \mathscr{F}_\nu)_{\nu\in-\mathbb{N}_0}$ appear in the following guise

$$Y_n := X_\nu \quad \text{and} \quad \mathscr{G}_n := \mathscr{F}_\nu \quad \text{mit} \quad n = |\nu|,\ \nu \in -\mathbb{N}_0,$$

where $(\mathscr{G}_n)_{n\in\mathbb{N}_0}$ is decreasing

$$\mathscr{G}_0 \supset \mathscr{G}_1 \supset \mathscr{G}_2 \supset \cdots \supset \mathscr{G}_\infty = \bigcap_{n\in\mathbb{N}_0} \mathscr{G}_n$$

and the random variables Y_n satisfy

$$Y_{n+1} \leqslant \mathbb{E}(Y_n \mid \mathscr{G}_{n+1}).$$

Typical example (as it appears in applications): $(X_t, \mathscr{F}_t)_{t\geqslant 0}$ is a continuous-time martingale[3] and $t_n \downarrow t$. Then $(X_{t_n}, \mathscr{F}_{t_n})_n = (Y_n, \mathscr{G}_n)_n$ is a backwards mg.

43.5 Theorem (MCT for backwards mg). *Let $(X_\nu, \mathscr{F}_\nu)_{\nu\in-\mathbb{N}_0}$ be a backwards mg. The limit $X_{-\infty} = \lim_{\nu\to-\infty} X_\nu$ exists a.s. and in L^1.*
In particular, $X_{-\infty} = \mathbb{E}(X_\nu \mid \mathscr{F}_{-\infty})$ a.s. for all $\nu \in -\mathbb{N}_0$.

⚠ The »backwards« condition implies L^1-boundedness (and much more).

Proof. 1° Since a backwards mg has a right end-point X_0, we see that

$$\forall \nu \in -\mathbb{N}_0 : \mathbb{E}X_\nu^+ \leqslant \mathbb{E}X_0^+ \implies \sup_{\nu\in-\mathbb{N}_0} \mathbb{E}X_\nu^+ \leqslant \mathbb{E}X_0^+ < \infty,$$

i.e. the MCT (Theorem 40.3) applies and shows $X_\nu \to X_{-\infty}$ a.s.

2° Since $X_\nu = \mathbb{E}(X_0 \mid \mathscr{F}_\nu)$, any backwards mg is uniformly integrable, and so we have L^1-convergence from Vitali's Theorem 42.6.

3° Fix $F \in \mathscr{F}_{-\infty}$ and note that $F \in \mathscr{F}_\mu$ for all $\mu \in -\mathbb{N}_0$. For any fixed $\nu \in -\mathbb{N}_0$

$$\int_F X_{-\infty}\, d\mathbb{P} \overset{L^1\text{-conv.}}{=} \lim_{\mu\to-\infty} \int_F X_\mu\, d\mathbb{P} \underset{\text{mg}}{\overset{\mathscr{F}_\mu\subset\mathscr{F}_\nu}{=}} \lim_{\mu\to-\infty} \int_F X_\nu\, d\mathbb{P} = \int_F X_\nu\, d\mathbb{P}. \qquad \square$$

Uniform integrability of a backwards sub-martingale is a bit more difficult.

43.6 Theorem (MCT for backwards sub-mg). *Let $(X_\nu, \mathscr{F}_\nu)_{\nu\in-\mathbb{N}_0}$ be a backwards sub-martingale.*

a) $X_{-\infty} = \lim_{\nu\to-\infty} X_\nu \in \mathbb{R}$ *exists a.s.*

b) *If* $\sup_{\nu\in-\mathbb{N}_0} \mathbb{E}|X_\nu| < \infty$, *then*

$$X_{-\infty} = L^1\text{-}\lim_{\nu\to-\infty} X_\nu \quad \textit{and} \quad X_{-\infty} \leqslant \mathbb{E}(X_\nu \mid \mathscr{F}_{-\infty}) \ \textit{for all} \ \nu \in -\mathbb{N}_0.$$

(3) The definition (§38.4) of a (sub-, super-)mg easily extends to the index set $I = [0,\infty)$, see also Def. 46.1 below.

Proof. a) and the second part of b) can be shown as in Theorem 43.5.

For the L^1-convergence we use Vitali's Theorem 42.6. Thus, we have to show that $(X_\nu)_{\nu\in-\mathbb{N}_0}$ is uniformly integrable.

Note that $\mathbb{E}X_\mu$ decreases as $\mu\downarrow-\infty$. Thus,

$$\sup_{\nu\in-\mathbb{N}_0}\mathbb{E}|X_\nu|<\infty \iff \lim_{\mu\to-\infty}\mathbb{E}X_\mu>-\infty. \tag{43.4}$$

The direction »$\Rightarrow$« is clear, the direction »$\Leftarrow$« follows from

$$|\mathbb{E}X_\nu|\leqslant\mathbb{E}|X_\nu|=2\mathbb{E}X_\nu^+-\mathbb{E}X_\nu\overset{\text{sub-mg}}{\leqslant}2\mathbb{E}X_0^+-\mathbb{E}X_\nu\leqslant2\mathbb{E}X_0^+-\lim_{\mu\to-\infty}\mathbb{E}X_\mu$$

as $\mu\mapsto\mathbb{E}X_\mu$ is decreasing. Thus, we get

$$\forall\epsilon>0\quad\exists\mu=\mu_\epsilon\quad\forall\nu\leqslant\mu:\quad\mathbb{E}X_\nu\geqslant\mathbb{E}X_\mu-\epsilon,$$

and the uniform integrability of $(X_\nu)_{\nu\in-\mathbb{N}_0}$ follows almost literally as in Corollary 43.3 (where X_μ takes the role of X_∞). □

44 Basic inequalities

In this chapter we discuss the basic inequalities for discrete (sub/super-) martingales $(X_n,\mathscr{F}_n)_{n\in\mathbb{N}}$.

44.1 Lemma. *Let $(X_n,\mathscr{F}_n)_{n\in\mathbb{N}_0}$ be a sub-mg. For all $n\in\mathbb{N}_0$ and $r>0$*

$$\mathbb{P}\Big(\max_{i\leqslant n}X_i\geqslant r\Big)\leqslant\frac{1}{r}\int_{\{\max_{i\leqslant n}X_i\geqslant r\}}X_n^+\,d\mathbb{P}\leqslant\frac{1}{r}\mathbb{E}(X_n^+). \tag{44.1}$$

Proof. Define a stopping time

$$T:=\min\{0\leqslant i\leqslant n\mid X_i\geqslant r\},\quad\min\emptyset:=\infty.$$

From the very definition we see that for $r>0$

$$\max_{i\leqslant n}X_i(\omega)\geqslant r\iff T(\omega)\leqslant n\iff T(\omega)\leqslant n\ \&\ X_T^+(\omega)=X_{n\wedge T}^+(\omega)\geqslant r.$$

By the Markov inequality

$$\begin{aligned}\mathbb{P}\Big(\max_{i\leqslant n}X_i\geqslant r\Big)&=\mathbb{P}\big(T\leqslant n,\ X_{n\wedge T}^+\geqslant r\big)\\&\leqslant\frac{1}{r}\int_{\{T\leqslant n\}}X_{n\wedge T}^+\,d\mathbb{P}\\&\overset{\S}{\leqslant}\frac{1}{r}\int_{\{T\leqslant n\}}X_n^+\,d\mathbb{P}\\&\leqslant\frac{1}{r}\int X_n^+\,d\mathbb{P}.\end{aligned}$$

In the step marked with § we use that $(X_n^+)_{n\in\mathbb{N}_0}$ is a sub-martingale (Thm. 38.8.d) in combination with the optional stopping theorem (Theorem 39.9.d) where we use $S \rightsquigarrow n \wedge T$ and $T \rightsquigarrow n$) and the fact that $\{T \leqslant n\} \in \mathscr{F}_{T\wedge n}$. □

44.2 Corollary. *Let $(X_n, \mathscr{F}_n)_{n\in\mathbb{N}_0} \subset L^p$, $p \geqslant 1$, be a martingale or a positive sub-martingale. For all $n \in \mathbb{N}_0$ and $r > 0$*

$$\mathbb{P}\left(\max_{i\leqslant n}|X_i| \geqslant r\right) \leqslant \frac{1}{r^p}\int_{\{\max_{i\leqslant n}|X_i|\geqslant r\}} |X_n|^p \,d\mathbb{P} \tag{44.2}$$

$$\leqslant \frac{1}{r^p}\mathbb{E}(|X_n|^p). \tag{44.3}$$

Proof. If $(X_n)_{n\in\mathbb{N}_0}$ is a martingale (or a positive sub-martingale), the conditional Jensen inequality shows that $(|X_n|^p)_{n\in\mathbb{N}_0}$ is a sub-martingale. Now we can apply Lemma 44.1. □

The following inequality was used by Kolmogorov in connection with his proof of the SLLN.

44.3 Corollary (Kolmogorov's inequality). *Let $(\xi_n)_{n\in\mathbb{N}} \subset L^2(\mathbb{P})$ be independent rv such that $\mathbb{E}\xi_n = 0$ and $\sigma_n^2 = \mathbb{V}\xi_n$. Set $X_n := \xi_1 + \dots + \xi_n$, $X_0 := 0$. Then*

$$\mathbb{P}\left(\max_{i\leqslant n}|X_i| \geqslant r\right) \leqslant \frac{1}{r^2}\sum_{i=1}^{n}\sigma_i^2 \quad \forall n \in \mathbb{N}_0,\ r > 0.$$

Proof. X_n is a mg for $\mathscr{F}_n := \sigma(\xi_1, \dots, \xi_n)$ and we have

$$\mathbb{E}X_n^2 = \mathbb{V}X_n = \sum_{i=1}^{n}\mathbb{V}\xi_i = \sum_{i=1}^{n}\sigma_i^2.$$ □

In analysis, estimates of the form (44.3) are often called »weak-type $(p,1)$-inequalities« since on the right-hand side there is an L^p-norm and on the left side a distribution function; in contrast, an inequality $\|\dots\|_q \leqslant C\|\dots\|_p$ would be a »strong (p,q)-inequality«. Let us see how to get from a weak $(1,p)$ a strong (p,p)-inequality.

44.4 Lemma. *Let $X, Y \geqslant 0$ be random variables such that*

$$\mathbb{P}(X \geqslant r) \leqslant \frac{1}{r}\int_{\{X\geqslant r\}} Y \,d\mathbb{P} \quad \forall r > 0. \tag{44.4}$$

This implies that for all $p > 1$ and $\frac{1}{p} + \frac{1}{q} = 1$

$$\mathbb{E}(X^p) \leqslant q^p\,\mathbb{E}(Y^p). \tag{44.5}$$

Proof. Let $p^{-1} + q^{-1} = 1$ and assume that $\mathbb{E}(X^p) < \infty$. By the layer-cake formula

(Theorem 16.5) and the elementary equality $\mathbb{1}_{\{X\geqslant r\}}(\omega) = \mathbb{1}_{[0,X(\omega)]}(r)$ we get

$$\begin{aligned}
\mathbb{E}(X^p) &= \int_0^\infty p r^{p-1}\,\mathbb{P}(X > r)\,dr \\
&\overset{(44.4)}{\leqslant} \int_0^\infty p r^{p-1}\left(\frac{1}{r}\int \mathbb{1}_{\{X\geqslant r\}}\cdot Y\,d\mathbb{P}\right)dr \\
&\overset{\text{Tonelli}}{\leqslant} \iint_0^\infty p r^{p-2}\,\mathbb{1}_{[0,X]}(r)\cdot Y\,dr\,d\mathbb{P} \\
&= \iint_0^X p r^{p-2}\,dr\,Y\,d\mathbb{P} \\
&= \frac{p}{p-1}\int X^{p-1}\,Y\,d\mathbb{P} \\
&\overset{\text{Hölder}}{\leqslant} \frac{p}{p-1}\left[\mathbb{E}\left(Y^p\right)\right]^{1/p}\left[\mathbb{E}\left(X^{q(p-1)}\right)\right]^{1/q} \\
&\leqslant q\,[\mathbb{E}(Y^p)]^{1/p}\,[\mathbb{E}(X^p)]^{1-1/p}.
\end{aligned}$$

We can now rearrange this inequality to get $[\mathbb{E}(X^p)]^{1/p} \leqslant q\,[\mathbb{E}(Y^p)]^{1/p}$.

If $\mathbb{E}(X^p) = \infty$, then we use $X \wedge n$, $n \in \mathbb{N}$, in place of X. The estimate (44.4) remains true for $X \wedge n$:

$$\{X \wedge n \geqslant r\} = \begin{cases} \{X \geqslant r\}, & r \leqslant n, \\ \emptyset, & r > n. \end{cases}$$

Therefore, our previous calculation yields

$$\mathbb{E}(X^p) \overset{\text{BL}}{=} \sup_{n\in\mathbb{N}} \mathbb{E}[(X \wedge n)^p] \leqslant \sup_{n\in\mathbb{N}} q^p\,\mathbb{E}(Y^p) = q^p\,\mathbb{E}(Y^p). \qquad \square$$

Now we combine Lemma 44.1 und 44.4. We use the following **standard notation**

$$X_n^* := \sup_{0\leqslant i\leqslant n} |X_i| \quad \text{and} \quad X^* := \sup_{n\in\mathbb{N}_0} X_n^* = \sup_{i\in\mathbb{N}_0} |X_i|.$$

44.5 Theorem (Doob; L^p maximal inequality). *Let $(X_n, \mathscr{F}_n)_{n\in\mathbb{N}_0} \subset L^p$, $p > 1$, be a martingale or a positive sub-martingale, and assume that it is L^p-bounded: $\sup_{n\in\mathbb{N}_0} \mathbb{E}(|X_n|^p) < \infty$.*

a) $X^* \in L^p$;

b) $X_\infty = \lim_{n\to\infty} X_n$ *exists in* L^p *and a.s.;*

c) $\mathbb{E}(X^{*p}) \leqslant q^p \sup_{n\in\mathbb{N}_0} \mathbb{E}(|X_n|^p) = q^p\,\mathbb{E}(|X_\infty|^p)$.

Proof. From Corollary 44.2 with $p = 1$ we see that

$$\mathbb{P}(X_n^* \geqslant r) \leqslant \frac{1}{r}\int_{\{X_n^*\geqslant r\}} |X_n|\,d\mathbb{P} \quad \forall n \in \mathbb{N},\ r > 0,$$

and by Lemma 44.4

$$\mathbb{E}(X_n^{*p}) \leqslant q^p\, \mathbb{E}(|X_n|^p) \leqslant q^p \sup_{i\in\mathbb{N}_0} \mathbb{E}(|X_i|^p).$$

A routine application of BL yields

$$\mathbb{E}(X^{*p}) = \sup_{n\in\mathbb{N}_0} \mathbb{E}(X_n^{*p}) \leqslant q^p \sup_{n\in\mathbb{N}_0} \mathbb{E}(|X_n|^p) < \infty$$

from which we infer that $X^* \in L^p$ or, equivalently $(X^*)^p \in L^1$. Since L^p-boundedness implies L^1-boundedness, we can use the MCT (Corollary 40.4) and see that

$$\exists X_\infty = \lim_{n\to\infty} X_n \quad \text{a.e.}$$

Moreover,

$$|X_\infty - X_n|^p \leqslant (|X_\infty| + |X_n|)^p \leqslant (2X^*)^p \in L^1,$$

and by dominated convergence

$$|X_\infty - X_n|^p \xrightarrow[n\to\infty]{} 0 \quad \text{in } L^1 \iff X_n \xrightarrow[n\to\infty]{} X_\infty \quad \text{in } L^p.$$

In particular, $\lim_{n\to\infty} \mathbb{E}(X_n^p) = \mathbb{E}(X_\infty^p)$. Since $(|X_n|^p)_{n\geqslant 0}$ is a sub-mg

$$\mathbb{E}(X_n^p) \leqslant \mathbb{E}(X_{n+1}^p) \leqslant \cdots \implies \mathbb{E}X_\infty^p = \lim_{n\to\infty} \mathbb{E}X_n^p = \sup_{n\in\mathbb{N}_0} \mathbb{E}X_n^p. \qquad \square$$

Azuma's inequality

This is a so-called »concentration of measure« inequality for super-martingales with bounded increments.

44.6 Theorem (Azuma 1967). *Let $(X_n, \mathscr{F}_n)_{n\in\mathbb{N}_0}$ be a super-martingale such that $|X_n - X_{n-1}| \leqslant c_n$ for a sequence $(c_n)_{n\in\mathbb{N}} \subset [0,\infty)$. Set $d_n^2 := c_1^2 + \cdots + c_n^2$. Then*

$$\mathbb{P}(X_n - X_0 \geqslant t) \leqslant \exp\left[-\frac{t^2}{2d_n^2}\right], \quad t \geqslant 0,\ n\in\mathbb{N}. \tag{44.6}$$

If $(X_n)_{n\in\mathbb{N}_0}$ is a martingale, one also has

$$\mathbb{P}(|X_n - X_0| \geqslant t) \leqslant 2\exp\left[-\frac{t^2}{2d_n^2}\right], \quad t \geqslant 0,\ n\in\mathbb{N}. \tag{44.7}$$

Proof. The function $e_t(x) := e^{tx}$, $x\in\mathbb{R}$, $t\geqslant 0$, is convex, see Fig. 44.1.

$$e^{tx} \leqslant \frac{e^{tc} - e^{-tc}}{2c}\, x + \frac{1}{2}\left(e^{tc} + e^{-tc}\right), \quad \forall x \in [-c, c].$$

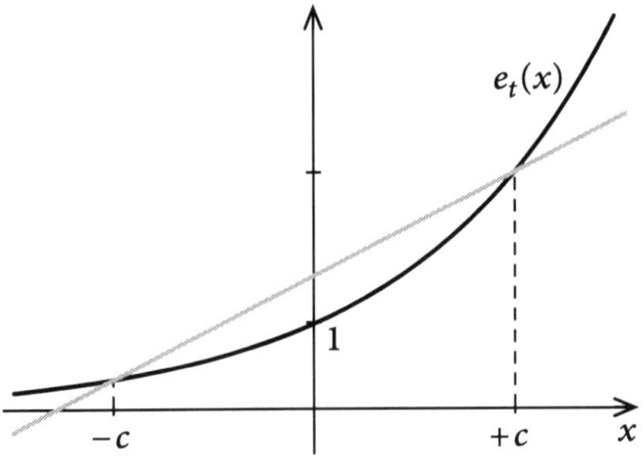

Fig. 44.1. The graph of $x \mapsto e^{tx}$ is in the interval $[-c, c]$ below the line connecting $(-c, e^{-tc})$ and (c, e^{tc})

If we set $x = (X_n - X_{n-1})$ and $c = c_n$, and take conditional expectations $\mathbb{E}(\cdots \mid \mathscr{F}_{n-1})$ on both sides, we get

$$\mathbb{E}\left(e^{t(X_n - X_{n-1})} \mid \mathscr{F}_{n-1}\right) \leqslant \underbrace{\frac{e^{tc_n} - e^{-tc_n}}{2c_n}}_{\geqslant 0} \underbrace{\mathbb{E}(X_n - X_{n-1} \mid \mathscr{F}_{n-1})}_{\leqslant 0 \text{ b/o super-mg}} + \frac{e^{tc_n} + e^{-tc_n}}{2}$$
$$\leqslant \frac{e^{tc_n} + e^{-tc_n}}{2} = \cosh(tc_n).$$

Observe that the cosh satisfies

$$\frac{1}{2}(e^{-y} + e^{y}) = \sum_{k=0}^{\infty} \frac{y^{2k}}{(2k)!} \leqslant \sum_{k=0}^{\infty} \frac{y^{2k}}{2^k\, k!} = e^{y^2/2}.$$

This implies that

$$\mathbb{E}\left(e^{t(X_n - X_{n-1})} \mid \mathscr{F}_{n-1}\right) \leqslant e^{t^2 c_n^2/2}.$$

Thus, with a telescoping sum argument and the tower property

$$\begin{aligned}
\mathbb{E}e^{t(X_n - X_0)} &= \mathbb{E}\left(\prod_{i=1}^{n} e^{t(X_i - X_{i-1})}\right) \\
&\overset{\text{tower}}{\underset{\text{pull}}{=}} \mathbb{E}\left(\prod_{i=1}^{n-1} e^{t(X_i - X_{i-1})} \mathbb{E}\left[e^{t(X_n - X_{n-1})} \mid \mathscr{F}_{n-1}\right]\right) \\
&\leqslant \mathbb{E}\left(\prod_{i=1}^{n-1} e^{t(X_i - X_{i-1})}\right) e^{t^2 c_n^2/2} \\
&\leqslant \cdots \leqslant e^{t^2 d_n^2/2}.
\end{aligned}$$

Using the Markov inequality for $t, \xi \geqslant 0$ gives

$$\begin{aligned}\mathbb{P}(X_n - X_0 \geqslant t) = \mathbb{P}\left(e^{\xi(X_n - X_0)} \geqslant e^{t\xi}\right) &\leqslant e^{-t\xi}\mathbb{E}e^{\xi(X_n - X_0)} \\ &\leqslant e^{-t\xi + \xi^2 d_n^2/2} \qquad (44.8)\\ &= e^{-\frac{1}{2}t^2/d_n^2 + \frac{1}{2}(\xi d_n - t/d_n)^2}.\end{aligned}$$

The right-hand side becomes minimal at $\xi = t/d_n^2$, and (44.6) follows.

If $(X_n)_{n\in\mathbb{N}_0}$ is a martingale, then (44.6) holds for the super-mg $(-X_n)_{n\in\mathbb{N}_0}$, i.e.

$$\mathbb{P}(X_n - X_0 \leqslant -t) \leqslant \exp\left[-\frac{t^2}{2d_n^2}\right], \quad t \geqslant 0,\ n \in \mathbb{N}.$$

Add this to (44.6) to obtain (44.7). □

45 Martingale proofs of some classical results

In this chapter we will prove some classical results of probability theory using martingale techniques.

Kolmogorov's 0-1*–law*

From Theorem 43.1 we know that all uniformly integrable martingales are of the form $X_n = \mathbb{E}(X \mid \mathscr{F}_n)$. We will now study the convergence »along a filtration«.

45.1 Theorem (Lévy 1935; Lévy's upward theorem). *Let* $X \in L^1(\mathscr{A})$, $(\mathscr{F}_n)_{n\in\mathbb{N}_0}$ *a filtration and* $\mathscr{F}_\infty = \sigma(\mathscr{F}_n, n \in \mathbb{N}_0)$. *Then*

$$\lim_{n\to\infty} \mathbb{E}(X \mid \mathscr{F}_n) = \mathbb{E}(X \mid \mathscr{F}_\infty) \quad \textit{a.s. and in } L^1. \qquad (45.1)$$

Proof. Define $X_n := \mathbb{E}(X \mid \mathscr{F}_n)$, $n \in \mathbb{N}_0$, and $X_\infty := \mathbb{E}(X \mid \mathscr{F}_\infty)$. We know from Lemma 42.2.b) that $(X_n, \mathscr{F}_n)_{n\in\mathbb{N}_0}$ is a ui martingale. By Theorem 43.1, the limit $Y := \lim_{n\to\infty} X_n \in L^1(\mathscr{F}_\infty)$ exists a.s. and in L^1.

We will show that $X_\infty = Y$ a.s.

$$\mathbb{E}(Y \mid \mathscr{F}_n) \overset{\S 43.1}{=} X_n \overset{\text{def}}{=} \mathbb{E}(X \mid \mathscr{F}_n) \overset{\text{tower}}{=} \mathbb{E}\Big(\underbrace{\mathbb{E}[X \mid \mathscr{F}_\infty]}_{=X_\infty} \mid \mathscr{F}_n\Big).$$

Consequently,

$$\int_F Y\,d\mathbb{P} = \int_F X_\infty\,d\mathbb{P} \quad \forall F \in \mathscr{F}_n,\ n \in \mathbb{N}_0, \ \text{ hence } \forall F \in \bigcup_{n\in\mathbb{N}_0} \mathscr{F}_n.$$

Since $\bigcup_{n\in\mathbb{N}_0} \mathscr{F}_n$ is a $\cap$-stable generator of $\mathscr{F}_\infty$, we get (Lemma 34.9)

$$Y = \mathbb{E}(X_\infty \mid \mathscr{F}_\infty) \overset{\mathscr{F}_\infty\text{-mble}}{=} X_\infty \quad \text{a.s.}$$ □

We will now apply Theorem 45.1 to the random variable $X = \mathbb{1}_F$ where $F \in \mathscr{F}_\infty$.

45.2 Corollary (Lévy 1935; 0-1–law). *Assume that $(\mathscr{F}_n)_{n\in\mathbb{N}_0}$ is a filtration. If $F \in \mathscr{F}_\infty = \sigma(\mathscr{F}_n, n \in \mathbb{N}_0)$, then*

$$\lim_{n\to\infty} \mathbb{P}(F \mid \mathscr{F}_n) = \mathbb{1}_F \quad a.s.$$

Note that $\mathbb{1}_F \in \{0, 1\}$ which explains the name »0-1–law«.

We can recover Kolomogorov's zero-one law (Theorem 31.8) as a special case of Lévy's zero-one law.

45.3 Corollary (Kolmogorov's 0-1–law). *Let $(\xi_n)_{n\in\mathbb{N}}$ be independent rv and set $\mathscr{T}_n := \sigma(\xi_n, \xi_{n+1}, \dots)$ and $\mathscr{T}_\infty := \bigcap_{n\in\mathbb{N}} \mathscr{T}_n$. If $F \in \mathscr{T}_\infty$, then $\mathbb{P}(F) = 0$ or $\mathbb{P}(F) = 1$.*

Proof. Since $(\xi_n)_{n\in\mathbb{N}}$ is a sequence of independent random variables, the natural filtration $\mathscr{F}_n := \sigma(\xi_1, \dots, \xi_n)$ is independent of the tail σ-algebra $\mathscr{T}_{n+1}$, hence $\mathscr{F}_n \perp\!\!\!\perp \mathscr{T}_\infty$. This means that $\mathbb{P}(F \mid \mathscr{F}_n) = \mathbb{P}(F)$ for all $F \in \mathscr{T}_\infty$ and $n \in \mathbb{N}$. Since $\mathscr{T}_\infty \subset \mathscr{F}_\infty$, the claim follows from

$$\mathbb{P}(F) \overset{\perp\!\!\!\perp}{=} \lim_{n\to\infty} \mathbb{P}(F \mid \mathscr{F}_n) \overset{\S 45.2}{=} \mathbb{1}_F \quad \text{a.s.} \qquad \square$$

Kolmogorov's L^1*-SLLN*

Let $\mathscr{G}_0 \supset \mathscr{G}_{-1} \supset \mathscr{G}_{-2} \supset \cdots \supset \mathscr{G}_{-\infty} := \bigcap_{\nu\in-\mathbb{N}_0} \mathscr{G}_\nu$ be a downwards directed filtration and $X \in L^1(\mathscr{A})$. Obviously, $\mathbb{E}(X \mid \mathscr{G}_\nu)$, $\nu \in -\mathbb{N}_0$ is a backwards martingale.

45.4 Theorem (Lévy's downward theorem). *Let $X \in L^1(\mathscr{A})$ and $(\mathscr{G}_\nu)_{\nu\in-\mathbb{N}_0}$ be a filtration. Then*

$$\lim_{\nu\to-\infty} \mathbb{E}(X \mid \mathscr{G}_\nu) = \mathbb{E}(X \mid \mathscr{G}_{-\infty}) \quad \textit{a.s. and in } L^1.$$

Proof. We apply the backwards MCT 43.5 to $X_\nu := \mathbb{E}(X \mid \mathscr{G}_\nu)$. There is an $\mathscr{G}_{-\infty}$-measurable random variable such that

$$X_{-\infty} = \lim_{\nu\to-\infty} X_\nu \quad \text{a.s. and in } L^1.$$

Moreover,

$$X_{-\infty} \overset{\S 43.5}{=} \mathbb{E}(X_0 \mid \mathscr{G}_{-\infty}) \overset{\text{def}}{=} \mathbb{E}\Big(\mathbb{E}[X \mid \mathscr{G}_0] \mid \mathscr{G}_{-\infty}\Big) \overset{\text{tower}}{=} \mathbb{E}\Big(X \mid \mathscr{G}_{-\infty}\Big). \qquad \square$$

Lévy's downward theorem yields an elegant proof of Kolmogorov's L^1-SLLN – with the bonus of L^1-convergence.

45.5 Theorem (Kolmogorov 1933; L^1-SLLN). *Let $(\xi_n)_{n\in\mathbb{N}} \subset L^1(\mathscr{A})$ be iid ran-*

dom variables. Then

$$\lim_{n\to\infty}\frac{1}{n}(\xi_1+\cdots+\xi_n)=\mathbb{E}\xi_1 \quad \textit{a.s. and in } \mathrm{L}^1$$

Proof. Set $S_n := \xi_1+\cdots+\xi_n$. We have

$$S_n=\mathbb{E}(S_n \mid S_n)=\sum_{i=1}^{n}\mathbb{E}(\xi_i \mid S_n)=n\,\mathbb{E}(\xi_1 \mid S_n)$$

and for $\mathscr{F}_{-n}:=\sigma(S_n,\xi_{n+1},\xi_{n+2}\dots)$

$$X_{-n}=\frac{1}{n}S_n=\mathbb{E}(\xi_1 \mid S_n)=\mathbb{E}(\xi_1 \mid S_n,\xi_{n+1},\xi_{n+2},\dots)=\mathbb{E}(X_{-1} \mid \mathscr{F}_{-n}).$$

This shows that $(X_{-n},\mathscr{F}_{-n})_{n\in\mathbb{N}}=(X_\nu,\mathscr{F}_\nu)_{\nu\in-\mathbb{N}}$ is a backwards martingale.

By Theorem 45.4

$$L=\lim_{n\to\infty}X_{-n}=\lim_{n\to\infty}\frac{S_n}{n}=\lim_{n\to\infty}\mathbb{E}(\xi_1 \mid \mathscr{F}_{-n})$$

in L^1 and a.s. Since

$$L=\underbrace{\lim_{n\to\infty}\frac{\xi_1+\cdots+\xi_k}{n}}_{=0,\ k\text{ fixed}}+\lim_{n\to\infty}\frac{\xi_{k+1}+\xi_{k+2}+\cdots+\xi_n}{n},$$

we see that the rv L is for every $k\in\mathbb{N}$ measurable w.r.t. $\mathscr{T}_{k+1}:=\sigma(\xi_{k+1},\xi_{k+2},\dots)$, i.e. it is $\mathscr{T}_\infty=\bigcap_{k\in\mathbb{N}}\mathscr{T}_k$-measurable.

By Kolmogorov's 0-1–law $\mathbb{P}(L=x)\in\{0,1\}$, i.e. L is a.s. constant:

$$L\overset{\text{a.s. cst.}}{=}\mathbb{E}L=\mathbb{E}\xi_1 \quad \text{a.s.}$$

□

The Hewitt–Savage 0-1*–law*

A sequence $(\xi_n)_{n\in\mathbb{N}}$ of real-valued random variables is said to be **exchangeable** if for every **finite** permutation $\pi:\{1,\dots,k\}\to\{1,\dots,k\}$ the random variables

$$\mathbb{X}=(\xi_1,\dots,\xi_k,\xi_{k+1},\dots) \quad\text{and}\quad \pi\mathbb{X}=(\xi_{\pi(1)},\dots,\xi_{\pi(k)},\xi_{k+1},\dots)$$

have the same probability distribution. For example, iid sequences are exchangeable. Define the canonical and the tail σ-algebras

$$\mathscr{F}_n:=\sigma(\xi_1,\dots,\xi_n) \quad\text{and}\quad \mathscr{T}_n:=\sigma(\xi_n,\xi_{n+1},\dots),$$

along with $\mathscr{F}_\infty:=\sigma\left(\bigcup_{n=1}^\infty \mathscr{F}_n\right)$ and $\mathscr{T}_\infty:=\bigcap_{n=1}^\infty \mathscr{T}_n$.

Using the factorization lemma (Lemma 7.16) we know that every set $F\in\mathscr{F}_\infty$ is of the form $F=\{\mathbb{X}\in B\}$ for a suitable Borel set $B\in\mathscr{B}(\mathbb{R}^{\mathbb{N}})=\mathscr{B}(\mathbb{R})^{\otimes\mathbb{N}}$. The set F is said to be **symmetric** if

$$\mathbb{1}_F=\mathbb{1}_B(\mathbb{X})=\mathbb{1}_B(\pi\mathbb{X}) \quad \text{for all finite permutations.}$$

It is not difficult to see that the family of symmetric set is a σ-algebra ✎.

45.6 Theorem (Hewitt–Savage). *Let $\mathbb{X} = (\xi_n)_{n\in\mathbb{N}}$ be exchangeable and independent, and $\mathscr{F}_n$, $\mathscr{T}_n$ be as above. If $F \in \mathscr{F}_\infty$ is a symmetric set, then $\mathbb{P}(F) \in \{0,1\}$.*

Proof. Since $F \in \mathscr{F}_\infty$, there is some $B \in \mathscr{B}(\mathbb{R})^{\otimes\mathbb{N}}$ such that $\mathbb{1}_F = \mathbb{1}_B(\mathbb{X})$.

Define $\mathbb{X}_n := (\xi_1,\dots,\xi_n)$ and $\mathbb{X}'_n := (\xi_{n+1},\dots,\xi_{2n})$. Since F is symmetric and $\mathbb{X}$ is exchangeable, we see that

$$\begin{aligned}
&\mathbb{E}(\mathbb{1}_F \mid \mathbb{X}_n) - \mathbb{1}_F \\
&= \mathbb{E}(\mathbb{1}_B(\mathbb{X}_n, \mathbb{X}'_n, \xi_{2n+1},\dots) \mid \mathbb{X}_n) - \mathbb{1}_B(\mathbb{X}_n, \mathbb{X}'_n, \xi_{2n+1},\dots) \\
&\sim \mathbb{E}(\mathbb{1}_B(\mathbb{X}'_n, \mathbb{X}_n, \xi_{2n+1},\dots) \mid \mathbb{X}'_n) - \mathbb{1}_B(\mathbb{X}'_n, \mathbb{X}_n, \xi_{2n+1},\dots) \\
&= \mathbb{E}(\mathbb{1}_F \mid \mathbb{X}'_n) - \mathbb{1}_F.
\end{aligned} \tag{45.2}$$

We have

$$\begin{aligned}
\mathbb{E}\big|\mathbb{1}_F - \mathbb{E}(\mathbb{1}_F \mid \mathscr{T}_n)\big| \quad &\leqslant \quad \mathbb{E}\big|\mathbb{1}_F - \mathbb{E}(\mathbb{1}_F \mid \mathbb{X}'_n)\big| + \mathbb{E}\big|\mathbb{E}(\mathbb{1}_F \mid \mathbb{X}'_n) - \mathbb{E}(\mathbb{1}_F \mid \mathscr{T}_n)\big| \\
&\overset{(45.2)}{=} \quad \mathbb{E}\big|\mathbb{1}_F - \mathbb{E}(\mathbb{1}_F \mid \mathbb{X}_n)\big| + \mathbb{E}\big|\mathbb{E}(\mathbb{1}_F \mid \mathbb{X}'_n) - \mathbb{E}(\mathbb{1}_F \mid \mathscr{T}_n)\big|
\end{aligned}$$

and by the tower property, observing that $\sigma(\mathbb{X}'_n) \subset \mathscr{T}_n$,

$$\begin{aligned}
&= \quad \mathbb{E}\big|\mathbb{1}_F - \mathbb{E}(\mathbb{1}_F \mid \mathbb{X}_n)\big| + \mathbb{E}\Big|\mathbb{E}\Big[\mathbb{E}(\mathbb{1}_F \mid \mathbb{X}'_n) - \mathbb{1}_F \mid \mathscr{T}_n\Big]\Big| \\
&\overset{\S 34.6.\text{d})}{\leqslant} \quad \mathbb{E}\big|\mathbb{1}_F - \mathbb{E}(\mathbb{1}_F \mid \mathbb{X}_n)\big| + \mathbb{E}\big|\mathbb{E}(\mathbb{1}_F \mid \mathbb{X}'_n) - \mathbb{1}_F\big| \\
&\overset{(45.2)}{=} \quad 2\mathbb{E}\big|\mathbb{E}(\mathbb{1}_F \mid \mathbb{X}_n) - \mathbb{1}_F\big|.
\end{aligned}$$

The left-hand side tends to $\mathbb{E}|\mathbb{1}_F - \mathbb{E}(\mathbb{1}_F \mid \mathscr{T}_n)| \to \mathbb{E}|\mathbb{1}_F - \mathbb{E}(\mathbb{1}_F \mid \mathscr{T}_\infty)|$ by Lévy's downward theorem (Theorem 45.4), while the right-hand side tends to zero by Lévy's upwards theorem (Theorem 45.1). Thus, $\mathbb{1}_F = \mathbb{E}(\mathbb{1}_F \mid \mathscr{T}_\infty)$ and, by Kolmogorov's 0-1–law (Corollary 45.3) $\mathbb{E}(\mathbb{1}_F \mid \mathscr{T}_\infty) = \mathbb{E}\mathbb{1}_F$ finishing the proof. □

💬 Replacing in the proof of the SLLN (Theorem 45.5) the Kolmogorov 0-1–law by the Hewitt–Savage 0-1–law, we see that the proof still holds for exchangeable independent random variables.

Convergence of sums of independent random variables

Recall the definition and some equivalent characterizations of convergence in distribution:

$$X_n \xrightarrow[n\to\infty]{d} X \overset{\text{def}}{\iff} \forall f \in C_b(\mathbb{R}) : \lim_{n\to\infty} \mathbb{E} f(X_n) = \mathbb{E} f(X) \tag{45.3}$$

$$\iff \forall \theta \in \mathbb{R} : \lim_{n\to\infty} \mathbb{E} e^{i\theta X_n} = \mathbb{E} e^{i\theta X} \tag{45.4}$$

$$\iff \forall \epsilon > 0 : \lim_{n\to\infty} \sup_{|\theta|\leqslant\epsilon} \left| \mathbb{E} e^{i\theta X_n} - \mathbb{E} e^{i\theta X} \right| = 0 \tag{45.5}$$

$$\iff \forall x \in \mathbb{R}, \mathbb{P}(X = x) = 0 : \lim_{n\to\infty} \mathbb{P}(X_n \leqslant x) = \mathbb{P}(X \leqslant x). \tag{45.6}$$

Our aim is the following result (compare this with Theorem 33.2).

45.7 Theorem (Lévy). *Let $(\xi_n)_{n\in\mathbb{N}}$ be independent random variables. For the series $\sum_{n=1}^{\infty} \xi_n$ it holds that*

$$\textit{a.s. convergence} \iff \mathbb{P}\textit{-convergence} \iff d\textit{-convergence.}$$

The implications »⇒« are clear; in order to show »d-convergence ⇒ a.s.-convergence« we need a lemma from analysis.

45.8 Lemma. *Let $(x_n)_{n\in\mathbb{N}} \subset \mathbb{R}$ be a sequence of numbers such that $\lim_{n\to\infty} e^{i\theta x_n}$ exists for all $\theta \in (-\epsilon, \epsilon)$. Then $\lim_{n\to\infty} x_n \in \mathbb{R}$ exists.*

Proof. Set $L(\theta) := \lim_{n\to\infty} e^{i\theta x_n}$. For all $r < \epsilon$

$$\begin{aligned} \frac{1}{r}\int_{-r}^{r} L(\theta)\,d\theta \overset{\text{DCT}}{=} \lim_{n\to\infty} \frac{1}{r}\int_{-r}^{r} e^{i\theta x_n}\,d\theta &= \lim_{n\to\infty} \frac{e^{irx_n} - e^{-irx_n}}{irx_n} \\ &= \lim_{n\to\infty} \frac{2\sin(rx_n)}{rx_n}. \end{aligned} \tag{45.7}$$

Since $L(0) = 1$, the integral on the lhS is for small $r < \epsilon$ not zero, thus $(x_n)_{n\in\mathbb{N}}$ is bounded. Set $x := \liminf_{n\to\infty} x_n$ and $x' := \limsup_{n\to\infty} x_n$.

If we apply the equalities (45.7) to the subsequences whose limits are x and x', we get

$$\forall r < \epsilon : \frac{\sin(rx)}{rx} = \frac{\sin(rx')}{rx'} \implies |x| = |x'|$$

since $y^{-1}\sin y$ is even and bijective for $0 < y < 1$.

Finally, since the limit $\lim_{n\to\infty} e^{i\theta x_n}$ exists, we see that $x = x'$. □

Proof of Theorem 45.7. Define $X_n := \xi_1 + \dots + \xi_n$ so that $X = \sum_{n=1}^{\infty} \xi_n$. We have to show $X_n \xrightarrow{d} X \implies X_n \xrightarrow{\text{a.s.}} X$.

By assumption, cf. (45.5), the limit $\lim_{n\to\infty} \mathbb{E} e^{i\theta X_n} = \mathbb{E} e^{i\theta X}$ exists locally uniformly. This means that

$$\inf_{|\theta|<\epsilon} |\phi_{X_n}(\theta)| > c > 0 \quad \text{for all } n \geqslant N(c, \epsilon).$$

Moreover, since $\phi_X(\theta) = \mathbb{E}e^{i\theta X}$ is continuous and $\phi_X(0) = 1$, we get

$$\inf_{|\theta|<\epsilon} |\phi_X(\theta)| > 2c > 0 \quad \text{for some } c \in (0, 1/2) \text{ and small } \epsilon = \epsilon(c).$$

Consider the martingale $\left(e^{i\theta X_n}/\phi_{X_n}(\theta), \mathscr{F}_n\right)_{n\geqslant N}$, $\mathscr{F}_n := \sigma(\xi_1, \dots, \xi_n)$. We have

$$\forall |\theta| < \epsilon \quad \forall n \geqslant N = N(c, \epsilon) : \quad \left|\frac{e^{i\theta X_n}}{\phi_{X_n}(\theta)}\right| \leqslant \frac{1}{c} \implies \sup_{n\geqslant N} \mathbb{E}\left|\frac{e^{i\theta X_n}}{\phi_{X_n}(\theta)}\right| < \infty.$$

The martingale convergence theorem (Theorem 40.3) gives $\lim_{n\to\infty} e^{i\theta X_n}/\phi_{X_n}(\theta)$ a.s. By assumption, $\lim_{n\to\infty} \phi_{X_n}(\theta) = \phi_X(\theta)$, thus $\lim_{n\to\infty} e^{i\theta X_n}$ a.s.

Now Lemma 45.8 shows $\lim_{n\to\infty} X_n \in \mathbb{R}$, finishing the proof. □

Khintchine's inequalities

The following inequality plays an important role in harmonic analysis and interpolation theory.

45.9 Theorem (Khintchine–Rademacher inequalities). *Let $(c_n)_{n\in\mathbb{N}} \subset \mathbb{R}$ be a sequence of real numbers and $(\xi_n)_{n\in\mathbb{N}}$ iid Bernoulli random variables such that $\mathbb{P}(\xi_1 = \pm 1) = \frac{1}{2}$. For all $0 < p < \infty$*

$$\gamma_p\left(\sum_{i=1}^n c_i^2\right)^{p/2} \leqslant \mathbb{E}\left(\left|\sum_{i=1}^n c_i\xi_i\right|^p\right) \leqslant \Gamma_p\left(\sum_{i=1}^n c_i^2\right)^{p/2}, \quad n \in \mathbb{N}. \tag{45.8}$$

The constants $0 < \gamma_p \leqslant \Gamma_p < \infty$ depend only on p.

45.10 Remark. If $p = 2$, we get from $\mathbb{E}\xi_i = 0$ that $\mathbb{V}\xi_i = \mathbb{E}\xi_i^2$, and by Bienaymé's identity

$$\mathbb{E}\left(\left|\sum_{i=1}^n c_i\xi_i\right|^2\right) = \mathbb{V}\left(\sum_{i=1}^n c_i\xi_i\right) = \sum_{i=1}^n \mathbb{V}(c_i\xi_i) = \sum_{i=1}^n c_i^2.$$

Thus, (45.8) shows that the L^p-norm (quasi-norm[1] if $0 < p < 1$) for $0 < p < \infty$ and the L^2-norm of $\sum_{i=1}^n c_i\xi_i$ are equivalent!

Proof of Theorem 45.9. Upper estimate. $S_0 := 0$, $S_n := \xi_1 + \dots + \xi_n$ is a martingale, and the martingale transform $X := c \bullet S$, $c = (c_i)_{i\in\mathbb{N}}$ is also a martingale.

Azuma's inequality (Theorem 44.6) yields

$$\mathbb{P}(|X_n| \geqslant t) \leqslant 2e^{-t^2/2d_n^2}, \quad d_n^2 = c_1^2 + \dots + c_n^2,$$

(1) A quasi-norm has all properties of a norm except the triangle inequality which holds only with a constant: $\|a + b\| \leqslant \gamma(\|a\| + \|b\|)$

and a direct calculation gives for all $0 < p < \infty$

$$\begin{aligned}
\mathbb{E}(|X_n|^p) &\overset{\S 16.5}{=} p\int_0^\infty t^{p-1}\mathbb{P}(|X_n| > t)\,dt \\
&\leqslant 2p\int_0^\infty t^{p-1}e^{-t^2/2d_n^2}\,dt \\
&\overset{s=t^2/2d_n^2}{=} d_n^p\, p2^{p/2}\int_0^\infty s^{p/2-1}e^{-s}\,ds \\
&= p2^{p/2}\Gamma(p/2)\left(d_n^2\right)^{p/2},
\end{aligned}$$

i.e. $\Gamma_p := p2^{p/2}\Gamma(p/2)$.

Lower estimate. The following duality trick is due to Littlewood (1930):

$$\mathbb{E}\left(X_n^2\right) = \mathbb{V}X_n \overset{\perp\!\!\!\perp}{=} \sum_{i=1}^n \mathbb{V}(c_i\xi_i) = \sum_{i=1}^n c_i^2$$

i.e. the lower estimate is just

$$\gamma_p\|X_n\|_2^p \leqslant \|X_n\|_p^p.$$

This inequality is obvious if $p \geqslant 2$ – since the function $p \mapsto \|X\|_p$ is increasing. If $p \in (0,2)$, we write

$$2 = p\alpha + 4\beta \quad \text{with suitable} \quad \alpha + \beta = 1,\ \alpha,\beta \in (0,1)$$

and use Hölder's inequality in conjunction with the upper estimate (marked below by §):

$$\begin{aligned}
\mathbb{E}\left(|X_n|^2\right) = \mathbb{E}\left(|X_n|^{p\alpha+4\beta}\right) &\overset{\text{Hölder}}{\leqslant} \left\{\mathbb{E}\left(|X_n|^p\right)\right\}^\alpha \left\{\mathbb{E}\left(|X_n|^4\right)\right\}^\beta \\
&\overset{\S}{\leqslant} \left\{\mathbb{E}\left(|X_n|^p\right)\right\}^\alpha \Gamma_4^\beta \left\{\mathbb{E}\left(|X_n|^2\right)\right\}^{2\beta}.
\end{aligned}$$

Rearranging gives

$$\mathbb{E}\left(|X_n|^2\right) \leqslant \Gamma_4^{\beta/(1-2\beta)} \{\mathbb{E}(|X_n|^p)\}^{\alpha/(1-2\beta)}.$$

Since $p\alpha + 4\beta = 2$, we get $\frac{\alpha}{1-2\beta} = \frac{2}{p}$ or $\frac{\beta}{1-2\beta} = \frac{2}{p} - 1$. This proves the lower estimate with $\gamma_p = \Gamma_4^{1-2/p}$ for $0 < p < 2$. □

45.11 Remark. In the 1960s and 1970s the following generalization of Khintchine's inequality was found. It involves the **adapted quadratic variation** or **square function** or **square bracket** of a martingale

$$[X]_\infty := X_0^2 + \sum_{i=1}^\infty (X_i - X_{i-1})^2 \in [0,\infty].$$

Theorem (Burkholder–Davis–Gundy inequalities). *Let* $(X_n, \mathscr{F}_n)_{n\in\mathbb{N}_0}$, $X_n \in L^p$, $p \in [1,\infty)$, *be a martingale. Then*

$$c_p \mathbb{E}\left([X]_\infty^{p/2}\right) \leqslant \mathbb{E}\left(\sup_{i\geqslant 0}|X_i|^p\right) \leqslant C_p \mathbb{E}\left([X]_\infty^{p/2}\right), \quad n \in \mathbb{N}. \tag{45.9}$$

The constants $0 < c_p \leqslant C_p < \infty$ *depend only on* p.

The proof is quite complicated, see e.g. [MaPs, Kap. 10].

46 Martingales in continuous time

The Definition 38.4 of a(sub/super-)martingale easily transfers to the index set $\mathbb{T} = [0,\infty)$.

46.1 Definition. A family of random variables $X = (X_t, \mathscr{F}_t)_{t\geqslant 0}$ which is adapted and integrable, i.e. X_t is $\mathscr{F}_t$-measurable and $\mathbb{E}|X_t| < \infty$, is called a

martingale (mg) if $\forall s \leqslant t: \quad \mathbb{E}(X_t \mid \mathscr{F}_s) = X_s$;
sub-martingale (sub-mg) if $\forall s \leqslant t: \quad \mathbb{E}(X_t \mid \mathscr{F}_s) \geqslant X_s$;
super-martingale (super-mg) if $\forall s \leqslant t: \quad \mathbb{E}(X_t \mid \mathscr{F}_s) \leqslant X_s$.

46.2 Remark. a) X is a mg $\iff$ $\pm X$ are sub-mg. Thus, it is usually enough to consider sub-martingales.

b) Note that

$$(X_t, \mathscr{F}_t)_{t\geqslant 0} \text{ is a sub-mg} \iff \begin{cases} (X_{t_n}, \mathscr{F}_{t_n})_{n\in\mathbb{N}_0} \text{ is a discrete sub-mg} \\ \forall (t_n)_n \subset [0,\infty),\ 0 \leqslant t_0 \leqslant t_1 \leqslant t_2 \leqslant \dots \end{cases}$$

c) The observation b) allows us to transfer many theorems on discrete martingales to the general setting by the following scheme:
Let $Q_n = \{q_1 < q_2 < \dots < q_n\} \subset \mathbb{Q}^+$ and $\bigcup_{n\in\mathbb{N}} Q_n = \mathbb{Q}^+$. Then

1° State a result for the discrete sub-martingale $(X_q, \mathscr{F}_q)_{q\in Q_n}$
2° Transfer the result to the discrete sub-martingale $(X_q, \mathscr{F}_q)_{q\in\mathbb{Q}}$.
3° Try to use density arguments to extend the result to $(X_t, \mathscr{F}_t)_{t\geqslant 0}$. This often requires additional regularity properties of $t \mapsto X_t$.

Example 1: Number of downcrossings of the sub-mg $(X_q)_{q\in Q_n\cap[0,t_0]}$ over $[a,b]$. Wlog we may assume that $t_0 \in Q_n$, otherwise we can add it. By Lemma 40.2

$$\mathbb{E}\left[D(X_q, q \in Q_n \cap [0,t_0], [a,b])\right] \leqslant \frac{1}{b-a}\mathbb{E}\left[(X_{t_0} - b)^+\right].$$

Now use Beppo Levi and $Q_n \uparrow \mathbb{Q}^+$ to get

$$\mathbb{E}\left[D(X_q, q \in \mathbb{Q}^+ \cap [0,t_0], [a,b])\right] \leqslant \frac{1}{b-a}\mathbb{E}\left[(X_{t_0} - b)^+\right].$$

If $D(X_q, q \in \mathbb{Q}^+ \cap [0,t_0], [a,b]) = D(X_t, t \in [0,t_0], [a,b])$ – e.g. if $t \mapsto X_t$ is (right- or left-) continuous, we have

$$\mathbb{E}\Big[D(X_t, t \in [0,t_0], [a,b])\Big] \leqslant \frac{1}{b-a}\mathbb{E}\Big[(X_{t_0} - b)^+\Big]. \tag{46.1}$$

Example 2: Doob's maximal L^p inequality for the mg or positive sub-mg $(X_q)_{q \in Q_n}$. The stopped process $(X_{q \wedge t_0})_{q \in Q_n}$ is a (positive sub-) martingale, and by Theorem 44.5 we have for $1 < p < \infty$ and $p^{-1} + q^{-1} = 1$

$$\mathbb{E}\left(\sup_{r \in Q_n} |X_{r \wedge t_0}|^p\right) \overset{44.5.c)}{\leqslant} q^p \mathbb{E}(|X_{t_0}|^p).$$

Now we use Beppo Levi and $Q_n \uparrow \mathbb{Q}^+$ to get

$$\mathbb{E}\left(\sup_{r \in \mathbb{Q}^+} |X_{r \wedge t_0}|^p\right) = \sup_{n \in \mathbb{N}} \mathbb{E}\left(\sup_{r \in Q_n} |X_{r \wedge t_0}|^p\right) \leqslant q^p \mathbb{E}(|X_{t_0}|^p).$$

If $\sup_{r \in \mathbb{Q}^+} |X_{r \wedge t_0}|^p = \sup_{t \geqslant 0} |X_{r \wedge t_0}|^p$ – e.g. if $t \mapsto X_t$ is (right- or left-) continuous, we have

$$\forall p \in (1,\infty),\ q = \frac{p}{p-1}: \quad \mathbb{E}\left(\sup_{t \leqslant t_0} |X_t|^p\right) \leqslant q^p \mathbb{E}(|X_{t_0}|^p). \tag{46.2}$$

46.3 Example. a) Let $(X_t, \mathscr{F}_t)_{t \geqslant 0}$ be a sub-mg and $a \in \mathbb{R}$. Then $(X_t \vee a, \mathscr{F}_t)_{t \geqslant 0}$ is a sub-mg.

Indeed: $\phi(x) := x \vee a$ is an increasing and convex function. Moreover, $\phi(X_t)$ is $\mathscr{F}_t$-measurable, and $\mathbb{E}\phi(X_t) \leqslant \mathbb{E}|X_t| + |a| < \infty$. Finally

$$\phi(X_s) \underset{X\text{ sub-mg}}{\overset{\phi\text{ increasing}}{\leqslant}} \phi(\mathbb{E}(X_t \mid \mathscr{F}_s)) \overset{\text{cJensen}}{\leqslant} \mathbb{E}(\phi(X_t) \mid \mathscr{F}_s).$$

b) Let $(X_t, \mathscr{F}_t)_{t \geqslant 0} \subset L^p$ be mg for some $p \in [1,\infty)$. Then $(|X_t|^p, \mathscr{F}_t)_{t \geqslant 0}$ is a sub-martingale.

Indeed: $\phi(x) = |x|^p$, obviously $\phi(X_t) \in L^1(\mathscr{F}_t)$, and

$$\phi(X_s) \underset{\text{mg}}{=} \phi(\mathbb{E}(X_t \mid \mathscr{F}_s)) \overset{\text{cJensen}}{\leqslant} \mathbb{E}(\phi(X_t) \mid \mathscr{F}_s).$$

The following Lemma is often used in applications.

46.4 Lemma. *Let $(X_t, \mathscr{F}_t)_{t \geqslant 0}$ be a sub-mg. Then $(X_t \vee a, \mathscr{F}_t)_{t \in [0,t_0]}$, $t_0 \geqslant 0$, $a \in \mathbb{R}$, is a uniformly integrable sub-martingale.*

Proof. Example 46.3.a) shows that we have a sub-martingale. Let us check »ui«: We have for all $t \leqslant t_0$

$$0 \leqslant X_t \vee a - a \leqslant \mathbb{E}(X_{t_0} \vee a - a \mid \mathscr{F}_t) =: Y_t$$
$$\implies |X_t \vee a| \leqslant (X_t \vee a - a) + |a| \leqslant Y_t + |a|$$

Because of Lemma 42.2.b) $(Y_t)_{t\leqslant t_0}$ is ui, thus for $R > 2|a|$, i.e. $R-|a| > R/2$,

$$\begin{aligned}\int_{|X_t\vee a|>R} |X_t \vee a|\,d\mathbb{P} &\leqslant \int_{Y_t>R/2} (Y_t+|a|)\,d\mathbb{P}\\ &= \int_{Y_t>R/2} Y_t\,d\mathbb{P} + |a|\mathbb{P}(Y_t > R/2)\\ &\leqslant \sup_{t\leqslant t_0}\int_{Y_t>R/2} Y_t\,d\mathbb{P} + \frac{2|a|}{R}\sup_{t\leqslant t_0}\mathbb{E}Y_t \xrightarrow[R\to\infty]{\text{ui}} 0.\end{aligned}$$ □

46.5 Corollary. *If $(X_t,\mathscr{F}_t)_{t\geqslant 0}\subset L^p$, $p\in[1,\infty)$ is a martingale, then the process $(|X_t|^p,\mathscr{F}_t)_{t\in[0,t_0]}$ is a uniformly integrable sub-martingale.*

Proof. The sub-mg property follows from Example 46.2, uniform integrability from Lemma 46.3. □

The following result is the key ingredient for the regularization of continuous-time sub-martingales.

46.6 Theorem. *Let $(X_t,\mathscr{F}_t)_{t\geqslant 0}$ be a sub-martingale. There is a measurable set $\Omega_0\subset\Omega$, $\mathbb{P}(\Omega_0)=1$, such that*

$$\forall\omega\in\Omega_0\quad\forall t\geqslant 0:\quad \exists\lim_{q\uparrow t>0} X_q(\omega)\in\mathbb{R}\quad\textit{and}\quad\exists\lim_{q\downarrow t\geqslant 0} X_q(\omega)\in\mathbb{R}.$$

Proof. We follow the strategy which we have already used in the proof of the MCT (Theorem 40.3): Since $\mathbb{E}|X_n|<\infty$,

$$\Omega_n^c := \bigcup_{a<b,a,b\in\mathbb{Q}}\left\{D(X_q, q\in\mathbb{Q}\cap[0,n],[a,b])=\infty\right\} \implies \mathbb{P}(\Omega_n)=1.$$

Thus,

$$\begin{aligned}&\left\{\omega \mid \lim_{q\downarrow t} X_q(\omega)\ \text{does not exist in } \overline{\mathbb{R}} \text{ for some } t\in[0,n]\right\}\\ &\qquad\subset\left\{\omega \mid \exists t\in[0,n]: \liminf_{q\downarrow t} X_q(\omega) < \limsup_{q\downarrow t} X_q(\omega)\right\}\subset\Omega_n^c.\end{aligned}$$

Using the scheme from Remark 46.2.c) for the inequality (44.1) we get

$$\mathbb{P}\left(\sup_{q\in\mathbb{Q}\cap[0,n]} X_q \geqslant r\right)\leqslant\frac{1}{r}\mathbb{E}X_n^+<\infty \implies \sup_{q\in\mathbb{Q}\cap[0,n]} X_q<\infty \text{ a.s.}$$

The same argument for $-X$ gives $\inf_{q\in\mathbb{Q}\cap[0,n]} X_q > -\infty$ a.s. Set

$$\Omega_0 := \bigcap_{n\in\mathbb{N}}\Omega_n\cap\left\{\sup_{q\in\mathbb{Q}\cap[0,n]} X_q<\infty\right\}\cap\left\{\inf_{q\in\mathbb{Q}\cap[0,n]} X_q>-\infty\right\}.$$

Since the intersection of countably many sets with full probability has again full probability, we have $\mathbb{P}(\Omega_0)=1$, and the assertion follows. □

46.7 Definition. Let $(X_t, \mathscr{F}_t)_{t\geqslant 0}$ be an adapted process.

a) $X_{t-}(\omega) := \limsup_{\mathbb{Q}^+\ni q\uparrow t>0} X_q(\omega)$ and $X_{t+}(\omega) := \limsup_{\mathbb{Q}^+\ni q\downarrow t\geqslant 0} X_q(\omega)$.

b) $\mathscr{F}_{t+} := \bigcap_{u>t} \mathscr{F}_u = \bigcap_{q>t, q\in\mathbb{Q}^+} \mathscr{F}_q$

c) $(\mathscr{F}_t)_{t\geqslant 0}$ is **right-continuous** if $\mathscr{F}_t = \mathscr{F}_{t+}$.

46.8 Theorem. *Let $(X_t, \mathscr{F}_t)_{t\geqslant 0}$ be a sub-martingale.*

a) $\mathbb{E}|X_{t+}| < \infty$.

b) $X_t \leqslant \mathbb{E}(X_{t+} \mid \mathscr{F}_t)$.

c) *$(X_{t+}, \mathscr{F}_{t+})$ is a sub-martingale.*

d) *$t \mapsto X_{t+}(\omega)$, $\omega \in \Omega_0$ (from Thm. 46.6), is right-continuous.*

e) *If $t \mapsto \mathbb{E}X_t$ is right-continuous, then $X_t = \mathbb{E}(X_{t+} \mid \mathscr{F}_t)$*

Before we look at the proof of Theorem 46.8, let us record the following result for martingales. In this case, $t \mapsto \mathbb{E}X_t = \mathbb{E}X_0 = \text{const}$ is trivially continuous:

46.9 Corollary. *If $(X_t, \mathscr{F}_t)_{t\geqslant 0}$ is a martingale, then $(X_{t+}, \mathscr{F}_{t+})_{t\geqslant 0}$ is a martingale.*

Proof of Theorem 46.8. Throughout the proof, q_n and r_n are rational numbers.
a), b) For $\mathbb{Q}^+ \ni q_n \downarrow t$ $(X_{q_n}, \mathscr{F}_{q_n})_{n\in\mathbb{N}}$ is a backwards sub-martingale (see also the remark before Theorem 43.5). Moreover,

$$\mathbb{E}X_t \overset{\text{sub-mg}}{\leqslant} \mathbb{E}X_{q_{n+1}} \leqslant \mathbb{E}X_{q_n} \implies \lim_{n\to\infty} \mathbb{E}X_{q_n} > -\infty$$

i.e. we are in the setting of Theorem 43.6: $X_{q_n} \xrightarrow[n\to\infty]{\text{a.s.},L^1} X_{t+}$. Thus,

$$X_{t+} \in L^1, \quad X_t \leqslant \mathbb{E}(X_{q_n} \mid \mathscr{F}_t), \quad \text{and} \quad \mathbb{E}(X_{q_n} \mid \mathscr{F}_t) \xrightarrow[n\to\infty]{} \mathbb{E}(X_{t+} \mid \mathscr{F}_t)$$

in L^1 and, for a subsequence, a.s.
c) Let $s < t$ and $s < r_n < t$, $r_n \downarrow s$. Then

$$X_{r_n} \overset{\text{sub-mg}}{\leqslant} \mathbb{E}(X_t \mid \mathscr{F}_{r_n}) \overset{\text{b)}}{\leqslant} \mathbb{E}\left[\mathbb{E}(X_{t+} \mid \mathscr{F}_t) \mid \mathscr{F}_{r_n}\right] \overset{\text{tower}}{=} \mathbb{E}(X_{t+} \mid \mathscr{F}_{r_n})$$

As $n \to \infty$, the left side converges to X_{s+} (see Theorem 46.6 and the definition of X_{s+}) while Lévy's downward theorem (Theorem 45.4) shows that the right-hand side converges to $\mathbb{E}(X_{t+} \mid \mathscr{F}_{s+})$.
d) Let $\omega \in \Omega_0$ and $t_n \downarrow t$. Pick $q_n > t_n$ such that $|X_{t_n+}(\omega) - X_{q_n}(\omega)| \leqslant \frac{1}{n}$ and $q_n \to t_n$. Thus,

$$|X_{t+}(\omega) - X_{t_n+}(\omega)| \leqslant \underbrace{|X_{t+}(\omega) - X_{q_n}(\omega)|}_{\to 0 \text{ b/o } \S 46.6} + |X_{q_n}(\omega) - X_{t_n+}(\omega)| \to 0.$$

e) Let $t \mapsto \mathbb{E}X_t$ be right-continuous.

$$X_t \leqslant \mathbb{E}(X_{t+} \mid \mathscr{F}_t) \iff \forall F^c \in \mathscr{F}_t : \quad \int_{F^c} X_t \, d\mathbb{P} \leqslant \int_{F^c} X_{t+} \, d\mathbb{P}$$

Thus, $\forall F \in \mathscr{F}_t$,

$$\int_F X_t \, d\mathbb{P} = \mathbb{E}X_t - \int_{F^c} X_t \, d\mathbb{P} \overset{\text{right-cts}}{=} \mathbb{E}X_{t+} - \int_{F^c} X_t \, d\mathbb{P}$$
$$\geqslant \mathbb{E}X_{t+} - \int_{F^c} X_{t+} \, d\mathbb{P} = \int_F X_{t+} \, d\mathbb{P}. \qquad \square$$

For the next theorem we need that $\mathscr{F}_t$ contains all measurable null sets $\mathscr{N} = \{N \in \mathscr{A} \mid \mathbb{P}(N) = 0\}$. This is done by **augmentation**, see also Definition 5.6.

$$\overline{\mathscr{F}_t} := \sigma(\mathscr{F}_t, \mathscr{N}) \overset{✎}{=} \{F \triangle N := (F \setminus N) \cup (N \setminus A) \mid F \in \mathscr{F}_t,\ N \in \mathscr{N}\} \tag{46.3}$$

$$\overset{✎}{=} \{F^* \in \mathscr{A} \mid \exists F, G \in \mathscr{F}_t : F \subset F^* \subset G \ \&\ \mathbb{P}(G \setminus F) = 0\}. \tag{46.4}$$

46.10 Theorem. *Let $(X_t, \mathscr{F}_t)_{t \geqslant 0}$ be a* **right-continuous** *sub-martingale (i.e. the paths $t \mapsto X_t(\omega)$ are right-continous for all ω).*

a) *$(X_t, \mathscr{F}_{t+})_{t \geqslant 0}$ is a sub-mg.*
b) *$(X_t, \overline{\mathscr{F}_{t+}})_{t \geqslant 0}$ is a sub-mg for $\overline{\mathscr{F}_{t+}} := \sigma(\mathscr{F}_{t+}, \mathscr{N})$.*
c) *$t \mapsto X_t(\omega)$ is for $\omega \in \Omega_0$ (from Thm. 46.6) is* **càdlàg**.[1]

⚠ The notation $\overline{\mathscr{F}_{t+}}$ is ambiguous since it is, a priori, not clear if we can interchange the operations »augmentation« and »make the filtration right-continuous«. One can show, however, that

$$\sigma(\mathscr{F}_{t+}, \mathscr{N}) \overset{\text{def}}{=} \sigma\Big(\bigcap\nolimits_{u>t} \mathscr{F}_u, \mathscr{N}\Big) = \bigcap\nolimits_{u>t} \sigma(\mathscr{F}_u, \mathscr{N}).$$

Proof. a) For $s < t$ we have, by the right-continuity and Theorem 46.8.c)

$$\mathbb{E}(X_t \mid \mathscr{F}_{s+}) \overset{X_t = X_{t+}}{=} \mathbb{E}(X_{t+} \mid \mathscr{F}_{s+}) \overset{\S 46.8.c)}{\geqslant} X_{s+} = X_s.$$

b) If $F^* \in \overline{\mathscr{F}_s}$, then $F^* = F \triangle N$ and $\int_{F^*} \cdots = \int_F \ldots$, so

$$\int_{F^*} X_s \, d\mathbb{P} = \int_F X_s \, d\mathbb{P} \leqslant \int_F X_t \, d\mathbb{P} = \int_{F^*} X_t \, d\mathbb{P}.$$

c) Let $\omega \in \Omega_0$. We know that $\lim_{\mathbb{Q} \ni q \uparrow t} X_q(\omega) = X_{t-}(\omega)$ exists. Let

$$t_1 < t_2 < t_3 < \ldots \uparrow t \quad \text{and} \quad t_1 < q_1 < t_2 < q_2 < \ldots \uparrow t$$

[1] This is a French acronym *continue à droite, limites finies à gauche*, i.e. right-continuous, finite left-hand limits.

By right-continunity we can take $q_n \in \mathbb{Q}$ such that $|X_{t_n}(\omega) - X_{q_n}(\omega)| < \frac{1}{n}$. Hence,

$$|X_{t-}(\omega) - X_{t_n}(\omega)| \leqslant |X_{t-}(\omega) - X_{q_n}(\omega)| + |X_{t_n}(\omega) - X_{q_n}(\omega)| \xrightarrow[n\to\infty]{} 0 + 0. \qquad \square$$

A **modification** of a process $(X_t)_{t\geqslant 0}$ is any process $(X'_t)_{t\geqslant 0}$ on the same probability space such that $\mathbb{P}(X_t = X'_t) = 1$ for all t.

⚠ There is the stronger notion of **indistinguishable** processes, i.e. one wants that $\mathbb{P}(X_t = X'_t\ \forall t) = 1$; mind the position of the »$\forall t$«.

46.11 Theorem. *Let $(X_t, \mathscr{F}_t)_{t\geqslant 0}$ be a sub-martingale such that $\mathscr{F}_t = \mathscr{F}_{t+}$ (right-continuous filtration) and $\mathscr{N} \subset \mathscr{F}_t$. If $t \mapsto \mathbb{E}X_t$ is right-continuous, then X has a modification which has only càdlàg paths.*

Proof. Let Ω_0 and X_{t+} be as in Theorem 46.6 and Definition 46.7, and

$$X'_t(\omega) := \begin{cases} X_{t+}(\omega), & \omega \in \Omega_0, \\ 0, & \omega \notin \Omega_0. \end{cases}$$

Then we have $X'_t = X_t = X_{t+}$ a.s. (Theorem 46.8.e)), X'_t is $\mathscr{F}_t$-measurable since $\Omega_0 \in \mathscr{F}_t$, and $(X'_t, \mathscr{F}_t)_{t\geqslant 0}$ is a sub-martingale (Theorem 46.8.c)). Moreover, $t \mapsto X'_t$ is right-continuous (Theorem 46.8.d)) and $t \mapsto X'_t$ is càdlàg (Theorem 46.10.c)). $\square$

💬 In many situations, e.g. if the Filtration $(\mathscr{F}_t)_{t\geqslant 0}$ comes from a Markov process, the augmented filtration $(\overline{\mathscr{F}_t})_{t\geqslant 0}$ is automatically right-continuous, see e.g. [BM, Theorem 6.21] (for Brownian motion or Lévy processes) and [9, Theorem 4, p. 61] (for Markov processes).

One of the most important martingale theorems is Doob's **optional stopping/sampling** theorem. Compare the following with the definitions and results from §39 and Theorem 43.4.

46.12 Definition. Let $(\mathscr{F}_t)_{t\geqslant 0}$ be a filtration. A **stopping time** is a random time $T : \Omega \to [0, \infty]$ such that

$$\{T \leqslant t\} \in \mathscr{F}_t \quad \forall t \geqslant 0. \tag{46.5}$$

Each stopping time T generates a σ-algebra

$$\mathscr{F}_T = \{F \in \mathscr{F}_\infty \mid F \cap \{T \leqslant t\} \in \mathscr{F}_t\ \forall t \geqslant 0\} \tag{46.6}$$

46.13 Remark (and Example). Let S, T be stopping times for the same filtration $(\mathscr{F}_t)_{t\geqslant 0}$.

a) (46.5) $\iff \forall t : \{T > t\} \in \mathscr{F}_t$. *Indeed:* $\{T \leqslant t\}^c = \{T > t\}$.

b) $\{T < t\} = \bigcup_{n\in\mathbb{N}} \underbrace{\{T \leqslant t - \tfrac{1}{n}\}}_{\in \mathscr{F}_{t-\frac{1}{n}} \subset \mathscr{F}_t} \in \mathscr{F}_t$.

c) $\{s \leqslant T < t\} = \{T < t\} \setminus \{T < s\} \in \mathscr{F}_t$.

d) The following results are proved as in the discrete case ✎.

- $\mathscr{F}_T$ is a σ-algebra (it contains all »information« on $X_t, t \leqslant T$);
- $T + a$, $a \geqslant 0$, is a stopping time;
- $S \leqslant T \implies \mathscr{F}_S \subset \mathscr{F}_T$;
- $S \wedge T$ is a stopping time and $\mathscr{F}_{S\wedge T} = \mathscr{F}_S \cap \mathscr{F}_T$.

e) $\{S < T\}, \{S \leqslant T\} \in \mathscr{F}_S \cap \mathscr{F}_T$. *Indeed:* $\{S < T\} \in \mathscr{F}_T$ as

$$\begin{aligned}\{S < T\} \cap \{T \leqslant t\} &= \bigcup_{q \in (0,t)\cap\mathbb{Q}} \{S < q < T\} \cap \{T \leqslant t\} \\ &= \bigcup_{q \in (0,t)\cap\mathbb{Q}} \underbrace{\{S < q\}}_{\in \mathscr{F}_q \subset \mathscr{F}_t} \cap \underbrace{\{T \leqslant q\}^c}_{\in \mathscr{F}_q \subset \mathscr{F}_t} \cap \{T \leqslant t\} \in \mathscr{F}_t\end{aligned}$$

and $\{S < T\} \in \mathscr{F}_S$ since we have

$$\begin{aligned}\{S < T\} \cap \{S \leqslant t\} &= \{S < T\} \cap \{S \leqslant t\} \cap \big(\{T \leqslant t\} \cup \{T > t\}\big) \\ &= \big(\{S < T\} \cap \{S \leqslant t\} \cap \{T \leqslant t\}\big) \cup \big(\{S < T\} \cap \{S \leqslant t\} \cap \{T > t\}\big) \\ &= \underbrace{\big(\{S < T\} \cap \{T \leqslant t\}\big)}_{\in \mathscr{F}_t \text{ see above}} \cup \big(\{S \leqslant t\} \cap \{t < T\}\big) \in \mathscr{F}_t.\end{aligned}$$

Changing the roles of S and T gives $\{S \leqslant T\} = \{S > T\}^c \in \mathscr{F}_S \cap \mathscr{F}_T$.

f) Let T be a stopping time. Then $T^n := (\lfloor 2^n T \rfloor + 1)/2^n$ is a discrete stopping time such that $T^n \downarrow T$. Note that $\{T^n = \infty\} = \{T = \infty\}$.

Indeed: $\{T^n = i2^{-n}\} = \{(i-1)2^{-n} \leqslant T < i2^{-n}\} \overset{c)}{\in} \mathscr{F}_{i2^{-n}}$.

Thus, $\{T^n \leqslant t\} \in \mathscr{F}_{t^*} \subset \mathscr{F}_t$ where t^* is the largest dyadic number less or equal to t.

g) We will see later (Lemma 60.2, if $t \mapsto X_t$ is continuous) that typical examples of stopping times are the

first entrance time into a closed set F $\quad \tau_F^\circ := \inf\{t \geqslant 0 \mid X_t \in F\}$,

first hitting time of an open set U $\quad \tau_U := \inf\{t > 0 \mid X_t \in U\}$.

h) $\forall t \geqslant 0 : \{T \leqslant t\} \in \mathscr{F}_{t+} \iff \forall t \geqslant 0 : \{T < t\} \in \mathscr{F}_t$.

Indeed: »$\Rightarrow$« follows from $\{T < t\} = \bigcup_{n=1}^\infty \{T \leqslant t - 1/n\}$ and the observation that $\{T \leqslant t - 1/n\} \in \mathscr{F}_{(t-1/n)+} \subset \mathscr{F}_t$.

Conversely, »$\Leftarrow$« follows from $\{T \leqslant t\} = \bigcap_{n=k}^\infty \{T < t + 1/n\} \in \mathscr{F}_{t+1/k}$ since we have $\{T < t + 1/n\} \in \mathscr{F}_{t+1/n} \subset \mathscr{F}_{t+1/k}$ for all $n \geqslant k \in \mathbb{N}$. Thus, $\{T \leqslant t\} \in \bigcap_{k=1}^\infty \mathscr{F}_{t+1/k} = \mathscr{F}_{t+}$.

46.14 Theorem (Optional stopping. Doob 1953). *Let $(X_t, \mathscr{F}_t)_{t\geqslant 0}$ be a sub-mg with càdlàg paths and $S \leqslant T$ be $\mathscr{F}_t$ stopping times. For all $k \in \mathbb{N}$ it holds that*

$X_{S\wedge k}, X_{T\wedge k} \in L^1(\mathbb{P})$ *and*

$$X_{S\wedge k} \leqslant \mathbb{E}(X_{T\wedge k} \mid \mathscr{F}_{S\wedge k}). \tag{46.7}$$

Moreover, if

a) *either* S,T *are bounded, i.e. if* $\mathbb{P}(T \leqslant K) = 1$ *for some constant* K,
b) *or* $\mathbb{P}(T < \infty) = 1$, $X_S, X_T \in L^1(\mathbb{P})$ *and* $\lim_{k\to\infty} \mathbb{E}\left[|X_k| \mathbb{1}_{\{T>k\}}\right] = 0$,

then

$$X_S \leqslant \mathbb{E}(X_T \mid \mathscr{F}_S). \tag{46.8}$$

Proof. As in Remark 46.13.f) we approximate the stopping times $S \leqslant T$ by dyadic stopping times $S^n \leqslant T^n$. Applying Theorem 39.6 and 39.9.b) with $S \rightsquigarrow S^n \wedge k$ and $T \rightsquigarrow T^n \wedge k$ we see that

$$X_{S^n\wedge k}, X_{T^n\wedge k} \in L^1(\mathbb{P}) \quad \text{and} \quad X_{S^n\wedge k} \leqslant \mathbb{E}\left[X_{T^n\wedge k} \,\middle|\, \mathscr{F}_{S^n\wedge k}\right].$$

The tower property shows because of $\mathscr{F}_{S\wedge k} \subset \mathscr{F}_{S^n\wedge k}$

$$\mathbb{E}\left[X_{S^n\wedge k} \,\middle|\, \mathscr{F}_{S\wedge k}\right] \leqslant \mathbb{E}\left[X_{T^n\wedge k} \,\middle|\, \mathscr{F}_{S\wedge k}\right]. \tag{46.9}$$

By Theorem 39.9 we see that $(X_{S^n\wedge k}, \mathscr{F}_{S^n\wedge k})_{n\geqslant 1}$ and $(X_{T^n\wedge k}, \mathscr{F}_{T^n\wedge k})_{n\geqslant 1}$ are backwards sub-martingales.

Since $\mathbb{E}X_{T^n\wedge k} \geqslant \mathbb{E}X_0 > -\infty$, the proof of Theorem 43.6.b) shows that the sub-martingales are uniformly integrable. Therefore,

$$X_{S^n\wedge k} \xrightarrow[n\to\infty]{\text{a.s., } L^1} X_{S\wedge k} \quad \text{and} \quad X_{T^n\wedge k} \xrightarrow[n\to\infty]{\text{a.s., } L^1} X_{T\wedge k}.$$

This implies that we have for all $F \in \mathscr{F}_{S\wedge k}$

$$\int_F X_{S\wedge k}\, d\mathbb{P} = \lim_{n\to\infty} \int_F X_{S^n\wedge k}\, d\mathbb{P} \overset{(46.9)}{\leqslant} \lim_{n\to\infty} \int_F X_{T^n\wedge k}\, d\mathbb{P} = \int_F X_{T\wedge k}\, d\mathbb{P}.$$

This proves (46.7), and (46.8) in the case a) where S, T are bounded.

Assume now that we are in the Case b). For all $F \in \mathscr{F}_S$

$$\begin{aligned} F \cap \{S \leqslant k\} \cap \{S \wedge k \leqslant t\} &= F \cap \{S \leqslant k\} \cap \{S \leqslant t\} \\ &= F \cap \{S \leqslant k \wedge t\} \in \mathscr{F}_{k\wedge t} \subset \mathscr{F}_t, \end{aligned}$$

which means that $F \cap \{S \leqslant k\} \in \mathscr{F}_{S\wedge k}$. Since $S \leqslant T < \infty$ we have

$$\int_{F\cap\{S>k\}} |X_k|\, d\mathbb{P} \leqslant \int_{\{S>k\}} |X_k|\, d\mathbb{P} \leqslant \int_{\{T>k\}} |X_k|\, d\mathbb{P} \xrightarrow[k\to\infty]{\text{b)}} 0.$$

By assumption, $X_S, X_T \in L^1(\mathbb{P})$. Using (46.7) we see that

$$\begin{aligned}
\int_F X_{S\wedge k}\,d\mathbb{P} &= \overbrace{\int_{F\cap\{S\leqslant k\}} \underbrace{X_{S\wedge k}}_{=X_S}\,d\mathbb{P}}^{\text{I}} + \overbrace{\int_{F\cap\{S>k\}} X_k\,d\mathbb{P}}^{\text{II}} \\
&\overset{\text{sub-mg}}{\leqslant} \int_{F\cap\{S\leqslant k\}} X_{T\wedge k}\,d\mathbb{P} + \int_{F\cap\{S>k\}} X_k\,d\mathbb{P} \\
&= \int_{F\cap\{T\leqslant k\}} \overbrace{X_{T\wedge k}}^{=X_T}\,d\mathbb{P} + \int_{F\cap\{S\leqslant k<T\}} \overbrace{X_{T\wedge k}}^{=X_k}\,d\mathbb{P} + \int_{F\cap\{S>k\}} X_k\,d\mathbb{P} \\
&= \underbrace{\int_{F\cap\{T\leqslant k\}} \overbrace{X_{T\wedge k}}^{=X_T}\,d\mathbb{P}}_{\text{III}} + \underbrace{\int_{F\cap\{T>k\}} X_k\,d\mathbb{P}}_{\text{IV}}.
\end{aligned}$$

By dominated convergence, we see that

$$\text{I} \xrightarrow[k\to\infty]{} \int_F X_S\,d\mathbb{P} \quad\text{and}\quad \text{III} \xrightarrow[k\to\infty]{} \int_F X_T\,d\mathbb{P},$$

while II, IV $\to 0$ because of our assumption b). This proves (46.8). □

46.15 Corollary. *Let $(X_t, \mathscr{F}_t)_{t\geqslant 0}$, $X_t \in L^1(\mathbb{P})$, be an adapted process with càdlàg paths. $(X_t, \mathscr{F}_t)_{t\geqslant 0}$ is sub-martingale if, and only if,*

$$\mathbb{E}X_S \leqslant \mathbb{E}X_T \quad \forall \text{ bounded } \mathscr{F}_t \text{ stopping times } S \leqslant T. \tag{46.10}$$

Proof. »⇒« If $(X_t, \mathscr{F}_t)_{t\geqslant 0}$ is a sub-martingale, we can use Theorem 46.14.

»⇐« Assume that (46.10) holds, let $s < t$, pick $F \in \mathscr{F}_s$, and define a random time $\rho := t\mathbb{1}_{F^c} + s\mathbb{1}_F$.

Since ρ is a bounded stopping time ✍ we can use (46.10) with $S := \rho$ and $T := t$, and we get

$$\mathbb{E}\big(X_s\mathbb{1}_F\big) + \mathbb{E}\big(X_t\mathbb{1}_{F^c}\big) = \mathbb{E}X_\rho \leqslant \mathbb{E}X_t.$$

If we rearrange this inequality, we see that $\mathbb{E}\big(X_s\mathbb{1}_F\big) \leqslant \mathbb{E}\big(X_t\mathbb{1}_F\big)$ for all $s < t$ and $F \in \mathscr{F}_s$, hence $X_s \leqslant \mathbb{E}(X_t \mid \mathscr{F}_s)$. □

46.16 Corollary. *Let $(X_t, \mathscr{F}_t)_{t\geqslant 0}$ be a uniformly integrable càdlàg sub-martingale. If $S \leqslant T$ are stopping times with $\mathbb{P}(T < \infty) = 1$, then Theorem 46.14.b) holds.*

Proof. Uniform integrability implies $\sup_{t\geqslant 0} \mathbb{E}|X(t)| \leqslant c < \infty$. Since $(X_t^+)_{t\geqslant 0}$ is a sub-martingale, we have for all $k \geqslant 0$

$$\begin{aligned}\mathbb{E}|X(S \wedge k)| &= 2\mathbb{E}X^+(S \wedge k) - \mathbb{E}X(S \wedge k) \\ &\overset{(46.7)}{\leqslant} 2\mathbb{E}X^+(k) - \mathbb{E}X(0) \\ &\leqslant 2\mathbb{E}|X(k)| + \mathbb{E}|X(0)| \leqslant 3c.\end{aligned}$$

By Fatou's lemma

$$\mathbb{E}|X(S)| \leqslant \liminf_{k\to\infty} \mathbb{E}|X(S \wedge k)| \leqslant 3c \quad \text{and similarly} \quad \mathbb{E}|X(T)| \leqslant 3c.$$

Thus,

$$\begin{aligned}\mathbb{E}\left[|X(k)|\mathbb{1}_{\{T>k\}}\right] &= \mathbb{E}\left[|X(k)|\left(\mathbb{1}_{\{T>k\}\cap\{|X(k)|>R\}} + \mathbb{1}_{\{T>k\}\cap\{|X(k)|\leqslant R\}}\right)\right] \\ &\leqslant \sup_{n\in\mathbb{N}} \mathbb{E}\left[|X(n)|\mathbb{1}_{\{|X(n)|>R\}}\right] + R\,\mathbb{P}(T > k) \\ &\xrightarrow[k\to\infty]{} \sup_{n\in\mathbb{N}} \mathbb{E}\left[|X(n)|\mathbb{1}_{\{|X(n)|>R\}}\right] \xrightarrow[R\to\infty]{\text{ui}} 0.\end{aligned}$$

□

X

Poisson Processes

47 Two special probability distributions

As before, $(\Omega, \mathscr{A}, \mathbb{P})$ is a probability space.

47.1 Definition. A random variable $\sigma : \Omega \to [0,\infty)$ is said to be **exponentially distributed** or **exponential** with parameter $\lambda > 0$ (notation: $\sigma \sim \mathsf{Exp}(\lambda)$) if

$$\mathbb{P}(\sigma \in ds) = \lambda e^{-\lambda s} \mathbb{1}_{[0,\infty)}(s)\,ds. \tag{47.1}$$

Let us collect a few properties of exponential random variables.

47.2 Remark (properties of $\mathsf{Exp}(\lambda)$). Assume that $\sigma \sim \mathsf{Exp}(\lambda)$ with parameter $\lambda > 0$.

a) Since $\int_x^\infty \lambda e^{-\lambda s}\,ds = e^{-\lambda x}$, we can (47.1) also write as

$$\mathbb{P}(\sigma > x) = \mathbb{P}(\sigma \geqslant x) = e^{-\lambda x} \quad \forall x \geqslant 0.$$

b) $\mathbb{E}\sigma = \lambda^{-1}$ and $\mathbb{V}\sigma = \lambda^{-2}$ ✍.

c) The characteristic function is $\phi_\sigma(\xi) = \mathbb{E}e^{i\xi\sigma} = \dfrac{\lambda}{\lambda - i\xi}$, $\xi \in \mathbb{R}$.

Indeed: $$\int_0^\infty e^{i\xi s} \lambda e^{-\lambda s}\,ds = \lambda \int_0^\infty e^{-(\lambda - i\xi)s}\,ds = \frac{\lambda}{\lambda - i\xi}.$$

For positive random variables, the **Laplace transformation** $\mathbb{E}e^{-t\sigma} = \lambda/(\lambda + t)$, $t \geqslant 0$ is used instead of the characteristic function; note that there is no convergence problem as $e^{-t\sigma} \leqslant 1$. Formally, the Laplace transform can be obtained from the c.f. if we set $\xi = it$, which can be used to show that the Laplace transform characterizes (the law of) a positive random variable, see also the appendix to this chapter.

d) Let $\sigma_1, \dots, \sigma_n \sim \mathsf{Exp}(\lambda)$ be iid copies of $\sigma \sim \mathsf{Exp}(\lambda)$. The sum $\sigma_1 + \dots + \sigma_n$ is a Gamma-distributed random variable $\sim \Gamma_{n,1/\lambda}$.

Recall that $X \sim \Gamma_{n,1/\lambda}$, if $\mathbb{P}(X \in dx) = \frac{\lambda^n}{\Gamma(n)} x^{n-1} e^{-\lambda x} \mathbb{1}_{(0,\infty)}(x)\,dx$.

Let us calculate the characteristic function of X:

$$\begin{aligned}
\mathbb{E}e^{i\xi X} &= \frac{\lambda^n}{\Gamma(n)}\int_0^\infty x^{n-1}e^{-\lambda x}e^{ix\xi}\,dx \\
&= \frac{\lambda^n}{i^{n-1}\Gamma(n)}\int_0^\infty e^{-\lambda x}\frac{d^{n-1}}{d\xi^{n-1}}e^{ix\xi}\,dx \\
&= \frac{\lambda^{n-1}}{i^{n-1}\Gamma(n)}\frac{d^{n-1}}{d\xi^{n-1}}\int_0^\infty \lambda e^{-\lambda x}e^{ix\xi}\,dx \\
&\overset{c)}{=} \frac{\lambda^{n-1}}{i^{n-1}\Gamma(n)}\frac{d^{n-1}}{d\xi^{n-1}}\frac{\lambda}{\lambda-i\xi} \\
&= \frac{\lambda^n}{(\lambda-i\xi)^n}.
\end{aligned}$$

On the other hand, since the rvs $\sigma_k \sim \mathsf{Exp}(\lambda)$ are iid,

$$\mathbb{E}e^{i\xi(\sigma_1+\cdots+\sigma_n)} = \left[\mathbb{E}e^{i\xi\sigma}\right]^n \overset{c)}{=} \left[\frac{\lambda}{\lambda-i\xi}\right]^n.$$

Because of the uniqueness of the characteristic function we conclude that $\sigma_1+\cdots+\sigma_n \sim X \sim \Gamma_{n,1/\lambda}$.

⚠ We could have derived this result by a direct calculation (using the convolution), but as soon as $n \geqslant 3$, these calculations are quite cumbersome.

e) **Lack of memory property.** For all $a, x \geqslant 0$ we have

$$\begin{aligned}
\mathbb{P}(\sigma > x+a \mid \sigma > a) &= \frac{\mathbb{P}(\sigma > a+x, \sigma > a)}{\mathbb{P}(\sigma > a)} \\
&= \frac{\mathbb{P}(\sigma > a+x)}{\mathbb{P}(\sigma > a)} = \frac{e^{-\lambda(a+x)}}{e^{-\lambda a}} \\
&= e^{-\lambda x} = \mathbb{P}(\sigma > x).
\end{aligned}$$

This has an interesting interpretation: If σ is the life time of some device, having survived a years does not influence its chance to live another x years, i.e. the device has »forgotten« its age. Moreover,

$$\mathbb{P}(\sigma > a+x) = \mathbb{P}(\sigma > x)\mathbb{P}(\sigma > a) \quad \forall a, x \geqslant 0. \tag{47.2}$$

We will see soon that (47.2) characterizes the exponential law.

47.3 Lemma (Cauchy's functional equation). *Let $\phi : [0,\infty) \to [0,\infty)$ be a right-continuous function which satisfies the following functional relation:*

$$\phi(s+t) = \phi(s)\phi(t) \quad \text{for all} \quad s, t \geqslant 0.$$

Then $\phi(t) = \phi(1)^t = e^{-\lambda t}$ with parameter $\lambda = -\log\phi(1)$.

Proof. We distinguish between two cases.

Case 1: If $\phi(a) = 0$ for some $a \geqslant 0$, then we have for all $t \geqslant 0$

$$\phi(a+t) = \phi(a)\phi(t) = 0 \implies \phi|_{[a,\infty)} \equiv 0.$$

Moreover, we see for all $n \in \mathbb{N}$ that

$$0 = \phi(a) = \left[\phi\left(\tfrac{a}{n}\right)\right]^n \implies \phi\left(\tfrac{a}{n}\right) = 0.$$

Since ϕ is right-continuous, we get that $\phi|_{[0,\infty)} \equiv 0$, and so we have $\phi(t) = \phi(1)^t$.

Case 2: If $\phi(1) \neq 0$, then we define $f(t) := \phi(t)\phi(1)^{-t}$. Observe that

$$f(s+t) = \phi(s+t)\phi(1)^{-(s+t)} = \phi(s)\phi(1)^{-s}\phi(t)\phi(1)^{-t} = f(s)f(t)$$

as well as $f(1) = 1$. Applying the functional relation k times yields for all $k, n \in \mathbb{N}$

$$f\left(\tfrac{k}{n}\right) = \left[f\left(\tfrac{1}{n}\right)\right]^k.$$

A similar calculation (with $k = n$) reveals that

$$\left[f\left(\tfrac{1}{n}\right)\right]^k = \left[f\left(\tfrac{1}{n}\right)\right]^{n\frac{k}{n}} = \left[f\left(\tfrac{n}{n}\right)\right]^{\frac{k}{n}} = \left[f(1)\right]^{\frac{k}{n}} = 1,$$

hence $f|_{\mathbb{Q}_+} \equiv 1$. Since f inherits the right-continuity of ϕ, we see that $f \equiv 1$ or $\phi(t) = [\phi(1)]^t$. □

47.4 Corollary. *The relation (47.2) uniquely characterizes the exponential law:* $\mathbb{P}(\sigma > t) = \mathbb{P}(\sigma > 1)^t$ *and* $\lambda = -\log \mathbb{P}(\sigma > 1)$.

47.5 Remark. a) Lemma 47.3 can be used to show that for a right-continuous f

$$f(s+t) = f(s) + f(t) \implies f(s) = cs,$$

just take $\phi(s) = e^{f(s)}$.

b) We can relax the assumption »ϕ is right-continuous« in Lemma 47.3 by »ϕ is a $\mathbb{R}$-valued function which is bounded in some non-empty open interval«. For the proof and further discussions we refer to the classic monograph by Aczel [1, Chapter 2.1].

We will need a further special distribution.

47.6 Definition. A random variable $N : \Omega \to \mathbb{N}_0$ is said to be **Poisson distributed** with parameter $\lambda > 0$ (notation: $N \sim \mathsf{Poi}(\lambda)$) if

$$\mathbb{P}(N = k) = e^{-\lambda}\frac{\lambda^k}{k!}, \quad k = 0, 1, 2, \ldots \tag{47.3}$$

47.7 Remark (properties of $\mathsf{Poi}(\lambda)$). Let $N \sim \mathsf{Poi}(\lambda)$ with parameter $\lambda > 0$.

a) We have $\mathbb{E}N = \mathbb{V}N = \lambda$ and $\mathbb{E}e^{i\xi N} = \exp\left[-\lambda(1-e^{i\xi})\right]$ for all $\xi \in \mathbb{R}$.

We start with the calculation of the characteristic function. We have

$$\begin{aligned}\mathbb{E}e^{i\xi N} &= \sum_{k=0}^{\infty} e^{i\xi k}\frac{\lambda^k}{k!}e^{-\lambda} = e^{-\lambda}\sum_{k=0}^{\infty}\frac{(e^{i\xi}\lambda)^k}{k!} \\ &= e^{-\lambda}e^{\lambda e^{i\xi}} = e^{-\lambda(1-e^{i\xi})}.\end{aligned}$$

From this we can get $\mathbb{E}N$ and $\mathbb{E}(N^2)$, hence $\mathbb{V}N = \mathbb{E}(N^2)-(\mathbb{E}N)^2$, by differentiation: Replace in the above calculation $e^{i\xi} \rightsquigarrow (1-s)$ to see

$$\mathbb{E}\left((1-s)^N\right) = e^{-\lambda(1-e^{i\xi})}\Big|_{e^{i\xi}=1-s} = e^{-\lambda s}, \quad s \in [0,1].$$

This substitution simplifies the following calculations (cf. §27.5.g))

$$\mathbb{E}N = -\mathbb{E}\left[\frac{d}{ds}(1-s)^N\right]\Big|_{s=0} = -\frac{d}{ds}\mathbb{E}\left[(1-s)^N\right]\Big|_{s=0} = -\frac{d}{ds}e^{-\lambda s}\Big|_{s=0} = \lambda$$

and

$$\begin{aligned}\mathbb{E}[N(N-1)] = \mathbb{E}\left[\frac{d^2}{ds^2}(1-s)^N\right]\Big|_{s=0} &= \frac{d^2}{ds^2}\mathbb{E}\left[(1-s)^N\right]\Big|_{s=0} \\ &= \frac{d^2}{ds^2}e^{-\lambda s}\Big|_{s=0} = \lambda^2.\end{aligned}$$

Thus, $\mathbb{V}N = [\mathbb{E}(N^2)-\mathbb{E}N] + \mathbb{E}N - (\mathbb{E}N)^2 = \lambda^2+\lambda-\lambda^2 = \lambda$.

b) If $X \sim \mathsf{Poi}(\lambda)$ and $Y \sim \mathsf{Poi}(\mu)$ are independent, then $X+Y \sim \mathsf{Poi}(\mu+\lambda)$. This follows from

$$\mathbb{E}e^{i\xi(X+Y)} \overset{\perp\!\!\!\perp}{=} \mathbb{E}e^{i\xi X}\mathbb{E}e^{i\xi Y} \overset{a)}{=} e^{-\lambda(1-e^{i\xi})}e^{-\mu(1-e^{i\xi})} = e^{-(\lambda+\mu)(1-e^{i\xi})}.$$

c) If $X_i \sim \mathsf{Poi}(\lambda_i)$ are independent random variables and $N = X_1+\dots+X_n$, then $N \sim \mathsf{Poi}(\lambda)$ with $\lambda = \sum_{i=1}^n \lambda_i$; moreover,

$$\mathbb{P}(X_1 = x_1,\dots,X_{n-1} = x_{n-1} \mid N = x) = \binom{x}{x_1,x_2,\dots,x_n}\prod_{i=1}^n\left(\frac{\lambda_i}{\lambda}\right)^{x_i}$$

where $x_1+\dots+x_n = x$. The first assertion follows from b), the second can be directly checked:

$$\begin{aligned}&\mathbb{P}(X_1 = x_1,\dots,X_{n-1} = x_{n-1} \mid N = x) \\ &= \frac{\mathbb{P}(X_1 = x_1,\dots,X_{n-1} = x_{n-1}, N = x)}{\mathbb{P}(N = x)} \\ &= \frac{\mathbb{P}(X_1 = x_1,\dots,X_{n-1} = x_{n-1}, X_n = x_n)}{\mathbb{P}(N = x)} \\ &\underset{\mathsf{Poi}}{\overset{\perp\!\!\!\perp}{=}} \frac{\prod_{i=1}^n e^{-\lambda_i}\lambda_i^{x_i}/x_i!}{e^{-\lambda}\lambda^x/x!} \\ &= \frac{x!}{x_1!\cdot\ldots\cdot x_n!}\prod_{i=1}^n\frac{\lambda_i^{x_i}}{\lambda^{x_i}}.\end{aligned}$$

💬 This is the **multinomial distribution**, cf. Example 22.8.b). We will come back to this in Theorem 50.5.

d) There is a partial converse to c). Assume that $N = X_1 + \dots + X_n \sim \mathsf{Poi}(\lambda)$ and that

$$\mathbb{P}(X_1 = x_1, \dots, X_{n-1} = x_{n-1} \mid N = x) = \binom{x}{x_1, \dots, x_n} p_1^{x_1} \cdot \ldots \cdot p_n^{x_n}$$

where $x = x_1 + \dots + x_n$, $x_i \in \mathbb{N}_0$ and $p_1 + \dots + p_n = 1$, $p_i \in (0,1)$.

In this case $X_1, \dots, X_n$ are independent and $X_i \sim \mathsf{Poi}(p_i \lambda)$. This follows from

$$\begin{aligned}
&\mathbb{P}(X_1 = x_1, \dots, X_{n-1} = x_{n-1}, X_n = x_n) \\
&\quad = \mathbb{P}(X_1 = x_1, \dots, X_{n-1} = x_{n-1}, N = x) \\
&\quad = \mathbb{P}(N = x)\mathbb{P}(X_1 = x_1, \dots, X_{n-1} = x_{n-1} \mid N = x) \\
&\quad = e^{-\lambda} \frac{\lambda^x}{x!} \binom{x}{x_1, \dots, x_n} p_1^{x_1} \cdot \ldots \cdot p_n^{x_n} = \prod_{i=1}^n e^{-\lambda p_i} \frac{(\lambda p_i)^{x_i}}{x_i!}.
\end{aligned}$$

💬 There are further variants of such converses, e.g.

$$\left.\begin{aligned} X + Y \sim \mathsf{Poi} \\ \phi_{X+Y}(\xi) = \phi_X(\xi)\phi_Y(\xi) \end{aligned}\right\} \implies X, Y \sim \mathsf{Poi}$$

(due to Raikov, see Lukacs [19, S. 243f.]) or (cf. Moran [21])

$$\left.\begin{aligned} X \perp\!\!\!\perp Y,\ X, Y \in \mathbb{N}_0 \\ \mathbb{P}(X = i \mid X + Y = n) = \binom{n}{i} p_n^i (1 - p_n)^{n-i} \\ \exists i_0 : \mathbb{P}(X = i_0) \cdot \mathbb{P}(Y = i_0) > 0 \end{aligned}\right\} \implies X, Y \sim \mathsf{Poi},\ p_n = p.$$

47.8 Theorem (countable additivity). *Let $X_i \sim \mathsf{Poi}(\lambda_i)$, $i \in \mathbb{N}$, be independent random variables. Then*

$$\sigma := \sum_{i=1}^{\infty} \lambda_i \begin{cases} < \infty \\ = \infty \end{cases} \iff S := \sum_{i=1}^{\infty} X_i \begin{cases} < \infty & a.s. \\ = \infty & a.s. \end{cases}$$

If $\sigma < \infty$, then $S \sim \mathsf{Poi}(\sigma)$.

Proof. Define $S_n := X_1 + \dots + X_n$ and $\sigma_n := \lambda_1 + \dots + \lambda_n$. We already know from Remark 47.7.b) that $S_n \sim \mathsf{Poi}(\sigma_n)$.

»⇒« We have, using the continuity of measures,

$$\begin{aligned}\mathbb{P}(S \leqslant m) &\overset{S_n\uparrow S}{=} \mathbb{P}\left(\bigcap_{n=1}^{\infty}\{S_n \leqslant m\}\right)\\ &= \lim_{n\to\infty} \mathbb{P}(S_n \leqslant m)\\ &= \lim_{n\to\infty} \sum_{k=0}^{m} e^{-\sigma_n}\frac{\sigma_n^k}{k!}\\ &= \sum_{k=0}^{m} e^{-\sigma}\frac{\sigma^k}{k!}\begin{cases}\in(0,\infty), & \text{if } \sigma<\infty,\\ =0, & \text{if } \sigma=\infty.\end{cases}\end{aligned}$$

Thus, $\sigma < \infty$ implies that $S \sim \mathsf{Poi}(\sigma)$ and $\mathbb{P}(S<\infty)=1$.

If $\sigma = \infty$, then $\mathbb{P}(S \leqslant m) = 0$ for all m, hence $\mathbb{P}(S > m) = 1$ for all m, and so $\mathbb{P}(S=\infty)=1$.

»⇐« Since $S_n \uparrow S$ a.s., we can use Beppo Levi to get

$$\mathbb{E}e^{-\xi S} \overset{BL}{=} \lim_{n\to\infty} \mathbb{E}e^{-\xi S_n} \overset{\mathsf{Poi}}{=} \lim_{n\to\infty} e^{-\sigma_n(1-\exp[-\xi])} = e^{-\sigma(1-\exp[-\xi])}.$$

This shows that the limit $\lim_{n\to\infty}\sigma_n \in [0,\infty]$ exists; thus,

$$\sigma=\infty \iff e^{-\sigma(1-\exp[-\xi])}=0 \iff \mathbb{E}e^{-\xi S}=0 \iff \mathbb{P}(S=\infty)=1.$$ ◻

*Appendix: Uniqueness of the Laplace transform**

We begin with a lemma from analysis which is helpful in order to show that finite measures are uniquely determined through their Laplace transforms.

Let μ be a measure on $([0,\infty),\mathscr{B}[0,\infty))$. Its **Laplace transform** is given by $\mathscr{L}[\mu](\lambda) = \int_{[0,\infty)} e^{-\lambda t}\,\mu(dt)$. If X is a rv with law μ, then $\mathscr{L}[\mu](\lambda) = \mathbb{E}e^{-\lambda X}$.

47.9 Lemma. *We have for all $t, x \geqslant 0$*

$$\lim_{\lambda\to\infty} e^{-\lambda t}\sum_{k\leqslant\lambda x}\frac{(\lambda t)^k}{k!} = \mathbb{1}_{[0,x)}(t) + \frac{1}{2}\mathbb{1}_{\{x\}}(t). \tag{47.4}$$

Proof. Let us rewrite (47.4) in probabilistic terms: If X is Poisson distributed with parameter λt, (47.4) means that

$$\lim_{\lambda\to\infty}\mathbb{P}(X\leqslant\lambda x) = \mathbb{1}_{[0,x)}(t) + \frac{1}{2}\mathbb{1}_{\{x\}}(t).$$

We have $\mathbb{E}X = \mathbb{V}X = \lambda t$. If $X_1,\dots,X_n \sim \mathsf{Poi}(\lambda t/n)$, then $S_n := X_1+\dots+X_n \sim \mathsf{Poi}(\lambda t)$. Without loss of generality we may assume that $\lambda = n$ is an integer. By the central limit theorem (Theorem 28.6), the sequence $(S_\lambda - \lambda t)/\sqrt{\lambda t}$ converges weakly to a standard normal random variable $G \sim \mathsf{N}(0,1)$. As $X \sim S_\lambda$,

$$\mathbb{P}(X\leqslant\lambda x) = \mathbb{P}\left(\frac{S_\lambda-\lambda t}{\sqrt{\lambda t}} \leqslant \sqrt{\lambda}\frac{x-t}{\sqrt{t}}\right) \xrightarrow{\lambda\to\infty} \begin{cases}\mathbb{P}(G<\infty)=1, & \text{if } x>t,\\ \mathbb{P}(G\leqslant 0)=\frac{1}{2}, & \text{if } x=t,\\ \mathbb{P}(G=-\infty)=0, & \text{if } x<t,\end{cases}$$

and the claim follows. ◻

47.10 Theorem. *A measure μ on $([0,\infty),\mathscr{B}[0,\infty))$ is finite if, and only if, $\mathscr{L}[\mu](0+)<\infty$. Any finite measure μ is uniquely determined by its Laplace transform.*

Proof. The first part of the assertion follows from monotone convergence since we have $\mu[0,\infty)=\int_{[0,\infty)} 1\,d\mu=\lim_{\lambda\to 0}\int_{[0,\infty)} e^{-\lambda t}\,\mu(dt)$.

For the uniqueness part we use first the differentiability lemma for parameter dependent integrals (Theorem 12.2) to get

$$(-1)^k\mathscr{L}^{(k)}[\mu](\lambda)=\int_{[0,\infty)} e^{-\lambda t}\,t^k\,\mu(dt).$$

Therefore,

$$\begin{aligned}\sum_{k\leqslant\lambda x}(-1)^k\mathscr{L}^{(k)}[\mu](\lambda)\frac{\lambda^k}{k!}&=\sum_{k\leqslant\lambda x}\int_{[0,\infty)}\frac{(\lambda t)^k}{k!}e^{-\lambda t}\,\mu(dt)\\&=\int_{[0,\infty)}\sum_{k\leqslant\lambda x}\frac{(\lambda t)^k}{k!}e^{-\lambda t}\,\mu(dt),\end{aligned}$$

and we conclude with Lemma 47.9 and dominated convergence that

$$\begin{aligned}\lim_{\lambda\to\infty}\sum_{k\leqslant\lambda x}(-1)^k\mathscr{L}^{(k)}[\mu](\lambda)\frac{\lambda^k}{k!}&=\int_{[0,\infty)}\left(\mathbb{1}_{[0,x)}(t)+\frac{1}{2}\mathbb{1}_{\{x\}}(t)\right)\mu(dt)\\&=\mu[0,x)+\frac{1}{2}\mu\{x\}.\end{aligned}\tag{47.5}$$

This shows that μ can be recovered from (all derivatives of) $\mathscr{L}[\mu]$. □

48 The Poisson process

In this chapter we will encounter one of the simplest stochastic processes in continuous time: The Poisson process. It is often used to model random arrivals, queues etc. Although we have already seen examples of stochastic processes – martingales in discrete and continuous time – let us finally give a formal definition of a stochastic process.

48.1 Definition. Let $(\Omega,\mathscr{A},\mathbb{P})$ be a probability space, $(E,\mathscr{E})$ a measurable space and $\mathbb{T}$ an (totally) ordered index set. A **(stochastic) process** $X=(X_t)_{t\in\mathbb{T}}$ is a family of E-valued random variables $X_t:\Omega\to E$.

A **filtration** $(\mathscr{F}_t)_{t\in\mathbb{T}}$ is a family of σ-algebras such that $\mathscr{F}_s\subset\mathscr{F}_t$ if $s\leqslant t$. A process is **adapted** to the filtration if each X_t is $\mathscr{F}_t$-measurable. Notation: $(X_t,\mathscr{F}_t)_{t\in\mathbb{T}}$. The filtration $\mathscr{F}_t^X:=\sigma(X_s,s\leqslant t)$ is the **natural** or **canonical filtration.**

- $\mathbb{T}$ is usually interpreted as »time«;
- E is called the **state space**;
- X is said to be discrete if $\mathbb{T}$ or E is a discrete set.

In the language of Definition 48.1 a discrete martingale $(X_n, \mathscr{F}_n)_{n\in\mathbb{N}_0}$ is a real-valued, adapted discrete stochastic process with time set $\mathbb{T} = \mathbb{N}_0$; similarly, a continuous-time martingale $(X_t, \mathscr{F}_t)_{t\geqslant 0}$ is a stochastic process.

48.2 Definition. Let $(\sigma_k)_{k\in\mathbb{N}}$ be iid copies of $\sigma \sim \mathsf{Exp}(\lambda)$, $\lambda > 0$ and $\tau_0 := 0$, $\tau_k := \sigma_1 + \dots + \sigma_k$. The process $(N_t)_{t\geqslant 0}$ defined by[(1)]

$$N_t(\omega) := \sum_{k=1}^{\infty} \mathbb{1}_{(0,t]}(\tau_k(\omega)), \quad t > 0, \quad N_0(\omega) := 0, \tag{48.1}$$

is called a **Poisson process** with **rate** (or **intensity**) λ. Notation: $\mathrm{PP}(\lambda)$.

48.3 Remark. a) If we interpret

- σ_k as waiting time between the arrival of customers $k-1$ and k,
- $\tau_k = \sigma_1 + \dots + \sigma_k$ as the arrival time of customer no. k; $\tau_0 := 0$,

then N_t is the number of customers in the queue at time $t \geqslant 0$. The graph of a path $t \mapsto N_t(\omega)$ is shown in Fig. 48.1. Each path $t \mapsto N_t(\omega)$ is a **càdlàg** (right

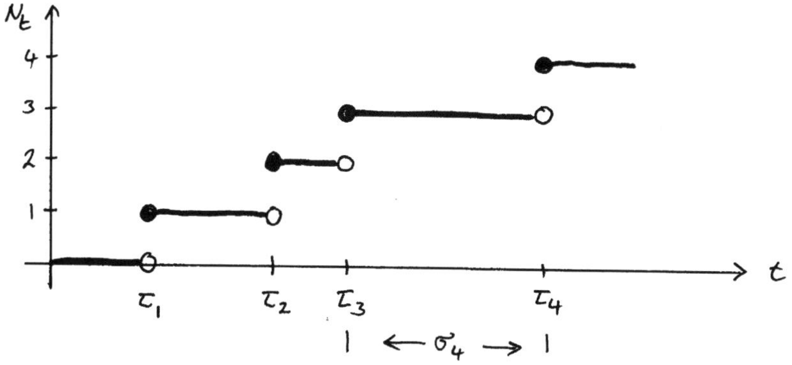

Fig. 48.1. A typical path of a Poisson process $\mathsf{Poi}(\lambda)$.

continuous, finite left limits) function.

b) We have the following important relation

$$\{N_t \leqslant n-1\} = \underbrace{\{N_t < n\}}_{\text{less than } n \text{ customers in the queue at time } t} = \overbrace{\{\tau_n > t\}}^{\text{customer no. } n \text{ arrives after time } t}.$$

Directly from the definition (48.1) we see that

$$N_t(\omega) = \#\big\{k \in \mathbb{N} \mid \tau_k(\omega) \leqslant t\big\}. \tag{48.2}$$

(1) An expression of the type $\sum_1^0$ (lower index larger than upper index) is said to be an **empty sum** and it is defined to be 0.

Notice that $\mathbb{P}(\tau_n < \infty) = 1$. This follows from the above equality since

$$\mathbb{P}(\tau_n = \infty) = \lim_{t\to\infty} \mathbb{P}(\tau_n > t) = \lim_{t\to\infty} \mathbb{P}(N_t < n) = \lim_{t\to\infty} e^{-t\lambda} \sum_{k=0}^{n-1} \frac{(t\lambda)^k}{k!} = 0$$

(here we use that $\lambda > 0$, otherwise $\mathbb{P}(\tau_1 = \infty) = 1$).

c) We can see $N_t(\omega) = N(\omega, (0,t])$ as a **random measure**. Observe that

$$\mathbb{1}_{(0,t]}(\tau_k) = \delta_{\tau_k}((0,t]) \implies N_t = N((0,t]) = \sum_{k=1}^{\infty} \delta_{\tau_k}((0,t]),$$

$$N(\omega, A) = \sum_{k=1}^{\infty} \delta_{\tau_k(\omega)}(A), \quad A \in \mathscr{B}[0,\infty). \tag{48.3}$$

Therefore,

$$\int_0^\infty f(t)\,dN_t = \int_0^\infty f(t)\,N(dt) = \sum_{k=1}^{\infty} f(\tau_k)$$

for all positive measurable functions $f : [0,\infty) \to [0,\infty)$ or for all measurable $f : [0,\infty) \to \mathbb{R}$ such that $\sum_{k=1}^{\infty} |f(\tau_k)| < \infty$ a.s.

In particular, $\int_{(a,b]} dN_t = \sum_{k=1}^{\infty} \mathbb{1}_{(a,b]}(\tau_k) = N_b - N_a$ is the number of arrivals in the time interval $(a,b]$.

The reason why $(N_t)_{t\geqslant 0}$ is said to be a »Poisson« process is that each rv is Poisson distributed: $N_t \sim \mathsf{Poi}(\lambda t)$. In order to show this, we need an auxiliary result.

48.4 Theorem (Campbell formula). *Let $(N_t)_{t\geqslant 0}$ be a PP(λ). For all functions $f \in \mathcal{L}_b^0[0,\infty)$ with compact support and $s \geqslant 0$ it holds that*

$$\mathbb{E} e^{i\int_0^\infty f(t+s)\,dN_t} = e^{-\lambda\int_0^\infty (1-\exp[if(t+s)])\,dt}. \tag{48.4}$$

Proof. By assumption, $\tau_k = \sigma_1 + \cdots + \sigma_k$ with iid $\sigma_k \sim \mathsf{Exp}(\lambda)$. Therefore,

$$\begin{aligned}
\phi(s) :=& \mathbb{E}\exp\left(i\int_0^\infty f(s+t)\,dN_t\right) \\
=& \mathbb{E}\exp\left(i\sum_{k=1}^{\infty} f(s+\sigma_1+\cdots+\sigma_k)\right) \\
\overset{\text{iid}}{=}& \int_0^\infty \underbrace{\mathbb{E}\exp\left(i\sum_{k=2}^{\infty} f(s+x+\sigma_2+\cdots+\sigma_k)\right)}_{=\phi(s+x)} \cdot \underbrace{e^{if(s+x)}}_{=:\gamma(s+x)} \underbrace{\mathbb{P}(\sigma_1 \in dx)}_{=\lambda e^{-\lambda x}\,dx} \\
=& \lambda \int_0^\infty \phi(s+x)\gamma(s+x)e^{-\lambda x}\,dx \\
=& \lambda e^{\lambda s} \int_s^\infty \gamma(t)\phi(t)e^{-\lambda t}\,dt.
\end{aligned}$$

This can be rewritten in the following form

$$e^{-\lambda s}\phi(s) = \lambda \int_s^\infty (e^{-\lambda t}\phi(t))\gamma(t)\,dt$$

and this integral equation leads to a differential equation

$$\phi'(t) - \lambda\phi(t) = -\lambda\phi(t)\gamma(t)$$

with »initial« condition $\phi(\infty) = 1$ – here we use that $\operatorname{supp} f$ is compact. This ODE can be solved in the usual way

$$\frac{\phi'(t)}{\phi(t)} = \lambda(1-\gamma(t)) \xrightarrow{\int_s^\infty \dots dt} \log\phi(\infty) - \log\phi(s) = \lambda \int_s^\infty (1-\gamma(t))\,dt,$$

i.e. $\phi(s) = e^{-\lambda\int_0^\infty (1-\gamma(t+s))\,dt}$ with $\gamma(t+s) = e^{if(t+s)}$. This is **some** solution to our integral equation, but we need to know that it is the only solution.

⚠ We may have lost information (resulting in additional solutions) as we pass from the integral equation to the differential equation.

⚠ Although our ODE is of the form $y' = F(x,y)$, the function F is not continuous, so standard uniqueness results for ODEs do not (directly) apply.

Let us show the uniqueness of the solution to the integral equation. To do so, write

$$e^{-\lambda s}\phi(s) = \phi(0) - \lambda \int_0^s (\phi(t)e^{-\lambda t})\gamma(t)\,dt \tag{48.5}$$

where we use that $\phi(0) = \mathbb{E}\exp\left(i\int_0^\infty f(t)\,dN_t\right)$ exists. We continue our proof after a brief (technical) interlude. ▮▮

48.5 Lemma (Gronwall). *Let $u, a, b : [0,\infty) \to [0,\infty)$ be positive measurable functions such that*

$$u(s) \leqslant a(s) + \int_0^s b(t)u(t)\,dt \quad \forall s \geqslant 0. \tag{48.6}$$

Then u is bounded by

$$u(s) \leqslant a(s) + \int_0^s a(t)b(t)\exp\left(\int_t^s b(r)\,dr\right)dt \quad \forall s \geqslant 0. \tag{48.7}$$

If $a(s) \equiv \alpha$ and $b(s) \equiv \beta$, this simplifies to

$$u(s) \leqslant \alpha e^{\beta s} \quad \forall s \geqslant 0. \tag{48.8}$$

Proof. If we set $y(s) := \int_0^s b(t)u(t)\,dt$, then (48.6) entails

$$y'(s) - b(s)y(s) \leqslant a(s)b(s) \quad \text{Lebesgue a.e.} \tag{48.9}$$

We define $z(s) := y(s)\exp\left(-\int_0^s b(t)\,dt\right)$, differentiate and use (48.9) to get

$$z'(s) \leqslant a(s)b(s)\exp\left(-\int_0^s b(t)\,dt\right) \quad \text{Lebesgue a.e.}$$

If we integrate this expression and observe that $z(0) = y(0) = 0$, we see

$$z(s) \leqslant \int_0^s a(t)b(t)\exp\left(-\int_0^t b(r)\,dr\right)dt,$$

and so

$$y(s) \leqslant \int_0^s a(t)b(t)\exp\left(\int_t^s b(r)\,dr\right)dt \quad \forall s > 0.$$

This yields (48.7) since $u(s) \leqslant a(s) + y(s)$. □

Conclusion of the proof of Theorem 48.4. Assume that (48.5) admits two solutions ϕ and ψ. Since $\phi(0) = \psi(0)$, we can subtract these solutions from each other and obtain

$$\underbrace{\left|e^{-\lambda s}[\phi(s) - \psi(s)]\right|}_{=u(s) \text{ in §48.5}} \leqslant \lambda \int_0^s \underbrace{\left|[\phi(t) - \psi(t)]e^{-\lambda t}\right|}_{=u(t) \text{ in §48.5}} |\gamma(t)|\,dt.$$

Applying Gronwall's lemma §48.5 shows that $u \equiv 0$ or $\phi \equiv \psi$. □

We will need one further concept.

48.6 Definition. A stochastic process $(X_t)_{t\geqslant 0}$ taking values in $\mathbb{R}^d$ is called a **Lévy process**, if

$$X_0 = 0 \quad \text{a.s.} \tag{L0}$$

$$X_t - X_s \sim X_{t-s} - X_0 \quad \forall s \leqslant t \tag{L1}$$

$$(X_{t_k} - X_{t_{k-1}})_{k=1}^n \text{ independent } \forall n \in \mathbb{N},\ \forall t_0 = 0 < t_1 < \cdots < t_n \tag{L2}$$

$$\mathbb{P}\left(t \mapsto X_t \text{ is càdlàg}\right) = 1. \tag{L3}$$

- Property (L1) is often called **stationary increments** and (L2) is called **independent increments**.
- Arguably the most important Lévy process is Brownian motion, see Ch. 58ff.

48.7 Theorem. *A Poisson process $(N_t)_{t\geqslant 0}$ is a Lévy process. Moreover,*

$$\mathbb{E}e^{i\xi N_t} = e^{-t\lambda(1-\exp[i\xi])} \tag{48.10}$$

and, in particular, $N_t \sim \mathsf{Poi}(\lambda t)$.

Proof. Fix $t_0 = 0 < t_1 < \cdots < t_n$ and $\xi_1, \dots, \xi_n \in \mathbb{R}$, and define

$$f(t) := \sum_{k=1}^{n} \xi_k \mathbb{1}_{(t_{k-1},t_k]}(t) \implies \int f(t)\,\mathrm{d}N_t = \sum_{k=1}^{n} \xi_k (N_{t_k} - N_{t_{k-1}})$$

(recall that N_t is a measure, see Remark 48.3.c)). A simple direct calculation (✍, compare with (27.9)) shows that

$$1 - e^{if(t)} = 1 - e^{i\sum_{k=1}^{n} \xi_k \mathbb{1}_{(t_{k-1},t_k]}(t)} = \sum_{k=1}^{n} (1 - e^{i\xi_k}) \mathbb{1}_{(t_{k-1},t_k]}(t),$$

and so

$$\int \left(1 - e^{if(t)}\right) \mathrm{d}t = \sum_{k=1}^{n} \left(1 - e^{i\xi_k}\right)(t_k - t_{k-1}).$$

Using Theorem 48.4 gives

$$\begin{aligned} \mathbb{E} e^{i\sum_{k=1}^{n} \xi_k (N_{t_k} - N_{t_{k-1}})} &= e^{-\lambda \sum_{k=1}^{n} (1-\exp[i\xi_k])(t_k - t_{k-1})} \\ &= \prod_{k=1}^{n} e^{-\lambda (t_k - t_{k-1})(1-\exp[i\xi_k])}. \end{aligned} \tag{48.11}$$

We can now check the properties (L0)–(L3).

1° Pick $n = 2$, $t_2 = t$, $t_1 = s$, $\xi_2 = \xi$, $\xi_1 = 0$.

This yields $\mathbb{E} e^{i\xi(N_t - N_s)} = e^{-\lambda(t-s)(1-\exp[i\xi])}$ proving (48.10) and (L1).

2° Inserting the result of 1° on the rhS of (48.11) shows

$$\mathbb{E} e^{i\sum_{k=1}^{n} \xi_k (N_{t_k} - N_{t_{k-1}})} = \prod_{k=1}^{n} \mathbb{E} e^{i\xi_k (N_{t_k} - N_{t_{k-1}})},$$

and we get (L2) using Kac's theorem (Corollary 27.8).

3° The properties (L0) and (L3) are trivial. □

There is also the following converse to Theorem 48.7.

48.8 Theorem. *A nontrivial*[2] *Lévy process $(N_t)_{t \geqslant 0}$ which changes its value only by jumps $\Delta N_t := N_t - N_{t-} \in \{0, 1\}$ is a Poisson process.*

Proof. We define the »jump times«

$$T_0 := 0, \quad T_1 := \inf\{t > T_0 \mid \Delta N_t = 1\}, \quad \dots \quad T_{k+1} := \inf\{t > T_k \mid \Delta N_t = 1\}$$

and observe that

$$\{T_k > t\} = \{N_t < k\} = \{N_t \leqslant k - 1\}, \quad t \geqslant 0,\ k \in \mathbb{N}.$$

In particular, T_k is a stopping time for the filtration $\mathscr{F}_t^N = \sigma(N_r, r \leqslant t)$.

[2] This means that $\mathbb{P}(N_t \neq 0) > 0$.

1° Since $t \mapsto N_t(\omega)$ is càdlàg, the sequence $(T_k(\omega))_{k\in\mathbb{N}}$ cannot have condensation points in compact intervals – otherwise N_t would explode.

2° For any $s, t > 0$ we have

$$\begin{aligned}
\mathbb{P}(T_1 > t+s) &= \mathbb{P}(T_1 > t+s,\ T_1 > t) \\
&= \mathbb{P}(N_{t+s} = 0,\ N_t = 0) \\
&= \mathbb{P}(N_{t+s} - N_t = 0,\ N_t = 0) \\
&\overset{\text{(L2)}}{=} \mathbb{P}(N_{t+s} - N_t = 0) \cdot \mathbb{P}(N_t = 0) \\
&\overset{\text{(L1)}}{=} \mathbb{P}(N_s = 0) \cdot \mathbb{P}(N_t = 0) \\
&= \mathbb{P}(T_1 > s) \cdot \mathbb{P}(T_1 > t)
\end{aligned}$$

and, by Corollary 47.4, this implies that $T_1 \sim \mathsf{Exp}(\lambda)$. In particular, $T_1 < \infty$ a.s.

3° Assume that $T_k^n(\omega) \downarrow T_k(\omega) = \inf_{n\in\mathbb{N}} T_k^n(\omega)$. Using the continuity of measures,

$$\{T_k \geqslant t\} = \bigcap_{n\in\mathbb{N}} \{T_k^n \geqslant t\}^{(3)} \implies \mathbb{P}(T_k \geqslant t) = \lim_{n\to\infty} \mathbb{P}(T_k^n \geqslant t).$$

Similarly, we get $\mathbb{P}(T_k < s) = \lim_{n\to\infty} \mathbb{P}(T_k^n < s)$ – mind the strict reverse inequality inside the set.

4° Define for $n \in \mathbb{N}$, $k \in \mathbb{N}$ the dyadic approximations $T_k^n := \dfrac{\lfloor 2^n T_k \rfloor + 1}{2^n}$, see Remark 46.13.f). For every $i \in \mathbb{N}$ we have

$$\begin{aligned}
\left\{T_k^n = \frac{i}{2^n}\right\} &= \left\{\frac{i-1}{2^n} \leqslant T_k < \frac{i}{2^n}\right\} \\
&= \underbrace{\left\{N_{i/2^n-} - N_{(i-1)/2^n-} = 1\right\}}_{\text{jump } \#k \text{ in } [(i-1)2^{-n}, i2^{-n})} \cap \underbrace{\left\{N_{(i-1)/2^n-} - N_0 = k-1\right\}}_{k-1 \text{ jumps in } [0,(i-1)2^{-n})}.
\end{aligned}$$

Using (L2) and $N_{t-1/\ell} \uparrow N_{t-}$ as $\ell \to \infty$, we see

$$\left\{T_k^n = \frac{i}{2^n}\right\} = \bigcup_{\ell=1}^{\infty} \left\{N_{i/2^n-1/\ell} - N_{(i-1)/2^n-1/\ell} = 1\right\} \cap \left\{N_{(i-1)/2^n-1/\ell} - N_0 = k-1\right\}$$
$$\perp\!\!\!\perp\ N_t - N_s \quad \forall s \geqslant i2^{-n}.$$

5° We will now show that $T_{k+1} - T_k \perp\!\!\!\perp T_1, \dots, T_k$. Assume that we already know that $T_1, \dots, T_k < \infty$ a.s. Let $t_1 < \dots < t_k$, $s > 0$.

(3) Indeed, $\omega \in \{T_k \geqslant t\} \iff \inf_{n\in\mathbb{N}} T_k^n(\omega) \geqslant t \iff \forall n : T_k^n(\omega) \geqslant t \iff \omega \in \bigcap_{n\in\mathbb{N}} \{T_k^n \geqslant t\}$.

$$\begin{aligned}
&\mathbb{P}(T_1 \geqslant t_1, \dots, T_k \geqslant t_k, \underbrace{T_{k+1} - T_k > s}_{=\{T_{k+1}-s>T_k\}}) \\
&\overset{3^\circ}{=} \lim_{n\to\infty} \mathbb{P}(T_1^n \geqslant t_1, \dots, T_k^n \geqslant t_k, \underbrace{T_{k+1} - s > T_k^n}_{=\left\{N_{T_k^n+s}-N_{T_k^n}=0\right\}}) \\
&= \lim_{n\to\infty} \sum_{i_\ell \geqslant t_\ell 2^n, \ell=1,\dots,k} \mathbb{P}\Big(\bigcap_{\ell=1}^k \{T_\ell^n = i_\ell 2^{-n}\} \cap \{N_{i_k 2^{-n}+s} - N_{i_k 2^{-n}} = 0\}\Big) \\
&\overset{(L2)}{\underset{4^\circ}{=}} \lim_{n\to\infty} \sum_{i_\ell \geqslant t_\ell 2^n, \ell=1,\dots,k} \mathbb{P}\Big(\bigcap_{\ell=1}^k \{T_\ell^n = i_\ell 2^{-n}\}\Big) \underbrace{\mathbb{P}\Big(N_{i_k 2^{-n}+s} - N_{i_k 2^{-n}} = 0\Big)}_{=\mathbb{P}(N_s=0)=\mathbb{P}(T_1>s),\ (L1)} \\
&= \mathbb{P}(T_1 \geqslant t_1, \dots, T_k \geqslant t_k) \cdot \mathbb{P}(T_1 > s).
\end{aligned}$$

If we set $t_1 = t_2 = \dots = t_k = 0$ and observe that $\{T_\ell \geqslant 0\} = \Omega$, then we get that $\mathbb{P}(T_{k+1} - T_k > s) = \mathbb{P}(T_1 > s)$; this proves, in particular, that $T_{k+1} < \infty$ a.s. Moreover, we conclude that

$$\mathbb{P}(T_1 \geqslant t_1, \dots, T_k \geqslant t_k, T_{k+1} - T_k > s) = \mathbb{P}(T_1 \geqslant t_1, \dots, T_k \geqslant t_k)\mathbb{P}(T_{k+1} - T_k > s);$$

this proves independence (Theorem 25.9.c)).

6° The random variables $(T_k - T_{k-1})_{k\in\mathbb{N}}$ are iid $\mathsf{Exp}(\lambda)$-rv. From 5° and 2° we know that $T_k - T_{k-1} \overset{5^\circ}{\sim} T_1 \overset{2^\circ}{\sim} \mathsf{Exp}(\lambda)$. Independence follows recursively:

$$\begin{aligned}
&\mathbb{P}\Big(T_{k+1} - T_k \geqslant s_{k+1}, \underbrace{T_k - T_{k-1} \geqslant s_k, \dots, T_1 - T_0 \geqslant s_1}_{\sigma(T_1,\dots,T_k)\text{-measurable}}\Big) \\
&\quad\overset{5^\circ}{=} \mathbb{P}\Big(T_{k+1} - T_k \geqslant s_{k+1}\Big) \cdot \mathbb{P}\Big(T_k - T_{k-1} \geqslant s_k, \dots, T_1 - T_0 \geqslant s_1\Big) \\
&\quad= \dots = \prod_{\ell=0}^k \mathbb{P}\Big(T_{\ell+1} - T_\ell \geqslant s_{\ell+1}\Big).
\end{aligned}$$

Thus, the differences $\sigma_k := T_k - T_{k-1}$, $k \in \mathbb{N}$, are the waiting times of a Poisson process. □

⚠ The proof of Theorem 48.8 shows a bit more: Since we can express N_t for $t \in (0, T_n]$ through $T_1, \dots, T_n$ – and vice versa – we get

$$\sigma(N_t, t \leqslant T_n) \overset{\text{def}}{=} \sigma(N_{t\wedge T_n}, t \geqslant 0) = \sigma(T_1, \dots, T_n). \tag{48.12}$$

48.9 Corollary. *A Poisson process* $\mathsf{PP}(\lambda)$ *has the* **strong Markov property**: *For each jump time* T_n, $N' := (N'_t)_{t\geqslant 0}$, $N'_t := N_{T_n+t} - N_{T_n}$, *is a Poisson process* $\mathsf{PP}(\lambda)$ *and* $N' \perp\!\!\!\perp \sigma(N_t, t \leqslant T_n)$.

Corollary 48.9 follows immediately from the observation preceding the statement and the fact that $T_{n+k} - T_n$ is the kth jump time T'_k of the process N', see Fig. 48.2.

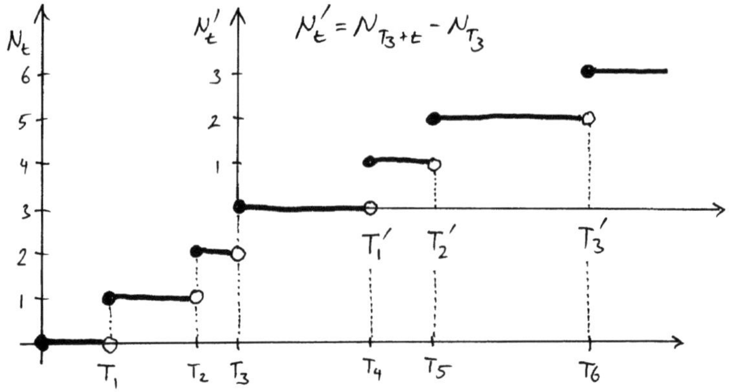

Fig. 48.2. The strong Markov property for the jump time T_3.

48.10 Corollary. *Every Lévy process* $N = (N_t)_{t\geqslant 0}$ *such that* $N_t \sim \mathsf{Poi}(\lambda t)$ *is a Poisson process* $PP(\lambda)$.

Proof. Since for all $s \leqslant t$, $N_t - N_s \sim N_{t-s} \in \mathbb{N}_0$, the paths $t \mapsto N_t$ are increasing càdlàg step functions, and $\Delta N_t = N_t - N_{t-} \in \mathbb{N}_0$. We have to show $\Delta N_t \in \{0,1\}$. Write $T_1, T_2, \ldots$ for the jump times.

As in the proof of Theorem 48.8, Step 1° we see that the sequence $(T_k(\omega))_{k\in\mathbb{N}}$ is discrete, and so

$$\sup_{r\leqslant t} \Delta N_r(\omega) = \lim_{n\to\infty} \max_{1\leqslant k\leqslant n} \left(N_{\frac{kt}{n}}(\omega) - N_{\frac{(k-1)t}{n}}(\omega) \right).$$

Note that for every $n \in \mathbb{N}$

$$\begin{aligned}
\mathbb{P}\left(\sup_{r\leqslant t} \Delta N_r \leqslant 1\right) &\geqslant \mathbb{P}\left(\max_{1\leqslant k\leqslant n} \left(N_{\frac{kt}{n}} - N_{\frac{(k-1)t}{n}}\right) \leqslant 1\right) \\
&= \mathbb{P}\left(N_{\frac{t}{n}} - N_0 \leqslant 1, \ldots, N_{\frac{kt}{n}} - N_{\frac{(k-1)t}{n}} \leqslant 1, \ldots, N_{\frac{nt}{n}} - N_{\frac{(n-1)t}{n}} \leqslant 1\right) \\
&\overset{(L2)}{=} \prod_{k=1}^{n} \mathbb{P}\left(N_{\frac{kt}{n}} - N_{\frac{(k-1)t}{n}} \leqslant 1\right) \\
&\overset{(L1)}{=} \mathbb{P}\left(N_{\frac{t}{n}} \leqslant 1\right)^n \\
&= \Big[\underbrace{e^{-\lambda\frac{t}{n}}}_{=\mathbb{P}\left(N_{\frac{t}{n}}=0\right)} + \underbrace{\frac{\lambda t}{n} e^{-\lambda\frac{t}{n}}}_{=\mathbb{P}\left(N_{\frac{t}{n}}=1\right)} \Big]^n \xrightarrow[n\to\infty]{} 1.
\end{aligned}$$

□

49 PPs, Markov processes and martingales

In this chapter, $N = (N_t)_{t \geqslant 0}$ is a Poisson process and $\mathscr{F}_t := \mathscr{F}_t^N = \sigma(N_r, r \leqslant t)$ is the natural filtration. Our first aim is to extend Corollary 48.9 and the independent increments property (L2).

49.1 Lemma. *Property (L2) of N, i.e. »$N_{t_k} - N_{t_{k-1}}$ are independent rv for any $t_0 = 0 < t_1 < \dots < t_n$« is equivalent to*

$$\forall s \leqslant t: \quad N_t - N_s \perp\!\!\!\perp \mathscr{F}_s^N. \tag{L2'}$$

Proof. (L2')⇒(L2): Pick any $t_0 = 0 < t_1 < \dots < t_n$ and $\xi_1, \dots, \xi_n \in \mathbb{R}$, and note that

$$\begin{aligned}
\mathbb{E}\left(e^{i\sum_{k=1}^n \xi_k (N_{t_k} - N_{t_{k-1}})}\right) &\overset{\text{tower}}{=} \mathbb{E}\left(\mathbb{E}\left[e^{i\sum_{k=1}^n \xi_k (N_{t_k} - N_{t_{k-1}})} \,\Big|\, \mathscr{F}_{t_{n-1}}^N\right]\right) \\
&\overset{\text{pull}}{=} \mathbb{E}\left(\mathbb{E}\left[e^{i\xi_n (N_{t_n} - N_{t_{n-1}})} \,\Big|\, \mathscr{F}_{t_{n-1}}^N\right] e^{i\sum_{k=1}^{n-1} \xi_k (N_{t_k} - N_{t_{k-1}})}\right) \\
&\overset{(L2')}{=} \mathbb{E}\left(e^{i\xi_n (N_{t_n} - N_{t_{n-1}})}\right) \cdot \mathbb{E}\left(e^{i\sum_{k=1}^{n-1} \xi_k (N_{t_k} - N_{t_{k-1}})}\right) \\
&= \dots = \prod_{k=1}^n \mathbb{E}\left(e^{i\xi_k (N_{t_k} - N_{t_{k-1}})}\right).
\end{aligned}$$

(L2)⇒(L2'): Let $0 = s_0 < s_1 < \dots < s_n \leqslant s < t$. Since the random variables $(N_t - N_s)$ and $(N_{s_k} - N_{s_{k-1}})$, $k = 1, \dots, n$ are independent, we conclude that

$$N_t - N_s \perp\!\!\!\perp \left(N_{s_1}, \dots, N_{s_k}, \dots N_{s_n}\right), \quad N_{s_k} = \sum_{i=1}^k \left(N_{s_i} - N_{s_{i-1}}\right).$$

This means that

$$N_t - N_s \perp\!\!\!\perp \bigcup_{n, 0 < s_1 < \dots < s_n \leqslant s} \sigma\left(N_{s_1}, \dots, N_{s_n}\right)$$

and, since the family on the right is ∩-stable, Theorem 25.5 shows that

$$N_t - N_s \perp\!\!\!\perp \sigma\left(\bigcup_{n, 0 < s_1 < \dots < s_n \leqslant s} \sigma\left(N_{s_1}, \dots, N_{s_n}\right)\right) \overset{✎}{=} \mathscr{F}_s^N. \qquad \square$$

We can now show Corollary 48.9 for general stopping times. Recall from Definition 46.12 that $T : \Omega \to [0, \infty]$ is a stopping time if $\{T \leqslant t\} \in \mathscr{F}_t$ for all $t \geqslant 0$; please refer to Remark 46.13 for essential properties.

49.2 Example. a) The time T_k of the kth jump of a Poisson process N is a stopping time for the natural filtration: $\{T_k > t\} = \{N_t < k\} \in \mathscr{F}_t^N$.

b) The random time $S := \inf\{t \geqslant a \mid N_t = N_{t-a}\}$ is the end-point of the first interval of length a without any jump; S is a stopping time for the natural filtration.

Indeed: $S(\omega)$ happens **after** some jump time $T_n(\omega)$; thus,

$$S = T_n + a \iff T_1 \leqslant a, T_2 - T_1 \leqslant a, \dots, T_n - T_{n-1} \leqslant a, T_{n+1} - T_n > a$$

and, by Remark 46.13 e),

$$\begin{aligned}\{S \leqslant t\} &= \bigcup_{n=0}^{\infty} \overbrace{\bigcap_{k=1}^{n}}^{n=0 \rightsquigarrow \bigcap_1^0 \cdots := \Omega} \{T_k - T_{k-1} \leqslant a\} \cap \{T_{n+1} - T_n > a\} \cap \{T_n + a \leqslant t\} \\ &= \bigcup_{n=0}^{\infty} \bigcap_{k=1}^{n} \underbrace{\{T_k \leqslant T_{k-1} + a\}}_{\in \mathscr{F}_{T_k} \subset \mathscr{F}_{T_n+a}} \cap \underbrace{\{T_{n+1} > T_n + a\}}_{\in \mathscr{F}_{T_n+a}} \cap \{T_n + a \leqslant t\} \in \mathscr{F}_t.\end{aligned}$$

The last step relies on the definition of $\mathscr{F}_{T_n+a}$: If $F \in \mathscr{F}_{T_n+a}$, then we have $F \cap \{T_n + a \leqslant t\} \in \mathscr{F}_t$.

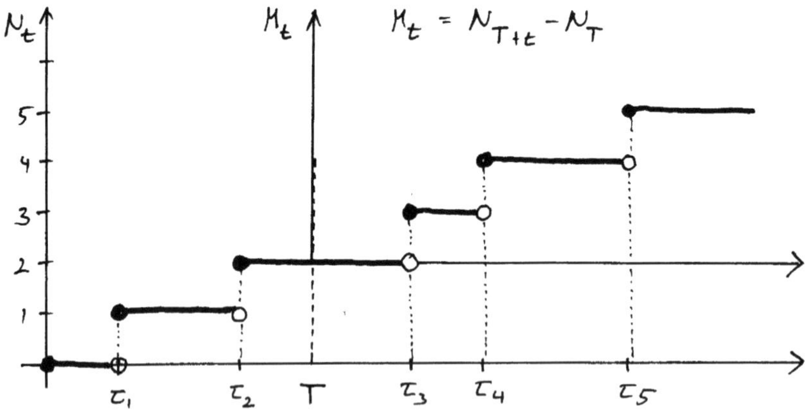

Fig. 49.1. The strong Markov property for a general stopping time T.

49.3 Theorem (Strong Markov property – SMP). *Let* N *be a Poisson process* PP(λ) *and* T *a stopping time such that* $\mathbb{P}(T < \infty) > 0$. *The process* $M = (M_t)_{t\geqslant 0}$, $M_t := N_{T+t} - N_T$, *is again a Poisson process* PP(λ) *on the space* $\{T < \infty\}$ *with* $\mathscr{A} \cap \{T < \infty\}$ *and* $\mathbb{P}'(F) := \mathbb{P}(F \cap \{T < \infty\})/\mathbb{P}(T < \infty) = \mathbb{P}(F \mid \{T < \infty\})$; *moreover,* M *is conditionally on* $\{T < \infty\}$ *independent of* $\mathscr{F}_T$.
Amendment: $(N_{T+t} - N_T)_{t\geqslant 0}$ *is conditionally on* $\{T < \infty\}$ *independent of* $(N_t)_{t\leqslant T}$.

Proof. It is clear that M has càdlàg paths which are either constant or have jumps of size 1. In view of Theorem 48.8 it is enough to show that M is a Lévy process, i.e. a càdlàg process with stationary and independent increments. Pick $0 = t_0 < t_1 < \cdots < t_n$, $\xi_1, \ldots, \xi_n \in \mathbb{R}$, $F \in \mathscr{F}_T$ and consider the dyadic approximation $T^m = (\lfloor T 2^m \rfloor + 1)2^{-m}$ (cf. Remark 46.13.f)). Using the independent and stationary increments property of N we get

$$
\begin{aligned}
&\mathbb{E}\Big(\mathbb{1}_{F\cap\{T<\infty\}} e^{i\sum_{k=1}^{n}\xi_k(M_{t_k}-M_{t_{k-1}})}\Big)\\
&= \lim_{m\to\infty}\mathbb{E}\Big(\mathbb{1}_{F\cap\{T<\infty\}} e^{i\sum_{k=1}^{n}\xi_k(N_{t_k+T^m}-N_{t_{k-1}+T^m})}\Big)\\
&= \lim_{m\to\infty}\sum_{j=0}^{\infty}\mathbb{E}\Big(\mathbb{1}_{F\cap\{T<\infty\}\cap\{T^m=j2^{-m}\}} e^{i\sum_{k=1}^{n}\xi_k(N_{t_k+T^m}-N_{t_{k-1}+T^m})}\Big)\\
&= \lim_{m\to\infty}\sum_{j=0}^{\infty}\mathbb{E}\Big(\underbrace{\mathbb{1}_{F\cap\{(j-1)2^{-m}\leqslant T<j2^{-m}\}}}_{\in\mathscr{F}_{j2^{-m}}} e^{i\sum_{k=1}^{n}\xi_k(N_{t_k+j2^{-m}}-N_{t_{k-1}+j2^{-m}})}\Big)\\
&\overset{(L2')}{\underset{(L2)}{=}} \lim_{m\to\infty}\sum_{j=0}^{\infty}\mathbb{E}\Big(\mathbb{1}_{F\cap\{(j-1)2^{-m}\leqslant T<j2^{-m}\}}\Big)\prod_{k=1}^{n}\mathbb{E}\Big(e^{i\xi_k(N_{t_k+j2^{-m}}-N_{t_{k-1}+j2^{-m}})}\Big)\\
&\overset{(L1)}{=} \lim_{m\to\infty}\sum_{j=0}^{\infty}\mathbb{E}\Big(\mathbb{1}_{F\cap\{(j-1)2^{-m}\leqslant T<j2^{-m}\}}\Big)\prod_{k=1}^{n}\mathbb{E}\Big(e^{i\xi_k N_{t_k-t_{k-1}}}\Big)\\
&= \mathbb{E}\Big(\mathbb{1}_{F\cap\{T<\infty\}}\Big)\prod_{k=1}^{n}\mathbb{E}\Big(e^{i\xi_k N_{t_k-t_{k-1}}}\Big).
\end{aligned}
$$

From this identity we conclude the following:

1° Take $F=\Omega$, $m=1$ and $t_1=t$. Dividing both sides by $\mathbb{P}(T<\infty)$, we see that $\mathbb{P}'(M_t\in d\omega)=\mathbb{P}(N_t\in d\omega)=\mathsf{Poi}(t\lambda)$.

2° Take $F=\Omega$ and divide by $\mathbb{P}(T<\infty)$. This shows that the random variables $M_{t_k}-M_{t_{k-1}}$ are independent and, under $\mathbb{P}'$, distributed like $M_{t_k-t_{k-1}}$. This shows (L2) and (L1) for M under $\mathbb{P}'$. Since (L0), (L3) are trivially satisfied, we see that M is a Lévy process, hence a PP(λ) by Theorem 48.8 or Corollary 48.10.

3° Take any $F\in\mathscr{F}_T$. Then we see, as in the proof of Lemma 49.1, that we have $F\cap\{T<\infty\}\perp\!\!\!\perp\mathscr{F}_\infty^M$, since the σ-algebra $\mathscr{F}_\infty^M$ is generated by the increments $M_{t_k}-M_{t_{k-1}}$, $0=t_0<t_1<\cdots<t_m<\infty$.

In order to see the amendment, we observe that

$$\{N_{t\wedge T}\geqslant k\}=\{T_k\leqslant t\wedge T\}\in\mathscr{F}_{t\wedge T}\subset\mathscr{F}_T,$$

see Remark 46.13.d) & e), shows that $N_{t\wedge T}$ is $\mathscr{F}_T$-measurable. □

Since the natural filtration $\mathscr{F}_t^N$ of a Poisson process is determined by the jump times $T_1, T_2, \dots$, every stopping time S is a function of the jump times, i.e. it is an educated guess that there are not too many stopping times apart from the jump times, see however Example 49.2.b).

49.4 Theorem. *Let* N *be a Poisson process* PP(λ), *denote its jump times by* T_k, *and let S be a further stopping time. Then*

$$T_{k-1} \leqslant S < T_k \implies S = T_{k-1}.$$

Proof. Let $T_0 = 0$ and assume that $k = 1$. Since $\mathbb{P}(S < T_1 < \infty) = 1$, we see from Theorem 49.3 that

$$\{T_1 - S > t\} = \{N_{S+t} - N_S = 0\} \perp\!\!\!\perp \mathscr{F}_S^N$$

and

$$\mathbb{P}(N_{S+t} - N_S = 0) = \mathbb{P}(N_t = 0) = e^{-\lambda t}.$$

Therefore, $T_1 - S \sim \mathsf{Exp}(\lambda)$, i.e. $\mathbb{E}(T_1 - S) = 1/\lambda$.

On the other hand, $T_1 \sim \mathsf{Exp}(\lambda)$. Then, $\mathbb{E}T_1 = 1/\lambda$ and $\mathbb{E}S = \mathbb{E}T_1 - \mathbb{E}(T_1 - S) = 0$; thus, $S = 0$ a.s.

Now assume that $k \geqslant 2$. We can use again Theorem 49.3 to see that the process $M_t = N_{T_{k-1}+t} - N_{T_{k-1}}$ is a $PP(\lambda)$. Since the kth jump of N is the first jump of M, we see

$$\begin{aligned} T_k - T_{k-1} &\leftrightarrow T_1^M \quad \text{(stopping time for M)} \\ S - T_{k-1} &\leftrightarrow S^M \quad \text{(stopping time for M)} \end{aligned}$$

and from $k = 1$ we conclude that $S - T_{k-1} \equiv 0$ a.s. or $S = T_{k-1}$ a.s. □

We will now proceed to the characterization of the Poisson process with the help of martingales. The definition of a discrete-time martingale (Def. 38.4) literally transfers to the continuous-time setting (see also §46), i.e. a stochastic process $(X_t, \mathscr{F}_t)_{t\geqslant 0}$ is a martingale if $X_t \in L^1(\mathscr{F}_t)$ and $\mathbb{E}(X_t \mid \mathscr{F}_s) = X_s$ for all $s \leqslant t$.

49.5 Example. Let $(N_t)_{t\geqslant 0}$ be a Poisson process $PP(\lambda)$ and $\mathscr{F}_t := \mathscr{F}_t^N$.

a) $M_t := N_t - \lambda t$ is a martingale. Clearly, $M_t \in L^1(\mathscr{F}_t)$. *Plausibility*: A necessary condition for M to be a martingale is $\mathbb{E}M_t = \mathbb{E}M_0 = 0$; thus, $\mathbb{E}N_t = \lambda t$ indicates that $N_t - \lambda t$ is indeed a good candidate. Now

$$\begin{aligned} \mathbb{E}(N_t - \lambda t \mid \mathscr{F}_s) &= \mathbb{E}(\underbrace{N_t - N_s}_{\perp\!\!\!\perp \mathscr{F}_s} + \underbrace{N_s}_{\mathscr{F}_s\text{-mble}} - \lambda t \mid \mathscr{F}_s) \\ &= \mathbb{E}(N_t - N_s) + N_s - \lambda t = \lambda(t-s) + N_s - \lambda t = M_s. \end{aligned}$$

b) $X_t := (N_t - \lambda t)^2 - \lambda t$ is a martingale. Clearly, $M_t \in L^1(\mathscr{F}_t)$. *Plausibility*: We already know that $M_t := N_t - \lambda t$ is a martingale, i.e. M_t^2 is a sub-martingale and we need a compensator A_t (cf. §38.10ff). $\mathbb{E}(M_t^2 - A_t) = \mathbb{E}(M_0^2 - A_0) = 0$

and $\mathbb{E}M_t^2 = \mathbb{V}N_t = \lambda t$, so $A_t = \lambda t$ seems to be a good choice.

$$\begin{aligned}
\mathbb{E}(M_t^2 - \lambda t \mid \mathscr{F}_s) &= \mathbb{E}\Big((\underbrace{[M_t - M_s]}_{\perp\!\!\!\perp \mathscr{F}_s} + \underbrace{M_s}_{\mathscr{F}_s\text{-mble}})^2 - \lambda t \mid \mathscr{F}_s\Big) \\
&= \mathbb{E}\big([M_t - M_s]^2 \mid \mathscr{F}_s\big) + \mathbb{E}\big(2M_s(M_t - M_s) \mid \mathscr{F}_s\big) + \mathbb{E}\big(M_s^2 \mid \mathscr{F}_s\big) - \lambda t \\
&\overset{\text{(L2)}}{\underset{\text{pull}}{=}} \underbrace{\mathbb{E}\big([M_t - M_s]^2\big)}_{\overset{\text{(L1)}}{=}\mathbb{E}(M_{t-s}^2) = \lambda(t-s)} + 2M_s \underbrace{\mathbb{E}\big(M_t - M_s\big)}_{=0} + M_s^2 - \lambda t \\
&= M_s^2 - \lambda s.
\end{aligned}$$

c) $Y_t := X_t - M_t = (N_t - \lambda t)^2 - N_t$ is a martingale. This shows that both $t \mapsto \lambda t$ and $t \mapsto N_t$ »compensate« the sub-martingale M_t^2 to become a martingale.

⚠ This is not a contradiction to the uniqueness of the compensator (see §38.10.c)) since the process $t \mapsto N_t$ is not predictable (it is »only« adapted) while $t \mapsto \lambda t$ is predictable (it is even deterministic). Please note that we have not defined »predictability« in a continuous-time setting, but the concept is important in connection with stochastic integration, see [25, Chapter IV.5].

49.6 Lemma. *If $(M_t, \mathscr{F}_t)_{t\geqslant 0}$ is a càdlàg martingale and τ a stopping time, then $(M_{\tau\wedge t}, \mathscr{F}_t)_{t\geqslant 0}$ is a martingale.*

Standard proof. Apply the optional stopping theorem (Corollary 46.15): Write $X_t = M_{t\wedge\tau}$, fix $s \leqslant t$ and take $S = s \wedge \tau$ and $T = t \wedge \tau$. □

Alternative ad-hoc argument. The following argument works in the present context since $M_t = N_t - \lambda t$ is the difference of two monotone càdlàg processes. Define the approximations $\tau^m := (\lfloor 2^m \tau \rfloor + 1)/2^m$ and $t^m := (\lfloor 2^m t \rfloor + 1)/2^m$. Since $s^m = i2^{-m}$ with $i \in \mathbb{N}_0$, Example 49.5.a) shows that

$$(M_{i2^{-m}}, \mathscr{F}_{i2^{-m}})_{i\in\mathbb{N}_0}$$

is a discrete martingale and, by the discrete optional stopping theorem §39.4,

$$(M_{i2^{-m}\wedge\tau^m}, \mathscr{F}_{i2^{-m}})_{i\in\mathbb{N}_0}$$

is also a discrete martingale. Since $s \leqslant s^m$, we see that $F \in \mathscr{F}_s \subset \mathscr{F}_{s^m}$, and we get for all $s \leqslant t$

$$\begin{array}{ccc}
\mathbb{E}(M_{t^m\wedge\tau^m}\mathbb{1}_F) & = & \mathbb{E}(M_{s^m\wedge\tau^m}\mathbb{1}_F) \\
\Big\downarrow {\scriptstyle m\to\infty} & & \Big\downarrow {\scriptstyle m\to\infty} \\
\mathbb{E}(M_{t\wedge\tau}\mathbb{1}_F) & = & \mathbb{E}(M_{s\wedge\tau}\mathbb{1}_F).
\end{array} \tag{49.1}$$

This proves the claim if we can justify the limits appearing in (49.1).

The **standard argument** uses backwards martingales and uniform integrability (Theorem 43.5)

With our application in mind we can use that $M_t = N_t - \lambda t$ for montone and right-continuous $t \mapsto N_t$ and $t \mapsto \lambda t$ to derive the L^1-convergence with the help of the monotone convergence theorem. □

49.7 Theorem. *Assume that $(N_t)_{t\geqslant 0}$, is a counting process, i.e. $N_0 = 0$ and $t \mapsto N_t$ changes its value only through jumps of size 1.*

$$(N_t)_{t\geqslant 0} \text{ is a PP}(\lambda) \iff \widetilde{N}_t := N_t - \lambda t \text{ is a mg w.r.t. } \mathscr{F}_t^N.$$

Proof. The direction »⇒« is Example 49.5.a). The converse direction »⇐« is proved in several steps.

1° Let $(T_n)_{n\in\mathbb{N}}$ be the jump times of $(N_t)_{t\geqslant 0}$. Clearly, $N_{T_n} = n$ and $N_{T_n-} = n-1$. For every $f : \mathbb{N}_0 \to \mathbb{R}$ we have

$$\begin{aligned}\int_{(0,t]} [f(N_{u-}+1) - f(N_{u-})]\,dN_u &\overset{\text{def}}{=} \sum_{n:\, T_n\leqslant t} \left[f(N_{T_n-}+1) - f(N_{T_n-})\right] \\ &= \sum_{n:\, T_n\leqslant t} [f(n) - f(n-1)] \\ &= f(N_t) - f(0).\end{aligned}$$

2° Let $r < t$. It is easy to check ✍(use a case-by-case analysis)

$$\mathbb{1}_{(0,t]}(T_n) = 1 \iff T_n \leqslant t \iff N_{T_n\wedge t} - N_{T_{n-1}\wedge t} = 1$$

and, consequently,

$$\begin{aligned}\int_{(0,t]} [f(N_{u-}+1) - f(N_{u-})]\,dN_u &\overset{\text{def}}{=} \sum_{n=1}^{\infty} \left[f(N_{T_n-}+1) - f(N_{T_n-})\right] \mathbb{1}_{(0,t]}(T_n) \\ &= \sum_{n=1}^{\infty} [f(n) - f(n-1)]\left(N_{T_n\wedge t} - N_{T_{n-1}\wedge t}\right).\end{aligned}$$

Now we perform the following operations

- we use that $\int_{(r,t]} \cdots = \int_{(0,t]} \cdots - \int_{(0,r]} \cdots$;
- we apply the conditional expectation $\mathbb{E}(\cdots \mid \mathscr{F}_r)$ on both sides;
- and use linearity to see that $\sum_n \mathbb{E}(\cdots \mid \mathscr{F}_r) = \mathbb{E}(\sum_n \cdots \mid \mathscr{F}_r)$;
- finally, Lemma 49.6 applies: Since $M_t = N_t - \lambda t$, we have
$$\mathbb{E}\left(N_{T_n\wedge t} - N_{T_n\wedge r} \mid \mathscr{F}_r\right) = \lambda \mathbb{E}(T_n \wedge t - T_n \wedge r \mid \mathscr{F}_r).$$

Together, this yields

$$\begin{aligned}
&\mathbb{E}(f(N_t)-f(N_r)\mid\mathscr{F}_r)\\
&=\lambda\mathbb{E}\left(\sum_{n=1}^{\infty}[f(n)-f(n-1)](T_n\wedge t-T_{n-1}\wedge t-T_n\wedge r+T_{n-1}\wedge r)\;\middle|\;\mathscr{F}_r\right)\\
&=\lambda\mathbb{E}\left(\sum_{n=1}^{\infty}\int_{T_{n-1}\wedge t}^{T_n\wedge t}\underbrace{[f(N_{u-}+1)-f(N_{u-})]}_{T_{n-1}\wedge t<u\leqslant T_n\wedge t\implies N_{u-}=n-1}\,du-\int_{T_{n-1}\wedge r}^{T_n\wedge r}\dots du\;\middle|\;\mathscr{F}_r\right)\\
&=\lambda\mathbb{E}\left(\int_r^t\left[f(N_{u-}+1)-f(N_{u-})\right]du\;\middle|\;\mathscr{F}_r\right).
\end{aligned}$$

3° We can now take a specific f:

- let $f(n)=1-e^{-\xi n}$, $\xi>0$;
- multiply both sides with $e^{\xi N_r}\mathbb{1}_F$ for $F\in\mathscr{F}_r$ and use a »pull-in«;
- apply the expectation $\mathbb{E}(\dots)$ on both sides,

and we see that

$$\begin{aligned}
\mathbb{E}\left(\left(1-e^{-\xi(N_t-N_r)}\right)\mathbb{1}_F\right)&=\lambda\mathbb{E}\left(\int_r^t\left[e^{-\xi N_{u-}}-e^{-\xi(N_{u-}+1)}\right]e^{\xi N_r}\mathbb{1}_F\,du\right)\\
&=\lambda(1-e^{-\xi})\mathbb{E}\left(\int_r^t e^{-\xi(N_{u-}-N_r)}\mathbb{1}_F\,du\right).
\end{aligned}$$

For every ω, the set $\{u : N_{u-}(\omega)\neq N_u(\omega)\}$ is countable, hence a Lebesgue null set. Therefore, we can replace $N_{u-}(\omega)$ and we get, after an obvious re-arrangement,

$$\begin{aligned}
\mathbb{E}\left(e^{-\xi(N_t-N_r)}\mathbb{1}_F\right)&=\mathbb{P}(F)-\lambda(1-e^{-\xi})\mathbb{E}\left(\int_r^t e^{-\xi(N_u-N_r)}\mathbb{1}_F\,du\right)\\
&=\mathbb{P}(F)-\lambda(1-e^{-\xi})\int_r^t\mathbb{E}\left(e^{-\xi(N_u-N_r)}\mathbb{1}_F\right)du.
\end{aligned}$$

4° This integral equation has a unique solution – use Gronwall's Lemma §48.5 and argue as in the proof of Theorem 48.4 – which is

$$\mathbb{E}\left(e^{-\xi(N_t-N_r)}\mathbb{1}_F\right)=\mathbb{P}(F)e^{-(t-r)\lambda(1-\exp[-\xi])}\quad\forall r<t\ \forall F\in\mathscr{F}_r.$$

Consequently,

- take $F=\Omega$ to get $N_t-N_r\sim N_{t-r}\sim\mathsf{Poi}(\lambda(t-r))$;
- take any $F\in\mathscr{F}_r$ to get $N_t-N_r\perp\!\!\!\perp\mathscr{F}_r$.

Now we can use Corollary 48.10 to deduce that N is a PP(λ). □

50 Superposititition, thinning and colouring of PPs

We will now study some transformations of a Poisson process. Again, $(N_t)_{t\geqslant 0}$ is a PP(λ) and we write $(\tau_k)_{k\in\mathbb{N}}$ for its jump times and set $\tau_0 = 0$. Recall that

$$N_t = N(0,t] = \sum_{k\in\mathbb{N}} \delta_{\tau_k}((0,t])$$

which means that we may interpret N_t as a measure

$$N(B) = N(\omega, B) = \sum_{k\in\mathbb{N}} \delta_{\tau_k(\omega)}(B), \quad B \in \mathscr{B}[0,\infty). \tag{50.1}$$

50.1 Theorem. *Let* $B_1,\dots,B_n \in \mathscr{B}[0,\infty)$ *be disjoint sets of finite Lebesgue measure* $|B_k| < \infty$.

$$N(B_1),\dots,N(B_n) \text{ are independent rv and } N(B_k) \sim \mathsf{Poi}(\lambda|B_k|).$$

💬 Random measures such that $N(B_k)$ are independent random variables for disjoint sets $B_1,\dots,B_n$ are often called **independently scattered**.

Proof. As in the proof of Theorem 48.7 we pick $\xi_1,\dots,\xi_n \in \mathbb{R}$ and define

$$f(t) = \sum_{k=1}^{n} \xi_k \mathbb{1}_{B_k}(t) \implies 1 - e^{if(t)} = \sum_{k=1}^{n} (1 - e^{i\xi_k}) \mathbb{1}_{B_k}(t).$$

Campbell's formula (Theorem 48.4) yields

$$\mathbb{E} e^{i\sum_{k=1}^{n}\xi_k N(B_k)} = \mathbb{E} e^{i\int f(t)\,dN_t} = e^{-\lambda\int_0^\infty (1-e^{if(t)})\,dt} = \prod_{k=1}^{n} e^{-\lambda|B_k|(1-e^{i\xi_k})};$$

taking $n = 1$ shows that $N(B) \sim \mathsf{Poi}(\lambda|B|)$. For general n we see independence with Kac's theorem (Corollary 27.8). □

50.2 Corollary. *Let* $A, B \in \mathscr{B}[0,\infty)$ *and define* $\widetilde{N}(A) := N(A) - \lambda|A|$ *where* $|A|$ *stands for Lebesgue measure* Leb(A). *If* $|A|, |B| < \infty$,

$$\mathbb{E}\widetilde{N}(A) = 0 \quad \text{and} \quad \mathbb{E}\widetilde{N}(A)\widetilde{N}(B) = \lambda|A \cap B|.$$

Proof. If $N(A) \sim \mathsf{Poi}(\lambda|A|)$, then $\mathbb{E}N(A) = \lambda|A|$ or, equivalently, $\mathbb{E}\widetilde{N}(A) = 0$.

Since $N(A)$ (and $|A|$) is a measure, we see that $\widetilde{N}(A) = \widetilde{N}(A \setminus B) + \widetilde{N}(A \cap B)$. Therefore,

$$\begin{aligned}\mathbb{E}\widetilde{N}(A)\widetilde{N}(B) &= \mathbb{E}\big(\widetilde{N}(A\setminus B) + \widetilde{N}(A\cap B)\big)\big(\widetilde{N}(B\setminus A) + \widetilde{N}(A\cap B)\big)\\ &\overset{\S}{=} \mathbb{E}\widetilde{N}(A\cap B)^2\\ &= \mathbb{V}N(A\cap B) = \lambda|A\cap B|.\end{aligned}$$

In the step marked with »§« we use the fact that for disjoint sets $C \cap D = \emptyset$ we have $\widetilde{N}(C) \perp\!\!\!\perp \widetilde{N}(D)$, hence $\mathbb{E}\widetilde{N}(C)\widetilde{N}(D) = \mathbb{E}\widetilde{N}(C)\cdot\mathbb{E}\widetilde{N}(D) = 0$. □

💬 Corollary 50.2 shows that $A \mapsto \widetilde{N}(A)$ is a (signed) **random orthogonal measure**. This opens up new possibilities, in particular, an approach to stochastic integration, see [BARCA, Chapter 10].

50.3 Lemma. *Assume that N and M are independent Poisson processes*[(1)] *with parameter λ and μ, respectively. Then*

$$\mathbb{P}\Big(\text{N and M jump at the same time}\Big) = 0.$$

Proof. Denote by $\tau_1, \tau_2, \dots$ the jump times of N. We have

$$\sum_{t>0} \Delta N_t \Delta M_t = \sum_{k\in\mathbb{N}} \Delta M_{\tau_k} \underbrace{\Delta N_{\tau_k}}_{=1} = \sum_{k\in\mathbb{N}} \Delta M_{\tau_k}.$$

Since $\tau_k \perp\!\!\!\perp M$, we see that

$$\mathbb{E}\sum_{k\in\mathbb{N}} \Delta M_{\tau_k} \overset{\text{Tonelli}}{=} \sum_{k\in\mathbb{N}} \mathbb{E}(\Delta M_{\tau_k}) \overset{\perp\!\!\!\perp}{=} \sum_{k\in\mathbb{N}} \mathbb{E}\Big[\mathbb{E}(\Delta M_t)\Big|_{t=\tau_k}\Big] = 0;$$

for the last equality we use $\mathbb{E}M_{t-} = \lim_{s\uparrow t} \mathbb{E}M_s = \lim_{s\uparrow t} \mu s = \mu t = \mathbb{E}M_t$. □

50.4 Theorem (superposition). *Let N^i be independent $PP(\lambda_i)$, $i \in \mathbb{N}$, such that $\lambda := \sum_{i=1}^\infty \lambda_i < \infty$. The superposition $N_t := \sum_{i=1}^\infty N_t^i$ is a $PP(\lambda)$.*

Proof. From Theorem 47.8 we know that $N_t \sim \mathsf{Poi}(\lambda t)$, since $\lambda < \infty$. Because of $\sum_{i=1}^\infty N_t^i < \infty$ a.s., the path $t \mapsto N_t$ is a càdlàg, monotonically increasing step function, and with the help of Lemma 50.3 we conclude that $\Delta N_t \in \{0, 1\}$.

Moreover, all increments of all processes

$$N_{t_k}^i - N_{t_{k-1}}^i,\ N_t^i - N_s^i \quad t_0 = 0 < t_1 < \dots < t_n \leqslant s < t,\ n \in \mathbb{N},\ i \in \mathbb{N}$$

are independent, and so

$$N_t - N_s \perp\!\!\!\perp N_{t_k} - N_{t_{k-1}} = \sum_{i=1}^\infty (N_{t_k}^i - N_{t_{k-1}}^i) \quad \forall t_0 = 0 < t_1 < \dots < t_n \leqslant s.$$

Finally, $N_t - N_s = \sum_{i=1}^\infty (N_t^i - N_s^i) \sim \bigstar_{i=1}^\infty \mathsf{Poi}(\lambda_i(t-s)) = \mathsf{Poi}(\lambda(t-s))$, cf. Theorem 47.8.

Now we can apply Theorem 48.8 to finish the proof. □

Recall from Remark 47.7.d) (with $n = 2$ and $p_1 = p$, $p_2 = q = (1-p)$ that

$$N_t \sim \mathsf{Poi}(\lambda t)\ \&\ \mathbb{P}(M_t \in \bullet \mid N_t) \sim \mathsf{Bin}(N_t, p) \implies \begin{cases} M_t \sim \mathsf{Poi}(p\lambda t), \\ N_t - M_t \sim \mathsf{Poi}(q\lambda t), \\ M_t \perp\!\!\!\perp N_t - M_t. \end{cases}$$

(1) This means that $\sigma(N_t, t \geqslant 0) \perp\!\!\!\perp \sigma(M_t, t \geqslant 0)$.

Mark the jumps of a $PP(\lambda)$ in red »1« or green »0«, choosing the colour randomly according to an iid sequence $(\beta_k)_{k\in\mathbb{N}}$ of Bernoulli rv.

50.5 Theorem (thinning). *Let N be a $PP(\lambda)$ whose jump times are denoted by τ_k, and let $(\beta_n)_{n\in\mathbb{N}}$ a sequence of iid Bernoulli $\mathsf{Bin}(p)$-rv which is independent of N; set $q = 1-p$. The processes*

$$N'_t := \sum_{k=1}^{\infty} \mathbb{1}_{(0,t]}(\tau_k)\beta_k \quad \text{and} \quad N''_t := \sum_{k=1}^{\infty} \mathbb{1}_{(0,t]}(\tau_k)(1-\beta_k)$$

are independent Poisson processes $PP(p\lambda)$ and $PP(q\lambda)$, respectively.

Proof. Clearly, $t \mapsto N'_t$ and $t \mapsto N''_t$ are càdlàg counting processes. Define $S(n) := \beta_1 + \dots + \beta_n$. Obviously, $N'_t = S(N_t)$ is the number of successes in the sequence β_k for $k \leqslant N_t$, and $N''_t = N_t - S(N_t)$.

By Remark 47.7.d), $N'_t \perp\!\!\!\perp N''_t$, $N'_t \sim \mathsf{Poi}(p\lambda t)$ and $N''_t \sim \mathsf{Poi}(q\lambda t)$.

Fix $s < t$ and pick $F \in \sigma(S(N_u), N_u - S(N_u), u \leqslant s)$. In the next calculation we use the following observations

a) $N_t - N_s$ is independent of N_s, F and $S(n)$;
b) F is independent of $S(x+N_s) - S(N_s)$ and we have $N_t - N_s \sim N_{t-s}$;
c) N_s is independent of S and $S(x+k) - S(k) \sim S(x)$;
d) N_{t-s} is independent of S;
e) $N'_{t-s} = S(N_{t-s})$ and $N''_{t-s} = N_{t-s} - S(N_{t-s})$ are independent (see the beginning of this proof).

$$\begin{aligned}
&\mathbb{E}\left(e^{i\xi[S(N_t)-S(N_s)]} e^{i\eta[N_t - S(N_t) - (N_s - S(N_s))]} \mathbb{1}_F\right) \\
&= \mathbb{E}\left(e^{i(\xi-\eta)[S((N_t-N_s)+N_s)-S(N_s)]} e^{i\eta(N_t-N_s)} \mathbb{1}_F\right) \\
&\overset{a)}{=} \int \mathbb{E}\left(e^{i(\xi-\eta)[S(x+N_s)-S(N_s)]} e^{i\eta x} \mathbb{1}_F\right) \mathbb{P}((N_t - N_s) \in dx) \\
&\overset{b)}{=} \int \mathbb{E}\left(e^{i(\xi-\eta)[S(x+N_s)-S(N_s)]} e^{i\eta x}\right) \mathbb{E}(\mathbb{1}_F)\, \mathbb{P}(N_{t-s} \in dx) \\
&\overset{c)}{=} \int \mathbb{E}\left(e^{i(\xi-\eta)S(x)} e^{i\eta x}\right) \mathbb{P}(N_{t-s} \in dx)\, \mathbb{E}(\mathbb{1}_F) \\
&\overset{d)}{=} \mathbb{E}\left(e^{i\xi S(N_{t-s})} e^{i\eta[N_{t-s} - S(N_{t-s})]}\right) \mathbb{E}(\mathbb{1}_F) \\
&\overset{e)}{=} \mathbb{E}\left(e^{i\xi N'_{t-s}}\right) \mathbb{E}\left(e^{i\eta N''_{t-s}}\right) \mathbb{E}(\mathbb{1}_F).
\end{aligned}$$

1° Taking $F = \Omega$, $\eta = 0$, we conclude that $N'_t - N'_s \sim N'_{t-s}$.
2° Taking $F = \Omega$, $\xi = 0$, we conclude that $N''_t - N''_s \sim N''_{t-s}$.
3° 1° & 2° show that $N'_t - N'_s$, $N''_t - N''_s$, and $F \in \sigma(N'_r, N''_r, r \leqslant s)$ are independent. In particular, $N'_t - N'_s \perp\!\!\!\perp \sigma(N'_r, r \leqslant s)$ as well as $N''_t - N''_s \perp\!\!\!\perp \sigma(N''_r, r \leqslant s)$.
4° From 1° –3° we conclude with the help of Theorem 48.8 that both N' and N'' are Poisson processes with intensity $p\lambda$ and $q\lambda$, respectively.

5° Applying 3° repeatedly ✎ we see that for $0 = t_0 < t_1 < \cdots < t_n$ all random variables $N'_{t_k} - N'_{t_{k-1}}, N''_{t_\ell} - N''_{t_{\ell-1}}, k, \ell = 1, \dots, n$, are independent. This shows that

$$\underbrace{\sigma(N'_{t_k} - N'_{t_{k-1}}, 1 \leqslant k \leqslant n)}_{\overset{\S 38.2}{=} \sigma(N'_{t_k}, 1 \leqslant k \leqslant n)} \perp\!\!\!\perp \underbrace{\sigma(N''_{t_k} - N''_{t_{k-1}}, 1 \leqslant k \leqslant n)}_{\overset{\S 38.2}{=} \sigma(N''_{t_k}, 1 \leqslant k \leqslant n)};$$

we can now argue like (at the end of the proof) in Lemma 49.1 and see that $\sigma(N'_t, t \geqslant 0) \perp\!\!\!\perp \sigma(N''_t, t \geqslant 0)$. □

Let us finally discuss the so-called **compound Poisson process**.

50.6 Definition. Let $N = (N_t)_{t \geqslant 0}$ denote a Poisson process $PP(\lambda)$ and let $H = (H_n)_{n \in \mathbb{N}}$ be a sequence of iid random variables which is independent of N. The stochastic process defined by[2]

$$X_t := \sum_{n=1}^{N_t} H_n, \tag{50.2}$$

is called a **compound Poisson process** (cPP).

50.7 Remark. a) $t \mapsto X_t$ is a càdlàg step function satisfying

$$\Delta X_t = \sum_{n=1}^{N_t} H_n - \sum_{n=1}^{N_{t-}} H_n = \begin{cases} 0, & N_t = N_{t-} \\ H_n, & N_t = n \neq N_{t-} \end{cases} = (\Delta N_t) H_{N_t}.$$

This means that X has the same jump times as N, but the jump sizes H_n are now iid random. Clearly, $H_n \equiv 1 \implies X_t = N_t$. Notice that, by construction, the interarrival times $(\sigma_n)_{n \in \mathbb{N}} \perp\!\!\!\perp (H_n)_{n \in \mathbb{N}}$.

b) Let $\tau_n = \sigma_1 + \cdots + \sigma_n$ be the jump times of N. We may write X_t as

$$X_t = \sum_{n=1}^{\infty} H_n \mathbb{1}_{(0,t]}(\tau_n) = \sum_{n=1}^{\infty} H_n \delta_{\tau_n}((0,t]);$$

so we may define an integral »driven by« dX_t, see Remark 48.3.c).

c) We can calculate the characteristic function of X_t. Assume that $H_1 \sim \mu$ and

(2) By definition, the »empty sum« and »empty product« are $\sum_{n=1}^{0} := 0$ and $\prod_{n=1}^{0} := 1$.

set $\chi(\xi) := \mathbb{E}e^{i\xi H_1}$. Using the independence of N and $(H_n)_{n\in\mathbb{N}}$ we see[(2)]

$$\begin{aligned}\mathbb{E}e^{i\xi X_t} = \mathbb{E}e^{i\xi\sum_{n=1}^{N_t} H_n} &= \sum_{k=0}^{\infty} \mathbb{E}e^{i\xi\sum_{n=1}^{k} H_n}\mathbb{P}(N_t = k) \\ &\overset{\text{iid}}{=} \sum_{k=0}^{\infty}\prod_{n=1}^{k} \mathbb{E}e^{i\xi H_n}\frac{(t\lambda)^k}{k!}e^{-t\lambda} \\ &\overset{\text{iid}}{=} \sum_{k=0}^{\infty}\chi(\xi)^k\frac{(t\lambda)^k}{k!}e^{-t\lambda} \\ &= e^{-t\lambda}e^{t\lambda\chi(\xi)} \\ &= e^{-t\lambda(1-\chi(\xi))} \qquad (50.3) \\ &= e^{-t\lambda\int(1-\exp[ix\xi])\,\mu(dx)}. \qquad (50.4)\end{aligned}$$

We also have the analogue of Campbell's formula (Theorem 48.4) for a cPP. As for a Poisson process we see

$$\int_0^\infty g(t)\,dX_t = \sum_{n=1}^{\infty} g(\underbrace{\sigma_1 + \cdots + \sigma_n}_{\text{iid Exp}(\lambda)})H_n,$$

and we get

50.8 Theorem (Campbell formula for cPP). *Let $X = (X_t)_{t\geqslant 0}$ denote a cPP with jump times $\tau_n = \sigma_1 + \cdots + \sigma_n$ and iid jump heights $H_n \sim \mu$.*

$$\mathbb{E}\exp\left(i\int_0^\infty f(t+s)\,dX_t\right) = \exp\left(-\lambda\int_0^\infty\int\left(1-e^{iyf(s+t)}\right)\mu(dy)\,dt\right) \qquad (50.5)$$

holds for $s \geqslant 0$ and all measurable $f : [0,\infty) \to \mathbb{R}$ with compact support.

Proof. The argument parallels the proof of Theorem 48.4, but with a different integral equation. Here we get

$$\begin{aligned}\phi(s) &:= \mathbb{E}\exp\left(i\int_0^\infty f(s+t)\,dX_t\right) \\ &= \mathbb{E}\exp\left(i\sum_{k=1}^{\infty} f(s+\sigma_1+\cdots+\sigma_k)H_k\right) \\ &\overset{\text{iid}}{=} \int_0^\infty \underbrace{\mathbb{E}\exp\left(i\sum_{k=2}^{\infty} f(s+x+\sigma_2+\cdots+\sigma_k)H_k\right)}_{=\phi(s+x)}\underbrace{\mathbb{E}\exp(if(s+x)H_1)}_{=:\gamma(s+x)}\underbrace{\mathbb{P}(\sigma_1\in dx)}_{=\lambda e^{-\lambda x}\,dx} \\ &= \lambda\int_0^\infty \phi(s+x)\gamma(s+x)e^{-\lambda x}\,dx \\ &= \lambda e^{\lambda s}\int_s^\infty \gamma(t)\phi(t)e^{-\lambda t}\,dt.\end{aligned}$$

Now we use that $H_1 \sim \mu$, i.e.

$$\gamma(s+x) = \mathbb{E}e^{if(s+x)H_1} = \int e^{iyf(s+x)}\,\mu(dy). \qquad \square$$

If we take in (50.6) $H_n \equiv 1$, then $\mu = \delta_1$ and we recover (48.4). We can use (50.6) to show the counterpart of Theorem 48.7 for cPPs.

50.9 Corollary. *A compound Poisson process* $(X_t)_{t\geqslant 0}$ *is a Lévy process and*

$$\mathbb{E}e^{i\xi X_t} = e^{-t\int_{\mathbb{R}}(1-\exp[i\xi y])\,\nu(dy)} \tag{50.6}$$

for some finite measure ν.

Proof. In order to see that X is a Lévy process, we can argue as in Theorem 48.7; we have already calculated the characteristic function in Remark 50.7.b) where $\nu(dy) := \lambda\cdot\mu(dy)$. $\square$

We close this chapter with a variant of the strong law of large numbers.

50.10 Theorem (LLN). *If* $(N_t)_{t\geqslant 0}$ *is a* PP(λ), *then* $\lim_{t\to\infty}\frac{1}{t}N_t = \lambda$ *a.s.*

Proof. The increments $N_k - N_{k-1} \in L^1(\mathbb{P})$ are iid rv. Using Kolmogorov's SLLN (Theorem 32.1) we see that

$$\frac{N_n}{n} = \frac{1}{n}\sum_{k=1}^{n}(N_k - N_{k-1}) \xrightarrow[n\to\infty]{\text{a.s.}} \mathbb{E}N_1 = \lambda.$$

Since $t \mapsto N_t$ is monotone, we can use a »sandwiching« argument

$$\forall n \leqslant t \leqslant n+1 : \quad \frac{n}{n+1}\frac{N_n}{n} = \frac{N_n}{n+1} \leqslant \frac{N_t}{t} \leqslant \frac{N_{n+1}}{n} = \frac{n+1}{n}\cdot\frac{N_{n+1}}{n+1}.$$

As the left- and right-hand sides converge a.s. to λ, the claim follows. $\square$

XI
Markov Chains

51 Random walks on the lattice $\mathbb{Z}^d$

We have already studied in Theorem 31.9 the fluctuation behaviour of a simple random walk on $\mathbb{Z}$. Recall that this was a stochastic process $S_n = X_1 + \cdots + X_n$ with iid steps $X_k \sim \mathsf{Bin}(\frac{1}{2})$. In this chapter we want to extend this study to arbitrary dimensions and not necessarily symmetric steps. In the notation introduced in Definition 48.1 this means that we consider stochastic processes with state space $(\mathbb{Z}^d, \mathscr{P}(\mathbb{Z}^d))$ and time set $\mathbb{T} = \mathbb{N}_0$:

$$(\Omega, \mathscr{A}, \mathbb{P}, (S_n)_{n\in\mathbb{N}}, \mathbb{Z}^d, \mathscr{P}(\mathbb{Z}^d)).$$

The guiding idea is to interpret S_n as the position (at time n) of a drunkard in a »city« $\mathbb{Z}^d$ who randomly chooses his next step $X_{n+1} = S_{n+1} - S_n$, see Fig. 51.1; we assume that the steps are iid.

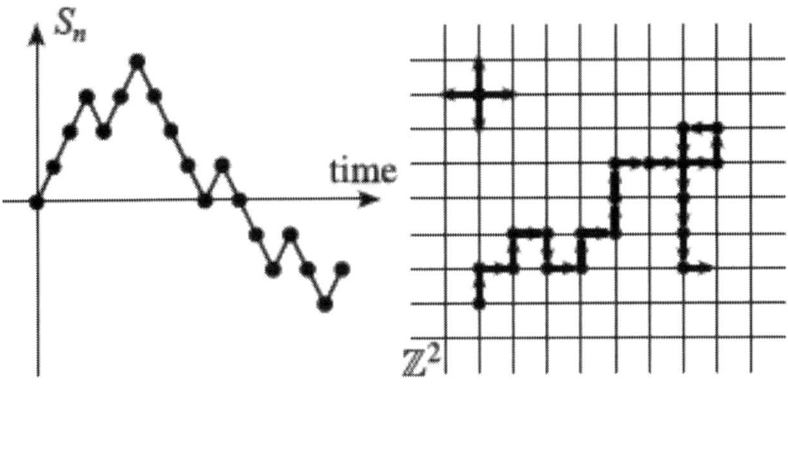

Fig. 51.1. A random walk in $d = 1$ and in New York ($d = 2$).

51.1 Definition. A **random walk** (RW) is a stochastic process of the form

$$S_n = \overbrace{S_0 + \underbrace{X_1 + \dots + X_n}_{\text{iid steps}}}^{\text{independent rvs}}, \quad n \in \mathbb{N}_0,$$

where the rvs $S_0, X_1, \dots, X_n$ are independent and the **steps** $X_1, \dots, X_n$ are iid.

A RW is said to be **simple** (SRW), if $S_0 \in \mathbb{Z}^d$ and $|X_1| = 1$; in particular, $\mathbb{P}(X_1 = e) = p_e$, $|e| = 1$, and $\sum_{|e|=1} p_e = 1$.

A SRW is said to be **symmetric**, if $p_e = 1/(2d)$ for all $|e| = 1$.

51.2 Remark. Here are some immediate consequences from the definition of a RW. As usual, let $n \geqslant m$,

a) Unless otherwise mentioned, we consider the natural (canonical) filtration $\mathscr{F}_n := \sigma(S_0, X_1, \dots, X_n) = \sigma(S_0, S_1, \dots, S_n)$ (✎ or Lemma 38.2).

b) $$S_n - S_m = X_n + \dots + X_{m+1} \rightsquigarrow \begin{cases} \perp\!\!\!\perp S_0, X_1, \dots, X_m \\ \perp\!\!\!\perp S_0 + X_1 + \dots + X_m = S_m \\ \sim X_1 + \dots + X_{n-m} = S_{n-m} - S_0 \end{cases}.$$

c) $S' := (S_{m+i} - S_m)_{i \in \mathbb{N}}$ is again a RW, $S'_0 = 0$ and $S' \perp\!\!\!\perp (S_i)_{0 \leqslant i \leqslant m}$, cf. Fig. 51.2. Note that the RWs S' and $S - S_0 = (S_n - S_0)_{n \in \mathbb{N}_0}$ have the same probability law since they are constructed from iid random steps. This means that they behave alike.

d) If $S_0, X_1 \in L^1$, then $(S_n - n\mathbb{E}X_1, \mathscr{F}_n)_{n \in \mathbb{N}_0}$ is a mg, cf. Example 49.5.a).

e) If $X_1 \in L^2$, then $((S_n - S_0 - n\mathbb{E}X_1)^2 - n\mathbb{V}X_1, \mathscr{F}_n)_{n \in \mathbb{N}_0}$ is a martingale, cf. Example 49.5.b).

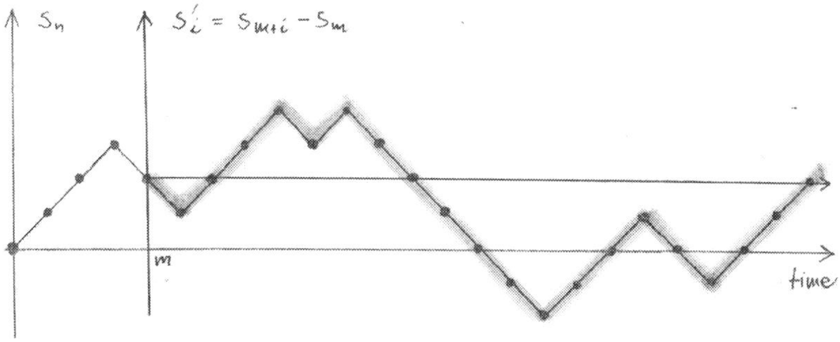

Fig. 51.2. The shift $S'_i := S_{m+i} - S_m$ gives us a new RW in a new coordinate grid; S' starts at 0 and it is independent of $(S_0, S_1, \dots, S_m)$.

We want to know how quickly a symmetric RW leaves an interval (in $d = 1$) or a region $\subset \mathbb{Z}^d$ – and whether it returns. For this, we need a few preparations.

51.3 Theorem (Wald's identities). *Let* $S_n = S_0 + X_1 + \cdots + X_n$ *be SRW and* T *a stopping time such that* $\mathbb{E}T < \infty$.

a) *If* $X_1 \in L^1$, *then*

$$S_T - S_0 \in L^1 \quad \text{and} \quad \mathbb{E}(S_T - S_0) = \mathbb{E}T \cdot \mathbb{E}X_1.$$

b) *If* $X_1 \in L^2$, *then*

$$S_T - S_0 \in L^2 \quad \text{and} \quad \mathbb{E}\left[(S_T - S_0 - T\cdot\mathbb{E}X_1)^2\right] = \mathbb{E}T \cdot \mathbb{V}X_1.$$

c) * *If* S *is a general RW (with possibly unbounded steps) with* $X_1 \in L^2$, *then* a) *and* b) *remain valid.*

Proof. a) The process $M_n := S_n - S_0 - n\mathbb{E}X_1$ is a martingale with bounded steps, i.e. it satisfies the assumptions of the optional sampling theorem §39.6.c). Thus,

$$M_T \in L^1, \quad \mathbb{E}M_T = \mathbb{E}M_0 = 0 \implies S_T \in L^1, \quad \mathbb{E}(S_T - S_0) = \mathbb{E}T \cdot \mathbb{E}X_1.$$

b) Similar to the proof of a), just use the mg $R_n := (S_n - S_0 - n\mathbb{E}X_1)^2 - n\mathbb{V}X_1$.

c)* If the steps of S are unbounded random variables, we consider first bounded stopping times $T_n := T \wedge n$ and use optional sampling in the form §39.6.a) to conclude that

$$\begin{aligned} \mathbb{E}(S_{T_n} - S_0) &= \mathbb{E}(T_n) \cdot \mathbb{E}X_1, \\ \mathbb{E}\left[(S_{T_n} - S_0 - T_n \cdot \mathbb{E}X_1)^2\right] &= \mathbb{E}(T_n) \cdot \mathbb{V}X_1. \end{aligned} \tag{51.1}$$

We want to take the limit $n \to \infty$. Define $M_n := S_n - S_0 - n\mathbb{E}X_1$; by optional stopping, $(M_{T_n})_{n\in\mathbb{N}}$ is a martingale. Thus, for all $n \geqslant m$

$$\mathbb{E}\left[(M_{T_n} - M_{T_m})^2\right] \overset{\S 41.3}{=} \mathbb{E}\left[M_{T_n}^2 - M_{T_m}^2\right] = \mathbb{E}(T_n - T_m) \cdot \mathbb{V}X_1 \xrightarrow[m,n\to\infty]{\text{DCT}} 0;$$

for the limit we use that $\mathbb{E}T < \infty$, i.e. $T \wedge n \uparrow T < \infty$ a.s. and T is an integrable majorant. Thus, $M_{T_n} \to M_T$ in both L^2 and L^1, which means that we can take limits in (51.1) to get the two Wald identities. □

51.4 Example (gambler's ruin). Let S be a one-dimensional simple RW starting at $S_0 = 0$ and with steps $X_i \sim p\delta_1 + q\delta_{-1}$ and $p + q = 1$.

a) Let $p = q = \frac{1}{2}$. When does S_n visit the position 1 for the first time? Define

$$T = T_1 := \inf\{n \in \mathbb{N} \mid S_n = 1\}$$

and check that it is a stopping time ✎. In this case $\mathbb{E}T < \infty$ **cannot hold** – even though $\mathbb{P}(T < \infty) = 1$ –, since otherwise we would have the following contradiction:

$$1 = \mathbb{E}S_T \overset{\S 51.3}{=} \mathbb{E}T \cdot \mathbb{E}X_1 = \mathbb{E}T \cdot 0 = 0.$$

b) Let $p = q = \frac{1}{2}$. We are interested in the first passage time of S_n at an upper or lower bound. If we interpret S_n as our fortune during play, the upper bound would be an aim we want to reach, the lower bound would be »ruin«: Let $a, b \in \mathbb{Z}$, $a < 0 < b$.

$$T = T(a,b) := \inf\{n \in \mathbb{N} : S_n = b \ \text{ or } \ S_n = a\} = T_a \wedge T_b.$$

We know from Theorem 31.9 or Example 39.5.c) that $\mathbb{P}(T < \infty) = 1$; let us assume for now that $\mathbb{E}T < \infty$ (cf. Theorem 51.5 for a proof). From Theorem 51.3.a) we get

$$0 = \mathbb{E}S_T = a\mathbb{P}(S_T = a) + b\mathbb{P}(S_T = b) \quad \text{and} \quad \mathbb{P}(S_T = a) + \mathbb{P}(S_T = b) = 1;$$

if we solve for $\mathbb{P}(S_T = a)$ and $\mathbb{P}(S_T = b)$ we get

$$\mathbb{P}(S_T = a) = \frac{b}{b-a} \quad \text{and} \quad \mathbb{P}(S_T = b) = \frac{-a}{b-a}.$$

Moreover, Theorem 51.3.b) shows

$$\underbrace{\mathbb{V}X_1}_{=1} \mathbb{E}T = \mathbb{E}S_T^2 = \mathbb{P}(S_T = a)a^2 + \mathbb{P}(S_T = b)b^2 = -ab.$$

c) Assume that $p \neq q$. We consider the martingale ✍ $M_n := (q/p)^{S_n}$.

Let $T = T(a,b)$ be the stopping time from b). Since martingales have constant expectations, we see

$$1 = \mathbb{E}\left(\frac{q}{p}\right)^{S_0} = \mathbb{E}\left(\frac{q}{p}\right)^{S_T} = \mathbb{P}(S_T = a)\left(\frac{q}{p}\right)^a + \mathbb{P}(S_T = b)\left(\frac{q}{p}\right)^b.$$

Since $\mathbb{P}(S_T = a) + \mathbb{P}(S_T = b) = 1$, we have 2 equations in 2 unknown variables, and we get

$$\mathbb{P}(S_T = a) = \frac{\left(\frac{q}{p}\right)^b - 1}{\left(\frac{q}{p}\right)^b - \left(\frac{q}{p}\right)^a} \quad \text{and} \quad \mathbb{P}(S_T = b) = \frac{1 - \left(\frac{q}{p}\right)^a}{\left(\frac{q}{p}\right)^b - \left(\frac{q}{p}\right)^a}.$$

If we combine this with the first Wald identity, $\mathbb{E}S_T = (p-q)\mathbb{E}T$, we can even work out $\mathbb{E}T$.

Let us now show that $\mathbb{E}T(a,b) < \infty$.

51.5 Theorem. *Assume that S, $S_0 = 0$, is a SRW on $\mathbb{Z}$, $a, b \in \mathbb{Z}$ with $a < 0 < b$ and $T(a,b) = \inf\{n \in \mathbb{N} : S_n \notin (a,b)\}$. If this is the case, then $\mathbb{E}T(a,b) < \infty$. For a symmetric SRW one has $\mathbb{E}T(a,b) = |a|b$.*

Proof. It is enough to show that $\mathbb{E}T(a,b) < \infty$; the formula $\mathbb{E}T(a,b) = |a|b$ follows from Example 51.4.b).

1° For $i \in (a,b) \cap \mathbb{Z}$ we have

$$\mathbb{P}(i + S_{b-a} \notin (a,b)) \geqslant p^{b-a},$$

since the SRW $i + S_n$ which starts at $i \in (a,b)$ can leave the interval (a,b) if it makes $b-a$ steps **to the right**, see Fig. 51.3.

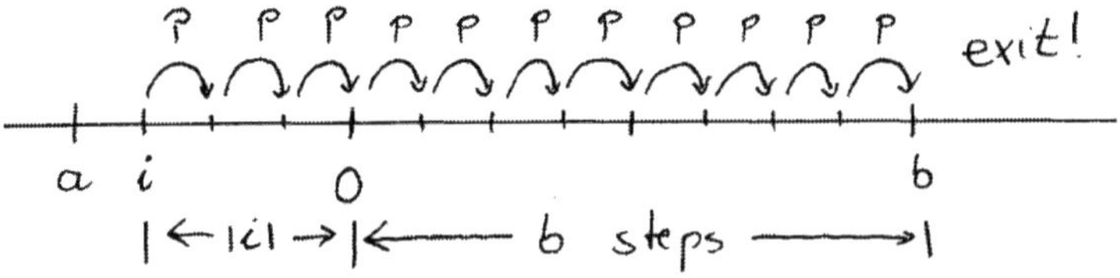

Fig. 51.3. The picture shows **one** possibility for a SRW to leave (a,b): Go $b-i$ steps to the right and note that $b-i \leqslant b-a = b+|a|$. Since there are other possibilities to leave (a,b), this is a lower bound of the exit probability.

2° Since the steps $(X_n)_{n\in\mathbb{N}}$ are iid rv, we have

$$S_{b-a} \perp\!\!\!\perp (X_{b-a+1} + \cdots + X_{2(b-a)}) = S_{2(b-a)} - S_{b-a} \sim S_{b-a} \tag{51.2}$$

and so

$$\begin{aligned}
&\mathbb{P}(T(a,b) > 2(b-a)) \\
&\quad\leqslant \mathbb{P}\Big(S_{b-a} \in (a,b), S_{2(b-a)} \in (a,b)\Big) \\
&\quad= \mathbb{E}\Big[\mathbb{1}_{(a,b)}(S_{b-a})\mathbb{1}_{(a,b)}((S_{2(b-a)} - S_{b-a}) + S_{b-a})\Big] \\
&\quad\overset{(51.2)}{=} \int \mathbb{E}\Big[\mathbb{1}_{(a,b)}(x)\mathbb{1}_{(a,b)}((S_{2(b-a)} - S_{b-a}) + x)\Big]\mathbb{P}(S_{b-a} \in dx) \\
&\quad\overset{(51.2)}{=} \int \mathbb{1}_{(a,b)}(x)\underbrace{\mathbb{E}\Big[\mathbb{1}_{(a,b)}(S_{b-a} + x)\Big]}_{=\mathbb{P}(x+S_{b-a}\in(a,b))}\mathbb{P}(S_{b-a} \in dx) \\
&\quad\overset{1^\circ}{\leqslant} (1-p^{b-a})\int \mathbb{1}_{(a,b)}(x)\,\mathbb{P}(S_{b-a} \in dx) \\
&\quad= (1-p^{b-a})\mathbb{P}(S_{b-a} \in (a,b)) \\
&\quad\leqslant (1-p^{b-a})^2.
\end{aligned}$$

3° If we iterate Step 2° we arrive at

$$\mathbb{P}(T(a,b) > n(b-a)) \leqslant (1-p^{b-a})^n, \quad n \in \mathbb{N}$$

which proves the claim since

$$\begin{aligned}
\mathbb{E}\frac{T(a,b)}{b-a} &\underset{\S16.5}{\overset{✎}{=}} \sum_{n=0}^{\infty} \mathbb{P}(T(a,b) > (b-a)n) \\
&\leqslant 1 + \sum_{n=1}^{\infty}(1-p^{b-a})^n = 1 + \frac{1-p^{b-a}}{p^{b-a}} = \frac{1}{p^{b-a}}.
\end{aligned}$$

□

We will now study the one-sided exit from an interval (a,∞), $a<0$, by letting $b\to\infty$. A simple calculation yields (see Example 51.4)

$$\lim_{b\to\infty}\mathbb{P}(S_{T(a,b)}=a)=\begin{cases}\left(\frac{p}{q}\right)^a=\left(\frac{q}{p}\right)^{|a|} & \text{if } p>q,\\ 1 & \text{if } p\leqslant q.\end{cases}$$

This is the probability that a player in a game with success probability p is ruined by the bank which is infinitely rich (aren't they all?); »ruin« means to loose at least EUR $|a|$.

51.6 Corollary. *Let* S, $S_0=0$ *be a SRW on* $\mathbb{Z}$ *and* $T_a=\inf\{n\in\mathbb{N}\mid S_n=a\}$. *If* $a<0$, *then*

$$\mathbb{P}(T_a<\infty)=\begin{cases}\left(\frac{p}{q}\right)^a=\left(\frac{q}{p}\right)^{|a|} & \text{if } p>q,\\ 1 & \text{if } p\leqslant q.\end{cases}$$

51.7 Corollary. *Let* S *be a SRW on* $\mathbb{Z}$. *It holds a.s. that*

$$\lim_{n\to\infty}\frac{S_n-S_0}{n}=p-q.$$

Proof. This follows from the SLLN (Theorem 32.1) as $\mathbb{E}X_1=p-q$. □

51.8 Remark. The case $q<p$ of Corollary 51.6 has a curious interpretation: If a gambler **who has an edge over the bank** plays with minimal possible wealth of $|a|=1$EUR against an infinitely rich bank, then he will not be ruined with probability of $1-\frac{q}{p}>0$. If this is the case, he will become infinitely rich, too. Indeed,

$$\mathbb{P}(S_n\to\infty\mid S_n\neq a\ \ \forall n)=1.$$

51.9 Theorem (Rekurrenz). *Let* S *be a symmetric SRW* ($p=q=\frac{1}{2}$) *on* $\mathbb{Z}$. *The walk* S *visits each* $x\in\mathbb{Z}$ *with probability* 1 *infinitely often.*

Proof. Assume that $S_0=0$. Clearly, $(S_n^a)_{n\in\mathbb{N}_0}$, $S_n^a:=S_n+a$ is for each $a\in\mathbb{Z}$ again a symmetric SRW such that $S_0^a=a$. Define $T_x^a:=\inf\{n\in\mathbb{N}\mid S_n^a=x\}$ and write $a\curvearrowright x\iff\mathbb{P}(T_x^a<\infty)=1$. Since $a,x\in\mathbb{Z}$ play symmetric roles, we get

$$\begin{aligned}\text{Corollary 51.6}&\implies a\curvearrowright x\quad\forall a\neq x\\&\implies x\curvearrowright a\quad\forall a\neq x\\&\implies x\curvearrowright a\curvearrowright x\quad\forall a\\&\implies x\curvearrowright x\curvearrowright x\curvearrowright\dots\end{aligned}$$

□

Let us study »recurrence« (i.e. infinite returns to one point) and »transience« (i.e. escape to infinity) in greater detail.

51.10 Definition. Let $(S_n)_{n\in\mathbb{N}_0}$, $S_0 = 0$, be a RW on $\mathbb{Z}^d$ (RW $\leftrightarrow$ we do not assume $|X_1| = 1$). The **first return time** to the starting position is the random time $T_0 := \inf\{n \in \mathbb{N} \mid S_n = 0\}$.

T_0 is a stopping time (Definition 39.1) and $T_0 = \infty$ means that »the RW does not return to $S_0 = 0$«. We a bit more **standard notation**

$u_n := \mathbb{P}(S_n = 0)$—the RW returns at time n to 0;
we have $u_0 = 1$ since $S_0 = 0$,

$f_n := \mathbb{P}(T_0 = n)$—the RW returns at time n **for the first time** to 0;
note that $f_0 = 0$; $(S_0, S_1, \dots, S_{T_0})$ is called an **excursion**.

51.11 Definition. A RW $(S_n)_{n\in\mathbb{N}_0}$, $S_0 = 0$, on $\mathbb{Z}^d$ (RW $\leftrightarrow$ we do not assume $|X_1| = 1$) is said to be

- **recurrent**, if $\sum_{n=1}^{\infty} f_n = 1$, i.e. $\mathbb{P}(T_0 < \infty) = 1$.
- **transient**, if $\sum_{n=1}^{\infty} f_n < 1$, i.e. $\mathbb{P}(T_0 < \infty) < 1$.

Roughly speaking, transience means that the return probability ρ satisfies $\rho = \mathbb{P}(T_0 < \infty) < 1$, i.e. we may return to 0, but n returns have the probability ρ^n; this is because the restarted RW $S_{T_0+i} - S_{T_0}$ is again a RW which is independent of $(S_0, \dots, S_{T_0})$ and behaves like the original RW.

51.12 Lemma. *A RW $(S_n)_{n\in\mathbb{N}_0}$, $S_0 = 0$, on $\mathbb{Z}^d$ (RW $\leftrightarrow$ we do not assume $|X_1| = 1$) is recurrent if, and only if, $\sum_{n=1}^{\infty} u_n = \infty$.*

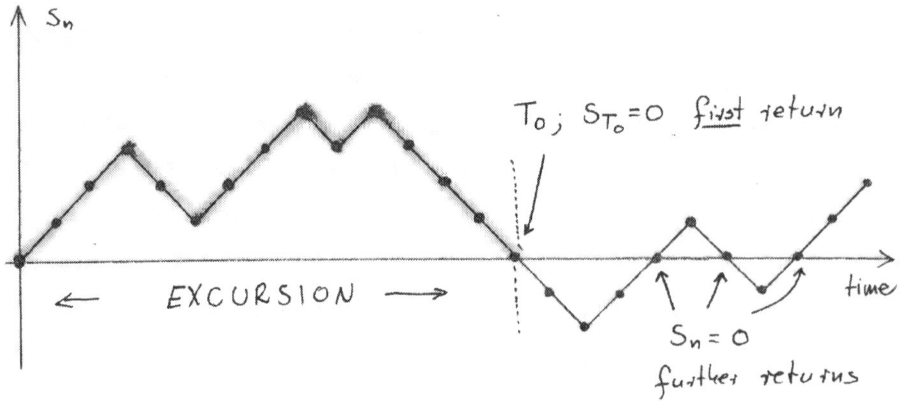

Fig. 51.4. The difference between $S_n = 0$ (being at 0 in time n) and $S_{T_0} = 0$ (being at 0 for the first time after starting from $S_0 = 0$).

Let us first understand the condition in Lemma 51.12. We have

$$\sum_{n=1}^{\infty} u_n = \sum_{n=1}^{\infty} \mathbb{E}\mathbb{1}_{\{S_n=0\}} \overset{\text{Tonelli}}{=} \mathbb{E}\sum_{n=1}^{\infty} \mathbb{1}_{\{S_n=0\}}$$

i.e. $\sum_{n=1}^{\infty} u_n$ counts the mean number of »visits« to 0.

Proof of Lemma 51.12. 1° We claim $u_n = \sum_{i=0}^{n} f_i u_{n-i},\ n \geqslant 1.$ This follows from

$$\begin{aligned}
\mathbb{P}(S_n = 0) &= \mathbb{P}(S_n = 0,\ T_0 \leqslant n) \\
&= \mathbb{P}\Big(\dot{\bigcup_{i=0}^{n}} \{T_0 = i\} \cap \{S_n = 0\}\Big) \\
&= \sum_{i=0}^{n} \mathbb{P}\big(\{T_0 = i\} \cap \{S_n = 0\}\big) \\
&= \sum_{i=0}^{n} \mathbb{P}\big(\{S_1 \neq 0, \dots, S_{i-1} \neq 0, S_i = 0\} \cap \{S_n - \underbrace{S_i}_{=0} = 0\}\big) \\
&\overset{\text{iid}}{=} \sum_{i=0}^{n} \mathbb{P}\big(\{S_1 \neq 0, \dots, S_{i-1} \neq 0, S_i = 0\}\big)\mathbb{P}\big(\{S_n - S_i = 0\}\big) \\
&\overset{\text{iid}}{=} \sum_{i=0}^{n} \mathbb{P}\big(\{T_0 = i\}\big)\mathbb{P}\big(\{S_{n-i} = 0\}\big) \\
&= \sum_{i=0}^{n} f_i u_{n-i}.
\end{aligned}$$

2° Consider for $x \in (-1, 1)$ the »probability generating functions«:

$$F(x) := \sum_{n=0}^{\infty} f_n x^n \quad \text{and} \quad U(x) := \sum_{n=0}^{\infty} u_n x^n \qquad (|x| < 1).$$

With the help of 1° we see that

$$\begin{aligned}
U(x) = \sum_{n=0}^{\infty} u_n x^n &= 1 + \sum_{n=1}^{\infty} \sum_{i=0}^{n} f_i u_{n-i} x^i x^{n-i} \\
&= 1 + \sum_{i=0}^{\infty} \sum_{n\geqslant 1, n=i}^{\infty} f_i x^i u_{n-i} x^{n-i} \\
&= 1 + \sum_{i=0}^{\infty} f_i x^i \sum_{k=0}^{\infty} u_k x^k = 1 + F(x)U(x).
\end{aligned}$$

Therefore,

$$F(x) = 1 - \frac{1}{U(x)} \implies \sum_{n=0}^{\infty} f_n = F(1) = \lim_{x\uparrow 1} F(x) = 1 - \lim_{x\uparrow 1} \frac{1}{U(x)}.$$

Thus, $F(1) = 1 \iff U(1-) = \infty$, as well as $F(1) < 1 \iff U(1-) < \infty$ with $U(1-) = \lim_{x\uparrow 1} U(x) = \sup_{x\in(0,1)} U(x)$. □

We will now apply the criterion for recurrence from Lemma 51.12 to a symmetric SRW.

51.13 Theorem (Pólya). *Let* $S = (S_n)_{n\in\mathbb{N}_0}$, $S_0 = 0$, *be a symmetric SRW on* $\mathbb{Z}^d$.

a) *S is recurrent, if* $d = 1, 2$.
b) *S is transient, if* $d \geqslant 3$.

Proof. We begin our analysis with a key observation: In order to return to $S_0 = 0$ we need an **even number of steps** since we have to retrace each of our »forward« steps in a »backwards« direction – and this holds in any dimension, see Fig. 51.5.

Fig. 51.5. A path $0 \rightsquigarrow 0$ needs an even number of steps.

A further essential tool is Stirling's formula $n! \approx \sqrt{2\pi n}(n/e)^n$.

1° Dimension $d = 1$. For $S_{2n} = 0$ we need n steps »forward« and n steps »backward« – but it does not matter when we make each step. If we pick the n positions for the forward steps within the total of $2n$ steps, we get

$$\mathbb{P}(S_{2n} = 0) = \binom{2n}{n}\left(\frac{1}{2}\right)^n\left(\frac{1}{2}\right)^n \approx \frac{\sqrt{2\pi 2n}}{\sqrt{2\pi n}\sqrt{2\pi n}} \frac{(2n)^{2n}}{e^{2n}} \frac{e^n}{n^n} \frac{e^n}{n^n}\left(\frac{1}{4}\right)^n = \frac{1}{\sqrt{\pi n}}.$$

Summing over $n \in \mathbb{N}$ yields $\sum_{n=1}^{\infty} u_{2n} \approx \sum_{n=1}^{\infty} (\pi n)^{-1/2} = \infty$, i.e. we are recurrent.

2° Dimension $d = 2$. Fig. 51.5 shows that we need for $S_{2n} = 0$

$$\alpha = \#\text{steps to the north} = \#\text{steps to the south}$$
$$\beta = \#\text{steps to the east} = \#\text{steps to the west}$$

such that $n = \alpha + \beta$. Since $2n = \alpha + \alpha + \beta + \beta$ we can use the multinomial coefficient to fix the directions within the $2n$ steps, and we get

$$u_{2n} = \sum_{\alpha+\beta=n} \binom{2n}{\alpha,\alpha,\beta,\beta}\left(\frac{1}{4}\right)^{2n} = \binom{2n}{n}\left(\frac{1}{4}\right)^{2n} \underbrace{\sum_{\alpha+\beta=n}\binom{n}{\alpha}\binom{n}{\beta}}_{=\binom{2n}{n},\ ✎} = \left[\frac{1}{4^n}\binom{2n}{n}\right]^2 \approx \frac{1}{\pi n}.$$

Again we have $\sum_{n=1}^{\infty} u_{2n} \approx \sum_{n=1}^{\infty} (\pi n)^{-1} = \infty$, i.e. recurrence.

3° Dimension $d \geqslant 3$. Without loss of generality we consider $d = 3$ only, higher dimensions follow the same pattern. In order to return to 0 we need an even number of steps $2n$ and we need the same number of steps »north–south«, »east–west« and »up–down«, say α, β and γ, respectively, such that $n = \alpha + \beta + \gamma$. Thus,

$$\begin{aligned}
u_{2n} &= \sum_{\alpha+\beta+\gamma=n} \binom{2n}{\alpha,\alpha,\beta,\beta,\gamma,\gamma} \left(\frac{1}{6}\right)^{2n} \\
&= \frac{1}{6^{2n}} \binom{2n}{n} \sum_{\alpha+\beta+\gamma=n} \binom{n}{\alpha,\beta,\gamma} \binom{n}{\alpha,\beta,\gamma} \\
&\leqslant \frac{1}{6^{2n}} \binom{2n}{n} \max_{\alpha+\beta+\gamma=n} \binom{n}{\alpha,\beta,\gamma} \underbrace{\sum_{\alpha+\beta+\gamma=n} \binom{n}{\alpha,\beta,\gamma}}_{=(1+1+1)^n = 3^n,\ \text{multinomial formula}} \\
&= \frac{3^n}{6^{2n}} \binom{2n}{n} \max_{\alpha+\beta+\gamma=n} \binom{n}{\alpha,\beta,\gamma}.
\end{aligned}$$

Assume that $n = 3m$ for some $m \in \mathbb{N}$. Then we have (see appendix)

$$\binom{3m}{\alpha,\beta,\gamma} \leqslant \binom{3m}{m,m,m}$$

and so

$$u_{6m} \leqslant \frac{3^{3m}}{6^{6m}} \binom{6m}{3m} \binom{3m}{m,m,m} \overset{\text{Stirling}}{\approx} \frac{1}{2\pi\sqrt{\pi}} \frac{1}{m\sqrt{m}}.$$

This shows that $\sum_{m=1}^{\infty} u_{6m} < \infty$. In order to deal with the still missing terms, we observe that

$$\underbrace{0 \rightsquigarrow * \rightsquigarrow \cdots \rightsquigarrow * \rightsquigarrow 0}_{6m-2\ \text{steps}} \overbrace{\rightsquigarrow * \rightsquigarrow 0}^{2\ \text{steps}} \implies \underbrace{0 \rightsquigarrow * \rightsquigarrow \cdots \rightsquigarrow * \rightsquigarrow * \rightsquigarrow 0}_{6m\ \text{steps}},$$

and so we get the following lower bounds

$$u_{6m} \geqslant \frac{1}{6} \cdot \frac{1}{6} \cdot u_{6m-2} \geqslant \left(\frac{1}{6} \cdot \frac{1}{6}\right)^2 \cdot u_{6m-4}.$$

Finally, we see

$$\begin{aligned}
\sum_{n=1}^{\infty} u_{2n} = \sum_{m=1}^{\infty} (u_{6m} + u_{6m-2} + u_{6m-4}) &\leqslant (1 + 6^2 + 6^4) \sum_{m=1}^{\infty} u_{6m} \\
&\approx \frac{1333}{2\pi\sqrt{\pi}} \sum_{m=1}^{\infty} \frac{1}{m\sqrt{m}} < \infty
\end{aligned}$$

which means that the RW is transient. ☐

*Appendix: Two combinatorial relations**

In the proof of Pólya's theorem we have used that

$$\sum_{\alpha+\beta=n}\binom{n}{\alpha}\binom{n}{\beta}=\binom{2n}{n} \quad\text{and}\quad \forall\alpha+\beta+\gamma=m:\quad \binom{3m}{\alpha,\beta,\gamma}\leqslant\binom{3m}{m,m,m}. \tag{51.3}$$

We want briefly indicate how one can prove these relations. The **first identity** is quite intuitive: Consider an $n\times n$ grid $\{0,1,2,\dots n\}\times\{0,1,2,\dots n\}$ and count the number N of paths from $(0,0)$ to (n,n) if we are only allowed to go »up« and »right«. Since a path is uniquely determined by the steps where we go »up«, we have to fix the n »up« steps within all $2n$ steps, and so

$$N=\binom{2n}{n}.$$

Another possibility to count the paths is to group them according to the place where a path passes through the minor diagonal $\{(i,n-i)\mid i=0,1,2,\dots n\}$, i.e. $(0,0)\to(i,n-i)\to(n,n)$. Each part of such a path has n steps and, fixing again the »up« steps, there are $\binom{n}{n-i}\binom{n}{i}$ such paths. Taking the sum yields

$$N=\sum_{i=0}^{n}\binom{n}{n-i}\binom{n}{i}=\sum_{\alpha+\beta=n}\binom{n}{\alpha}\binom{n}{\beta}$$

finishing the argument.

In order to see the **second relation**, we note that the claim is equivalent to showing $m!m!m!\leqslant\alpha!\beta!\gamma!$ for all $\alpha,\beta,\gamma\in\mathbb{N}_0$ such that $\alpha+\beta+\gamma=3m$. There is nothing to show if $\alpha=\beta=\gamma=m$. Assume that $\alpha>m>\beta$ and some $\gamma\in\mathbb{N}_0$ given by $\gamma=3m-\alpha-\beta$. We can decrease α by 1 and increase β by 1, leaving γ unchanged. We still have $\alpha-1\geqslant m\geqslant\beta+1$ as well as

$$(\alpha-1)!(\beta+1)!\gamma!\leqslant\alpha!\beta!\gamma! \iff (\alpha-1)!(\beta+1)!\leqslant\alpha!\beta! \iff \beta+1\leqslant\alpha.$$

This shows that we can decrease any index $>m$ by 1 and increase any index $<m$ by 1 reducing the r.h.S. $\alpha!\beta!\gamma!$ with each operation. Repeating this procedure finally leads to the minimum value $m!m!m!$.

52 Finite Markov chains

The steps of a random walk $S_n=X_1+\dots+X_n$ are iid random variables. In this and the following chapters we want to consider processes (still indexed by $\mathbb{N}_0$) such that the increment S_n-S_{n-1} may depend on $S_0,\dots,S_{n-1}$. In fact, we only assume a mild dependence on the last position S_{n-1} which means that the process has the Markov property.

Notation. Contrary to the random walk setting, the process is now called X, and we use $s,t,u\in\mathbb{N}_0$ to denote time and $i,j,k\in E$ to denote the states. We assume that $E=\{i(1),\dots,i(r)\}$, $r\in\mathbb{N}\cup\{\infty\}$, is countable and we often identify $i(m)\in E$ with $m\in\mathbb{N}\cup\{\infty\}$.

⚠ Note that $i(m)$ refers to the mth element of E while $X_t=i_t$ means that X_t attains some value $i_t\in E$, i.e. the subscript refers to time.

We need a few algebraic preparations.

52.1 Definition. Let $Q = (q_{ij})_{i,j\in E} \in \mathbb{R}^{r\times r}$ and $e = (e_i)_{i\in E} \in \mathbb{R}^r$ for $r \in \mathbb{N}\cup\{\infty\}$.

a) We write $Q \geqslant 0$, $e \geqslant 0$ (resp. $Q > 0$, $e > 0$) if all entries are $\geqslant 0$ (resp. > 0).
b) Q is said to be **stochastic**, if

$$Q \geqslant 0 \quad \text{and the row sums satisfy} \quad \sum_{j\in E} q_{ij} = 1 \quad \forall i \in E.$$

c) $\mu = (\mu_i)_{i\in E}$ is a **probability (vector/measure)**,(1) if $\mu \geqslant 0$ and $\sum_{i\in E}\mu_i = 1$.

52.2 Lemma. *Let $Q \in \mathbb{R}^{r\times r}$ and $f \in \mathbb{R}^r$. The following are equivalent*

a) *Q is stochastic*
b) 1. $f \geqslant 0 \implies Qf \geqslant 0$;
   2. $Q\mathbb{1} = \mathbb{1}$ *for* $\mathbb{1} = (1,\dots,1)^\top \in \mathbb{R}^r$.
c) *If μ is a probability vector, then μQ is a probability vector.*

Proof. a)⇒b) is obvious.

b)⇒a): Let $e_j = (0,\dots,0,\overset{j}{1},0\dots)^\top$. Then

$$q_{ij} = (Qe_j)_i \overset{\text{b)}}{\geqslant} 0 \quad \text{and} \quad (Q\mathbb{1})_i = \sum_{j\in E} q_{ij}1 = \sum_{j\in E} q_{ij} \overset{\text{b)}}{=} 1.$$

a)⇒c): $\mu' := \mu Q \implies \mu'_j = \sum_{i\in E}\mu_i q_{ij}$. Since Q is stochastic, we get $\mu'_j \geqslant 0$ and

$$\sum_{j\in E}\mu'_j = \sum_{j\in E}\sum_{i\in E}\mu_i q_{ij} = \sum_{i\in E}\mu_i \underbrace{\sum_{j\in E} q_{ij}}_{=1} = \sum_{i\in E}\mu_i = 1.$$

c)⇒a): The vector $\epsilon_i := (0,\dots,0,\overset{i}{1},0\dots)$ is a probability vector; by assumption, $\epsilon_i Q$ is again a probability vector, and so

$$(\epsilon_i Q)_j = q_{ij} \geqslant 0 \quad \text{and} \quad \sum_{j\in E} q_{ij} = \sum_{j\in E}(\epsilon_i Q)_j = 1.$$ □

52.3 Lemma. *Assume that $Q' = (q'_{ij})$, $Q'' = (q''_{ij}) \in \mathbb{R}^{r\times r}$ are stochastic matrices. The product $Q := Q'Q'' = (q_{ij})$ is again stochastic.*
Amendment: $Q'' > 0 \implies Q > 0$.

(1) We identify $(\mu_i)_{i\in E}$ with $\sum_{i\in E}\mu_i\delta_i$.

Proof. We have

$$\sum_{j\in E} q_{ij} = \sum_{j\in E}\sum_{k\in E} q'_{ik}q''_{kj} = \sum_{k\in E} q'_{ik} \underbrace{\sum_{j\in E} q''_{kj}}_{=1} = \sum_{k\in E} q'_{ik} = 1.$$

Fix $i \in E$. For the amendment we observe that $\sum_k q'_{ik} = 1$ implies that $q'_{ik(i)} > 0$ for some $k(i)$. The definition of the matrix product shows

$$q_{ij} = \sum_{k\in E} q'_{ik}q''_{kj} \geqslant q'_{ik(i)}q''_{k(i)j} > 0 \quad \text{for any } j \in E.$$ □

52.4 Definition. Let $X = (X_0, X_1, \dots, X_t)$, $t \in \mathbb{N}$, be a discrete stochastic process taking values in $E = \{i(1), \dots, i(r)\}$, $r \in \mathbb{N} \cup \{\infty\}$. Assume that $\mu = (\mu_i)_{i\in E}$ is a probability vector and $P(s) = (p_{ij}(s))_{ij}$, $s = 1, \dots, t$, stochastic $r \times r$-matices. X is called a **Markov chain** (MC) with **initial distribution** μ and **transition matrices** $P(1), \dots, P(t)$, if

$$\mathbb{P}(X_0 = i_0, \dots, X_t = i_t) = \mu_{i_0} p_{i_0 i_1}(1) \dots p_{i_{t-1} i_t}(t). \tag{52.1}$$

X is called a **homogeneous Markov chain** if $P(1) = \dots = P(t) = P$.
Notation: MC(P(t)), MC(P) or MC(μ, P(t)), MC(μ, P) to emphasize the initial distribution.

Let us check that (52.1) defines a probability measure on E^{t+1}, $t \in \mathbb{N}$:

$$\begin{aligned}
&\sum_{i_0\in E}\sum_{i_1\in E}\dots\sum_{i_t\in E} \mathbb{P}(X_0 = i_0, \dots X_t = i_t) \\
&\quad = \sum_{i_0\in E}\sum_{i_1\in E}\dots\sum_{i_t\in E} \mu_{i_0} p_{i_0 i_1}(1) \dots p_{i_{t-1} i_t}(t) \\
&\quad = \underbrace{\sum_{i_0\in E} \mu_{i_0}}_{=1} \underbrace{\sum_{i_1\in E} p_{i_0 i_1}(1)}_{=1} \dots \underbrace{\sum_{i_t\in E} p_{i_{t-1} i_t}(t)}_{=1} = 1.
\end{aligned}$$

In the same way we see that for $s < t$

$$\mathbb{P}(X_0 = i_0, \dots, X_s = i_s) = \mu_{i_0} p_{i_0 i_1}(1) \dots p_{i_{s-1} i_s}(s), \quad 1 \leqslant s \leqslant t \tag{52.2}$$

defines a probability measure on E^{s+1}, i.e. $(X_1, \dots, X_s)$ is also a MC(P(s)).

52.5 Lemma. *If X is a MC(μ, P(t)), then for all $s \leqslant t$*

$$\begin{aligned}
\mathbb{P}(X_s = j \mid X_0 = i_0, \dots, X_{s-2} = i_{s-2}, X_{s-1} = i) \\
= p_{ij}(s) = \mathbb{P}(X_s = j \mid X_{s-1} = i)
\end{aligned} \tag{52.3}$$

Proof. We have

$$\begin{aligned}
&\mathbb{P}(X_s = j \mid X_0 = i_0, \dots, X_{s-2} = i_{s-2}, X_{s-1} = i) \\
&\quad = \frac{\mathbb{P}(X_s = j, X_0 = i_0, \dots, X_{s-2} = i_{s-2}, X_{s-1} = i)}{\mathbb{P}(X_0 = i_0, \dots, X_{s-2} = i_{s-2}, X_{s-1} = i)} \\
&\quad = \frac{\mu_{i_0} p_{i_0 i_1}(1) \dots p_{i_{s-2},i}(s-1) p_{ij}(s)}{\mu_{i_0} p_{i_0 i_1}(1) \dots p_{i_{s-2},i}(s-1)} = p_{ij}(s).
\end{aligned}$$ □

💬 Recall from the chapter on conditional expectations §35 that the classical usage and the abstract usage of conditioning coincide (cf. Example 35.2): $\mathbb{P}(F \mid X = x) = \mathbb{P}(F \mid \{X = x\})$ if $\mathbb{P}(X = x) > 0$.

52.6 Remark. a) In view of the last remark, we see that formula (52.3) becomes

$$\mathbb{P}(X_s = j \mid X_0, \dots, X_{s-1}) \overset{\text{i)}}{=} p_{X_{s-1} j}(s) \overset{\text{ii)}}{=} \mathbb{P}(X_s = j \mid X_{s-1}). \tag{52.4}$$

The second equality ii) follows from the first equality i), if we apply conditional expectations $\mathbb{E}(\cdots \mid X_{s-1})$ on both sides of i) and use the tower property (on the left) and the $\sigma(X_{s-1})$-measurability on the right.

b) Let X be a $\mathrm{MC}(\mu, P)$ such that $P = \begin{pmatrix} \mu_1 \dots \mu_r \\ \cdots \\ \mu_1 \dots \mu_r \end{pmatrix}$ (in each row is $\mu^\top$). This implies that the rv $X_0, \dots, X_t$ are iid. Indeed, we see from the definition of a MC

$$\begin{aligned}
\mathbb{P}(X_0 = i_0, X_1 = i_1, \dots, X_t = i_t) &= \mu_{i_0} \mu_{i_1} \cdots \mu_{i_t} \\
&= \mathbb{P}(X_0 = i_0) \cdot \ldots \cdot \mathbb{P}(X_t = i_t) \quad \forall i_0, \dots, i_t \in E,
\end{aligned}$$

and independence follows with Theorem 25.9.

c) We can represent a homogenous MC with a directed (multi-)graph:

- $E = \{i(1), \dots, i(r)\}$ are the vertices;
- the directed edges have the weights p_{ij} and p_{ji}, respectively;
- $(X_0, \dots, X_t) = (i_0, i_1, \dots, i_t)$ is a path with t steps within the graph.

52.7 Lemma. *If X is a homogenous* $\mathrm{MC}(\mu, P)$, *then for all* $s, t \in \mathbb{N}$

$$\mathbb{P}(X_{s+t} = j \mid X_t = i) = p_{ij}^{(s)} \quad \text{where} \quad P^s = \underbrace{P \cdot \ldots \cdot P}_{s \text{ factors}} = (p_{ij}^{(s)})_{ij} \tag{52.5}$$

and

$$\mathbb{P}(X_s = j) = (\mu P^s)_j = \sum_{i \in E} \mu_i p_{ij}^{(s)}. \tag{52.6}$$

Proof. We have

$$\mathbb{P}(X_{s+t}=j\mid X_t=i)=\frac{\mathbb{P}(X_{s+t}=j,X_t=i)}{\mathbb{P}(X_t=i)}$$
$$=\frac{\mathbb{P}(X_{s+t}=j,\ X_{s+t-1}\in E,\dots,X_{t+1}\in E,X_t=i,\ X_{t-1}\in E,\dots,X_0\in E)}{\mathbb{P}(X_t=i,\ X_{t-1}\in E,\dots,X_0\in E)}$$

(the highlighted terms do not contribute, they are $=\Omega$), and so

$$\mathbb{P}(X_{s+t}=j\mid X_t=i)=\frac{\sum_{i_0,\dots,i_{t-1}\in E}\sum_{i_{t+1},\dots i_{s+t-1}\in E}\left(\mu_{i_0}p_{i_0i_1}\cdots p_{i_{t-1}i}\right)\left(p_{ii_{t+1}}\cdots p_{i_{s+t-1}j}\right)}{\sum_{i_0,\dots,i_{t-1}\in E}\mu_{i_0}p_{i_0i_1}\cdots p_{i_{t-1}i}}$$
$$=\sum_{i_{t+1},\dots i_{s+t-1}\in E}p_{ii_{t+1}}\cdots p_{i_{s+t-1}j}=p_{ij}^{(s)}.$$

(52.5)⇒(52.6): Set $t=0$ and sum over $i\in E$ to see

$$\mathbb{P}(X_s=j)=\sum_{i\in E}\mathbb{P}(X_s=j,X_0=i)$$
$$=\sum_{i\in E}\mathbb{P}(X_s=j\mid X_0=i)\mathbb{P}(X_0=i)=\sum_{i\in E}p_{ij}^{(s)}\mu_i.\qquad\square$$

Interpretation: $P_{ij}^s=p_{ij}^{(s)}$ is the probability, for $i\to j$ in s steps. Formula (52.5) means that this probability does not depend on the time t when we start, but only on the difference of time $s=(t+s)-t$; it does also not depend on the previous steps $X_0,X_1,\dots,X_{t-1}$, i.e. we could run our MC up to time t, stop and re-start at the last known position X_t.

Typical application: Let $s<t<u$:

$$\mathbb{P}(X_s=i,X_t=j)=\mathbb{P}(X_t=j\mid X_s=i)\mathbb{P}(X_s=i)=\overbrace{(\mu P^s)_i}^{\text{in } s \text{ steps } \mu\to i}\underbrace{p_{ij}^{(t-s)}}_{\text{in } t-s \text{ steps } i\to j};$$
$$\mathbb{P}(X_s=i,X_t=j,X_u=k)=(\mu P^s)_iP_{ij}^{t-s}P_{jk}^{u-t}.$$

Ergodic Markov chains and the law of large numbers

52.8 Definition. A stochastic matrix P is said to be **ergodic** if there is some $s\in\mathbb{N}$ such that $P^s>0$. A MC(P) is said to be **ergodic** if P is ergodic.

- An ergodic MC can reach every state in at most s steps. The corresponding incidence (or Levi) graph has no unconnected sub-graph.
- The ergodic theorem shows that an ergodic MC(μ,P) »forgets« the initial distribution μ if $t\gg1$. The notion »ergodic« is originally from physics (statistical mechanics).

52.9 Example. Here are a few typical examples for **non-ergodic** ⚠ MCs:

a) P is a sub-diagonal block matrix. Let $E = \{i(1), \dots, i(r)\} = \dot{\bigcup}_{m=1}^{n} E_m$ and assume

$$p_{ij} \neq 0 \iff (i,j) \in E_m \times E_{m+1} \quad \text{or} \quad (i,j) \in E_n \times E_1,$$

This means that one can only go from $E_1 \to E_2 \to \cdots \to E_n \to E_1$.

Consequence: Every power P^s has some entries $p_{ij}^{(s)} = 0$.

b) There is one isolated state j_0, i.e. $p_{ij_0} = 0$ for all $i \neq j_0$.

c) There are several isolated »islands«. Let $E = \dot{\bigcup}_{m=1}^{n} E_m$ and

$$p_{ij} \neq 0 \iff i,j \in E_m.$$

This means that P is a block diagonal matrix, the MC remains forever in the block of states where it started at time $t = 0$.

52.10 Definition. A probability vector $\pi = (\pi_{i(1)}, \dots, \pi_{i(r)})$ is said to be **stationary** (or **invariant**) for the $r \times r$ transition matrix P if $\pi P = \pi$.

52.11 Theorem (ergodic theorem for MC). *Let X be an ergodic* MC(P). *There exists a unique stationary distribution* π *and*

$$\lim_{t\to\infty} p_{ij}^{(t)} = \pi_j > 0 \quad \forall i,j \in E.$$

Proof. The proof proceeds in several steps.

1° A metric on the probability vectors. Let μ', μ'' be probability vectors. Then

$$d(\mu', \mu'') := \frac{1}{2} \sum_{i\in E} |\mu_i' - \mu_i''| \tag{52.7}$$

is a metric ✍ on the set of all probability vectors Π, and (Π, d) is a complete metric space; due to the normalization of d we have $d(\pi, 0) = \frac{1}{2}$.(2) Note that

$$0 = \sum_{i\in E} \mu_i' - \sum_{i\in E} \mu_i'' = \sum_{i\in E} (\mu_i' - \mu_i'') = \sum_{i\in E} (\mu_i' - \mu_i'')^+ - \sum_{i\in E} (\mu_i' - \mu_i'')^-$$

and so

$$\begin{aligned} d(\mu', \mu'') &= \frac{1}{2} \sum_{i\in E} |\mu_i' - \mu_i''| = \frac{1}{2} \sum_{i\in E} (\mu_i' - \mu_i'')^+ + (\mu_i' - \mu_i'')^- \\ &= \sum_{i\in E} (\mu_i' - \mu_i'')^+ = \sum_{i\in E,\, \mu_i' \geqslant \mu_i''} (\mu_i' - \mu_i'') \leqslant \sum_{i\in E} \mu_i' = 1. \end{aligned}$$

(2) If $r = |E| < \infty$ this is obvious since $E \simeq \mathbb{R}^r$ and d is the ℓ^1-norm. If $r = \infty$, we can use the Riesz–Fischer theorem (Theorem 14.10 and Example 14.14: $\ell^1(E) = L^1$ with counting measure). In order to see that the limit π of a sequence $(\mu^{(n)})_{n\in\mathbb{N}}$ is a probability vector, use the lower triangle inequality: $|1/2 - d(\pi, 0)| = |d(\mu^{(n)}, 0) - d(\pi, 0)| \leqslant d(\mu_n, \pi) \to 0$.

2° Let μ', μ'' be probability vectors and $Q = (q_{ij})$ a stochastic matrix. The probability vectors $\mu'Q$ and $\mu''Q$ (cf. Lemma 52.2) satisfy

$$d(\mu'Q, \mu''Q) \leqslant (1-\alpha)\, d(\mu', \mu''), \quad \alpha := \min_{i,j\in E} q_{ij}. \tag{52.8}$$

Note that $Q > 0$ implies that $\alpha \in (0,1)$.

Indeed: Set $J := \{j \in E \mid (\mu'Q)_j - (\mu''Q)_j > 0\}$. Since we have

$$\sum_{j\in E} (\mu'Q)_j = 1 = \sum_{j\in E} (\mu''Q)_j,$$

the set $E \setminus J \neq \emptyset$ and, say, $k \in E \setminus J$. This k satisfies

$$\sum_{j\in J} q_{ij} = \sum_{j\in E} q_{ij} - \sum_{j\in E\setminus J} q_{ij} \leqslant 1 - q_{ik} \leqslant 1 - \alpha.$$

Recall from the end of Step 1° that $d(\mu', \mu'') = \sum_{j\in E} (\mu'_j - \mu''_j)^+$. This and the definition of J show

$$\begin{aligned} d(\mu'Q, \mu''Q) &= \sum_{j\in E} ((\mu'Q)_j - (\mu''Q)_j)^+ \\ &= \sum_{j\in J} ((\mu'Q)_j - (\mu''Q)_j) \\ &= \sum_{j\in J} \sum_{i\in E} (\mu'_i - \mu''_i) q_{ij} \\ &\leqslant \sum_{i\in E} (\mu'_i - \mu''_i)^+ \sum_{j\in J} q_{ij} \\ &\overset{(52.8)}{\leqslant} (1-\alpha)\, d(\mu', \mu''). \end{aligned}$$

3° $\mu^{(t)} := \mu P^t$ is a d-Cauchy sequence. By assumption, there is a $w \in \mathbb{N}$ such that $Q := P^w > 0$; thus, $\alpha \in (0,1)$. In view of (52.8), we get for $t > w$ and any $u > 0$

$$\begin{aligned} d(\mu^{(t)}, \mu^{(t+u)}) = d(\mu P^t, \mu P^{t+u}) &\leqslant (1-\alpha)\, d(\mu P^{t-w}, \mu P^{t+u-w}) \\ &\leqslant \ldots \leqslant (1-\alpha)^m d(\mu P^{t-mw}, \mu P^{t+u-mw}) \end{aligned}$$

as long as $0 \leqslant t - mw < w$ for $m \in \mathbb{N}$.

Since $t \to \infty \implies m \to \infty \implies (1-\alpha)^m \to 0$, we conclude that

$$\lim_{t,u\to\infty} d(\mu^{(t)}, \mu^{(t+u)}) = 0 \implies \exists \pi = \lim_{t\to\infty} \mu^{(t)}.$$

This limit does not depend on the initial distribution μ. Moreover,

$$\pi P = \lim_{t\to\infty} \mu P^t P = \lim_{s\to\infty} \mu P^s = \pi.$$

4° Uniqueness of the limit. If π, π' are stationary distributions, then

$$d(\pi,\pi') \overset{\text{stationary}}{=} d(\pi P^w, \pi' P^w) \overset{(52.8)}{\leqslant} (1-\alpha) d(\pi,\pi') \overset{\alpha\in(0,1)}{\Longrightarrow} d(\pi,\pi') = 0.$$

Finally, we see that $\pi = \pi \underbrace{P^w}_{>0} > 0$ (✎, cf. amendement to §52.3). □

52.12 Remark. We can determine the speed of convergence in Theorem 52.11. Let π be stationary. With the calculation used in Step 3° of the proof of the ergodic theorem we find for $t > w$

$$d(\mu P^t, \pi) = d(\mu P^t, \pi P^t) \leqslant (1-\alpha)^m d(\mu P^{t-mw}, \pi P^{t-mw}) \leqslant (1-\alpha)^m.$$

Since $0 \leqslant t - mw < w$, we have

$$d(\mu P^t, \pi) \leqslant (1-\alpha)^{\frac{t}{w}-1} = \frac{1}{1-\alpha}\beta^t, \quad \beta := (1-\alpha)^{1/w} < 1.$$

This shows that $\mu P^t \to \pi$ **exponentially fast.**

We are now interested in the meaning of a stationary distribution. Let us define the number of visits of state i up to time t, and the number of consecutive visits to the states (i,j) up to time t:

$$\begin{aligned} v_i^t(\omega) &:= \#\{0 \leqslant s \leqslant t \mid X_s(\omega) = i\} = \sum_{s=0}^t \mathbb{1}_{\{i\}}(X_s(\omega)); \\ v_{ij}^t(\omega) &:= \#\{1 \leqslant s \leqslant t \mid X_{s-1}(\omega) = i,\ X_s(\omega) = j\} \\ &= \sum_{s=1}^t \mathbb{1}_{\{i\}}(X_{s-1}(\omega)) \mathbb{1}_{\{j\}}(X_s(\omega)). \end{aligned}$$

52.13 Theorem (LLN for MC). *Assume that X is an ergodic* MC(P) *with discrete state space and stationary distribution* π. *For every* $\epsilon > 0$ *it is true that*

$$\lim_{t\to\infty} \mathbb{P}\left(\left|\frac{v_i^t}{t} - \pi_i\right| \geqslant \epsilon\right) = 0, \quad i \in E,$$

$$\lim_{t\to\infty} \mathbb{P}\left(\left|\frac{v_{ij}^t}{t} - \pi_i p_{ij}\right| \geqslant \epsilon\right) = 0, \quad i,j \in E.$$

💬 This means that $\frac{1}{t} v_i^t \approx \pi_i$ is the proportion of »time« which an MC spends in state i.

Proof. (3) We have

$$\mathbb{E}\mathbb{1}_{\{i\}}(X_s) = \mathbb{P}(X_s = i) = \sum_{j\in E} \mu_j p_{ji}^{(s)} \xrightarrow[s\to\infty]{\S 52.11} \pi_i.$$

(3) You should compare the following proof with the proof of the WLLN §28.2

Summation over $s = 0, \ldots, t$ shows

$$\mathbb{E}\left[\frac{1}{t}v_i^t\right] = \mathbb{E}\left[\frac{1}{t}\sum_{s=0}^{t}\mathbb{1}_{\{i\}}(X_s)\right] \xrightarrow[t\to\infty]{} \pi_i. \tag{52.9}$$

Therefore, we can find for every $\epsilon > 0$ some $t(\epsilon)$ such that for all $t \geqslant t(\epsilon)$

$$\left\{\omega \mid \left|\frac{v_i^t(\omega)}{t} - \pi_i\right| \geqslant \epsilon\right\} \overset{(52.9)}{\subset} \left\{\omega \mid \left|\frac{v_i^t(\omega)}{t} - \frac{1}{t}\mathbb{E}v_i^t\right| \geqslant \frac{\epsilon}{2}\right\}.$$

The Chebyshev–Markov inequality now yields

$$\mathbb{P}\left\{\left|\frac{v_i^t}{t} - \frac{1}{t}\mathbb{E}v_i^t\right| \geqslant \frac{\epsilon}{2}\right\} = \mathbb{P}\left\{\left|v_i^t - \mathbb{E}v_i^t\right| \geqslant \frac{t\epsilon}{2}\right\} \leqslant \frac{4\mathbb{V}v_i^t}{\epsilon^2 t^2}.$$

We show that $\mathbb{V}v_i^t \leqslant Ct$ for some constant. Define $\mu_i^{(s)} := \mathbb{E}\mathbb{1}_{\{i\}}(X_s) = \mathbb{P}(X_s = i)$. From the very definition of the variance we get

$$\begin{aligned}
\mathbb{V}v_i^t &= \mathbb{E}\left[\left(\sum_{s=0}^{t}(\mathbb{1}_{\{i\}}(X_s) - \mu_i^{(s)})\right)^2\right] \\
&= \underbrace{\sum_{s=0}^{t}\underbrace{\mathbb{E}\left[(\mathbb{1}_{\{i\}}(X_s) - \mu_i^{(s)})^2\right]}_{\leqslant 1}}_{\leqslant t+1} + 2\sum_{s<u}\mathbb{E}\left[(\mathbb{1}_{\{i\}}(X_s) - \mu_i^{(s)})(\mathbb{1}_{\{i\}}(X_u) - \mu_i^{(u)})\right].
\end{aligned}$$

Moreover,

$$\begin{aligned}
\mathbb{E}\left[\left(\mathbb{1}_{\{i\}}(X_s) - \mu_i^{(s)}\right)\left(\mathbb{1}_{\{i\}}(X_u) - \mu_i^{(u)}\right)\right] &= \mathbb{E}\left[\mathbb{1}_{\{i\}}(X_s)\mathbb{1}_{\{i\}}(X_u) - \mu_i^{(s)}\mu_i^{(u)}\right] \\
&= \sum_{j\in E}\mu_j p_{ji}^{(s)} p_{ii}^{(u-s)} - \mu_i^{(s)}\mu_i^{(u)} \\
&=: R(s,u).
\end{aligned}$$

Using the ergodic theorem §52.11 and Remark 52.12 we see that for large $s \gg 1$

$$\begin{aligned}
p_{ji}^{(s)} &= \pi_i + \beta_{ji}^{(s)}, & |\beta_{ji}^{(s)}| &\leqslant c\lambda^s, \\
\mu_i^{(s)} &= \sum_{j\in E}\mu_j p_{ji}^{(s)}\pi_i + d_i^{(s)}, & |d_i^{(s)}| &\leqslant c\lambda^s,
\end{aligned}$$

and, thus,

$$\begin{aligned}
|R(s,u)| &= \left|\sum_{j\in E}\mu_j(\pi_i + \beta_{ji}^{(s)})(\pi_i + \beta_{ii}^{(u-s)}) - (\pi_i + d_i^{(s)})(\pi_i + d_i^{(u)})\right| \\
&\leqslant c'(\lambda^s + \lambda^u + \lambda^{u-s})
\end{aligned}$$

for some constant c'. From this we get $\sum_{s<u}|R(s,u)| \leqslant c''t$ and $\mathbb{V}v_i^t \leqslant Ct$.

In order to deal with v_{ij}^t, we can use the same approach, but we have now to estimate $\mathbb{V}v_{ij}^t$ which is only notationally more complicated. Here is a »replacement« table

$$\begin{aligned} v_i^t &\longrightarrow v_{ij}^t, \\ \mathbb{E}\mathbb{1}_{\{i\}}(X_s) &\longrightarrow \mathbb{E}\mathbb{1}_{\{i\}}(X_{s-1})\mathbb{1}_{\{j\}}(X_s), \\ \pi_i &\longrightarrow \pi_i p_{ij}; \end{aligned}$$

moreover, you should use that

$$\begin{aligned} \mathbb{E}\mathbb{1}_{\{i\}}(X_{s-1})\mathbb{1}_{\{j\}}(X_s) &= \mathbb{P}(X_{s-1} = i, X_s = j) \\ &= \mathbb{P}(X_s = j \mid X_{s-1} = i)\mathbb{P}(X_{s-1} = i) \\ &\overset{(52.5)}{\underset{(52.6)}{=}} \sum_{k\in E} \mu_k p_{ki}^{(s-1)} p_{ij}. \end{aligned}$$

After some lengthy calculations, one finally arrives at $\mathbb{V}v_{ij}^t \leqslant C't$. □

An application: Asymptotics of powers of positive matrices

We want to study the asymptotic behaviour of large powers of matrices A^t as $\mathbb{N} \ni t \to \infty$.

52.14 Definition. Let $A \in \mathbb{C}^{r\times r}$, $r < \infty$. By $\sigma(A) \subset \mathbb{C}$ we denote the **spectrum** of A. The **spectral radius** is $\rho(A) := \max\{|\lambda| \mid \lambda \in \sigma(A)\}$. If $\|x\|$ is a norm in $\mathbb{C}^r$, then $\|A\| := \sup_{x\neq 0} \|Ax\|/\|x\|$ is the corresponding matrix or operator norm.

It is easy to see that for any norm $\|\bullet\|$ in $\mathbb{C}^r$ we have

$$\rho(A) \leqslant \|A\| \quad \text{and} \quad \rho(A) = \lim_{t\to\infty} \|A^t\|^{1/t}. \tag{52.10}$$

The next theorem is a special case of the theorem of Perron and Frobenius on irreducible matrices.

52.15 Theorem (Perron–Frobenius). *Let Q be an $r \times r$-matrix, $r < \infty$, such that $Q > 0$. Then*

- $\lambda = \rho(Q) > 0$ *is an eigenvalue, and*
- *there are eigenvectors $Qe = \lambda e > 0$ and $Q^* f = \lambda f > 0$.*

Moreover, these properties uniquely characterize λ, and the corresponding eigenspace is one-dimensional.

Proof. The proof proceeds in several steps. Claims are highlighted .

1° We have $\rho(Q) > 0$. In fact, if $\rho(Q) = 0$, then $\sigma(Q) = \{0\}$ and Q is nilpotent, i.e. for some power $Q^k = 0$, and this contradicts $Q > 0$

2° ρ(Q) is an eigenvalue. We may assume $\rho(Q) = 1$, otherwise $Q \rightsquigarrow Q/\rho(Q)$.

Since $\rho(\theta Q) = \theta < 1$ for any $\theta \in (0,1)$, the Neumann series converges and we see

$$(\mathrm{id} - \theta Q)^{-1} = \sum_{n=0}^{\infty} \theta^n Q^n \geqslant \sum_{n=0}^{m} \theta^n Q^n \quad \forall m \in \mathbb{N}. \tag{52.11}$$

Assume that $\rho(Q) = 1 \notin \sigma(Q)$. This means that $(\mathrm{id} - Q)^{-1}$ exists, and we find with (52.11), letting $\theta \uparrow 1$,

$$(\mathrm{id} - Q)^{-1} \geqslant \sum_{n=0}^{m} Q^n \implies \lim_{n\to\infty} Q^n = 0.$$

In particular, we get $\rho(Q)^n = \rho(Q^n) \overset{(52.10)}{\leqslant} \|Q^n\| \to 0$ and we infer that $\rho(Q) < 1$; this is but impossible, and so $\rho(Q) = 1 \in \sigma(Q)$.

3° There is an eigenvector $e > 0$. Again we assume that $\rho(Q) = 1$. Since 1 is an eigenvalue, there is some $v \in \mathbb{C}^r$ such that $Qv = v$. Define

$$e := \widetilde{v} := (|v_1|, \dots, |v_r|)^\top.$$

We claim that $Qe = e$; in particular, this means that $e = Qe > 0$ (since $Q > 0$ and $e \geqslant 0$). Assume that $Qe \neq e$. A direct calculation yields

$$e = \widetilde{v} = \widetilde{Qv} \overset{\text{def}}{=} (|(Qv)_1|, \dots, |(Qv)_r|)^\top \leqslant Q\widetilde{v} = Qe.$$

This shows that $w := (Q - \mathrm{id})e \geqslant 0$ and, by assumption, $w \neq 0$. Using that $Q > 0$ we see

$$(Q - \mathrm{id})Qe = Qw > 0 \implies \exists \epsilon > 0 : Qw > \epsilon Qe.$$

Setting $z := Qe > 0$ we get

$$(Q - \mathrm{id})z = (Q - \mathrm{id})Qe \geqslant \epsilon z \iff Qz \geqslant (1+\epsilon)z.$$

We can rewrite this using the matrix $P := (1+\epsilon)^{-1}Q > 0$:

$$Pz \geqslant z \implies P^m z \geqslant z;$$

for this implication use $P^m = \mathrm{id} + \sum_{i=0}^{m-1} P^i(P - \mathrm{id})$ and the fact that $P^i(P - \mathrm{id}) \geqslant 0$. By construction, $\rho(P) = (1+\epsilon)^{-1} < 1$. Now (52.10) implies that

$$\|P^m\| \leqslant (\rho(P) + \epsilon)^m \quad \forall m \gg 1 \implies P^m \xrightarrow[m\to\infty]{} 0.$$

Thus, $0 = \lim_{m\to\infty} P^m z \geqslant z > 0$, which is absurd – thus, $Qe = e$.

4° The assertion for Q^* can be shown as in 3° .

5° Uniqueness. Assume that $Qe' = \lambda' e'$ for some λ' and $e' > 0$. This gives for the eigenvector $f > 0$ of Q^* with eigenvalue λ

$$\lambda' \langle e', f \rangle = \langle Qe', f \rangle = \langle e', Q^* f \rangle = \lambda \langle e', f \rangle \implies \lambda' = \lambda.$$

Thus, e' is an eigenvector for the eigenvalue λ.

If e, e' were **linearly independent**, then $e+\gamma e' \neq 0$ for any $\gamma \in \mathbb{R}$, and we may choose γ in such a way that $e+\gamma e'$ has at least one coordinate equal to 0. So,

$$\lambda(e+\gamma e') = Q(e+\gamma e') \xRightarrow{\text{as in } 3^\circ} \lambda(\widetilde{e+\gamma e'}) = Q(\widetilde{e+\gamma e'}) > 0$$

for the last estimate, we use that $Q > 0$ and $(\widetilde{e+\gamma e'}) \geqslant 0$. Since this is a contradiction, we see that $e = ce'$. In particular, the eigenspace is one-dimensional.

6° The assertions for f can be shown as in 5° . □

We will now combine the ergodic theorem (Theorem 52.11) and the Perron–Frobenius theorem to study the asymptotic behaviour of powers of matrices. We use the following **standard notation**

$$\alpha_n \approx \beta_n \iff \lim_{n\to\infty} \frac{\alpha_n}{\beta_n} = 1.$$

52.16 Corollary. *Let $Q = (q_{ij}) > 0$ be an $r \times r$-matrix, $r < \infty$. Then $q_{ij}^{(n)} \approx \lambda^n e_i f_j$ where $\lambda > 0$ and $e, f \in \mathbb{R}^r$ are as in Theorem 52.15.*

▲ λ is the largest eigenvalue of Q since $\lambda = \rho(Q)$ and $\lambda > 0$.

Proof. Using the Perron–Frobenius theorem we find $\lambda > 0$ and all eigenvectors $e, f \in \mathbb{R}^r$ such that

$$\sum_{i\in E} e_i = 1 \quad \text{and} \quad \langle e, f\rangle = \sum_{i\in E} e_i f_i = 1.$$

Since the eigenspaces are one-dimensional, this normalization uniquely fixes e and f. Set

$$p_{ij} = \frac{q_{ij}e_j}{\lambda e_i} \implies \sum_{j\in E} p_{ij} = \frac{(Qe)_i}{\lambda e_i} = 1.$$

This shows that the matrix $P = (p_{ij})$ is stochastic and > 0. Using

$$\sum_{i\in E} e_i f_i p_{ij} = \sum_{i\in E} e_i f_i \frac{q_{ij}e_j}{\lambda e_i} = \frac{1}{\lambda} e_j \sum_{i\in E} f_i q_{ij} = \frac{1}{\lambda} e_j (Q^* f)_j = e_j f_j$$

we see that $\pi \in \mathbb{R}^r$ such that $\pi_i = e_i f_i$ is the stationary distribution of P. So,

$$\begin{aligned} q_{ij}^{(n)} &= \sum_{i_1,\dots,i_{n-1}\in E} q_{ii_1} q_{i_1 i_2} \cdots q_{i_{n-2}i_{n-1}} q_{i_{n-1}j} \\ &= \lambda^n \sum_{i_1,\dots,i_{n-1}\in E} p_{ii_1} p_{i_1 i_2} \cdots p_{i_{n-2}i_{n-1}} p_{i_{n-1}j} e_i e_j^{-1} = \lambda^n e_i p_{ij}^{(n)} e_j^{-1}. \end{aligned}$$

The ergodic theorem now applies and shows that (exponentially fast):

$$\frac{q_{ij}^{(n)}}{\lambda^n} = e_i p_{ij}^{(n)} e_j^{-1} \xrightarrow[n\to\infty]{\text{Theorem 52.11}} e_i \pi_j e_j^{-1} = e_i f_j.$$ □

53 The scope of Markov chains*

We will give now a few typical examples of homogeneous Markov chains in discrete time and with discrete state space E, $\mathscr{E} = \mathscr{P}(E)$. We identify E as subset of $\mathbb{Z}$ assuming that $0 \in E$. As before, $r, s, t \in \mathbb{N}_0$ denote times, $i, j, k \in E$ states, $\mu = (\mu_i)_{i \in E}$ is the initial distribution and $P = (p_{ij})_{i,j \in E}$ the transition matrix.

53.1 Example (sequence of iid rv). Let $(X_s)_{s \in \mathbb{N}_0}$ be a sequence of iid rv with values in $E = \mathbb{N}_0$ and probability distribution $\mathbb{P}(X_s = i) = p_i$ where $p_i \geqslant 0$ and $\sum_{i \in \mathbb{N}_0} p_i = 1$. Because of independence, we see that for all $i_0, \dots, i_t, j \in E$

$$\begin{aligned}\mathbb{P}(X_{t+1} = j \mid X_0 = i_0, \dots, X_t = i_t) &= \mathbb{P}(X_{t+1} = j) = p_j \\ &= \mathbb{P}(X_{t+1} = j \mid X_t = i_t).\end{aligned}$$

This means that $(X_s)_{s \in \mathbb{N}}$ is a MC »with zero memory«, initial distribution $\mu = p$ and transition matrix

$$P = \begin{pmatrix} p_0 & p_1 & p_2 & p_3 & \cdots \\ p_0 & p_1 & p_2 & p_3 & \cdots \\ \vdots & \vdots & \vdots & \vdots & \\ p_0 & p_1 & p_2 & p_3 & \cdots \\ \vdots & \vdots & \vdots & \vdots & \end{pmatrix}.$$

53.2 Example (random walk on a lattice). Let $(\xi_s)_{s \in \mathbb{N}}$ be a sequence of iid rv taking values in $E = \mathbb{N}_0$ and probability distribution $\mathbb{P}(\xi_s = i) = p_i$ where $p_i \geqslant 0$ and $\sum_{i \in \mathbb{N}_0} p_i = 1$. We define the random walk

$$X_0 = i_0 \quad \text{and} \quad X_t = i_0 + \sum_{s=1}^{t} \xi_s.$$

Since $X_{t+1} = X_t + \xi_{t+1}$, we find for all $0 \leqslant i_0 \leqslant \dots \leqslant i_t \leqslant j$

$$\begin{aligned}\mathbb{P}(X_{t+1} = j \mid X_0 = i_0, \dots, X_t = i_t) &= \mathbb{P}(X_t + \xi_{t+1} = j \mid X_0 = i_0, \dots, X_t = i_t) \\ &\stackrel{\text{iid}}{=} \mathbb{P}(i_t + \xi_{t+1} = j) = p_{j-i_t} = p_{i_t j} \\ &= \mathbb{P}(X_{t+1} = j \mid X_t = i_t).\end{aligned}$$

This shows that $(X_s)_{s \in \mathbb{N}}$ is a MC with increasing paths, $\mu = \delta_{i_0}$ and

$$p_{ij} = \begin{cases} p_{j-i}, & i \leqslant j, \\ 0, & i > j, \end{cases} \quad \text{and} \quad P = \begin{pmatrix} p_0 & p_1 & p_2 & p_3 & \cdots \\ 0 & p_0 & p_1 & p_2 & \cdots \\ 0 & 0 & p_0 & p_1 & \cdots \\ 0 & 0 & 0 & p_0 & \cdots \\ \vdots & \vdots & \vdots & \vdots & \ddots \end{pmatrix}.$$

If $E = \mathbb{Z}$, we get a walk on $\mathbb{Z}$ with two-sided infinite transition matrix

$$P = \begin{pmatrix} \ddots & \vdots & \vdots & \vdots & \vdots & \\ \cdots & p_0 & p_1 & p_2 & p_3 & \cdots \\ \cdots & p_{-1} & p_0 & p_1 & p_2 & \cdots \\ \cdots & p_{-2} & p_{-1} & p_0 & p_1 & \cdots \\ & \vdots & \vdots & \vdots & \vdots & \ddots \end{pmatrix}.$$

53.3 Example (success runs). Consider iid Bernoulli experiments with success probability p and denote by X_t the length (so far) of the current run of successes after t trials; this means that X_t is the number of trials since the last failure (where the start $X_0 = 0$ is understood as a failure). The state space of this process is $E = \mathbb{N}_0$. By definition

$$p_{i,i+1} = \mathbb{P}(X_{t+1} = i+1 \mid X_t = i) = p \quad \text{and} \quad p_{i,0} = \mathbb{P}(X_{t+1} = 0 \mid X_t = i) = q,$$

while all other (conditional) probabilities are zero: $p_{ij} = 0$ if $j \neq 0, i+1$. The transition matrix of this MC is

$$P = \begin{pmatrix} q & p & 0 & 0 & 0 & 0 & \cdots \\ q & 0 & p & 0 & 0 & 0 & \cdots \\ q & 0 & 0 & p & 0 & 0 & \cdots \\ q & 0 & 0 & 0 & p & 0 & \cdots \\ \vdots & \vdots & & & & \ddots & \end{pmatrix}.$$

53.4 Example (homogeneous (lazy) SRW). Let $(X_s)_{s\in\mathbb{N}}$ be the RW on $E = \mathbb{Z}$ of Example 53.4, but assume that for $p, q, r \geqslant 0$ and $p + r + q = 1$

$$p_i = \begin{cases} p, & i = 1, \\ r, & i = 0, \\ q, & i = -1, \\ 0, & \text{else}, \end{cases} \quad \text{and} \quad P = \begin{pmatrix} \ddots & \ddots & \ddots & & & & 0 \\ & q & r & p & & & \\ & & q & r & p & & \\ & & & q & r & p & \\ 0 & & & & \ddots & \ddots & \ddots \end{pmatrix}.$$

If $r = 0$ this is the standard SRW on $\mathbb{Z}$, if $r > 0$ this is a »lazy« SRW.

53.5 Example (inhomogeneous (lazy) SRW). We are interested in the boundary behaviour. Therefore, we study three different state spaces. Let $(X_s)_{s\in\mathbb{N}}$ be the RW of Example 53.2.

a) Let $E = \mathbb{Z}$ (i.e. no boundary) and set for $p_i, q_i, r_i \geqslant 0$ with $p_i + r_i + q_i = 1$

$$p_{ij} = \begin{cases} p_i, & j = i+1, \\ r_i, & j = i, \\ q_i, & j = i-1, \\ 0, & \text{else}, \end{cases} \quad \text{and } P = \begin{pmatrix} \ddots & \ddots & \ddots & & & & 0 \\ & q_{-1} & r_{-1} & p_{-1} & & & \\ & & q_0 & r_0 & p_0 & & \\ & & & q_1 & r_1 & p_1 & \\ & 0 & & & \ddots & \ddots & \ddots \end{pmatrix}.$$

b) Let $E = \mathbb{N}_0$, i.e. $i = 0$ is a one-sided boundary. With p_{ij} as in Part a), we have

$$P = \begin{pmatrix} r_0 & p_0 & 0 & \cdots & & & \\ q_1 & r_1 & p_1 & 0 & \cdots & & \\ 0 & q_2 & r_2 & p_2 & 0 & \cdots & \\ \vdots & & \ddots & \ddots & \ddots & & \end{pmatrix} \quad \text{where} \quad \begin{matrix} p_0 + r_0 = 1, \\ p_i + r_i + q_i = 1, \ i > 0. \end{matrix}$$

If $r_0 = 1$ and $p_0 = 0$, the boundary $i = 0$ is **absorbing**, if $r_0 = 0$ and $p_0 = 1$, the boundary is **reflecting**.

c) Let $E = \{0, 1, \dots m\}$, i.e. both $i = 0$ and $i = m$ are boundaries. With p_{ij} as in Part a), we have

$$P = \begin{pmatrix} r_0 & p_0 & 0 & \cdots & \cdots & \cdots & 0 \\ q_1 & r_1 & p_1 & 0 & \cdots & \cdots & 0 \\ 0 & q_2 & r_2 & p_2 & 0 & \cdots & 0 \\ \vdots & & \ddots & \ddots & \ddots & & \\ \vdots & & & \ddots & \ddots & \ddots & \\ 0 & \cdots & \cdots & 0 & q_{m-1} & r_{m-1} & p_{m-1} \\ 0 & \cdots & \cdots & \cdots & 0 & q_m & r_m \end{pmatrix}, \quad \begin{matrix} p_0 + r_0 = 1 = q_m + r_m, \\ p_i + r_i + q_i = 1, \ \text{else}. \end{matrix}$$

The boundary classification is the same as in Part c).

53.6 Example (Ehrenfest urn). The Ehrenfest urn is a realization of the Markov chain from Example 53.5.c). It was proposed by Tatiana and Paul Ehrenfest to model the second law of thermodynamics. Let $E = \{0, 1, 2, \dots, 2n\}$ and assume that at time $t = 0$ $2n$ numbered balls are placed randomly into two containers, I and II. At each time $t \in \mathbb{N}$ we pick a number $\{1, 2, \dots, 2n\}$ at random, remove the corresponding ball from its current container and place it into the other container.

Denote by X_t the number of balls in container I at time t. By definition

$$p_{i,i+1} = \mathbb{P}(X_{t+1} = i + 1 \mid X_t = i) = \frac{2n - i}{2n},$$

since the ball must be in container II if we want to increase the number of balls in container I, and

$$p_{i,i-1} = \mathbb{P}(X_{t+1} = i - 1 \mid X_t = i) = \frac{i}{2n}.$$

If $|i - j| \neq 1$, then $p_{ij} = 0$. The boundaries at 0 and $2n$ are reflecting. This shows a tendency towards some kind of »equilibrium« where both containers have the same number of balls.

53.7 Example (recursive sequences). Let $E = \mathbb{Z}$ (with some modifications, we can also consider a general state space) and $f : E \times \mathbb{R} \to E$ be any function. Moreover, $(U_s)_{s \in \mathbb{N}}$ is a sequence of iid $\mathbb{R}$-valued random variables. For fixed $i_0 \in E$ we define

$$X_0 := i_0 \quad \text{and} \quad X_{t+1} := f(X_t, U_t), \quad t \in \mathbb{N}_0.$$

Observe that for all $i_0,\dots,i_t,j \in E$

$$\begin{aligned}\mathbb{P}(X_{t+1}=j \mid X_0=i_0,\dots,X_t=i_t) &= \mathbb{P}(f(X_t,U_t)=j \mid X_0=i_0,\dots,X_t=i_t)\\ &\overset{\text{iid}}{=} \mathbb{P}(f(i_t,U_t)=j) = p_{i_t j}\\ &= \mathbb{P}(X_{t+1}=j \mid X_t=i_t),\end{aligned}$$

and this shows that the sequence $(X_t)_{t\in\mathbb{N}}$ is a Markov chain.

Let us discuss a few special cases of the above construction.

a) $X_{t+1} = U_t + X_t = X_0 + \sum_{s=1}^{t+1} U_s$ where the steps U_s are iid $\mathbb{Z}$-valued rv – see Example 53.2.

b) (auto-regressive process of order 1). Let $\beta \in (-1,1)$ and set

$$X_0 := i_0 \quad\text{and}\quad X_{t+1} = \beta X_t + U_{t+1}, \quad t \in \mathbb{N}_0.$$

This model is most interesting, if $E = \mathbb{R}$. Of course, this requires some changes when we check the Markov property at the beginning of this example.

c) (storage model). At the end of each day $t \in \mathbb{N}$ we check the number of items X_t in our storage. There is a maximum number M of items that can be stored, and there is a critical minimum $m < M$ which is needed to run the business. We have the following strategy: If $X_t \leqslant m$, then we re-supply »immediately« (over night) to the maximum level M.

Assume that $X_0 = M$, U_t is the number of items sold during day t and we assume that the sequence $(U_s)_{s\in\mathbb{N}}$ is iid. This means that our state space is $E = \{0,1,2,\dots,M\}$, and we have

$$X_0 = M \quad\text{and}\quad X_{t+1} = \begin{cases}(M-U_{t+1})^+, & \text{if } X_t \leqslant m,\\ (X_t - U_{t+1})^+, & \text{if } m < X_t \leqslant M.\end{cases}$$

Thus $X_{t+1} = f(X_t, U_{t+1})$ for the function

$$f(i,x) = (M-x)^+ \mathbb{1}_{[0,m]}(i) + (i-x)^+ \mathbb{1}_{(m,M]}(i).$$

Assume that $\mathbb{P}(U_s = i) = p_i$ and define $p_{ij} := \mathbb{P}(f(i,U_{t+1}) = j)$. We have

- if $i \leqslant m$ and $j = 1,\dots,M$

$$p_{ij} = \mathbb{P}(f(i,U_{t+1})=j) = \mathbb{P}((M-U_{t+1})^+ = j) = \mathbb{P}(U_{t+1} = M-j) = p_{M-j}.$$

- if $i \leqslant m$ and $j = 0$

$$p_{i0} = \mathbb{P}(f(i,U_{t+1})=0) = \mathbb{P}(U_{t+1} \geqslant M) = \sum_{k=M}^{\infty} p_k.$$

- if $m+1 \leqslant i \leqslant M$

$$p_{ij} = \mathbb{P}((i - U_{t+1})^+ = j) = \begin{cases} 0, & \text{if } j > i, \\ \mathbb{P}(U_{t+1} \geqslant i) = \sum_{k=i}^\infty p_i, & \text{if } j = 0, \\ \mathbb{P}(U_{t+1} = i - j) = p_{i-j}, & \text{if } j = 1, \dots, i. \end{cases}$$

Overall we get the following transition matrix (initial state ↓ in the right-hand column)

$$P = \begin{pmatrix} \sum\limits_{k=M}^\infty p_k & p_{M-1} & p_{M-2} & \cdots & \cdots & \cdots & \cdots & p_0 \\ \vdots & \vdots & \vdots & & & & & \vdots \\ \sum\limits_{k=M}^\infty p_k & p_{M-1} & p_{M-2} & \cdots & \cdots & \cdots & \cdots & p_0 \\ \sum\limits_{k=m+1}^\infty p_k & p_m & p_{m-1} & \cdots & p_0 & 0 & \cdots & 0 \\ \sum\limits_{k=m+2}^\infty p_k & p_{m+1} & p_m & \cdots & p_1 & p_0 & \cdots & 0 \\ \vdots & & & & & \ddots & \ddots & \\ \sum\limits_{k=M}^\infty p_k & p_{M-1} & p_{M-2} & \cdots & \cdots & \cdots & p_1 & p_0 \end{pmatrix} \quad \begin{matrix} 0 \\ \vdots \\ m \\ m+1 \\ m+2 \\ \vdots \\ M \end{matrix}$$

d) The model discussed in Part c) is actually so general that we can use it to show the existence of a general MC with a given matrix P in discrete time and with discrete state space E, cf. also Lemma 54.1 below. Let us quickly sketch the argument. Assume that $E = \{0, 1, \dots, m\}$ is finite and that $P = (p_{ij})_{i,j\in E}$ is a given transition matrix. Construct an iid sequence $(U_s)_{s\in\mathbb{N}}$ of uniformly distributed random variables in $[0,1]$. We are going to construct a function satisfying

$$f : E \times [0,1] \to E, \quad \mathbb{P}(f(i, U_{t+1}) = j) = p_{ij}, \quad t \in \mathbb{N}_0,\ i, j \in E.$$

Indeed, define

$$f(i,x) = j \quad \text{if} \quad \sum_{k=0}^{j-1} p_{ik} \leqslant x < \sum_{k=0}^{j} p_{ik},$$

where we use the »empty sum convention« $\sum_0^{-1} := 0$. Since the U_s are uniformly distributed, we finally get

$$\mathbb{P}(f(i, U_{t+1}) = j) = \mathbb{P}\left(\sum_{k=0}^{j-1} p_{ik} \leqslant U_{t+1} < \sum_{k=0}^{j} p_{ik} \right) = p_{ij}.$$

53.8 Example (branching processes). A population (at generation $t \in \mathbb{N}_0$) consists of X_t individuals. In each generation, every individuum k has $\xi_k^{(t)} \in \mathbb{N}_0$ offspring, but the parent individuum dies; the sequences $\xi^{(0)} := (\xi_k^{(0)})_{k\in\mathbb{N}}$, $\xi^{(1)} := (\xi_k^{(1)})_{k\in\mathbb{N}}$, $\xi^{(t)} := (\xi_k^{(2)})_{k\in\mathbb{N}}$, ... are assumed to be independent sequences of independent random variables. The initial population is $X_0 = i_0$; once we have reached $X_{t+1} = 0$, the population is gone forever.

Using the »empty-sum convention« $\sum_1^0 := 0$, we see that

$$X_{t+1} = \sum_{k=1}^{X_t} \xi_k^{(t)} = f(X_n, \xi^{(t)}), \quad \xi^{(t)} = (\xi_k^{(t)})_{k\in\mathbb{N}};$$

this is a generalization of Example 53.7.

The process X is temporally homogeneous, if the law of $\xi_k^{(t)}$ does not depend on t; if $\xi_k^{(t)}$ does not depend on time t nor the state space variable k, we speak of a so-called **Galton–Watson** process. An example of a state-dependent, temporally homogeneous process is, e.g.

$$\mathbb{P}(\xi_k^{(t)} = j) = \begin{cases} e^{-\gamma k} q_j, & \text{if } j \in \mathbb{N}, \\ 1 - e^{-\gamma k}, & \text{if } j = 0, \end{cases}$$

where $\gamma > 0$ and $\sum_{j\in\mathbb{N}} q_j = 1$. Note that $\gamma > 0$ plays the role of a damping constant: If $k \gg 1$, less offspring is born.

53.9 Example (Wright–Fisher model). We want to study the fluctuation of gene frequencies in a population of fixed size 2N. We assume that there are two types of genes: g and G. Denote by X_t the number of g-type individuals at epoch t. In each time-step $t \to t+1$ we replace the whole population with 2N new individuals, and their gene type is decided by a Bernoulli experiment: We get type g or G with probability p_i or $q_i = 1 - p_i$, respectively, where $X_t = i$ is the number of individuals of type g in the old population.

This means that our state space is $E = \{0, 1, \dots, 2N\}$ and for $i, j \in E$ we have

$$\begin{aligned}\mathbb{P}(X_{t+1} = j \mid X_0 = i_0, \dots, X_{t-1} = i_{t-1}, X_t = i) &= \mathbb{P}(X_{t+1} = j \mid X_t = i) \\ &= \binom{2N}{j} p_i^j (1-p_i)^{2N-j}.\end{aligned}$$

The following different scenarios are usually considered.

a) Simple reproduction model: We set $p_i = i/(2N)$ and $q_i = (2N - i)/2N$, i.e. the probabilities are proportional to the existing gene pool. The states 0 and 2N are absorbing and, in the long run, the system ends up in one of these states.

b) Reproduction with mutation: When choosing a new generation, we allow for spontaneous mutations, namely

$$\mathbb{P}(g \rightsquigarrow G) = \gamma_1 \quad \text{and} \quad \mathbb{P}(G \rightsquigarrow g) = \gamma_2.$$

Thus, we get

$$p_i = \frac{i}{2N}(1-\gamma_1) + \frac{2N-i}{2N}\gamma_2 \quad \text{and} \quad q_i = \frac{i}{2N}\gamma_1 + \frac{2N-i}{2N}(1-\gamma_2).$$

In the long run we end up in a stationary equilibrium.

c) Reproduction with selection: We assume that the g-genes grow by a factor $(1+s)$ without mutation, i.e.

$$p_i = \frac{(1+s)i}{2N + si} \quad \text{and} \quad q_i = 1 - p_i.$$

54 The (strong) Markov property

We will now consider **infinite** Markov chains $X = (X_0, X_1, X_2, \dots)$ with transition matrix P and state space $E = \{i(1), \dots, i(r)\}$, $r \in \mathbb{N} \cup \{\infty\}$.[(1)] Since we have to construct infinitely many X_t, the existence of X is indeed an issue, and the naive approach of (52.2) does not work in this setting.

54.1 Lemma. *Let $E = \{i(1), \dots, i(r)\}$, $r \in \mathbb{N} \cup \{\infty\}$ be the state space and P a $r \times r$ stochastic matrix. There exists a probability space $(\Omega', \mathscr{A}', \mathbf{P}')$ such that for every $i \in E$ there is a $MC(\delta_i, P)$ with $\delta_i = (\underbrace{0, \dots, 0, 1}_{i}, 0, \dots)$ as initial distribution.*

Proof. We use $(\Omega', \mathscr{A}', \mathbf{P}') = ([0,1), \mathscr{B}[0,1), \lambda)$ and define on this space a sequence $(U_t)_{t\in\mathbb{N}}$ of iid uniformly distributed random variables. Recurseively, we define

$$X_0^i := i,$$

$$X_1^i := i(n) \quad \text{if} \quad \sum_{m=1}^{n-1} p_{i,i(m)} < U_1 \leqslant \sum_{m=1}^{n} p_{i,i(m)},$$

$$\vdots$$

$$X_{t+1}^i := i(n) \quad \text{if} \quad \sum_{m=1}^{n-1} p_{X_t^i,i(m)} < U_{t+1} \leqslant \sum_{m=1}^{n} p_{X_t^i,i(m)}.$$

With this construction we have $\mathbf{P}'(X_0^i = i) = 1$ and $\mathbf{P}'(X_{t+1}^i = j) = p_{i,j}$. Moreover,

$$\begin{aligned}
&\mathbf{P}'\left(X_{t+1}^i = i(n) \mid X_0^i = i_0, X_1^i = i_1, \dots, X_t^i = i_t\right) \\
&= \mathbf{P}'\left(\sum_{m=1}^{n-1} p_{X_t^i,i(m)} < U_{t+1} \leqslant \sum_{m=1}^{n} p_{X_t^i,i(m)} \mid X_0^i = i_0, X_1^i = i_1, \dots, X_t^i = i_t\right) \\
&= \mathbf{P}'\left(\sum_{m=1}^{n-1} p_{i_t,i(m)} < U_{t+1} \leqslant \sum_{m=1}^{n} p_{i_t,i(m)} \mid X_0^i = i_0, X_1^i = i_1, \dots, X_t^i = i_t\right) \\
&\overset{\text{iid}}{=} \mathbf{P}'\left(\sum_{m=1}^{n-1} p_{i_t,i(m)} < U_1 \leqslant \sum_{m=1}^{n} p_{i_t,i(m)}\right) = p_{i_t,i(n)}.
\end{aligned}$$

□

We can now construct a **canonical version** of the Markov chain. Set

$$\Omega := E^{\mathbb{N}_0}, \quad \omega = (\omega(t))_{t\in\mathbb{N}_0}, \quad X_t(\omega) := \omega(t), \quad \mathscr{A} := \mathscr{F}_\infty := \sigma(X_t, t \geqslant 0).$$

54.2 Theorem (canonical MC). *Let $E = \{i(1), \dots, i(r)\}$, $r \in \mathbb{N} \cup \{\infty\}$ and P be a stochastic $r \times r$-matrix. For every $i \in E$ there is a unique probability measure $\mathbb{P}^i$*

(1) ⚠ Note that $i(m)$ refers to the mth element of E while $X_t = i_t$ means that X_t attains some value $i_t \in E$, i.e. the subscript refers to time.

on $\Omega = E^{\mathbb{N}_0}$, $\mathscr{A} = \sigma(X_t, t \geqslant 0)$ *such that the coordinate projections* $X_t : \Omega \to E$, $X_t(\omega) = \omega(t)$ *are a* $MC(\delta_i, P)$.

Proof. We use the $MC(\delta_i, P)$ $(X_t^i)_{t\in\mathbb{N}_0}$ on $(\Omega', \mathscr{A}', \mathsf{P}')$ from Lemma 54.1. The map $\Psi : \Omega' \to \Omega$, $\omega' \mapsto \omega := (X_t^i(\omega'))_{t\in\mathbb{N}_0}$ is measurable, since each coordinate projection $X_t \circ \Psi : \Omega' \to E$, $\omega' \mapsto X_t^i(\omega')$ is measurable (see Theorem 16.9.b) and the construction of $\mathscr{A}$ §19.2 ✍).

Define $\mathbb{P}^i := \mathsf{P}' \circ \Psi^{-1}$. Then we have

$$\begin{aligned}\mathbb{P}^i(X_0 = i_0, X_1 = i_1, \dots, X_t = i_t) &= \mathsf{P}'(X_0^i = i_0, X_1^i = i_1, \dots, X_t^i = i_t)\\ &= \mathsf{P}'(X_0^i = i_0) \cdot p_{i_0,i_1} \cdot \dots \cdot p_{i_{t-1},i_t}\\ &= \mathbb{P}^i(X_0 = i_0) \cdot p_{i_0,i_1} \cdot \dots \cdot p_{i_{t-1},i_t}.\end{aligned}$$

In order to see the uniqueness of $\mathbb{P}^i$, assume that there is a further such measure $\mathbb{Q}^i$. By construction, $\mathbb{P}^i = \mathbb{Q}^i$ on all cylinder sets of the form

$$\{\omega \in \Omega \mid X_0(\omega) = i_0, X_1(\omega) = i_1, \dots, X_t(\omega) = i_t\}, \quad i_0, \dots, i_t \in E.$$

But this is a $\cap$-stable family which generates $\mathscr{F}_\infty$, hence $\mathbb{P}^i = \mathbb{Q}^i$ on $\mathscr{F}_\infty$. □

54.3 Definition. Let $X = (X_t)_{t\in\mathbb{N}_0}$ be a stochastic process on $(\Omega, \mathscr{A}, \mathbb{P})$ with natural filtration $\mathscr{F}_t^X := \sigma(X_s, s \leqslant t)$ and state space $(E, \mathscr{E})$. The process X is called a **Markov process** (MP), if it enjoys the following **Markov property** (MP):

$$\mathbb{P}(X_{t+1} \in B \mid \mathscr{F}_t^X) = \mathbb{P}(X_{t+1} \in B \mid X_t) \quad \forall t \in \mathbb{N}_0,\ B \in \mathscr{E}. \tag{54.1}$$

54.4 Lemma. *The Markov property* (54.1) *is equivalent to*

$$\mathbb{E}(f(X_{t+1}) \mid \mathscr{F}_t^X) = \mathbb{E}(f(X_{t+1}) \mid X_t) \quad \forall t \in \mathbb{N}_0,\ f : E \xrightarrow{\text{bdd, mble}} \mathbb{R}. \tag{54.2}$$

Proof. (54.2)⇒(54.1): Take $f = \mathbb{1}_B$.
(54.1)⇒(54.2): Because of the linearity of conditional expectation and the conditional Beppo Levi theorem, we can easily check the conditions of the monotone class theorem (Theorem 7.17) from which the claim follows. □

54.5 Example. A discrete $MC(\mu, P)$ X is a Markov process. Take the realization of X on $(\Omega, \mathscr{A}, \mathbb{P}^i)$. From Lemma 52.5 and (52.4) we know that

$$\mathbb{P}^i(X_{t+1} = j \mid \mathscr{F}_t^X) = p_{X_t, j} = \mathbb{P}^i(X_{t+1} = j \mid X_t) \quad \forall t \in \mathbb{N}_0,\ i, j \in E.$$

Thus, a MC enjoys the Markov property which takes the following form:

$$\mathbb{P}^i(X_{t+1} = j \mid \mathscr{F}_t^X) = \mathbb{P}^i(X_{t+1} = j \mid X_t) = \mathbb{P}^{X_t}(X_{t+1} = j) \quad \mathbb{P}^i\text{-a.s.} \tag{54.3}$$

Instead of $(X_0,X_1,X_2,\dots)$ we can consider the shifted Markov chain which starts at time t and position X_t, i.e. $(X'_0,X'_1,\dots)=(X_t,X_{t+1},\dots)$. We have

$$p_{X_t j}=\mathbb{P}^i(X'_1=j\mid X'_0=X_t) \tag{54.4}$$

i.e. X' is a $\mathrm{MC}(\mu P^t,P)$; note that $X_t\sim\mu P^t$.

Notation: If μ is a probability, we set $\mathbb{P}^\mu:=\sum_{i\in E}\mu_i\mathbb{P}^i$. The corresponding expectation is denoted by $\mathbb{E}^\mu$. Thus,

$$\mathbb{P}^\mu(X_{t+1}=j\mid\mathscr{F}_t^X)=\mathbb{P}^\mu(X_{t+1}=j\mid X_t)\overset{(54.4)}{=}\underset{\substack{\uparrow\\ \text{one step}}}{\overset{\text{new starting point}}{\overset{\downarrow}{\mathbb{P}^{X_t}}}}(X_1=j)$$

⚠ All equalities hold only $\mathbb{P}^\mu$-almost surely; this explains the »disappearance« of μ in the right-most term.

The following theorem is stated for a discrete MC with discrete state space $(E,\mathscr{E})$, and $\mathscr{E}=\mathscr{P}(E)$, but both statement and proof hold for general state spaces.

54.6 Theorem. *Let $X=(X_t)_{t\in\mathbb{N}_0}$ be a discrete $\mathrm{MC}(\mu,P)$ with a discrete state space $(E,\mathscr{E})$. For every $t\in\mathbb{N}_0$ the following ($\mathbb{P}^\mu$-a.s. holding) assertions are equivalent:*

a) $\mathbb{P}^\mu(X_{t+1}\in B\mid\mathscr{F}_t^X)=\mathbb{P}^\mu(X_{t+1}\in B\mid X_t)$ *for all* $B\in\mathscr{E}$;

a') $\mathbb{E}^\mu(f(X_{t+1})\mid\mathscr{F}_t^X)=\mathbb{E}^\mu(f(X_{t+1})\mid X_t)$ *for all* $f:E\xrightarrow{\text{bdd, mble}}\mathbb{R}$;

b) $\mathbb{P}^\mu(X_{t+1}\in B\mid\mathscr{F}_t^X)=\mathbb{P}^{X_t}(X_1\in B)$ *for all* $B\in\mathscr{E}$;

b') $\mathbb{E}^\mu(f(X_{t+1})\mid\mathscr{F}_t^X)=\mathbb{E}^{X_t}f(X_1)$ *for all* $f:E\xrightarrow{\text{bdd, mble}}\mathbb{R}$;

c) $\mathbb{P}^\mu(X_{t+1}\in B_1,\dots,X_{t+s}\in B_s\mid\mathscr{F}_t^X)=\mathbb{P}^{X_t}(X_1\in B_1,\dots,X_s\in B_s)$ *for all* $s\in\mathbb{N}_0$ *and* $B_1,\dots,B_s\in\mathscr{E}$;

d) $\mathbb{P}^\mu((X_{t+1},X_{t+2},\dots)\in\Gamma\mid\mathscr{F}_t^X)=\mathbb{P}^{X_t}((X_1,X_2,\dots)\in\Gamma)$ *for all* $\Gamma\in\mathscr{E}^{\otimes\mathbb{N}}$.

Proof. a)⇔a') and b)⇔b') follow with the technique of Lemma 54.4.

a)⇒b): This is the notation which we have introduced in Example 54.5.

b)⇒c): (IA) We use induction on s. Assume that c) holds for $1,\dots,s$.

Induction step $s \rightsquigarrow s+1$: Let $B_1,\dots,B_{s+1} \in \mathscr{E}$. Then

$$
\begin{aligned}
&\mathbb{P}^\mu(X_{t+1} \in B_1, X_{t+2} \in B_2, \dots, X_{t+s+1} \in B_{s+1} \mid \mathscr{F}_t^X)(\omega) \\
&\overset{\text{tower}}{=} \mathbb{E}^\mu\Big(\mathbb{E}^\mu\Big[\mathbb{1}_{B_1}(X_{t+1})\mathbb{1}_{B_2}(X_{t+2})\dots\mathbb{1}_{B_{s+1}}(X_{t+s+1}) \mid \mathscr{F}_{t+1}^X\Big]\Big|\mathscr{F}_t^X\Big)(\omega) \\
&\overset{\text{pull}}{=} \mathbb{E}^\mu\Big(\mathbb{1}_{B_1}(X_{t+1})\mathbb{E}^\mu\Big[\mathbb{1}_{B_2}(X_{t+2})\dots\mathbb{1}_{B_{s+1}}(X_{t+s+1}) \mid \mathscr{F}_{t+1}^X\Big]\Big|\mathscr{F}_t^X\Big)(\omega) \\
&\overset{\text{IA}}{=} \mathbb{E}^\mu\Big(\mathbb{1}_{B_1}(X_{t+1})\mathbb{E}^{X_{t+1}}\Big[\mathbb{1}_{B_2}(X_1)\dots\mathbb{1}_{B_{s+1}}(X_s)\Big]\Big|\mathscr{F}_t^X\Big)(\omega) \\
&\overset{\text{b')}}{=} \mathbb{E}^{X_t(\omega)}\Big(\mathbb{1}_{B_1}(X_1)\mathbb{E}^{X_1}\Big[\mathbb{1}_{B_2}(X_1)\dots\mathbb{1}_{B_{s+1}}(X_s)\Big]\Big) \\
&\overset{\text{IA}}{=} \mathbb{E}^{X_t(\omega)}\Big(\mathbb{1}_{B_1}(X_1)\mathbb{E}^{X_0}\Big[\mathbb{1}_{B_2}(X_2)\dots\mathbb{1}_{B_{s+1}}(X_{s+1}) \mid \mathscr{F}_1^X\Big]\Big) \\
&\overset{\text{pull}}{=} \mathbb{E}^{X_t(\omega)}\Big(\mathbb{E}^{X_t(\omega)}\Big[\mathbb{1}_{B_1}(X_1)\mathbb{1}_{B_2}(X_2)\dots\mathbb{1}_{B_{s+1}}(X_{s+1}) \mid \mathscr{F}_1^X\Big]\Big) \\
&\overset{\text{tower}}{=} \mathbb{E}^{X_t(\omega)}\Big(\mathbb{1}_{B_1}(X_1)\mathbb{1}_{B_2}(X_2)\dots\mathbb{1}_{B_{s+1}}(X_{s+1})\Big);
\end{aligned}
$$

observe that – under $\mathbb{P}^{X_t(\omega)}$ – we have $X_0 \equiv X_t(\omega)$ a.s.

c)⇒d): We can rewrite d) in an integral form:

$$
\int_F \mathbb{1}_\Gamma(X_{t+1}, X_{t+2}, \dots)\, d\mathbb{P}^\mu = \int_F \int \mathbb{1}_\Gamma(X_1, X_2, \dots)\, d\mathbb{P}^{X_t}\, d\mathbb{P}^\mu \quad \forall F \in \mathscr{F}_t^X;
$$

this shows that both sides of the equality are, for fixed F, measures in $\Gamma \in \mathscr{E}^{\otimes \mathbb{N}}$. If we take

$$
\Gamma = B_1 \times \dots \times B_s \times E \times E \dots
$$

then it is clear that d) is exactly c).

Since cylinder sets of this form are a ∩-stabile generator of $\mathscr{E}^{\otimes \mathbb{N}}$, see Step 1° in the proof of Theorem 19.4, the uniqueness of measures theorem tells us that this equality holds for all sets $\Gamma \in \mathscr{E}^{\otimes \mathbb{N}}$,

d)⇒a): Take $\Gamma = B \times E \times E \times \dots$ for any $B \in \mathscr{E}$. □

We will now discuss the strong Markov property. Recall from Definition 39.1 that a random time $\tau : \Omega \to \mathbb{N}_0 \cup \{\infty\}$ is a stopping time if

$$
\{\tau \leqslant t\} \in \mathscr{F}_t^X \quad \forall t \in \mathbb{N}_0.
$$

The σ-algebra generated by τ is

$$
\mathscr{F}_\tau^X = \Big\{F \in \mathscr{F}_\infty^X \mid F \cap \{\tau \leqslant t\} \in \mathscr{F}_t^X \;\; \forall t \in \mathbb{N}_0\Big\}
$$

where $\mathscr{F}_\infty^X = \sigma(X_t, t \in \mathbb{N}_0) = \sigma\big(\bigcup_{t \in \mathbb{N}_0} \mathscr{F}_t^X\big)$.

54.7 Definition. Let $X = (X_t)_{t \in \mathbb{N}_0}$ be a stochastic process with natural filtration $\mathscr{F}_t^X = \sigma(X_s, s \leqslant t)$ and state space $(E, \mathscr{E})$. The process X is called a **strong**

Markov process (SMP) if it enjoys the following **strong Markov property** (SMP): For every stopping time τ and all $t \in \mathbb{N}$, $B \in \mathscr{E}$ and $\mathbb{P}^\mu$-a.a. $\omega \in \{\tau < \infty\}$ it holds that

$$\mathbb{P}^\mu(X_{\tau+t} \in B \mid \mathscr{F}_\tau^X)(\omega) = \mathbb{P}^{X_\tau(\omega)}(X_t \in B) \tag{54.5}$$

Similar to Theorem 54.6, the next result remains valid for general state spaces – without changing the proof.

54.8 Theorem. *A discrete* MC(μ, P) *with discrete state space is a strong Markov process. Moreover, for any* $B_1, \dots, B_t \in \mathscr{E}$ *and every stopping time* τ *one has for* $\mathbb{P}^\mu$*-a.a.* $\omega \in \{\tau < \infty\}$

$$\mathbb{P}^\mu(X_{\tau+1} \in B_1, \dots, X_{\tau+t} \in B_t \mid \mathscr{F}_\tau^X)(\omega) = \mathbb{P}^{X_\tau(\omega)}(X_1 \in B_1, \dots, X_t \in B_t). \tag{54.6}$$

Proof. First, we note that the rv X_τ is $\mathscr{F}_\tau^X$-measurable: If $B \in \mathscr{E}$ and $t \in \mathbb{N}_0$, then

$$\begin{aligned}\{X_\tau \in B\} \cap \{\tau \leqslant t\} &= \bigcup_{s=0}^{t} \{X_s \in B\} \cap \{\tau = s\} \\ &= \bigcup_{s=0}^{t} \underbrace{\{X_s \in B\}}_{\in \mathscr{F}_s^X \subset \mathscr{F}_t^X} \cap \underbrace{\{\tau = s\}}_{\in \mathscr{F}_s^X} \in \mathscr{F}_t^X,\end{aligned}$$

i.e. $\{X_\tau \in B\} \in \mathscr{F}_\tau^X$. Since the map $i \mapsto \mathbb{P}^i(\cdot)$ is measurable,[(2)] the composition $\omega \mapsto \mathbb{P}^{X_\tau(\omega)}(\cdot)$ is $\mathscr{F}_\tau^X$-measurable.

We have to show for all $F \in \mathscr{F}_\tau^X$ that

$$\int_{F \cap \{\tau < \infty\}} \mathbb{1}_B(X_{\tau+t})\, d\mathbb{P}^\mu = \int_{F \cap \{\tau < \infty\}} \mathbb{P}^{X_\tau}(X_t \in B)\, d\mathbb{P}^\mu.$$

Observe that $\{\tau < \infty\} = \bigcup_{0 \leqslant s < \infty} \{\tau = s\}$ and we can use this to »freeze« τ and reduce the strong Markov property to the »normal« Markov property §54.6.b); this happens at the step marked with §. We use, in particular, that $F \cap \{\tau = s\} = \underbrace{(F \cap \{\tau \leqslant s\})}_{\in \mathscr{F}_s^X} \cap \underbrace{\{\tau = s\}}_{\in \mathscr{F}_s^X} \in \mathscr{F}_s^X$:

(2) For a MC with discrete state space this is always true, in general one has to assume this measurability.

$$\begin{aligned}
\int_{F\cap\{\tau<\infty\}} \mathbb{1}_B(X_{\tau+t})\,d\mathbb{P}^\mu &= \sum_{s=0}^\infty \int_{F\cap\{\tau=s\}} \mathbb{1}_B(X_{\tau+t})\,d\mathbb{P}^\mu \\
&= \sum_{s=0}^\infty \int_{F\cap\{\tau=s\}} \mathbb{1}_B(X_{s+t})\,d\mathbb{P}^\mu \\
&\overset{\S}{=} \sum_{s=0}^\infty \int_{F\cap\{\tau=s\}} \mathbb{P}^{X_s}(X_t\in B)\,d\mathbb{P}^\mu \\
&= \sum_{s=0}^\infty \int_{F\cap\{\tau=s\}} \mathbb{P}^{X_\tau}(X_t\in B)\,d\mathbb{P}^\mu \\
&= \int_{F\cap\{\tau<\infty\}} \mathbb{P}^{X_\tau}(X_t\in B)\,d\mathbb{P}^\mu.
\end{aligned}$$

The additional assertion follows from the above argument, if we replace $X_t \rightsquigarrow (X_1,\dots,X_t)$ and $B \rightsquigarrow B_1\times\cdots\times B_t$ and use §54.6.c) instead of §54.6.b). □

The following equivalent re-formulation of Theorem 54.8 should be compared with Theorem 49.3 (for the Poisson process). Note that in the Poisson case, the independent increments properties leads an even stronger assertion since we do not have to use conditioning on $\{X_\tau = i\}$.

54.9 Corollary. *Let X be a discrete $MC(\mu,P)$ with discrete state space $(E,\mathscr{E})$ and let τ be a stopping time. Conditional on $\{X_\tau = i\}\cap\{\tau<\infty\}$, the shifted process $X_{\tau+\bullet} = (X_{\tau+t})_{t\in\mathbb{N}_0}$ is a $MC(\delta_i,P)$ which is independent of $X_1,\dots,X_\tau$.*

Proof. Fix $F\in\mathscr{F}_\tau^X$ and $B_1,\dots,B_t\in\mathscr{E}$. We have

$$\begin{aligned}
&\mathbb{P}^\mu\Big(\{X_{\tau+1}\in B_1,\dots,X_{\tau+t}\in B_t\}\cap F \mid X_\tau=i,\tau<\infty\Big)\cdot\mathbb{P}^\mu\Big(X_\tau=i,\tau<\infty\Big) \\
&\quad= \mathbb{P}^\mu\Big(\{X_{\tau+1}\in B_1,\dots,X_{\tau+t}\in B_t\}\cap F\cap\{X_\tau=i\}\cap\{\tau<\infty\}\Big) \\
&\quad\overset{\text{tower}}{=} \mathbb{E}^\mu\Big[\mathbb{E}^\mu\Big(\mathbb{1}_{\{X_{\tau+1}\in B_1,\dots,X_{\tau+t}\in B_t\}}\mathbb{1}_F\mathbb{1}_{\{X_\tau=i\}}\mathbb{1}_{\{\tau<\infty\}}\mid\mathscr{F}_\tau^X\Big)\Big] \\
&\quad\overset{\text{pull}}{=} \mathbb{E}^\mu\Big[\mathbb{1}_F\mathbb{1}_{\{X_\tau=i\}}\mathbb{1}_{\{\tau<\infty\}}\mathbb{E}^\mu\Big(\mathbb{1}_{\{X_{\tau+1}\in B_1,\dots,X_{\tau+t}\in B_t\}}\mid\mathscr{F}_\tau^X\Big)\Big] \\
&\quad\overset{\text{SMP}}{=} \mathbb{E}^\mu\Big[\mathbb{1}_F\mathbb{1}_{\{X_\tau=i\}}\mathbb{1}_{\{\tau<\infty\}}\mathbb{E}^{X_\tau}\Big(\mathbb{1}_{\{X_1\in B_1,\dots,X_t\in B_t\}}\Big)\Big] \\
&\quad= \mathbb{E}^\mu\Big[\mathbb{1}_F\mathbb{1}_{\{X_\tau=i\}}\mathbb{1}_{\{\tau<\infty\}}\mathbb{E}^i\Big(\mathbb{1}_{\{X_1\in B_1,\dots,X_t\in B_t\}}\Big)\Big] \\
&\quad= \mathbb{P}^\mu\Big(F\cap\{X_\tau=i\}\cap\{\tau<\infty\}\Big)\cdot\mathbb{P}^i\Big(X_1\in B_1,\dots,X_t\in B_t\Big).
\end{aligned}$$

Dividing by $\mathbb{P}^\mu\Big(X_\tau=i,\tau<\infty\Big)$ yields

$$\begin{aligned}
&\mathbb{P}^\mu\Big(\{X_{\tau+1}\in B_1,\dots,X_{\tau+t}\in B_t\}\cap F\mid X_\tau=i,\tau<\infty\Big) \\
&\quad= \mathbb{P}^\mu\Big(F\mid X_\tau=i,\tau<\infty\Big)\cdot\mathbb{P}^i\Big(X_1\in B_1,\dots,X_t\in B_t\Big).
\end{aligned}$$

If we take $F = \Omega$, we see that $X_{\tau+\bullet}$ is a Markov chain under the conditional probability measure $\mathbb{P}^\mu(\cdots \mid X_\tau = i, \tau < \infty)$:

$$\mathbb{P}^\mu\Big(X_{\tau+1} \in B_1, \dots, X_{\tau+t} \in B_t \mid X_\tau = i, \tau < \infty\Big) = \mathbb{P}^i\Big(X_1 \in B_1, \dots, X_t \in B_t\Big).$$

Re-inserting this into the long calculation further up, we see the conditional independence of F and $X_{\tau+\bullet}$:

$$\begin{aligned}&\mathbb{P}^\mu\Big(\{X_{\tau+1} \in B_1, \dots, X_{\tau+t} \in B_t\} \cap F \mid X_\tau = i, \tau < \infty\Big)\\ &\quad = \mathbb{P}^\mu\Big(F \mid X_\tau = i, \tau < \infty\Big) \cdot \mathbb{P}^\mu\Big(X_{\tau+1} \in B_1, \dots, X_{\tau+t} \in B_t \mid X_\tau = i, \tau < \infty\Big). \qquad \square\end{aligned}$$

We close this chapter with two examples which illustrate the power of the strong Markov property.

54.10 Example (motion detector). Let X be a $MC(\mu, P)$ with discrete state space. We assume that we can observe X only if it changes its position. Denote the observable process by $(Z_t)_{t\in\mathbb{N}_0}$. It is given by

$$Z_t = X_{\sigma_t} \quad \text{and} \quad \left|\begin{aligned}\sigma_0 &= 0,\\ \sigma_1 &= \inf\{s > 0 \mid X_s \neq X_0\},\\ \sigma_{t+1} &= \inf\{s > \sigma_t \mid X_s \neq X_{\sigma_t}\}\\ &= \inf\{u > 0 \mid X_{\sigma_t+u} \neq X_{\sigma_t}\} + \sigma_t.\end{aligned}\right.$$

We are going to show that Z is itself a Markov chain.

Notice that $\sigma_{t+1} = \sigma_t + \widetilde{\sigma}_1$, where $\widetilde{\sigma}_1$ is defined for the MC $X_{\sigma_t+\bullet}$ as we have defined σ_1 for the MC X.

Let us assume, for simplicity, that $\sigma_t < \infty$ a.s., i.e. there are no »absorbing« states where X can get trapped. Since the σ_t's are stopping times ✍, we can use the strong MP to get

$$\begin{aligned}\mathbb{P}^\mu(Z_{t+1} = i_{t+1} \mid Z_0, \dots, Z_t) &= \mathbb{P}^\mu(X_{\sigma_{t+1}} = i_{t+1} \mid X_{\sigma_0}, \dots, X_{\sigma_t})\\ &= \mathbb{P}^{X_{\sigma_t}}(X_{\sigma_1} = i_{t+1}) =: \widetilde{p}_{X_{\sigma_t} i_{t+1}}.\end{aligned}$$

We want to find the explicit value of $\widetilde{p}_{ij} = \mathbb{P}^i(X_{\sigma_1} = j)$. Since

$$\mathbb{P}^i(X_{\sigma_1} = j, \sigma_1 = s) = \mathbb{P}^i(X_0 = i, \dots, X_{s-1} = i, X_s = j) = p_{ii}^{s-1} p_{ij},$$

we can sum over all $s \in \mathbb{N}$ and see for $i \neq j$:

$$\widetilde{p}_{ij} = \mathbb{P}^i(X_{\sigma_1} = j) = \frac{p_{ij}}{1 - p_{ii}} \quad \text{and} \quad \widetilde{p}_{ii} = 0.$$

In particular, Z is a $MC(\mu, \widetilde{P})$.

54.11 Example (observation window). Let X be a $MC(\mu, P)$ with discrete state space $(E, \mathscr{E})$. We assume that we can observe X only when $X_s \in W$ for some

$W \in \mathscr{E}$. We denote the observable process by $(Y_t)_{t\in\mathbb{N}_0}$. It is given by

$$Y_t = X_{\tau_t} \quad \text{and} \quad \left| \begin{aligned} \tau_0 &= \inf\{s \geqslant 0 \mid X_s \in W\}, \\ \tau_1 &= \inf\{s > \tau_0 \mid X_s \in W\}, \\ \tau_{t+1} &= \inf\{s > \tau_t \mid X_s \in W\} \\ &= \inf\{u > 0 \mid X_{\tau_t+u} \in W\} \end{aligned} \right.$$

The random times τ_t are stopping times ✍, and we have: $\tau_{t+1} = \tau_t + \widetilde{\tau}_1$ where $\widetilde{\tau}_1$ is defined using $X_{\tau_t+\bullet}$ as we have defined τ_1 for X; (⚠ $\widetilde{\tau}_0 = 0$ as $X_{\tau_t} \in W$). To keep things simple, we assume that $\tau_t < \infty$ a.s. Using Corollary 54.9 we see that for $i_1,\dots,i_{t+1} \in W$

$$\begin{aligned} &\mathbb{P}^\mu(Y_{t+1} = i_{t+1} \mid Y_t = i_t,\dots,Y_0 = i_0) \\ &\quad = \mathbb{P}^\mu(X_{\tau_{t+1}} = i_{t+1} \mid X_{\tau_t} = i_t,\dots,X_{\tau_0} = i_0) \\ &\quad = \mathbb{P}^{i_t}(X_{\tau_1} = i_{t+1}) = \bar{p}_{i_t i_{t+1}} \end{aligned}$$

which means that Y is a MC with transition matrix $\bar{P} = (\bar{p}_{ij})_{i,j\in W}$ and initial distribution $\bar{\mu} \sim X_{\tau_0}$. We will determine the explicit form of $\bar{p}_{ij}$ in the following chapter (Theorem 55.2).

55 Enter, hit, run and return

We continue studying the path behaviour of discrete Markov chains. Throughout this chapter, $X = (X_n)_{n\in\mathbb{N}_0}$ is a $MC(\mu, P)$ with state space $E = \{i(1),\dots,i(r)\}$, $r \in \mathbb{N} \cup \{\infty\}$, and $\mathscr{E} = \mathscr{P}(E)$.

Enter: Entrance times

The [**first**] **entrance time**[(1)] into a set $B \in \mathscr{B}$ is the random time

$$\tau_B^\circ = \tau_B^{X,\circ} = \inf\{t \in \mathbb{N}_0 \mid X_t \in B\}, \quad \inf\emptyset := \infty. \tag{55.1}$$

We are interested in the probability to enter B and the time it takes to get there:

$$h_i^B = \mathbb{P}^i(\tau_B^\circ < \infty) \quad \text{and} \quad k_i^B = \mathbb{E}^i \tau_B^\circ \overset{✍}{=} \sum_{t\in\mathbb{N}} \mathbb{P}^i(\tau_B^\circ \geqslant t). \tag{55.2}$$

55.1 Example. Consider the MC shown in Fig. 55.1.
Problem: Find $h_2^{\{4\}}$ and $k_2^{\{1,4\}}$ – these are the so-called »absorption probability« and mean »absorption time«, since X does not leave the states 1 and 4.
Solution: We use a difference equation. Obviously,

$$\begin{aligned} h_1^{\{4\}} &= 0, & k_1^{\{1,4\}} &= 0 \\ h_4^{\{4\}} &= 1, & k_4^{\{1,4\}} &= 0. \end{aligned}$$

(1) »first« is often dropped.

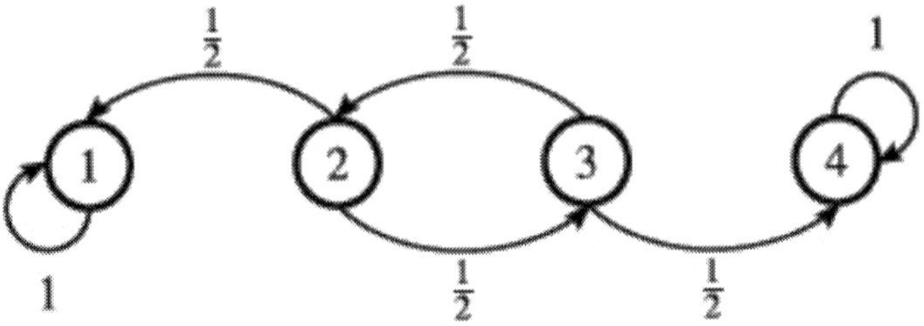

Fig. 55.1. An example of a MC with absorbing states 1 and 4.

If we make one step from the starting point $X_0 = 2$, then we get

$$h_2^{\{4\}} = \tfrac{1}{2}h_1^{\{4\}} + \tfrac{1}{2}h_3^{\{4\}} \qquad k_2^{\{1,4\}} = 1 + \tfrac{1}{2}k_1^{\{1,4\}} + \tfrac{1}{2}k_3^{\{1,4\}},$$

and, similarly, from $X_0 = 3$,

$$h_3^{\{4\}} = \tfrac{1}{2}h_2^{\{4\}} + \tfrac{1}{2}h_4^{\{4\}} \qquad k_3^{\{1,4\}} = 1 + \tfrac{1}{2}k_2^{\{1,4\}} + \tfrac{1}{2}k_4^{\{1,4\}}.$$

We can solve these two systems of 4 equations in 4 unknown variables each, and we see

$$h_2^{\{4\}} = \tfrac{1}{3} \qquad k_2^{\{1,4\}} = 2.$$

Example 55.1 is a special instance of the following more general result.

55.2 Theorem. *Let X be a* $\mathrm{MC}(\mu, P)$ *with discrete state space* $(E, \mathscr{E})$ *and* $B \in \mathscr{E}$. *The vector* $h^B := (h_i^B)_{i \in E} \geqslant 0$ *is the minimal (or smallest) positive solution of the linear system*

$$\begin{cases} h_i^B = 1, & i \in B, \\ h_i^B = \sum_{j \in E} p_{ij} h_j^B, & i \notin B; \end{cases} \tag{55.3}$$

»minimal« means that every solution $x = (x_i) \geqslant 0$ *of (55.3) satisfies* $h^B \leqslant x$.

Proof. h^B solves (55.3). If $i \in B$, then $h_i^B = 1$. If $i \notin B$, then

$$\begin{aligned}
h_i^B &= \mathbb{P}^i(\tau_B^\circ < \infty) \\
&\overset{i\notin B}{=} \mathbb{P}^i(1 \leqslant \tau_B^\circ < \infty) \\
&= \sum_{j\in E} \mathbb{P}^i(1 \leqslant \tau_B^\circ < \infty, X_1 = j) \\
&= \sum_{j\in E} \mathbb{P}^i(1 \leqslant \tau_B^\circ < \infty \mid X_1 = j) \cdot \mathbb{P}^i(X_1 = j) \\
&\overset{\text{MP}}{=} \sum_{j\in E} \mathbb{P}^{X_1}(0 \leqslant \tau_B^\circ < \infty)\Big|_{X_1=j} \mathbb{P}^i(X_1 = j) \\
&= \sum_{j\in E} p_{ij} h_j^B.
\end{aligned}$$

Minimality. Let $x = (x_i)_{i\in E} \geqslant 0$ be any solution of (55.3). We have

$$\begin{aligned}
x_i &= h_i^B = 1, && i \in B, \\
x_i &= \sum_{j\in E} p_{ij} x_j = \sum_{j\in B} p_{ij} \cdot 1 + \sum_{j\notin B} p_{ij} x_j, && i \notin B.
\end{aligned}$$

We can iterate these identities, plugging them into itself to get for $i \notin B$

$$\begin{aligned}
x_i &= \sum_{j\in B} p_{ij} + \sum_{j\notin B} p_{ij} \Bigg(\sum_{k\in B} p_{jk} + \sum_{k\notin B} p_{jk} x_k \Bigg) \\
&= \mathbb{P}^i(X_1 \in B) + \mathbb{P}^i(X_1 \notin B, X_2 \in B) + \underbrace{\sum_{j\notin B} \sum_{k\notin B} p_{ij} p_{jk} x_k}_{\geqslant 0}.
\end{aligned}$$

After a few more iterations we arrive at

$$\begin{aligned}
x_i &\geqslant \mathbb{P}^i(X_1 \in B) + \mathbb{P}^i(X_1 \notin B, X_2 \in B) + \cdots + \mathbb{P}^i(X_1 \notin B, \ldots, X_{t-1} \notin B, X_t \in B) \\
&= \mathbb{P}^i(\tau_B^\circ = 1) + \mathbb{P}^i(\tau_B^\circ = 2) + \cdots + \mathbb{P}^i(\tau_B^\circ = t) \\
&= \mathbb{P}^i(1 \leqslant \tau_B^\circ \leqslant t) \\
&\xrightarrow[t\to\infty]{} \mathbb{P}^i(1 \leqslant \tau_B^\circ < \infty) \overset{i\notin B}{=} \mathbb{P}^i(0 \leqslant \tau_B^\circ < \infty) =: h_i^B.
\end{aligned}$$

□

55.3 Example (birth-and-death process). We will now study a generalization of the *ruin problems* (Problem 51.4), which is also a simple model for a population: Let $E = \mathbb{N}_0$. The initial distribution μ indicates the probability that at time 0 there are $X_0 \in \mathbb{N}_0$ individuals alive; X_t is the size of the population at time t. Moreover,

$$\begin{aligned}
\mathbb{P}^\mu(X_{t+1} = i+1 \mid X_t = i) &= p_i = \text{probability of a birth}, \\
\mathbb{P}^\mu(X_{t+1} = i-1 \mid X_t = i) &= q_i = \text{probability of a death}.
\end{aligned}$$

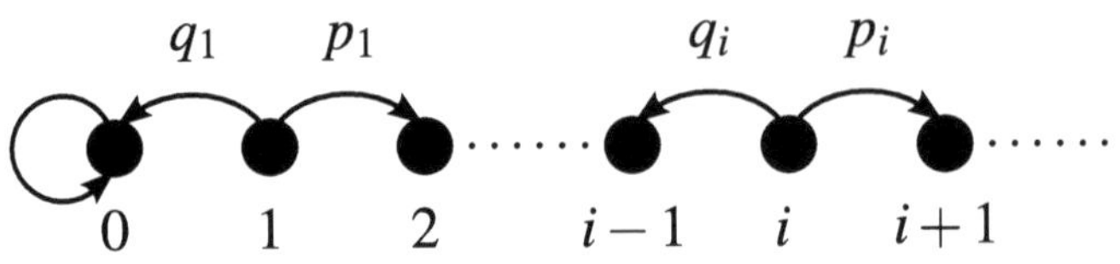

Fig. 55.2. The birth-and-death Markov chain.

Note that p_i and q_i depend on the current size i of the population. We are interested in the **extinction** of the population, i.e.

$$h_i := h_i^{\{0\}} = \mathbb{P}^i(\tau_{\{0\}}^\circ < \infty).$$

Using the »let's-make-a-step« argument, we get

$$h_0 = 1 \quad \text{and} \quad h_i = p_i h_{i+1} + q_i h_{i-1}.$$

Moreover, using that $p_i + q_i = 1$, this implies

$$p_i h_i + q_i h_i = h_i = p_i h_{i+1} + q_i h_{i-1} \implies h_i - h_{i+1} = \frac{q_i}{p_i}(h_{i-1} - h_i).$$

Iterating this step we arrive at

$$h_i - h_{i+1} = \frac{q_i}{p_i}(h_{i-1} - h_i) = \cdots = \underbrace{\frac{q_i q_{i-1} \cdots q_1}{p_i p_{i-1} \cdots p_1}}_{r_i}(h_0 - h_1).$$

Now we can sum over $i = 0, \dots, k-1$ and see

$$h_k = h_0 - (h_0 - h_k) \overset{h_0=1}{=} 1 - (1 + r_1 + \cdots + r_{k-1})(1 - h_1).$$

It remains to determine the value of $1 - h_1$. Set $r := 1 + \sum_{i=1}^\infty r_i$. Since $h = (h_k)_{k \in \mathbb{N}_0}$ is a probability vector, we have necessarily

$$\forall k: \quad h_k \geqslant 0 \implies 1 - r(1 - h_1) \geqslant 0. \tag{55.4}$$

Case 1: $r = \infty$. In this case we have $1 - h_1 = 0$ and $h_k = 1$.
Case 2: $r < \infty$. Since we have one degree or freedom – our system is underdetermined – we should take the minimal solution in (55.4) (compare this with Theorem 55.2), i.e.

$$h_1 = \frac{r-1}{r} = \frac{\sum_{i=1}^\infty r_i}{1 + \sum_{i=1}^\infty r_i} \implies h_k = \frac{\sum_{i=k}^\infty r_i}{1 + \sum_{i=1}^\infty r_i}.$$

⚠ Example 55.3 is a prime example illustrating the fact that one can use extremality properties (here: minimality) to replace a missing lateral condition (here: value of r).

55.4 Theorem. *Let X be a* $\mathrm{MC}(\mu, P)$ *with discrete state space* $(E, \mathscr{E})$ *and* $B \in \mathscr{E}$. *The vector* $k^B := (k_i^B)_{i\in E} \geqslant 0$ *is the minimal (or smallest) positive solution of the linear system*

$$\begin{cases} k_i^B = 0, & i \in B, \\ k_i^B = 1 + \sum_{j \notin B} p_{ij} k_j^B, & i \notin B. \end{cases} \tag{55.5}$$

Proof. The second equality in (55.5) shows that either all k_i^B, $i \notin B$, are finite or infinite. We assume first that k^B is finite.

k^B solves (55.5). If $X_0 = i \in B$, then $\tau_B^\circ = 0$ and $k_i^B = 0$. If $X_0 = i \notin B$, we find because of

$$\tau_B^\circ = \tau_B^{X,\circ} = 1 + \tau_B^{X_{1+\bullet},\circ} \qquad (\blacktriangle\ X_0 \notin B)$$
$$\mathbb{E}^i(\tau_B^\circ \mid X_1) = \mathbb{E}^i(1 + \tau_B^{X_{1+\bullet},\circ} \mid X_1) = 1 + \mathbb{E}^{X_1}\tau_B^\circ$$

(the rather intuitive last equality actually needs the SMP) and so

$$k_i^B = \mathbb{E}^i\tau_B^\circ = \sum_{j\in E}\mathbb{E}^i(\tau_B^\circ \mathbb{1}_{\{X_1=j\}}) = \sum_{j\in E}\overbrace{\mathbb{E}^i(\tau_B^\circ \mid X_1 = j)}^{=1+\mathbb{E}^j\tau_B^\circ=1+k_j^B}\overbrace{\mathbb{P}^i(X_1 = j)}^{=p_{ij}^{(1)}}$$
$$= 1 + \sum_{j\notin B} p_{ij}k_j^B.$$

Minimality. Assume that $y \geqslant 0$ is a solution of (55.5). By iteration, we see that $k_i^B = y_i = 0$ for $i \in B$ and for $i \notin B$

$$\begin{aligned}
y_i &= 1 + \sum_{j\notin B} p_{ij}y_j \\
&= 1 + \sum_{j\notin B} p_{ij}\Big(1 + \sum_{k\notin B} p_{jk}y_k\Big) \\
&= 1 + \sum_{j\notin B} p_{ij} + \sum_{j\notin B}\sum_{k\notin B} p_{ij}p_{jk}y_k \\
&= \mathbb{P}^i(\tau_B^\circ \geqslant 1) + \mathbb{P}^i(\tau_B^\circ \geqslant 2) + \sum_{j\notin B}\sum_{k\notin B} p_{ij}p_{jk}y_k \\
&= \cdots = \sum_{s=1}^t \mathbb{P}^i(\tau_B^\circ \geqslant s) + \sum_{j_1\notin B}\cdots\sum_{j_t\notin B} p_{ij_1}\cdots p_{j_{t-1}j_t}y_{j_t} \\
&\geqslant \sum_{s=1}^t \mathbb{P}^i(\tau_B^\circ \geqslant s) \xrightarrow[t\to\infty]{} k_i^B.
\end{aligned}$$

In the last line we use the (by now familiar) identity $\mathbb{E}T = \sum_{s=1}^\infty \mathbb{P}(T \geqslant s)$.

Infinite k^B. If one, hence all, $k_i^B = \infty$, $i \in B$, the above minimality argument still holds and shows that $k^B = \infty$ is the smallest positive solution to (55.5). □

Hit and return: The classification of states

Now we want to know, how often a MC returns to its starting position. For this it is useful to consider the **shifted** MC:

$$X = (X_0, X_1, X_2, \dots) \rightsquigarrow X \circ \theta_t := X_{t+\bullet} = (X_t, X_{t+1}, X_{t+2}, \dots).$$

We may also shift X by a random time τ, i.e. $X \circ \theta_\tau = (X_\tau, X_{\tau+1}, \dots)$.

55.5 Definition. The *kth hitting time* of a set $B \in \mathscr{E}$ is defined as

$$\tau_B = \tau_B^1 := \inf\{t \in \mathbb{N} \mid X_t \in B\},$$

$$\tau_B^{k+1} := \inf\{t > \tau_B^k \mid X_t \in B\} = \tau_B^k + \underbrace{\tau_B \circ \theta_{\tau_B^k}}_{\text{1st hitting time of the shifted chain}}$$

If $B = \{j\}$ is a singleton, one usually writes τ_j^k etc.

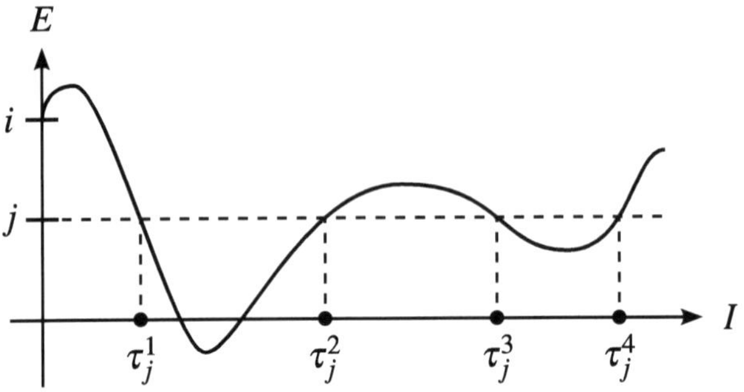

Fig. 55.3. The hitting times τ_j^k.

55.6 Definition. The **occupation time** (»number of visits«) and **hitting probability** of a set $B \in \mathscr{E}$ are defined as

$$v_B := \sup\{k \geqslant 1 \mid \tau_B^k < \infty\} = \sum_{t=0}^{\infty} \mathbb{1}_B(X_t),$$

$$r_{iB} := \mathbb{P}^i(\tau_B < \infty) = \mathbb{P}^i(v_B > 0),$$

$(\sup \emptyset := 0)$. If $B = \{j\}$ is a singleton, it is common to write v_j and r_{ij}.

55.7 Theorem. *Let* X *be a* MC(μ, P) *with discrete state space* (E, $\mathscr{E}$). *For all* $i, j \in E$

$$\mathbb{P}^i(v_j \geqslant k) = \mathbb{P}^i(\tau_j^k < \infty) = r_{ij} r_{jj}^{k-1}, \tag{55.6}$$

$$\mathbb{E}^i v_j = \frac{r_{ij}}{1 - r_{jj}}, \tag{55.7}$$

$$= \sum_{t=1}^{\infty} p_{ij}^{(t)}. \tag{55.8}$$

Proof. From Fig. 55.4 we can read off that $\tau_j^{k+1} = \tau_j^k + \tau_j^1 \circ \theta_{\tau_j^k}$. Therefore, the strong Markov property yields

$$\begin{aligned}
\mathbb{P}^i(\tau_j^{k+1} < \infty) &= \mathbb{P}^i(\tau_j^k < \infty,\ \tau_j \circ \theta_{\tau_j^k} < \infty) \\
&\overset{\text{tower}}{=} \mathbb{E}^i\left[\mathbb{E}^i\left(\mathbb{1}_{\{\tau_j^k < \infty\}} \mathbb{1}_{\{\tau_j \circ \theta_{\tau_j^k} < \infty\}} \mid \mathscr{F}_{\tau_j^k}\right)\right] \\
&\overset{\text{pull}}{=} \mathbb{E}^i\left[\mathbb{1}_{\{\tau_j^k < \infty\}} \mathbb{E}^i\left(\mathbb{1}_{\{\tau_j \circ \theta_{\tau_j^k} < \infty\}} \mid \mathscr{F}_{\tau_j^k}\right)\right] \\
&\overset{\text{SMP}}{=} \mathbb{E}^i\Big[\mathbb{1}_{\{\tau_j^k < \infty\}} \underbrace{\mathbb{E}^{X_{\tau_j^k}}}_{=\mathbb{E}^j} \mathbb{1}_{\{\tau_j < \infty\}}\Big] \\
&= \mathbb{P}^i(\tau_j^k < \infty) \cdot \mathbb{P}^j(\tau_j < \infty) \\
&= \mathbb{P}^i(\tau_j^k < \infty) \cdot r_{jj} = \cdots = \underbrace{\mathbb{P}^i(\tau_j < \infty)}_{=r_{ij}} \cdot r_{jj}^k.
\end{aligned} \tag{55.9}$$

Clearly,

$$\begin{aligned}
\tau_B^k < \infty &\iff \text{»there are } k \text{ or more visits to B«} \\
&\iff v_B \geqslant k
\end{aligned}$$

and this gives the fundamental relation[2] $\{\tau_B^k < \infty\} = \{v_B \geqslant k\}$, hence we see that $\mathbb{P}^i(\tau_B^k < \infty) = \mathbb{P}^i(v_B \geqslant k)$. Therefore, we find for the expected value

$$\mathbb{E}^i v_j = \sum_{k=1}^{\infty} \mathbb{P}^i(v_j \geqslant k) = \sum_{k=1}^{\infty} r_{ij} r_{jj}^{k-1} = \frac{r_{ij}}{1 - r_{jj}}$$

as well as

$$\mathbb{E}^i v_j = \mathbb{E}^i\left(\sum_{t=1}^{\infty} \mathbb{1}_{\{j\}}(X_t)\right) \overset{\text{BL}}{=} \sum_{t=1}^{\infty} \mathbb{E}^i \mathbb{1}_{\{j\}}(X_t) = \sum_{t=1}^{\infty} \overbrace{\mathbb{P}^i(X_t = j)}^{\overset{\text{def}}{=} p_{ij}^{(t)}}. \qquad \square$$

💬 If $i = j$, then (55.6) shows that $\mathbb{P}^i(v_i \geqslant k) = r_{ii}^k$, i.e. the number of visits v_i is under $\mathbb{P}^i$ **geometrically distributed** with mean $\mathbb{E}^i v_i = r_{ii}/(1 - r_{ii})$, cf. (55.7).

We will now classify the states according to the (expected) number of returns.

(2) Compare this with Remark 48.3.b).

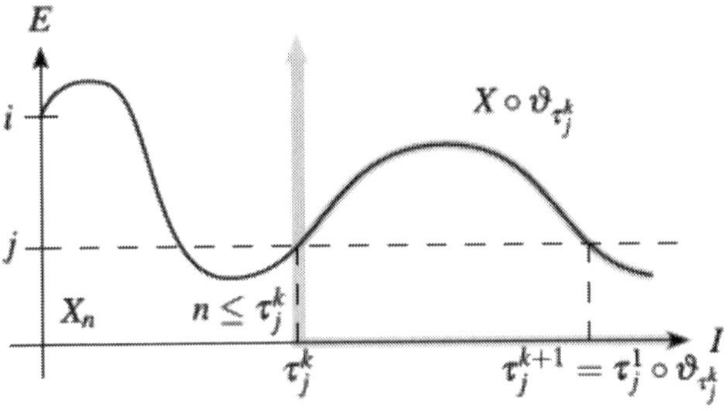

Fig. 55.4. The shifted chain $X \circ \theta_{\tau_j^k}$ and the shifted coordinate grid are highlighted in grey. Observe that the $k+1$st hitting time of X is the first hitting time of $X \circ \theta_{\tau_j^k}$.

55.8 Definition. Let X be a $MC(\mu, P)$ with discrete state space $(E, \mathscr{E})$. A state $i \in E$ is called **recurrent**, if $r_{ii} = 1$, and **transient**, if $r_{ii} < 1$.

55.9 Remark. a) Theorem 55.7 and Definition 55.8 show that we have for each $i \in E$ the following **dichotomy**:

$$\begin{aligned} \textit{either} \text{ recurrent:} \quad & r_{ii} = 1 \iff \mathbb{E}^i v_i = \infty \iff \mathbb{P}^i(v_B = \infty) = 1, \\ \textit{or} \text{ transient:} \quad & r_{ii} < 1 \iff \mathbb{E}^i v_i < \infty \iff \mathbb{P}^i(v_B = \infty) = 0 \\ & \qquad\qquad\qquad\qquad\quad \iff \mathbb{P}^i(v_B < \infty) = 1. \end{aligned}$$

⚠ This means that we are in the truly exceptional situation that

$$\mathbb{E}^i v_i = \infty \implies v_i = \infty \ \mathbb{P}^i\text{-a.s.}$$

b) Assume that $i \in E$ is transient. Then we have $\lim_{n\to\infty} p_{ii}^{(n)} = 0$, since $\mathbb{E}^i v_i < \infty$ shows that the series (55.8) converges.

c) Assume that $i \in E$ is transient. Then we have $\mathbb{E}^i \tau_i = \infty$. This follows from $1 > r_{ii} = \mathbb{P}^i(\tau_i < \infty) \implies \mathbb{P}^i(\tau_i = \infty) > 0$.

There is a simple connection between recurrence and stationary distributions.

55.10 Lemma. *Assume that* X *is a* $MC(\mu, P)$ *with discrete state space* $(E, \mathscr{E})$ *such that there exists a stationary distribution* π. *Every state* $j \in E$ *with* $\pi_j > 0$ *is recurrent.*

Proof. For every $t \in \mathbb{N}$ we have

$$0 < \pi_j = (\pi P^t)_j = \sum_{i \in E} \pi_i p_{ij}^{(t)}.$$

Summing over $t \in \mathbb{N}$ yields

$$\infty = \sum_{t=1}^{\infty} \sum_{i \in E} \pi_i p_{ij}^{(t)} = \sum_{i \in E} \pi_i \sum_{t=1}^{\infty} p_{ij}^{(t)} \overset{(55.7)}{\underset{(55.8)}{=}} \sum_{i \in E} \pi_i \frac{r_{ij}}{1 - r_{jj}} \leqslant \frac{1}{1 - r_{jj}}.$$

This shows that $r_{jj} = 1$. □

55.11 Definition. Let X be a MC(μ, P) with discrete state space $(E, \mathscr{E})$. The greatest common divisor

$$d_i := \gcd\{t \in \mathbb{N} \mid p_{ii}^{(t)} > 0\}$$

is called a **period** of the state $i \in E$. A state is **aperiodic**, if $d_i = 1$.

A typical example of a 2-periodic MC is the chain with state space $E = \{1, 2\}$ and transition matrix $P = \left(\begin{smallmatrix} 0 & 1 \\ 1 & 0 \end{smallmatrix}\right)$; this follows from the observation

$$\begin{pmatrix} 0 & 1 \\ 1 & 0 \end{pmatrix}^{2n} = \begin{pmatrix} 1 & 0 \\ 0 & 1 \end{pmatrix} \quad \text{and} \quad \begin{pmatrix} 0 & 1 \\ 1 & 0 \end{pmatrix}^{2n+1} = \begin{pmatrix} 0 & 1 \\ 1 & 0 \end{pmatrix}.$$

55.12 Lemma. *Let X be a MC(μ, P) with discrete state space $(E, \mathscr{E})$. For every $i \in E$ one has*

$$d_i = 1 \iff \exists t(i) \quad \forall t \geqslant t(i) : p_{ii}^{(t)} > 0.$$

Proof. Since there are at least 2 prime numbers among the $t \geqslant t(i)$, the direction »$\Leftarrow$« is clear.

Conversely, »$\Rightarrow$«, we consider the set $D_i = \{t \in \mathbb{N} \mid p_{ii}^{(t)} > 0\}$. From $P^{t+s} = P^t P^s$ we see that

$$p_{ii}^{(t+s)} = \sum_{j \in E} p_{ij}^{(s)} p_{ji}^{(t)} \geqslant p_{ii}^{(s)} p_{ii}^{(t)} > 0 \quad \forall s, t \in D_i,$$

and we get $t + s \in D_i$.

Since D_i is closed under addition and $\gcd(D_i) = 1$, the additive group generated by D_i is $\mathbb{Z}$, thus

$$\exists t_1, \ldots, t_n \in D_i \quad \exists z_1, \ldots, z_n \in \mathbb{Z} : 1 = \sum_{m=1}^{n} z_m t_m.$$

Define $N := t_1 \sum_{m=1}^{n} |z_m| t_m \in \mathbb{N}$. Every $t \geqslant N$ can be written as multiple of t_1 plus remainder $r \in \{0, 1, \ldots, t_1 - 1\}$:

$$t = N + k t_1 + r = k t_1 + \sum_{m=1}^{n} \underbrace{(t_1 |z_m| + r z_m)}_{\geqslant 0} t_m \in D_i.$$ □

55.13 Definition. A Markov chain is **irreducible**, if $r_{ij} > 0$ for all $i, j \in \mathrm{E}$.

Interpretation: Denote by $t(i,j)$ the number of steps needed to go from state i to state $j \neq i$. Because of (55.8) we know that for an irreducible chain $t(i,j) < \infty$ for all $i \neq j$. If we compare irreducibility with ergodicity (Definition 52.8) we see that an ergodic chain satisfies $\mathrm{P}^s > 0$ for some s, i.e. $\max_{i,j} t(i,j) \leqslant s$ – we can go from any i to any other j in a **fixed** number of steps.

55.14 Lemma. *Let* X *be a* $\mathrm{MC}(\mu, \mathrm{P})$ *with discrete state space* $(\mathrm{E}, \mathscr{E})$. *Define for* $i \in \mathrm{E}$ *the* ***recurrence class*** $\mathrm{E}_i := \{j \in \mathrm{E} \mid r_{ij} > 0\}$. *If* $i \in \mathrm{E}$ *is recurrent, then* $r_{jk} = 1$ *for all* $j, k \in \mathrm{E}_i$, *i.e. all states in* E_i *are recurrent.*

The set E_i contains all $j \in \mathrm{E}$ which can be reached from i in any number of steps and with positive probability. Sometimes one writes $i \to j \overset{\text{def}}{\iff} j \in \mathrm{E}_i$. By $i \leftrightarrow j$ we mean that both $i \to j$ and $j \to i$; this is an equivalence relation and the equivalence classes are called »communicating classes«; the states in a communicating class are mutually accessible.

Proof. Assume that i is recurrent and $j \in \mathrm{E}_i$. Using $\tau_i = \tau_j + \tau_i \circ \theta_{\tau_j}$ we see

$$\begin{aligned} 0 = \mathbb{P}^i(\tau_i = \infty) \quad &\geqslant \quad \mathbb{P}^i(\tau_j < \infty, \tau_i \circ \theta_{\tau_j} = \infty) \\ &\underset{\text{as (55.9)}}{\overset{\text{SMP}}{=}} \mathbb{P}^i(\tau_j < \infty) \cdot \underbrace{\mathbb{P}^j}_{X_{\tau_j} = j}(\tau_i = \infty) \quad = \quad \underbrace{r_{ij}}_{>0}(1 - r_{ji}). \end{aligned}$$

This shows that $r_{ji} = 1$. Since $r_{ij}, r_{ji} > 0$, there are $s, t \in \mathbb{N}$ with $p_{ij}^{(s)}, p_{ji}^{(t)} > 0$. Thus,

$$\begin{aligned} \mathbb{E}^j v_j \overset{(55.8)}{=} \sum_{u=1}^{\infty} p_{jj}^{(u)} \quad &\geqslant \quad \sum_{u=1}^{\infty} p_{jj}^{(t+u+s)} \\ &= \quad \sum_{u=1}^{\infty} \sum_{k,k' \in \mathrm{E}} p_{jk}^{(t)} p_{kk'}^{(u)} p_{k'j}^{(s)} \\ &\geqslant \quad \sum_{u=1}^{\infty} p_{ji}^{(t)} p_{ii}^{(u)} p_{ij}^{(s)} \qquad (55.10) \\ &= \quad p_{ji}^{(t)} p_{ij}^{(s)} \sum_{u=1}^{\infty} p_{ii}^{(u)} \\ &\overset{(55.8)}{=} \underbrace{p_{ji}^{(t)} p_{ij}^{(s)}}_{>0} \mathbb{E}^i v_i \underset{i \text{ recurrent}}{\overset{55.9.a)}{=}} \infty, \end{aligned}$$

and we conclude that j is recurrent, cf. Remark 55.9.a).

Now we can switch the roles $i \leftrightsquigarrow j$ and we see that $r_{ij} = 1$. Finally,

$$r_{jk} = \mathbb{P}^j(\tau_k < \infty) \geqslant \mathbb{P}^j(\tau_i < \infty, \ \tau_k \circ \theta_{\tau_i} < \infty) \underset{\text{as above}}{\overset{\text{SMP}}{=}} r_{ji} r_{ik} = 1.$$ □

55.15 Theorem. *Let X be an irreducible* MC(P) *with discrete state space.*

a) *All* $i \in E$ *are either recurrent or transient.*

b) $d_i = d$ *for all* $i \in E$.

c) *Every stationary distribution* π *satisfies* $\pi > 0$.

d) *If X is recurrent, then* $\mathbb{P}^\mu(\tau_i < \infty) = 1$ *for all* $i \in E$.

Proof. a) *Case 1:* There exists some recurrent $i \in E$. From Lemma 55.14 we know that all $j \in E_i$ are recurrent, but by irreducibility $E_i = E$, so all $j \in E$ are recurrent.

Case 2: There exists no recurrent $i \in E$. Then all $i \in E$ are transient.

b) Because of irreducibility, we find for all $i, j \in E$ numbers $s, t \in \mathbb{N}$ such that $p_{ij}^{(s)}, p_{ji}^{(t)} > 0$. The calculation in (55.10) shows that

$$p_{jj}^{(s+u+t)} \geqslant p_{ji}^{(t)} p_{ii}^{(u)} p_{ij}^{(s)}.$$

Recall that $a \mid b$ means »a is a divisor of b«. From the last estimate we get

$$\begin{aligned} u = 0 \implies p_{jj}^{(s+t)} > 0 \quad &\implies d_j \mid (s+t) \\ p_{ii}^{(u)} > 0 \implies p_{jj}^{(t+u+s)} > 0 &\implies d_j \mid (t+u+s) \end{aligned}$$

Consequently, d_j divides every u with $p_{ii}^{(u)} > 0$. By the definition of the greatest common divisor $d_i = \gcd\{u \mid p_{ii}^{(u)} > 0\}$, we see that $d_j \leqslant d_i$.

Changing the roles of $i \leftrightsquigarrow j$ we also get $d_i \leqslant d_j$, thus $d_i = d_j$.

c) Let π be a stationary distribution. Since $\pi \neq 0$, there is some i_0 with $\pi_{i_0} > 0$; because of irreducibility, there is some $t \in \mathbb{N}$ such that $p_{i_0 j}^{(t)} > 0$. This shows

$$\pi_j = (\pi P^t)_j = \sum_{i \in E} \pi_i p_{ij}^{(t)} \geqslant \pi_{i_0} p_{i_0 j}^{(t)} > 0.$$

d) Since all $i \in E$ are recurrent, we see from Lemma 55.14 that

$$\forall j, k \in E: \quad \mathbb{P}^j(\tau_k < \infty) = r_{jk} = 1;$$

therefore,

$$\mathbb{P}^\mu(\tau_k < \infty) = \sum_{j \in E} \mu_j \mathbb{P}^j(\tau_k < \infty) = \sum_{j \in E} \mu_j = 1.$$ □

Run: Coupling and (again) the ergodic theorem

Let $X' = (X'_t)_{t \in \mathbb{N}_0}$ and $Y' = (Y'_t)_{t \in \mathbb{N}_0}$ be two stochastic processes with state spaces $(E, \mathscr{E})$ and $(E', \mathscr{E}')$, respectively. We assume that the processes live on $(\Omega_1, \mathscr{A}_1, \mathbb{P}_1)$ and $(\Omega_2, \mathscr{A}_2, \mathbb{P}_2)$, respectively. A **coupling** is the realization of X', Y' on **one** probability space, that is an $E\times E'$-valued process (X,Y) on some $(\Omega,\mathscr{A},\mathbb{P})$, such that

$$(X_1,\dots,X_t)\sim(X'_1,\dots,X'_t)\quad\text{and}\quad(Y_1,\dots,Y_t)\sim(Y'_1,\dots,Y'_t)\quad\forall t\in\mathbb{N}.$$

This means that we prescribe the marginal distributions of (X,Y), but we do not specify the joint distribution of (X,Y) – please compare the situation with the proof of Theorem 28.1 – nor the initial distribution of X_0 and Y_0.

The simplest coupling procedure is **independent coupling** where we define $\Omega:=\Omega_1\times\Omega_2$ and $\mathbb{P}=\mathbb{P}_1\otimes\mathbb{P}_2$ and $X\perp\!\!\!\perp Y$. For our purposes this is sufficient.

55.16 Theorem (independent coupling). *Assume that X is a* $MC(\mu,P)$ *and Y is a* $MC(\nu,Q)$ *taking values in* $E=\{1,\dots,r\}$ *and* $E'=\{1,\dots,r'\}$, $r,r'\in\mathbb{N}\cup\{\infty\}$, *respectively; it is assumed that X and Y are independent, i.e.* $\mathscr{F}^X_\infty\perp\!\!\!\perp\mathscr{F}^Y_\infty$.

a) (X,Y) *is a* $MC(\mu\otimes\nu,P\otimes Q)$ *where*
$$\mu\otimes\nu_{(i,i')}=\mu_i\nu_{i'}\ \text{ and }\ P\otimes Q_{(i,i')(j,j')}:=p_{ij}q_{i'j'}.$$

b) *If X, Y are irreducible and aperiodic, then* (X,Y) *is irreducible and aperiodic.*

c) *If* π,π' *are stationary distributions of X, Y, then* $\pi\otimes\pi'$ *is stationary for* (X,Y).

d) *If X, Y are irreducible, aperiodic and have stationary distributions, then* (X,Y) *is recurrent.*

Proof. a) We have

$$\begin{aligned}
&\mathbb{P}^{\mu\otimes\nu}\big((X_t,Y_t)=(j,j')\mid(X_{t-1},Y_{t-1})=(i,i')\big)\\
&=\frac{\mathbb{P}^{\mu\otimes\nu}(X_{t-1}=i,X_t=j,Y_{t-1}=i',Y_t=j')}{\mathbb{P}^{\mu\otimes\nu}(X_{t-1}=i,Y_{t-1}=i')}\\
&\overset{\perp\!\!\!\perp}{=}\frac{\mathbb{P}^{\mu\otimes\nu}(X_{t-1}=i,X_t=j)\,\mathbb{P}^{\mu\otimes\nu}(Y_{t-1}=i',Y_t=j')}{\mathbb{P}^{\mu\otimes\nu}(X_{t-1}=i)\,\mathbb{P}^{\mu\otimes\nu}(Y_{t-1}=i')}\\
&=\mathbb{P}^{\mu}(X_t=j\mid X_{t-1}=i)\,\mathbb{P}^{\nu}(Y_t=j'\mid Y_{t-1}=i')=p_{ij}q_{i'j'}.
\end{aligned}$$

b) Assume that X is irreducible and aperiodic. Lemma 55.12 and the definition of irreducibility give

$$\begin{aligned}
&\text{irreducible:} && \forall i,j,k\in E\quad\exists s,t:\quad p^{(t)}_{ji},p^{(s)}_{ik}>0,\\
&\text{aperiodic:} && \forall u\gg 1: p^{(u)}_{ii}>0.
\end{aligned}$$

As in the proof of Theorem 55.15.b) we conclude that

$$p^{(t+u+s)}_{jk}\geqslant p^{(t)}_{ji}p^{(u)}_{ii}p^{(s)}_{ik}>0\implies p^{(T)}_{jk}>0\quad\forall T\gg 1. \tag{55.11}$$

A similar calculation works for Y, and so

$$P\otimes Q^T_{(j,j')(k,k')}=p^{(T)}_{jk}q^{(T)}_{j'k'}>0\quad\forall T\gg 1.$$

Since $(j,j'),(k,k') \in E \times E'$ are arbitrary, it follows that (X,Y) is aperiodic (see Lemma 55.12). Irreducibility follows with (55.7) from

$$\frac{r_{(i,j)(i',j')}}{1-r_{(j,j)(j',j')}} \overset{(55.7)}{=} \mathbb{E}^{(i,i')} v_{(j,j')} \overset{(55.8)}{=} \sum_{t=1}^{\infty} (P \otimes Q)^t_{(i,i')(j,j')} = \sum_{t=1}^{\infty} p^{(t)}_{ij} q^{(t)}_{i'j'} > 0.$$

c) We have ✍ $(\pi \otimes \pi')(P \otimes Q) = (\pi P) \otimes (\pi' Q) = \pi \otimes \pi'$.

d) From Part b), c) we know that (X,Y) is irreducible and has a stationary distribution $\pi \otimes \pi'$. Theorem 55.15 shows $\pi \otimes \pi' > 0$ and, according to Lemma 55.10, (X,Y) is recurrent. □

We will use, as in the proof of Theorem 52.11, a metric on the space of probability measures.[3]

$$d(\mu, \nu) := \sup_{B \in \mathscr{B}} |\mu(B) - \nu(B)|.$$

55.17 Lemma. *Let* $(X', \Omega_1, \mathscr{A}_1, \mathbb{P}_1)$ *be a* MC(μ, P) *and* $(Y', \Omega_2, \mathscr{A}_2, \mathbb{P}_2)$ *a* MC(ν, P) *both taking values in* $E = \{i(1), \dots, i(r)\}$, $r \in \mathbb{N} \cup \{\infty\}$. *Assume that the independent coupling* (X,Y) *is irreducible and recurrent. Then*

$$\lim_{t\to\infty} d\left(\mathbb{P}_1^{\mu}(X' \circ \theta_t \in \bullet), \mathbb{P}_2^{\nu}(Y' \circ \theta_t \in \bullet)\right) = 0. \tag{55.12}$$

55.18 Remark. Lemma 55.17 shows, in particular,

$$\left|\mathbb{P}_1^{\mu}(X' \circ \theta_t \in \Gamma) - \mathbb{P}_2^{\nu}(Y' \circ \theta_t \in \Gamma)\right| \xrightarrow[t\to\infty]{} 0 \quad \forall \Gamma \in \mathscr{B}^{\otimes \mathbb{N}_0}$$

and, if we take $\Gamma = \{j\} \times E \times E \times \dots$ and $\mu = \delta_i$, $\nu = \delta_{i'}$, then

$$\left|p^{(t)}_{ij} - p^{(t)}_{i'j}\right| = \left|\mathbb{P}_1^i(X'_t = j) - \mathbb{P}_2^{i'}(Y'_t = j)\right| \xrightarrow[t\to\infty]{} 0 \quad \forall i, i', j \in E.$$

Proof of Lemma 55.17. We have seen in Thm. 55.16 that (X,Y) is a MC$(\mu \otimes \nu, P \otimes P)$ on a suitable probability space $(\Omega, \mathscr{A}, \mathbb{P})$. Since (X,Y) is irreducible and recurrent, Theorem 55.15.d) shows that

$$\begin{aligned}\tau &:= \inf\{t > 0 \mid X_t = Y_t\} \\ &= \inf\{t > 0 \mid (X_t, Y_t) = (i,i) \text{ for some } i \in E\} < \infty.\end{aligned}$$

Define (see Fig. 55.5)

$$\widetilde{X}_t := \begin{cases} X_t & \text{for } \tau > t; \\ Y_t & \text{for } \tau \leqslant t. \end{cases}$$

[3] This is the so-called **total variation** distance. It coincides with the metric introduced in Chapter 52.

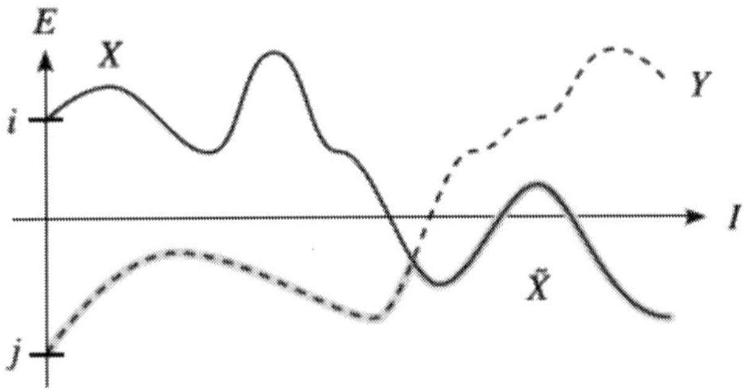

Fig. 55.5. The »mixed« path $\widetilde{X}$ obtained from joining the paths of X and Y. The highlighted portions of the paths of X and Y are removed.

The strong Markov property shows for all $s \leqslant t$ and $B_1, \dots, B_t \in \mathscr{E}$

$$\begin{aligned}
&\mathbb{P}^{\mu\otimes\nu}(\widetilde{X}_1 \in B_1, \dots, \widetilde{X}_t \in B_t, \tau = s) \\
&\quad= \mathbb{P}^{\mu\otimes\nu}(X_1 \in B_1, \dots, X_\tau \in B_\tau, Y_{\tau+1} \in B_{\tau+1}, \dots, Y_t \in B_t, \tau = s) \\
&\quad\overset{\text{SMP}}{=} \mathbb{E}^{\mu\otimes\nu}\left[\mathbb{1}_{B_1}(X_1)\cdots\mathbb{1}_{B_\tau}(X_\tau)\mathbb{1}_{\{\tau=s\}}\mathbb{P}^{(X_\tau,Y_\tau)}(Y_1 \in B_{s+1}, \dots, Y_{t-s} \in B_t)\right] \\
&\quad\overset{\S}{=} \mathbb{E}^{\mu\otimes\nu}\left[\mathbb{1}_{B_1}(X_1)\cdots\mathbb{1}_{B_\tau}(X_\tau)\mathbb{1}_{\{\tau=s\}}\mathbb{P}^{(X_\tau,Y_\tau)}(X_1 \in B_{s+1}, \dots, X_{t-s} \in B_t)\right] \\
&\quad= \cdots = \mathbb{P}^{\mu\otimes\nu}(X_1 \in B_1, \dots, X_t \in B_t, \tau = s).
\end{aligned}$$

in the step marked with § we use that $X_\tau = Y_\tau$ and P is the transition matrix of both X and Y.

Summing this equality in $s = 0, 1, 2, \dots$ reveals that $\widetilde{X} \sim X$. Finally,

$$\begin{aligned}
&|\mathbb{P}_1^\mu(X' \circ \theta_t \in \Gamma) - \mathbb{P}_2^\nu(Y' \circ \theta_t \in \Gamma)| \\
&\quad= |\mathbb{P}^{\mu\otimes\nu}(X \circ \theta_t \in \Gamma) - \mathbb{P}^{\mu\otimes\nu}(Y \circ \theta_t \in \Gamma)| \\
&\quad= |\mathbb{P}^{\mu\otimes\nu}(\widetilde{X} \circ \theta_t \in \Gamma) - \mathbb{P}^{\mu\otimes\nu}(Y \circ \theta_t \in \Gamma)| \\
&\quad\leqslant |\mathbb{P}^{\mu\otimes\nu}(\widetilde{X} \circ \theta_t \in \Gamma, t < \tau) - \mathbb{P}^{\mu\otimes\nu}(Y \circ \theta_t \in \Gamma, t < \tau)| \\
&\qquad+ \underbrace{|\mathbb{P}^{\mu\otimes\nu}(\widetilde{X} \circ \theta_t \in \Gamma, t \geqslant \tau) - \mathbb{P}^{\mu\otimes\nu}(Y \circ \theta_t \in \Gamma, t \geqslant \tau)|}_{=0 \text{ since } \widetilde{X}=Y \text{ if } t\geqslant\tau} \\
&\quad\leqslant 2\mathbb{P}^{\mu\otimes\nu}(\tau > t) \xrightarrow[t\to\infty]{\text{as } \tau<\infty \text{ a.s.}} 0.
\end{aligned}$$

□

The following theorem should be compared with Theorem 52.11.

55.19 Theorem (ergodic theorem). *Let X be an irreducible and aperiodic* MC(μ, P) *with discrete state space* $(E, \mathscr{E})$. *In this case one of the following alternatives* a) *or* b) *applies:*

a) X *has a unique stationary distribution* $\pi > 0$ *such that*

$$\lim_{t\to\infty} d\left(\mathbb{P}^{\mu}(X\circ\theta_t \in \bullet), \mathbb{P}^{\pi}(X\in\bullet)\right) = 0; \tag{55.13}$$

b) X *does not have a stationary distribution and*

$$\lim_{t\to\infty} p_{ij}^{(t)} = 0 \quad \forall i,j\in E. \tag{55.14}$$

Proof. 1° $\neg$(55.14) $\Rightarrow$ existence of a stationary distribution π. Assume that (55.14) does not hold, i.e.

$$\exists i_0, j_0 \in E: \quad \limsup_{t\to\infty} p_{i_0 j_0}^{(t)} > 0.$$

Since E is countable, we can repeatedly take sub-sequences and conclude – with a diagonal argument – that there is a sequence $(t_n)_{n\in\mathbb{N}}$ such that for each $i_0 \in E$

$$\forall j\in E \quad \exists \gamma_j \geqslant 0: \quad \lim_{n\to\infty} p_{i_0 j}^{(t_n)} = \gamma_j.$$

Since $\gamma_{j_0} > 0$, we get

$$0 < \sum_{j\in E}\gamma_j = \sum_{j\in E}\liminf_{n\to\infty} p_{i_0 j}^{(t_n)} \overset{\text{Fatou}}{\leqslant} \liminf_{n\to\infty} \underbrace{\sum_{j\in E} p_{i_0 j}^{(t_n)}}_{=1} = 1.$$

Assume that X, Y are independent MC(μ, P) and MC(ν, P), respectively, and consider the independent coupling (X, Y). By Theorem 55.16, the chain (X, Y) is an irreducible MC($\mu\otimes\nu$, P$\otimes$P).

Assume, for a moment, that (X, Y) is transient. Then we know

$$\sum_{t=1}^{\infty}\left(p_{ij}^{(t)}\right)^2 = \sum_{t=1}^{\infty}(P\otimes P)_{(ii)(jj)}^{t} < \infty \implies \text{(55.14) holds}$$

which is impossible. Thus, (X, Y) is recurrent and we can use Lemma 55.17 (and Remark 55.18) to see that

$$\lim_{t\to\infty}|p_{ij}^{(t)} - p_{i_0 j}^{(t)}| = 0 \quad \forall i,j\in E \implies \lim_{n\to\infty} p_{ij}^{(t_n)} = \gamma_j \quad \forall i,j\in E.$$

This qualifies $(\gamma_j)_{j\in E}$ as a serious candidate for π.

From $P^tP = P^{t+1} = PP^t$ we see that

$$\sum_{j\in E} p_{ij}^{(t)} p_{jk} = p_{ik}^{(t+1)} = \sum_{j\in E} p_{ij} p_{jk}^{(t)} \quad \forall i,k\in E. \tag{55.15}$$

Setting $t = t_n$ and letting $n \to \infty$ gives

$$\sum_{j\in E} \gamma_j p_{jk} = \sum_{j\in E} \lim_{n\to\infty} p_{ij}^{(t_n)} p_{jk} \overset{\text{Fatou}}{\leqslant} \liminf_{n\to\infty} \sum_{j\in E} p_{ij}^{(t_n)} p_{jk}$$
$$\overset{(55.15)}{=} \liminf_{n\to\infty} \sum_{j\in E} p_{ij} p_{jk}^{(t_n)}$$
$$\overset{\text{DCT}}{=} \sum_{j\in E} p_{ij} \liminf_{n\to\infty} p_{jk}^{(t_n)}$$
$$= \sum_{j\in E} p_{ij} \gamma_k = \gamma_k.$$

Now we sum over $k \in E$ and see that on both sides of the inequality we get $\sum_{k\in E} \gamma_k$ (this sum is finite!). This means that in the above equality there must be »=« as »<« would produce a contradiction. This shows that $\gamma P = \gamma$ and, therefore, $\pi := (\sum_{k\in E} \gamma_k)^{-1} \gamma$ is a stationary distribution.

2° Assume that X does not have a stationary distribution. In this case, 1° shows that (55.14) holds.

3° Assume that π is a stationary distribution of X. Theorem 55.15 shows that $\pi > 0$. Let Y be a MC(π, P) which is independent of X, and denote by (X, Y) the independent coupling (cf. Theorem 55.17). Since (X, Y) is irreducible and recurrent, we get

$$d(\mathbb{P}^\mu(X \circ \theta_t \in \bullet), \mathbb{P}^\pi(X \in \bullet)) \overset{\text{stationary}}{=} d(\mathbb{P}^\mu(X \circ \theta_t \in \bullet), \mathbb{P}^\pi(X \circ \theta_t \in \bullet));$$

(55.12) shows that the right-hand side converges to 0 as $t \to \infty$. Since the limit is unique, we conclude that π is unique. □

We can express the limit (55.13) using the hitting time $\tau_j := \tau_j^1$ (Def. 55.5).

55.20 Corollary. *Let X be a* MC(μ, P) *with discrete state space* (E, $\mathscr{E}$) *and assume that* $j \in E$ *is an aperiodic state. Then*

$$\lim_{t\to\infty} p_{ij}^{(t)} = \frac{\mathbb{P}^i(\tau_j < \infty)}{\mathbb{E}^j \tau_j} \tag{55.16}$$

Proof. 1° *Case 1:* $i = j$ *and* j *is transient.* Then, $\lim_{t\to\infty} p_{jj}^{(t)} = 0$ and $\mathbb{E}^j \tau_j = \infty$, see Remark 55.9.b),c), i.e. (55.16) holds.

2° *Case 2:* $i = j$ *and* j *is recurrent.* Define $E_j := \{k \in E \mid r_{jk} > 0\}$; the set E_j contains all states which can be reached from j. If we start in j, we may restrict X to E_j and we see that

- X is irreducible and recurrent (Lemma 55.14);
- X is aperiodic ($d_j = 1$ and Theorem 55.15.b)).

Theorem 55.19 shows that the $\lim_{t\to\infty} p_{jj}^{(t)}$ exists. In order to identify the limit, we define

$$v_j(t) := \sup\left\{k \geqslant 0 \mid \tau_j^k \leqslant t\right\} = \sum_{s=1}^{t} \mathbb{1}_{\{j\}}(X_s), \quad t \in \mathbb{N}.$$

Since the sequence $(\tau_j^t)_{t\in\mathbb{N}_0}$ is a random walk (w.r.t. the law $\mathbb{P}^j$ ✎)(4), we can use the SLLN and find

$$\frac{v_j(\tau_j^t)}{\tau_j^t} = \frac{t}{\tau_j^t} = \frac{1}{\frac{\tau_j^t}{t}} \xrightarrow[t\to\infty]{\text{a.s. under } \mathbb{P}^j} \frac{1}{\mathbb{E}^j \tau_j}.$$

From the very definition of $v_j(t)$ we see that $\tau_j^{v_j(t)} \leqslant t \leqslant \tau_j^{v_j(t)+1}$. Since $t \mapsto v_j(t)$ is monotone increasing, we get

$$\frac{v_j(t)}{t} \leqslant \frac{v_j(\tau_j^{v_j(t)+1})}{\tau_j^{v_j(t)}} = \frac{v_j(\tau_j^{v_j(t)+1})}{\tau_j^{v_j(t)+1}} \underbrace{\overbrace{\frac{\tau_j^{v_j(t)+1}}{v_j(t)+1} \frac{v_j(t)}{\tau_j^{v_j(t)}}}^{\to \mathbb{E}^j\tau_j/\mathbb{E}^j/\tau_j = 1} \frac{v_j(t)+1}{v_j(t)}}_{\to 1} \xrightarrow[t\to\infty]{\text{a.s.}} \frac{1}{\mathbb{E}^j\tau_j}.$$

The lower estimate follows with a similar calculation.

Since $v_j(t) \leqslant t$, we can use dominated convergence to see

$$\frac{1}{t}\sum_{s=1}^{t} p_{jj}^{(s)} \overset{(55.8)}{=} \frac{1}{t}\mathbb{E}^j v_j(t) = \mathbb{E}^j \frac{v_j(t)}{t} \xrightarrow[t\to\infty]{} \frac{1}{\mathbb{E}^j\tau_j};$$

since j is recurrent, $\mathbb{P}^j(\tau_j < \infty) = 1$ and $\mathbb{P}^j(\tau_j < \infty)/\mathbb{E}^j\tau_j = 1/\mathbb{E}^j\tau_j$.

The above calculation shows us the Cesàro (arithmetic mean) limit of the sequence $(p_{jj}^{(s)})_{s\in\mathbb{N}}$. Since the limit $\lim_{s\to\infty} p_{jj}^{(s)}$ exists, it must coincide with its Cesàro limit ✎, and the claim follows.

3° *Case 3: $i \neq j$.* We have

$$\begin{aligned} p_{ij}^{(t)} = \mathbb{P}^i(X_t = j) &= \mathbb{P}^i\left(X_t = j,\ \tau_j \leqslant t\right) \\ &= \mathbb{P}^i\left[(X \circ \theta_{\tau_j})_{t-\tau_j} = j,\ \tau_j \leqslant t\right] \\ &\overset{\text{SMP}}{=} \mathbb{E}^i\left[\mathbb{1}_{\{\tau_j \leqslant t\}} \mathbb{P}^{X_{\tau_j}}(X_{t-\tau_j} = j)\right]. \end{aligned}$$

Since $X_{\tau_j} = j$, we can use the first two steps 1° & 2° and find with dominated convergence

$$p_{ij}^{(t)} = \mathbb{E}^i\left[\mathbb{1}_{\{\tau_j \leqslant t\}} \underbrace{p_{jj}^{(t-\tau_j)}}_{\to 1/\mathbb{E}^j\tau_j}\right] \xrightarrow[t\to\infty]{} \frac{\mathbb{P}^i(\tau_j < \infty)}{\mathbb{E}^j\tau_j}. \qquad \square$$

(4) This is essentially the strong Markov property!

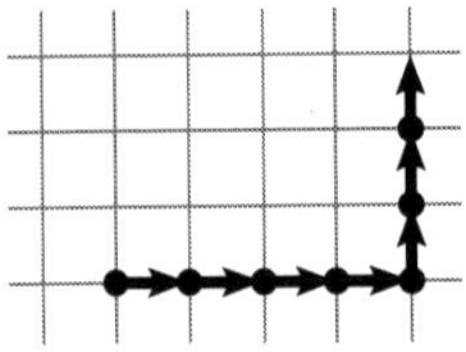

$$\begin{pmatrix} i_1 \\ i_2 \\ i_3 \\ \vdots \\ i_d \end{pmatrix} \xrightarrow[\text{steps}]{|j_1-i_1|} \begin{pmatrix} j_1 \\ i_2 \\ i_3 \\ \vdots \\ i_d \end{pmatrix} \xrightarrow[\text{steps}]{|j_2-i_2|} \begin{pmatrix} j_1 \\ j_2 \\ i_3 \\ \vdots \\ i_d \end{pmatrix} \xrightarrow[\text{steps}]{|j_3-i_3|} \begin{pmatrix} j_1 \\ j_2 \\ j_3 \\ \vdots \\ i_d \end{pmatrix} \longrightarrow \cdots \longrightarrow \begin{pmatrix} j_1 \\ j_2 \\ j_3 \\ \vdots \\ j_d \end{pmatrix}$$

Fig. 56.1. A path from $i = (i_1, \dots, i_d)$ to $j = (j_1, \dots, j_d)$ along the coordinate axes.

56 General random walks and recurrence

Let us return to the topic of Chapter 51 and study again the recurrence and transience of random walks on the lattice using methods from the chapters on Markov chains. In a second step we want to look at general random walks, i.e. random walks whose step distribution is arbitrary.

Recall that a RW is a stochastic process of the form $X_t = X_0 + \xi_1 + \cdots + \xi_t$ where the steps $(\xi_s)_{s\in\mathbb{N}}$ are iid random variables, which are independent of the starting point X_0. If $X_t \in \mathbb{Z}^d$ and $|\xi_s| = 1$ the RW is called a simple random walk (SRW). Since

$$\begin{aligned} \mathbb{P}(X_{t+1} = i + e \mid X_t = i) &= \mathbb{P}(X_1 = i + e \mid X_0 = i) \\ &= \mathbb{P}^i(X_1 = i + e) = p_e \end{aligned}$$

a SRW is a homogeneous Markov chain with state space $(\mathbb{Z}^d, \mathscr{P}(\mathbb{Z}^d))$ and transition matrix $P = (p_{ij})_{i,j\in\mathbb{Z}^d}$, $p_{i,j} = \begin{cases} p_e, & \text{if } j - i = e, \\ 0, & \text{else.} \end{cases}$ In fact, a SRW is also spatially homogeneous, since p_{ij} depends only on the difference $j - i$:

$$\mathbb{P}^i(X_t = j) = \mathbb{P}(i + \xi_1 + \cdots + \xi_t = j) = \mathbb{P}(\xi_1 + \cdots + \xi_t = j - i) = \mathbb{P}^0(X_t = j - i).$$

We have seen in Remark 55.9 that

$$X \quad \text{recurrent} \iff \sum_{t=1}^{\infty} p_{ii}^{(t)} \overset{(55.8)}{=} \mathbb{E}^i v_i = \infty \quad \forall i \in \mathbb{Z}^d,$$

$$X \quad \text{transient} \iff \sum_{t=1}^{\infty} p_{ii}^{(t)} \overset{(55.8)}{=} \mathbb{E}^i v_i < \infty \quad \forall i \in \mathbb{Z}^d.$$

56.1 Lemma. *A SRW is irreducible, i.e.* $r_{ij} = \mathbb{P}^i(\tau_j < \infty) > 0$, $i, j \in \mathbb{Z}^d$.

Proof. It is enough to find a path from $i \to j$ which has positive probability. This can be achieved if we move along the coordinate axes, cf. Fig. 56.1. The probability for this particular path is at least

$$\prod_{n=1}^{d} \min\{p_{e_n}, p_{-e_n}\}^{|j_n - i_n|} > 0, \quad e_n = (\underbrace{0, \dots, 0, 1}_{n}, 0 \dots)^\top \qquad \square$$

56.2 Remark. An important **consequence of Lemma 56.1** is that it is enough to consider $\mathbb{E}^0 v_0$ and $p_{00}^{(t)}$ for transience and recurrence, cf. Theorem 55.15. In view of Theorem 51.13 (Pólya's theorem) and the fact that $X_t = i + \xi_1 + \cdots + \xi_t$ is as good a random walk as $X_t' = \xi_1 + \cdots + X_t$ this is not surprising.

In §§51.11–51.13 we used $u_t = \mathbb{P}^0(X_t = 0)$ which is exactly $p_{00}^{(t)}$. Thus, we have checked in the convergence behaviour of

$$\mathbb{E}^i v_i\Big|_{i=0} = \frac{r_{ii}}{1-r_{ii}}\Big|_{i=0} = \sum_{t=1}^{\infty} p_{00}^{(t)}$$

in order to establish the recurrence/transience of a SRW.

We want to study now random walks $X_t = X_0 + \xi_1 + \cdots + \xi_t$, with general step distributions $\xi_1 \sim \mu$, i.e. random walks which do not necessarily live on a lattice $\mathbb{Z}^d$ (or any other additive group). We have seen in Theorem 31.9 that a symmetric SRW on $\mathbb{Z}$ fluctuates, i.e. $\liminf_{t\to\infty} X_t = -\infty$ and $\limsup_{t\to\infty} X_t = +\infty$ a.s.; a SRW with $p \neq q$ tends to $\pm\infty$: This is the SLLN as $p \neq q \iff \mathbb{E}\xi_1 \neq 0$ and $\frac{1}{t}\mathbb{E}X_t \to \mathbb{E}\xi_1$ shows that $\mathbb{E}X_t \to \pm\infty$, depending on the sign of $\mathbb{E}\xi_1$. This result still holds in the general case.

56.3 Theorem. *Let $X_t = \xi_1 + \cdots + \xi_t$, $X_0 = 0$ be a (general) RW on $\mathbb{R}$. With probability 1 one of the following three alternatives holds.*

a) $\lim_{t\to\infty} X_t = \infty$.

b) $\lim_{t\to\infty} X_t = -\infty$.

c) $\begin{cases} X_t \equiv 0 & (\xi_1 \sim \delta_0), \\ -\infty = \liminf\limits_{t\to\infty} X_t < \limsup\limits_{t\to\infty} X_t = \infty & (\xi_1 \nsim \delta_0). \end{cases}$

If $\mathbb{E}|\xi_1| < \infty$, then a)–c) correspond to the cases $\mathbb{E}\xi_1 > 0$, $\mathbb{E}\xi_1 < 0$ and $\mathbb{E}\xi_1 = 0$. The directions $\mathbb{E}\xi_1 > 0 \Rightarrow$ a) and $\mathbb{E}\xi_1 > 0 \Rightarrow$ b) are easy – this is just the SLLN – but all other implications need more effort.

Proof. Exactly as in the proof of Theorem 31.9 we see that the event $\{X = \pm\infty\}$ for the random variable $X := \limsup_{t\to\infty} X_t$ is terminal. By Kolmogorov's 0-1 law (Theorem 31.8) we get that $\mathbb{P}(X = \pm\infty) = 0$ or $\mathbb{P}(X = \pm\infty) = 1$. Indeed,

$$\big\{X = \pm\infty\big\} = \big\{\limsup_{t\to\infty}(\xi_s + \cdots + \xi_t) = \pm\infty\big\} \in \sigma(\xi_s, \xi_{s+1}, \dots) \quad \forall s \in \mathbb{N},$$

hence, $\{X = \pm\infty\} \in \mathscr{T}_\infty = \bigcap_{s\in\mathbb{N}} \sigma(\xi_s, \xi_{s+1}, \dots)$.

Assume that $|X| < \infty$ a.s. Since the steps are iid, $(X_{t+1} - \xi_1)_{t\geqslant 0} \sim (X_t)_{t\geqslant 0}$, and so $X \sim X - \xi_1$. Since $|X| < \infty$, this entails that $\xi_1 = 0$. The argument is the same as in Case 2 of the proof of Theorem 31.9:

$$\mathbb{E}e^{i\eta X} = \mathbb{E}e^{i\eta(X-\xi_1)} \overset{\perp\!\!\!\perp}{=} \mathbb{E}e^{i\eta X}\mathbb{E}e^{-i\eta\xi_1} \overset{\perp\!\!\!\perp}{=} \mathbb{E}e^{i\eta X}\phi_{\xi_1}(\eta).$$

Since $\phi_X(\eta) = \mathbb{E}e^{i\eta X}$ is (uniformly) continuous in a neighbourhood of $\eta = 0$ and

$\phi_X(0) = 1$, we see for sufficiently small $\epsilon > 0$ that $\phi_X|_{[-\epsilon,\epsilon]} \neq 0$; hence we can divide and conclude that $\phi_{\xi_1}|_{[-\epsilon,\epsilon]} = 1$. Lemma 56.4 now shows that $\xi_1 \equiv 0$ which is ruled out. Therefore, $\mathbb{P}(X = \pm\infty) = 1$.

The same argument applies to $X' := \liminf_{t\to\infty} X_t$, and we see from this that $\mathbb{P}(X' = \pm\infty) = 1$.

Finally, as $X' \leqslant X$, the case $X' = +\infty$ and $X = -\infty$ is not possible; since a), b) deal with $X = X' = +\infty$ and $X = X' = -\infty$, respectively, we see that the alternatives a)–c) cover all cases. □

56.4 Lemma. *Let* $Y : \Omega \to \mathbb{R}^d$ *be a random variable whose characteristic function* $\phi_Y(\eta) = \mathbb{E}e^{i\langle\eta, Y\rangle}$ *is constant on some neighbourhood of* $\eta = 0$*, then* $\phi_Y \equiv 1$ *and* $Y \equiv 0$ *a.s.*

Proof. Assume that $\phi_Y|_{B_\epsilon(0)} \equiv 1$, and fix $|\eta| < 1$ and $\xi \in \mathbb{R}^d$. We have

$$\left|\phi_Y(\xi+\eta) - \phi_Y(\xi)\right|^2 = \left|\mathbb{E}\left[e^{i\langle\xi, Y\rangle}(e^{i\langle\eta, Y\rangle} - 1)\right]\right|^2 \leqslant \mathbb{E}\left[\left|e^{i\langle\eta, Y\rangle} - 1\right|^2\right]$$

and since $|e^{ix} - 1|^2 = (e^{ix} - 1)(e^{-ix} - 1) = 2 - 2\,\mathrm{Re}\,e^{ix}$, we see with Fubini

$$\left|\phi_Y(\xi+\eta) - \phi_Y(\xi)\right|^2 \leqslant 2\mathbb{E}\left[1 - \mathrm{Re}\,e^{i\langle\eta, Y\rangle}\right] = 2(1 - \mathrm{Re}\,\phi_Y(\eta)).$$

The right-hand side is 0 for all $|\eta| < \epsilon$, and for $|\xi| < \epsilon$, the above estimate shows that $\phi_Y|_{B_{2\epsilon}(0)} \equiv 1$; by iteration we get $\phi_Y \equiv 1$. Because of the uniqueness of the characteristic function we see that $Y \sim \delta_0$, i.e. $Y \equiv 0$ a.s. □

We can now study random walks on $\mathbb{R}^d$ with general step distributions. In order to keep things simple, we assume that $X_0 = 0$.

56.5 Lemma (strong Markov property (SMP) for the RW). *Let* $X_t = \xi_1 + \cdots + \xi_t$*,* $X_0 = 0$*, be a RW on* $\mathbb{R}^d$ *and* $\mathscr{F}_t = \sigma(\xi_1, \dots, \xi_t) = \sigma(X_1, \dots, X_t)$ *its natural filtration. Moreover, assume that* τ *is a stopping time. Then*

$$\mathbb{P}(\{X_{t+\tau} - X_\tau \in B\} \cap F) = \mathbb{P}(X_t \in B) \cdot \mathbb{P}(F) \tag{56.1}$$
$$\forall t \in \mathbb{N},\ F \in \mathscr{F}_\tau \cap \{\tau < \infty\},\ B \in \mathscr{B}(\mathbb{R}^d);$$

$$\mathbb{P}\left(\bigcap_{i=1}^n \{X_{t_i+\tau} - X_\tau \in B_i\} \cap F\right) = \mathbb{P}\left(\bigcap_{i=1}^n \{X_{t_i} \in B_i\}\right) \cdot \mathbb{P}(F) \tag{56.2}$$
$$\forall t_1 < \cdots < t_n,\ F \in \mathscr{F}_\tau \cap \{\tau < \infty\},\ B_1, \dots, B_n \in \mathscr{B}(\mathbb{R}^d);$$

$$\mathbb{P}(\{X_{\cdot+\tau} - X_\tau \in \Gamma\} \cap F) = \mathbb{P}(X_\cdot \in \Gamma) \cdot \mathbb{P}(F) \tag{56.3}$$
$$\forall F \in \mathscr{F}_\tau \cap \{\tau < \infty\},\ \Gamma \in \mathscr{B}(\mathbb{R}^d)^{\otimes\mathbb{N}}.$$

Proof. Compare the following arguments with the proof of Theorem 54.6 and Corollary 54.9.

Clearly, (56.3)⇒(56.2)⇒(56.1). Moreover, (56.2)⇒(56.3) follows in the same way as we have shown §54.6.c)⇒§54.6.d). It remains to show (56.1) and (56.2).

Proof of (56.1): Let $F \in \mathscr{F}_\tau \cap \{\tau < \infty\}$, $B \in \mathscr{B}(\mathbb{R}^d)$ and $t \geqslant 0$. Because of the definition of $\mathscr{F}_\tau$, we know that $F \cap \{\tau = s\} = F \cap \{\tau \leqslant s\} \cap \{\tau \leqslant s-1\} \in \mathscr{F}_s$, hence

$$\begin{aligned}
\mathbb{P}(\{X_{t+\tau} - X_\tau \in B\} \cap F) &= \sum_{s=0}^{\infty} \mathbb{P}(\underbrace{\{X_{t+s} - X_s \in B\}}_{\substack{=\xi_{s+1}+\dots+\xi_{t+s} \perp\!\!\!\perp \mathscr{F}_s \\ \sim \xi_1+\dots+\xi_t}} \cap \underbrace{\{\tau = s\} \cap F}_{\in \mathscr{F}_s}) \\
&\overset{\perp\!\!\!\perp}{=} \sum_{s=0}^{\infty} \mathbb{P}(X_t \in B) \cdot \mathbb{P}(\{\tau = s\} \cap F) \\
&= \mathbb{P}(X_t \in B) \cdot \mathbb{P}(F).
\end{aligned}$$

Proof of (56.2): As before, we see that

$$\begin{aligned}
\mathbb{P}\Big(\bigcap_{i=1}^{n} \{X_{t_i+\tau} - X_\tau \in B_i\} \cap F\Big) &= \sum_{s=0}^{\infty} \mathbb{P}\Big(\bigcap_{i=1}^{n} \{X_{t_i+s} - X_s \in B_i\}\Big) \cdot \mathbb{P}\big(\{\tau = s\} \cap F\big) \\
&\overset{?!}{=} \sum_{s=0}^{\infty} \mathbb{P}\Big(\bigcap_{i=1}^{n} \{X_{t_i} \in B_i\}\Big) \cdot \mathbb{P}\big(\{\tau = s\} \cap F\big),
\end{aligned}$$

only the equality marked by »?!« needs an argument. Without loss of generality we assume that $n = 2$ – larger n increase only the notational complexity – and $t = t_1 < t_2 = t'$. We have

$$\begin{aligned}
&\mathbb{P}\big(X_{t+s} - X_s \in B,\ X_{t'+s} - X_s \in B'\big) \\
&= \mathbb{P}\big(X_{t+s} - X_s \in B,\ (X_{t'+s} - X_{t+s}) + (X_{t+s} - X_s) \in B'\big) \\
&\overset{\perp\!\!\!\perp}{=} \int \mathbb{P}\big(\underbrace{X_{t+s} - X_s}_{\sim X_t} \in B,\ y + \underbrace{(X_{t+s} - X_s)}_{\sim X_t} \in B'\big)\, \mathbb{P}\big(\underbrace{(X_{t'+s} - X_{t+s})}_{\sim X_{t'} - X_t} \in dy\big) \\
&= \int \mathbb{P}\big(X_t \in B,\ y + X_t \in B'\big)\, \mathbb{P}\big((X_{t'} - X_t) \in dy\big) \\
&\overset{\perp\!\!\!\perp}{=} \mathbb{P}\big(X_t \in B, X_{t'} - X_t + X_t \in B'\big).
\end{aligned}$$

□

Let us now study recurrence and transience of general random walks. We need the following notation (compare this with the notation used in Thm. 52.13).

56.6 Definition. Let $X_t = \xi_1 + \dots + \xi_t$, $X_0 = 0$ be a RW on $\mathbb{R}^d$. We define the

a) *occupation measure* $\quad B \mapsto v(B) = \sum_{t=0}^{\infty} \mathbb{1}_B(X_t);$

b) *mean occupation measure* $\quad B \mapsto \mathbb{E}v(B) = \sum_{t=0}^{\infty} \mathbb{P}(X_t \in B);$

c) *accessible points* $\quad \mathcal{A} = \bigcap_{\epsilon>0} \big\{x \in \mathbb{R}^d \mid \mathbb{E}v(B_\epsilon(x)) > 0\big\};$

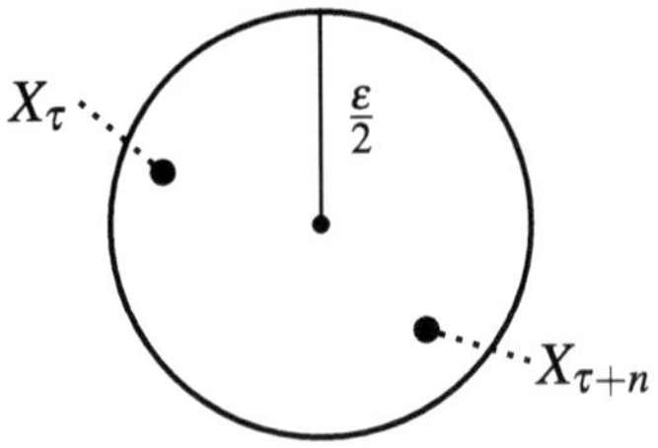

Fig. 56.2. The distance of X_τ and any other $X_{t+\tau} \in B_{\epsilon/2}(y_k)$.

mean recurrence set $\mathcal{M} = \bigcap_{\epsilon>0} \{x \in \mathbb{R}^d \mid \mathbb{E}v(B_\epsilon(x)) = \infty\}$;

recurrence set $\mathcal{R} = \bigcap_{\epsilon>0} \{x \in \mathbb{R}^d \mid v(B_\epsilon(x)) = \infty \text{ a.s.}\}$.

⚠ We cannot directly apply the results of Chapter 55 since the random walk X does not (necessarily) have a discrete state space.

💬 It is clear that we always have $\mathcal{R} \subset \mathcal{M} \subset \mathcal{A}$.

The following result extends Remark 55.9 and Theorem 55.15, as well as Theorem 51.13.

56.7 Theorem (recurrence–transience dichotomy). *For an arbitrary RW on $\mathbb{R}^d$, $X_t = \xi_1 + \cdots + \xi_t$, $X_0 = 0$, the following alternative holds:*

a) *X is recurrent: $\mathbb{P}(\lim_{t\to\infty} |X_t| = \infty) < 1$. In this case, $\mathcal{R} \neq \emptyset$ and $\mathcal{R} = \mathcal{M} = \mathcal{A}$ is a non-trivial closed sub-group of the group $(\mathbb{R}^d, +)$.*

b) *X is transient: $\mathbb{P}(\lim_{t\to\infty} |X_t| = \infty) = 1$. In this case, $\mathcal{R} = \emptyset$ and $\mathcal{R} = \mathcal{M} = \emptyset$.*

Proof. Let us first discuss the consequences of $\mathbb{P}(\lim_{t\to\infty} |X_t| = \infty) < 1$ resp. $= 1$.

1° Assume that $\mathbb{P}(\lim_{t\to\infty} |X_t| = \infty) < 1$. This entails that, with probability > 0, there is a bounded subsequence, i.e.

$$\exists r > 0 : \quad \mathbb{P}(X_t \in B_r(0) \text{ for } \infty \text{ many } t) > 0.$$

For every $\epsilon > 0$ we can cover $B_r(0) = \bigcup_{k=1}^N B_{\epsilon/2}(y_k)$ by finitely many balls, and we see that

$$\exists k_0 : \quad \mathbb{P}(X_t \in B_{\epsilon/2}(y_k) \text{ for } \infty \text{ many } t) > 0$$

The event $\{X_t \in B \text{ infinitely often}\}$ is symmetric, since it does not depend on finite perturbations of the sequence of steps $(\xi_s)_{s\in\mathbb{N}}$. Thus, by Theorem 45.6, $\mathbb{P}(X_t \in B_{\epsilon/2}(y_{k_0}) \text{ for } \infty \text{ many } t) = 1$; in particular, $\tau = \inf\{t \mid X_t \in B_{\epsilon/2}(x_{k_0})\}$ is an a.s. finite stopping time. Since $X_\tau \in B_{\epsilon/2}(y_{k_0})$, we can read off Fig. 56.2 that $X_{\tau+t} \in B_{\epsilon/2}(y_{k_0}) \implies |X_{\tau+t} - X_\tau| \leqslant \epsilon$; thus,

$$\begin{aligned}
1 &= \mathbb{P}(X_t \in B_{\epsilon/2}(y_{k_0}) \text{ for } \infty \text{ many } t)\\
&= \mathbb{P}(X_{\tau+t} \in B_{\epsilon/2}(y_{k_0}) \text{ for } \infty \text{ many } t)\\
&\overset{\text{Fig. 56.2}}{\leqslant} \mathbb{P}(|X_{\tau+t} - X_\tau| < \epsilon \text{ for } \infty \text{ many } t)\\
&\overset{\text{SMP §56.5}}{=} \mathbb{P}(|X_t| < \epsilon \text{ for } \infty \text{ many } t).
\end{aligned}$$

This proves that $0 \in \mathcal{R}$, hence $\mathcal{R} \neq \emptyset$ as claimed in a).

2° Now let $\mathbb{P}\left(\lim_{t\to\infty} |X_t| = \infty\right) = 1$. In this case, we define the sets

$$A_t := \left\{|X_t| < r,\ |X_{t+s}| \geqslant r,\ \ \forall s \geqslant u\right\}.$$

Note that the conditions $|X_t| < r$ & $|X_{t+s} - X_t| \geqslant 2r\ \ \forall s \geqslant u$ imply that we are in A_t – just observe that $|X_{t+s}| \geqslant |X_{t+s} - X_t| - |X_t| > 2r - r = r$ – and so

$$\begin{aligned}
\mathbb{P}(A_t) &\geqslant \mathbb{P}\left(|X_t| < r,\ |X_{t+s} - X_t| \geqslant 2r,\ \ \forall s \geqslant u\right)\\
&\overset{\text{SMP §56.5}}{=} \mathbb{P}(|X_t| < r)\underbrace{\mathbb{P}(|X_s| \geqslant 2r,\ \ \forall s \geqslant u)}_{>0}.
\end{aligned}$$

We have constructed the sets A_t in such a way that each ω can appear at most in $u \in \mathbb{N}$ many of the A_t. Indeed,

$$\begin{aligned}
\omega \in A_1 &\implies |X_1(\omega)| < r \quad \text{and} \quad |X_{1+u}(\omega)| \geqslant r,\ |X_{2+u}(\omega)| \geqslant r, \dots\\
\omega \in A_2 &\implies |X_2(\omega)| < r \quad \text{and} \quad |X_{2+u}(\omega)| \geqslant r,\ |X_{3+u}(\omega)| \geqslant r, \dots\\
&\ \ \vdots\\
\omega \in A_u &\implies |X_u(\omega)| < r \quad \text{and} \quad |X_{2u}(\omega)| \geqslant r,\ |X_{2u+1}(\omega)| \geqslant r, \dots\\
\omega \in A_{u+1} &\implies |X_{u+1}(\omega)| < r \quad \text{and} \quad |X_{2u+1}(\omega)| \geqslant r,\ |X_{2u+2}(\omega)| \geqslant r, \dots
\end{aligned}$$

i.e. if $\omega \in A_1 \cap \dots \cap A_u$, then we have $\omega \notin A_{u+1} \cup A_{u+2} \cup \dots$ This shows

$$\begin{aligned}
u \geqslant \sum_{t=1}^{\infty} \mathbb{1}_{A_t} &\implies u \geqslant \sum_{t=1}^{\infty} \mathbb{P}(A_t) \geqslant \sum_{t=1}^{\infty} \mathbb{P}(|X_t| < r)\underbrace{\mathbb{P}(|X_s| \geqslant 2r,\ \ \forall s \geqslant u)}_{>0}\\
&\implies \mathbb{E}v(B_r(0)) = \sum_{t=1}^{\infty} \mathbb{P}(|X_t| < r) < \infty,
\end{aligned}$$

and since $r > 0$ is arbitrary, we see that $0 \notin \mathcal{M}$.

Observe that $v(B_r(x)) = \sum_t \mathbb{1}_{B_r(x)}(X_t) = \sum_t \mathbb{1}_{B_r(0)}(X_t - x)$. If we use the RW $X_t - x$ instead of X_t, the previous argument shows that $x \notin \mathcal{M}$ for any $x \in \mathbb{R}^d$, thus, $\mathcal{M} = \emptyset$.

Since $\mathcal{R} \subset \mathcal{M} = \emptyset$ we see that $\mathcal{R} = \emptyset$ as claimed in b).

3° Each of the pairs $\mathbb{P}(\lim_{t\to\infty}|X_t| = \infty) = 1$ vs. $\mathbb{P}(\lim_{t\to\infty}|X_t| = \infty) < 1$ and $\mathcal{R} = \emptyset$ vs. $\mathcal{R} \neq \emptyset$ are genuine alternatives. Therefore, the steps 1° and 2° show that

$$\text{a)} \iff \mathcal{R} \neq \emptyset \iff \mathbb{P}(\lim_{t\to\infty}|X_t| = \infty) < 1,$$
$$\text{b)} \iff \mathcal{R} = \emptyset \iff \mathbb{P}(\lim_{t\to\infty}|X_t| = \infty) = 1.$$

We still have to show the remaining claims in a) and b).

4° $\mathcal{R}$ is a closed set. If $\mathcal{R} = \emptyset$, then there is nothing to show. Assume that $\mathcal{R} \neq \emptyset$. We will prove that $\mathcal{R}^c$ is an open set. If $x \in \mathcal{R}^c$, we know that

$$\exists \epsilon > 0 : \quad \mathbb{P}(v(B_\epsilon(x)) < \infty) > 0 \implies \forall y \in B_{\epsilon/2}(x) : \quad \mathbb{P}(v(B_{\epsilon/2}(y)) < \infty) > 0$$

since $B_{\epsilon/2}(y) \subset B_\epsilon(x)$ for all $|y - x| < \epsilon/2$ and $v(\cdot)$ is monotone. This shows that $B_{\epsilon/2}(x) \subset \mathcal{R}^c$, i.e. $\mathcal{R}^c$ is open.

5° If $\mathcal{R} \neq 0$, then $x \in \mathcal{A},\ y \in \mathcal{R} \implies y - x \in \mathcal{R}$. Observe that

$$y - x \notin \mathcal{R} \implies \exists \epsilon' > 0,\ N \geqslant 1 : \quad \mathbb{P}(|X_t - (y-x)| \geqslant 2\epsilon',\ \forall t \geqslant N) > 0,$$
$$x \in \mathcal{A} \implies \forall \epsilon > 0\ \exists s \in \mathbb{N} : \quad \mathbb{P}(|X_s - x| < \epsilon) > 0.$$

Similar to the argument used at the beginning of Step 2° , we find for any $u \in \mathbb{N}$ and $\epsilon = \epsilon'$

$$\begin{aligned}
&\mathbb{P}(|X_t - y| \geqslant \epsilon,\ \forall t \geqslant u + s) \\
&\quad\geqslant \mathbb{P}(|X_t - X_s - (y-x)| \geqslant 2\epsilon,\ |X_s - x| < \epsilon,\ \forall t \geqslant u+s) \\
&\quad\overset{\text{SMP}}{\underset{56.5}{=}} \mathbb{P}(|X_{t-s} - (y-x)| \geqslant 2\epsilon,\ \forall t \geqslant u+s) \cdot \mathbb{P}(|X_s - x| < \epsilon) > 0.
\end{aligned}$$

This implies that $y \notin \mathcal{R}$. Thus, we have shown that for all $x \in \mathcal{A}$

$$y - x \notin \mathcal{R} \implies y \notin \mathcal{R}$$

which is equivalent to saying that $y \in \mathcal{R} \implies y - x \in \mathcal{R}$.

6° If $\mathcal{R} \neq 0$, then $\mathcal{R} \neq \emptyset$ is a group. Let $r, q \in \mathcal{R} \subset \mathcal{A}$ and pick x, y in 5° as follows:

$$x = y = r \overset{5^\circ}{\implies} y - x = r - r = 0 \in \mathcal{R},$$
$$x = r,\ y = 0 \implies y - x = 0 - r = -r \in \mathcal{R},$$
$$x = -r,\ y = q \implies y - x = q + r \in \mathcal{R}.$$

7° If $\mathcal{R} \neq 0$, then we have $\emptyset \neq \mathcal{R} = \mathcal{A}$. Let $a \in \mathcal{A}$ and pick x, y in 5° as follows:

$$x = a,\ y = 0 \overset{5^\circ}{\implies} -a \in \mathcal{R} \overset{6^\circ}{\implies} a \in \mathcal{R} \implies \mathcal{A} \subset \mathcal{R} \overset{3^\circ}{\implies} \mathcal{A} = \mathcal{R}. \qquad \square$$

The assertion of Theorem 56.7 is theoretically most satisfying but it is not a readily checked criterium. Therefore, we want to obtain easily applicable tests for recurrence and transience.

56.8 Lemma (scaling). *Let* $X_t = \xi_1 + \cdots + \xi_t$, $X_0 = 0$, *be a RW on* $\mathbb{R}^d$. *For every* $\epsilon > 0$ *and* $r \geqslant 1$ *one has*

$$\sum_{t=0}^{\infty} \mathbb{P}(|X_t| \leqslant r\epsilon) \leqslant c r^d \sum_{t=0}^{\infty} \mathbb{P}(|X_t| \leqslant \epsilon). \tag{56.4}$$

Proof. We cover the ball $B_{r\epsilon}(0) = \bigcup_{k=1}^{N} B_{\epsilon/2}(y_k)$ with finitely many smaller balls. This requires approximately

$$\frac{\text{volume}(B_{r\epsilon}(0))}{\text{volume}(B_{\epsilon/2}(0))} \approx \frac{(r\epsilon)^d}{(\epsilon/2)^d} \approx r^d$$

balls, i.e. we have $N \leqslant cr^d$ for some constant $c < \infty$. Define stopping times ✍

$$\tau_k := \inf\left\{t \mid X_t \in B_{\epsilon/2}(y_k)\right\}.$$

Using the subadditivity of $\mathbb{P}(\cdot)$ we get (cf. Fig. 56.2)

$$\begin{aligned}
\sum_{t=0}^{\infty} \mathbb{P}(|X_t| \leqslant r\epsilon) &\leqslant \sum_{k=1}^{N}\sum_{t=0}^{\infty} \mathbb{P}(X_t \in B_{\epsilon/2}(y_k)) \\
&= \sum_{k=1}^{N}\sum_{t=0}^{\infty} \mathbb{P}(X_t \in B_{\epsilon/2}(y_k),\ \underbrace{t \geqslant \tau_k}_{X_t \in B_{\epsilon/2}(y_k) \text{ not possible if } t < \tau_k}) \\
&= \sum_{k=1}^{N}\sum_{s=0}^{\infty} \mathbb{P}(X_{s+\tau_k} \in B_{\epsilon/2}(y_k),\ \tau_k < \infty) \\
&\overset{\text{Fig. 56.2}}{\leqslant} \sum_{k=1}^{N}\sum_{s=0}^{\infty} \mathbb{P}(|X_{s+\tau_k} - X_{\tau_k}| < \epsilon,\ \tau_k < \infty) \\
&\overset{\text{SMP §56.5}}{=} \underbrace{\sum_{k=1}^{N} \mathbb{P}(\tau_k < \infty)}_{\leqslant N \leqslant cr^d} \sum_{s=0}^{\infty} \mathbb{P}(|X_s| < \epsilon).
\end{aligned}$$

□

56.9 Theorem (recurrence if $d = 1, 2$). *A RW* $X_t = \xi_1 + \cdots + \xi_t$, $X_0 = 0$, *on* $\mathbb{R}^d$ *is recurrent, if*

a) $d = 1$ *and* $\frac{1}{t} X_t \xrightarrow[n\to\infty]{\mathbb{P}} 0$ *(WLLN for the steps* ξ_t*).*

b) $d = 2$ *and* $\frac{1}{\sqrt{t}} X_t \xrightarrow[n\to\infty]{d} G \sim N(0, \Gamma)$ *(CLT for the steps* ξ_t*).*

💬 A sufficient condition for the WLLN is $\mathbb{E}|\xi_1| < \infty$ and $\mathbb{E}\xi_1 = 0$, cf. Theorem 32.1; note that the SLLN implies the WLLN.

💬 A sufficient condition for the CLT (also for $d = 2$) is $\mathbb{E}|\xi_1|^2 < \infty$ and $\mathbb{E}\xi_1 = 0$, cf. Theorem 28.6 for $d = 1$; the case $d = 2$ can be reduced to $d = 1$ using Definition 36.1 and the Cramér–Wold device [WT, Korollar 9.19], considering the one-dimensional Gaussian rv $\langle \ell, (X_t^{(1)}, X_t^{(2)}) \rangle$ for all $\ell \in \mathbb{R}^2$.

Proof of Theorem 56.9. For any $\epsilon > 0$ and $r \geqslant 1$ we find with Lemma 56.8

$$\begin{aligned}\sum_{n=0}^{\infty} \mathbb{P}(|X_n| \leqslant \epsilon) &\geqslant \gamma r^{-d} \sum_{n=0}^{\infty} \mathbb{P}(|X_n| \leqslant \epsilon r)\\ &= \gamma \sum_{n=0}^{\infty} \int_n^{n+1} r^{-d}\, \mathbb{P}(|X_{\lfloor t \rfloor}| \leqslant \epsilon r)\, dt\\ &= \gamma \int_0^{\infty} \mathbb{P}(|X_{\lfloor t r^d \rfloor}| \leqslant \epsilon r)\, dt,\end{aligned}$$

where $\lfloor s \rfloor$ denotes the integer part of $s \geqslant 0$.

a) Since $\lim_{r\to\infty} \mathbb{P}(|X_{\lfloor tr \rfloor}| \leqslant \epsilon r) = 1$, we can use Fatou's Lemma to deduce

$$\begin{aligned}\sum_{n=0}^{\infty} \mathbb{P}(|X_n| \leqslant \epsilon) &\geqslant \liminf_{r\to\infty} \gamma \int_0^{\infty} \mathbb{P}(|X_{\lfloor tr \rfloor}| \leqslant \epsilon r)\, dt\\ &\geqslant \gamma \int_0^{\infty} \underbrace{\liminf_{r\to\infty} \mathbb{P}(|X_{\lfloor tr \rfloor}| \leqslant \epsilon r)}_{=1}\, dt = \infty.\end{aligned}$$

This gives that $0 \in \mathcal{M}$ (the mean recurrence set), hence $\mathcal{M} \neq \emptyset$, and Theorem 56.7 proves recurrence.

b) For $c \leqslant 1$ and any 2-dimensional centered Gaussian rv with covariance matrix Γ we have (see Theorem 36.2)

$$\begin{aligned}\mathbb{P}(|G| \leqslant c) &= \int_{|y|\leqslant c} \frac{1}{2\pi\sqrt{\det\Gamma}} e^{-\frac{1}{2}\langle y, \Gamma^{-1} y\rangle}\, dy\\ &\underset{dy = c^2\,dx}{\overset{y=cx}{=}} c^2 \int_{|x|\leqslant 1} \frac{1}{2\pi\sqrt{\det\Gamma}} e^{-\frac{c^2}{2}\langle x, \Gamma^{-1} x\rangle}\, dx\\ &\overset{|c|\leqslant 1}{\geqslant} \gamma' c^2.\end{aligned}$$

Yet another application of Fatou's lemma yields

$$\begin{aligned}\sum_{n=0}^{\infty} \mathbb{P}(|X_n| \leqslant \epsilon) &\geqslant \liminf_{r\to\infty} \gamma \int_0^{\infty} \mathbb{P}(|X_{\lfloor tr^2 \rfloor}| \leqslant \epsilon r)\, dt\\ &\geqslant \gamma \int_0^{\infty} \liminf_{r\to\infty} \mathbb{P}\left(\frac{|X_{\lfloor tr^2 \rfloor}|}{r\sqrt{t}} \leqslant \frac{\epsilon}{\sqrt{t}}\right) dt\\ &= \gamma \int_0^{\infty} \mathbb{P}\left(|G| \leqslant \frac{\epsilon}{\sqrt{t}}\right) dt\\ &\geqslant \gamma \int_1^{\infty} \mathbb{P}\left(|G| \leqslant \frac{\epsilon}{\sqrt{t}}\right) dt\\ &\geqslant \gamma\gamma'\epsilon \int_1^{\infty} \frac{dt}{t} = \infty.\end{aligned}$$

Again we see that $0 \in \mathcal{M}$, hence $\mathcal{M} \neq \emptyset$, and we conclude recurrence from Theorem 56.7. □

Let us return to the remark following the statement of Theorem 56.3.

56.10 Corollary (Chung–Fuchs theorem). *Let $X_t = \xi_1 + \cdots + \xi_t$, $X_0 = 0$, be a RW on $\mathbb{R}$ with integrable steps $\mathbb{E}|\xi_1| < \infty$. With probability 1 one of the following alternatives holds:*

a) $\mathbb{E}\xi_1 > 0 \iff \lim_{t\to\infty} X_t = \infty$.
b) $\mathbb{E}\xi_1 < 0 \iff \lim_{t\to\infty} X_t = -\infty$.
c) $\mathbb{E}\xi_1 = 0 \iff \begin{cases} X_t \equiv 0 & (\xi_1 \sim \delta_0) \\ -\infty = \liminf\limits_{t\to\infty} X_t < \limsup\limits_{t\to\infty} X_t = \infty & (\xi_1 \nsim \delta_0) \end{cases}$

Proof. We know from Theorem 56.3 that »$\lim_{t\to\infty} X_t = \infty$ a.s.«, »$\lim_{t\to\infty} X_t = -\infty$ a.s.«, »$\liminf_{t\to\infty} = -\infty < \infty = \limsup_{t\to\infty}$ a.s.« are mutually exclusive alternatives, and so are »$\mathbb{E}\xi_1 > 0$«, »$\mathbb{E}\xi_1 < 0$« and »$\mathbb{E}\xi_1 = 0$«. Therefore it is enough to show, in each case, »$\Longrightarrow$«.

a) & b) follow from the SLLN as $\frac{1}{t}\mathbb{E}X_t \to \mathbb{E}\xi_1$.

c) follows from Theorem 56.9.a) and Theorem 56.7.a): If $\xi_1 \neq 0$, then X_t does not stay put at 0; since $\mathcal{A} = \mathcal{M} = \mathcal{R}$, we see that $\mathcal{R} \supsetneq \{0\}$, i.e. the group $\mathcal{R}$ is unbounded, and we infer that the lower and upper limits must be $-\infty$ and $+\infty$, respectively. □

57 The Chung–Fuchs criterium

We want to derive a very handy criterium for recurrence and transience of a RW in terms of the characteristic function of the steps. Recall from Chapter 27 that the characteristic function of a random variable $Y \sim \mu$ on $\mathbb{R}^d$ is the function

$$\check{\mu}(\eta) := \mathbb{E}e^{i\langle \eta, Y\rangle} = \int_{\mathbb{R}^d} e^{i\langle \eta, y\rangle}\,\mu(dy).$$

If $\mu(dx) = m(x)\,dx$, then we identify $\check{m}(\eta) = \check{\mu}(\eta)$. Recall also that $\check{\mu}$ determines μ, hence Y, uniquely. We will need the following useful auxiliary result.

57.1 Lemma. *Let μ and ν be two probability measures on $\mathbb{R}^d$.*

a) $\int \check{\mu}\,d\nu = \int \check{\nu}\,d\mu$ *(Plancherel's identity)*
b) $\widecheck{\mu * \nu} = \check{\mu}\cdot\check{\nu}$ *(convolution theorem)*

Proof. a) Since $|\check{\mu}|, |\check{\nu}| \leqslant 1$, we can use Fubini's theorem and get

$$\begin{aligned}\int \check{\mu}(\eta)\,\nu(d\eta) &= \iint e^{i\langle y,\eta\rangle}\,\mu(dy)\,\nu(d\eta) \\ &= \iint e^{i\langle y,\eta\rangle}\,\nu(d\eta)\,\mu(dy) = \int \check{\nu}(y)\,\mu(dy).\end{aligned}$$

b) The convolution satisfies

$$\begin{aligned}\widecheck{\mu * \nu}(\eta) &= \int e^{i\langle y,\eta\rangle}\,\mu * \nu(dy)\\ &\overset{\S 17.7}{=} \iint e^{i\langle (x+y),\eta\rangle}\,\mu(dx)\,\nu(dy) = \check{\mu}(\eta)\check{\nu}(\eta);\end{aligned}$$

alternatively, combine Theorem 25.14 and the easy direction »⇒« of Corollary 27.8. □

57.2 Theorem (Chung–Fuchs test). *Let* $X_t = \xi_1 + \cdots + \xi_t$, $X_0 = 0$, *be a RW on* $\mathbb{R}^d$ *whose steps satisfy* $\xi_1 \sim \mu$; X *is transient if, and only if,*

$$\exists \epsilon > 0 : \quad \sup_{r<1} \int_{B_\epsilon(0)} \operatorname{Re} \frac{1}{1 - r\check{\mu}(\eta)}\,d\eta < \infty. \tag{57.1}$$

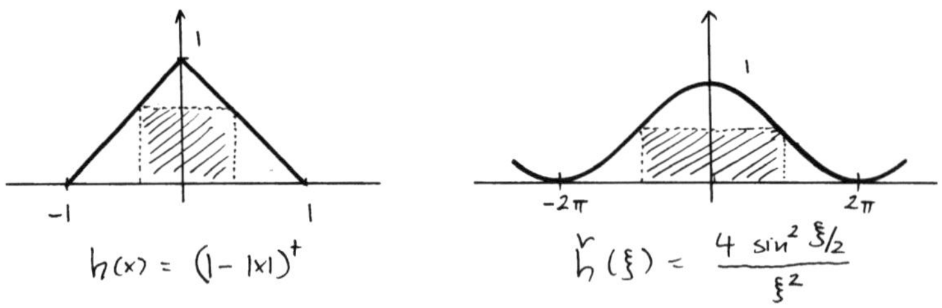

Fig. 57.1. The triangular distribution (tent-function) $h(x) = (1 - |x|)^+$ in $d = 1$ and íts characteristic function $\check{h}(\xi)$. Note that both functions are, in a neighbourhood of zero, strictly positive and that we may inscribe a rectangle below the graph. This propertiy makes the pair $(h, \check{h})$ almost as special as the normal distribution and its characteristic function.

Proof. 1° Let $h(x) = (1 - |x|)^+$ be a tent function in $\mathbb{R}$. The d-dimensional tent function (it is a »pyramid«) and its characteristic funciton is given by

$$H(x) = \prod_{j=1}^{d} h(x_j), \qquad x = (x_1, \dots, x_d) \in \mathbb{R}^d,$$

$$\check{H}(\sigma) = \prod_{j=1}^{d} \check{h}(\sigma_j), \qquad \sigma = (\sigma_1, \dots, \sigma_d) \in \mathbb{R}^d.$$

Since $X_t \sim \mu^{*t} = \mu * \cdots * \mu$ (t factors), we have

$$\begin{aligned}\int \check{H}(\tfrac{1}{a}\sigma)\,\mathbb{P}(X_t \in d\sigma) &\overset{\text{✎}}{=} \int a^d\,\widecheck{H(a\cdot)}(\sigma)\,\mu^{*t}(d\sigma)\\ &\overset{\S 57.1}{=} \int a^d\,H(ax)\,\check{\mu}(x)^t\,dx.\end{aligned}$$

Now we multiply both sides by r^t $(r < 1)$, sum over $t \in \mathbb{N}_0$, where we use $\mu^{*0} = \delta_0$;

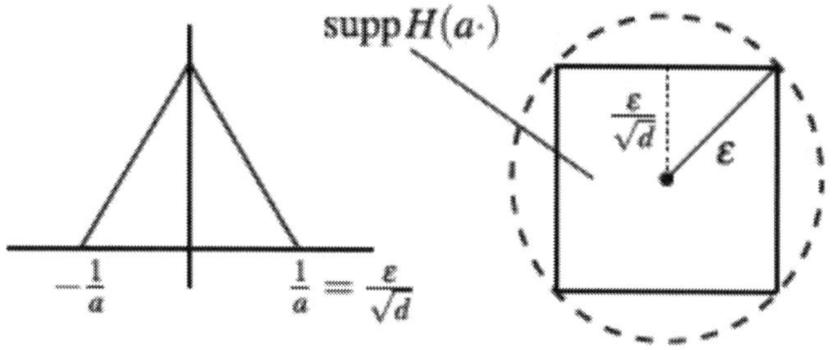

Fig. 57.2. Support of the tent function in $d=1$ and $d>1$.

because of the uniform convergence of the series we may interchange summation and integration.

$$\int \check{\mathrm{H}}(\tfrac{1}{a}\sigma)\sum_{t=0}^{\infty} r^t\mu^{*t}(d\sigma) = \int a^d\,\mathrm{H}(ax)\sum_{t=0}^{\infty}\underbrace{r^t\check{\mu}(x)^t}_{|r\check{\mu}|\leqslant r<1} dx$$
$$= a^d\int\frac{\mathrm{H}(ax)}{1-r\check{\mu}(x)}\,dx.$$

Since the expression on the left is real-valued, we may take the real part on the right. Using »Re $\int = \int$Re« we see that

$$\int \check{\mathrm{H}}(\tfrac{1}{a}\sigma)\sum_{t=0}^{\infty} r^t\mu^{*t}(d\sigma) = a^d\int \mathrm{Re}\,\frac{\mathrm{H}(ax)}{1-r\check{\mu}(x)}\,dx. \tag{57.2}$$

2° We show (57.1)⇒transience. Pick $a=\sqrt{d}/\epsilon$. Observe – see Fig. 57.1 and Fig. 57.2 – that we have $\mathbb{1}_{\mathrm{B}_a(0)}(\sigma)\leqslant\gamma'\check{\mathrm{H}}(\frac{1}{a}\sigma)$ and $|\mathrm{H}(ax)|\leqslant\mathbb{1}_{\mathrm{B}_\epsilon(0)}(x)$ since the support satisfies $\operatorname{supp}\mathrm{H}(ax)\subset[-1/a,1/a]^d$. This shows

$$\begin{aligned}\sum_{t=0}^{\infty}\mathbb{P}(|X_t|<a) = \sum_{t=0}^{\infty}\mu^{*t}(\mathrm{B}_a(0)) &= \sum_{t=0}^{\infty}\int\mathbb{1}_{\mathrm{B}_a(0)}(\sigma)\,\mu^{*t}(d\sigma)\\
&\overset{\text{BL}}{\leqslant}\gamma'\sup_{r<1}\sum_{t=0}^{\infty}\int\check{\mathrm{H}}(\tfrac{1}{a}\sigma)\,r^t\mu^{*t}(d\sigma)\\
&\overset{\text{BL}}{=}\gamma'\sup_{r<1}\int\check{\mathrm{H}}(\tfrac{1}{a}\sigma)\sum_{t=0}^{\infty}r^t\mu^{*t}(d\sigma)\\
&\overset{(57.2)}{=}\gamma''\sup_{r<1}a^d\int\mathrm{Re}\,\frac{\mathrm{H}(ax)}{1-r\check{\mu}(x)}\,dx\\
&\leqslant\gamma''\sup_{r<1}a^d\int_{\mathrm{B}_\epsilon(0)}\mathrm{Re}\,\frac{1}{1-r\check{\mu}(x)}\,dx<\infty.\end{aligned}$$

This shows that $\mathcal{M}=\emptyset$, i.e. X is transient (Theorem 56.7).

3° We show transience⟹(57.1). It is easy to see that

$$\check{\check{h}}(x) = 2\pi h(x), \quad x \in \mathbb{R}.$$

Therefore, we can use (57.2) (with $\mathrm{H} \rightsquigarrow \check{\mathrm{H}}$) and find

$$(2\pi)^d \int \mathrm{H}(\tfrac{1}{a}\sigma) \sum_{t=0}^{\infty} r^t \mu^{*t}(d\sigma) = a^d \int \operatorname{Re} \frac{\check{\mathrm{H}}(ax)}{1 - r\check{\mu}(x)}\, dx. \tag{57.2'}$$

Since $\gamma \mathbb{1}_{\mathrm{B}_\epsilon(0)}(x) \leqslant \check{\mathrm{H}}(\frac{1}{\epsilon}x)$ for sufficiently small ϵ (✍, see Fig. 57.1), we have for $a = 1/\epsilon$

$$\begin{aligned}
\sup_{r<1} \int_{\mathrm{B}_\epsilon(0)} \operatorname{Re} \frac{1}{1 - r\check{\mu}(x)}\, dx \quad &\leqslant \quad \frac{1}{\gamma} \sup_{r<1} \int \operatorname{Re} \frac{\check{\mathrm{H}}(\frac{1}{\epsilon}x)}{1 - r\check{\mu}(x)}\, dx \\
&\overset{(57.2')}{\underset{r=1}{\leqslant}} \gamma' \int \epsilon^d \mathrm{H}(\epsilon\sigma) \sum_{t=0}^{\infty} \mu^{*t}(d\sigma) \\
&\leqslant \gamma' \sum_{t=0}^{\infty} \mu^{*t}(\mathrm{B}_{2\sqrt{d}/\epsilon}(0)) \overset{\mathcal{M}=\emptyset}{\underset{\text{transient}}{<}} \infty.
\end{aligned}$$

In the last line we use that $\operatorname{supp} \mathrm{H}(\epsilon\cdot) \subset [-1/\epsilon, 1/\epsilon]^d \subset \overline{\mathrm{B}_{\sqrt{d}/\epsilon}(0)} \subset \mathrm{B}_{2\sqrt{d}/\epsilon}(0)$, see Fig. 57.2. □

An important special case are symmetric RWs. A random variable $\xi_1 \sim \mu$ is called **symmetric**, if

$$\xi_1 \sim -\xi_1 \iff \mu(\mathrm{B}) = \mu(-\mathrm{B}), \quad -\mathrm{B} = \{-b \mid b \in \mathrm{B}\}.$$

⚠ symmetric ($\xi_1 \sim -\xi_1$) and **rotationally** symmetric ($\xi_1 \sim \mathrm{R}\xi_1$ for any rotation $\mathrm{R} \in \mathbb{R}^{d\times d}$) are different properties if $d \geqslant 2$.

💬 the RW $\mathrm{X}_t = \xi_1 + \cdots + \xi_t$, $\mathrm{X}_0 = 0$, is said to be symmetric, if ξ_1 is symmetric.

57.3 Lemma. *A rv $\xi_1 \sim \mu$ is symmetric if, and only if, $\check{\mu}(\eta) \in \mathbb{R}$.*

Proof. This is Theorem 27.5.c). □

57.4 Corollary. *Let $\mathrm{X}_t = \xi_1 + \cdots + \xi_t$, $\mathrm{X}_0 = 0$, be a symmetric RW on $\mathbb{R}^d$. The random walk X is transient if, and only if,*

$$\exists \epsilon > 0 : \quad \int_{\mathrm{B}_\epsilon(0)} \frac{d\eta}{1 - \check{\mu}(\eta)} < \infty. \tag{57.3}$$

Proof. We compare (57.1) and (57.3): For every $\epsilon > 0$ this gives

$$\begin{aligned}
\sup_{r<1} \int_{\mathrm{B}_\epsilon(0)} \operatorname{Re} \frac{1}{1 - r\check{\mu}(\eta)}\, d\eta &\overset{\text{symm.}}{\underset{\check{\mu}\in\mathbb{R}}{=}} \sup_{r<1} \int_{\mathrm{B}_\epsilon(0)} \frac{1}{1 - r\check{\mu}(\eta)}\, d\eta \\
&\overset{\text{BL}}{=} \int_{\mathrm{B}_\epsilon(0)} \frac{1}{1 - \check{\mu}(\eta)}\, d\eta.
\end{aligned}$$

□

Here are two further criteria which are useful for applications.

57.5 Corollary. *Let $X_t = \xi_1 + \cdots + \xi_t$, $X_0 = 0$, be a RW on $\mathbb{R}^d$ and let $\epsilon > 0$ be fixed.*

$$\int_{B_\epsilon(0)} \operatorname{Re} \frac{1}{1-\check{\mu}(\eta)}\, d\eta = \infty \implies \textit{recurrence;} \tag{57.4}$$

$$\int_{B_\epsilon(0)} \frac{1}{1-\operatorname{Re}\check{\mu}(\eta)}\, d\eta < \infty \implies \textit{transience.} \tag{57.5}$$

Proof. Proof of (57.4) We use (57.1). Let $r_n \uparrow 1$ be a sequence and apply Fatou's lemma

$$\begin{aligned}
\sup_{r<1} \int_{B_\epsilon(0)} \operatorname{Re} \frac{1}{1-r\check{\mu}(\eta)}\, d\eta &\geqslant \sup_{n\in\mathbb{N}} \int_{B_\epsilon(0)} \operatorname{Re} \frac{1}{1-r_n\check{\mu}(\eta)}\, d\eta \\
&\geqslant \liminf_{n\to\infty} \int_{B_\epsilon(0)} \operatorname{Re} \frac{1}{1-r_n\check{\mu}(\eta)}\, d\eta \\
&\geqslant \int_{B_\epsilon(0)} \liminf_{n\to\infty} \operatorname{Re} \frac{1}{1-r_n\check{\mu}(\eta)}\, d\eta \\
&= \int_{B_\epsilon(0)} \operatorname{Re} \frac{1}{1-\check{\mu}(\eta)}\, d\eta = \infty.
\end{aligned}$$

Proof of (57.5) Since $\eta \mapsto \operatorname{Re}\check{\mu}(\eta)$ is continuous and $\check{\mu}(0) = 1$, we know that $\operatorname{Re}\check{\mu}(\eta) \geqslant 0$ for all $\eta \in B_\epsilon(0)$ and some $\epsilon > 0$ ✍. Therefore, we get for $r < 1$

$$\begin{aligned}
\int_{B_\epsilon(0)} \operatorname{Re} \frac{1}{1-r\check{\mu}(\eta)}\, d\eta &\overset{\operatorname{Re}\frac{1}{z} \leqslant \frac{1}{\operatorname{Re} z}}{\leqslant} \int_{B_\epsilon(0)} \frac{1}{1-r\operatorname{Re}\check{\mu}(\eta)}\, d\eta \\
&\leqslant \int_{B_\epsilon(0)} \frac{1}{1-\operatorname{Re}\check{\mu}(\eta)}\, d\eta < \infty.
\end{aligned}$$

□

Let ξ_1 be an $\mathbb{R}^d$-valued random variable and ξ_1' be an independent copy of ξ_1. We call $\widetilde{\xi}_1 := \xi_1 - \xi_1'$ the **symmetrization** of ξ_1; if $\xi_1 \sim \mu$, then

$$\widetilde{\xi}_1 = \xi_1 - \xi_1' \sim \xi_1' - \xi_1 = -\widetilde{\xi}_1$$

as well as

$$\begin{aligned}
\mathbb{E}\left(e^{i\langle\eta,\widetilde{\xi}_1\rangle}\right) &= \mathbb{E}\left(e^{i\langle\eta,\xi_1\rangle} e^{-i\langle\eta,\xi_1'\rangle}\right) \\
&\overset{\perp\!\!\!\perp}{=} \mathbb{E}\left(e^{i\langle\eta,\xi_1\rangle}\right) \cdot \mathbb{E}\left(e^{-i\langle\eta,\xi_1'\rangle}\right) \\
&\overset{\xi_1\sim\xi_1'}{=} \mathbb{E}\left(e^{i\langle\eta,\xi_1\rangle}\right) \cdot \mathbb{E}\left(e^{-i\langle\eta,\xi_1\rangle}\right) \\
&= \check{\mu}(\eta) \cdot \overline{\check{\mu}(\eta)} = |\check{\mu}(\eta)|^2.
\end{aligned}$$

57.6 Corollary. *Let $X_0 = 0$, $X_t = \xi_1 + \cdots + \xi_t$, be a random walk on $\mathbb{R}^d$, $X'_0 = 0$, $X'_t := \xi'_1 + \cdots + \xi'_t$, an independent copy, and $\widetilde{X}_t := X_t - X'_t$ the symmetrization. Then*

$$X_t \text{ recurrent} \implies \widetilde{X}_t \text{ recurrent},$$

or, equivalently,

$$\widetilde{X}_t \text{ transient} \implies X_t \text{ transient}.$$

Proof. 1° Define $S_t := X_{2t}$. One step of this RW is $S_1 = \xi_1 + \xi_2$, and we see that

$$\mathbb{E}e^{i\langle\eta, S_1\rangle} = \mathbb{E}e^{i\langle\eta,\xi_1\rangle}e^{i\langle\eta,\xi_2\rangle} \overset{\text{iid}}{=} \mathbb{E}e^{i\langle\eta,\xi_1\rangle}\mathbb{E}e^{i\langle\eta,\xi_1\rangle} = \left(\check{\mu}(\eta)\right)^2.$$

2° The characteristic function of $\widetilde{X}_1 = \xi_1 - \xi'_1$ – this is the step of the RW $\widetilde{X}_t$ –, is

$$\mathbb{E}e^{i\langle\eta,\widetilde{X}_1\rangle} = \mathbb{E}e^{i\langle\eta,\xi_1\rangle}e^{-i\langle\eta,\xi'_1\rangle} \overset{\text{iid}}{=} \mathbb{E}e^{i\langle\eta,\xi_1\rangle}\overline{\mathbb{E}e^{i\langle\eta,\xi_1\rangle}} = \left|\check{\mu}(\eta)\right|^2.$$

3° We show $\widetilde{X}_t$ transient $\Rightarrow S_t = X_{2t}$ transient. Use (57.1) and estimate the integrands in the following way:

$$\underbrace{\operatorname{Re}\frac{1}{1 - r\check{\mu}^2(\eta)}}_{\text{integrand in (57.1) for } S_t = X_{2t}} \overset{\operatorname{Re}\frac{1}{z} \leqslant \frac{1}{\operatorname{Re} z}}{\leqslant} \frac{1}{1 - r\operatorname{Re}\check{\mu}^2(\eta)} \overset{\operatorname{Re} z \leqslant |z|}{\leqslant} \underbrace{\frac{1}{1 - r|\mu(\eta)|^2}}_{\text{integrand in (57.1) for } \widetilde{X}_t}.$$

Integrating this chain of inequalities over $B_\epsilon(0)$ shows that the right-hand side, hence the left-hand side, is finite, proving the claim.

4° $\quad X_{2t}$ transient $\overset{\S 56.7}{\Longrightarrow} |X_{2t}| \xrightarrow[t\to\infty]{\text{a.s.}} \infty \Longrightarrow |X_{2t+1}| \xrightarrow[t\to\infty]{\text{a.s.}} \infty.$

The last implication uses that the shifted random walk $X_{2t+1} - X_1$ behaves (stochastically) like X_{2t} and that $\mathbb{P}(|X_1| < \infty) = 1$. Since both $|X_{2t}| \to \infty$ and $|X_{2t+1}| \to \infty$, we conclude that $|X_t| \to \infty$ (✍, this is a purely deterministic argument) and we see with Theorem 56.7 that $(X_t)_{t\in\mathbb{N}_0}$ is transient. □

We will finally show that any random walk is transient, if it is **genuinely** d-dimensional for some $d \geqslant 3$. We have to exclude that a RW lives in a $(d-1)$-dimensional subspace. This is the rationale of the next definition.

57.7 Definition. A random variable ξ_1 with values in $\mathbb{R}^d$ is called **genuinely** d**-dimensional**, if

$$\forall \theta \neq 0: \quad \mathbb{P}(\langle\theta, \xi_1\rangle \neq 0) > 0.$$

We need the following auxiliary estimate.

57.8 Lemma. *One has $\frac{1}{4}t^2 \leqslant 1 - \cos t \leqslant \frac{1}{2}t^2$ for all $|t| \leqslant 1$.*

Proof. Since $t \mapsto \sin(t)/t$ is even, it is enough to consider $0 \leqslant t \leqslant 1$. Because of the concavity of $\sin t$ on $[0,1]$ (cf. Fig. 57.3) we see that

$$\frac{\sin 1}{1} \leqslant \frac{\sin t}{t} \leqslant 1 \iff \sin 1 \cdot t \leqslant \sin t \leqslant t$$
$$\implies \sin 1 \int t\,dt \leqslant \int \sin t\,dt \leqslant \int t\,dt.$$

This implies $\frac{\sin 1}{2} t^2 \leqslant 1 - \cos t \leqslant \frac{1}{2} t^2$ and the assertion follows. □

57.9 Theorem. *Let $d \geqslant 3$. Every genuinely d-dimensional RW (i.e. $\xi_1 \sim \mu$ is genuinely d-dimensional) is transient.*

Proof. 1° We use (57.5), i.e. we have to show that for some $\epsilon > 0$

$$\int_{B_\epsilon(0)} \frac{1}{1 - \operatorname{Re} \check{\mu}(\eta)}\, d\eta < \infty.$$

To do so, we first bound the denominator of the integrand. Using that μ is a probability measure, we find

$$\begin{aligned}
1 - \operatorname{Re} \check{\mu}(\eta) &= \int \big(1 - \cos\langle y, \eta\rangle\big)\, \mu(dy) \\
&\geqslant \frac{1}{4} \int_{|\langle y, \eta\rangle| \leqslant 1} |\langle y, \eta\rangle|^2\, \mu(dy) \\
&\underset{\theta \in \mathbb{S}^{d-1}}{\overset{\eta = r\theta}{\geqslant}} \frac{1}{4} r^2 \int_{|\langle y, \theta\rangle| \leqslant 1/r} |\langle y, \theta\rangle|^2\, \mu(dy) \\
&\geqslant \frac{C}{4} r^2 \qquad \forall r \leqslant r_0.
\end{aligned}$$

The bound $C = \inf_{\theta \in \mathbb{S}^{d-1}} \inf_{r \leqslant r_0} \int_{|\langle y, \theta\rangle| \leqslant 1/r} |\langle y, \theta\rangle|^2\, \mu(dy) > 0$ will be shown in the last step.

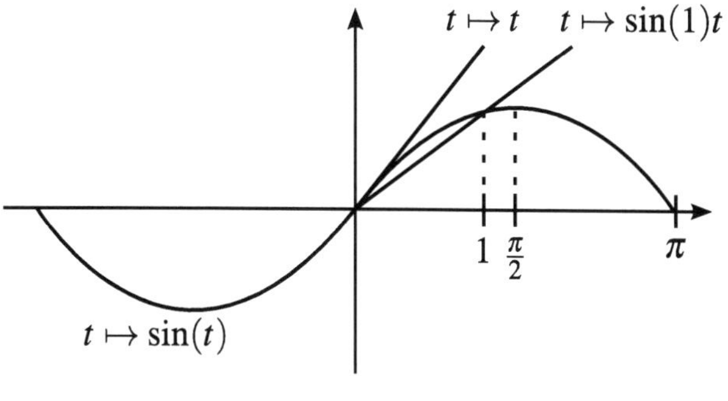

Fig. 57.3. For all $t \in [0,1]$ we have $\sin 1 \cdot t \leqslant \sin t \leqslant t$.

2° If we use spherical coordinates in (57.5) and bound the Jacobian by r^{d-1}, we see that

$$\begin{aligned}\int_{B_\epsilon(0)} \frac{1}{1-\operatorname{Re}\check{\mu}(\eta)}\,d\eta &\leqslant \int_0^\epsilon \int_{\mathbb{S}^{d-1}} \frac{1}{1-\operatorname{Re}\check{\mu}(r\theta)}\,r^{d-1}\,dr\,d\theta \\ &\overset{1^\circ}{\leqslant} \int_0^\epsilon r^{d-1}\frac{4}{C}r^{-2}\,dr \int_{\mathbb{S}^{d-1}} d\theta \\ &= \frac{C'}{4}\int_0^\epsilon r^{d-3}\,dr < \infty.\end{aligned}$$

3° Assume that $C = 0$. Then there are sequences $r_n \downarrow 0$ and $\theta_n \in \mathbb{S}^{d-1}$ such that

$$\int_{|\langle y,\theta_n\rangle|\leqslant 1/r_n} |\langle y,\theta_n\rangle|^2\,\mu(dy) \leqslant \frac{1}{n}.$$

Since $\mathbb{S}^{d-1}$ is compact, we may assume that $\theta_n \to \theta \in \mathbb{S}^{d-1}$, otherwise we could extract a suitable subsequence. By Fatou's lemma,

$$\begin{aligned}0 &= \liminf_{n\to\infty} \int_{|\langle y,\theta_n\rangle|\leqslant \frac{1}{r_n}} |\langle y,\theta_n\rangle|^2\,\mu(dy) \\ &\geqslant \int \liminf_{n\to\infty}\left(\mathbb{1}_{\{|\langle y,\theta_n\rangle|\leqslant \frac{1}{r_n}\}}|\langle y,\theta_n\rangle|^2\right)\mu(dy) \\ &= \int |\langle y,\theta\rangle|^2\,\mu(dy) > 0\end{aligned}$$

since ξ_1 is, by assumption, genuinely d-dimensional. Since this is a contradiction, we have $C > 0$. □

57.10 Remark. In fact, the criterium (57.4) is *necessary and sufficient*, i.e. the following **Spitzer test** holds:

$$\exists \epsilon > 0 : \quad \int_{B_\epsilon(0)} \operatorname{Re}\frac{1}{1-\check{\mu}(\eta)}\,d\eta \begin{cases} = \infty & \Longrightarrow \text{Rekurrenz,} \\ < \infty & \Longrightarrow \text{Transienz.}\end{cases} \tag{57.4'}$$

This improvement of the Chung–Fuchs criterium (Theorem 56.7) is surprisingly difficult to prove. The »simple« direction (57.4) follows with Fatou's lemma. The converse (due to Ornstein and Stone, 1969) is significantly more difficult and it requires some new ideas and techniques. A proof can be found in the – not easily accessible – paper by Port & Stone [24].

XII
Brownian Motion

58 First steps towards Brownian motion

Let $S_n := X_1 + \ldots + X_n$ be a symmetric random walk (RW), i.e. the steps $(X_k)_{k\in\mathbb{N}}$ are iid symmetric random variables; we assume that $\mathbb{E}X_1 = 0$ and $\mathbb{V}X_1 = \sigma^2 > 0$.

We interpret $n = 1, 2, \ldots$ as units of »time«, i.e. we have one step per time unit. Let us scale the time and make ever more steps per unit of time. To do so, we interpolate the RW

$$\xi_t^n := S_{\lfloor nt\rfloor} + (nt - \lfloor nt\rfloor)X_{\lfloor nt\rfloor+1}, \quad t \in [0,1];$$

$\lfloor x\rfloor = \max\{k \in \mathbb{Z} \mid k \leqslant x\}$ denotes the integer part of $x \geqslant 0$.

If we scale only time, but not the position, the limit will not exist. In fact,

$$\begin{aligned}
\mathbb{E}\xi_t^n &= \mathbb{E}S_{\lfloor nt\rfloor} + (nt - \lfloor nt\rfloor)\mathbb{E}X_{\lfloor nt\rfloor+1}\\
\mathbb{V}\xi_t^n &= \mathbb{V}S_{\lfloor nt\rfloor} + (nt - \lfloor nt\rfloor)^2\mathbb{V}X_{\lfloor nt\rfloor+1}\\
&= \lfloor nt\rfloor\sigma^2 + \underbrace{(nt - \lfloor nt\rfloor)^2}_{\leqslant 1}\sigma^2.
\end{aligned}$$

The last few lines indicate that we should scale the position with the factor $1/\sqrt{n}$

$$B_t^n := \frac{1}{\sqrt{n}}\,\xi_t^n = \frac{1}{\sqrt{n}}\left(S_{\lfloor nt\rfloor} + (nt - \lfloor nt\rfloor)X_{\lfloor nt\rfloor+1}\right).$$

Thus,

$$\mathbb{V}B_t^n = \left(\frac{\lfloor nt\rfloor}{n} + \frac{(nt - \lfloor nt\rfloor)^2}{n}\right)\sigma^2 \xrightarrow[n\to\infty]{} t\sigma^2.$$

Setting $t = i/n$, we get

$$B_{i/n}^n = \frac{1}{\sqrt{n}}\,S_i$$

and choosing $i = i(n)$ in such a way that $i/n = t = \text{const.}$, yields

$$B_{i/n}^n = \sqrt{t}\,\frac{S_i}{\sqrt{i}} \xrightarrow[n\to\infty]{\text{CLT}} \sqrt{t}\,G \quad \text{for} \quad G \sim N(0,\sigma^2).$$

In fact, the convergence holds for the finite-dimensional distributions ✍ and even for the whole path $t \mapsto B_t^n$; the last property is quite deep, a proof can be found in Schilling/Partzsch [BM, Theorem 14.5]:

58.1 Theorem (invariance theorem; Donsker 1951). *Let $(B_t^n)_{t\in[0,1]}$ be as above, $\sigma^2 = 1$ and $\Phi : (C([0,1],\mathbb{R}), \|\cdot\|_\infty) \to \mathbb{R}$ be any uniformly continuous and bounded functional. Then*

$$\lim_{n\to\infty} \mathbb{E}\Phi(B_\bullet^n) = \mathbb{E}\Phi(B_\bullet)$$

where $B = (B_t)_{t\in[0,1]}$ is a ***Brownian motion****.*

58.2 Definition. Let $(\Omega, \mathscr{A}, \mathbb{P})$ be any probability space. A (**d-dimensional**) **Brownian motion** is a stochastic process $B = (B_t)_{t\geqslant 0}$ indexed by $[0,\infty)$ taking values in $\mathbb{R}^d$ such that for $0 \leqslant s < t < \infty$

$$B_0(\omega) = 0 \quad \forall \omega \in \Omega; \tag{B0}$$

$$B_t - B_s \sim B_{t-s} - B_0; \tag{B1}$$

$$B_t - B_s \perp\!\!\!\perp \mathscr{F}_s^B := \sigma(B_r, r \leqslant s); \tag{B2}$$

$$B_t - B_s \sim \mathsf{N}(0, t-s)^{\otimes d}; \tag{B3}$$

$$t \mapsto B_t(\omega) \quad \text{is continuous } \forall \omega \in \Omega. \tag{B4}$$

Notation: We use BM [BM($\mathbb{R}^d$)] to denote a [d-dimensional] Brownian motion.

💬 $B = (B_t)_t$ is also called a **Brownian motion**, if

- the index set is not $[0,\infty)$ but $[0,T)$ or $[0,T]$;
- the starting point is not 0 but x;
- we replace the σ-algebras $\mathscr{F}_t^B$ by $\mathscr{F}_t \supset \mathscr{F}_t^B$. Note that
  $$\forall t \geqslant 0 : \mathscr{F}_t \supset \mathscr{F}_t^B \iff \forall t \geqslant 0 : B_t \text{ is } \mathscr{F}_t\text{-measurable}.$$
  The last property means that $(B_t)_{t\geqslant 0}$ is **adepted** to $(\mathscr{F}_t)_{t\geqslant 0}$. We call $(B_t, \mathscr{F}_t)_t$ a **Brownian motion with filtration**.

💬 We often write $B(t,\omega)$ and $B(t)$ instead of $B_t(\omega)$ and B_t.

💬 B is a Lévy process, cf. Definition 48.6.

⚠ We can change the »$\forall\omega \in \Omega$« in the properties (B0), (B4) to »for $\mathbb{P}$-almost all $\omega \in \Omega$«. In this case, we should make sure that all (measurable) null sets are contained in $\mathscr{F}_0$.

58.3 Remark. A one-dimensional Brownian motion BM($\mathbb{R}$) has the following properties

a) $\mathbb{E}B_t = 0$ and $\mathbb{V}B_t = t$ and $\operatorname{Cov}(B_s, B_t) = \mathbb{E}B_tB_s = \min(s,t)$;
b) $\mathbb{E}\exp(i\xi B_t) = \exp(-\frac{t}{2}\xi^2)$;
c) $t \mapsto B_t$ ist extremely irregular;

d) The property (B2) is equivalent to the so-called **independent increments** property (the proof is identical to the proof of Lemma 49.1):

$$\forall n \in \mathbb{N},\ 0 = t_0 < t_1 < \cdots < t_n\ :\ (\mathrm{B}(t_i) - \mathrm{B}(t_{i-1}))_{i=1,\dots,n} \text{ independent.} \qquad (\mathrm{B}2')$$

59 Existence of Brownian motion

We want to show that $\mathrm{BM}(\mathbb{R})$ and $\mathrm{BM}(\mathbb{R}^d)$ exist. We begin with a few preparations.

59.1 Lemma. *Let $G \sim \mathsf{N}(m, C)$ be a d-dimensional Gaussian vector and $A \in \mathbb{R}^{d\times d}$. Then $AG \sim \mathsf{N}(Am, ACA^\top)$.*

Proof. We have by Theorem 36.2

$$\mathbb{E}e^{i\langle\xi, AG\rangle} = \mathbb{E}e^{i\langle A^\top\xi, G\rangle} = e^{i\langle A^\top\xi, m\rangle - \frac{1}{2}\langle A^\top\xi, CA^\top\xi\rangle} = e^{i\langle\xi, Am\rangle - \frac{1}{2}\langle\xi, ACA^\top\xi\rangle}. \qquad \square$$

59.2 Lemma. *Let $X, \Gamma \sim \mathsf{N}(0, t)$. Then*

$$X \perp\!\!\!\perp \Gamma \implies X - \Gamma \perp\!\!\!\perp X + \Gamma \quad \textit{and} \quad X \pm \Gamma \sim \mathsf{N}(0, 2t).$$

Proof. We have

$$X \perp\!\!\!\perp \Gamma \implies \begin{pmatrix} X \\ \Gamma \end{pmatrix} \sim \text{normal} \overset{\S 59.1}{\implies} \begin{pmatrix} X - \Gamma \\ X + \Gamma \end{pmatrix} = \begin{pmatrix} 1 & -1 \\ 1 & 1 \end{pmatrix} \begin{pmatrix} X \\ \Gamma \end{pmatrix} \sim \text{normal.}$$

Thus, $X \pm \Gamma$ is normal and

$$\mathbb{E}(X \pm \Gamma) = \mathbb{E}X \pm \mathbb{E}\Gamma = 0 \quad \text{and} \quad \mathbb{V}(X \pm \Gamma) \overset{X \perp\!\!\!\perp \Gamma}{=} \mathbb{V}X + \mathbb{V}\Gamma = 2t.$$

By Corollary 36.3

$$\operatorname{Cov}(X - \Gamma, X + \Gamma) = \mathbb{E}(X - \Gamma)(X + \Gamma) = \mathbb{E}(X^2 - \Gamma^2) = t - t = 0$$

implies that $X - \Gamma \perp\!\!\!\perp X + \Gamma$. $\square$

59.3 Lemma. *Assume that X', X'', Z are rv with $X' \perp\!\!\!\perp X''$ and $(X', X'') \perp\!\!\!\perp Z$. Then X', X'', Z are independent.*

Proof. We use Kac's theorem (Corollary 27.8). Fix $\xi, \eta, \zeta \in \mathbb{R}^d$. Then

$$\begin{aligned} \mathbb{E}e^{i\xi X' + i\eta X'' + i\zeta Z} &\overset{(X', X'') \perp\!\!\!\perp Z}{=} \mathbb{E}e^{i\xi X' + i\eta X''}\,\mathbb{E}e^{i\zeta Z} \\ &\overset{X' \perp\!\!\!\perp X''}{=} \mathbb{E}e^{i\xi X'}\,\mathbb{E}e^{i\eta X''}\,\mathbb{E}e^{i\zeta Z}. \end{aligned} \qquad \square$$

59.4 Lemma. *Let* $G_n \sim N(0, t_n)$. *Then*

$$G_n \xrightarrow[n\to\infty]{d} G \iff t_n \xrightarrow[n\to\infty]{} t;$$

in this case $G \sim N(0, t)$.

Proof. We have for all $\xi \in \mathbb{R}$

$$G_n \xrightarrow{d} G \iff \mathbb{E}e^{i\xi G_n} \to \mathbb{E}e^{i\xi G} \iff e^{-\frac{1}{2}t_n\xi^2} \to \mathbb{E}e^{i\xi G} \iff t_n \to t.$$

In particular, $\mathbb{E}e^{i\xi G} = e^{-\frac{1}{2}t\xi^2}$. □

59.5 Remark. We explain now the motivation for the construction of Brownian motion. Assume that $(B_t)_{t\geqslant 0}$ is a BM($\mathbb{R}$) and fix $0 \leqslant s < t < u$. Then

$$\begin{pmatrix} B_s \\ B_t \\ B_u \end{pmatrix} = \begin{pmatrix} 1 & 0 & 0 \\ 1 & 1 & 0 \\ 1 & 1 & 1 \end{pmatrix} \begin{pmatrix} B_s \\ B_t - B_s \\ B_u - B_t \end{pmatrix}$$

is Gaussian, as $(B_s, B_t - B_s, B_u - B_t)$ is Gaussian (Example 58.3.d) and Lemma 59.1). Now we can use Theorem 36.4 and we see

$$\mathbb{P}(B_t \in \cdot \mid B_s = x, B_u = y) = N(m, \sigma^2),$$
$$m = \frac{(u-t)x + (t-s)y}{u-s}, \quad \sigma^2 = \frac{(u-t)(t-s)}{u-s}.$$

An alternative direct proof of this is given in [BM, Section 3.4, pp. 31–32] or [BM, Problem 3.8, p. 37]. This relation is best understood in the form of a sketch (Fig. 59.1).

This leads to the following algorithm for the construction of a Brownian path $W(t)$ for $t \in [0,1]$:

1° Pick $W(0) = 0$ and $W(1) \sim N(0,1)$.

2° Assume that we have already constructed $W(k2^{-n})$, $k = 0, 1, \dots, 2^n$ such that

$$W(k2^{-n}) - W((k-1)2^{-n}) \sim N(0, 2^{-n}) \text{ are iid rvs.}$$

3° Pick iid $\Gamma_{2^n+k} \sim N(0, \frac{1}{4}2^{-n})$, $k = 0, 1, \dots, 2^n - 1$ and set

$$W(\ell 2^{-(n+1)}) := \begin{cases} W(k2^{-n}) & \text{if } \ell = 2k, \\ \frac{1}{2}(W(k2^{-n}) + W((k+1)2^{-n})) + \Gamma_{2^n+k} & \text{if } \ell = 2k+1. \end{cases}$$

4° Define

$$W_{2^n}(t, \omega) := \text{linear interpolation of } (W(k2^{-n}, \omega), k = 0, 1, \dots, 2^n).$$

Note that the refined increments $W(\ell 2^{-(n+1)}) - W((\ell-1)2^{-(n+1)})$ are again independent. This can be seen from the following argument:

- Fix n. By assumption $W(k2^{-n}) - W((k-1)2^{-n})$, $k = 1, \dots, 2^n$ are independent.

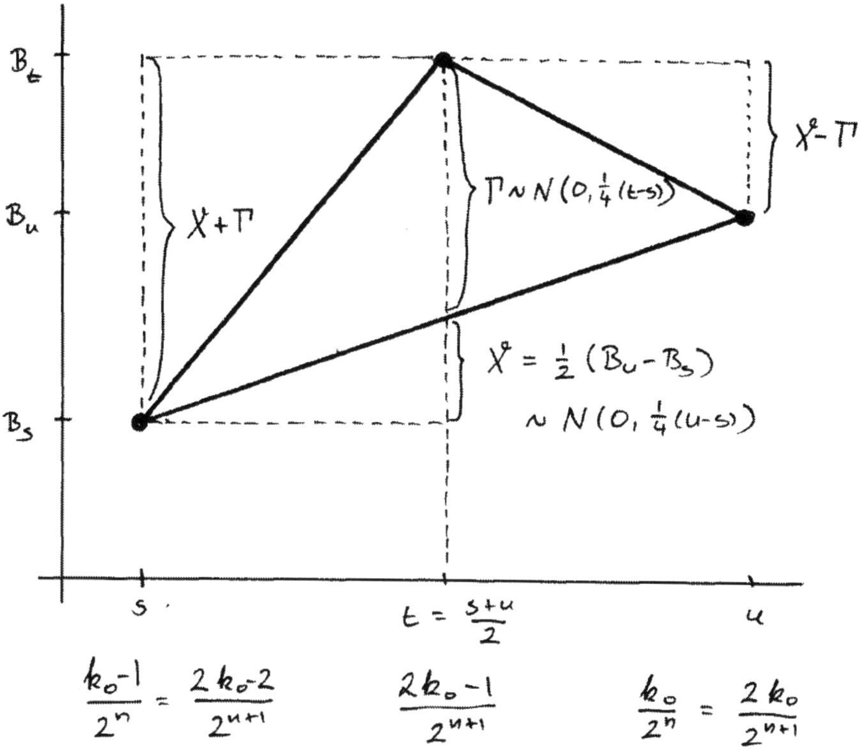

Fig. 59.1. Γ is independent of B_s, B_t

- Pick k_0 and $\Gamma = \Gamma_{2^n+k_0} \sim \mathsf{N}(0, \frac{1}{4}2^{-n})$ independent of everything else and observe that

$$X := \frac{1}{2}(W(k_0 2^{-n}) - W((k_0-1)2^{-n})) \sim \mathsf{N}(0, \tfrac{1}{4}2^{-n}) \overset{\S 59.2}{\Longrightarrow} X+\Gamma \perp\!\!\!\perp X-\Gamma$$

- Since $X \pm \Gamma \perp\!\!\!\perp W(k2^{-n}) - W((k-1)2^{-n}), k \neq k_0$, we get from Lemma 59.3

$$\left.\begin{aligned} &X+\Gamma = W((2k_0-1)2^{-(n+1)}) - W((2k_0-2)2^{-(n+1)}),\\ &X-\Gamma = W(2k_0 2^{-(n+1)}) - W((2k_0-1)2^{-(n+1)}),\\ &W(k2^{-n}) - W((k-1)2^{-n}),\ \forall k \neq k_0 \end{aligned}\right\}\quad \text{are independent}$$

- now repeat this with a different k_0.

59.6 Theorem (Lévy 1940). *The series*

$$W(t,\omega) := \sum_{n=0}^{\infty}\Big(W_{2^{n+1}}(t,\omega) - W_{2^n}(t,\omega)\Big) + W_1(t,\omega), \quad t \in [0,1],$$

converges a.s. uniformly. In particular $(W(t))_{t\in[0,1]}$ *is a* $\mathrm{BM}(\mathbb{R})$.

Proof. Set $\Delta_n(t,\omega) := W_{2^{n+1}}(t,\omega) - W_{2^n}(t,\omega)$. By construction,

$$\Delta_n\big((2k-1)2^{-n-1},\omega\big) = \Gamma_{2^n+(k-1)}(\omega) \sim \mathsf{N}(0,\tfrac{1}{4}2^{-n}) \text{ iid} \quad k=1,2,\dots,2^n.$$

Thus,

$$\begin{aligned}
&\mathbb{P}\left(\max_{1\leqslant k\leqslant 2^n}\left|\Delta_n\big((2k-1)2^{-n-1}\big)\right| > \frac{x_n}{\sqrt{2^{n+2}}}\right)\\
&\leqslant 2^n\mathbb{P}\left(\left|\sqrt{2^{n+2}}\,\Delta_n\big(2^{-n-1}\big)\right| > x_n\right)\\
&= \frac{2\cdot 2^n}{\sqrt{2\pi}}\int_{x_n}^{\infty} e^{-r^2/2}\,dr\\
&\leqslant \frac{2^{n+1}}{\sqrt{2\pi}}\int_{x_n}^{\infty}\frac{r}{x_n}e^{-r^2/2}\,dr\\
&= \frac{2^{n+1}}{x_n\sqrt{2\pi}}e^{-x_n^2/2}.
\end{aligned}$$

Choose $c>1$ and $x_n := c\sqrt{2n\log 2}$. Then

$$\begin{aligned}
\sum_{n=1}^{\infty}\mathbb{P}\left(\max_{1\leqslant k\leqslant 2^n}\left|\Delta_n\big((2k-1)2^{-n-1}\big)\right| > \frac{x_n}{\sqrt{2^{n+2}}}\right) &\leqslant \sum_{n=1}^{\infty}\frac{2^{n+1}}{c\sqrt{2\pi}}e^{-c^2\log 2^n}\\
&= \frac{2}{c\sqrt{2\pi}}\sum_{n=1}^{\infty}2^{-(c^2-1)n} < \infty.
\end{aligned}$$

By Borel–Cantelli (Lemma 29.6 or Theorem 31.1),

$$\exists\Omega_0\subset\Omega,\ \mathbb{P}(\Omega_0)=1 \quad \forall\omega\in\Omega_0 \quad \exists N(\omega)\in\mathbb{N} \quad \forall n\geqslant N(\omega):$$
$$\max_{1\leqslant k\leqslant 2^n}\left|\Delta_n\big((2k-1)2^{-n-1}\big)\right| \leqslant c\sqrt{\frac{n\log 2}{2^{n+1}}}.$$

$\Delta_n(t)$ is the distance between the polygonal arcs $W_{2^{n+1}}(t)$ and $W_{2^n}(t)$; clearly, the maximum is attained at one of the midpoints of the intervals $[(k-1)2^{-n}, k2^{-n}]$, $k=1,\dots,2^n$, see Figure 59.1. Thus

$$\sup_{0\leqslant t\leqslant 1}\left|W_{2^{n+1}}(t,\omega) - W_{2^n}(t,\omega)\right| \leqslant \max_{1\leqslant k\leqslant 2^n}\left|\Delta_n\big((2k-1)2^{-n-1},\omega\big)\right| \leqslant c\sqrt{\frac{n\log 2}{2^{n+1}}}$$

and so

$$W(t,\omega) := \lim_{N\to\infty} W_{2^N}(t,\omega) = \sum_{n=0}^{\infty}\Big(W_{2^{n+1}}(t,\omega) - W_{2^n}(t,\omega)\Big) + W_1(t,\omega)$$

exists for all $\omega\in\Omega_0$ uniformly in $t\in[0,1]$.

Therefore, $t\mapsto W(t,\omega)$, $\omega\in\Omega_0$, inherits the continuity of the polygonal arcs $t\mapsto W_{2^n}(t,\omega)$. Set

$$\widetilde{W}(t,\omega) := \begin{cases} W(t,\omega), & \omega\in\Omega_0,\\ 0, & \omega\notin\Omega_0.\end{cases}$$

By construction, we find for all $0 \leqslant i \leqslant k \leqslant 2^n$, $\mathbb{P}$-almost surely

$$\begin{aligned}\widetilde{W}(k2^{-n}) - \widetilde{W}(i2^{-n}) &= W_{2^n}(k2^{-n}) - W_{2^n}(i2^{-n}) \\ &= \sum_{\ell=i+1}^{k} \left(W_{2^n}(\ell 2^{-n}) - W_{2^n}((\ell-1)2^{-n})\right) \overset{\text{iid}}{\sim} \mathsf{N}(0,(k-i)2^{-n}).\end{aligned}$$

Since $t \mapsto \widetilde{W}(t)$ is continuous and the dyadic numbers are dense in $[0,1]$, we conclude that the increments $\widetilde{W}(t_i) - \widetilde{W}(t_{i-1})$, $0 = t_0 < t_1 < \cdots < t_N \leqslant 1$ are independent $\mathsf{N}(0, t_i - t_{i-1})$ distributed random variables (Lemma 59.4; independence is preserved under $\mathbb{P}$-limits). This shows that $(\widetilde{W}(t))_{t\in[0,1]}$ is a BM. □

Now it is easy to construct a real-valued Brownian motion for all $t \geqslant 0$.

59.7 Corollary. *Let $(W^k(t))_{t\in[0,1]}$, $k \in \mathbb{N}_0$, be independent* BM($\mathbb{R}$) *on the same probability space. Then*

$$B(t) := \begin{cases} W^0(t), & t \in [0,1); \\ W^0(1) + W^1(t-1), & t \in [1,2); \\ \sum_{i=0}^{k-1} W^i(1) + W^k(t-k), & t \in [k,k+1),\ k \geqslant 2; \end{cases} \tag{59.1}$$

is a BM($\mathbb{R}$) *indexed by $t \in [0,\infty)$.*

Proof. Let B^k be a copy of the process W from Theorem 59.6 on the probability space $(\Omega_k, \mathscr{A}_k, \mathbb{P}_k)$. Define, on the product space $(\Omega, \mathscr{A}, \mathbb{P}) = \bigotimes_{k\geqslant 0}(\Omega_k, \mathscr{A}_k, \mathbb{P}_k)$, $\omega = (\omega_0, \omega_1, \omega_2, \ldots)$, processes $W^k(\omega) := B^k(\omega_k)$; because of the product construction, these are independent Brownian motions on $(\Omega, \mathscr{A}, \mathbb{P})$.

Let us check (B0)–(B4) for the process $(B(t))_{t\geqslant 0}$ defined by (59.1). The properties (B0) and (B4) are obvious. Let $s < t$ such that $s \in [\ell, \ell+1), t \in [m, m+1)$ where $\ell \leqslant m$. Then

$$\begin{aligned}B(t) - B(s) &= \sum_{i=0}^{m-1} W^i(1) + W^m(t-m) - \sum_{i=0}^{\ell-1} W^i(1) - W^\ell(s-\ell) \\ &= \sum_{i=\ell}^{m-1} W^i(1) + W^m(t-m) - W^\ell(s-\ell) \\ &= \begin{cases} W^m(t-m) - W^m(s-m) \sim \mathsf{N}(0, t-s), & \ell = m, \\ \underbrace{\left(W^\ell(1) - W^\ell(s-\ell)\right) + \sum_{i=\ell+1}^{m-1} W^i(1) + W^m(t-m)}, & \ell < m. \end{cases} \\ &\overset{\perp\!\!\!\perp}{\sim} \mathsf{N}(0, 1-s+\ell) \star \mathsf{N}(0,1)^{\star m-\ell-1} \star \mathsf{N}(0, t-m) = \mathsf{N}(0, t-s)\end{aligned}$$

This proves (B1). We show (B2) only for two increments $B(u)-B(t)$ and $B(t)-B(s)$ where $s<t<u$. As before, $s\in[\ell,\ell+1), t\in[m,m+1)$ and $u\in[n,n+1)$ where $\ell\leqslant m\leqslant n$. Then

$$B(u)-B(t)=\begin{cases} W^m(u-m)-W^m(t-m), & m=n,\\ \left(W^m(1)-W^m(t-m)\right)+\sum_{i=m+1}^{n-1}W^i(1)+W^n(u-n), & m<n.\end{cases}$$

By assumption, the random variables appearing in the representation of the increments $B(u)-B(t)$ and $B(t)-B(s)$ are independent which means that the two increments are independent. The case of finitely many, not necessarily adjacent, increments is similar. □

We want to construct $BM(\mathbb{R}^d)$. For this the following characterization of BM is useful.

59.8 Theorem. *Let $(W_t)_{t\geqslant 0}$, $W_0=0$, be a stochastic process which is adapted to the filtration $(\mathscr{F}_t)_{t\geqslant 0}$. W is a $BM(\mathbb{R}^d)$ if, and only if, it has continuous paths and*

$$\forall\xi\in\mathbb{R}^d,\ s<t:\quad \mathbb{E}\left[e^{i\langle\xi,W_t-W_s\rangle}\mid\mathscr{F}_s\right]=e^{-\frac12(t-s)|\xi|^2}. \tag{59.2}$$

Proof. Assume that (59.2) holds. Take expectations in (59.2) and set $s=0$ to get

$$\mathbb{E}e^{i\langle\xi,W_t-W_s\rangle}=e^{-\frac12(t-s)|\xi|^2}\quad\text{and}\quad \mathbb{E}e^{i\langle\xi,W_t\rangle}=e^{-\frac12 t|\xi|^2}. \tag{59.3}$$

This shows that $W_t-W_s\sim\mathsf{N}(0,t-s)^{\otimes d}$, i.e. (B3), (B1). Moreover, (59.2) reads

$$\forall\xi\in\mathbb{R}^d,\ s<t:\ \mathbb{E}\left[e^{i\langle\xi,W_t-W_s\rangle}\mid\mathscr{F}_s\right]=\mathbb{E}e^{i\langle\xi,W_t-W_s\rangle}. \tag{59.4}$$

Let $F\in\mathscr{F}_s$, multiply (59.4) with $\mathbb{1}_F$ and take expectations. This gives

$$\begin{aligned}\mathbb{E}\left(\mathbb{1}_F e^{i\langle\xi,W_t-W_s\rangle}\right) &= \mathbb{E}\left(\mathbb{E}\left[\mathbb{1}_F e^{i\langle\xi,W_t-W_s\rangle}\mid\mathscr{F}_s\right]\right)\\ &\overset{\text{pull}}{=} \mathbb{E}\left(\mathbb{1}_F\mathbb{E}\left[e^{i\langle\xi,W_t-W_s\rangle}\mid\mathscr{F}_s\right]\right)\\ &\overset{(59.4)}{=} \mathbb{E}e^{i\langle\xi,W_t-W_s\rangle}\cdot\mathbb{E}\mathbb{1}_F.\end{aligned}$$

From this we conclude that $W_t-W_s\perp\!\!\!\perp\mathscr{F}_s$ (Example 27.9.a)), hence (B2).

Since (B0), (B4) are assumed, it follows that W is a BM.

Conversely, let W be a BM for the filtration $\mathscr{F}_t$. Then

$$\mathbb{E}\left[e^{i\langle\xi,W_t-W_s\rangle}\mid\mathscr{F}_s\right]\overset{(B2)}{=}\mathbb{E}\left[e^{i\langle\xi,W_t-W_s\rangle}\right]\overset{(B1)}{=}\mathbb{E}\left[e^{i\langle\xi,W_{t-s}\rangle}\right]\overset{(B3)}{=}e^{-\frac12(t-s)|\xi|^2}.$$ □

59.9 Corollary. a) *Let $(B_t,\mathscr{F}_t)_t$, $B_t=(B_t^1,\dots,B_t^d)$, be a $BM(\mathbb{R}^d)$. Then the coordinate processes B^i are $BM(\mathbb{R})$, which are independent for their natural filtrations.*

b) *Conversely, if* B^i, $i = 1\dots,d$ *are independent one-dimensional* BM($\mathbb{R}$), *then* $B = (B^1,\dots,B^d)$ *is a* BM($\mathbb{R}^d$).

Proof. a) Take in Theorem 59.8 $\xi = \theta e_i$ where e_i = ith unit vector. Then B^i is a BM($\mathbb{R}$) for the original filtration.

In order to see independence, we have to switch to the natural filtrations $\mathscr{F}_t^i := \sigma(B_s^i, s \leqslant t)$. Set $t_0 = 0$. We have to show that $\mathscr{F}_\infty^1,\dots,\mathscr{F}_\infty^d$ are independent. Since $\mathscr{F}_\infty^i$ is generated by the $\cap$-stable system $\mathscr{G}_\infty^i$ (as defined below), it is enough to show that

$$\mathscr{G}_\infty^i := \bigcup_{\substack{0\leqslant t_1<\dots<t_n<\infty\\ n\in\mathbb{N}}} \sigma\big((B_{t_1}^i,\dots,B_{t_n}^i)\big),\quad i=1,\dots,d,\quad \text{independent}$$

$$\iff \quad \forall t_1<\dots<t_n : \{(B_{t_1}^i,\dots,B_{t_n}^i), i=1,\dots,d\}\quad \text{independent}$$

$$\Longleftarrow \quad \forall t_1<\dots<t_n : \{(B_{t_1}^i - B_{t_0}^i,\dots,B_{t_n}^i - B_{t_{n-1}}^i), i=1,\dots,d\}\quad \text{independent}$$

$$\Longleftarrow \quad \forall t_1<\dots<t_n, \forall i=1,\dots,d : B_{t_k}^i - B_{t_{k-1}}^i \quad \text{independent.}$$

In order to see the independence of the increments (across time and dimension) we fix $\xi_k = (\xi_k^1,\dots,\xi_k^d) \in \mathbb{R}^d$ and $t_0 = 0 \leqslant t_1 < \dots < t_n$. Using (B2) for $(B_t)_{t\geqslant 0}$ we get

$$\begin{aligned}
\mathbb{E}\left[\exp\left(i\sum_{k=1}^n\sum_{j=1}^d \xi_k^j\big(B_{t_k}^j - B_{t_{k-1}}^j\big)\right)\right] &= \mathbb{E}\left[\exp\left(i\sum_{k=1}^n \langle \xi_k, B_{t_k} - B_{t_{k-1}}\rangle\right)\right]\\
&\overset{\text{(B2)}}{=} \prod_{k=1}^n \mathbb{E}\Big(\exp\big(i\langle \xi_k, B_{t_k} - B_{t_{k-1}}\rangle\big)\Big)\\
&= \prod_{k=1}^n e^{-\frac{1}{2}(t_k-t_{k-1})|\xi_k|^2}\\
&= \prod_{k=1}^n\prod_{j=1}^d e^{-\frac{1}{2}(t_k-t_{k-1})(\xi_k^j)^2}\\
&\overset{\text{(B3)}}{=} \prod_{k=1}^n\prod_{j=1}^d \mathbb{E}e^{i\xi_k^j(B_{t_k}^j - B_{t_{k-1}}^j)}.
\end{aligned}$$

b) For the converse note that $B_0 = 0$, $B_t - B_s \sim N(0, t-s)^{\otimes d}$, which gives stationary; continuity of $t \mapsto B_t$ is clear. Finally, for $s < t$

$$B_t^i - B_s^i \perp\!\!\!\perp \mathscr{F}_s^i \quad (\text{b/o BM}(\mathbb{R})),$$

$$B^k \perp\!\!\!\perp \mathscr{F}_\infty^i,\ i \neq k \quad (\text{b/o } B^1,\dots,B^d \text{ independent}),$$

and we see with a short calculation ✍ that $B_t - B_s \perp\!\!\!\perp \sigma(\mathscr{F}_s^1,\dots,\mathscr{F}_s^d)$. □

60 BM as a martingale

Throughout this chapter $B_t = B(t) = (B^1(t), \dots, B^d(t))$ is a $BM(\mathbb{R}^d)$ with filtration $\mathscr{F}_t$.

60.1 Example. Let $(B_t, \mathscr{F}_t)_{t \geqslant 0}$, be a $BM(\mathbb{R}^d)$ with filtration $\mathscr{F}_t$. In most of the examples below, the integrability and adaptedness are clear, and we do not check them.

a) $(B_t)_{t \geqslant 0}$ is a martingale with respect to $\mathscr{F}_t$.

Indeed: Let $0 \leqslant s \leqslant t$. Using (B2), we get

$$\mathbb{E}(B_t \mid \mathscr{F}_s) = \mathbb{E}(B_t - B_s \mid \mathscr{F}_s) + \mathbb{E}(B_s \mid \mathscr{F}_s) = \mathbb{E}(B_t - B_s) + B_s = B_s.$$

b) $M_t := |B_t|^2$ is a positive sub-martingale with respect to $\mathscr{F}_t$.

Indeed: For all $0 \leqslant s < t$ we can use the (conditional) Jensen inequality to get

$$\mathbb{E}(M_t \mid \mathscr{F}_s) = \sum_{i=1}^d \mathbb{E}((B_t^i)^2 \mid \mathscr{F}_s) \geqslant \sum_{i=1}^d \mathbb{E}(B_t^i \mid \mathscr{F}_s)^2 = \sum_{i=1}^d (B_s^i)^2 = M_s.$$

c) $M_t := |B_t|^2 - d \cdot t$ is a martingale with respect to $\mathscr{F}_t$.

Indeed: Since $|B_t|^2 - d \cdot t = \sum_{i=1}^d \left((B_t^i)^2 - t\right)$ it is enough to consider $d = 1$. For $s < t$ we see

$$\begin{aligned}
\mathbb{E}\left[B_t^2 - t \,\middle|\, \mathscr{F}_s\right] &= \mathbb{E}\left[\left((B_t - B_s) + B_s\right)^2 - t \,\middle|\, \mathscr{F}_s\right] \\
&= \mathbb{E}\left[(B_t - B_s)^2 + 2B_s(B_t - B_s) + B_s^2 - t \,\middle|\, \mathscr{F}_s\right] \\
&\overset{B_t - B_s \perp\!\!\!\perp \mathscr{F}_s}{=} \underbrace{\mathbb{E}\left[(B_t - B_s)^2\right]}_{=t-s} + 2B_s \underbrace{\mathbb{E}\left[B_t - B_s\right]}_{=0} + B_s^2 - t \\
&= B_s^2 - s.
\end{aligned}$$

💬 Let $d = 1$. The term t appearing in $B_t^2 - t$ has several meanings: It is the compensator (making the sub-martingale B_t^2 into a martingale, cf. Theorem 38.10 for the discrete case) as well as the quadratic variation (see Theorem 61.2). A famous theorem due to Lévy characterizes Brownian motion among all martingales with continuous paths: BM is the only L^2-martingale such that $B_t^2 - t$ is again a martingale, see [BM, Theorem 9.12, 18.5] or [25, Theorem IV.3.6].

d) $M^\xi(t) := e^{i\langle \xi, B(t)\rangle + \frac{t}{2}|\xi|^2}$ is for all $\xi \in \mathbb{R}^d$ a complex-valued martingale with respect to $\mathscr{F}_t$.

Indeed: For $0 \leqslant s < t$ we have, cf. Part a),

$$\begin{aligned}
\mathbb{E}(M^{\xi}(t) \mid \mathscr{F}_s) &= e^{\frac{t}{2}|\xi|^2}\, \mathbb{E}(e^{i\langle \xi, B(t)-B(s)\rangle} \mid \mathscr{F}_s)\, e^{i\langle \xi, B(s)\rangle} \\
&\overset{B_t - B_s \perp\!\!\!\perp \mathscr{F}_s}{\underset{B_t-B_s \sim B_{t-s}}{=}} e^{\frac{t}{2}|\xi|^2}\, \mathbb{E}(e^{i\langle \xi, B(t-s)\rangle}) \cdot e^{i\langle \xi, B(s)\rangle} \\
&= e^{\frac{t}{2}|\xi|^2}\, e^{-\frac{t-s}{2}|\xi|^2}\, e^{i\langle \xi, B(s)\rangle} \\
&= M^{\xi}(s).
\end{aligned}$$

e) $M^{\zeta}(t) := e^{\langle \zeta, B(t)\rangle - \frac{t}{2}\zeta^2}$, $\zeta^2 = \sum_{k=1}^{d} \zeta_k^2$, is for all $\zeta \in \mathbb{C}^d$ a complex-valued martingale with respect to $\mathscr{F}_t$.
Indeed: Since $\mathbb{E}e^{|\zeta||B(t)|} < \infty$ for all $\zeta \in \mathbb{C}^d$ ✎, we can use the argument of Part d).

We can learn a lot on the behaviour of BM using **stopping times**. Recall that

$$\tau : \Omega \to [0,\infty] \text{ stopping time} \iff \forall t : \{\tau \leqslant t\} \in \mathscr{F}_t.$$

As announced in Remark 46.13.g) we will now show that entrance and hitting times are stopping times.

60.2 Lemma. *Let $(X_t, \mathscr{F}_t)_{t \geqslant 0}$ be a BM($\mathbb{R}$) or any other adapted process with continuous paths (for all ω, otherwise we need that $\mathscr{F}_0$ contains all null sets).*

a) *The **first entrance time into a closed set** F $\tau_F^\circ := \inf\{t \geqslant 0 \mid X_t \in F\}$ is a stopping time.*

b) *The **first hitting time of an open set** U $\tau_U := \inf\{t > 0 \mid X_t \in U\}$ satisfies $\{\tau_U < t\} \in \mathscr{F}_t$ and $\{\tau_U \leqslant t\} \in \mathscr{F}_{t+}$.*

Proof. a) The distance $x \mapsto d(x, F) := \inf_{y \in F}|x - y|$ of x and F is (Lipschitz) continuous in x ✎. We want to show that

$$\left\{\tau_F^\circ \leqslant t\right\} = \left\{\omega \in \Omega \mid \inf_{r \in \mathbb{Q}\cap[0,t]} d(X_r(\omega), F) = 0\right\} =: \Omega_t.$$

Since $\Omega_t \in \mathscr{F}_t$, we see that τ_F° is a stopping time.
»⊃«: Assume that $\omega \in \Omega_t$. There is a sequence $(r_i)_{i \in \mathbb{N}} \subset \mathbb{Q}^+ \cap [0,t]$ and some $s \leqslant t$ such that

$$r_i \xrightarrow[i\to\infty]{} s, \quad d(X(r_i,\omega), F) \xrightarrow[i\to\infty]{} 0 \quad \text{and} \quad X(r_i,\omega) \xrightarrow[i\to\infty]{} X(s,\omega).$$

Since $d(x, F)$ is continuous, we get $d(X(s,\omega), F) = 0$, i.e. $X(s,\omega) \in F$, as F is closed. Thus, $\tau_F^\circ(\omega) \leqslant s \leqslant t$.
»⊂«: Let $\omega \notin \Omega_t$. Then

$$\inf_{\mathbb{Q}^+ \ni r \leqslant t} d(X(r,\omega), F) \geqslant \delta > 0 \overset{\text{cts.}}{\underset{\text{paths}}{\Longrightarrow}} X(s,\omega) \notin F,\ \forall s \leqslant t \overset{\text{F closed}}{\Longrightarrow} \tau_F^\circ(\omega) > t.$$

b) We claim for all $t > 0$

$$\left\{\tau_U < t\right\} = \bigcup_{\mathbb{Q}^+ \ni r < t} \underbrace{\left\{X(r) \in U\right\}}_{\in \mathscr{F}_r \subset \mathscr{F}_t} \in \mathscr{F}_t.$$

»⊂«: We have $\tau_U(\omega) < t \implies \exists s < t : X(s,\omega) \in U$. Since the paths are continuous[1] and U is open, we see that

$$\exists s < r < t,\ r \in \mathbb{Q} : X(r,\omega) \in U \implies \omega \in \bigcup_{\mathbb{Q}^+ \ni r < t} \{X(r) \in U\}.$$

»⊃«: Assume that $\omega \in \{X(r) \in U\}$ for some $r \in (0,t) \cap \mathbb{Q}$. Clearly, this shows that $\tau_U(\omega) \leqslant r < t$.

Finally,

$$\{\tau_U \leqslant t\} = \bigcap_{n \geqslant k} \underbrace{\left\{\tau_U < t + \tfrac{1}{n}\right\}}_{\in \mathscr{F}_{t+1/k}} \in \mathscr{F}_{t+\frac{1}{k}} \quad \text{for all } k \in \mathbb{N}$$

and this shows that $\{\tau_U \leqslant t\} \in \bigcap_{k=1}^{\infty} \mathscr{F}_{t+\frac{1}{k}} = \mathscr{F}_{t+}$. □

60.3 Example (first passage time). Let $(B_t)_{t \geqslant 0}$ be a BM($\mathbb{R}$) and $\tau_b := \tau^\circ_{\{b\}}$ the **first passage time**. We have

$$\sup_{t \geqslant 0} B_t \geqslant \sup_{n \geqslant 1} B_n = \sup_{n \geqslant 1}(X_1 + X_2 + \cdots + X_n)$$

where $X_i = B_i - B_{i-1}$ are iid standard normal rvs. Exactly as in Theorem 31.9 (cf. also Theorem 56.3) we get

$$S := \limsup_{n \to \infty}(X_1 + X_2 + \cdots + X_n) = X_1 + \limsup_{n \to \infty}(X_2 + \cdots + X_n) =: X_1 + S'.$$

Since the rvs X_i are iid, we know that $S \sim S'$ and $X_1 \perp\!\!\!\perp S'$. Since the random variable $X_1 \sim \mathsf{N}(0,1)$ is not trivial, we conclude exactly as in §31.9 that $S = \infty$, the only thing that changes is that we have to replace in (31.3) the c.f. $\cos\xi$ of a Bernoulli rv by the c.f. $e^{-\frac{1}{2}\xi^2}$ of the standard normal rv X_1. By symmetry, we see that $\liminf_{n \to \infty} X_n = -\infty$, and we get

$$\mathbb{P}\left(-\infty = \liminf_{t \to \infty} B_t < \limsup_{t \to \infty} B_t = +\infty\right) = 1, \tag{60.1}$$

hence

$$\mathbb{P}(\tau_b < \infty) = 1 \quad \forall b \in \mathbb{R}; \tag{60.2}$$

(you should compare this argument with Theorem 31.9 and 56.3).

The continuity of the sample paths and (60.1) imply that the (random) set $\{t \geqslant 0 \mid B_t(\omega) = b\}$ is a.s. unbounded.

In particular, a one-dimensional Brownian motion is **point recurrent**, i.e. it returns to each level $b \in \mathbb{R}$ time and again.

We can now prove the »continuous-time« version of Wald's identities (cf. Theorem 51.3).

[1] right-continuous is enough for this conclusion

60.4 Theorem (Wald's identities. Wald 1944). *Let $(B_t, \mathscr{F}_t)_{t\geqslant 0}$ be BM($\mathbb{R}$) and assume that τ is an $\mathscr{F}_t$ stopping time. If $\mathbb{E}\tau < \infty$, then $B_\tau \in L^2(\mathbb{P})$ and we have*

$$\mathbb{E}B_\tau = 0 \quad and \quad \mathbb{E}B_\tau^2 = \mathbb{E}\tau.$$

Proof. By optional stopping $(B_{\tau\wedge t})_{t\geqslant 0}$ is an $(\mathscr{F}_{\tau\wedge t})_{t\geqslant 0}$-martingale; therefore, its expectation is constant: $\mathbb{E}B_{\tau\wedge t} = \mathbb{E}B_0 = 0$.
Doob's maximal L^p-inequality (46.2) (for $p = 2$) gives

$$\mathbb{E}\left[B_{\tau\wedge t}^2\right] \leqslant \mathbb{E}\left[\sup_{s\leqslant t} B_s^2\right] \leqslant 4\mathbb{E}\left[B_t^2\right] = 4t,$$

hence $B_{\tau\wedge t} \in L^2(\mathbb{P})$.

Again by optional stopping (Theorem 46.14) and Example 60.1.c) we see that $M_t := B_{\tau\wedge t}^2 - \tau\wedge t$ is a martingale, and so

$$\mathbb{E}\left[M_t^2\right] = \mathbb{E}\left[M_0^2\right] = 0 \quad \text{or} \quad \mathbb{E}\left[B_{\tau\wedge t}^2\right] = \mathbb{E}(\tau\wedge t).$$

By the martingale property we see

$$\mathbb{E}\left[B_{\tau\wedge t}B_{\tau\wedge s}\right] = \mathbb{E}\left[B_{\tau\wedge s}^2\right] \quad \forall s \leqslant t,$$

hence

$$\mathbb{E}\left[(B_{\tau\wedge t} - B_{\tau\wedge s})^2\right] = \mathbb{E}\left[B_{\tau\wedge t}^2 - B_{\tau\wedge s}^2\right] = \mathbb{E}(\tau\wedge t - \tau\wedge s) \xrightarrow[s,t\to\infty]{\text{DCT}} 0.$$

This means that $(B_{\tau\wedge t})_{t\geqslant 0}$ is an L^2-Cauchy sequence which converges in L^2 and, for a subsequence, a.s. Since $\tau < \infty$ a.s. and since $t \mapsto B_t$ is continuous, we see that $L^2\text{-}\lim_{t\to\infty} B_{\tau\wedge t} = B_\tau$ (in order to identify the limit, use a subsequence argument). In particular,

$$\infty > \mathbb{E}\left[B_\tau^2\right] = \lim_{t\to\infty}\mathbb{E}\left[B_{\tau\wedge t}^2\right] = \lim_{t\to\infty}\mathbb{E}[\tau\wedge t] \overset{\text{BL}}{=} \mathbb{E}\tau.$$

Since L^2-convergence implies L^1-convergence, we get

$$\mathbb{E}B_\tau = \lim_{t\to\infty}\mathbb{E}B_{t\wedge\tau} = 0. \qquad \square$$

The following corollaries are typical applications of Wald's identities (compare with Example 51.4.b)).

60.5 Corollary. *Let $(B_t)_{t\geqslant 0}$ be a BM($\mathbb{R}$) and $\tau := \inf\{t \geqslant 0 \mid B_t \notin (-a,b)\} = \tau^\circ_{(-a,b)^c}$, $a, b \geqslant 0$. Then*

$$\mathbb{P}(B_\tau = -a) = \frac{b}{a+b}, \quad \mathbb{P}(B_\tau = b) = \frac{a}{a+b} \quad and \quad \mathbb{E}\tau = ab. \tag{60.3}$$

Proof. Observe that $t\wedge\tau$ is a bounded stopping time. Moreover,

$$B_{t\wedge\tau} \in [-a,b] \implies |B_{t\wedge\tau}| \leqslant a \vee b,$$

and so

$$\mathbb{E}(t\wedge\tau) \overset{\S 60.4}{=} \mathbb{E}\left(B^2_{t\wedge\tau}\right) \leqslant a^2 \vee b^2.$$

By monotone convergence, $\mathbb{E}\tau \leqslant a^2 \vee b^2 < \infty$. Using again Theorem 60.4 and the fact that $\mathbb{P}(\tau<\infty)=1$, we get

$$-a\,\mathbb{P}(B_\tau=-a)+b\,\mathbb{P}(B_\tau=b)=\mathbb{E}B_\tau=0,$$
$$\mathbb{P}(B_\tau=-a)+\mathbb{P}(B_\tau=b)=1.$$

Solving this system of equations yields

$$\mathbb{P}(B_\tau=-a)=\frac{b}{a+b} \quad\text{and}\quad \mathbb{P}(B_\tau=b)=\frac{a}{a+b}.$$

The second identity from Theorem 60.4 gives

$$\mathbb{E}\tau=\mathbb{E}\left(B^2_\tau\right)=a^2\,\mathbb{P}(B_\tau=-a)+b^2\,\mathbb{P}(B_\tau=-b)=ab. \qquad \square$$

⚠ The following result is disturbing (but not unexpected; see Example 51.4.a)): BM($\mathbb{R}$) hits 1 infinitely often with probability one – but the average time Brownian motion needs to get there is infinite.

60.6 Corollary. *Let $(B_t)_{t\geqslant 0}$ be a BM($\mathbb{R}$) and $\tau_1=\inf\{t\geqslant 0 \mid B_t=1\}$ be the first passage time of the level 1. Then $\mathbb{E}\tau_1=\infty$.*

Proof. Example 60.3 shows that $\mathbb{P}(\tau_1<\infty)=1$. By definition, $B_{\tau_1}=1$, therefore $\mathbb{E}B_{\tau_1}=\mathbb{E}B^2_{\tau_1}=1$. In view of Wald's identities (Theorem 60.4) this is only possible if $\mathbb{E}\tau_1=\infty$. $\square$

61 How regular is a BM?

Let $B=(B_t)_{t\geqslant 0}$ be a BM($\mathbb{R}$). We are interested in the regularity of the (**sample**) **path** $t\mapsto B_t(\omega)=B(t,\omega)$.

61.1 Definition. Let $f:[0,\infty)\to\mathbb{R}$ be a function. For every $p>0$ and any finite **partition** $\Pi=\{0=t_0\leqslant t_1<\dots<t_n=t\}$ of $[0,t]$ we define the p-**variational sum** as

$$S_p(f,\Pi):=\sum_{t_i,t_{i-1}\in\Pi}|f(t_i)-f(t_{i-1})|^p. \tag{61.1}$$

The **total variation** of a function f is defined as

$$\|f\|_{TV[0,t]}:=\sup\left\{S_1(f,\Pi):\Pi \text{ any finite partition of } [0,t]\right\}. \tag{61.2}$$

Clearly, $S_p(f,\Pi)$ is a measure of the oscillations of the function f, and we are mainly interested in the limit $|\Pi|:=\max_i|t_i-t_{i-1}|\to 0$.

61.2 Theorem (quadratic variation). *Let $(B_t)_{t\geqslant 0}$ be BM($\mathbb{R}$) and $(\Pi_n)_{n\geqslant 1}$ be any sequence of finite partitions of $[0,t]$ satisfying $\lim_{n\to\infty}|\Pi_n| = 0$. Then the mean-square limit exists:*

$$\operatorname{var}_2(B;t) = L^2(\mathbb{P})\text{-}\lim_{n\to\infty} S_2(B,\Pi_n) = t.$$

Proof. Let $\Pi = \{t_0 = 0 < t_1 < \ldots < t_n = t\}$ be some partition of $[0,t]$.

$$\mathbb{E}S_2(B,\Pi) = \sum_{i=1}^n \underbrace{\mathbb{E}\left[(B(t_i) - B(t_{i-1}))^2\right]}_{\stackrel{(B1)}{=}\mathbb{E}[B(t_i - t_{i-1})^2]} = \sum_{i=1}^n (t_i - t_{i-1}) = t.$$

Therefore,

$$\mathbb{E}\left[\left(S_2(B,\Pi) - t\right)^2\right] = \mathbb{V}\left[S_2(B,\Pi)\right] = \mathbb{V}\left[\sum_{i=1}^n \left(B(t_i) - B(t_{i-1})\right)^2\right].$$

By (B2) the random variables $(B(t_i) - B(t_{i-1}))^2$, $i = 1,\ldots,n$, are independent with mean zero. Using Bienaymés identity we find

$$\begin{aligned}
\mathbb{E}\left[\left(S_2(B,\Pi) - t\right)^2\right] &\stackrel{(B2)}{=} \sum_{i=1}^n \mathbb{V}\left[\left(B(t_i) - B(t_{i-1})\right)^2\right] \\
&= \sum_{i=1}^n \mathbb{E}\left[\left(\left(B(t_i) - B(t_{i-1})\right)^2 - (t_i - t_{i-1})\right)^2\right] \\
&\stackrel{(B1)}{=} \sum_{i=1}^n \mathbb{E}\left[\left(B(t_i - t_{i-1})^2 - (t_i - t_{i-1})\right)^2\right] \\
&= \sum_{i=1}^n (t_i - t_{i-1})^2 \underbrace{\mathbb{E}\left[\left(B(1)^2 - 1\right)^2\right]}_{=2} \\
&\leqslant 2|\Pi| \sum_{i=1}^n (t_i - t_{i-1}) = 2|\Pi|\, t \xrightarrow[|\Pi|\to 0]{} 0.
\end{aligned}$$

□

61.3 Corollary. *For a BM($\mathbb{R}$) we have $\mathbb{P}\left(\forall t : \|B_\bullet\|_{TV[0,t]} = \infty\right) = 1$.*

Proof. Let Π_n be any sequence of partitions of $[0,t]$ with $|\Pi_n| \to 0$. Then

$$\begin{aligned}
&\sum_{t_{i-1},t_i\in\Pi_n} \left(B(t_i) - B(t_{i-1})\right)^2 \\
&\quad\leqslant \max_{t_{i-1},t_i\in\Pi_n} |B(t_i) - B(t_{i-1})| \sum_{t_{i-1},t_i\in\Pi_n} |B(t_i) - B(t_{i-1})| \\
&\quad\leqslant \max_{t_{i-1},t_i\in\Pi_n} |B(t_i) - B(t_{i-1})| \cdot \|B_\bullet\|_{TV[0,t]} \\
&\quad\xrightarrow[|\Pi_n|\to 0]{\text{b/o uniform continuity}} 0 \cdot \|B_\bullet\|_{TV[0,t]},
\end{aligned}$$

since $s \mapsto B(s)$ is continuous on the compact interval $[0,t]$, hence uniformly continuous.

The left-hand side converges (for a subsequence) almost surely to t. Thus: $\mathbb{P}(\|B_\bullet\|_{TV[0,t]} < \infty) = 0$ for any $t > 0$.

Since $t \mapsto \|f\|_{TV[0,t]}$ is increasing, we get

$$\mathbb{P}\Big(\exists t \in (0,\infty) : \|B_\bullet\|_{TV[0,t]} < \infty\Big) = \mathbb{P}\Big(\exists q \in \mathbb{Q}^+ : \|B_\bullet\|_{TV[0,q]} < \infty\Big) = 0. \qquad \square$$

We can use Theorem 61.2 to learn more about the irregularity of a Brownian path.

Recall that a function $f : [0,\infty) \to \mathbb{R}$ is called **(locally) Hölder continuous** of order $\alpha > 0$, if for every interval $I \subset [0,\infty)$

$$\exists c = c(f,\alpha,I) \quad \forall s,t \in I : \quad |f(t) - f(s)| \leqslant c|t-s|^\alpha \tag{61.3}$$

61.4 Corollary. *Almost all Brownian paths are nowhere (locally) Hölder continuous of order $\alpha > \frac{1}{2}$.*

On the upside, one can show that almost all paths are Hölder continuous for any index $\alpha < \frac{1}{2}$; see [BM, Theorem 10.1]. With even more care, one can see that the modulus of continuity of a Brownian motion is of order $\sqrt{2t|\log t|}$; see [BM, Theorem 10.6].

Proof. Fix some interval $[a,b] \subset [0,\infty)$, $a < b$ rational, $\alpha > \frac{1}{2}$ and assume that (61.3) holds for $f(t) = B(t,\omega)$ and some constant $c = c(\omega,\alpha,[a,b])$.

Then we find for all finite partitions Π of $[a,b]$

$$\begin{aligned} S_2(B(\cdot,\omega),\Pi) &= \sum_{t_{i-1},t_i \in \Pi} \Big(B(t_i,\omega) - B(t_{i-1},\omega)\Big)^2 \\ &\leqslant c(\omega)^2 \sum_{t_{i-1},t_i \in \Pi} (t_i - t_{i-1})^{2\alpha} \leqslant c(\omega)^2 (b-a)|\Pi|^{2\alpha-1}. \end{aligned}$$

By Theorem 61.2, the left-hand side converges almost surely to $b-a$ for some subsequence Π_n with $|\Pi_n| \to 0$. Since $2\alpha - 1 > 0$, the right-hand side tends to 0, and we have a contradiction.

This shows that (61.3) can only hold on a $\mathbb{P}$-null set $N_{a,b} \subset \Omega$. But then $\bigcup_{0 \leqslant a<b, a,b\in\mathbb{Q}} N_{a,b}$ is still an exceptional (i.e., null) set, which is the same for all intervals. $\square$

We are going to show that almost all paths of Brownian motion are nowhere differentiable.

61.5 Theorem (Paley, Wiener, Zygmund 1931). *Let* B *be* $\mathrm{BM}(\mathbb{R})$. *Then the paths* $t \mapsto B_t(\omega)$ *are for almost all* $\omega \in \Omega$ *nowhere differentiable.*

Proof. (*Dvoretzky, Erdös, Kakutani 1961*). Define for every $n \in \mathbb{N}$

$$\mathrm{A}_n := \{\omega \in \Omega \mid \mathrm{B}(\cdot,\omega) \text{ nowhere differentiable in } [0,n)\}.$$

It is not clear if the set A_n is measurable. We will show that $\Omega \setminus \mathrm{A}_n \subset \mathrm{N}_n$ where $\mathbb{P}(\mathrm{N}_n) = 0$ is a measurable null set.

Assume that a function f is differentiable at $t_0 \in [0,n)$. Then

$$\exists \delta > 0 \ \exists \mathrm{L} > 0 \ \forall t \in \mathrm{B}_\delta(t_0): \quad |f(t) - f(t_0)| \leqslant \mathrm{L} \cdot |t - t_0|.$$

Consider for sufficiently large $k \in \mathbb{N}$ the grid $\left\{\frac{i}{k} \mid 1 \leqslant i \leqslant nk\right\}$. Then there exists a smallest index $i = i(k)$ such that

$$t_0 \leqslant \frac{i}{k} \quad \text{and} \quad \frac{i}{k}, \dots, \frac{i+3}{k} \in \mathrm{B}_\delta(t_0).$$

For $\ell = i+1, i+2, i+3$ we get therefore

$$\begin{aligned} \left|f\left(\tfrac{\ell}{k}\right) - f\left(\tfrac{\ell-1}{k}\right)\right| &\leqslant \left|f\left(\tfrac{\ell}{k}\right) - f(t_0)\right| + \left|f(t_0) - f\left(\tfrac{\ell-1}{k}\right)\right| \\ &\leqslant \mathrm{L} \cdot \left(\left|\tfrac{\ell}{k} - t_0\right| + \left|\tfrac{\ell-1}{k} - t_0\right|\right) \\ &\leqslant \mathrm{L} \cdot \left(\tfrac{4}{k} + \tfrac{3}{k}\right) = \tfrac{7\mathrm{L}}{k}. \end{aligned}$$

If $f(t) = \mathrm{B}(t,\omega)$ is a Brownian path, this implies that for the sets

$$\mathrm{C}_m^{\mathrm{L}} := \bigcap_{k=m}^{\infty} \bigcup_{i=1}^{kn} \bigcap_{\ell=i+1}^{i+3} \left\{\left|\mathrm{B}\left(\tfrac{\ell}{k}\right) - \mathrm{B}\left(\tfrac{\ell-1}{k}\right)\right| \leqslant \tfrac{7\mathrm{L}}{k}\right\}$$

we have

$$\Omega \setminus \mathrm{A}_n \subset \bigcup_{\mathrm{L}=1}^{\infty} \bigcup_{m=1}^{\infty} \mathrm{C}_m^{\mathrm{L}}.$$

Our assertion follows if we can show that $\mathbb{P}(\mathrm{C}_m^{\mathrm{L}}) = 0$ for all $m, \mathrm{L} \in \mathbb{N}$.

$$\begin{aligned} \mathbb{P}(\mathrm{C}_m^{\mathrm{L}}) &\leqslant \mathbb{P}\left(\bigcup_{i=1}^{kn} \bigcap_{\ell=i+1}^{i+3} \left\{\left|\mathrm{B}\left(\tfrac{\ell}{k}\right) - \mathrm{B}\left(\tfrac{\ell-1}{k}\right)\right| \leqslant \tfrac{7\mathrm{L}}{k}\right\}\right) \\ &\leqslant \sum_{i=1}^{kn} \mathbb{P}\left(\bigcap_{\ell=i+1}^{i+3} \left\{\left|\mathrm{B}\left(\tfrac{\ell}{k}\right) - \mathrm{B}\left(\tfrac{\ell-1}{k}\right)\right| \leqslant \tfrac{7\mathrm{L}}{k}\right\}\right) \\ &\overset{\text{(B2)}}{=} \sum_{i=1}^{kn} \prod_{\ell=i+1}^{i+3} \mathbb{P}\left(\left\{\left|\mathrm{B}\left(\tfrac{\ell}{k}\right) - \mathrm{B}\left(\tfrac{\ell-1}{k}\right)\right| \leqslant \tfrac{7\mathrm{L}}{k}\right\}\right) \\ &\overset{\text{(B1)}}{=} kn\,\mathbb{P}\left(\left\{\left|\mathrm{B}\left(\tfrac{1}{k}\right)\right| \leqslant \tfrac{7\mathrm{L}}{k}\right\}\right)^3 \\ &\leqslant kn\left(\frac{c}{\sqrt{k}}\right)^3 \xrightarrow{k\to\infty} 0. \end{aligned}$$

For the last estimate we use that

$$\mathbb{P}\left(\left|B\left(\tfrac{1}{k}\right)\right| \leqslant \tfrac{7L}{k}\right) = \sqrt{\frac{1}{2\pi\frac{1}{k}}} \int_{-7L/k}^{7L/k} \underbrace{e^{-x^2/(2\frac{1}{k})}}_{\leqslant 1} dx$$

$$\leqslant \frac{\sqrt{k}}{\sqrt{2\pi}} 14L \frac{1}{k} = \frac{14L}{\sqrt{2\pi}} \frac{1}{\sqrt{k}}. \qquad \square$$

Theorem 61.5 is devastating: There is no chance to define a stochastic integral $\int_0^\infty f(t)\,dB_t$ in a classical sense by, say, $\int_0^\infty f(t)\,\dot{B}_t\,dt$, since the derivative $\dot{B}_t$ does not exist.

Recall the notion of a **Riemann–Stieltjes Integral**. Let $\alpha, f : [a,b] \to \mathbb{R}$ be functions, and set

$$\int_a^b f(t)\,d\alpha(t) := \lim_{|\Pi|\to 0} \sum_{t_i,t_{i-1}\in\Pi} f(s_i)(\alpha(t_i) - \alpha(t_{i-1})) \tag{61.4}$$

(provided that the limit exists) where $\Pi = \{t_0 = a < t_1 < \cdots < t_n = b\}$ is a partition of $[a,b]$ and $s_i \in [t_{i-1}, t_i]$. This is a generalization of the Riemann integral (take $\alpha(t) = t$) and it is in line with the Lebesgue integral induced by the measure $\mu(s,t] := \alpha(t) - \alpha(s)$ and a right-continuous distribution function α.

We also need the **Banach–Steinhaus theorem** from functional analysis, e.g. Rudin [28, Thm. 5.8].

61.6 Theorem (Banach–Steinhaus). *Let $(\mathcal{X}, \|\cdot\|_{\mathcal{X}})$, $(\mathcal{Y}, \|\cdot\|_{\mathcal{Y}})$ be Banach spaces and $S^\Pi : \mathcal{X} \to \mathcal{Y}$ be a family of continuous linear maps indexed by Π from an arbitrary index set. Then*

$$\forall x \in \mathcal{X}: \quad \sup_{\Pi} \|S^\Pi[x]\|_{\mathcal{Y}} < \infty \implies \sup_{\Pi} \underbrace{\sup_{\|x\|_{\mathcal{X}}\leqslant 1} \|S^\Pi[x]\|_{\mathcal{Y}}}_{\text{operator norm } \|S^\Pi\|_{\mathcal{X}\to\mathcal{Y}}} < \infty.$$

61.7 Lemma. *Let $\alpha : [a,b] \to \mathbb{R}$. If $A[f] := \int_a^b f(t)\,d\alpha(t)$ exists for all $f \in C[a,b]$, then $\|\alpha\|_{TV[a,b]} < \infty$.*

Proof. We apply the Banach–Steinhaus theorem for the Banach spaces

$$(\mathcal{X}, \|\cdot\|_{\mathcal{X}}) = (C[a,b], \|\cdot\|_\infty) \quad \text{and} \quad (\mathcal{Y}, \|\cdot\|_{\mathcal{Y}}) = (\mathbb{R}, |\cdot|).$$

We take the family of all finite partitions $\Pi = \{t_0 = a < t_1 < \ldots < t_n = b\}$ of $[a,b]$ as index set and define linear maps $S^\Pi : \mathcal{X} \to \mathcal{Y}$ by

$$S^\Pi[f] := \sum_{t_i,t_{i-1}\in\Pi} f(t_{i-1})(\alpha(t_i) - \alpha(t_{i-1})).$$

Because of

$$|S^\Pi[f]| \leqslant \|f\|_\infty \sum_{t_i,t_{i-1}\in\Pi} |\alpha(t_i) - \alpha(t_{i-1}| = c_\alpha \|f\|_\infty$$

we see that the maps S^Π are bounded.

For the saw-tooth function

$$f_\Pi(s) := \begin{cases} \operatorname{sgn}(\alpha(t_i) - \alpha(t_{i-1})), & \text{if } s = t_{i-1},\ i = 1, \dots, n, \\ \operatorname{sgn}(\alpha(t_n) - \alpha(t_{n-1})), & \text{if } s = t_n, \\ \text{piecewise linear}, & \text{in-between}, \end{cases}$$

we find

$$S^\Pi[f_\Pi] = \underbrace{\sum_{t_i, t_{i-1} \in \Pi} |\alpha(t_i) - \alpha(t_{i-1})|}_{= S_1(\alpha;\Pi)} \leqslant \sup_{\|f\|_\infty \leqslant 1} \left|S^\Pi[f]\right|.$$

By assumption $\lim_{|\Pi| \to 0} S^\Pi[f] = \int_a^b f(t)\, d\alpha(t) < \infty$ for all $f \in C[a,b]$. Therefore, $\sup_\Pi \left|S^\Pi[f]\right| < \infty$ and, by the Banach–Steinhaus theorem,

$$\|\alpha\|_{TV[a,b]} = \sup_\Pi S_1(\alpha, \Pi) = \sup_\Pi S^\Pi[f_\Pi] \leqslant \sup_\Pi \sup_{\|f\|_\infty \leqslant 1} \left|S^\Pi[f]\right| < \infty. \qquad \square$$

61.8 Remark. This means that $\int_0^t f(s)\, dB_s(\omega)$ cannot be defined for all (deterministic) continuous integrands f as a Riemann–Stieltjes integral.

Like the Riemann integral, the Riemann-Stieltjes integral coincides with the Lebesgue integral for continuous functions f. One can construct the measure explicitly from α. This means that also Lebesgue theory will not be a way out of our problem.

The **real way out** was discovered by K. Itô who proposed to change the limit in the definition of (61.4) to an L^2 or even a $\mathbb{P}$-limit. Technically, this is possible, since BM($\mathbb{R}$) is a martingale (Example 60.1).

62 The Markov Property (MP)

We begin with an important application of Theorem 59.8.

62.1 Theorem. *Let* $B = (B_t, \mathscr{F}_t)_{t \geqslant 0}$ *be a* BM($\mathbb{R}^d$) *and fix* $s > 0$. *Set* $W_t := B_{t+s} - B_s$, $\mathscr{F}_t^W := \sigma(B_{r+s} - B_s, r \leqslant t)$ *and* $W := (W_t, \mathscr{F}_t^W)_{t \geqslant 0}$. *Then* W *is a* BM($\mathbb{R}^d$) *and* $W \perp\!\!\!\perp (B_t)_{0 \leqslant t \leqslant s}$.

Proof. We begin with the proof of the independence.[1] Fix $s > 0$ and $m, n \in \mathbb{N}$, and pick times $t_n > \cdots > t_1 > t_0 = s = s_m > \cdots > s_1 > s_0 = 0$. Because of (B2′), the random variables

$$B(t_k) - B(t_{k-1}),\ B(s_i) - B(s_{i-1}), \quad 1 \leqslant k \leqslant n,\ 1 \leqslant i \leqslant m,$$

[1] Compare this argument with the proof of Theorem 50.5, Step 5°.

are independent. Since we can represent each $W(t_k)$ and $B(s_j)$ from t- resp. s-increments (which preserves independence, Corollary 25.7), we see that

$$(W(t_n),\dots,W(t_1),W(t_0)) \perp\!\!\!\perp (B(s_m),\dots,B(s_1),B(s_0))$$

and, therefore,

$$\bigcup_{\substack{s<t_1<\dots<t_n\\ n\in\mathbb{N}}} \sigma(W(t_n),\dots,W(t_0)) \;\perp\!\!\!\perp\; \bigcup_{\substack{s_1<\dots<s_m\leqslant s\\ m\in\mathbb{N}}} \sigma(B(s_m),\dots,B(s_0)).$$

Since both families are $\cap$-stable ✎ generators of $\mathscr{F}_\infty^W$ and $\mathscr{F}_s^B$, respectively, the independence $\mathscr{F}_\infty^W \perp\!\!\!\perp \mathscr{F}_s^B$ follows from Theorem 25.5.

We will now show that W is a BM($\mathbb{R}^d$). Clearly, $W_0 = 0$ and $t \mapsto W_t$ is continuous; so we can use Lemma 59.8. In the step marked by »§« we use that $B_{u+s} - B_{t+s} \perp\!\!\!\perp \mathscr{F}_{t+s}^B$:

$$\begin{aligned}\mathbb{E}\left[e^{i\langle\xi,W_u-W_t\rangle} \mid \mathscr{F}_t^W\right] &= \mathbb{E}\left[e^{i\langle\xi,B_{u+s}-B_{t+s}\rangle} \mid \sigma(B_{r+s}-B_s, r\leqslant t)\right]\\ &\overset{\S}{=} \mathbb{E}\left[e^{i\langle\xi,B_{u+s}-B_{t+s}\rangle}\right]\\ &= \mathbb{E}e^{i\langle\xi,B_{u-t}\rangle}\\ &= e^{-\frac{1}{2}(u-t)|\xi|^2}.\end{aligned}$$

□

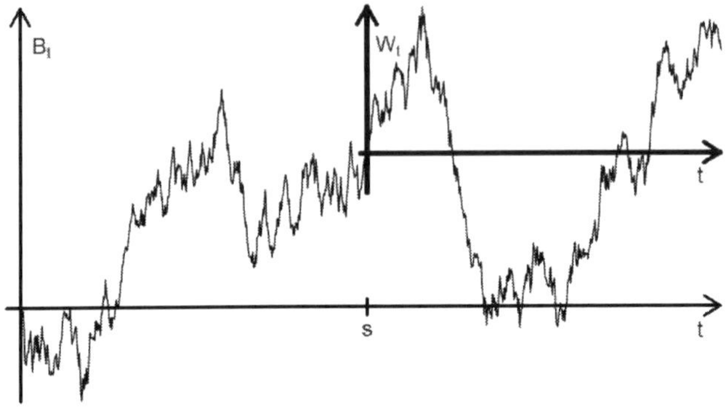

Fig. 62.1. Splitting a Brownian path into the »past« $(B_t)_{t\leqslant s}$ and the »future« $(W_t + y)_{t\geqslant 0}$. $W + y$ is a BM independent of the past, starting at y.

62.2 Remark. a) Theorem 62.1 allows us to split a Brownian path into two independent pieces

$$B(t+s) = B(t+s) - B(s) + B(s) = W(t) + y\big|_{y=B(s)}.$$

Observe that $W(t) + y$ is a Brownian motion started at $y \in \mathbb{R}^d$.

We can interpret Theorem 62.1 also as a **renewal property**:

$$B_0 = 0 \xrightarrow{\quad t+s \quad \text{seconds} \quad} B_{t+s} = x$$

$$\iff B_0 = 0 \xrightarrow{s \text{ seconds}} B_s = y = W_0 + y \xrightarrow[\text{using W}]{t \text{ seconds}} W_t + y = x.$$

This situation is shown in Figure 62.1.

b) The finite-dimensional distributions (fdd) are the measures

$$\mathbb{P}^x(B_{t_1} \in A_1, \dots, B_{t_n} \in A_n) := \mathbb{P}(B_{t_1} + x \in A_1, \dots, B_{t_n} + x \in A_n) \tag{62.1}$$

where $0 \leqslant t_1 < \dots < t_n$ and $A_1, \dots, A_n \in \mathscr{B}(\mathbb{R}^d)$.

We write $\mathbb{E}^x = \int \cdots d\mathbb{P}^x$. Clearly, $\mathbb{P}^0 = \mathbb{P}$ and $\mathbb{E}^0 = \mathbb{E}$.

Interpretation: $\mathbb{P}^x(B_s \in A)$ denotes the probability that a Brownian particle starts at time $t = 0$ at the point x and travels in s units of time into the set A. This notation is compatible with our use of $\mathbb{P}^i$ in the chapter on Markov chains.

c) The finite dimensional distributions (62.1) determine the $\mathbb{P}^x$ uniquely (this is due to Kolmogorov's theorem, [BM, Appendix A.1, pp. 357–363]). Thus, $\mathbb{P}^x$ is a well-defined measure on $(\Omega, \mathscr{F}_\infty^B)$. Moreover, $x \mapsto \mathbb{E}^x u(B_t) = \mathbb{E}^0 u(B_t + x)$ is measurable for any bounded Borel function u.

We will need the following formulae for conditional expectations which generalize Proposition 34.12.c) and Theorem 35.13. A proof is in [BM, Appendix A.2, pp. 363–364].

62.3 Lemma. *Assume that $\mathscr{X} \perp\!\!\!\perp \mathscr{Y} \subset \mathscr{A}$ are independent σ-algebras. If X is an $\mathscr{X}/\mathscr{B}(\mathbb{R}^d)$ and Y a $\mathscr{Y}/\mathscr{B}(\mathbb{R}^d)$-measurable random variable, then*

$$\mathbb{E}\big[\Phi(X,Y) \mid \mathscr{X}\big] = \mathbb{E}\Phi(x,Y)\big|_{x=X} = \mathbb{E}\big[\Phi(X,Y) \mid X\big] \tag{62.2}$$

for all bounded Borel measurable functions $\Phi : \mathbb{R}^d \times \mathbb{R}^d \to \mathbb{R}$.

If $\Psi : \mathbb{R}^d \times \Omega \to \mathbb{R}$ is bounded and $\mathscr{B}(\mathbb{R}^d) \otimes \mathscr{Y}/\mathscr{B}(\mathbb{R})$-measurable, then

$$\mathbb{E}\big[\Psi(X(\cdot),\cdot) \mid \mathscr{X}\big] = \mathbb{E}\Psi(x,\cdot)\big|_{x=X} = \mathbb{E}\big[\Psi(X(\cdot),\cdot) \mid X\big]. \tag{62.3}$$

62.4 Theorem (Markov property). *Let $(B_t, \mathscr{F}_t)_{t \geqslant 0}$ be a BM$(\mathbb{R}^d)$. Then*

$$\mathbb{E}\big[f(B_{t+s}) \mid \mathscr{F}_s\big] = \mathbb{E}\big[f(B_t + x)\big]\Big|_{x=B_s} = \mathbb{E}^{B_s}\big[f(B_t)\big] \tag{62.4}$$

holds for all bounded measurable functions $f : \mathbb{R}^d \to \mathbb{R}$ and

$$\mathbb{E}\big[\Psi(B_{\bullet+s}) \mid \mathscr{F}_s\big] = \mathbb{E}\big[\Psi(B_\bullet + x)\big]\Big|_{x=B_s} = \mathbb{E}^{B_s}\big[\Psi(B_\bullet)\big] \tag{62.5}$$

holds for all functionals $\Psi : C[0,\infty) \to \mathbb{R}$ which are $\mathscr{B}(C[0,\infty))/\mathscr{B}(\mathbb{R})$-measurable and bounded (and may depend on the whole Brownian path).

Proof. Write $f(B_{t+s}) = f(B_s + B_{t+s} - B_s) = f(B_s + (B_{t+s} - B_s))$.

The equality (62.4) follows from (B2) if we use (62.2) with $X = B_s$, $Y = B_{t+s} - B_s$, $\mathscr{X} = \mathscr{F}_s$ and $\Phi(x, y) = f(x+y)$.

The equality (62.5) follows from Theorem 62.1 if we take in (62.3) $\mathscr{Y} = \mathscr{F}_\infty^W$, $W_t := B_{t+s} - B_s$, $\mathscr{X} = \mathscr{F}_s$ and $X = B_s$. □

62.5 Remark. a) ⚠⚠ It is helpful to write (62.4) in integral form and with all the ω:

$$\begin{aligned}\mathbb{E}\left[f(B_{t+s}) \mid \mathscr{F}_s\right](\omega) &= \int_\Omega f(B_t(\omega') + B_s(\omega))\,\mathbb{P}(d\omega') \\ &= \int_\Omega f(B_t(\omega'))\,\mathbb{P}^{B_s(\omega)}(d\omega') \\ &= \mathbb{E}^{B_s(\omega)}\left[f(B_t)\right].\end{aligned}$$

b) Typical examples of functionals Ψ appearing in (62.5) are:

$$\Psi(B_\bullet) = \max_{0 \leqslant t \leqslant 1} B_t \quad \text{or} \quad \Psi(B_\bullet) = \int_0^1 B_s\,ds \quad \text{or} \quad \Psi(B_\bullet) = \mathbb{1}_{\{\tau_U \leqslant t\}}$$

where τ_U is the first hitting time of an open set U.

c) In general, any d-dimensional, $\mathscr{F}_t$ adapted stochastic process $(X_t)_{t \geqslant 0}$ with (right-)continuous paths is called a **Markov process** if

$$\mathbb{E}\left[f(X_{t+s}) \mid \mathscr{F}_s\right] = \mathbb{E}\left[f(X_{t+s}) \mid X_s\right] \quad \forall s, t \geqslant 0,\ f : \mathbb{R}^d \xrightarrow[\text{mble.}]{\text{bdd.}} \mathbb{R}. \tag{62.4a}$$

From the factorization lemma (Lemma 7.16) we know that there exists some measurable function $g^f_{s,t+s} : \mathbb{R}^d \to \mathbb{R}$ such that

$$\mathbb{E}\left[f(X_{t+s}) \mid X_s\right] = g^f_{s,t+s}(X_s).$$

Let us assume that $g^f_{s,t+s}(x) = g^f_t(x)$, i.e. it **depends only on the difference** $t = (t+s) - s$. In this case we have a **homogeneous Markov process**, and we write

$$\begin{gathered}\mathbb{E}\left[f(X_{t+s}) \mid \mathscr{F}_s\right] = \mathbb{E}^{X_s}\left[f(X_t)\right] \\ \text{with} \quad \mathbb{E}^x\left[f(X_t)\right] := g^f_t(x).\end{gathered} \tag{62.4b}$$

This is the case for Brownian motion. In fact, (62.4b)⇒(62.4a). Indeed, apply $\mathbb{E}[\cdots \mid X_s]$ to both sides of (62.4b):

$$\begin{aligned}\mathbb{E}\left[f(X_{t+s}) \mid X_s\right] &\overset{\text{tower}}{=} \mathbb{E}\left[\mathbb{E}\left[f(X_{t+s}) \mid \mathscr{F}_s\right] \mid X_s\right] \\ &\overset{(62.4b)}{=} \mathbb{E}\left[\mathbb{E}^{X_s}\left[f(X_t)\right] \mid X_s\right] \\ &\overset{\text{pull}}{=} \mathbb{E}^{X_s}\left[f(X_t)\right] \\ &\overset{(62.4b)}{=} \mathbb{E}\left[f(X_{t+s}) \mid \mathscr{F}_s\right],\end{aligned}$$

and we get (62.4a).

d) Markov processes are often used in the literature. In fact, the Markov property (62.4a), (62.4b) can be seen as a means to obtain the finite dimensional distributions from one-dimensional distributions. Let us illustrate this for three points of time.

Fix $0 \leqslant s < t < u$ and let $\phi, \psi, \chi : \mathbb{R}^d \to \mathbb{R}$ arbitrary bounded measurable functions. Then we have

$$\mathbb{E}\big[\phi(B_s)\,\psi(B_t)\,\chi(B_u)\big] = \iiint\limits_{\mathbb{R}^d\times\mathbb{R}^d\times\mathbb{R}^d} \phi(x)\psi(y)\chi(z)\,\mathbb{P}(B_s \in dx, B_t \in dy, B_u \in dz).$$

Let us calculate the same expectation in a different way:

$$\begin{aligned}
&\mathbb{E}\big[\phi(B_s)\,\psi(B_t)\,\chi(B_u)\big]\\
&= \mathbb{E}\big[\phi(B_s)\,\psi(B_t)\mathbb{E}\big(\chi(B_u)\mid\mathscr{F}_t\big)\big]\\
&\overset{(62.4)}{=} \mathbb{E}\big[\phi(B_s)\,\psi(B_t)\mathbb{E}^{B_t}\chi(B_{u-t})\big]\\
&\overset{(62.4)}{=} \mathbb{E}\big[\phi(B_s)\,\underbrace{\mathbb{E}^{B_s}\big(\psi(B_{t-s})\mathbb{E}^{B_{t-s}}\chi(B_{u-t})\big)}_{=\int\psi(y)\int\chi(z)\,\mathbb{P}^y(B_{u-t}\in dz)\,\mathbb{P}^{B_s}(B_{t-s}\in dy)}\big]\\
&= \int\limits_{\mathbb{R}^d}\phi(x)\Big(\int\limits_{\mathbb{R}^d}\psi(y)\int\limits_{\mathbb{R}^d}\chi(z)\,\mathbb{P}^y(B_{u-t}\in dz)\,\mathbb{P}^x(B_{t-s}\in dy)\Big)\mathbb{P}(B_s\in dx).
\end{aligned}$$

Comparing these intgrals and using $\phi = \mathbb{1}_A$, $\psi = \mathbb{1}_B$, $\chi = \mathbb{1}_C$ gives

$$\mathbb{P}(B_s \in dx, B_t \in dy, B_u \in dz) = \mathbb{P}^y(B_{u-t} \in dz)\mathbb{P}^x(B_{t-s} \in dy)\mathbb{P}(B_s \in dx).$$

If we iterate the calculation from Remark 62.5.d) we arrive at the following **Chapman–Kolmogorov relations**.

62.6 Theorem. *Let $(B_t)_{t\geqslant 0}$ be a BM($\mathbb{R}^d$) (or a general Markov process), $x_0 = 0$ and $t_0 = 0 < t_1 < \ldots < t_n$. Then*

$$\mathbb{P}^0(B_{t_1} \in dx_1, \ldots, B_{t_n} \in dx_n) = \prod_{i=1}^n \mathbb{P}^{x_{i-1}}(B_{t_i - t_{i-1}} \in dx_i). \tag{62.6}$$

63 Strong Markov Property (SMP)

The MP (62.4) remains valid if we replace s by a stopping time $\tau(\omega)$.

⚠ We allow $\tau = \infty$, so B_τ needs to be defined also in this case. In general, if

$(X_t)_{t\geqslant 0}$ is a stochastic process and τ a stopping time, we set

$$X_\tau(\omega) := \begin{cases} X_{\tau(\omega)}(\omega) & \text{if } \tau(\omega) < \infty, \\ 0 & \text{if } \tau(\omega) = \infty. \end{cases}$$

Recall the approximation of a stopping time τ through dyadic stopping times, cf. Remark 46.13.f):

$$\tau_n := \frac{\lfloor 2^n \tau \rfloor + 1}{2^n} \downarrow \tau.$$

We also need a slightly larger σ-algebra than $\mathscr{F}_\tau$ from (46.6), namely

$$\mathscr{F}_{\tau+} = \left\{ F \in \mathscr{F}_\infty \mid F \cap \{\tau \leqslant t\} \in \mathscr{F}_{t+} \ \forall t \geqslant 0 \right\}. \tag{63.1}$$

⚠ Please **note the** »+« in the definition.

63.1 Lemma. *Let τ be an $\mathscr{F}_t$ stopping time and $\tau_n \downarrow \tau$ the dyadic approximation. We have $F \cap \{(k-1)2^{-n} \leqslant \tau < k2^{-n}\} = F \cap \{\tau_n = k2^{-n}\} \in \mathscr{F}_{k/2^n}$ for all $F \in \mathscr{F}_{\tau+}$.*

Proof. Let $F \in \mathscr{F}_{\tau+}$. We have

$$\begin{aligned} F \cap \{\tau_n = k2^{-n}\} &= F \cap \{(k-1)2^{-n} \leqslant \tau < k2^{-n}\} \\ &= F \cap \{\tau < k2^{-n}\} \cap \{\tau < (k-1)2^{-n}\}^c \\ &= \underbrace{\bigcup_{i\geqslant 1} \left(F \cap \left\{ \tau \leqslant k2^{-n} - \tfrac{1}{i} \right\} \right)}_{\in \mathscr{F}_{(k/2^n - 1/i)+} \subset \mathscr{F}_{k/2^n}} \cap \underbrace{\{\tau < (k-1)2^{-n}\}^c}_{\in \mathscr{F}_{k/2^n}} \in \mathscr{F}_{k/2^n}. \end{aligned}$$ □

The following theorem is the random-time analogue of Theorem 62.1

63.2 Theorem. *Let $(B_t, \mathscr{F}_t)_{t\geqslant 0}$ be a $\mathrm{BM}(\mathbb{R}^d)$ and $\sigma < \infty$ a finite stopping time. Then $(W_t)_{t\geqslant 0}$, $W_t := B_{\sigma+t} - B_\sigma$, is again a $\mathrm{BM}(\mathbb{R}^d)$ which is independent of $\mathscr{F}_{\sigma+}$.*

Proof. By Lemma 63.1, $\sigma_n := (\lfloor 2^n \sigma \rfloor + 1)/2^n$ is a decreasing sequence of stopping times such that $\inf_{n\geqslant 1} \sigma_n = \sigma$. For all $0 \leqslant s < t$, $\xi \in \mathbb{R}^d$ and all $F \in \mathscr{F}_{\sigma+}$ we find by the continuity of the sample paths

$$\begin{aligned}
&\mathbb{E}\Big[e^{i\langle\xi,\, B(\sigma+t)-B(\sigma+s)\rangle}\mathbb{1}_F\Big]\\
&= \lim_{n\to\infty}\mathbb{E}\Big[e^{i\langle\xi,\, B(\sigma_n+t)-B(\sigma_n+s)\rangle}\mathbb{1}_F\Big]\\
&= \lim_{n\to\infty}\sum_{k=1}^{\infty}\mathbb{E}\Big[e^{i\langle\xi,\, B(k2^{-n}+t)-B(k2^{-n}+s)\rangle}\mathbb{1}_{\{\sigma_n=k2^{-n}\}}\cdot\mathbb{1}_F\Big]\\
&= \lim_{n\to\infty}\sum_{k=1}^{\infty}\mathbb{E}\Big[\overbrace{e^{i\langle\xi,\, B(k2^{-n}+t)-B(k2^{-n}+s)\rangle}}^{\perp\!\!\!\perp\, \mathscr{F}_{k2^{-n}},\ \sim B(t-s)\text{ by (B2), (B1)}}\underbrace{\mathbb{1}_{\{(k-1)2^{-n}\leqslant\sigma<k2^{-n}\}}\mathbb{1}_F}_{\in\mathscr{F}_{k2^{-n}}\text{ as }F\in\mathscr{F}_{\sigma+}}\Big]\\
&= \lim_{n\to\infty}\sum_{k=1}^{\infty}\mathbb{E}\Big[e^{i\langle\xi,\, B(t-s)\rangle}\Big]\mathbb{P}\Big(\{(k-1)2^{-n}\leqslant\sigma<k2^{-n}\}\cap F\Big)\\
&= \mathbb{E}\Big[e^{i\langle\xi,\, B(t-s)\rangle}\Big]\mathbb{P}(F).
\end{aligned}$$

In the last equality we use $\dot{\bigcup}_{k=1}^{\infty}\{(k-1)2^{-n}\leqslant\sigma<k2^{-n}\}=\{\sigma<\infty\}=\Omega$ for all $n\geqslant 1$. Taking, in particular, $F=\Omega$ we deduce

$$\mathbb{E}\Big[e^{i\langle\xi,\, B(\sigma+t)-B(\sigma+s)\rangle}\Big]=\mathbb{E}\Big[e^{i\langle\xi,\, B(t-s)\rangle}\Big].$$

Finitely many increments are treated in the same way:
Let $t_0=0<t_1<\ldots<t_N$ and $\xi_1,\ldots,\xi_N\in\mathbb{R}^d$ and $F\in\mathscr{F}_{\sigma+}$. Then we get

$$\begin{aligned}
\mathbb{E}\Big[e^{i\sum_{n=1}^{N}\langle\xi_n,\, B(\sigma+t_n)-B(\sigma+t_{n-1})\rangle}\mathbb{1}_F\Big] &= \prod_{n=1}^{N}\mathbb{E}\Big[e^{i\langle\xi_n,\, B(t_n-t_{n-1})\rangle}\Big]\mathbb{P}(F)\\
&= \prod_{n=1}^{N}\mathbb{E}\Big[e^{i\langle\xi_n,\, B(\sigma+t_n)-B(\sigma+t_{n-1})\rangle}\Big]\mathbb{P}(F).
\end{aligned}$$

From this we conclude that $W(t_n)-W(t_{n-1})=B(\sigma+t_n)-B(\sigma+t_{n-1})$ are independent random variables with law $\mathsf{N}(0,t_n-t_{n-1})$; thus we have a BM.

Moreover, all increments are independent of $F\in\mathscr{F}_{\sigma+}$ under $\mathbb{P}$. As in Theorem 62.1, this gives that $(B_{\sigma+t_1}-B_\sigma,\ldots,B_{\sigma+t_N}-B_{\sigma+t_{N-1}})\perp\!\!\!\perp\mathscr{F}_{\sigma+}$ and $\mathscr{F}_\infty^W:=\sigma(B_{\sigma+t}-B_\sigma,t\geqslant 0)\perp\!\!\!\perp\mathscr{F}_{\sigma+}$ under $\mathbb{P}$. □

We can now use (62.2), (62.3) to deduce from Theorem 63.2 the following consequences:

63.3 Theorem (Strong Markov property). *Let $(B_t)_{t\geqslant 0}$ be a $\mathrm{BM}(\mathbb{R}^d)$ and σ be a stopping time. Then we have for all $t\geqslant 0$ and bounded and measurable functions $f:\mathbb{R}^d\to\mathbb{R}$*

$$\begin{aligned}
\mathbb{E}\Big[f(B_{t+\sigma})\mid\mathscr{F}_{\sigma+}\Big](\omega)\mathbb{1}_{\{\sigma<\infty\}}(\omega) &= \mathbb{E}\Big[f(B_t+x)\Big]\Big|_{x=B_\sigma(\omega)}\mathbb{1}_{\{\sigma<\infty\}}(\omega)\\
&= \mathbb{E}^{B_\sigma(\omega)}\Big[f(B_t)\Big]\mathbb{1}_{\{\sigma<\infty\}}(\omega).
\end{aligned}\tag{63.2}$$

For all bounded $\mathscr{B}(C)/\mathscr{B}(\mathbb{R})$-measurable functionals $\Psi : C[0,\infty) \to \mathbb{R}$, which may depend on a whole Brownian path, one has

$$\begin{aligned}\mathbb{E}\Big[\Psi(B_{\bullet+\sigma}) \mid \mathscr{F}_{\sigma+}\Big](\omega)\mathbb{1}_{\{\sigma<\infty\}}(\omega) &= \mathbb{E}\Big[\Psi(B_\bullet + x)\Big]\Big|_{x=B_\sigma(\omega)} \mathbb{1}_{\{\sigma<\infty\}}(\omega) \\ &= \mathbb{E}^{B_\sigma(\omega)}\Big[\Psi(B_\bullet)\Big]\mathbb{1}_{\{\sigma<\infty\}}(\omega).\end{aligned} \tag{63.3}$$

Proof. We use Theorem 63.2 and (62.2) with $\mathscr{Y} = \mathscr{F}_\infty^W$, $W_t = B_{t+\sigma\wedge n} - B_{\sigma\wedge n}$ and $\mathscr{X} = \mathscr{F}_{(\sigma\wedge n)+}$, $X = B_{\sigma\wedge n}$ to get

$$\mathbb{E}\Big[f(B_{t+\sigma\wedge n}) \mid \mathscr{F}_{(\sigma\wedge n)+}\Big] = \mathbb{E}\Big[f(W_t + x)\Big]_{x=B_{\sigma\wedge n}} = \mathbb{E}^{B_{\sigma\wedge n}}\Big[f(W_t)\Big]$$

and we get from this

$$\mathbb{E}\Big[f(B_{t+\sigma\wedge n}) \mid \mathscr{F}_{(\sigma\wedge n)+}\Big]\mathbb{1}_{\{\sigma\leqslant n\}} = \mathbb{E}^{B_{\sigma\wedge n}}\Big[f(W_t)\Big]\mathbb{1}_{\{\sigma\leqslant n\}} = \mathbb{E}^{B_\sigma}\Big[f(W_t)\Big]\mathbb{1}_{\{\sigma\leqslant n\}}.$$

Since $\mathbb{1}_{\{\sigma\leqslant n\}}$ is $\mathscr{F}_\sigma \cap \mathscr{F}_n \overset{\S 46.13.d)}{=} \mathscr{F}_{\sigma\wedge n}$ measurable, we can use a pull-in argument on the left-hand side. Since $f(B_{t+\sigma\wedge n})\mathbb{1}_{\{\sigma\leqslant n\}} = f(B_{t+\sigma})\mathbb{1}_{\{\sigma\leqslant n\}}$ we arrive, after a pull-out, at

$$\mathbb{E}\Big[f(B_{t+\sigma}) \mid \mathscr{F}_{(\sigma\wedge n)+}\Big]\mathbb{1}_{\{\sigma\leqslant n\}} = \mathbb{E}^{B_\sigma}\Big[f(W_t)\Big]\mathbb{1}_{\{\sigma\leqslant n\}}.$$

We can now use Lemma 63.4 (below) to replace $\mathscr{F}_{(\sigma\wedge n)+}$ by $\mathscr{F}_{\sigma+}$ and let $n \to \infty$ to get (63.2).

The equality (63.3) follows similarly. □

In the proof of Theorem 63.3 we use the following auxiliary result.

63.4 Lemma*. *Let $Z : \Omega \to \mathbb{R}$ be an integrable random variable, $(\mathscr{F}_t)_{t\geqslant 0}$ a filtration and σ a stopping time. Then*

$$\mathbb{E}\Big(Z \mid \mathscr{F}_{(\sigma\wedge n)+}\Big)\mathbb{1}_{\{\sigma\leqslant n\}} = \mathbb{E}(Z \mid \mathscr{F}_{\sigma+})\mathbb{1}_{\{\sigma\leqslant n\}} \quad a.s.$$

Proof. Since σ is also a stopping time for the filtration $(\mathscr{F}_{t+})_{t\geqslant 0}$, we can use the very definition of $\mathscr{F}_{\tau+}$ for $\tau = \sigma\wedge n$ and $\tau = \sigma$ to conclude from Remark 46.13.d) that

$$\mathscr{F}_{(\sigma\wedge n)+} = \mathscr{F}_{\sigma+} \cap \mathscr{F}_{n+}.$$

Since both $\mathbb{E}\big(Z \mid \mathscr{F}_{(\sigma\wedge n)+}\big)$ and $\mathbb{E}(Z \mid \mathscr{F}_{\sigma+})$ are $\mathscr{F}_{\sigma+}$-measurable, it is enough to show that

$$\forall F \in \mathscr{F}_{\sigma+} : \quad \int_F \mathbb{E}\Big(Z \mid \mathscr{F}_{(\sigma\wedge n)+}\Big)\mathbb{1}_{\{\sigma\leqslant n\}}\,d\mathbb{P} = \int_F \mathbb{E}(Z \mid \mathscr{F}_{\sigma+})\mathbb{1}_{\{\sigma\leqslant n\}}\,d\mathbb{P}.$$

If $F \in \mathscr{F}_{\sigma+}$ the very definition of this σ-algebra shows $F \cap \{\sigma \leqslant n\} \in \mathscr{F}_{n+}$; on the other hand $\{\sigma \leqslant n\} \in \mathscr{F}_{\sigma+}$, hence $F\cap\{\sigma \leqslant n\} \in \mathscr{F}_{\sigma+}$, and we get $F\cap\{\sigma \leqslant n\} \in \mathscr{F}_{(\sigma\wedge n)+}$

and $F \cap \{\sigma \leqslant n\} \in \mathscr{F}_{\sigma+}$. Therefore,

$$\begin{aligned}\int_F \mathbb{E}\left(Z \mid \mathscr{F}_{(\sigma\wedge n)+}\right) \mathbb{1}_{\{\sigma\leqslant n\}}\, d\mathbb{P} &= \int_{F\cap\{\sigma\leqslant n\}} \mathbb{E}\left(Z \mid \mathscr{F}_{(\sigma\wedge n)+}\right) d\mathbb{P}\\ &= \int_{F\cap\{\sigma\leqslant n\}} Z\, d\mathbb{P}\\ &= \int_{F\cap\{\sigma\leqslant n\}} \mathbb{E}(Z \mid \mathscr{F}_{\sigma+})\, d\mathbb{P}\\ &= \int_F \mathbb{E}(Z \mid \mathscr{F}_{\sigma+}) \mathbb{1}_{\{\sigma\leqslant n\}}\, d\mathbb{P},\end{aligned}$$

and the claim follows. □

We are now going to consider a somewhat delicate special case:

- $\tau = \tau(B_\bullet)$ is a stopping time which can be expressed as a functional of a Brownian path, e.g. a first hitting time;
- σ is a stopping time such that $\sigma \leqslant \tau$;
- Set $\tau' = \tau(B_{\bullet+\sigma})$ the stopping time τ for the shifted process $B_{\bullet+\sigma}$, which is the remaining time, counting from σ, until the event described by τ happens;
- $W_\bullet := B_{\bullet+\sigma} - B_\sigma$; then $\tau' = \tau(W_\bullet + B_\sigma)$, and the functionals $f(B_\tau)$ and $f(W_{\tau'} + B_\sigma)$ have the same distribution.

Thus, (63.3) implies the following result.

63.5 Corollary. *Let $(B_t)_{t\geqslant 0}$ be a BM($\mathbb{R}^d$), $\tau = \tau(B_\bullet)$ a first hitting time and $\sigma \leqslant \tau$ a further stopping time. Set $\tau' = \tau(B_{\bullet+\sigma})$ and $W_\bullet := B_{\bullet+\sigma} - B_\sigma$. Then*

$$\mathbb{E}\left[f(B_\tau) \mid \mathscr{F}_{\sigma+}\right](\omega) = \mathbb{E}\left[f(W_{\tau'} + x)\right]\Big|_{x=B_\sigma(\omega)} = \mathbb{E}^{B_\sigma(\omega)} f(W_{\tau'}) \tag{63.4}$$

holds for $\mathbb{P}$-a.a. $\omega \in \{\tau < \infty\}$ and all bounded, measurable $f : \mathbb{R}^d \to \mathbb{R}$.

⚠ Observe the usual abuse of notation: B and W appearing in (63.4) are usually identified, and the same letter, say B, is used. In this case τ' becomes τ.

64 The Reflection Principle

In our closing chapter we discuss one of the most famous applications of the strong Markov property.

- Let $(B_t)_{t\geqslant 0}$ be a one-dimensional BM.
- Let $\tau_b := \tau^\circ_{\{b\}} = \inf\{t \geqslant 0 \mid B_t = b\}$ be the first passage time at level b.
- Assume that $\tau_b < t$. Stop B at τ_b and start anew from $B_{\tau_b} = b$.

Since $W_s + b = B_{\tau_b+s} - B_{\tau_b} + b = B_{\tau_b+s}$ is again a BM($\mathbb{R}$), we see that

the probabilities to be at time t above or below b are the same.

$B_{\tau_b} = b$ and τ_b is $\mathscr{F}_{\tau_b}^B$-measurable (✍, but also use your intuition: The event $\tau_b < u$ can be decided if we know $(B_{t\wedge\tau_b})_{t\geqslant 0}$), so

$$\begin{aligned}\mathbb{P}(\tau_b \leqslant t,\, B_t < b) &= \mathbb{P}(\underbrace{\{\tau_b \leqslant t\}}_{\in \mathscr{F}_{\tau_b}^B} \cap \underbrace{\{B_{\tau_b+(t-\tau_b)} - B_{\tau_b} < 0\}}_{\in \mathscr{F}_\infty^W \perp\!\!\!\perp \mathscr{F}_{\tau_b}^B,\ \sim W_{t-\tau_b}}) \\ &= \mathbb{P}(\{\tau_b \leqslant t\} \cap \{W_{t-\tau_b} < 0\}).\end{aligned}$$

Since τ_b and W are independent, we can treat τ_b, resp., $t - \tau_b$ like a fixed time and use the symmetry of W in the form $W_{t-\tau_b} \sim -W_{t-\tau_b}$. Thus,

$$\begin{aligned}\mathbb{P}(\tau_b \leqslant t,\, B_t < b) &= \mathbb{P}(\{\tau_b \leqslant t\} \cap \{W_{t-\tau_b} > 0\}) \\ &= \mathbb{P}(\{\tau_b \leqslant t\} \cap \{B_{\tau_b+(t-\tau_b)} - B_{\tau_b} > 0\}) \\ &= \mathbb{P}(\tau_b \leqslant t,\, B_t > b).\end{aligned}$$

From this we get

$$\begin{aligned}\mathbb{P}(\tau_b \leqslant t) &= \mathbb{P}(\tau_b \leqslant t,\, B_t \geqslant b) + \mathbb{P}(\tau_b \leqslant t,\, B_t < b) \\ &= \mathbb{P}(\tau_b \leqslant t,\, B_t \geqslant b) + \mathbb{P}(\tau_b \leqslant t,\, B_t > b) \\ &= \mathbb{P}(B_t \geqslant b) + \mathbb{P}(B_t > b) \\ &= 2\mathbb{P}(B_t \geqslant b).\end{aligned}$$

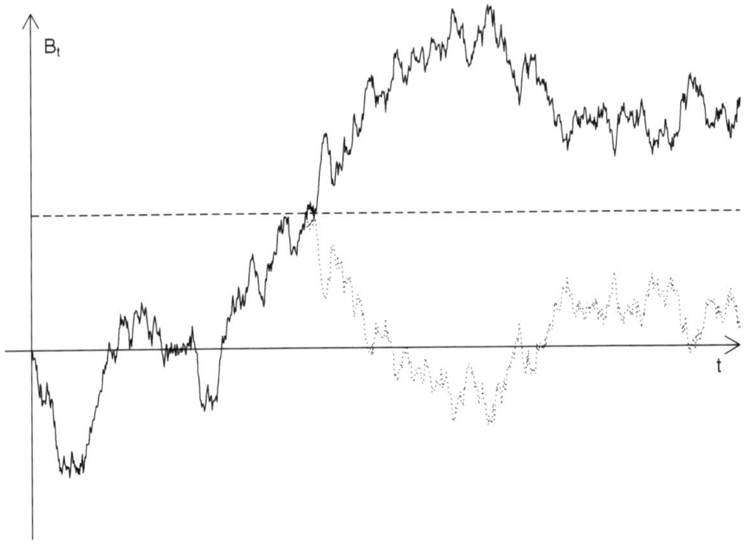

Fig. 64.1. Brownian motion reaching a level b and the reflection of its path at the level b (shown in grey).

By the definition of τ_b,

$$\{\tau_b \leqslant t\} = \{M_t \geqslant b\} \quad \text{where} \quad M_t := \sup_{s \leqslant t} B_s.$$

Therefore, for all $b \geqslant 0$,

$$\begin{aligned}
\mathbb{P}(M_t \geqslant b) &= \mathbb{P}(\tau_b \leqslant t) \\
&= 2\mathbb{P}(B_t \geqslant b) \\
&= \mathbb{P}(B_t \geqslant b) + \mathbb{P}(-B_t \leqslant -b) \qquad (64.1) \\
&\overset{B_t \sim -B_t}{=} \mathbb{P}(B_t \geqslant b) + \mathbb{P}(B_t \leqslant -b) \\
&= \mathbb{P}(|B_t| \geqslant b).
\end{aligned}$$

From this we can calculate the probability distributions of various functionals of a Brownian motion.

64.1 Theorem (Lévy 1939). *Let $(B_t)_{t \geqslant 0}$ be a BM($\mathbb{R}$), $b \in \mathbb{R}$, and set*

$$\tau_b = \inf\{t \geqslant 0 \mid B_t = b\}, \quad M_t := \sup_{s \leqslant t} B_s, \quad \textit{and} \quad m_t := \inf_{s \leqslant t} B_s.$$

Then,

$$M_t \sim |B_t| \sim -m_t \sim M_t - B_t \sim B_t - m_t \sim \sqrt{\frac{2}{\pi t}}\, e^{-x^2/(2t)} \mathbb{1}_{[0,\infty)}(x)\, dx, \qquad (64.2)$$

and

$$\tau_b \sim \frac{|b|}{\sqrt{2\pi t^3}} e^{-b^2/(2t)} \mathbb{1}_{[0,\infty)}(t)\, dt. \qquad (64.3)$$

Proof. 1° The equalities (64.1) immediately show that $M_t \sim |B_t|$.
Since $-B$ is agan a BM, we see that $m_t = \inf_{s \leqslant t} B_s = -\sup_{s \leqslant t}(-B_s) \sim -M_t$.

2° Time-reversal: $(B_t)_{t \geqslant 0}$ BM $\Longrightarrow \forall a > 0 : (B_a - B_{a-s})_{s \in [0,a]}$ BM .
Indeed, if $0 = t_0 < t_1 < \cdots < t_n \leqslant a$, then

$$(B_a - B_{a-t_k}) - (B_a - B_{a-t_{k-1}}) = (B_{a-t_{k-1}} - B_{a-t_k}), \quad k = 1 \ldots, n,$$

are independent and distributed like $B_{t_k - t_{k-1}}$. Moreover, $t \mapsto (B_a - B_{a-t})$ is continuous, and $B_a - B_{a-0} = 0$.

3° Using the time-reversed Brownian motion with $a = t$, we see that

$$M_t - B_t = \sup_{s \leqslant t}(B_s - B_t) = \sup_{s \leqslant t}(B_{t-s} - B_t) \sim \sup_{s \leqslant t} B_s = M_t$$

and, by symmetry, $M_t - B_t \sim B_t - m_t$.

4° The formula (64.3) follows for $b > 0$ from $\{\tau_b \geqslant t\} = \{M_t \leqslant b\}$ if we differentiate the following equality in t:

$$\mathbb{P}(\tau_b \geqslant t) = \mathbb{P}(M_t \leqslant b) = \sqrt{\frac{2}{\pi t}} \int_0^b e^{-x^2/(2t)}\,dx \overset{x=y\sqrt{t}}{=} \sqrt{\frac{2}{\pi}} \int_0^{b/\sqrt{t}} e^{-y^2/2}\,dy.$$

For $b < 0$ we use again the symmetry of a Brownian motion. □

⚠ From (64.2) we know that $M_t \sim |B_t|$ for each $t \geqslant 0$. However, the finite-dimensional distributions of $(M_t)_{t\geqslant 0}$ and $(|B_t|)_{t\geqslant 0}$ – hence the law of the processes M and |B| – are different. (Note that both M and |B| are not Markov processes, otherwise the one-dimensional distributions would already determine the fdd, cf. Theorem 62.6.) Here it is obvious from the behaviour of the paths: $t \mapsto M_t$ is a.s. increasing, $t \mapsto |B_t|$ is positive but not necessarily monotone.

⚠ We derived (64.1) with the Markov property of a Brownian motion, i.e. the fact that $\mathscr{F}_s^B$ and $\mathscr{F}_\infty^W$, $W_t = B_{t+s} - B_s$ are independent. Often the following problematic argument can be found:

$$\begin{aligned}\mathbb{P}(\tau_b \leqslant t,\, B_t < b) &= \mathbb{E}\Big(\mathbb{1}_{\{\tau_b\leqslant t\}}\mathbb{E}\big[\mathbb{1}_{(-\infty,0)}(B_t - b) \mid \mathscr{F}_{\tau_b}^B\big]\Big)\\ &\overset{?!}{=} \mathbb{E}\Big(\mathbb{1}_{\{\tau_b\leqslant t\}}\mathbb{E}^{B_{\tau_b}}\big[\mathbb{1}_{(-\infty,0)}(B_{t-\tau_b} - b)\big]\Big)\\ &= \mathbb{E}\Big(\mathbb{1}_{\{\tau_b\leqslant t\}}\mathbb{E}^{b}\big[\mathbb{1}_{(-\infty,0)}(B_{t-\tau_b} - b)\big]\Big)\\ &= \mathbb{E}\Big(\mathbb{1}_{\{\tau_b\leqslant t\}}\underbrace{\mathbb{E}\big[\mathbb{1}_{(-\infty,0)}(B_{t-\tau_b})\big]}_{=1/2}\Big).\end{aligned}$$

The problem appears in the step marked by »?!«: Because of the shift caused by the strong Markov property, we get $t - \tau_b$, but we do not know whether $t - \tau_b$ is positive – and if it were, our calculation would certainly not be covered by (63.2). Please note that the indicator function is outside the expectation. Below, we'll see how one can fix this.

A rigorous argument needs the following **stronger version of the SMP**. Please note carefully the dependence of the random times on the fixed path ω in the expression below.

64.2 Theorem. *Let $(B_t)_{t\geqslant 0}$ be a $\mathrm{BM}(\mathbb{R}^d)$, τ an $\mathscr{F}_t^B$ stopping time and $\eta \geqslant \tau$ where η is an $\mathscr{F}_{\tau+}^B$-measurable random time. Then we have for all $\omega \in \{\eta < \infty\}$ and all measurable, bounded $f : \mathbb{R}^d \to \mathbb{R}$*

$$\begin{aligned}\mathbb{E}\big[f(B_\eta) \mid \mathscr{F}_{\tau+}^B\big](\omega) &= \mathbb{E}^{B_\tau(\omega)}\big[f(B_{\eta(\omega)-\tau(\omega)}(\cdot))\big] \qquad (64.4)\\ &= \int f(B_{\eta(\omega)-\tau(\omega)}(\omega'))\,\mathbb{P}^{B_\tau(\omega)}(d\omega').\end{aligned}$$

Proof. Either look it up in [BM, Thm. 6.11], or try a proof by yourself: Approximate τ and η as in the proof of Theorem 63.2 and observe that the approximations preserve the order $\tau \leqslant \eta$. □

The fact that τ is a stopping time and η is an $\mathscr{F}_{\tau+}$-measurable random time with $\eta \geqslant \tau$ looks artificial. Typical examples of such pairs (τ, η) are $(\tau, \tau + t)$, $(\tau, \tau \vee t)$ or $(\sigma \wedge t, t)$ where σ is a further stopping time.

The proof of the reflection principle uses only the strong Markov property and the symmetry of a Brownian motion. Therefore, we can rephrase the reflection principle in the following more general version. Let $(B(t), \mathscr{F}_t)_{t\geqslant 0}$ be a Brownian motion and τ an a.s. finite stopping time. We consider the process

$$W(t,\omega) := \begin{cases} B(t,\omega), & \text{if } 0 \leqslant t < \tau(\omega) \leqslant +\infty, \\ 2B(\tau(\omega),\omega) - B(t,\omega), & \text{if } \tau(\omega) \leqslant t < \infty, \end{cases} \tag{64.5}$$

as shown in Fig. 64.2.

The trajectories of the processes B and W coincide on $[0, \tau(\omega))$. On $[\tau(\omega), \infty)$ we have $W(t,\omega) = B(\tau(\omega),\omega) - \big(B(t,\omega) - B(\tau(\omega),\omega)\big)$, i.e. every trajectory of W is the reflection of the original trajectory with respect to the axis $y = B(\tau(\omega), \omega)$.

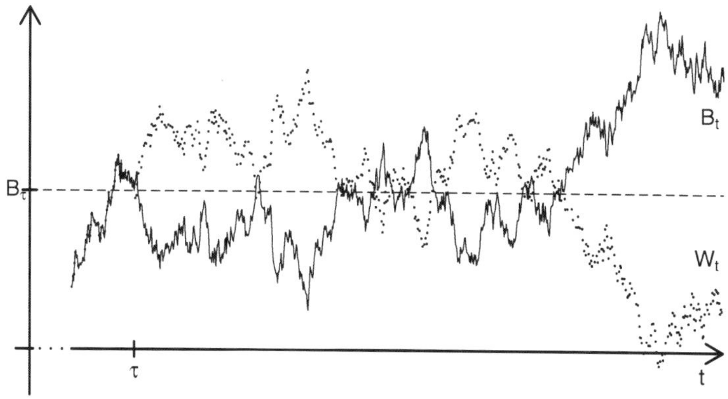

Fig. 64.2. Brownian motion reaching a level $y = B_\tau$ for the first time, and the reflection of its path at that level.

From Theorem 63.2 we know that $(B(\tau) - B(\tau + t))_{t\geqslant 0}$ is a Brownian motion which is independent of the stopped process $(B(t \wedge \tau))_{t\geqslant 0}$. Therefore it is plausible that the process W, which is a concatenation of these two processes, is again a Brownian motion. This is indeed so; the following theorem is proved in [BM, Theorem 6.12].

64.3 Theorem (no proof). *Let $(B(t), \mathscr{F}_t)_{t\geqslant 0}$ be a $BM(\mathbb{R}^d)$ and let $(W(t))_{t\geqslant 0}$ the process given by (64.5). Then $(W(t))_{t\geqslant 0}$ is again a Brownian motion.*

One could even randomize this construction by constructing paths from the »excursions« (the arcs away from $y = B_\tau$) and flipping them with equal probability over or below the line $y = B_\tau$ with an independent Bernoulli random variable; we will still get a Brownian motion.

Bibliography

[MIMS] Schilling, R.L.: *Measures, Integrals and Martingales.* Cambridge University Press, Cambridge 2017 (2nd ed).

[CEX] Schilling, R.L., Kühn, R.: *Counterexamples in Measure and Integration.* Cambridge University Press, Cambridge 2021.

[MI] Schilling, R.L.: *Maß und Integral.* De Gruyter, Berlin 2014.

[WT] Schilling, R.L.: *Wahrscheinlichkeit.* De Gruyter, Berlin 2017.

[MaPs] Schilling, R.L.: *Martingale und Prozesse.* De Gruyter, Berlin 2018.

[BM] Schilling, R.L., Partzsch, L.: *Brownian Motion. An Introduction to Stochastic Processes.* De Gruyter, Berlin 2014 (2nd ed).

[BARCA] Khoshnevisan, D., Schilling. R.L.: *From Lévy-type Processes to Parabolic SPDEs.* Birkhäuser, Cham 2016.

[1] Aczél, J.: *Lectures on Functional Equations and Their Applications.* Academic Press, New York (NY) 1966.

[2] Alt, H.-W.: *Linear Functional Analysis. An Application-Oriented Introduction.* Springer, Cham 2016 [translation of the 6th German Edition 2012].

[3] Artin, E.: *The Gamma Function.* Holt, Rinehart and Winston, New York 1964 [reprinted by Dover, Mineola 2015. German original: Die Gamma-Funktion, Hamburg 1931].

[4] Bauer, H.: *Wahrscheinlichkeitstheorie.* De Gruyer, Berlin 1991 (4. Auflage) [translation: *Probability Theory.* De Gruyter, Berlin 1996].

[5] Bayes, T.: An Essay towards Solving a Problem in the Doctrine of Chances. *Philosophical Transactions of the Royal Society* **53** (1763) 370–418 [reprinted as: *Versuch zur Lösung eines Problems der Wahrscheinlichkeitsrechnung.* Engelmann, Leipzig 1908].

[6] Borel, E.: *Éléments de la théorie des probabilités.* Hermann, Paris 1909.

[7] Bru, M.-F., Bru, B.: *Les jeux d l'infini et du hasard 1,2.* Presses universitaires de Franche-Comté, Bescançon 2018.

[8] Chung, K.-L.: *A Course in Probability Theory.* Academic Press, San Diego (CA) 1974 (2nd ed).

[9] Chung, K.-L.: *Lectures from Markov Processes to Brownian Motion.* Springer, New York 1982 [2nd edition appeared as Chung, K.-L., Walsh, J.B.: *Markov processes, Brownian motion, and time symmetry,* Springer, Berlin 2004].

[10] Diaconis, P., Skyrms, B.: *Ten Great Ideas About Chance.* Princeton University Press, Princeton (NJ) 2018.

[11] Etemadi, N.: An elementary proof of the strong law of large numbers, *Zeitschrift für Wahrscheinlichkeitstheorie und verwandte Gebiete* **55** (1981) 119–122.

[12] Etemadi, N.: On some classical results in probability theory. *Sankhya A* **47** (1985) 215–221.

[13] Honderich, T. (ed.): *The Oxford Companion to Philosophy.* Oxford University Press, Oxford 1995.

[14] Kato, T.: *Perturbation Theory for Linear Operators.* Springer, Berlin 1980 (2nd ed).

[15] Kolmogoroff, A.: *Grundbegriffe der Wahrscheinlichkeitsrechnung.* Ergebnisse der Mathematik und ihrer Grenzgebiete Bd. II, Heft 3, Springer, Berlin 1933.

[16] Lang, S.: *Linear Algebra.* Springer, New York 1987 (3rd ed).

[17] Laplace, P.S.: Mémoire sur la probabilité des causes par les événements. *Savants étranges* **6** (1774) 621–656 [reprinted as: Memoir on the Probability of the Causes of Events. *Statistical Science* **1** (1986) 359–378].

[18] Laplace, P.S.: *Théorie analytique des probabilités.* Courcier, Paris 1812.

[19] Lukacs, E.: *Characteristic Functions.* Hafner, New York 1970.

[20] McGrayne, S.B.: *The Theory that Would Not Die.* Yale University Press, New Haven (CT) 2011.

[21] Moran, P.A.P.: A characteristic property of the Poisson distribution. *Proceedings of the Cambridge Philosophical Society* **48** (1951) 206–207.

[22] Olver, F.W.J. (et al.): *NIST Handbook of Mathematical Functions.* Cambridge University Press, Cambridge 2010 (free online access: `http://dlmf.nist.gov/`).

[23] Parthasarathy, K.R.: *Probability Measures on Metric Spaces.* AMS Chelesa, Providence (RI) 2005.

[24] Port, S.C., Stone, C.J.: Potential Theory of Random Walks on Abelian Groups. *Acta Mathematica* **122** (1969) 19–114.

[25] Revuz, D., Yor, M.: *Continuous Martingales and Brownian Motion.* Springer, Berlin 1999 (3rd ed).

[26] Rosenhouse, J.: *The Monty Hall Problem.* Oxford University Press, Oxford 2009.

[27] Rudin, W.: *Principles of Mathematical Analysis.* Mc-Graw Hill, New York 1976 (3rd ed) [German Edition: Analysis 4. Aufl. Oldenbourg, München 2009].

[28] Rudin, W.: *Real and Complex Analysis*. Mc-Graw Hill, New York 1986 (3rd ed) [German Edition: Reelle und Komplexe Analysis 2. Aufl. Oldenbourg, München 2009].

[29] Savage, L.J.: *The Foundations of Statistics*. Dover, New York 1972 (2nd edn., 1st edn Wiley 1954).

[30] Schneider, I. (ed.): *Die Entwicklung der Wahrscheinlichkeitstheorie von den Anfängen bis 1933. Einführung und Texte* [The development of probability theory from the begginnings until 1933. Introduction and sources. In German]. Wissenschaftliche Buchgesellschaft, Darmstadt 1988.

[31] Stromberg, K.: The Banach–Tarski paradox. *American Mathematical Monthly* **86** (1979) 151–161.

[32] Whittaker, E.T., Watson, G.N.: *A Course of Modern Analysis*. Cambridge University Press, Cambridge 1973 [reprint of the 4th ed. 1927].

Name and subject index

Printed in Great Britain
by Amazon

79274001R00258